Emerging Synthesis Techniques for Luminescent Materials

Ratnesh Tiwari
Bhilai Institute of Technology, India

Vikas Dubey
Bhilai Institute of Technology, India

Sanjay J. Dhoble
Rashtrasant Tukadoji Maharaj Nagpur University, India

A volume in the Advances in Chemical and Materials Engineering (ACME) Book Series

IGI Global
DISSEMINATOR OF KNOWLEDGE

Published in the United States of America by
IGI Global
Engineering Science Reference (an imprint of IGI Global)
701 E. Chocolate Avenue
Hershey PA, USA 17033
Tel: 717-533-8845
Fax: 717-533-8661
E-mail: cust@igi-global.com
Web site: http://www.igi-global.com

Library of Congress Cataloging-in-Publication Data

Names: Tiwari, Ratnesh, 1982- editor. | Dubey, Vikas, 1985- editor. | Dhoble, Sanjay J., 1967- editor.
Title: Emerging synthesis techniques for luminescent materials / Ratnesh Tiwari, Vikas Dubey, and Sanjay J. Dhoble, editors.
Description: Hershey, PA : Engineering Science Reference, [2018] | Includes bibliographical references.
Identifiers: LCCN 2017036323| ISBN 9781522551706 (h/c) | ISBN 9781522551713 (eISBN)
Subjects: LCSH: Optoelectronic devices--Materials. | Electroluminescent devices--Materials. | Phosphors.
Classification: LCC TK7871.68 .E455 2018 | DDC 621.3815/20284--dc23 LC record available at https://lccn.loc.gov/2017036323

This book is published in the IGI Global book series Advances in Chemical and Materials Engineering (ACME) (ISSN: 2327-5448; eISSN: 2327-5456)

British Cataloguing in Publication Data
A Cataloguing in Publication record for this book is available from the British Library.

Advances in Chemical and Materials Engineering (ACME) Book Series

J. Paulo Davim
University of Aveiro, Portugal

ISSN:2327-5448
EISSN:2327-5456

Mission

The cross disciplinary approach of chemical and materials engineering is rapidly growing as it applies to the study of educational, scientific and industrial research activities by solving complex chemical problems using computational techniques and statistical methods.

The **Advances in Chemical and Materials Engineering (ACME) Book Series** provides research on the recent advances throughout computational and statistical methods of analysis and modeling. This series brings together collaboration between chemists, engineers, statisticians, and computer scientists and offers a wealth of knowledge and useful tools to academics, practitioners, and professionals through high quality publications.

Coverage

- Biomaterials
- Multifuncional and Smart Materials
- Thermo-Chemical Treatments
- Toxicity of Materials
- Nanomaterials
- Electrochemical and Corrosion
- Ductility and Crack-Resistance
- Wear of Materials
- Numerical Techniques
- Fracture Mechanics

IGI Global is currently accepting manuscripts for publication within this series. To submit a proposal for a volume in this series, please contact our Acquisition Editors at Acquisitions@igi-global.com or visit: http://www.igi-global.com/publish/.

The Advances in Chemical and Materials Engineering (ACME) Book Series (ISSN 2327-5448) is published by IGI Global, 701 E. Chocolate Avenue, Hershey, PA 17033-1240, USA, www.igi-global.com. This series is composed of titles available for purchase individually; each title is edited to be contextually exclusive from any other title within the series. For pricing and ordering information please visit http://www.igi-global.com/book-series/advances-chemical-materials-engineering/73687. Postmaster: Send all address changes to above address.

Titles in this Series

For a list of additional titles in this series, please visit: www.igi-global.com/book-series

Advanced Solid Catalysts for Renewable Energy Production
Sergio González-Cortés (Oxford University, UK) and Freddy Emilio Imbert (Univsersidad de Los Andes, Venezuela)
Engineering Science Reference • copyright 2018 • 520pp • H/C (ISBN: 9781522539032) • US $225.00 (our price)

Production, Properties, and Applications of High Temperature Coatings
Amir Hossein Pakseresht (University of Tehran, Iran & Materials and Energy Research Center, Iran)
Engineering Science Reference • copyright 2018 • 557pp • H/C (ISBN: 9781522541943) • US $235.00 (our price)

Energetic Materials Research, Applications, and New Technologies
Rene Francisco Boschi Goncalves (Federal University of Para, Brazil) José Atilio Fritz Fidel Rocco (Aeronautics Institute of Technology, Brazil) and Koshun Iha (Aeronautics Institute of Technology, Brazil)
Engineering Science Reference • copyright 2018 • 367pp • H/C (ISBN: 9781522529033) • US $225.00 (our price)

Chemical Compound Structures and the Higher Dimension of Molecules Emerging Research and Opportunities
Gennadiy Vladimirovich Zhizhin (Russian Academy of Natural Sciences, Russia)
Engineering Science Reference • copyright 2018 • 221pp • H/C (ISBN: 9781522541080) • US $175.00 (our price)

Emergent Research on Polymeric and Composite Materials
R. Somashekar (University of Mysore, India) and Thejas Urs G. (University of Mysore, India)
Engineering Science Reference • copyright 2018 • 341pp • H/C (ISBN: 9781522530237) • US $235.00 (our price)

3D Printing and Its Impact on the Production of Fully Functional Components Emerging Research and Opportunities
Petar Kocovic (Union – Nikola Tesla University, Serbia)
Engineering Science Reference • copyright 2017 • 115pp • H/C (ISBN: 9781522522898) • US $150.00 (our price)

Advanced Applications of Supercritical Fluids in Energy Systems
Lin Chen (Tohoku University, Japan & Japan Society for the Promotion of Science (JSPS), Japan) and Yuhiro Iwamoto (Nagoya Institute of Technology, Japan)
Engineering Science Reference • copyright 2017 • 682pp • H/C (ISBN: 9781522520474) • US $235.00 (our price)

Innovative Applications of Mo(W)-Based Catalysts in the Petroleum and Chemical Industry Emerging Research and Opportunities
Hui Ge (Chinese Academy of Sciences, China) Xingchen Liu (Chinese Academy of Sciences, China) Shanmin Wang (Oak Ridge National Laboratory, USA) Tao Yang (China University of Petroleum, China) and Xiaodong Wen (Synfuels China, China)
Engineering Science Reference • copyright 2017 • 162pp • H/C (ISBN: 9781522522744) • US $135.00 (our price)

IGI Global
DISSEMINATOR OF KNOWLEDGE

701 East Chocolate Avenue, Hershey, PA 17033, USA
Tel: 717-533-8845 x100 • Fax: 717-533-8661
E-Mail: cust@igi-global.com • www.igi-global.com

Table of Contents

Detailed Table of Contents

The growing demand for energy-efficient materials needs advancement in existing approaches. Photon upconversion is an effective way to overcome this issue. Novel upconverting materials show their potential in the advancement of energy conservation and lighting technology. In this contest, lanthanide-activated inorganic luminescent materials play a key role. The unique features of these materials like non-toxic nature, narrow multicolour emissions, long luminescence lifetime, high signal/noise ratio, and good chemical durability make them more promising. In this chapter, an overview of photon upconversion process in various lanthanide doped/codoped materials is given. Introductory idea about upconversion mechanism, criteria of material selection, novel synthesis roots, reports on upconversion-emitting materials, and emergent applications of these materials are presented.

The upconversion (UC) of the rare earth doped MY_2O_4 (M=Ba, Sr, Ca) has been extensively investigated due to their potential applications in many fields, such as color display, high density memories, optical data storage, sensor and energy solar cell, etc. Many series of them, especially the Er^{3+}, Yb^{3+} doped MY_2O_4 were studied in this chapter, due to the thermal and mechanical toughness, high optical transmittance from the ultraviolet to the infrared regions, and a low nonlinear refractive index compared to the other commercial laser glasses. The energy transfer (ET) mechanism of rare earth doped phosphor plays an important role in the upconversion process.

Section 3
Emerging Synthesis Techniques of Nano-Materials for Solid State Lighting

In this section, emerging syntheses are reported for solid state lighting phosphors, and their applications for white light emitting diodes and various synthesis techniques are reported.

Phosphor materials are an integral and important part in the white light generation in LEDs and have participated in the global warming reduction significantly. In this chapter, the authors focus on different types of phosphor materials belonging to a family of apatites, fluoroapatites, silicates, oxides, phosphates, and borates with different crystal structures. The detailed investigations on their crystal chemical structure, synthesis, and photoluminescence properties are briefly discussed. The improved optical properties make

Preface

This book provides information about emerging synthesis techniques of luminescent materials and phosphors has assumed ever more significance in the overall scheme of scientific progress. This book is aimed at providing a sound introduction to the dosimetry, upconversion, solid state lighting and OLED's, Mechanoluminescence and Electroluminescence phosphors for scientist and academicians. We hope that it will beneficial for society also.

Phosphors are solid luminescent materials that emit photons when excited by an external energy source. Luminescence continues to play a major technological role for mankind. Almost all modern phosphors are synthesized by solid-state reactions at high temperatures. Updated versions of these techniques are presented in this book along with other techniques such as sol–gel and combustion synthesis, hydrothermal, chemical route, molten salt etc. that have been developed over the past few decades.

The past few decades have seen spectacular developments in research on luminescence. There has been phenomenal growth in the subject and significant progress has been made. Rare earth ion-activated phosphors have numerous applications in the display, lighting, and medical industries. In recent years, the luminescent properties of phosphate materials have been widely investigated for their many advantages, such as excellent thermal and chemical stability, and the development of optical devices based on rare earth (RE) ion-doped materials has proven to be one of the most interesting fields of research.

Rare earth-activated phosphors are of interest due to their unusual stability and valuable luminescent applications such as phosphors for lamps, color TV screens, long lasting devices, laser hosts, scintillators, and pigments. The energy transfer phenomenon has been studied extensively in inorganic phosphors, crystals, solutions, and glasses. Hence, in order to contribute to such knowledge, an attempt has been made in this book to consider efficient phosphors based on rare earth-activated phosphates, to study their luminescence properties, and to explore potential new materials and applications.

Solid-state lighting using light-emitting diodes (LED) and phosphor material to generate white light is the current research focus in the lighting industry. Solid-state lighting technology has several advantages over conventional fluorescent lamps, such as reduced power consumption, compactness, efficient light output, and longer lifetime. Solid-state lighting will have its impact in reducing global electricity consumption. White light-emitting diodes can save about 80% in energy terms and do not require any harmful ingredients, in contrast to conventional light sources, such as incandescence light bulbs and luminescent tubes. White LEDs thus have a great potential to replace the latter and are considered to represent the next generation of solid-state lighting devices.

It emphasis the quality of research and explore new phenomenon occurs in day to day activities in the field of opto-electronics.

Preface

Emerging synthesis techniques of luminescent materials has assumed great importance during the last few decades because of excellent luminescent materials have been developed for lighting and display with greater performance in its brightness, color rendering, efficiency, and lifetime.

In recent years a lot of research is going on the display devices applications. therefore in this field got a Nobel prize in year 2014 (Physics and Chemistry).

The present book give the impact of the photoluminescence and study of rare earth doped phosphors for fluorescent lamps, TV tubes, electroluminescent display devices, high quality image intensifiers and SSL (Solid State Lightening), LED (Light Emitting Diode) and LASER materials Application. Our basic aim to produce the WLED (White light emitting diode) for single host and single dopant and to remove the costly tri – band phosphor by single one. Also it gives information about prediction of earth quake lights, mechanoluminscence sensors, TL-Dosimeters, OSL (optically stimulated luminescence), medical imaging phosphors etc.

It provide the great helps to young researchers, scientists, society, industries and medical field for future instruments fabrication and its applications.

Chapter 1: This chapter describes the basic concepts of swift heavy ions and their modifications. The basic requirements for Photoluminescence and thermoluminescence are also described in brief. The procedure for the calculation of different trapping parameters via different approaches is discussed. Finally, the procedure to be used for the dosimetric applications is also discussed for these ion irradiated materials. This discussion will benefit the present researcher at a basic level and help them with the future aspects of the ion irradiated phosphor materials in the nanodimension.

Chapter 2: The shape of a thermoluminescence (TL) glow curve has fundamental importance for calculating the characteristic parameters of trap levels within the band gap. TL analysis are mostly based on the three-parameter general order kinetics model. The parameters are activation energy, order of kinetics and frequency factor. Peak shape method is one of the most prominent methods for extracting the activation energy from a TL curve. An overview of different peak shape methods along with an alternative approach formulated directly from basic TL equations is presented here. Generally, peak shape method requires prior knowledge of order of kinetics to determine activation energy which creates a difficulty due to the non-uniqueness of symmetry factor for a particular value of order of kinetics. A modified version of peak shape method which is free from this constraint is discussed here. Activation energies from experimental curves of tremolite and actinolite are estimated using peak shape method. Limitation of peak shape method for saturated TL peaks with heavy retrapping is also discussed.

Chapter 3: The present chapter provides the useful information about synthesis and characterization of dysprosium doped oxide and fluoride based phosphors such as $SrGd_2O_4$, $CaSO_4$ and CaF_2. Various techniques viz. acid-recrystallization, chemical co-precipitation and homogenous precipitation cum auto-combustion methods were adopted to synthesize these phosphors for large scale production. All the prepared phosphors were characterized by X-ray diffraction analysis, scanning electron microscopy and transmission electron microscopy techniques. The thermoluminescence (TL) studies were performed after different irradiation sources such as gamma rays, thermal neutrons and low energy ions (H, Ar and N), respectively. Linear dose responses were observed in a wide range of doses for all the samples. Various trapping parameters namely order of kinetics, activation energy and frequency factors were calculated by using computerized glow curve deconvolution (CGCD) method.

Chapter 4: Growing demand of energy efficient materials need advancement in existing approaches. Photon upconversion is an effective way to overcome this issue. Novel upconverting materials show their potentiality in the advancement of energy conservation and lighting technology. In this contest, lanthanide activated inorganic luminescent materials play key role. The unique features of these materials like non-toxic nature, narrow multicolour emissions, long luminescence lifetime, high signal/noise ratio, and good chemical durability make them more promising. Herein, an overview of photon upconversion process in various lanthanide doped/codoped materials is given. Introductory idea about upconversion mechanism, criteria of material's selection, novel synthesis roots, reports on upconversion emitting materials, and emergent applications of these materials are presented.

Chapter 5: The upconversion (UC) of the rare earth doped MY_2O_4 (M=Ba, Sr, Ca) has been extensively investigated due to their potential applications in many fields, such as color display, high density memories, optical data storage, sensor and energy solar cell, etc. Many series of them, especially the Er^{3+}, Yb^{3+} doped MY_2O_4 were studied in this work, due to the thermal and mechanical toughness, high optical transmittance from the ultraviolet to the infrared regions, and a low nonlinear refractive index compared to the other commercial laser glasses. Especially energy transfer (ET) mechanism of rare earth doped phosphor gives in details it plays an important role in the upconversion process. All prepared phosphor with variable concentration of Er^{3+} (1, 2 and 3 mol%) at fixed concentration if Yb^{3+} (0.5mol%) was studied by photoluminescence analysis and it is found that the excitation spectra of prepared phosphor shows sharp excitation centred at 980nm laser excitation peak. The emission spectra with variable concentration of Er^{3+} shows strong peaks at 545nm and 594nm, the dominant peak at 545nm due to the ($^2H_{11/2}$,$^4S_{3/2}$) transition and weaker transition at 594nm with peak associated to 613nm. Spectrophotometric determinations of peaks are evaluated by Commission Internationale de I'Eclairage (CIE) technique. These upconverted emissions are attributed to a two photon process. The excitation wavelength dependence on the upconverted luminescence together with its time evolution after infrared pulsed excitation suggest that energy transfer upconversion processes are responsible for the upconversion luminescence. The upconversion mechanisms were studied in detail through laser power dependence. The cross relaxation process in Er^{3+} was also investigated.

Chapter 6: Phosphor materials become integral and important part in the white light generation in LEDs and participating in the global warming reduction significantly. In this chapter we mainly emphasis on different type of phosphor materials belongs to a family of apatites, fluoroapatites, silicates, oxides, phosphates and borates with different crystal structures. The detailed investigations on their crystal chemical structure, synthesis and photoluminescence properties are briefly discussed. The improved optical properties makes the phosphor potential candidate for smart panel displays and white light emitting diodes (including solid state lighting). The crystal structure have a great influence on the chemical and luminescence properties of any phosphor, hence change of activator ions (Eu^{2+}, Ce^{3+}, Mn^{2+} and Tb^{3+}) concentrations a great change can be achieved in the phosphor performances. In this book we have tried our best to correlate the structure-compositions-property of the phosphor materials with special emphasis on white LEDs.

Chapter 7: This chapter consists of Systematic synthesis of magnesium silicate nanoparticles by solution combustion method using ODH fuel. Compared to other fuels ODH has the better advantages of obtaining crystalline nanomaterials without post calcinations. By doping different rare earth ions to magnesium silicate, the product can be used for luminescent applications. Synthesis of Magnesium ferrite nanoparticle by molten salt method followed by the synthesis of polymer composite in the form of fibers by the process of electrospinning.

Chapter 8: Hydrothermal synthesis is easy, portable, less-hazardous and low cost synthesis method. Various researchers across the globe are worked on the synthesis of different materials via this route. Practically, fluorides are difficult to synthesize due to their hygroscopic nature by conventional methods. But, the hydrothermal synthesis is used to prepare several compositionally optimized fluoride based materials using closed-system physical and chemical processes in an aqueous solution at low temperatures and pressures. The silent features of hydrothermal method over conventional methods of materials processing, are crystallization of materials, crystal growth, in the processing of a wide range of materials not only the bulk crystals, but fine particles with a controlled size and morphology. Therefore, in order to place its numerous recent development, past and current research needs to club in a one article. This review is a recent clocking updates for synthesis, materials and their applications.

Chapter 9: Considering the imminent global energy-crisis and inefficient energy utilization, energy efficient organic light emitting diodes (OLEDs) considered as one of the most competitive candidates for new-generation smart displays and particularly for future energy-saving lighting sources. Recently much effort has been made to generate white OLEDs comprising of both fluorescent and phosphorescent materials since it offers more advantages. Full color displays white OLEDs requires all primary colors red, green and blue. In recent decades numerous phenanthroimidazole based blue, green, yellow and red emitting materials have been developed for efficient OLEDs. In organic electronics considerable interest is shown on Phenanthroimidazole, due to ease in fluorophore modification at N1 and C2 position. On the other hand, these materials have the rigid skeleton, show high carrier mobilities, high thermal stability, as well as good film forming ability, leading to promising applications in high-performance OLEDs. This book chapter focuses on the design and synthesis of phenanthroimidazole based blue, green, yellow and red fluorescent materials and their applications in OLEDs. At first, some nondoped blue, green, and yellow fluorescent materials endowed with unipolar and bipolar-transporting abilities are comprehensively studied. Then attention has been paint for typical blue, green, yellow, orange, and red PhOLEDs of phenanthroimidazole based fluorophores as host materials are briefly presented. The molecular design concept, general synthetic routes for PI materials, thermal stabilities, and the applications of fluorophores in fluorescent OLEDs and host materials in PhOLEDs are reviewed.

Chapter 10: The chapter reports synthesis and characterization of europium doped $Y_4Al_2O_9$ phosphor for display and dosimetric applications. The europium activated $Y_4Al_2O_9$ (YAM) phosphor synthesized via solid state reaction method. Synthesized phosphors were characterized by powder X-ray diffraction (PXRD) techniques, scanning electron microscopy (SEM) technique and transmission electron microscopy (TEM) technique. Particle size calculated from TEM analysis and crystallite size was calculated by Scherer's formula. All synthesized phosphor for different concentration of europium ion were studied by photo, thermo and mechanoluminescence study. It is found that for photoluminescence analysis of Eu^{3+} doped phosphor have prominent spectra in red region and electric dipole transition ($^5D_0 \rightarrow {}^7F_2$) dominant over the magnetic dipole transition ($^5D_0 \rightarrow {}^7F_1$) due to non-centro symmetry between rare earth ions. Broad excitation spectra found for photoluminescence study.

Chapter 11: Over the past few decades, many efforts have been put to optimize the properties of electroluminescent devices such as Electroluminescence (EL) panel, Organic Light emitting Diode (OLED), Flat Panel Display, EL lamps etc. In this chapter, we provide a comprehensive review of the state-of-the-art research activities related to EL phenomena, the principle of electroluminescence, different types of EL, fabrication of devices, studies on novel electroluminescent materials, their characteristics and potential applications. We began with a historical background of Electroluminescence, description of its structure, working principle and parameters with mechanism. We discussed the experiments determining the electrical, optical and physical properties of the powder as well as thin film EL cells. The recent progress on the improvement of their characteristics and finding novel structures, such as the latest achievements in using various semiconductor nanostructures in polymer matrix as emitting layer in EL devices summarized.

As this book covers relatively wide areas and numerous contents connected to complex scientific issues, errors and omissions may be unavoidable due to the limited knowledge and competence of the authors, therefore we sincerely appreciate the criticism and comments from the readers.

We feel confident it will be highly valued for its documentation of the important field of emerging synthesis techniques of luminescent materials and its practical applications.

Ratnesh Tiwari
Bhilai Institute of Technology, India

Vikas Dubey
Bhilai Institute of Technology, India

S. J. Dhoble
Rashtrasant Tukadoji Maharaj Nagpur University, India

Acknowledgment

The editors would like to acknowledge the help of all the people involved in this project and, more specifically, to the authors and reviewers that took part in the review process. Without their support, this book would not have become a reality.

First, the editors would like to thank each one of the authors, Hendrik C Swart, Sudipta Som, Somrita Dutta, Vijay Kumar, Sukhamoy Bhattacharyya, Partha Sarathi Majumdar, Jyoti Singh, Mahesh S. Bhadane, Sanjay Daga Dhole, Jairam Manam, Shailendra Satish Dahiwale,Anurag Pandey, Surya Prakash Tiwari, Viresh Dutta, Vinod Kumar, Deepak Kumar, Shefali Verma, Kartikey Verma, Neha Dubey, Ravi Shrivastava; Jagjeet Kaur Saluja; Rituraj Chandrakar, Sivakumar Vaidyanathan, Jairam Tagare, R. Marikumar, S. Kasturi, S. Satish Kumar, Ramachandra Naik, Revathi V, S. C Prashantha, H Nagabhushana, Nilesh Sureshrao Ugemuge, Shubha Tripathi,Manish Kumar Mishra,Vikas Kumar Jain and Vikas Lahariya, Meera Ramrakhiani, for their contributions. Our sincere gratitude goes to the chapter's authors who contributed their time and expertise to this book.

Editors are very thankful to Late Dr. B.P. Chandra (Ex-vice-chancellor Pt. R.S.U. Raipur) for motivating us.

Second, the editors wish to acknowledge the valuable contributions of the reviewers Dr. Yogita Parganiha, Dr. Pradeep Mullick, regarding the improvement of quality, coherence, and content presentation of chapters. Most of the authors also served as referees; we highly appreciate their double task.

Ratnesh Tiwari
Bhilai Institute of Technology, India

Vikas Dubey
Bhilai Institute of Technology, India

S. J. Dhoble
Rashtrasant Tukadoji Maharaj Nagpur University, India

Section 1

Emerging Synthesis Techniques of Thermoluminescence Dosimetry Materials and Its Applications

In this section, luminescence materials are synthesized via different synthesis routes and its thermoluminescence properties are investigated. Some theoretical approaches are adopted for glow peak.

Chapter 1
Swift Heavy Ion Synthesis and Modifications of Nanophosphors for Dosimetric Application:
Effect of Swift Heavy Ion Irradiation

Sudipta Som
National Taiwan University, Taiwan

Somrita Dutta
National Chiao Tung University, Taiwan

Vijay Kumar
Chandigarh University, India

Hendrik C. Swart
University of the Free State, South Africa

ABSTRACT

Recently, there has been a huge demand for the experimental and theoretical efforts for the dosimetry of the ionizing radiation due to its diagnostic and therapeutic purposes. Different ionizing radiations include low ionizing radiation such as x-rays, γ-rays and densely ionizing radiation swift heavy ions (SHIs). When SHIs pass through a material, intense electronic excitations occur along the ion trajectory due to the inhomogeneous distribution of the energy inside the material. This internal disorder creates a large number of defects in the materials. The characterization of such defects provides material's technological importance for different practical applications in radiation dosimetry. Different luminescence behavior including thermoluminescence (TL) is the well-established and very sensitive technique for the characterization of the structural disorder and defects in solids. Therefore, this chapter deals with the basic concept of swift heavy ion and ion-induced modification for dosimetric applications.

DOI: 10.4018/978-1-5225-5170-6.ch001

INTRODUCTION

- **Swift Heavy Ion**: Swift means very fast, i.e. having very high velocity. It is well known that if any object with a mass m traverses with velocity v, it will gain the kinetic energy indicated as $1/2mv^2$. As the energy of the substance is directly proportional to the square of the velocity, the word swift indicates that it is having an elevated energy. Therefore, highly energetic heavy ions are generally known as Swift Heavy Ions. The present chapter deals with swift heavy ion induced synthesis and modification of phosphor materials mainly for dosimetric applications. This chapter will describe the history and the types of the radiation in brief along with its interaction with matter. Based on this discussion, this chapter will introduce different luminescence phenomena. Various simulation procedures will be revealed to obtain an idea of the suitability of the materials for dosimetric appliances.

Before going into the details, the outset of some prologue to theindispensable concepts and terminology of the radiation physics is necessary to be first mentioned. This interesting subject starts with the discovery of X-rays by Wilhelm K. Rontgen in the 1890s along with the foundation of radium by the Curies and radioactivity by Henri Becquerel. From then, X-rays havebecome handy paraphernalia in the practice of medicine, diagnostic radiography within a brief duration and hence setting some kind of record for the hasty implementation of innovative machinery forindustrial applications. However, the chronological advancement of the knowledge of radiological physics draws the research attention since then and has become an imperative partof the materials physics.

The radiation falls on an object and can excite and ionizethe atoms i.e., carries sufficient energy tofree electrons from atoms or moleculesof matter with which they interact are known as ionizing radiation. It is well known that at least 4- 25eV energy isrequired to craft a valenceelectron free from the atom and thereby the radiations which carry energies greater than the above-mentioned energy will be named as "ionizing." Ionizing radiation can be classified in two different ways as directly or indirectly ionizing radiation based on the nature of the radiation that creates the ionizing consequence.

- **Directly Ionizing:** If any radiation of charged massive particle carries adequate kinetic energy to ionize target atoms unswervingly by elementary interaction through the Coulomb force, these ionizing radiation will be categorized as directly ionizing radiation. It includes electrons,protons, nuclei, muons, pions, and so on. In general, alpha particles and beta particles defined as the Helium nuclei evicted from atomic nuclei, and electrons ejected usually at relativistic speeds respectively are also in this category.
- **Indirectly Ionizing:** The electrically neutral substances when falling on an object, indirect ionization occurs due to secondary ionizations. Therefore these substances do not interact sturdily with the target. The neutron radiation is one of the examples of indirectly ionizing radiation.

These ionizing radiations have many militaries, industrial, and medical uses along with the appliances in materials science. Therefore, the present chapter deals with the basic understandings of ionizing radiation and its application for thedosimetric purpose.

BACKGROUND

The highly energetic cosmic rays (Figure 1) constitute a serious dosimetric problem for aircrew and space dosimetry. The component used in various subsystems of the spacecraft near the lower earth orbit gets exposed to these high energy heavy ion radiations. (Vij et al., 2010)This necessitates the concurrent development for the measurement of doses/ fluences of such energetic particles with great precision of accuracy especially while dealing with human beings. Recently, there is a huge demand for experimental and theoretical efforts for the dosimetry of the ionizing radiation due to its diagnostic and therapeutic purposes.(Manjunatha et al., 2013) However, the ability to emit visible light radiation when suitably excited makes the phosphor materials suitable for TL dosimeter (TLD).

These TLD materials may be in the form of chips, single crystals or microcrystalline size powder. There are a number of commercially available TLDs under different trade names.The most popular names are LiF:Mg,Ti (TLD-100), LiF:Mg,Cu,P (TLD-700H), CaSO4:Dy (TLD-900), CaF_2:Dy (TLD-200), and Al_2O_3:C (TLD-500).(Salah, 2011) But, efforts are still being made by either improving the TL

Figure 1. Cosmic ray, spacecraft and heavy ion dosimeter
https://en.wikipedia.org/wiki/Space_station,https://www.space.com/32644-cosmic-rays.html, http://ndtteste.sell.ecer.com/iz6bf2be1-radiometer-x-ray-pipeline-crawler-personal-dosimeter-ndt-x-ray-equipment-images.html

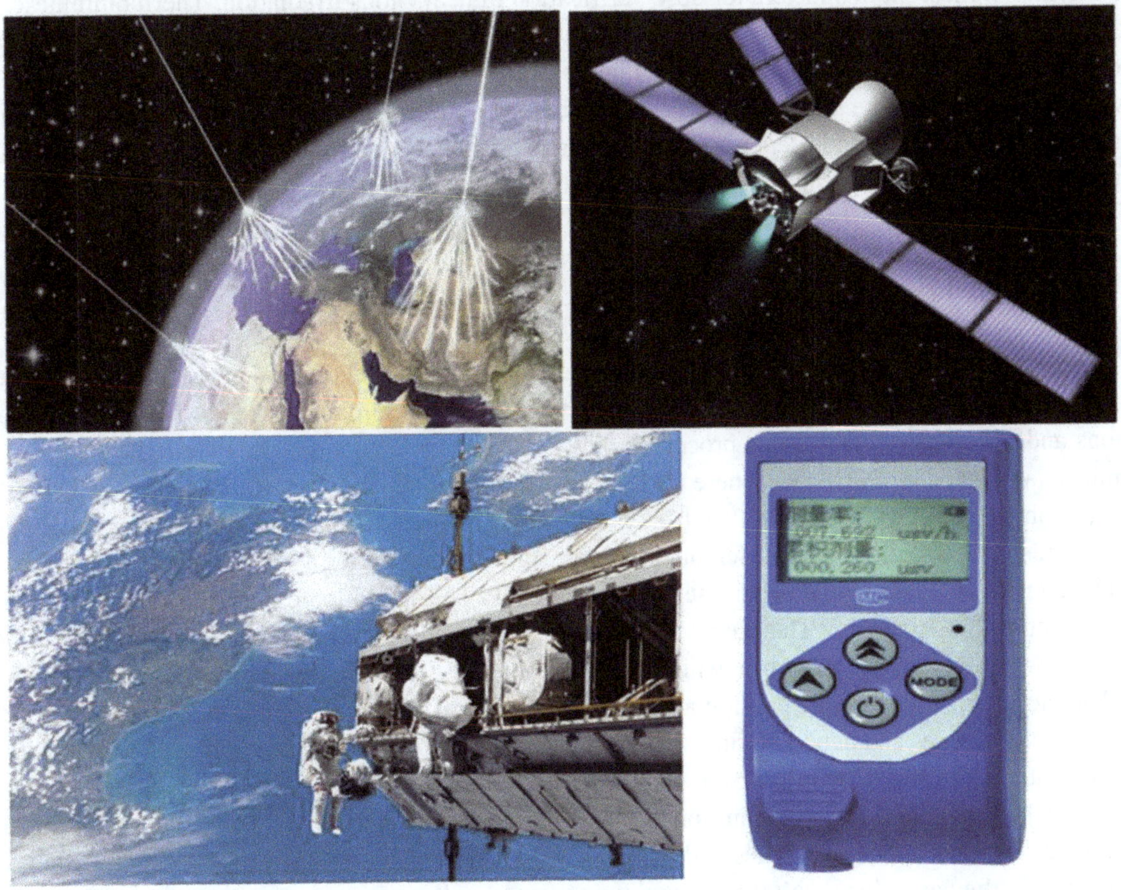

characteristics of these phosphors by preparing them using different methods or by doping with different impurities or developing some new ones. However, all these phosphors have 'dose-ranges' depending on their 'sensitivity' and 'response characteristics' (linearity and saturation) to high-energy radiations. At low doses, very low signal to noise makes it difficult to estimate doses, while at higher doses the saturation of the TL signal makes it difficult. For example, Hasan et al.have suggested the use of the high-temperature peak of $CaSO_4$: Dy for estimating such high doses of γ-rays, but it is also not very accurate and convenient as there is not much variation in TL intensity with doses, undesired changes in the glowing structure also makes it difficult.(Hasan,Kitis, Charalambou, 1985) Moreover, the black body radiation from the heating plate may add some inaccuracies at higher temperatures.

Recently, heavy ion beams are widely used in space dosimetry applications, radiotherapy and diagnostic.(Salah, 2011) The cancer treatment with such highly energetic ion beams has several advantages in comparison with the conventional irradiation with photons. The physical advantage is a significantly more favorable dose deposition profile. Whereas with conventional irradiation the dose deposition decreases in proportion to the penetration depth, with ion beams it gradually increases, and then declines rapidly beyond a sharply defined maximum known as the Bragg peak near the end of therange of ion beam. Therefore, the penetration depths of those ions can be projected to have a maximum absorption at the tumors positions. However, the dose of those energetic ions needs to be measured with great precision and accuracy, especially, while dealing with human beings. This has triggered investigations to use thermoluminescent dosimeters for dose verification in heavy ions irradiation. Thermoluminescent materials exhibit differences in dose response between sparsely ionizing radiation like X-ray or γ-ray and densely ionizing radiation like heavy charged particles (HCP). This is due to different spatial dose distributions, which cause saturation effects in the case of densely ionizing radiation.(Salah, 2011) Ion beam effects on the materials depend on the ion energy, fluence, and ion species. When energetic ion penetrates through the material it loses its kinetic energy in two modes:

1. Nuclear and
2. Electronic stopping.

The nuclear stopping is dominant at low energies (keV) and the energy lost in this process is called nuclear energy loss. The other mode of energy loss is by exciting or ionizing the atom by inelastic collisions and the energy spent in this process is called as electronic energy loss. Electronic stopping is dominant at high energies (MeV), where the displacement of atoms due to elastic collisions is insignificant.(Salah, 2011) There are several studies on different TLD materials aimed at the understanding of their TL response/ efficiencies to different ion beams irradiations.(Gnaser, 1999) However, most of these TLD materials are not capable of precisely detecting HCP. The saturation effect is the major problem, which occurs at relatively low fluences. Moreover, there is a significant variation in the TL glow curves structure with an increase in fluence, which is undesirable for the use in dosimetry.

Therefore, this chapter will provide a basic introduction of swift heavy ion, thermoluminescence, dosimetric properties and nanomaterials. The intrinsic properties of swift heavy ion and the application for the materials synthesis will be discussed in details. The description of nanomaterials synthesis will also be elaborated on. The basic definition of thermoluminescence and defect related relaxation will be discussed. The dosimetric properties and its effect will also be given in this chapter. Finally, the relation between the heavy ions with the thermoluminescence will be evaluated on the basis of dosimetry applications.

SWIFT HEAVY ION AND RELATED PHENOMENA

Ion Beam Interaction

The research field of ion beam physics started more than a hundred years back with the discovery of Electron beams and beams of positive ions by Thomson in 1897 and Goldstein in 1902 respectively. Accelerated ion beams are classified into two broad categories depending upon the energy imparted to the materials:

- Low-energy and
- High-energy heavy ion beams

Low energy ion beams have an energy ranging from a few keV to a few hundreds of keV, which lose energy to the material (up to 10 keV/ amu) through elastic collisions with the nucleus or atoms of the material. This type of energy loss is called ‘nuclear energy loss’ $(dE/dx)_n$.(Hodgson, Balgin, Pal, Neri, Hammer, 1980) The high energy heavy ions are known as ‘swift heavy ions (SHI)’ have energy ranging from a few tens of MeV to a few-GeV and lose energy to the target (≥ 1 MeV/amu) predominantly via inelastic collisions with the target material.(Nastasi, Mayer, Hirvonen, 1996) This type of energy loss is called ‘electronic energy loss’ $(dE/dx)_e$. (Avasthi, Mehta, 2011)

The effect of SHI is different in different classes of materials. It is more prominent in insulators than in metals and semiconductors.(Mohanta, Mishra, Choudhury,2004) Different models were proposed to explain the effect of SHI in materials, the Coulomb explosion (Dufour, Khomenkov, Rizza,Toulemonde,2012) and thermal spike model.(Krauser, Zollondz, Weidinger, Trautmann, 2003)

In the Coulomb explosion model, the target atoms are ionized by electronic excitation during the passage of SHI through the materials. As a result, a cylinder of positive ions is created along the path of ions. These positive ions are mutually repulsive and explode radially due to theconversion of electrostatic energy to coherent radial atomic movements under Coulomb forces until the ions are screened by conduction electrons. As a result, ion tracks are formed along the path of the projectile ion due to radial Coulomb explosion. This Coulomb explosion process is predominant in insulators where there are little conduction electrons (Figure2).

According to the thermal spike model, SHI transmits its energy to the electrons of target materials as kinetic energy via inelastic collision. This kinetic energy is transmitted to the lattice by electron-phonon interactions, resulting in the increase of lattice temperature above the melting point of the material, followed by rapid quenching. This leads to the formation of an ‘amorphous track’ along the path of the projectile ion when the melt solidifies.(Schiwietz, Staufenbiel, Roth, Czerski, Grande, 2005) The passage of SHI through a material is shown in Figure 3.

Intense electronic energy deposition is responsible for the modification of the material properties. It can produce columnar defect, point defect and even amorphous tracks along the path of its motion. SHI is also used to produce materials with novel properties, which cannot be generated by any other means. Using ion beam irradiation, one can create new defects as well as to anneal out pre-existing defects in materials. This remarkable ability of SHI provides unique opportunity to engineer structural, optical and optoelectronic properties of phosphor materials. Recently, heavy ion beam irradiations are extensively used in radiotherapy and diagnostic for the dosimetric purpose.

For both Figure2 and 3 the sky-blue dots indicate the target materials, the orange dot implies the SHI. The heated zone is shown as yellow and red.

Ion Beam Assisted Synthesis

Recently, nanotechnology has become one of the fastest emerging innovative aspectsin science and engineering. The topiccrops up from the unification of materials science, physics, chemistry, biology, and electronicsto constructnovel well-designed systems of nanoscale proportions generally allied with dimensions in the range of 0.1 to 100 nm. However, the synthesis of metals nanoparticles along with various semiconductors materials with the help of irradiation technology is now well recognized and the reactionmechanism is realistically implicit. Among all the approaches, 'top-down' methodologyis still now dynamically encouraged by the scientists. With the help of the concept of this approach, the electron beamnanolithography and reactive ion etching are used to fabricate different nanodevicesnowadays via the deposition of the energy of radiation on thematerials. Ion track membrane is another illustration of the formation nano-sized cylindrical structure by ionbeam on the plastic film which is potentialfor chemical detectors and biosensors. (McKinlay, 1981) Ionizing radiation such as gamma radiation and electron beam has been used widely in industryfor crosslinking of polymer, polymer blend and composites also. This section describes that the radiation technique is essential to thefabrication of nanostructures with

Figure 2. Schematic representation of the Coulomb Explosion model
Aumayr, Facsko, El-Said, Trautmann, Schleberger, 2011

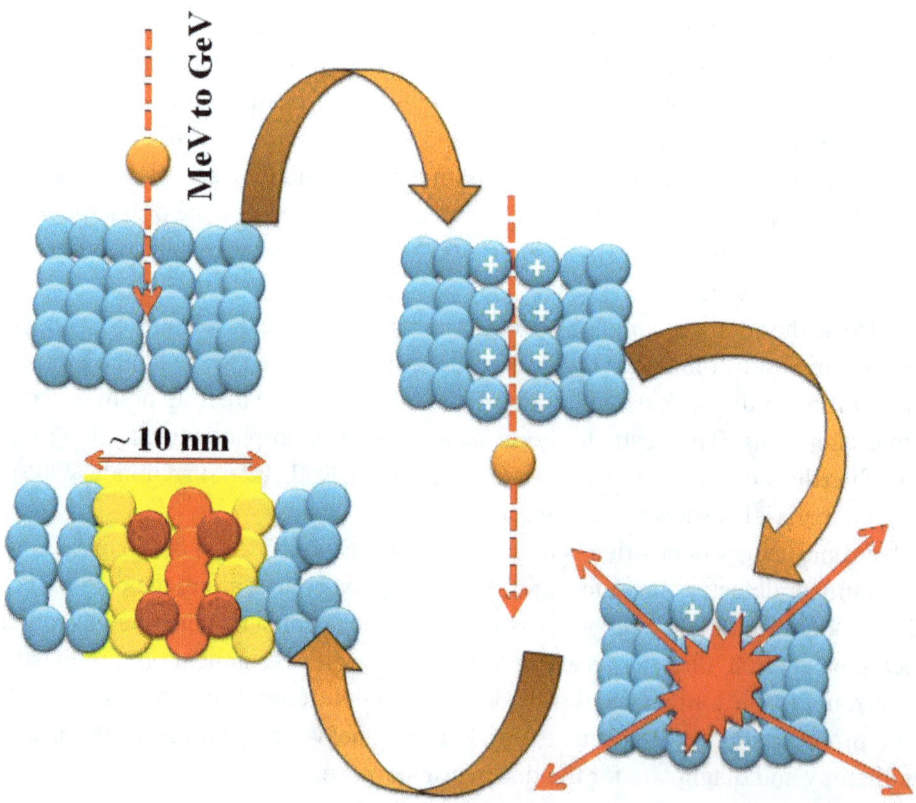

Figure 3. Schematic representation of the Thermal spike model
Schiwietz, Staufenbiel, Roth, Czerski, Grande, 2005

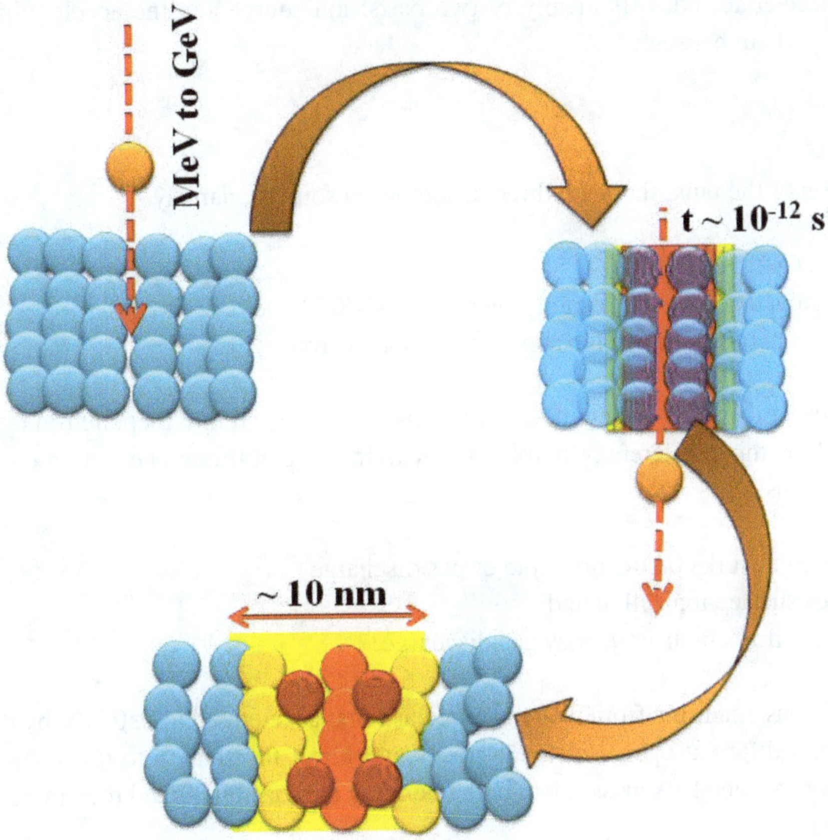

high resolution and a highaspect ratio owing to the fact that the radiation beams can be focused on a few nanometers or less and scanned withquite high speed.(Kellerer, 1985) This discussion indicates the ability of the swift heavy ions for the synthesis of different nanomaterials including phosphors.

Swift Heavy Ion Beam Irradiation

Swift heavy ion can be generated from the Pelletron Accelerator. In the present chapter, a 15 UD Pelletron Accelerator located at IUAC New Delhi is described for the production of theswift heavy ion beam. (Kraft, Schattat, Bolse, 2002) The schematic diagram of the IUAC Pelletron and its different parts are given in Figure 4.

Characteristics

- Tandem electrostatic accelerator of Van de Graff type (Avasthi, 2002) in vertical configuration
- It can go upto a maximum terminal voltage of 16 MV and
- It is capable of accelerating any ion from proton to uranium upto energy of around 200 MeV

Pelletron Accelerator

The pelletron accelerator consists mainly of two parts: ion source and the accelerating column with many auxiliary parts in between.

Ion Source

In the top portion of the tank, there are three different ion sources, namely

- R.F. source (ALPHATROSS)
- Source of Negative Ions by Cesium Sputtering (SNICS) and
- the direct extraction negative ion source (Duoplasmatron)

These sources can produce different negative ions to be injected into thepelletron tank. These three ion sources work on three different principles and with the help of these; one can produce almost every type of negative ions:

- Duoplasmatron works on the principle of gas discharge
- SNICS uses surface ionization and
- R.F. source utilizes high-frequency gas discharge.

The negative ions emerging from the ion source are first accelerated to 250 keV by the high voltage deck potential and different optical elements focus and inject the same into the vertical accelerating tube. The injector magnet does mass selection by bending the ions by 90°and then injects them into the accelerating tube.

Accelerating Column

The singly ionized negative ions then follow a vertically downward path and get accelerated through the accelerating tube path. It consists of a vertical insulating cylindrical tank of height 26.5 m and diameter 5.5 m, filled with SF_6 (Sulphur hexafluoride) at high pressure (>200 psi) as an insulating gas. The tank consists of the high voltage accelerating terminal of height 3.18 m and diameter 1.52 m. The terminal is connected to the tank vertically through ceramic titanium tubes called accelerating tubes. A potential gradient is maintained through these tubes from high voltage to ground, from the top of the tank to the terminal as well as from the terminal to the bottom of the tank. The insulating column which supports the high potential terminal consists of thirty 1 MV modules, 15 on either side of the terminal. The upper portion of the column is referred to as low energy section and the portion below the terminal as the high energy section. The shorter section with no potential gradient, commonly known as the Dead Section, is provided each, in the low and high energy sections for equipment housing. Both are provided with an electron trap and a sputter ion pump. The low energy dead section (LEDS) is also provided with an electrostatic quadrupole lens while the high energy dead section (HEDS) is equipped with a second foil stripper assembly. A shorting rod system is also provided for temporarily shorting selected column modules without entering the pressure vessel. Two insulating shafts run one from each ground end to the terminal and are used to drive four 400 cycles per second (cps) generators, which provide power for the equipment like heater lenses, pumps, foil changer etc., housed in the column, dead sections and terminal. The charging of high voltage terminal to 15 MV is done by using the pelletron charging chains. (Avasthi, 2002)

There are two independent subsystems with one charging chain in each so that each chain is required to supply 100 μA current. Negative ions from the ion source are focussed by different optical instruments and then injected by the injector magnet into accelerator with some small energy (~300 keV) provided by the negative potential deck at the ion source. In the terminal, negative ions are stripped of a few electrons by the positive ions which are again accelerated as they proceed towards the bottom of the tank at the ground potential.

The beam energy gained by the ion in the accelerating tube at the end of the tube is determined by,

$$E_{beam} = (q + 1)V_T \tag{1}$$

where V_T is the terminal voltage, q is the number of positive charges on the ion after stripping.(Avasthi, 2002)

These high energy ions were analyzed to the required energy with the help of 90°bending magnet known as the analyzer magnet and directed to the desired experimental area with the help of a switching magnet which can deflect the beam into any one of the seven beamlines in the beam hall for the experiments.

Generally, the irradiation of the samples is accomplished via the formation of thin films and pellet. Stopping and Range of Ions in Matter (SRIM) computer program (Ziegler, Biersack, Ziegler, 2008) is generally used to obtainthe range of the ions inside the materials. And the pellet width and thin film width is made greater or less than the above-mentioned range for ion irradiation or implantation, respectively. Some special arrangement is also be made for the powder sample for which the range is very less and it is quite impossible to make a pellet. The whole procedure along with the special arrangement for the ion irradiation is described below.

Special Arrangements for Irradiation of Powder Sample

- An aluminum strip with five holes of diameter 1 cm and inner thickness 0.5 mm are generally used.
- One side of the strip is covered with the silver foil. This layer acts here as a conducting layer. If any secondary electron will be produced during irradiation it will be absorbed by this layer and thus help in irradiation process. A suitable quantity of powder depending on the projectile range is put into the holes on the other side of the strip and is gently pressed with a solid cylinder of thesamediameter so that it does not fall out from the strip as shown in Figure 5. In this figure, the schematic diagram of an Al strip with silver foil is shown and this stripe is attached to the ladder.
- The strip is then carefully mounted on the sample holder having four faces fitted on the rectangular ladder of a scattering chamber connected to a beam line.
- The four faces are indicated as A, B, C, D and in this figure, only C side is shown. A circular quartz plate is shown in the ladder where a square is marked to indicate the ion irradiation area.
- This indication helps to mark the ion irradiation area on the TV screen and we can easily put our sample within this area for irradiation.

Figure 4. Schematic diagram of 15 UD Pelletron Accelerator
Avasthi, 2002

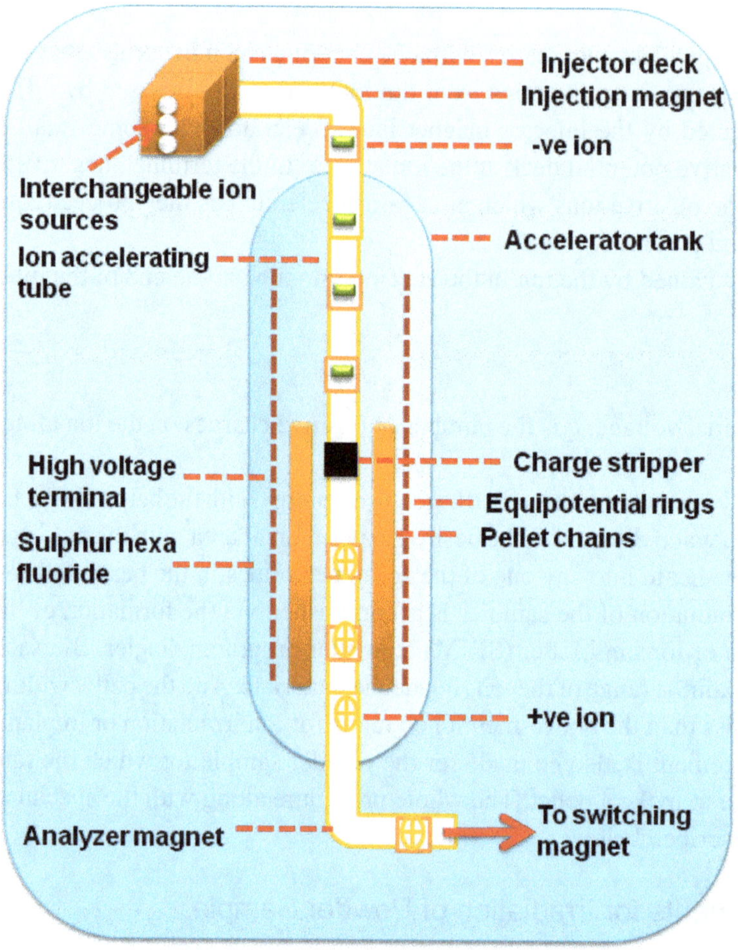

The ladder can be moved up-down or rotated about the vertical axis in order to bring the samples into the path of theion beam for irradiation. A pressure of 10^{-6} Torr was maintained inside the chamber during ion irradiation. After irradiating a sample to the desired fluence, the beam was turned off and a new sample was brought into the position for irradiation.The idea about the fluence range and different species of SHI will be clear from the previous literature.The present research group has also reported the same in aliterature on the comparative study of luminescence modifications of Y_2O_3:Tb^{3+} phosphor persuaded by 150 MeV Ni^{7+}, 120 MeV Ag^{9+} and 110 MeV Au^{8+} swift heavy ions in the fluence range 1×10^{11}- 1×10^{13} ions/cm^2. (Som et al., 2014a)

Gamma Irradiation Facility

Along with the swift heavy ion irradiation, a short description of gamma- ion irradiation is also necessary for this chapter. A different model for gamma irradiation can be observed and in this chapter, The Gamma Chamber (model GC-1200 made by BRIT, Mumbai) installed at IUAC, New Delhi is described for Gamma irradiation of prepared phosphors. In the same literature discussed before(Som et al., 2014),

Figure 5. Special arrangement for powder sample irradiation
Som et al., 2014a

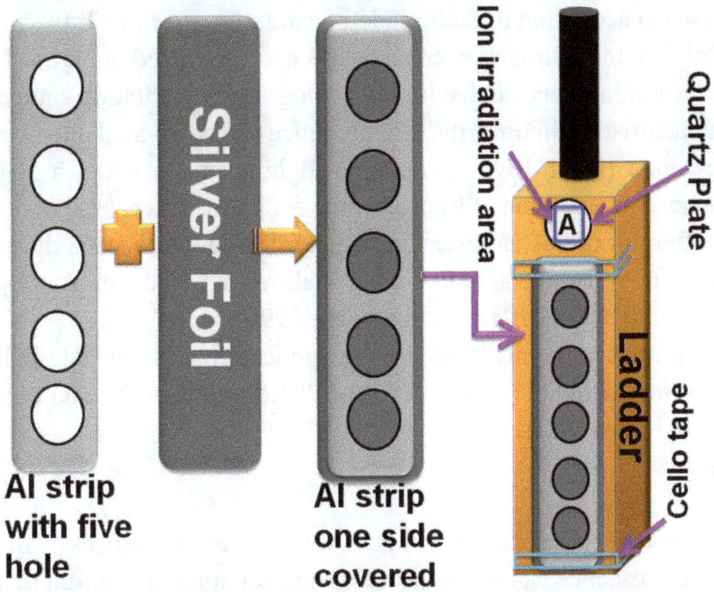

the present authors used gamma-ray for the excitation of phosphor materials to get the idea about the TL behavior of the materials. In that work, the present research group used the GC-1200 unit designed to house up to 185 TBq (5000Ci) of theCobalt-60 source.The lead shielding in this instrument is adequate enough to bring down the radiation leakage level on the exterior of the unit well below the accepted standards. It offers a dose rate of about 7.5KGy/ hr. The lead shielding brings down the radiation leakage level on the exterior of the chamber, well below the accepted standards. In that work, the dose for γ-ray was chosen as 5 kGy. The chamber consists of the following parts:

- Source and cylindrical source cage
- Lead Flask as Biological shield
- Central drawer with sample chamber
- Mechanical driver system
- External cabinet

Now the clear understanding of the phosphor application necessitates the deeper knowledge of luminescence phenomena which is the backbone of the present chapter.

Luminescence

'Luminescence', the name originates from the Latin word lumen that means light. The term "luminescenz" was first used in 1888 by the great German physicist, Eilhardt Wiedemann. An excitation source kicks a ground state electron to a higher excited state and when it returns to its original state if it will emit a part of the given energy as photon then this process is known as luminescence. It is often described as 'cold light' to distinguish it from incandescent light emission. In the last two centuries, more observations and discoveries were made including luminescence from rare earth and transition metal ions doped inorganic materials. These materials were given the name 'phosphors' that absorb, store energy and lastly convert it into thelight.(Blasse, Grabmaier,1994)Luminescent materials can be excited by different types of energy and are therefore classified according to their mode of excitation given in Table 1.

The applicability of different luminescence materials is summarized in Figure 6.(Ronda, 2008) This figure displays different luminescence behavior depending on the excitation sources.

On the basis of characteristics lifetime, the luminescence can be divided into two categories: fluorescence and phosphorescence. (Kitai, 1993) Emission of light during or within $\tau_c \sim 10^{-8}$ sec of excitation is known as Fluorescence and emission of light after the excitation has ceased ($\tau_c > 10^{-8}$ sec) is known as Phosphorescence or Afterglow. Phosphorescence differs from fluorescence in that the phosphorescence is always associated with trapping of carriers at rather shallow traps and their subsequent thermal release prior to their radiative combination.(Blasse, Grabmaier, 1994)

The present research work is mainly based on the application of solid state lighting and dosimetry. These two applications are mainly the consequence of Photoluminescence and thermoluminescence.

Photoluminescence

In luminescence, if the source of excitation will be a photon then this process will be known as photoluminescence. Photoluminescence can be described as an excitation of electron to a higher energy state by absorbing photons (electromagnetic radiation) and then return to a lower energy state accompanied by the emission of a photon in the visible region. The photoluminescence can be recorded by recording two types of spectral excitation and emission (Figure7).

In the case of an excitation spectrum, the wavelength of theexcitation source is varied and the intensity of the emitted light at a fixed emission wavelength is measured as a function of the excitation wavelength. The excitation spectrum gives information about the possible excited states of the phosphor and hence the value of the band gap of thephosphor. For excitation, a different light source can be used(Blasse, Grabmaier, 1994)

- Incandescent Lamps for visible and infrared radiation
- Arc lamp such as Xenon lamp, Mercury Lamp, Deuterium lamp for UV/ visible/ infrared radiation.
- Lasers for different wavelength regions

Table 1. Luminescence Phenomena(Ronda, 2008)

Phenomena	Cause of Excitation
Radioluminescence	Nuclear radiations i.e. γ- rays, β-particles, α-particles etc.
Cathodoluminescence	Electrons beam
Electroluminescence	Electric field
Chemiluminescence	Chemical energy
Triboluminescence	Mechanical energy
Sonoluminescence	Sound waves
Bioluminescence	Biochemical energy
Thermoluminescence	Any type of radiation followed by the release of energy on heating the sample

Figure 6. Different luminescence phenomena
Ronda, 2008

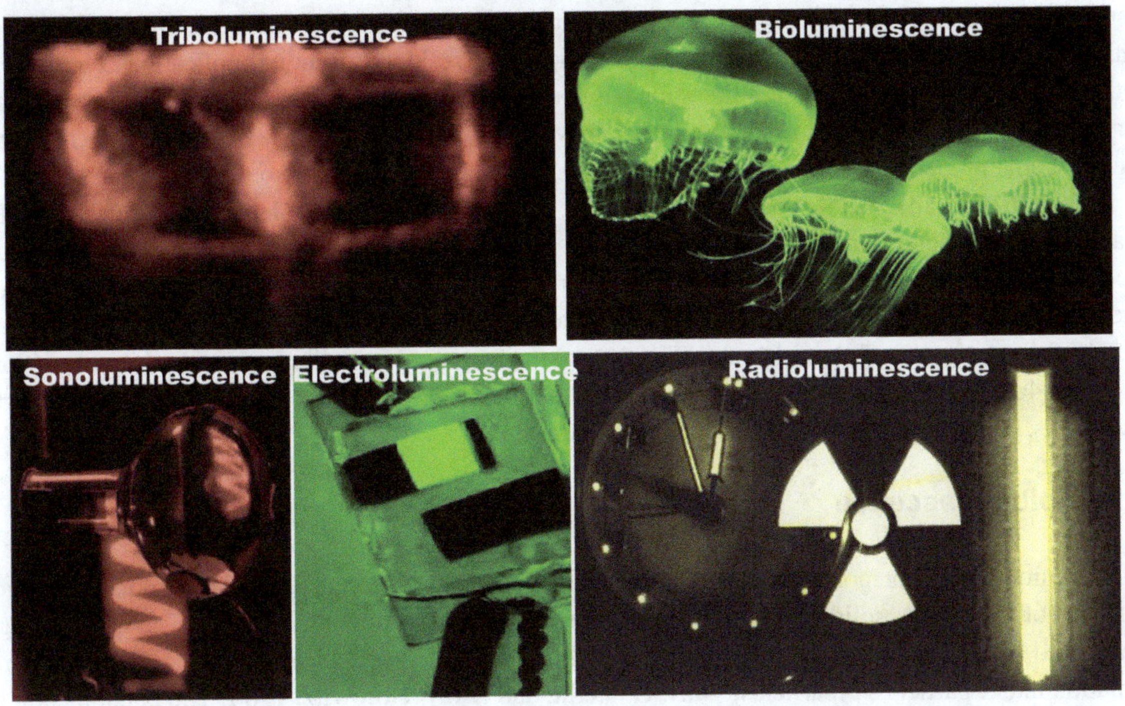

Figure 7. Schematic diagram of the photoluminescence phenomena
Ronda, 2008

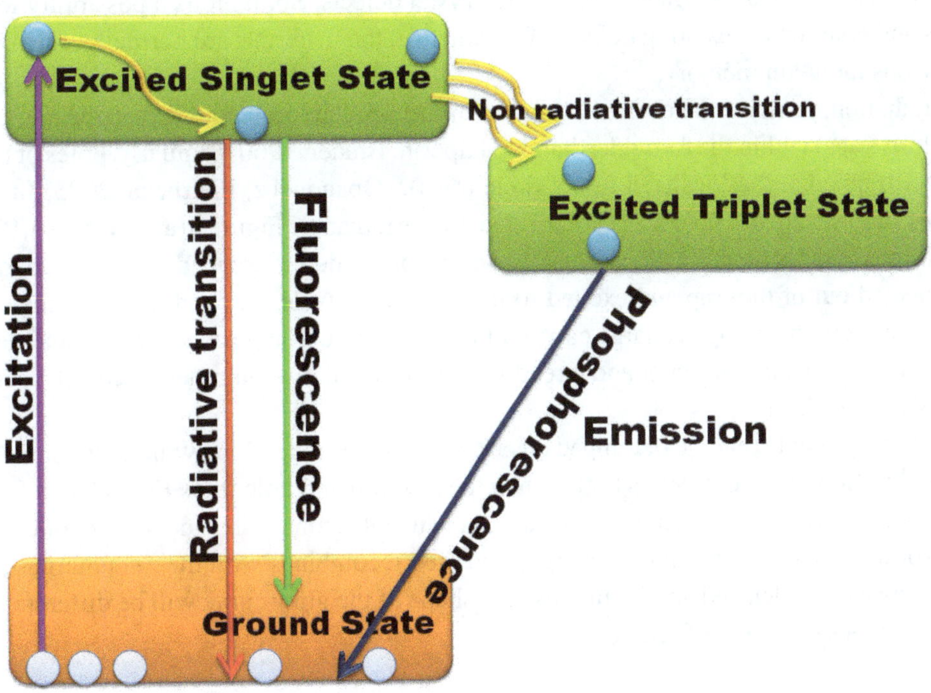

The emission spectrum gives the information about the radiative transition of the electron to a lower excited state or ground state from a higher energy state and also provides information on the spectral distribution of the light emitted by a sample. This process is shown in Figure4.Depending on the color emission, the materials are categorized in different display and solid state lighting applications.The present author's group has published a lot of literature based on photoluminescence emission and their transition characteristics. In one research work, the authors have reported the defect correlated self-quenching and spectroscopic properties of calcium titanate ($CaTiO_3$) phosphors based on photoluminescence. (Som et al. 2014b) In this paper, Eu^{3+} doped phosphors were illuminated via ultraviolet light which showed the characteristic red photoluminescence corresponding to the $^5D_0 \rightarrow ^7F_J$ transitions of Eu^{3+}. In another paper, the authors have presented the photometric characterization of Eu^{3+}/Dy^{3+}coactivated yttrium oxysulfide phosphor for near white light emission.This paper also comprised the photoluminescence phenomena based on the energy transfer mechanism. The decay kinetics was also well explained in this paper.(Som et al. 2014c) These two papers will provide a basic understanding of photoluminescence, rare earth transitions, decay kinetics and energy transfer phenomena.

Thermoluminescence

Thermoluminescence or more aptly called thermally stimulated luminescence (TSL) is the phenomenon of emission of light from a solid which has been previously exposed to ionizing radiation. The stored energy is released under conditions of increasing temperature. Excitation is achieved by exposure to ionizing radiation like α-particles, β-particles, γ-ray, UV light, and X-ray.(Horowitz, 1984) The phenomenon of thermoluminescence can be explained on the basis of the band theory of solids. (McKinlay, 1981) The forbidden energy gap of the solid contains many localized levels corresponding to various defects such as vacant lattice sites, interstitial atoms, and impurity atoms. The electron and hole traps can already be present in the material prior to irradiation, due to crystal defects. So, there is a possibility for trapping of electrons and holes on the existing defects. To avoid this, the materials are normally annealed to erase out the previous radiation memory.

After irradiation, some of the excited electrons in the conduction band get trapped at the localized levels (electron traps) within the forbidden band. (Halperin, Braner, 1960) Similarly, holes of the valence band also get trapped at the localized levels (hole traps). (Opanowicz, Pietrucha, 2003) In this way a large number of electron and hole traps are formed after irradiation Figure8 (a). (Watnabe,1951)

When the material is heated either the electron or hole which is trapped with less energy will be thermally ejected out of the trap and excited to the conduction/valence band, move in the crystal and finally recombines with a trapped charge carrier of opposite sign. The place where recombination takes place is called a recombination center and if recombination is radiative, luminescence will occur Figure8 (b) & (c).(Watnabe,1951)

It may also be possible that the detrapped electrons and holes recombine with their counterpart at an entirely new site (luminescence center) other than the electron and hole traps Figure8 (d). In general, a detrapped charge carrier may become retrapped from which it again excites into the conduction/valence band by absorbing energy E (activation energy) and then recombines with their counterpart, so the luminescence emission is delayed and in this case the shape of the glow curve will be different to the case when retrapping does not occur Figure8 (e).

Figure 8. Schematic representation of the TL phenomenon
Watnabe,1951

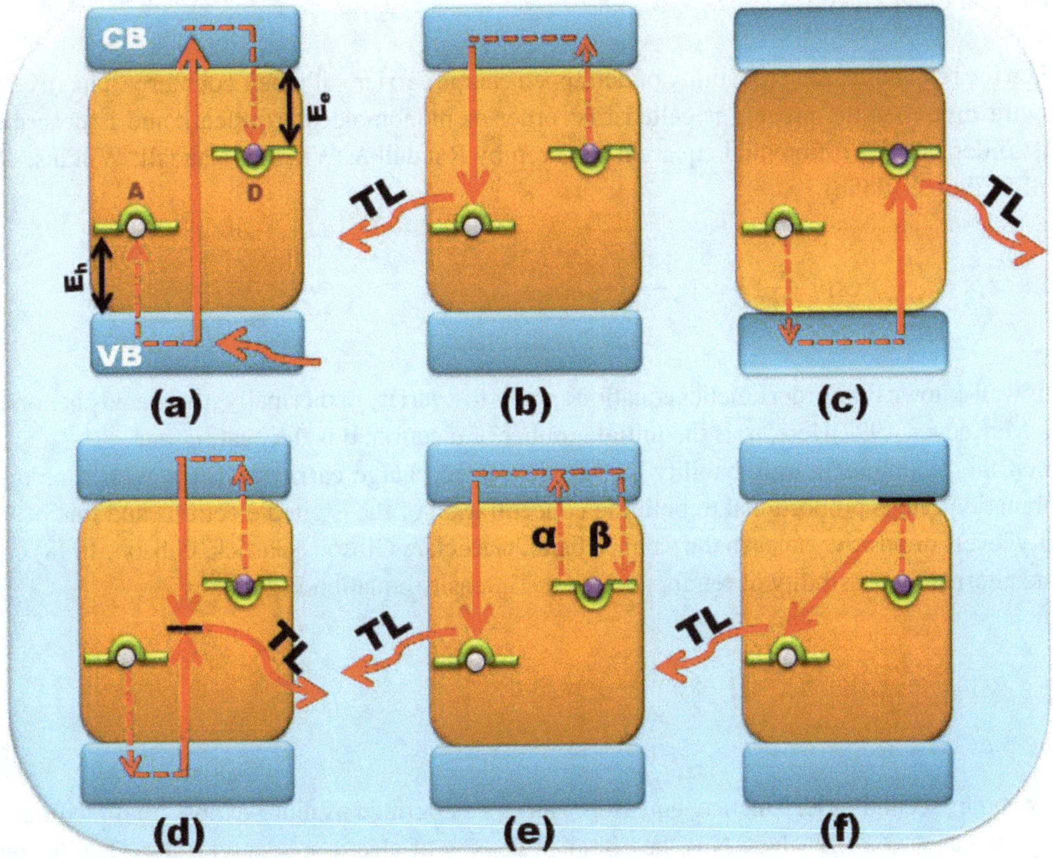

It is also possible that detrapped charge carriers recombine directly with their counterpart without exciting into conduction/valence band. This type of recombination is called tunneling Figure8 (f).(Watnabe,1951)

The various peaks observed in a TL spectrum may be attributed to trap levels associated with the intrinsic and extrinsic defects present in the material.(Horowitz, 1984) The probability (p) per unit time for thermal excitation from the trap is exponentially dependent upon temperature according to the relation:

$$p = s \exp\left\{\frac{-E}{kT}\right\} \tag{2}$$

where s is a constant known as frequency factor with dimensions of reciprocal time and E is the activation energy or depth of the trap.

The exact equations of the TL process depend on the mechanism of recombination of detrapped charge carriers with their counterpart. On the basis of this, thermoluminescence phenomena can be studied under the three types of kinetics:

- First order kinetics (slow retrapping)
- Second order kinetics (fast retrapping)
- General order kinetics

When the recombination probability of detrapped charge carrier with their counterpart has a constant value with time then the process is called first order or monomolecular process and is described by the first order kinetics differential equation as given by Randall & Wilkins:(Randall, Wilkins, 1945a; Randall, Wilkins, 1945b)

$$I = n_0 s \exp[-(\frac{E}{KT})] \exp[-(\frac{s}{q}) \int \exp(-\frac{E}{KT}).dT] \tag{3}$$

This well-known First order kinetics equation applies to a variety of thermally stimulated phenomenon. (Chen, McKeever, 1997)Here, n_0 is the initial number of electron, q is the heating rate.

When the recombination probability for the detrapped charge carriers change with time and the recombination rate is proportional to both the concentration of the excited electrons and that of vacant impurity levels or positive holes in the valence band, Garlick & Gibson,(Garlick, Gibson, 1948) considered the alternative possibility of retrapping and the intensity equation is given by:

$$I = n_0^2 s' \exp(-\frac{E}{KT})[1 + n_0(\frac{s'}{q}) \int_{T_0}^{T} \exp(-\frac{E}{KT}) dT]^{-2} \tag{4}$$

This is the 'second order kinetics' equation. Here, s' is defined as a pre-exponential factor and can be expressed as $s' = s / N$ where N is the effective density of electron. Second order TL glow peak is wider and more symmetric than a first order TL glow peak. It is because of the fact that in a second order kinetics, a significant number of released electrons are retrapped before they recombine, in this way giving rise to a delay in luminescence emission and a spreading out of the emission over a wider temperature range.

The general order kinetics equation was given by May & Partridge (May, Partridge, 1964) to describe nearly all the experimental TL glow peaks. The first and second order kinetics equations are just the special case of this general order equation given by:

$$I = \frac{-dn}{dt} = s' n^b e^{-E/kT} \tag{5}$$

where b is the kinetic order, which can have values other than 1 or 2 and $s' = s/n_0^{b-1}$. Solving this equation for $b \neq 1$ by assuming a linear heating rate q gives:

$$I = s n_0 \exp(-\frac{E}{KT})[(b-1)(\frac{s}{q}) \int_{T_0}^{T} \exp(-\frac{E}{KT}) dT + 1]^{-\frac{b}{b-1}} \tag{6}$$

This equation includes the second order case (b = 2) but is not valid for b = 1.

Farrington Daniels first proposed the use of thermoluminescence phenomenon as a practical research tool in several different fields of application such as dating in archeology and geological objects, dosimetry such as Radiation dosimetry, Personnel dosimetry, Environmental Dosimetry, Medical dosimetry etc. and in the identification of defect levels inside the materials.(Chen, McKeever, 1997)

Trapping Parameters

To get the idea about the dosimetric behavior of the materials, the idea about the trapping parameters is required. (Opanowicz, 2007) Trapping parameters of trap states are investigated by avariety of methods based on the observation of thermoluminescence glow curve.(Sunta, Ayta, Chubaci,Watanabe, 2001; Haering, Adams, 1960) A thermoluminescence glow curve may consist of several peaks. The positions, shapes, and intensities of the glow peaks depend on various parameters of the trapping states, which are responsible for the TL phenomenon. (Opanowicz, 2000)The most important parameter to be determined are the trap depth or activation energy (E), which is the thermal energy required to liberate the trapped electrons or holes, the frequency factor (s) which represents the product of the number of times an electron hits the wall and the wall reflection coefficient, treating the trap as a potential well(Opanowicz, 2007) and the order of kinetics (b) obeyed by the TL process. Various experimental methods are currently used for determining trapping parameters and are classified according to the way in which the glow curve data are utilized. (Mckeever, 1985) These methods are:

Computerized Glow Curve Deconvolution

Computerized glow curve deconvolution technique is an efficient technique for the evaluation of trapping parameters. When the glow curves are composite in nature, deconvolution of the peaks is needed. Computerized glow curve deconvolution is widely used for studying the TL mechanism as well as for TL dosimetry. (Mckeever, 1985) In the present studies, the TLanal computer program (Chung, Choe, Lee, Kim, Chang, 2005) was used for glow curve deconvolution. This is a quick and efficient computer program in order to resolve the peaks from the composite TL glow curve.

This computer program is based on Halperin and Braner equations which describe the flow of the charges between the various energy levels during a trap emptying by thermal heating. Chung et al.(Chung, Choe, Lee, Kim, 2007) derived the equation (7) for TL intensity by applying some physical assumptions to Halperin and Braner equations. Assuming the free carrier concentration in the conduction band (n_c) to be very small compared to trap carrier concentration (n) and also the rate of change of n_c to be much smaller than the rate of change of n, the TL intensity can be expressed as:

$$I(t) = -\alpha \frac{dn}{dt} = \alpha \frac{n^2 p}{n + R(N - n)} \tag{7}$$

where R is the ratio of the retrap and recombination coefficient, N is the concentration of the available electron traps of depth E below the conduction band and α is a scaling factor of the detecting system. Here, p = s exp(−E/kT)is defined as the probability per unit time of the release of an electron from the trap and sis the frequency factor for the escape from the traps to the conduction band. In the case of the linear temperature rise, equation (7) is called the general one trap (GOT) equation. Although this method is based on the GOT model, there is no restriction in the temperature profile; this program can be used in the case where the temperature changes linearly as well as arbitrarily. The software works for the deconvolution of the TL glow curves having mixed order, first-order, second-order and general order kinetics. It is shown that this method/software is efficient and fast in generating TL glow peaks and can be adopted into a numerical curve fitting.(Chung, Choe, Lee, Kim, 2007)

The kinetic parameters of trap levels in maximum γ-dose/ion fluence irradiated phosphors were determined for each deconvoluted peak by this computer program. The theoretically generated glow curves were fitted with the experimental glow curves and the quality of fitting was checked by calculating the figure of merit (FOM) for each fitting defined by

$$FOM = \frac{\sum TL_{Exp} - TL_{The}}{\sum TL_{The}} \tag{8}$$

Here TL_{Exp} and TL_{The} represent the TL intensity of experimental and theoretical glow curves respectively. The summation extends over all the available experimental data points. The quality of fitting and choice of an appropriate number of peaks was refined by repeating the process of fitting in order to get the minimum figure of merit (FOM) with aminimum number of possible peaks. The fits were considered adequate when the FOM values were observed below 5%, with most actually being below 2%.(Chung, Choe, Lee, Kim, 2007) In the present studies, FOM was found around 1% which confirms a very good agreement between theoretically generated and experimentally recorded glow curves.

GLOW CURVE SHAPE METHODS

The various methods based on the shape of the glow curve were used to calculate the trapping parameters of the thermoluminescence glow peaks. (Furetta,2003)

Order of Kinetics

According to Chen, (Chen, 1969a) the following shape parameters are required to determine the order of kinetics: total half intensity width ($\omega = T_2 - T_1$), high-temperature half width ($\delta = T_2 - T_m$) and low-temperature half width ($\tau = T_m - T_1$). Here T_m is the peak temperature, T_1 and T_2 are the temperatures on either side of T_m corresponding to half peak intensity as shown in Figure 9. Chen introduced the symmetry factor (μ_g) to differentiate between first and second order TL glow peaks:

$$\mu_g = \delta / \omega = (T_2 - T_m) / (T_2 - T_1) \tag{9}$$

$\mu_g = 0.42$ for first order kinetics

$\mu_g = 0.52$ for second order kinetics

Figure 9. Schematic representation of a thermoluminescence glow peak (Chen, 1969a)

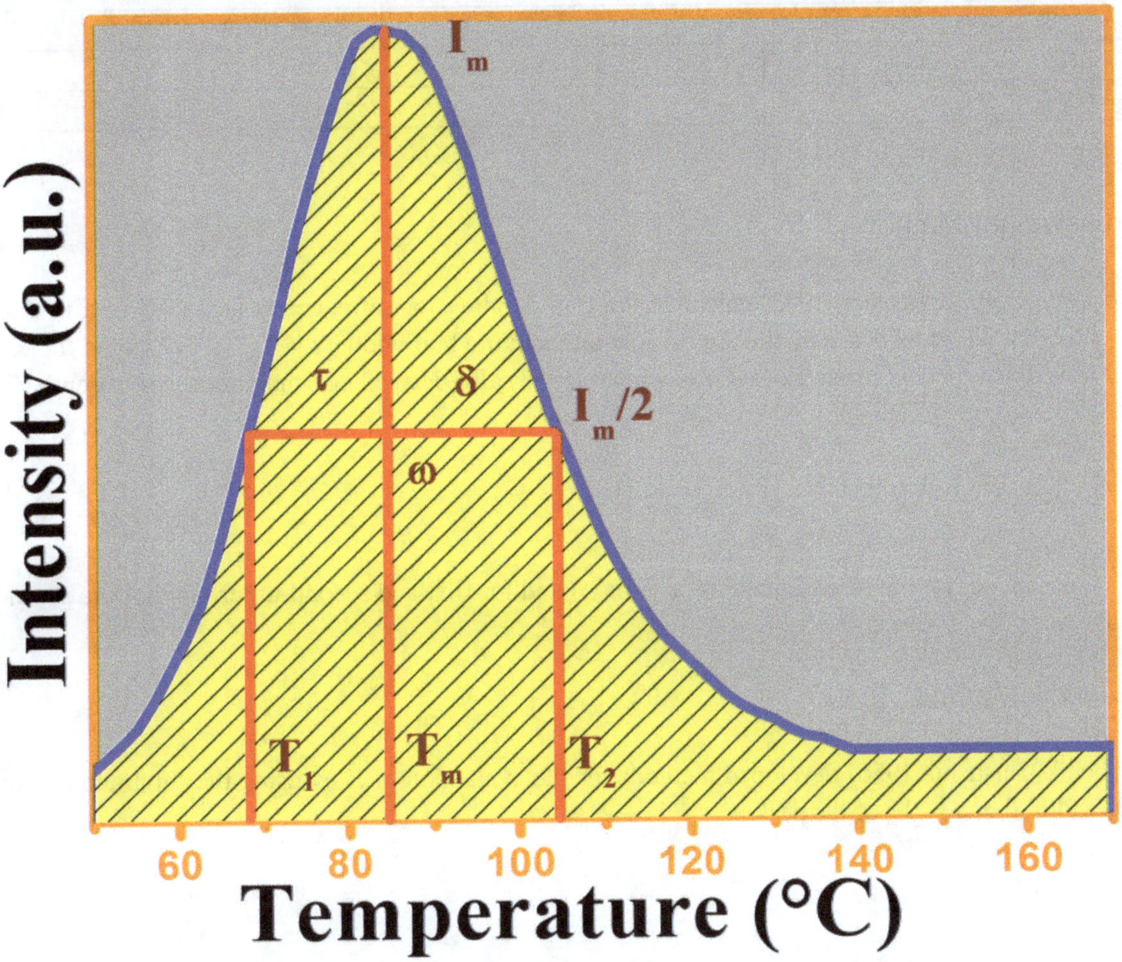

Trap Depth

Chen Method

Chen (Chen, 1969b) has given three equations each for first, second and general order peaks, to evaluate trap depth by using either the 'full width of the peak at its half-height' $\omega = T_2 - T_1$, its 'low temperature half width' $\tau = T_m - T_1$ or its 'high temperature half width' $\delta = T_2 - T_m$. A general formula for E was given by:

$$E = c_\gamma (k\, T_m^{\;2}/\gamma) - b_\gamma (2\, k\, T_m) \qquad (10)$$

where γ is τ, δ or ω. The values of constants used in Chen's equation for first, second and general order kinetics are given in Table 2.

Table 2. Values of constants used in Chen's equation

First Order			Second Order		
$c_\tau = 1.51$	$c_\delta = 0.976$	$c_\omega = 2.52$	$c_\tau = 1.81$	$c_\delta = 1.71$	$c_\omega = 3.54$
$b_\tau = 1.58$	$b_\delta = 0$	$b_\omega = 1$	$b_\tau = 2.0$	$b_\delta = 0$	$b_\omega = 1.0$

Grossweiner Method

Grossweiner (Grossweiner, 1953) calculate the trap depth (E) and frequency factor (s) of a first and second order TL glow peaks by the use of 'low-temperature half width' $\tau = T_m - T_1$ based on the Randall & Wilkins model.Grossweiner (Grossweiner, 1953) also derived a formula for the trap depth (E):

$$E_\gamma = c_\gamma \frac{kT_m T_1}{\gamma} \tag{11}$$

The constants c_γ for the three equations (τ, δ or ω) and for first and second-order kinetics are given in Table 3.

Lushchik Method

Lushchik (Lushchik,1956) developed methods for the evaluation of E by using the 'high-temperature half width' $\delta = T_2 - T_m$ of the peak:

$$E_\gamma = c_\gamma \frac{kT_m^2}{\gamma} \tag{12}$$

where c_γ for the three equations (τ, δ or ω) and for first and second-order kinetics are given in Table 3.

Frequency Factor

After knowing the parameters E and b, frequency factor (s) was calculated by Chen and Winer equation (Chen, Winer, 1970)

$$\frac{qE}{kT_m^2} = s\left[1 + (b-1)\frac{2kT_m}{E}\right]\exp\left(\frac{-E}{kT_m}\right) \tag{13}$$

Table 3. Values of constants used in Grossweiner and Luschik equations

Order	Grossweiner			Luschik		
First	c_τ=1.41	c_δ=1.0152	c_ω=2.4171	c_τ=1.3556	c_δ=0.976	c_ω=2.3238
Second	c_τ=1.68	c_δ=1.8313	c_ω=3.5217	c_τ=1.5651	c_δ=1.706	c_ω=3.2808

Here q is the heating rate and b are the order of kinetics and equals 1 for 1st order kinetics and 2 for 2nd order kinetics.(Horowitz, Rosenkrantz, Mahajna, Yosian, 1996)

For example, the present author's group has reported earlierthe optimization and characterization of the trap level distribution after the γ- ray irradiationundoped/codoped $CaMoO_4$ phosphors.In this literature, the effect of rare earth ions for the shift in glow peak temperature was discussed. Different trapping parameters such as trap depth, order of kinetics and frequency factor were summarized and the suitability of the material for dosimetric application was reported.(Dutta, Som, Sharma, 2013)

Dosimetric Behavior

The variation in the intensity of glow peaks with ion irradiation indicates the suitability of the materials for thedosimetric purpose. The TL intensity of the phosphor generally increases linearly with ion irradiation and then saturates. The intensity variation can be explained by the track interaction model.[44]The recombination occurs entirely within the tracksat low ion fluences. However, the non-radiative competitive centers in the intermediate regionintercept the electrons escaping from the tracks. Therefore, the TL intensity is proportional to the number of ion beam tracks. At higher ion fluences, the distance between the two tracks decreases. Hence, the electrons escaping the track can reach the neighboring track. It enhancesthe recombination of the luminescent centers and thereby increases the TL intensity. When the ion fluencesfurther increase, the tracks overlap each other. Therefore, the effective ionization is reduced. It results in sublinearity of the TL intensity and finally saturation. The effect of nanomaterials for this linearity will be checked. However, if any TL response is linear throughout a wide region, that materials can be suitable for dosimetry application. The TL fading is also checked for these materials for atleast 60 days. Less fading over 60 days for the irradiated phosphor will be potential for practical appliances. The linear TL response, maximum sensitivity and less fading of the ion irradiated phosphor shows its suitability as effective TL dosimeter for swift heavy ion irradiations.

FUTURE RESEARCH DIRECTIONS

Current research trends demonstrate an utter concern towards the modifications of materials induced by swift heavy ions owing to its unique ability to engineer and modify the materials in a controlled fashion such as physical, chemical, structural, and optical properties. Electronic excitations are fashioned by theseionizing irradiations to change the materials' properties. In thecase of phosphor materials, this high energy deposition and heat dissemination during ion irradiation may construct new colorcenters, point defects and sometimes the reorganization of the trapping and luminescent centers are observed which results in the modification of luminescence properties. Therefore swift heavy ion is always important for phosphor materials. Moreover, the agglomerated phosphor particles can be destructed via the ion irradiation to form nanoparticles which will be abreakthroughin the research field.

CONCLUSION

This chapter describes the basic concepts of swift heavy ions and their modifications. The basic requirements for Photoluminescence and thermoluminescence are also described in brief. The procedure for the calculation of different trapping parameters via different approaches is discussed. Finally, the procedure to be used for thedosimetric applications is also discussed for these ion irradiated materials. This discussion will benefit the present researcher at a basic level and help them with the future aspects of the ion irradiated phosphor materials in the nanodimension.

ACKNOWLEDGMENT

Authors are highly thankful to Dr. S.P. Lochab and the Pelletron group of Inter University Accelerator Center, New Delhi, India for providing the ion beam and other facilities. Financial support from the University of the Free State, Bloemfontein, South Africa is also acknowledged.

REFERENCES

Aumayr, F., Facsko, S., El-Said, A. S., Trautmann, C., & Schleberger, M. (2011).Single ion induced surface nanostructures: a comparison between slow highly charged and swift heavy ions. J. Phys.: Condens. Matter, 23, 393001 (23pp). doi:10.1088/0953-8984/23/39/393001

Avasthi, D. K. (2002). Some interesting aspects of swift heavy ions in materials science. *Current Science*, *78*, 1297–1302.

Avasthi, D. K., & Mehta, G. K. (2011). *Swift Heavy Ions for Materials Engineering and Nanostructuring* (Vol. 145). Springer Series in Materials Science. doi:10.1007/978-94-007-1229-4

Blasse, G., & Grabmaier, B. C. (1994). *Luminescent materials.* Berlin: Spring-Verlag.

Chen, R. (1969). On the Calculation of Activation Energies and Frequency Factors from Glow Curves. *Journal of Applied Physics*, *40*(2), 570–585. doi:10.1063/1.1657437

Chen, R. (1969). Glow Curves with General Order Kinetics. *Journal of the Electrochemical Society*, *116*(9), 1254–1257. doi:10.1149/1.2412291

Chen, R., & McKeever, S. W. S. (1997). *Theory of Thermoluminescence and Related Phenomena*. Singapore: World Scientific. doi:10.1142/2781

Chen, R., & Winer, S. A. A. (1970). Effects of Various Heating Rates on Glow Curves. *Journal of Applied Physics*, *40*(13), 5227–5228. doi:10.1063/1.1658652

Chung, K. S., Choe, H. S., Lee, J. I., & Kim, J. L. (2007). A new method for the numerical analysis of thermoluminescence glow curve. *Radiation Measurements*, *42*(4-5), 731–734. doi:10.1016/j.radmeas.2007.02.028

Chung, K. S., Choe, H. S., Lee, J. I., Kim, J. L., & Chang, S. Y. (2005). A computer program for the deconvolution of thermoluminescence glow curves. *Radiation Protection Dosimetry*, *115*(1-4), 343–349. doi:10.1093/rpd/nci073 PMID:16381744

Dufour, C., Khomenkov, V., Rizza, G., & Toulemonde, M. (2012). Ion-matter interaction: The three-dimensional version of the thermal spike model. Application to nanoparticle irradiation with swift heavy ions. *Journal of Physics. D, Applied Physics*, *45*(6), 065302. doi:10.1088/0022-3727/45/6/065302

Dutta, S., Som, S., & Sharma, S. K. (2013). Optimization and characterization of trap level distribution in γ-irradiated doped/codoped $CaMoO_4$ phosphors. *Physica B, Condensed Matter*, *417*, 39–45. doi:10.1016/j.physb.2013.02.024

Furetta, C. (2003). *Handbook of Thermoluminescence* (2nd ed.). Singapore: World Scientific Publishing Co. Pte. Ltd. doi:10.1142/5167

Garlick, G. F. J., & Gibson, A. F. (1948). The Electron Trap Mechanism of Luminescence in Sulphide and Silicate Phosphors. *Proceedings of the Physical Society*, *60*(6), 574–590. doi:10.1088/0959-5309/60/6/308

Gnaser, H. (1999). *Low-Energy Ion Irradiation of Solid Surfaces*. Berlin: Springer-Verlag.

Grossweiner, I. (1953). A Note on the Analysis of First-Order Glow Curves. *Journal of Applied Physics*, *24*(10), 1306–1307. doi:10.1063/1.1721152

Haering, R. R., & Adams, E. N. (1960). Theory and Application of Thermally Stimulated Currents in Photoconductors. *Physical Review*, *117*(2), 451–454. doi:10.1103/PhysRev.117.451

Halperin, A., & Braner, A. A. (1960). Evaluation of Thermal Activation Energies from Glow Curves. *Physical Review*, *117*(2), 408–415. doi:10.1103/PhysRev.117.408

Hasan, F., Kitis, G., & Charalambou, S. (1985). Regenerated thermoluminescence in $CaSO_4$-Dy. *Journal of Physics. C. Solid State Physics*, *18*(8), 1743–1751. doi:10.1088/0022-3719/18/8/020

Hodgson, R. T., Balgin, J. E. E., Pal, R., Neri, J. M., & Hammer, D. A. (1980). Ion beam annealing of semiconductors. *Applied Physics Letters*, *37*(2), 187–188. doi:10.1063/1.91819

Horowitz, Y. S. (1984). *Thermoluminescence & Thermoluminescent Dosimetry*. Boca Raton, FL: CRC Press.

Horowitz, Y. S., Rosenkrantz, M., Mahajna, S., & Yosian, D. (1996). The track interaction model for alpha particle induced thermoluminescence supralinearity: Dependence of the supralinearity on the vector properties of the alpha particle radiation field. *Journal of Physics. D, Applied Physics*, *29*(1), 205–217. doi:10.1088/0022-3727/29/1/031

Kellerer, A. (1985). Fundamentals of microdosimetry. In K. R. Kase, B. E. Bjarngard, & F. H. Attix (Eds.), *The Dosimetry of lonuinsRadiation* (Vol. 1). New York: Academic Press.

Kitai, H. (1993). *Solid State Luminescence: Theory, Materials & Devices*. London: Chapman & Hall. doi:10.1007/978-94-011-1522-3

Kraft, S., Schattat, B., Bolse, W., Klaumünzer, S., Harbsmeier, F., Kulinska, A., & Löffl, A. (2002). Ion beam mixing of ZnO/SiO_2 and Sb/Ni/Si interfaces under swift heavy ion irradiation. *Journal of Applied Physics*, *91*(3), 1129–1134. doi:10.1063/1.1425439

Krauser, J., Zollondz, J. H., Weidinger, A., & Trautmann, C. (2003). Conductivity of nanometer-sized ion tracks in diamond-like carbon films. *Journal of Applied Physics*, *94*(3), 1959–1964. doi:10.1063/1.1587263

Lushchik, B. (1956). The Investigation of Trapping Centers in Crystals by the Method of Thermal Bleaching. *Soviet Physics, JETP*, *3*, 390–399.

Manjunatha, C., Sunitha, D. V., Nagabhushana, H., Singh, F., Sharma, S. C., Chakradhar, R. P. S., & Nagabhushana, B. M. (2013). Thermoluminescence properties of 100 MeV Si^{7+} swift heavy ions and UV irradiated $CdSiO_3$: Ce^{3+} nano phosphor. *Journal of Luminescence*, *134*, 358–368. doi:10.1016/j.jlumin.2012.08.020

May, E., & Partridge, J. A. (1964). Thermoluminescent Kinetics of Alpha-Irradiated Alkali Halides. *The Journal of Chemical Physics*, *40*(5), 1401–1408. doi:10.1063/1.1725324

Mckeever, S. W. S. (1985). *Thermoluminescence in solids*. Cambridge, UK: Cambridge University Press. doi:10.1017/CBO9780511564994

McKinlay, F. (1981). *Thermoluminescence Dosimetry. Mishawaka*. CRC Press.

Mohanta, D., Mishra, N. C., & Choudhury, A. (2004). SHI-induced grain growth and grain fragmentation effects in polymer-embedded CdS quantum dot systems. *Materials Letters*, *58*(29), 3694–3699. doi:10.1016/j.matlet.2004.06.061

Nastasi, M., Mayer, J. W., & Hirvonen, J. K. (1996). *Ion-Solid Interactions*. Cambridge, UK: Cambridge University Press. doi:10.1017/CBO9780511565007

Opanowicz, A. (2000). Determination of the recombination coefficient and the density of deep traps from the simultaneously measured thermally stimulated conductivity and luminescence. *Journal of Physics. D, Applied Physics*, *33*(14), 1635–1642. doi:10.1088/0022-3727/33/14/301

Opanowicz, A. (2007). Analysis of thermally stimulated luminescence and conductivity without quasi-equilibrium approximation. *Journal of Physics. D, Applied Physics*, *40*(16), 4980–4990. doi:10.1088/0022-3727/40/16/034

Opanowicz, A., & Pietrucha, P. (2003). Heating-rate method for determination of kinetic parameters from thermally stimulated conductivity and luminescence. *Journal of Applied Physics*, *93*(2), 957–967. doi:10.1063/1.1530713

Randall, J. T., & Wilkins, M. H. F. (1945a). Phosphorescence and Electron Traps. I. The Study of Trap Distributions. *Proc. Roy. Soc.*, *184*(999), 365–389. doi:10.1098/rspa.1945.0024

Randall, J. T., & Wilkins, M. H. F. (1945b). Phosphorescence and Electron Traps. II. The Study of Trap Distributions. *Proc. Roy. Soc.*, *184*(999), 390–407. doi:10.1098/rspa.1945.0025

Ronda, C. (2008). *Luminescence: from theory to application*. Weinheim: Wiley-VCH publication.

Salah, N. (2011). Nanocrystalline materials for the dosimetry of heavy charged particles: A review. *Radiation Physics and Chemistry*, *80*(1), 1–10. doi:10.1016/j.radphyschem.2010.08.003

Schiwietz, G., Staufenbiel, F., Roth, M., Czerski, K., & Grande, P. L. (2005). *Electrons as a tool for short-time track characterization*. Aschaenburg.

Som, S., Dutta, S., Chowdhury, M., Kumar, V., Kumar, V., Swart, H. C., & Sharma, S. K. (2014). A comparative investigation on ion impact parameters and TL response of Y_2O_3: Tb^{3+}nano phosphor exposed to swift heavy ions for space dosimetry. *Journal of Alloys and Compounds*, *589*, 5–18. doi:10.1016/j.jallcom.2013.11.182

Som, S., Kunti, A. K., Kumar, V., Kumar, V., Dutta, S., Chowdhury, M., ... Swart, H. C. (2014). Defect correlated fluorescent quenching and electron-phonon coupling in the spectral transition of Eu^{3+} in $CaTiO_3$ for red emission in display application. *Journal of Applied Physics*, *115*(19), 193101. doi:10.1063/1.4876316

Som, S., Mitra, P., Kumar, V., Kumar, V., Terblans, J. J., Swart, H. C., & Sharma, S. K. (2014). The energy transfer phenomena and color tunability in Y_2O_2S: Eu^{3+}/ Dy^{3+} micro-fibers for white emission in solid-state lighting applications. *Dalton Transactions (Cambridge, England)*, *43*(26), 9860–9871. doi:10.1039/C4DT00349G PMID:24710323

Sunta, M., Ayta, W. E. F., Chubaci, J. F. D., & Watanabe, S. (2001). Dependence of quasi-equilibrium on heating rate and its use in the study of thermoluminescent materials. *Journal of Physics. D, Applied Physics*, *34*(22), 3285–3295. doi:10.1088/0022-3727/34/22/313

Vij, A., Kumar, R., Chawla, A. K., Lochab, S. P., Chandra, R., & Singh, N. (2010). Swift heavy ion induced synthesis and enhanced photoluminescence of SrS: Ce nanoparticles. *Optical Materials*, *33*(1), 58–62. doi:10.1016/j.optmat.2010.07.024

Watnabe, S. (1951). Properties of a $CaSO_4$-Mn Phosphor under Vacuum Ultraviolet Excitation. *Physical Review*, *83*(4), 785–791. doi:10.1103/PhysRev.83.785

Ziegler, J. F., Biersack, J. P., & Ziegler, M. D. (2008). *SRIM- The Stopping and Range of Ions in Matter*. Lulu Press Co.

KEY TERMS AND DEFINITIONS

Ionizing Radiation: The radiation that carries sufficient energy to free electrons from atoms or molecules of matter with which they interact is known as ionizing radiation.

Phosphor: The luminescent materials that can emit light after the absorption of a photon are generally known as phosphor.

Photoluminescence: It is the light emission from any form of matter after the absorption of electromagnetic radiation (i.e., photon).

Radiation Dosimetry: The measurement procedure for the dose of ionizing radiation is known as radiation dosimetry.

Swift Heavy Ion: Highly energetic heavy ions are known as swift heavy ions.

Thermoluminescence: It is a type of luminescence in which the formerly absorbed energy from electromagnetic radiation or different ionizing radiation from the materials is re-emitted as light after the introduction of heating to the material. This phenomenon is completely diverse from that of black body radiation.

Chapter 2
An Overview on Peak Shape Method for Thermoluminescence Glow Curve Analysis:
Application on Tremolite and Actinolite Glow Peaks

Sukhamoy Bhattacharyya
Acharya Prafulla Chandra College, India

Partha Sarathi Majumdar
Acharya Prafulla Chandra College, India

ABSTRACT

The shape of a thermoluminescence (TL) glow curve has fundamental importance for calculating the characteristic parameters of trap levels within the band gap. TL analysis are mostly based on the three-parameter general order kinetics model. The parameters are activation energy, order of kinetics, and frequency factor. Peak shape method is one of the most prominent methods for extracting the activation energy from a TL curve. An overview of different peak shape methods along with an alternative approach formulated directly from basic TL equations is presented in this chapter. Generally, peak shape method requires prior knowledge of order of kinetics to determine activation energy which creates a difficulty due to the non-uniqueness of symmetry factor for a particular value of order of kinetics. A modified version of peak shape method which is free from this constraint is discussed here. Activation energies from experimental curves of tremolite and actinolite are estimated using peak shape method. Limitation of peak shape method for saturated TL peaks with heavy retrapping is also discussed.

INTRODUCTION

Thermoluminescence (TL) is a phenomenon of light emission when a material (a semiconductor or an insulator) is heated after being exposed to some ionizing radiation such as X-ray, γ-ray, UV light etc. Impurities and structural defects present in the solids create localized energy levels in the band

DOI: 10.4018/978-1-5225-5170-6.ch002

gap. Energy levels nearer to the conduction band (CB) are potential electron traps whereas those close to valence band (VB) are potential hole traps functioning as recombination centres. In practice, a TL material should have the traps deep enough to be stable at room temperature so that the emission is predominantly in the visible region. Over the last few decades many TL materials have been investigated and developed. The most important application of TL is in the field of radiation dosimetry, including environmental aspects. Moreover, TL has also got very important application in dating where it plays a role complementary to radiocarbon dating.

The theoretical studies of thermoluminescence are based on some models which are mainly phenomenological. The simplest and most widely used theory of TL analysis is based on three important parameters *viz.* i) activation energy (E), ii) frequency factor (s) and iii) order of kinetics (b). These parameters are also important in indicating the applicability and usefulness of a TL material. These parameters may be extracted by analysing the TL glow curve which is the plot of intensity (I) of emitted light vs. temperature (T).

One of the most successful models for TL analysis is the General Order Kinetics (GOK) model (May & Partridge, 1964; Chen, 1969a) where TL intensity is directly expressed as a function of temperature and involves E, s and b. The GOK model yields the first order equation of Randall & Wilkins (1945) for $b \to 1$ and the second order equation of Garlick & Gibson (1948) for $b = 2$. Moreover, based on the band picture of solids, there are other models also such as One Trap One Recombination centre (OTOR), Non-interactive Multitrap System (NMTS), Interactive Multitrap System (IMTS) etc. (Pagonis, Kitis & Furetta, 2006). These models are mathematically expressed in terms of differential equations governing the trafficking of charge carriers. In OTOR model, the existence of one trap and one recombination centre is assumed within band gap whereas in NMTS and IMTS models, multiple trap levels including thermally disconnected deep traps (TDDT) are also considered.

There are various techniques to determine the trapping parameters of a TL curve (Chen & Pagonis, 2011; Chen & McKeever, 1997). The most commonly used methods are (i) Peak Shape (PS) method, (ii) Various Heating Rates (VHR) method, (iii) Initial Rise (IR) method, (iv) Area method, (v) Computerised Glow Curve Deconvolution (CGCD) technique etc. This review focuses on the applicability of peak shape method for determining the activation energy from a TL glow curve. Peak shape method is widely used in TL studies and is primarily based on the geometric shape of the glow curve which is governed by the trapping parameters. Along with theoretical considerations, experimental TL curves of tremolite and actinolite materials belonging to silicate minerals group that are found in abundance in the crust of earth have also been studied in the light of peak shape method for extracting the activation energies that are useful to interpret the TL mechanism in these materials. The plan of the article is as follows: a brief review of the prominent works on peak shape method is discussed in section 2; the mathematical formulation of peak shape method is elaborated in section 3; the application of peak shape method to analyse the experimental TL curves of tremolite and actinolite is given in section 4; the limitations of peak shape method in case of saturated TL peaks with heavy retrapping discussed in section 5 and concluding remarks are summarized in section 6.

A BRIEF REVIEW ON PEAK SHAPE METHOD

An elementary model to analyse a TL process may be given on the basis of band theory of solids. The presence of electron traps and recombination centres within the band gap facilitate the process. A schematic diagram considering one such trap and one recombination centre (RC) is shown in Figure 1. Here

Figure 1. Schematic diagram of TL process on the basis of band theory of solids

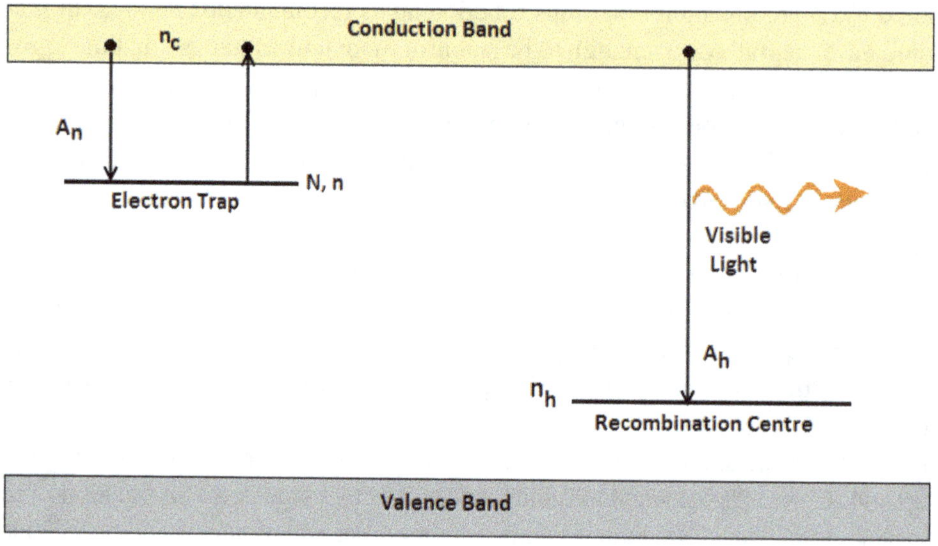

n, n_c and n_h are the instantaneous concentrations (m^{-3}) of filled traps, CB electrons and trapped holes in RC respectively. N (m^{-3}) is the total concentration of active traps. A_h is the recombination probability (m^3s^{-1}) of an electron with a hole in RC and A_n is the retrapping probability of an electron in electron trap. The basic charge neutrality condition is given by (Pagonis et al., 2006)

$$n_h = n + n_c \tag{1}$$

The intensity (I) of the emitted light is equal to the rate of recombination of electrons with holes in the recombination centres. A typical glow curve is shown in Figure 2 where T_m is the peak temperature and T_1, T_2 are the half intensity temperatures $\left(T_1 < T_m < T_2\right)$. The well known peak shape parameters are defined as: $\tau = T_m - T_1$, $\delta = T_2 - T_m$ and $\omega = T_2 - T_1$. The quantity $\mu_g = \dfrac{\delta}{\omega}$ is known as the symmetry factor which plays a crucial role in peak shape method. For a first order peak, $\mu_g \simeq 0.42$ and for second order peak $\mu_g \simeq 0.52$. So the value of μ_g may serve as an indicator of the order of kinetics of a TL process.

The pioneering work on the peak shape method to determine activation energy was proposed by Grosswiener (1953) where the peak position and peak shape parameter τ were employed to calculate the trap depth of first-order glow curves. According to Grosswiener (1953), the peak shape relation is

$$E = 1.51\frac{kT_mT_1}{\tau} \tag{2}$$

Here k is the Boltzmann constant. This relation was applied to an experimental TL glow curve of ice (Grosswiener 1953). Lushchik (1956) also proposed a formula for peak shape method which may be written as

Figure 2. A sample TL glow curve with peak shape parameters

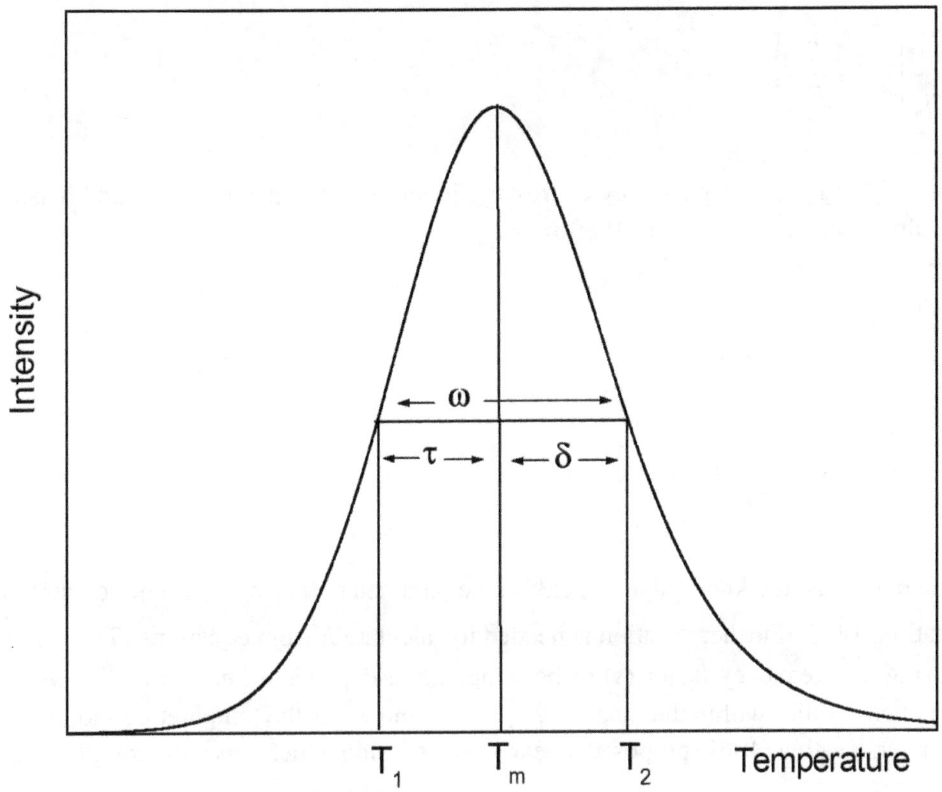

$$E = \frac{kT_m^2}{\delta} \tag{3}$$

Among the initial theories, the most relevant work in peak shape method was due to Halperin & Braner (1960) for the evaluation of thermal activation energies from glow curves of excited crystals. Following this methodology, in a subsequent work, Halperin, Braner, Ben-Zvi & Kristianpoller (1960) computed activation energies from various glow peaks of NaCl and KCl single crystals. According to them (Halperin & Braner, 1960; Halperin et al., 1960), the activation energy in terms of the peak shape parameter δ and μ_g may be written as

$$E = \frac{q}{\delta} kT_m^2 \left(1 - \frac{2kT_m}{E}\right) \tag{4}$$

where the parameter q depends on the shape of the glow peak and on the kinetics of the process. For a first order process $(b = 1)$, q is given by

$$q_{(1)} = 1.72 \frac{\mu_g}{1 - \mu_g}\left(1 - 3.16 \frac{kT_m}{E}\right) \tag{5}$$

whereas, an approximated expression of q for most of the second order $\left(b=2\right)$ TL processes is

$$q_{(2)}=\frac{2\mu_g}{1-\mu_g}\left(1-\frac{4kT_m}{E}\right) \tag{6}$$

Equations (5,6) may be simplified by expressing in terms of the parameter τ and hence E may be given by (Halperin et al., 1960; Chen, 1969b)

$$E=1.72\frac{kT_m^2}{\tau}\left(1-5.16\frac{kT_m}{E}\right)\text{ for } b=1$$

$$\text{and } E=\frac{2kT_m^2}{\tau}\left(1-6\frac{kT_m}{E}\right)\text{ for } b=2 \tag{7}$$

It may be noted that the knowledge of peak shape parameters (*i.e.* δ, τ) are not enough to estimate E from equations (4, 7). Further iteration is needed to calculate E from equations (7).

Considering the frequency factor (s) to be temperature dependent *i.e.* $s=s''T^a$, where the temperature exponent 'a' lies within the range $-2\leq a\leq 2$ and s'' is the temperature independent part of frequency factor, Keating (1961) proposed a peak shape relation which was valid for all values of a. The relation holds good, as reported by Keating (1961), for $10<\frac{E}{kT_m}<35$ and $0.75<\frac{\delta}{\tau}<0.9$.

The most widely used formula of peak shape method was proposed by Chen (1969a). Considering all the previous theories discussed above and by analysing a large number of numerical TL peaks, Chen presented a set of peak shape relations which may be summarized as (Chen 1969a)

$$E_\alpha=c_\alpha\frac{kT_m^2}{\alpha}-b_\alpha\left(2kT_m\right) \tag{8}$$

where c_α, b_α are constants and α stands for τ, δ and ω. The coefficients c_α and b_α (Chen 1969a) are given in Table 1 in terms of μ_g for fractional intensity $x=\frac{I}{I_m}=\frac{1}{2}$, I_m being the peak intensity.

Though most of the peak shape formulas deal with half intensity points $\left(x=\frac{1}{2}\right)$, a few accounts are available in literature where other fractional intensity points are also available. Fleming (1968) reported the temperatures on both rising and falling side of the curve at which the fractional intensity $x=\frac{1}{2},\frac{2}{3}$ and $\frac{4}{5}$ for both first and second order kinetics, considering the trap depth within 0.1 to 1.5 eV. Christodoulides (1985) presented peak shape expressions of E corresponding to first order kinetics where TL

Table 1. Coefficients from equation (8) for $x = \frac{1}{2}$ (Chen, 1969a,b). Here $\mu_g' = \mu_g - 0.42$.

Kinetic Order	Coefficient	τ	δ	ω
1st	c_α	1.51	0.976	2.52
	b_α	1.58	0.0	1.0
2nd	c_α	1.81	1.71	3.54
	b_α	2.0	0.0	1.0
General	c_α	$1.51 + 3.0\mu_g'$	$0.976 + 7.3\mu_g'$	$2.52 + 10.2\mu_g'$
	b_α	$1.58 + 4.2\mu_g'$	0.0	1.0

intensity is $\frac{1}{4}, \frac{1}{2}$ and $\frac{3}{4}$ of the maximum intensity. Mazumdar, Singh & Gartia (1988) extended the work of Christodoulides (1985) to determine *E* for non-first order kinetics considering fractional intensity $x = \frac{1}{2}, \frac{2}{3}$ and $\frac{4}{5}$.

MATHEMATICAL FORMULATION OF PEAK SHAPE METHOD

The peak shape formulas discussed in the previous section have been obtained empirically by considering a large number of TL glow curves. However, it is possible to develop a mathematical foundation of peak shape method from the fundamental TL equations which are based on the band theory of solids. According to OTOR model, the basic differential equations governing the TL phenomenon are (Pagonis et al., 2006)

$$I = -\frac{dn_h}{dt} = A_h n_c n_h \tag{9}$$

$$-\frac{dn}{dt} = nse^{-\frac{E}{kT}} - A_n n_c \left(N - n\right) \tag{10}$$

$$-\frac{dn}{dt} = nse^{-\frac{E}{kT}} - A_n n_c \left(N - n\right) - A_h n_c n_h \tag{11}$$

Here s is the frequency factor as mentioned before. To obtain an analytical solution of these differential equations, the following Quasi-Equilibrium (QE) conditions (Pagonis et al., 2006) are introduced:

$$\left|\frac{dn_c}{dt}\right| \ll \left|\frac{dn}{dt}\right|$$

and $$\left|\frac{dn_c}{dt}\right| \ll \left|\frac{dn_h}{dt}\right| \tag{12}$$

Since the electron density in the conduction band (n_c) remains almost stationary, $\left|\frac{dn_c}{dt}\right| \simeq 0$. So, from equation (11), it is obtained that

$$n_c \simeq \frac{nse^{-E/kT}}{A_n\left(N-n\right)+A_h n_h} \tag{13}$$

Therefore, the expression for TL intensity in OTOR model (using QE conditions) becomes

$$I = -\frac{dn_h}{dt} = \frac{A_h n_h nse^{-E/kT}}{A_n\left(N-n\right)+A_h n_h} \tag{14}$$

This equation is also known as Generalised One Trap (GOT) equation in literature (Levy, 1985). Now, according to the following considerations

1. The probability of recombination with a hole in RC is much greater than the probability of retrapping an electron *i.e.* $A_h n_h \gg A_n\left(N-n\right)$, and
2. Density of electrons in the conduction band is negligible as compared to trapped electrons *i.e.* $n_h \approx n$ with $n_c \ll n$

equation (14) reduces to

$$I \simeq -\frac{dn}{dt} = nse^{-E/kT} \tag{15}$$

which is the famous first order kinetics ($b = 1$) equation of Randall & Wilkins (1945).

Again, assuming equal probability of recombination and retrapping (*i.e.* $A_h = A_n$) along with $n_h \approx n$, equation (14) yields

$$I \simeq -\frac{dn}{dt} = \frac{s}{n} n^2 e^{-E/kT} \tag{16}$$

Equation (16) is the well known equation of Garlick & Gibson (1948) for second order kinetics.

The second order equation (16) may be derived from equation (14) in an alternative way by using the following assumptions

1. Retrapping dominates over recombination i.e. $A_n \left(N - n\right) \gg A_h n_h$,
2. Traps are far from saturation i.e. $n \ll N$, and
3. $n_h \approx n$

In this case, the TL intensity is expressed as

$$I \simeq -\frac{dn}{dt} = \frac{sA_h}{NA_n} n^2 e^{-E/kT} \tag{17}$$

Both the equations (16) and (17) can be recast in the same form by putting $s' = \frac{s}{N}$ or $\frac{sA_h}{NA_n}$ respectively.

It may be noted that all the observed TL processes cannot be classified in a straightforward way under first order or second order kinetics. The value of order of kinetics (*b*) might also lie between 1 and 2. To encompass such cases, May & Partridge (1964) proposed a general order kinetic model (GOK) which may be expressed as

$$I \simeq -\frac{dn}{dt} = \frac{s}{N^{b-1}} n^b e^{-E/kT} \tag{18}$$

Corresponding to the well known linear heating scheme $T = T_0 + \beta T$, β being the heating rate and T_0 the initial temperature, the TL intensity (*I*) in GOK model takes the form (Bhattacharyya, Karmakar, Majumdar & Singh, 2016)

$$I(T) = N f^b s e^{-E/kT} \left[1 + (b-1) s \frac{f^{b-1}}{\beta} \int_{T_0}^{T} e^{-E/kT'} dT' \right]^{-\frac{b}{b-1}} \tag{19}$$

Here $f = \frac{n_0}{N}$ is the filling ratio and *n* is the initial concentration of trapped electrons. For $b \to 1$, using the standard limit

$$\lim_{z \to \infty} \left(1+z\right)^{1/z} = e$$

equation (19) reduces to the first order equation of Randall & Wilkins (1945)

$$I(T) = n_0 s e^{-E/kT} \exp\left[-\frac{s}{\beta}\int_{T_0}^{T} e^{-E/kT'} dT'\right] \tag{20}$$

Equation (20) corresponds to the first order equation (15) mentioned before. For $b = 2$, equation (19) reduces to the second order equation of Garlick & Gibson (1948)

$$I(T) = s\frac{n_0^{\;2}}{N} e^{-E/kT}\left[1+\frac{s'}{\beta}\int_{T_0}^{T} e^{-E/kT'} dT'\right]^{-2} \tag{21}$$

Equation (21) corresponds to the second order equation (16). The TL peak position (T_m) may be obtained from the maxima conditions which are given as

$$\frac{\beta E}{kT_m^{\;2}} = s e^{-E/kT} \text{ for } b = 1$$

$$\text{and } \frac{\beta E}{kT_m^{\;2}} = b\, s f^{b-1} e^{-E/kT}\left[1+(b-1)\, s\frac{f^{b-1}}{\beta}\int_{T_0}^{T} e^{-E/kT'} dT'\right]^{-1} \text{ for } b \neq 1 \tag{22}$$

The temperature integral $J\left(T,T_0\right) = \int_{T_0}^{T} e^{-E/kT'} dT'$ occurring in equations (19 – 22) cannot be evaluated analytically in a closed form. However, In the limit $T_0 \to 0$, $J\left(T,T_0\right)$ is approximated as (Abramowitz & Stegun, 1965)

$$J\left(T,T_0\right) \simeq \int_{0}^{T} e^{-E/kT'} dT' = \frac{E}{ku} E_2(u) \tag{23}$$

with $u = \dfrac{E}{kT}$ where $E_2(u)$ is the 2nd exponential integral (Abramowitz & Stegun, 1965). Eliminating $s\dfrac{f^{b-1}}{\beta}$ from equations (19) and (22), it may be written that

$$\frac{I}{I_m} = e^{u_m - u}\left[1 - \frac{b-1}{b}F\left(u, u_m\right)\right]^{-\frac{b}{b-1}} \text{ for } b \neq 1$$

$$\text{and } \frac{I}{I_m} = \exp\left[u_m - u + F\left(u, u_m\right)\right]^{-\frac{b}{b-1}} \text{ for } b = 1 \tag{24}$$

$$\text{where } F\left(u, u_m\right) = u_m^{\ 2} e^{u_m}\left[\frac{E_2\left(u_m\right)}{u_m} - \frac{E_2\left(u\right)}{u}\right] \tag{25}$$

Here, I_m is the peak intensity corresponding to peak temperature T_m and $u_m = \frac{E}{kT_m}$. Let T_x be the temperature for which the fractional intensity is $\frac{I}{I_m} = x$. The nomenclature $T_x = T_x^-$ for $T_x < T_m$ (rising side of the peak) and $T_x = T_x^+$ for $T_x > T_m$ (falling side of the peak) is adopted here. For any particular value of *x*, T_x^- and T_x^+ can be calculated by solving equation (24) numerically. For the range of values $10 < u_m < 100$, good linear relations may be obtained between the following pairs of variables:

1. u_m and $\frac{u_x^-}{u_x^- - u_m}$
2. u_m and $\frac{u_x^+}{u_m - u_x^+}$
3. u_m and $\frac{u_x^+ u_x^-}{u_m\left(u_m - u_x^+\right)}$

where $u_x^{\pm} = \frac{E}{kT_x^{\pm}}$. Conventionally, $x = \frac{1}{2}$ and corresponding values of *E* are denoted as E_τ, E_δ and E_ω. From these linear relations, the activation energies corresponding to any arbitrary value of *x* are expressed as

$$E_{\tau x} = C_{\tau x}\frac{kT_m^2}{\tau_x} - D_{\tau x}kT_m \tag{26}$$

$$E_{\delta x} = C_{\delta x} \frac{kT_m^2}{\tau_x} - D_{\delta x} kT_m \tag{27}$$

$$E_{\omega x} = C_{\omega x} \frac{kT_m^2}{\tau_x} - D_{\omega x} kT_m \tag{28}$$

The peak shape coefficients C_{ix} and D_{ix} $\left(i = \tau, \delta, \omega\right)$ depend on the order of kinetics. From equations (26 – 28) it is evident that any arbitrary point (other than the conventional half intensity points) may be used to derive activation energy by peak shape method. Since the analysis of a complex TL curve becomes complicated due to the overlapping of several component glow peaks each of which is characterized by own kinetic parameters, it might be convenient to use the glow intensity of an isolated peak close to its maximum intensity where interference from neighbouring peaks will be practically minimum (Fleming, 1968). Thus, the values of $x = \frac{2}{3}$ and $\frac{4}{5}$, apart from the conventional value $\frac{1}{2}$, may be very useful (Fleming, 1968; Singh et al., 2012a) .

By using the technique of non-linear regression, C_{ix} and D_{ix} may be expressed as a quadratic function of *b* for a particular value of *x* that may be given by

$$C_{ix} = C_{ix}^{(0)} + C_{ix}^{(0)} b + C_{ix}^{(0)} b^2 \tag{29}$$

$$D_{ix} = D_{ix}^{(0)} + D_{ix}^{(0)} b + D_{ix}^{(0)} b^2 \tag{30}$$

In Table 2, the peak shape coefficients (Singh et al., 2012a) are given corresponding to different values of fractional intensity (*x*). The suffices from *C*'s and *D*'s have been dropped in the table as the parameters are mentioned explicitly.

The peak shape relations given by equations (26 – 28) have been applied to determine the activation energies (E_τ, E_δ and E_ω) from some numerically simulated TL peaks. The results are exhibited in Table 3 which shows that the estimated values are in good agreement with input value of *E*.

It is evident that in order to apply the equations (29,30) to a TL peak, prior knowledge of *b* is essential. The value of *b* may be obtained from the symmetry factor $\left(\mu_{gx}\right)$. For any arbitrary value of *x*, μ_{gx} may be expressed as

$$\mu_{gx} = \frac{T_x^+ - T_m}{T_x^+ - T_x^-} = \frac{\left(u_m - u_x^+\right) u_x^-}{\left(u_x^- - u_x^+\right) u_m} \tag{31}$$

Table 2. Peak shape coefficients (Singh et al., 2012a) corresponding to equations (26 – 30) for different values of fractional intensity (x)

$\frac{I}{I_m}(=x)$	i	$C^{(0)}$	$C^{(1)}$	$C^{(2)}$	$D^{(0)}$	$D^{(1)}$	$D^{(2)}$
$\frac{1}{2}$	τ	1.019	0.504	– 0.066	– 1.059	– 1.217	0.109
	δ	0.105	0.926	– 0.048	0.154	– 0.205	– 0.128
	ω	1.124	1.427	– 0.113	– 0.902	– 0.346	– 0.061
$\frac{2}{3}$	τ	0.684	0.426	– 0.055	– 0.720	– 1.210	0.098
	δ	0.146	0.683	– 0.048	0.184	– 0.432	– 0.094
	ω	0.830	1.108	– 0.103	– 0.529	– 0.607	– 0.029
$\frac{4}{5}$	τ	0.449	0.342	– 0.043	– 0.480	– 1.184	0.085
	δ	0.153	0.487	– 0.041	0.180	– 0.606	– 0.062
	ω	0.602	0.829	– 0.084	– 0.293	0.777	– 0.006

Equation (31) shows that along with $u_x^{\pm}$ which depends on *b* following equation (24), μ_{gx} has also a dependence on u_m which is different for different TL peaks. Therefore, for different TL peaks corresponding to same *b*, μ_{gx} might be of different values *i.e.* a non-uniqueness of the symmetry factor is inherent in the shapes of TL curves.

For experimental TL peaks, the activation energy (*E*) is unknown for which u_m cannot be determined. However, for experimental curves, u_m usually lies between 20 and 40. For $20 < u_m < 40$, μ_g varies from 0.426 to 0.415 for $b = 1$ and from 0.526 to 0.514 for $b = 2$. The dependence of μ_g on *b* is much stronger than the dependence on u_m. A rough estimate of the variation of μ_g is about 9 – 20% for the variation of *b* from 1 to 2 whereas it is around 5% when u_m changes from 20 to 40. Singh et al. (2012a) proposed a relation between μ_g and *b* corresponding to fractional intensity $x = \frac{1}{2}$ for $20 < u_m < 40$ as

$$\mu_{g_{\frac{1}{2}}} = 0.245 + 0.186\,b - 0.024\,b^2 \tag{32}$$

Similar relations for any arbitrary *x* with $20 < u_m < 40$ may be derived by using equation (31). For $x = \frac{2}{3}$ and $\frac{4}{5}$, the respective relations are given by

Table 3. Activation energies of synthetic TL curves determined from Peak shape coefficients (Singh et al., 2012a) given in Table 2 for different values of fractional intensity (x). The input values are: E = 1.0 eV, $s = 1.0 \times 10^{12}\ s^{-1}$. *The linear heating rate is* $\beta = 1.0\ Ks^{-1}$

Input Value of *b*	*x*	Activation Energy (eV)		
		E_τ	E_δ	E_ω
1.0	$\frac{1}{2}$	0.998	0.998	0.998
	$\frac{2}{3}$	0.998	0.997	0.998
	$\frac{4}{5}$	0.996	0.996	0.996
1.5	$\frac{1}{2}$	0.999	1.000	1.000
	$\frac{2}{3}$	0.999	1.001	1.000
	$\frac{4}{5}$	0.998	0.999	0.999
2.0	$\frac{1}{2}$	1.001	1.002	1.003
	$\frac{2}{3}$	1.002	1.004	1.003
	$\frac{4}{5}$	1.000	1.002	1.001

$$\mu_{g_{\frac{2}{3}}} = 0.319 + 0.138\,b - 0.021\,b^2 \tag{33}$$

and $$\mu_{g_{\frac{4}{5}}} = 0.361 + 0.103\,b - 0.016\,b^2 \tag{34}$$

The values of *b* has been estimated for the synthetic peaks mentioned before in Table 3 and the results are displayed in Table 4. It may be noted that, because of the dependence of μ_g on u_m, some error will

Table 4. Order of kinetics determined from μ_g by using equations (32 – 34) for different values of fractional intensity (x). The input values are: E = 1.0 eV, $s = 1.0 \times 10^{12}\ s^{-1}$. The linear heating rate is $\beta = 1.0\ Ks^{-1}$

Input Value of *b*	T_m (K)	*x*	μ_g	Output Value of *b*
1.0	384.57	$\frac{1}{2}$	0.418	1.08
		$\frac{2}{3}$	0.437	1.01
		$\frac{4}{5}$	0.453	1.07
1.5	384.22	$\frac{1}{2}$	0.477	1.56
		$\frac{2}{3}$	0.482	1.54
		$\frac{4}{5}$	0.487	1.64
2.0	383.88	$\frac{1}{2}$	0.518	1.97
		$\frac{2}{3}$	0.513	2.03
		$\frac{4}{5}$	0.510	2.19

be inherent in the value of *b* estimated in this way (Singh et al., 2012a), as is evident from Table 4. This discussion is consistent with the work of Chen (1969a) also from where it is evident that μ_g is not unique for a particular value of *b*. Considering a wide range of values of *E* and *s*, Chen (1969a) has shown that for a specific value of *b*, μ_g varies within a range of $\pm 7\%$. So, in the absence of knowledge of *E* and *s*, the estimated values of *b* will have some uncertainty. In order to avoid this problem, the original peak shape formulas of Chen included instead of *b* (1969a).

Kitis & Pagonis (2007) further expressed the peak shape formulas of Chen (1969a) for $x = \frac{1}{2}$ in terms of order of kinetics and modified the coefficients accordingly. The relations are:

$$E_\tau = C_\tau b\left[b^{1/(b-1)} - 1\right]\frac{kT_m^2}{\tau} - \left[\frac{b^{1/(b-1)}}{b^{1/(b-1)} - 1}\right](2kT_m) = c_\tau \frac{kT_m^2}{\tau} - b_\tau(2kT_m) \tag{35}$$

$$E_\delta = C_\delta b\frac{kT_m^2}{\tau} = c_\delta \frac{kT_m^2}{\tau} - b_\delta(2kT_m) \tag{36}$$

$$E_\omega = C_\omega b^{1/(b-1)}\frac{kT_m^2}{\omega} - 2kT_m = c_\omega \frac{kT_m^2}{\tau} - b_\omega(2kT_m) \tag{37}$$

Here, $b_\delta = 0$ and $b_\omega = 1$. The coefficients C_τ, C_δ and C_ω are evaluated by simulation. The net values of the coefficients c_τ, c_δ, c_ω and b_τ are listed in Table 5 for $b = 1, 1.5$ and 2. These coefficients are in good agreement with the values given in Table 2 corresponding to $x = \frac{1}{2}$ for respective values of b. Kitis & Pagonis (2007) have also presented a simple relation between μ_g and b considering a large number of glow curves which is given as

$$\mu_g = \left[\frac{1}{1+(b-1)\Delta m}\right]^{-1/(b-1)} \tag{38}$$

where $\Delta m = \frac{2}{u_m}$. This follows for $x = \frac{1}{2}$ from equation (31) by retaining the first two terms in the asymptotic expression (Abramowitz & Stegun, 1965) of $E_2(u_m)$, $E_2(u_x^+)$ and $E_2(u_x^-)$ following equations (24) and (25). However, from experimental point of view, equation (38) has no additional advantage over equation (32) to determine b from μ_g since u_m is unknown. Instead, equation (32) is more convenient to use.

Table 5. Coefficients from equations (35 – 37) corresponding to $x = \frac{1}{2}$ *(Kitis & Pagonis,2007)*

Order of Kinetics	c_τ	c_δ	c_ω	b_τ
1.0	1.50	0.98	2.48	1.58
1.5	1.66	1.37	3.03	1.80
2.0	1.80	1.74	3.52	2.0

Peak Shape Method without Prior Knowledge of Order of Kinetics

In the previous section, it is seen that to determine E it is essential to have prior knowledge of b which may be estimated from μ_g following equation (32) or (38). On the other hand, as is already discussed, μ_g has a weak but considerable dependence on u_m apart from the strong dependence on b. As a result, b cannot be determined accurately from a knowledge of μ_g only. Singh et al. (2012a) has shown that if μ_g is estimated from b without considering the effect of u_m, an error as high as 20% may occur in the value of μ_g. Singh, Karmakar, Singh & Singh (2012b) suggested a modified version of peak shape method to remove this difficulty where the dependence of μ_g on both b and u_m were considered. In this method, b is calculated in an iterative manner and hence determination of E does not require any prior knowledge of b. This is to be noted that only E_ω can be determined in this method.

To a good approximation, for a particular value of u_m, E_ω may be expressed as a function of μ_g as (Singh et al., 2012b)

$$E_\omega = \frac{kT_m^2}{\omega}\left(C_0 + C_1\mu_g + C_2\mu_g^2\right) + kT_m\left(D_0 + D_1\mu_g + D_2\mu_g^2\right) \tag{39}$$

The coefficients C_i, D_i $(i = 1, 2, 3)$ will vary with $u_m = \dfrac{E}{kT_m}$. Starting from a suitable guess value of u_m, E is estimated iteratively (Singh et al., 2012b). This final value of E can be used to determine b following the work of Singh, Karmakar & Singh (2013). Unlike the conventional method, this modified peak shape method enables simultaneous evaluation of E and b. In Table 6, the results derived from a number of synthetic peaks by applying this version of peak shape method (Singh et al., 2012b; Singh et al., 2013) are listed.

Table 6. Simultaneous determination of activation energy and order of kinetics from peak shape method (Singh et al., 2012b; Singh et al., 2013). Input values to generate the peaks are also listed. The linear heating rate is $\beta = 1.0\ Ks^{-1}$

Input Values for Peak Generation				Output Values	
E (eV)	**s (s⁻¹)**	**b**	**T_m (K)**	**E_ω (eV)**	**b**
1.0	1.0×10^{12}	1.0	384.6	0.997	1.001
		1.5	384.2	0.995	1.498
		2.0	383.9	0.997	2.001
		2.5	383.5	0.998	2.504

Peak Shape Method in Mixed Order Kinetics Model

In mixed order kinetics (MOK) model the concept of thermally disconnected deep traps comes into play. The charge neutrality condition is modified as

$$n_h = n + n_c + m \tag{40}$$

where *m* is the concentration of electrons trapped at deep traps. The expression for TL intensity in MOK model is given by (Kitis, Chen & Pagonis, 2008; Singh, Singh & Mazumdar, 1998)

$$I = -\frac{dn}{dt} = s_1' n(n+m) e^{-E/kT} \tag{41}$$

where $s_1' = \frac{sA_h}{nA_n}$. A parameter $\alpha = \frac{n_0}{n_0 + m}$ is defined (n_0 being the initial concentration of trapped electrons) in this approach through which the effect of symmetry factor and hence the order of kinetics is incorporated in the calculation of activation energy. This MOK model is a kind of intermediation between the first and second order cases: $\alpha \to 0$ *i.e.* $n_0 \ll m$ leads to $b = 1$ and $\alpha \to 1$ *i.e.* $n_0 \gg m$ leads to $b = 2$ (Yossian & Horowitz, 1997). Singh et al. (1998) expressed the peak shape relations for *E* corresponding to $x = \frac{1}{2}$ in a similar form given by equations (26 – 28) except for the fact that the coefficients C_i, D_i are expressed as quadratic functions of α instead of *b* as were given in equations (29,30). Thus, the coefficients C_i, D_i for MOK model are expressed as (Singh et al., 1998)

$$C_{ix} = C_{ix}^{(0)} + C_{ix}^{(0)}\alpha + C_{ix}^{(0)}\alpha^2 \tag{42}$$

$$D_{ix} = D_{ix}^{(0)} + D_{ix}^{(0)}\alpha + D_{ix}^{(0)}\alpha^2 \tag{43}$$

According to Kitis et al. (2008), the activation energies from peak shape method for MOK model may be expressed as

$$E_\tau = C_\tau \left(\frac{1-\mu_g''}{\mu_g''}\right)\left(\frac{F_m + \alpha}{F_m}\right)\frac{kT_m^2}{\tau} \tag{44}$$

$$E_\delta = C_\delta \left(\frac{F_m + \alpha}{F_m}\right)\frac{kT_m^2}{\delta} \tag{45}$$

$$E_{\omega} = C_{\omega}\left(\frac{1}{\mu''_{g}}\right)\left(\frac{F_{m}+\alpha}{F_{m}}\right)\frac{kT_{m}^{2}}{\omega} \tag{46}$$

where $F_{m} = \alpha\left(\frac{n_{m}+m}{n_{m}}\right)$. $n_{m} = \int_{T_{m}}^{\infty} I(T)dT$ is the density of trapped electrons at peak temperature. $\mu''_{g} = \frac{n_{m}}{n_{0}} = \frac{1-\alpha}{F_{m}-\alpha}$ is known as the integral symmetry factor which does not differ much from the conventional symmetry factor μ_{g}. The coefficients C_{τ}, C_{δ} and C_{ω} are determined by simulation (Kitis et al., 2008; Chen, Kristianpoller & Davidson, 1981).

ANALYSIS OF TL GLOW CURVES OF TREMOLITE AND ACTINOLITE

In this section, a demonstration of the peak shape method has been made by analysing experimental TL curves (Singh, 1989) of tremolite and actinolite materials. Both of these materials belong to silicate group of minerals. Tremolite is a member of the amphibole group of silicate minerals containing double silicate chain. Actinolite is an intermediate member in a solid-solution series between magnesium-rich tremolite and iron-rich ferro-actinolite. Actinolites are also manifested in nature in the form of jade gemstone. The TL behaviour of tremolites and actinolites have not been explored exhaustively though they have been under investigation for many decades. Northup & Lee (1940) included tremolite in the list of thermoluminescent materials. They reported that tremolite is a white fibrous material that exhibits a faint bluish white glow. Jain & Mitra (1977) reported green hornblende (tremolite) showed little TL sensitivity. Melo, Valerio & Caldas (2004) studied the TL characteristics of mineral samples acquired as jade from different parts of the world. In a subsequent work, they have also reported the TL dosimetric properties of the Brazilian actinolite irradiated by high dose of gamma rays (Melo, Valerio & Caldas, 2008). In the present report, TL glow peaks (Singh, 1989) of tremolite [$Ca_2Mg_5(Si_8O_{22})(OH)_2$)] and actinolite [$Ca_2Fe_2^{+5}(Si_8O_{22})(OH)_2$)] are investigated to extract the activation energy according to the peak shape method.

The TL behaviour of tremolite were recorded (Singh, 1989) after γ-ray excitation with a dose of 4.2 KGy and the resulting glow curve is shown in Figure 3. The glow curve in this case exhibits three distinct peaks and the second peak is the most intense. The glow curve of actinolite is also recorded (Singh, 1989) with a γ-ray dose of 4.2 KGy and is shown in Figure 3. The TL pattern for actinolite is rather complex exhibiting four peaks. The present analysis deals with all the peaks of tremolite and the first peak of actinolite. The respective peak positions and other parameters of both the materials obtained by curve fitting technique (Singh, 1989) are summarised in Table 7. The first two peaks of tremolite are presented in Figure 4 and the third one in Figure 5. The first peak of actinolite is displayed in Figure 5.

The consistency of the peak positions displayed in Table 7 has been checked by using the curve fitted values of *s* (Table 7) and equation (22). The required input value of *b* is estimated by using equations (32 – 34) from symmetry factor μ_{gx} which was calculated according to equation (31) for fractional intensities $x = \frac{1}{2}, \frac{2}{3}$ and $\frac{4}{5}$. In the next step, the activation energies $\left(E_{\tau}, E_{\delta}, E_{\omega}\right)$ of tremolite and ac-

Figure 3. Experimental glow curves of tremolite and actinolite (Singh, 1989)

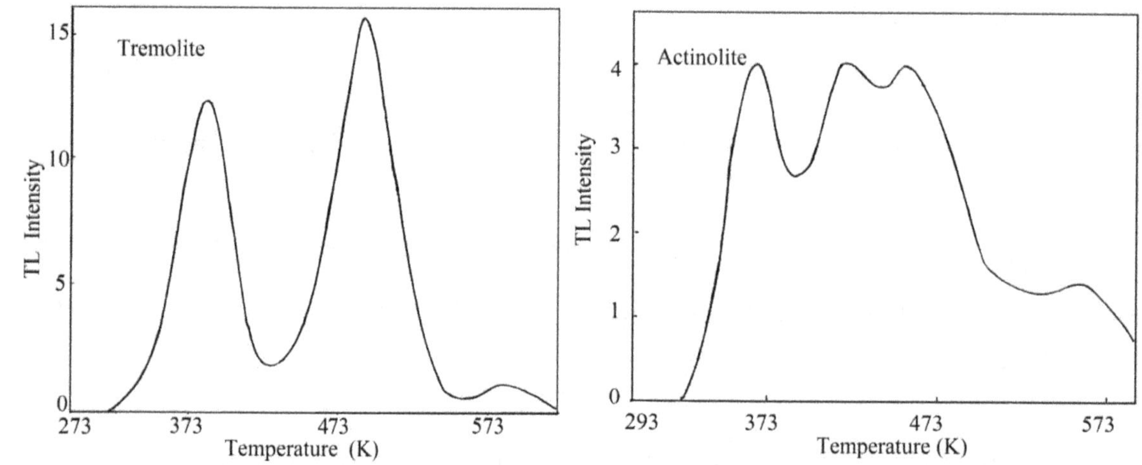

Figure 4. First and second glow peak of tremolite (Singh, 1989)] after being resolved from the composite peak given in Figure 3

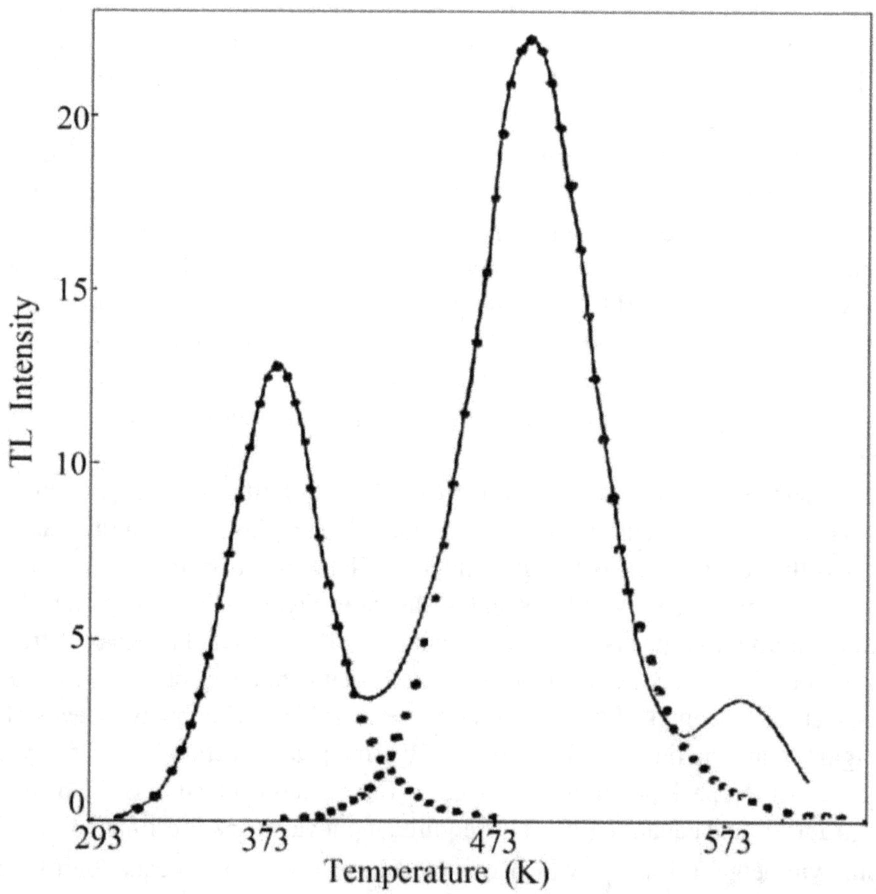

Figure 5. The third glow peak of tremolite and the first peak of actinolite (Singh, 1989)

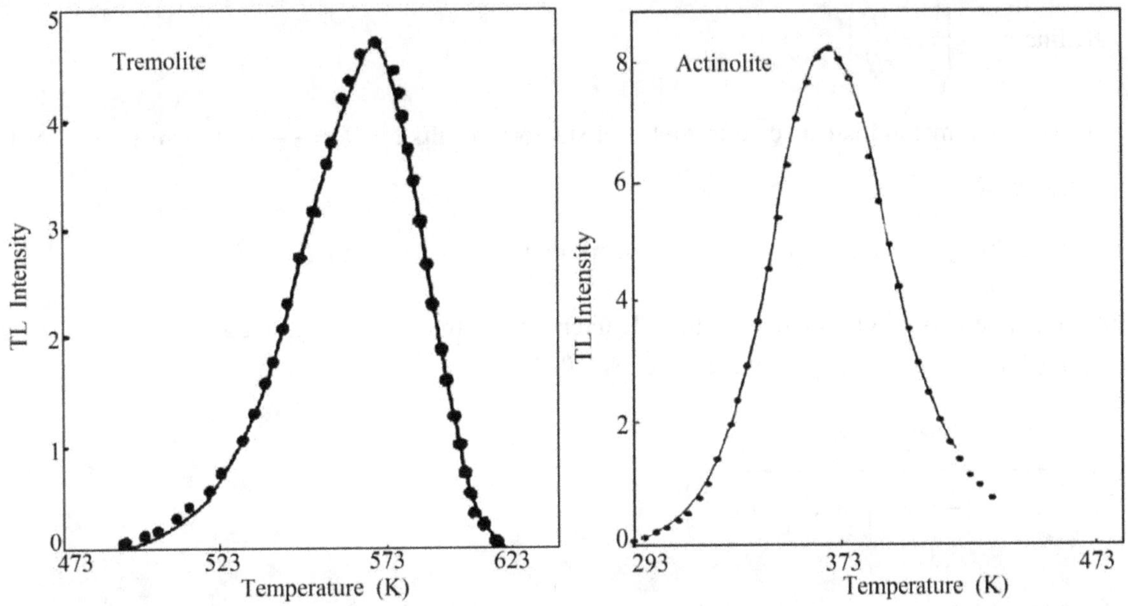

tinolite are determined using the equations (26 – 28) with the set of coefficients listed in Table 2. The results obtained are given in Table 8. The activation energies obtained from different fractional intensities corresponding to different shape parameters are not only in good agreement among themselves, but also are consistent with the curve fitted values (Singh, 1989).

The activation energies of TL peaks of tremolite and actinolite have also been determined by using equations (7) and (8) corresponding to the peak shape methods of Halperin et al. (1960) and Chen (1969a) respectively. The coefficients given in Table 1 are used while adopting the method of Chen (1969a). The results are displayed in Table 9. It is evident that the activation energies determined from Chen's method (1969a) is in good agreement with the other results given in Table 7 and Table 8 but those from the method of Halperin et al. (1960) yield some error.

LIMITATION OF PEAK SHAPE METHOD: SATURATED TL PEAKS WITH HEAVY RETRAPPING

Though peak shape method, as discussed so far, is a very successful and widely used technique for analysing experimental and numerical TL peaks, it has a limitation in case of saturated TL peaks with heavy retrapping. 'Heavy retrapping' condition is realised when the retrapping probability $\left(A_n\right)$ of the electrons is greater than the recombination probability $\left(A_h\right)$ *i.e.* $R = \frac{A_n}{A_h} > 1$. However, peak shape method works well in case of saturated peaks when the retrapping probability is negligible. For saturated TL peaks, the electron traps are completely filled *i.e.* $n_0 = N$. The following cases may be specified:

1. Peak shape method is expected to yield satisfactory results for $R \leq 1$, irrespective of the value of filling ratio $\left(f = \frac{n_0}{N}\right)$.

2. Peak shape method is expected to yield satisfactory results for $f = \frac{n_0}{N} \ll 1$, irrespective of the value of *R*.

3. Applicability of peak shape method is uncertain when $\frac{n_0}{N} \simeq 1$ and $R > 1$.

In the framework of OTOR model, the TL intensity for the case of saturated TL peaks with high retrapping $\left(R > 1\right)$ is given by (Kitis & Vlachos, 2013)

$$I = \frac{NRs\,e^{-E/kT}}{(1-R)^2 W\left[-1, e^{-Z_2}\right]^2 + W\left[-1, e^{-Z_2}\right]} \tag{47}$$

If the recombination process dominates over retrapping $\left(R < 1\right)$ for saturated peaks, this expression becomes

$$I = \frac{NRs\,e^{-E/kT}}{(1-R)^2 W\left[e^{Z_1}\right] + W\left[e^{Z_1}\right]^2} \tag{48}$$

where $Z_1 = \frac{1}{C} - \ln(C) + \frac{s}{(1-R)\beta}\int_{T_0}^{T} e^{-E/kT'}dT'$ (49)

Table 7. Curve fitted results of peak positions and other kinetic parameters determined from TL glow curves of tremolite and actinolite (Singh, 1989)

Sample	Type of Irradiation	Heating Rate (Ks^{-1})	Peak no.	Curve Fitted Parameters			
				T_m (K)	*E* (eV)	*s* (s^{-1})	*b*
Tremolite	γ-ray	0.667	1	377	0.85	9.5×10^9	2.0
			2	487	1.13	1.7×10^{10}	2.0
			3	573	1.40	6.8×10^{10}	1.0
Actinolite	γ-ray	0.667	1	369	0.70	3.0×10^8	2.0

Table 8. Activation energies of tremolite and actinolite determined by using the peak shape coefficients (Singh et al., 2012a) from equations (26 – 30) given in Table 2

Material	T_m (K)	x	μ_{gx}	b	Mean b	Activation Energy (eV)			
						E_τ	E_δ	E_ω	Mean E
Tremolite	377.7	$\frac{1}{2}$	0.521	2.00	2.10	0.851	0.852	0.852	0.872
		$\frac{2}{3}$	0.515	2.08		0.861	0.876	0.869	
		$\frac{4}{5}$	0.511	2.23		0.879	0.910	0.896	
	487.1	$\frac{1}{2}$	0.520	1.99	2.10	1.129	1.127	1.129	1.158
		$\frac{2}{3}$	0.515	2.08		1.144	1.165	1.155	
		$\frac{4}{5}$	0.511	2.23		1.169	1.211	1.191	
	573.0	$\frac{1}{2}$	0.419	1.09	1.07	1.425	1.499	1.458	1.441
		$\frac{2}{3}$	0.438	1.02		1.405	1.418	1.411	
		$\frac{4}{5}$	0.454	1.09		1.432	1.473	1.452	
Actinolite	369.1	$\frac{1}{2}$	0.524	2.03	2.15	0.703	0.711	0.708	0.726
		$\frac{2}{3}$	0.518	2.14		0.713	0.736	0.725	
		$\frac{4}{5}$	0.513	2.29		0.730	0.763	0.747	

Table 9. Activation energies of tremolite and actinolite determined by using the peak shape methods of Halperin et al. (1960) and of Chen (1969a) for $x = \frac{1}{2}$

Material	T_m (K)	μ_{gx}	Activation Energy (eV)				
			Halperin et al. [11]		Chen [2]		
			E_τ		E_τ	E_δ	E_ω
Tremolite	377.7	0.521	0.823		0.848	0.851	0.853
	487.1	0.520	1.104		1.126	1.125	1.131
	573.0	0.419	1.467		1.400	1.385	1.406
Actinolite	369.1	0.524	0.633		0.697	0.713	0.708

$$Z_2 = \frac{1}{C} - \ln(C) + \frac{s}{(1-R)\beta}\int_{T_0}^{T} e^{-E/kT'} dT' \quad (50)$$

and $$C = \frac{n_0 \mid 1-R \mid}{N R} \quad (51)$$

$W\left[e^{Z_1}\right]$ and $W\left[-1, e^{-Z_2}\right]$ are the principal and the second branches of the Lambert *W*-function respectively (Corless, Gonnet, Hare, Jeffrey & Knuth, 1996; Corless, Jeffrey & Knuth, 1997).

It should be noted that saturated TL peaks with heavy retrapping have not yet been observed experimentally. Some glow curves are generated numerically for such cases ($R > 1$) corresponding to both saturated and unsaturated cases following equation (47) and the peak shape method is applied on these curves. Figure 6 displays the numerically simulated glow peaks corresponding to $R = 30$ for both unsaturated ($f = 0.1$) and saturated ($f = 1.0$) cases. Similar glow peaks may be simulated for heavy recombination cases following equation (48). The results displayed in Table 10 show that the activation energies derived from peak shape method deviates from the input value for saturated TL peaks with heavy retrapping whereas those corresponding to unsaturated peaks, the results are quite in agreement.

CONCLUDING REMARKS

Peak shape method is apparently the simplest of different methods of determination of activation energy from a TL glow curve. However, there are some difficulties in the application of this method both from theoretical and experimental points of view. Most of the variants of peak shape method require prior knowledge of order of kinetics. The order of kinetics may be determined from the symmetry factor of the peaks but the symmetry factor has a dependence on $\frac{E}{kT_m}$ also. Because of this non-uniqueness of symmetry factor for a particular value of order of kinetics, an error gets introduced in the estimation

Table 10. Activation energies determined by peak shape method applied on numerically simulated unsaturated and saturated TL peaks with heavy retrapping for $x = \frac{1}{2}$. *The input values are:* $E = 1.0\,eV$; $n_0 = 10^{10}\ cm^{-3},\ s = 10^{12}\ s^{-1};\ \beta = 1.0\ Ks^{-1}$

Filling Ratio (*f*)	*R*	T_m (K)	μ_g	Effective *b*	Activation Energy (eV)		
					E_τ	E_δ	E_ω
0.1	3	451.13	0.522	2.01	1.045	1.039	1.043
	30	489.48	0.525	2.04	1.043	1.042	1.044
	300	534.78	0.527	2.01	1.046	1.053	1.051
1.0	3	416.86	0.527	2.07	0.885	0.883	0.885
	30	453.18	0.520	1.99	0.810	0.827	0.820
	300	492.44	0.522	2.01	0.810	0.832	0.823

Figure 6. Unsaturated and saturated synthetic TL glow peaks with heavy retrapping ($R > 1$ *)*

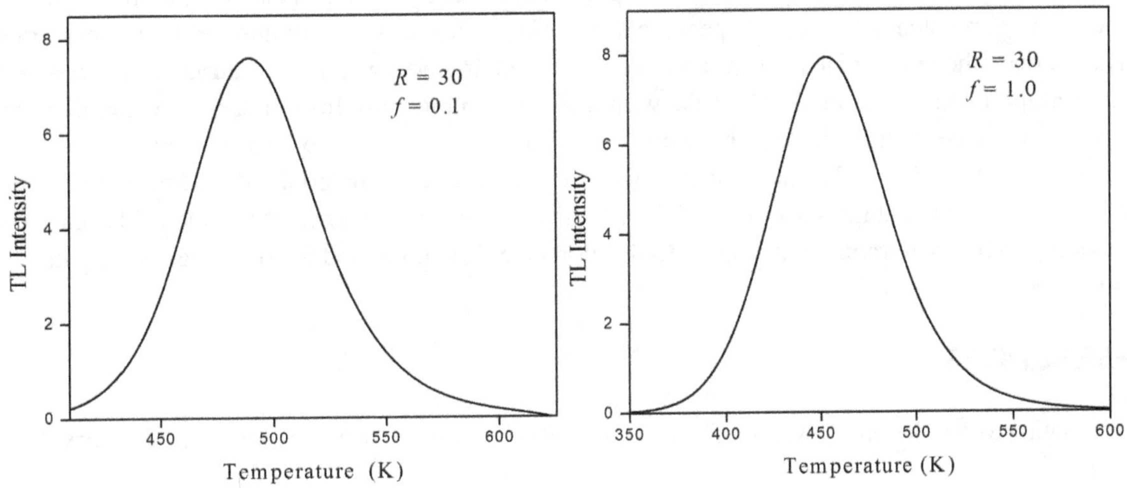

of order of kinetics and hence the accuracy of activation energy determined thereof is also compromised. In fact the error in estimating activation energy estimated this way may mount up to 20%. To avoid this problem, one has to perform an independent experiment such as decay of TL to have a precise estimation of order of kinetics. In the present work it has been shown that the difficulty arising due to the non-uniqueness of the symmetry factor may be overcome by employing a modified peak shape method that does not require prior knowledge of the order of kinetics. A new aspect is seen where simultaneous determination of activation energy and order of kinetics is possible by using this modified peak shape method.

From experimental point of view, the occurrence of an isolated TL peak is very rare. The peak under observation might be affected both on high and low temperature sides due to overlapping with satellite peaks. To avoid this difficulty, it might be more useful if the peak shape method is applied to not only the half-intensity points but also to other points in the upper portion of the peak where the points might be less affected due to the presence of neighbouring peaks. The symmetry factor should be checked not only for fractional intensity $x = \frac{1}{2}$ but also for other values such as $\frac{2}{3}, \frac{4}{5}$ and order of kinetics should be calculated accordingly. The consistency of these values of order of kinetics has to be checked before applying the peak shape method to experimental TL peaks.

The peak shape method has been applied on glow peaks of γ-irradiated tremolite and actinolite materials. Though tremolite-actinolite system belongs to highly complex mineral group in respect of their chemical composition and presence of impurities, their TL pattern is rather simple. TL curve of tremolite, being irradiated in room temperature by γ-ray exhibits three peaks while that of actinolite shows at least four peaks. As per as sensitivity is concerned, the TL response of tremolite is about five times more than that of actinolite. The present work addresses all three peaks of tremolite and only the first peak of actinolite. Analysis of all these TL peaks have been performed by different prominent versions of peak shape method. Except for the method of Halperin et al. (1960), all other results are quite consistent among themselves and also in reasonable agreement with the results obtained by curve fitting method.

In case of saturated TL peaks with heavy retrapping, the activation energies cannot be determined accurately in conventional peak shape method, though such peaks are yet to be observed experimentally. This shows a limitation of the applicability of peak shape method. The peak shape method applied on numerically simulated saturated TL peaks corresponding to heavy retrapping yields a considerable deviation from the input value of activation energy, though it works well for unsaturated TL peaks with heavy retrapping and also for all TL peaks with heavy recombination. In summary, the simplicity and effectiveness of peak shape method have made it a very useful tool for extracting activation energies corresponding to different TL mechanisms inherent in luminescent materials. It will be a challenging task for both experimentalists and theoreticians to detect practical evidence of saturated TL peaks corresponding to heavy retrapping and hence to develop an efficient method for studying such peaks.

REFERENCES

Abramowitz, M., & Stegun, I. A. (Eds.). (1965). *Handbook of mathematical functions*. New York, USA: Dover.

Bhattacharyya, S., Karmakar, M., Majumdar, P. S., & Singh, S. D. (2016). On the thermoluminescence glow curve analysis recorded under hyperbolic heating profile. *Int. J. Lumin. Appl., 6*, 194.

Chen, R. (1969). Glow curves with general order kinetics. *Journal of the Electrochemical Society, 116*(9), 1254. doi:10.1149/1.2412291

Chen, R. (1969). On the calculation of activation energies and frequency factors from glow curves. *Journal of Applied Physics, 40*(2), 570–585. doi:10.1063/1.1657437

Chen, R., Kristianpoller, N., Davidson, Z., & Visocekas, R. (1981). Mixed first and second order kinetics in thermally stimulated processes. *Journal of Luminescence, 23*(3-4), 293–303. doi:10.1016/0022-2313(81)90135-6

Chen, R., & McKeever, S. W. S. (1997). *Theory of thermoluminescence and related phenomena*. Singapore: World Scientific Publishing. doi:10.1142/2781

Chen, R., & Pagonis, V. (2011). *Thermally and Optically Stimulated Luminescence: A Simulation Approach*. Chichester, UK: Wiley. doi:10.1002/9781119993766

Christodoulides, C. (1985). Determination of activation energies by using the widths of peaks of thermoluminescence and thermally stimulated depolarisation currents. *Journal of Physics. D, Applied Physics*, *18*(8), 1501–1510. doi:10.1088/0022-3727/18/8/015

Corless, R. M., Gonnet, G. H., Hare, D. E. G., Jeffrey, D. J., & Knuth, D. E. (1996). On the Lambert W function. *Advances in Computational Mathematics*, *5*(1), 329–359. doi:10.1007/BF02124750

Corless, R. M., Jeffrey, D. J., & Knuth, D. E. (1997). A sequence of series for the Lambert W Function. In *ISSAC '97: Proceedings of the 1997 International Symposium on Symbolic and Algebraic Computation*, (pp. 197). New York: ACM Press. 10.1145/258726.258783

Fleming, R. J. (1968). Tabulation of glow-curve characteristics. *Canadian Journal of Physics*, *46*(14), 1569–1573. doi:10.1139/p68-483

Garlick, G. F. J., & Gibson, A. F. (1948). The electron trap mechanism of luminescence in sulphide and silicate phosphors. *Proceedings of the Physical Society*, *40*(6), 574–590. doi:10.1088/0959-5309/60/6/308

Grossweiner, L. I. (1953). A Note on the Analysis of First Order Glow Curves. *Journal of Applied Physics*, *24*(10), 1306–1307. doi:10.1063/1.1721152

Halperin, A., & Braner, A. A. (1960). Evaluation of thermal activation energies from glow curves. *Physical Review*, *117*(2), 408–415. doi:10.1103/PhysRev.117.408

Halperin, A., Braner, A. A., Ben-Zvi, A., & Kristianpoller, N. (1960). Thermal Activation Energies in NaCl and KC1 Crystals. *Physical Review*, *117*(2), 416–422. doi:10.1103/PhysRev.117.416

Jain, V. K., & Mitra, S. (1977). Thermoluminescence studies of some silicate minerals. *Thermochimica Acta*, *18*(2), 241–244. doi:10.1016/0040-6031(77)80024-5

Keating, P. N. (1961). Thermally Stimulated Emission and Conductivity Peaks in the Case of Temperature Dependent Trapping Cross Sections. *Proceedings of the Physical Society*, *78*(6), 1408–1415. doi:10.1088/0370-1328/78/6/343

Kitis, G., Chen, R., & Pagonis, V. (2008). Thermoluminescence glow-peak shape methods based on mixed order kinetics. [a]. *Physica Status Solidi*, *205*(5), 1181–1189. doi:10.1002/pssa.200723470

Kitis, G., & Pagonis, V. (2007). Peak shape methods for general order thermoluminescence glow-peaks: A reappraisal. *Nucl. Instm. Methods B*, *262*(2), 313–322. doi:10.1016/j.nimb.2007.05.027

Kitis, G., & Vlachos, N. D. (2013). General semi-analytical expressions for TL, OSL and other luminescence stimulation modes derived from the OTOR model using the Lambert W-function. *Radiation Measurements*, *48*, 47–54. doi:10.1016/j.radmeas.2012.09.006

Levy, P. W. (1985). Thermoluminescence kinetics in materials exposed to the low doses applicable to dating and dosimetry. *Nuclear Tracks*, *10*, 547.

Luschik, C. B. (1956). The Investigation of Trapping Centres in Crystals by the Method of Thermal Bleaching. *Soviet Physics, JETP*, *3*, 390.

May, C. E., & Partridge, J. A. (1964). Thermoluminescent kinetics of alpha-irradiated alkali halides. *The Journal of Chemical Physics*, *40*(5), 1401–1409. doi:10.1063/1.1725324

Mazumdar, P. S., Singh, S. J., & Gartia, R. K. (1988). The determination of activation energy from the shape of a thermoluminescence peak. *Journal of Physics. D, Applied Physics*, *21*(5), 815–819. doi:10.1088/0022-3727/21/5/023

Melo, A. P., Valerio, M. E. G., & Caldas, L. V. E. (2004). Thermoluminescent characteristics of mineral samples acquired as jade. *Nucl. Instm. Methods B*, *218*, 198–201. doi:10.1016/j.nimb.2003.12.058

Melo, A. P., Valerio, M. E. G., & Caldas, L. V. E. (2007). Thermoluminescent characteristics of Actinolite-Teflon composites for gamma high-doses. *Physica Status Solidi*, *4*(c), 1052–1055. doi:10.1002/pssc.200673854

Northup, M. A., & Lee, O. I. (1940). Experiments on the Thermoluminescence of Some Common and Unusual Minerals. *Journal of the Optical Society of America*, *30*(5), 206. doi:10.1364/JOSA.30.000206

Pagonis, V., Kitis, G., & Furetta, C. (2006). *Numerical and Practical Exercises in Thermoluminescence*. New York: Springer.

Randall, J. T., & Wilkins, M. H. F. (1945). Phosphorescence and electron traps. II. the interpretation of long-period phosphorescence. *Proceedings of the Royal Society of London. Series A, Mathematical and Physical Sciences*, *184*(999), 390–407. doi:10.1098/rspa.1945.0025

Singh, S. J., Karmakar, M., Bhattacharya, M., Singh, S. D., Singh, W. S., & Azharuddin, S. (2012). On the reappraisal of the peak shape method for general order thermoluminescence. *Indian Journal of Physics*, *86*(2), 113–116. doi:10.100712648-012-0013-6

Singh, S. J., Karmakar, M., & Singh, S. D. (2013). On the determination of the order of kinetics in thermoluminescence by peak-shape method. *Rad. Eff. Def. Sol.*, *168*(5), 352–357. doi:10.1080/10420150.2013.771358

Singh, S. J., Karmakar, M., Singh, W. S., & Singh, S. D. (2012). Modified peak shape method for the determination of activation energy in thermoluminescence. *Physica Scripta*, *86*(3), 035702. doi:10.1088/0031-8949/86/03/035702

Singh, T. S. C. (1989). *Thermally stimulated luminescence of some solids* (Unpublished doctoral dissertation). Manipur University, Imphal, India.

Singh, W. S., Singh, S. D., & Mazumdar, P. S. (1998). On the mixed order kinetics of the thermoluminescence glow peak. *Journal of Physics Condensed Matter*, *10*(22), 4937–4946. doi:10.1088/0953-8984/10/22/018

Yossian, D., & Horowitz, Y. S. (1997). Mixed-order and general-order kinetics applied to synthetic glow peaks and to peak 5 in Lif:Mg,Ti (TLD-100). *Radiation Measurements*, *27*(3), 465–471. doi:10.1016/S1350-4487(97)00007-3

Chapter 3
Phosphors for Various Dosimetry Applications Derived by Different Synthesis Routes

Jyoti Singh
Galgotias University, India

Mahesh S. Bhadane
Savitribai Phule Pune University, India

Vikas Dubey
Bhilai Institute of Technology, India

Sanjay Daga Dhole
Savitribai Phule Pune University, India

Jairam Manam
IIT(ISM) Dhanbad, India

Shailendra Satish Dahiwale
Savitribai Phule Pune University, India

ABSTRACT

The chapter provides useful information about synthesis and characterization of dysprosium doped oxide and fluoride-based phosphors such as $SrGd_2O_4$, $CaSO_4$, and CaF_2. Various techniques (e.g., acid-recrystallization, chemical co-precipitation, and homogenous precipitation cum auto-combustion methods) were adopted to synthesize these phosphors for large-scale production. All the prepared phosphors were characterized by x-ray diffraction analysis, scanning electron microscopy, and transmission electron microscopy techniques. The thermoluminescence (TL) studies were performed after different irradiation sources such as gamma rays, thermal neutrons, and low energy ions (H, Ar, and N), respectively. Linear dose responses were observed in a wide range of doses for all the samples. Various trapping parameters, namely order of kinetics, activation energy, and frequency factors, were calculated by using computerized glow curve deconvolution (CGCD) method.

DOI: 10.4018/978-1-5225-5170-6.ch003

INTRODUCTION

The phenomenon of thermoluminescence (TL) takes place when the radiation incident on insulating crystals and a fraction of deposited energy is stored at the defect centres or colour centres in the crystal lattice. After heating the respective crystal, this stored energy is released in the form of visible light (Bull, 1985, pp 105-113). Under certain limitations, the radiation dose absorbed through the crystal is directly proportional to the amount of light emitted. In the radiation dosimetry, the TL phosphors are utilized as a passive dosimeter for monitoring the integrated radiation doses obtained from various sources (Bull, 1985, pp 105-113). Linear dose response, low fading, radiation resistant, simple TL profile, less dependence on the energy of irradiation, high chemical robustness, and high sensitivity are essential requirements for materials used for dosimetric applications (Jayasudha et al., 2017; Gupta et al., 2016; Manam et al., 2010; Reddy et al., 2016). An appropriate combination of host-dopant and optimization of luminescent centres incorporated in the host lattice plays an important role in making sensitive phosphors because a single host does not hold all the characteristics of TLD phosphors very often (Jayasudha et al., 2017; Gupta et al., 2016; Manam et al., 2010; Reddy et al., 2016; Kumar et al., 2016; Wang et al., 2005).

TL responses of rare-earth doped oxide and fluoride based matrices are extensively studied due to their high chemical stability, wide range radiation dose response, less fading and high sensitivity to very small doses (Gupta et al., 2016; Wanget et al., 2005; Manam & Das, 2008; Dubey et al., 2016; Sun et al., 2014; Lakshmanan, 2001). These materials are mainly applicable in radiation dosimetry for environmental and personnel monitoring. A lot of researches have been done to develop such synthetic materials which exhibit all the characteristics of a good TLD material (Jayasudha et al., 2017; Gupta et al., 2016; Manam et al., 2010; Reddy et al., 2016; Kumar et al., 2016; Wang et al., 2005; Manam & Das, 2008; Dubey et al., 2016; Sun et al., 2014). On the basis of these researches, it is found that mixed alkali/alkaline oxide and sulphate matrices constitute a class of good TL phosphors and their TL performances improved when an appropriate amount of activators are incorporated (Manam & Das, 2008; Dubey et al., 2016; Sun et al., 2014; Lakshmanan, 2001; Salah et al., 2006; Shinsho et al., 2008; Salah N., 2008; Bhadane et al., 2016; Wagiran et al., 2012; Mondragon et al., 2008; Lakshmanan et al., 2008). As a novel luminescent host, the structural and thermoluminescence properties of SrY_2O_4 (Dubey et al., 2016) and $BaGd_2O_4$ (Sun et al., 2014) have been extensively investigated. However, no research work has been claimed so far concerning to TL studies of $SrGd_2O_4$: Dy^{3+} host synthesized via homogeneous precipitation cum auto-combustion method.

Additionally, it is worthwhile to mention that Dy doped $CaSO_4$ (having a tissue no. ~15.3) and CaF_2 (having a tissue no. ~16.3) is an admired phosphor for their excellent dosimetric applications. In the past couple of decades, a considerable amount of work is involved in search of development in $CaSO_4$: Dy and CaF_2: Dy phosphor materials (Lakshmanan, 2001; Salah et al., 2006; Shinsho et al., 2008; Salah N., 2008; Bhadane et al., 2016; Wagiran et al., 2012; Mondragon et al., 2008; Lakshmanan et al., 2008).

In India, $CaSO_4$: Dy is used as a dosimetry material for personnel as well as environmental applications. $CaSO_4$: Dy is one of the highly sensitive microphosphor for gamma dosimetry as compared to nano and has been studied by a various group (Salah et al., 2006; Shinsho et al., 2008; Lakshmanan et al., 2008) using different synthesis methods. $CaSO_4$: Dy and other microcrystalline phosphors are widely been used as a thermoluminescence (TL) dosimetric materials and are useful for low as well as high dose measurements (Madhusoodanan et al., 1999). Overall, $CaSO_4$ is the most interesting phosphor activated with Dysprosium, because of the considerable stability, sensitivity, repeatability, etc. It also enhances the luminescence efficiency and defects population of the phosphor. However, $CaSO_4$:Dy is also reported on the basis of mixed field dosimetry (Ayyangar et al., 1974) for gamma as well as neutron and found an additive and no permanent damage to the γ- ray sensitivity observed after irradiating maximum thermal neutron fluence. Shortly, the neutron detectors or dosimetry is not easy as the dosimetry of gamma radiations and other radiations because of the numerous direct and indirect energy transfer processes involved and the variation of the reaction cross sections with energy (Marrale et al., 2012; Patra et al., 2015). Neutron dosimetry is vital a device which is used in a numerous field such as nuclear medicine, nuclear power, particle physics, magnetism field, etc (Shani G., 2013). Specially, thermal neutron has less energy (~0.25 eV) and favorable uses for medical purposes plus nuclear research laboratories, reactors, and industries (Triolo et al., 2007; Bhadane et al., 2016). Thermal neutron response of every TL materials depends on the capture cross section of the elements of TL materials for the thermal neutron. Basically, the thermal neutron interacts with phosphor materials primarily through scattering and absorption processes i.e. Elastic scattering, Inelastic scattering, Non-elastic scattering, and Capture process. However, in the present case, inelastic and non-elastic scattering processes are most probable as they form (n, α) and (n, γ) reactions which produces prompt α and γ particles responsible to have increase dosimetric change due to thermal neutrons (Yazici et al., 2002).

Another side, calcium fluoride (CaF_2) has got attention because it has a wide-spread energy band gap values in the range of 10 to 11 eV. Therefore, the probability of the electron re-trapped or de-trapped range is high, whenever ionizing radiation interacts with phosphors. The TL properties of CaF_2 can be increased by adding different dopants such as Dy, Eu, Cu (Bhadane et al., 2016; Wulf & Gniadecka, 1996; Zahedifar et al., 2012; Bos & Dielhof, 1991; Campos & Lima, 1987; Wagiran et al., 2012), Mn (Mondragon, 2008), Tm (Deilhof et al., 1988) etc. Among all, Dy doped calcium fluoride (here after referred as CaF_2: Dy) is one of the most sensitive phosphors, and is commercially available as a TLD-200 from Harshaw Chemical Company (Bhadane et al., 2017). Several research groups have studied the TL properties of CaF_2: Dy nanophosphor for dosimetric applications, especially for beta, gamma, electron, UV, neutron, low energy (Fain & Montret, 1980) and swift heavy ions respectively (Yamashita et al., 1971; Rajurkar et al., 1994; Kumar et al., 2008; Bos, 2001; Azorin, Furetta & Scacco, 1993; Salah, Sahare & Rupasov, 2007; Kitis, Gomez-Ros & Tuyn, 1998; Pagonis, Kitis & Furetta, 2006; Graser, 1999; Sunitha et al., Singh, Manam & Singh, 2017; Jain, 2011; Avasthi & Mehta, 2011; Gnaser, 1999). On the other hand, using low energy ion irradiation, not much work has been carried out for dosimetric applications.

SYNTHESIS ROUTES

Generally, there are many types of synthesis routes to prepare the different types of the sample for various applications. Among them, one of the key routes to prepare the microparticles is acid recrystallization and for the preparation of nanoparticles, we have employed chemical co-precipitation method and homogeneous precipitation cum auto-combustion method, respectively.

Acid Re-Crystallization Method

Yamashita et al. illustrated the acid re-crystallization method was the well-known method for the preparation of microparticles for different application purposes. To prepare $CaSO_4$: Dy microcrystalline phosphor for thermal neutron dosimetry, the following procedure has been adopted:

- **Apparatus Required:** Heating Mantle, 1 or 2-liter size rounded bottom flask, Cylindrical condensed system, 2 conical flasks, 10 mm diameter rubber pipe which was easily connected with condenser and conical flasks (high quality), double distilled or de-ionized water etc.
  - Firstly, we have to take AR graded $CaSO_4.2H_2O$ and an appropriate amount of Dy_2O_3 (Sigma Aldrich) as an activator and dissolved in to calculated percentage concentration of H_2SO_4 (Acid) in a rounded bottom flask which is connected to a sealed condensed system.
  - The acid was distilled out in a flask connected to the condenser as shown in Figure 1.
  - Heat to the acid solution till boiling point reach of H_2SO_4, after the complete evaporation of the acid, was getting the recrystallized $CaSO_4$: Dy (0.1 mol%) impurity in the form of wet powder (i.e. phosphor).
  - After cooling the round bottom flask, the $CaSO_4$: Dy phosphor thus obtained in microcrystalline form.
  - Such a phosphor was washed repeatedly with double distilled water to remove the acidic parameter and dried in an oven at 100 °C-120 °C for 5h.
  - Finally, the phosphors was collected in powder form, crushed it (using the pestle mortar) to obtain the particle size ranging from 150 to 250 μm.

Therefore, this microcrystalline powder was required heat treatments to use as TLD phosphors.

Chemical Co-Precipitation Method

This is very easy and time consuming synthesis route for the preparation of nanoparticles. By using this method, CaF_2: Dy nanophosphors for low energy ion dosimetry are synthesized. The procedure is as follows:

Figure 1. Complete set up of Acid re-crystallization Method

1. Primarily, some of the apparatus is required such as beakers/ conical flask, burette, magnetic needle, stirrer machine etc.
2. For the preparation of CaF_2: Dy nanophosphor, we have to take a appropriate amount of $CaCl_2$ and NH_4F were separately dissolved in every 20 ml doubled distilled water with continuous stirring for 30 min.
3. After this, 0.1 mol % Dy was added into $CaCl_2$ solution under the continuous stirring process of 30 min.
4. Then 50 ml ethanol was added to the solution and stirred again for 20 min.
5. Further, NH_4F solution was added drop wise in the above solution results in white precipitate as shown in Figure 2.
6. After 30 min, the white precipitate was filtered, washed with double distilled water and ethanol several times to remove un-reacted precursors and reaction by products.
7. Then, it was dried at 120 °C for 5 h and finally compound was ready for further processes.

Figure 2. Set up of Chemical co-precipitation Method

For the Low Energy Ion Beam (LEIB) irradiation purposes, pellet form phosphor material was required. Therefore, by using the hydraulic pressure around 6 to 7 ton load, a very thin (thickness less than 0.5 cm & diameter is 1 cm) CaF_2: Dy nanophosphor pellets were prepared. These pellets were irradiated by Low Energy Ion Beam Facility (LEIBF) of Ar, H, and N ions in the fluence range from 1×10^{13} to 5×10^{14} ion/cm^2 at a beam current ~600 nA in vacuum ~10^{-6} torr.

Homogeneous Precipitation cum Auto-Combustion Method

The powder samples of $SrGd_2O_4$: Dy samples were prepared by homogeneous precipitation cum auto-combustion method after controlled heating at 1200°C for 3h in air ambience. The procedure is as follows:

1. Firstly, the stoichiometric amounts of starting materials $Sr(NO_3)_2$ (Otto Chemie Pvt. Limited 99%), $Gd(NO_3)_3.6H_2O$ (Otto Chemie Pvt. Limited, 99.9%) and $Dy(NO_3)_3.xH_2O$ (Sigma-Aldrich, 99.9%) were dissolved in the excess of ethanol.
2. The above mixture taken in a glass beaker was stirred well for 30 min under heating at 80°C to get a homogeneous clear solution, after that, the quantity of urea (H_2NCONH_2) along with the boric acid (little amount works as a flux) was added.
3. The molar ratio of the reactants $Sr(NO_3)_2$: $Gd(NO_3)_3$: $Dy(NO_3)_3.xH_2O$: $(NH_2)_2CO$ was taken 1: 2(1-x): 2x: 6.667 (where, x= 3 mol%).
4. Afterward, the obtained solution was continuously stirred at 100°C for at least 3h until the liquid totally evaporated and white powders was left.
5. In the next step, the powder was ground properly and then introduced into a muffle furnace at 1200°C for 3h.

The samples prepared by auto-combustion method show the pure $SrGd_2O_4$ phase than the samples prepared by simple combustion route.

CHARACTERIZATION TECHNIQUES

All the samples were characterized by the following characterization techniques for analysis purpose:

X-ray diffraction pattern was studied for structural analysis at room temperature for micro and nano-phosphor samples using Cu-target (Cu-K_α= 1.54 Å) on Bruker AXS D8 Advance X-ray Diffractometer, to confirm the formation of the structure. SEM and FESEM images were obtained using Scanning Electron Microscope, "JEOL JSM-6360A (FEI, Netherlands) and field-emission scanning electron microscope (FESEM Supra-55, Germany), respectively and TEM images were obtained on "TECHNAI G2 20 (FEI, Netherlands)" operated at 200 kV to observe the morphology. TL glow curves measurements were recorded for all the samples after irradiation of thermal neutron, low energy ions and gamma rays on a Nucleonix TLD Reader (Model 1009I) taking ~5 mg of sample each time; at a constant heating rate of 5 °C/s. The as-synthesized $SrGd_2O_4$: Dy samples were exposed with γ-rays obtained from ^{60}Co target in the dose ranging from 80 Gy to 2 kGy at room temperature. However, the irradiation procedure of thermal neutron and low ion beams are as follows:

Irradiation Facilities

^{252}Cf Thermal Neutron Chamber and Sample Irradiation Process

^{252}Cf based thermal neutron source having a flux of 10^6 n/cm^2/second and half life 2.645 years was used for irradiation purposes (Campos & Lima, 1987). The neutron spectrum is having an energy ~0.025 eV which corresponds to thermal neutrons. The schematic of the ^{252}Cf based thermal neutron facility is shown in Figure 3. The ^{252}Cf irradiator presents cylindrical shape with a coaxial stainless steel filter fitted in the center of the cylindrical chamber. The ^{252}Cf is covered by paraffin wax as a moderator which filters out the high energy neutrons and converts to thermal neutrons. The irradiation of the sample was done in the outer shell of the ring which gets constant flux of thermal neutrons. The sample rod was used to keep and take out the samples from irradiation chamber. Moderator tank of paraffin wax is fully surrounded by borated paraffin to shield the neutrons. It is also covered by thin lead shielding to attenuate the unwanted gamma rays emitted from the ^{252}Cf source and still remains at the sample position with small amount. Consequently, to avoid these gamma contamination during thermal neutron irradiation, set of experiments were performed to estimate the actual contribution of the gamma dose. Here, $CaSO_4$:Dy phosphor covered with aluminum foil was kept in a 5 mm cylindrical plastic vial filled up with the

Figure 3. Schematic of the cylindrical type ^{252}Cf moderated thermal neutron irradiation chamber

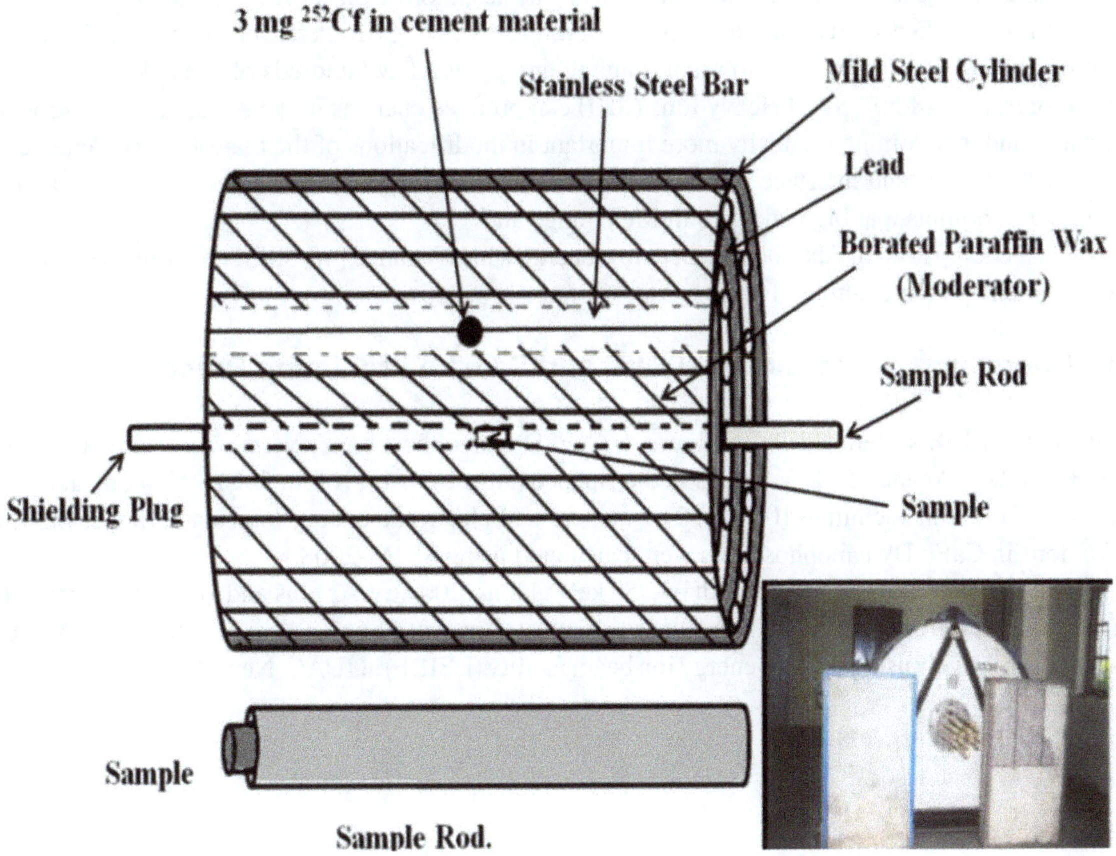

boric powder. The boric powder was used to stop the thermal neutron and aluminum foil was used to stop the alpha particles produced from boron. Additionally, a pure $CaSO_4$: Dy sample was kept for irradiation to see the cumulative dose effects. The dose separation way was adopted to estimate the pure thermal neutrons.

Ion Beams

Low and high energy ion beams have numerous uses mostly in material science, nuclear physics, medical sciences and biological sciences. Ion beam techniques are very prevailing tools in materials modification technology due to their high flexibility and the possibility to control exactly technological important parameters like energy, fluence, beam size and position. According to the effect responsible for the change of surface properties, the modification of materials by energetic ions may be classified in the following way: Doping, Ion beam mixing, Use of radiation defects, Ion beam irradiation etc. According to different applications of ion beams, they are useful and applied for most of scientific, technology and research area such as: Engineering of materials, Ion beam mixing and multilayer, Ion beam induced adhesion and stitching, Electronic devices, Polymers and so on.

Swift Heavy Ions (SHI)

A Pelletron called high energy ion beam accelerator (tandem Van de Graaff type accelerator), especially used for the basic and applied research in nuclear physics, atomic physics, materials science, biosciences and other allied fields available at IUAC, New Delhi. A 15UD Pelletron accelerator is capable of delivering any ion from proton to uranium up to energy of a few hundreds of MeV depending upon the nature of ion [48-50]. Swift Heavy Ions (SHI) can produce changes in the solid, thin film surfaces, interfaces and is becoming gradually more important in modifications of the materials. The interaction of ions with surface and interface results in the modifications of structural characteristics and applied loss $(dE/dx)_e$ dominant at high energies in MeV range and

Nuclear energy loss $(dE/dx)_n$ dominant at low energies in keV range [48-50]. Therefore, in the present case a low energy ion beam has been used for dosimetric purposes.

Low Energy Ion Beam Facility (LEIB) and Sample Irradiation Process

By using the LEIB system shown in Figure 4, CaF_2:Dy nanophosphors pellets were irradiated by 100 keV H, 200 keV Ar and 350 keV N ions in the fluence range from 1×10^{13} to 5×10^{14} ion/cm^2 at a beam current ~600 nA in vacuum ~10^{-6} torr. The nuclear and electronic energy losses and ranges of H, Ar and N ions in CaF_2: Dy nanophosphors were calculated using SRIM-2008.

CaF_2:Dy pellets were irradiated with by 350 keV N ions, 200 keV Ar ions and 100 keV H ions at the fluences varying from 1×10^{13} to 5×10^{14} ions cm^{-2} at a current of 660 nAcm^{-2}, 500 nAcm^{-1} and 630 nAcm^{-1} respectively using the low-energy ion beam facility (LEIBF) at IUAC, New Delhi (Gnaser, 1999).

Figure 4. Low Energy Ion Beam Facility a) Main irradiation chamber system and b) sample ladder inserted into the chamber

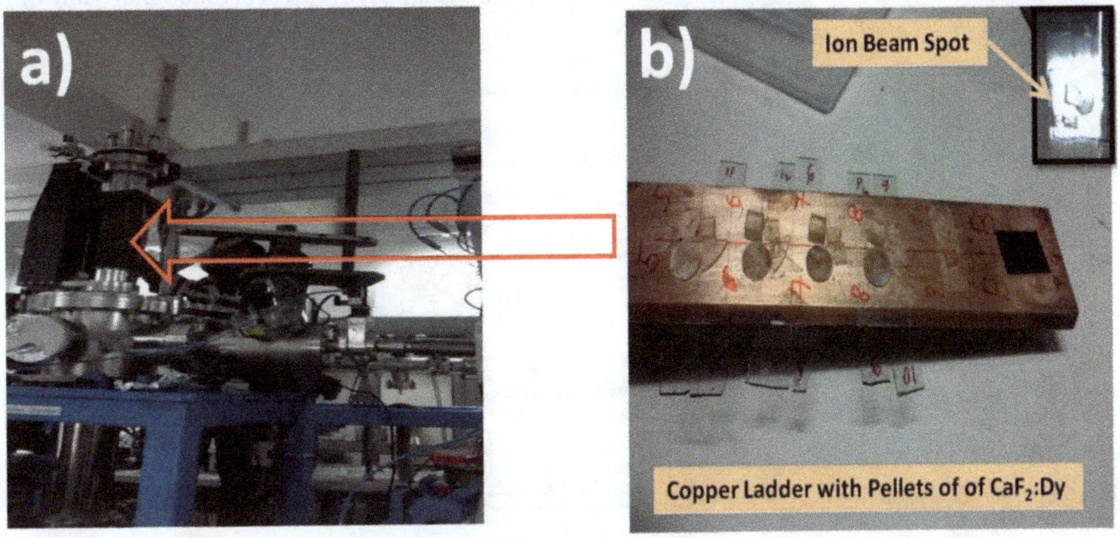

INTERACTION MECHANIMS

Figure 5, shows the schematic illustration of the interaction of ion beam with nanophosphor. The conduction band (CB) of the electronic state and the valence band (VB) are empty when there is no interaction of ion beam with nanophosphor. When ion beam interacts with CaF_2: Dy producing substantial energy transfer across the interface which results in electron gets excited and trapped to metastable state from VB via CB. Usually, when ion beam interacts with CaF_2: Dy, the traps and recombination centre (RC) between the forbidden zone get filled up. A sufficient amount of temperature (heat) is needed to remove them. When sufficient energy is given, the peak glows.

Bos et al. have explained detail mechanism based on energy band model. electron-hole trappings and their electronic transition in a TL material. Figure 5 shows in case (I), the rest ion source; means (off mode) non-interacts with pellets have not been produced any effect on the sample. Hence, the conduction band (CB) of the electronic state is as it is and the valence band (VB) empty. In (II) case, the energetic ion beams are interacted or penetrated through the interface of CaF_2: Dy pellets, produce substantial energy transfer across the interface which results electron gets excited and trapped to metastable state from VB via CB. Traditionally, when ion beam interacts with the CaF_2: Dy pellets; that time traps and recombination centre (RC) between the forbidden zone get filled up. Hence, a store energy in the form of traps was required a sufficient amount of temperature (heat) to removed them in case (III) and it gives glow (IV). As can be seen in Figure 5 case (V), the CaF_2: Dy pellets was mounted on copper ladder serially and different types of ion beam interacts with the samples. The practical layout of sample mounted copper ladder along with CaF_2: Dy pellets and ion beam spot shows in inset image of case (V). The ion energy has been chosen in such a way that the maximum nuclear energy loss on the sample interface and its complete range as calculated by the SRIM code (stopping and range of ions in matter).

Figure 5. The Schematic design for Low Energy Ion Beam interacts with the phosphor (I), (II), (III), (IV) and (V) along with the simple energy level scheme consisting of OTOR centre in Thermoluminescence models. Copper ladder with CaF_2: Dy pellets mounted on it and inset represents the inner view of the ion irradiation beam spot

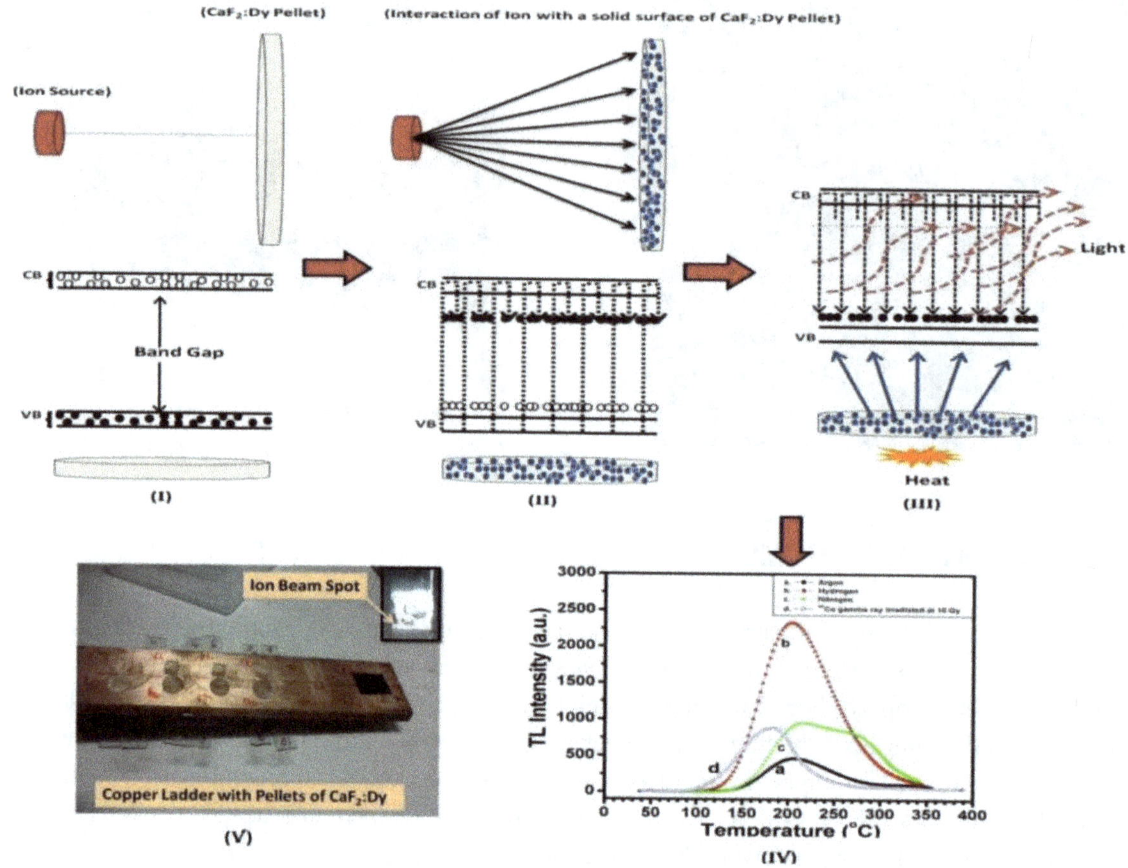

RESULTS AND DISCUSSION

$CaSO_4$: Dy Microphopshor for Thermal Neutron Dosimetry

Structural and Morphological Characterizations

XRD pattern of the $CaSO_4$: Dy phosphor irradiated with thermal neutrons at different fluences from 6×10^{11} n/cm^2 to 52×10^{11} n/cm^2 as shown in Figure 6. From the XRD results, it is observed that, peak intensity decreases with increase in the neutron fluence without any peak shift. This reduction in peak intensity is mainly due to the thermal neutrons, which produces prompt alpha particles through (n, α) reaction and predominantly responsible to induce defects in the phosphor. If neutron fluence increases the concentration of defects also increase which eventually further decreases the periodic lattice arrangement of the phosphor results into a decrease in the intensity of the XRD peaks. Figure 6 also shows the XRD

Figure 6. X-ray diffraction pattern of microcrytstalline $CaSO_4$: Dy irradiated with different thermal neutron fluences

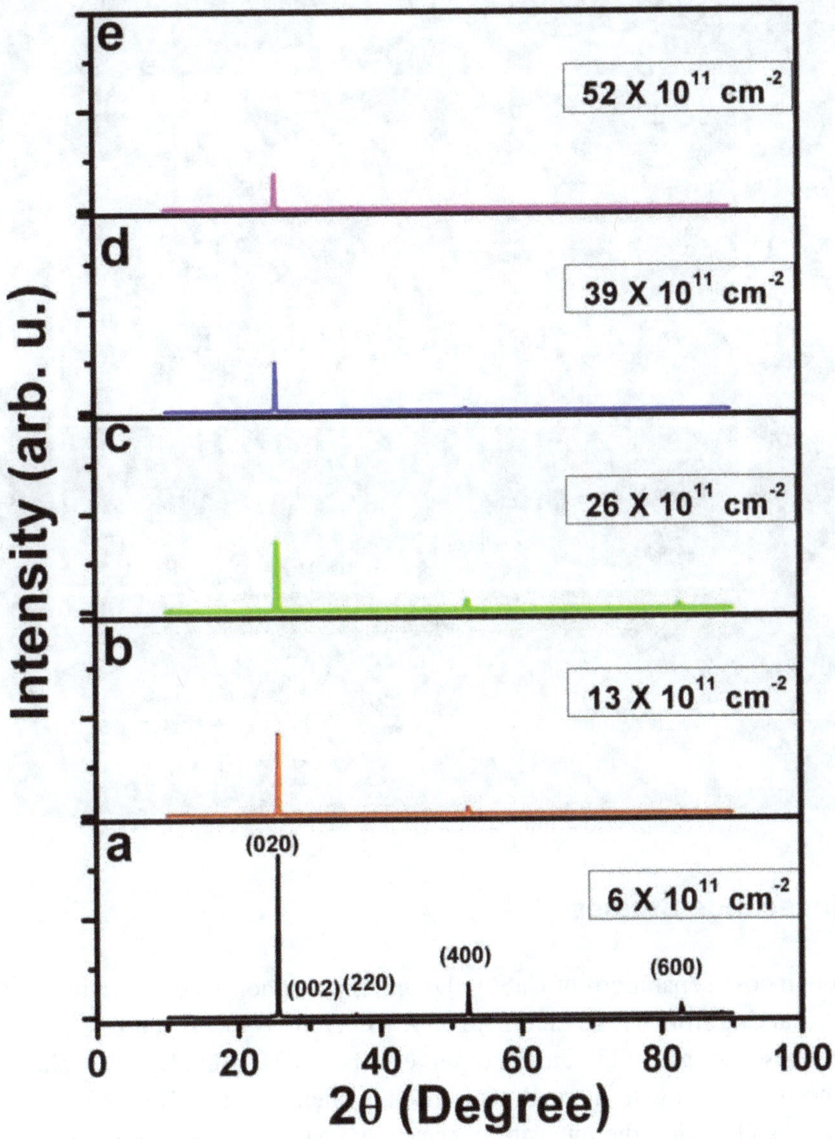

patterns of as prepared $CaSO_4$: Dy phosphor and is compared with the standard JCPDS Card No. 37-1496. From the data as well as sample, it is found that the structure is orthorhombic with phase a=6.993, b=7.001 and c=6.241 for $CaSO_4$: Dy phosphor. Moreover, size, shape and overall morphology of the micro particles have been examined using the SEM analysis which is shown in Figure 7. Corresponding particle distribution is also plotted in the inset of image. It is observed from particle distributions that the highest number of particle is having the size in the range of 200-250 μm.

Figure 7. SEM image of $CaSO_4$: Dy microphosphor

Thermoluminescence Studies

TL spectra (through dose separation) of $CaSO_4$: Dy microphosphor irradiated with thermal neutron at different fluences varying from 6×10^{11} n/cm^2 to 60×10^{11} n/cm^2 (glow spectra's a-f) is shown in Figure 8. The thermoluminescence glow (TL) did not observe for pristine $CaSO_4$: Dy microphosphor. But, when the microphosphor irradiated with thermal neutrons at a fluence of 6×10^{11} ^{n}/cm^2, the TL glow curve observed at 240 °C and further the intensity of the glow peak increases with increase in the neutron fluence. However, there is no change in the position and shape of the TL peak.

For the dosimetric purposes, the phosphor should have numerous significant properties to select a best possible TL phosphor such as relatively single glow curve, high TL efficiency and intensity glow curve should appear after 170° C (means deeper level traps), good linearity of the TL signal in the specific useful range of radiation dose, long term stability (i.e. low fading), reproducibility, effective atomic number (Z_{eff}) close to that of the biological tissue etc (Bhadane et al., 2017). $CaSO_4$: Dy microphosphor is very sensitive and also exhibits single major spectrum at 240° C for thermal neutron irradiation. Moreover, the higher temperature glow curve is advantageous since higher temperature peak has lesser fading. Therefore, $CaSO_4$: Dy microphosphor is a good material for TLD.

Figure 8. TL glow spectra of $CaSO_4$: Dy microphosphor irradiated with thermal neutrons at fluences from 6×10^{11} n/cm² to 60×10^{11} n/cm²

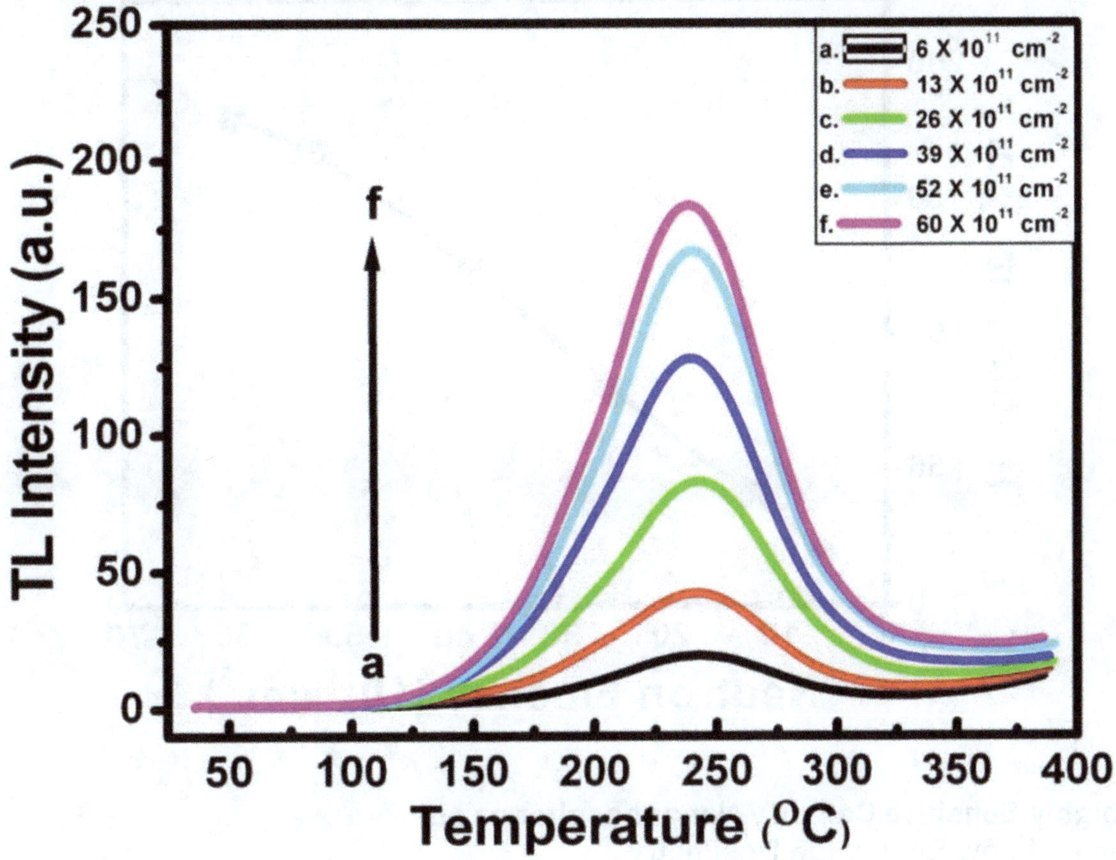

Response

The behavior of TL response with respect to thermal neutron fluence is shown in Figure 9. The linear response is computed as TL signal, normalized to the mass of the $CaSO_4$: Dy microphosphor powder samples. Primarily, the assessment of the TL sensitivity from the slope (using the regression square fitting i.e. R^2 fitting) of the linear curve which is nearly equal to one (=0.987). Therefore, response of $CaSO_4$: Dy microphosphpr linearly increases with thermal neutron fluence. Beyond the thermal neutron fluence of 52×10^{11} n/cm², the TL response of the microphosphor saturates.

Due to increase in the thermal neutron fluence a saturation effects occur because the distances between neighboring TC(trapping centre)/LC(luminescent centre) complex decrease and the TC/LC complex begins to merge and overlap. The overlapping regions do not contribute to additional TL since they do not result in additional trapping charge carriers due to the full occupancy of the available trapping centers and luminescence centers. In case of microparticles, the surface to volume ratio is large as compared to bulk which results in a higher surface barrier energy of the microparticles. This also explains the high TL sensitivity of the microphosphor (Bhadane et al., 2016).

Figure 9. TL response of $CaSO_4$: Dy microphosphor irradiated with thermal neutrons at different fluences from 6 × 10^{11} n/cm² to 60 × 10^{11} n/cm.²

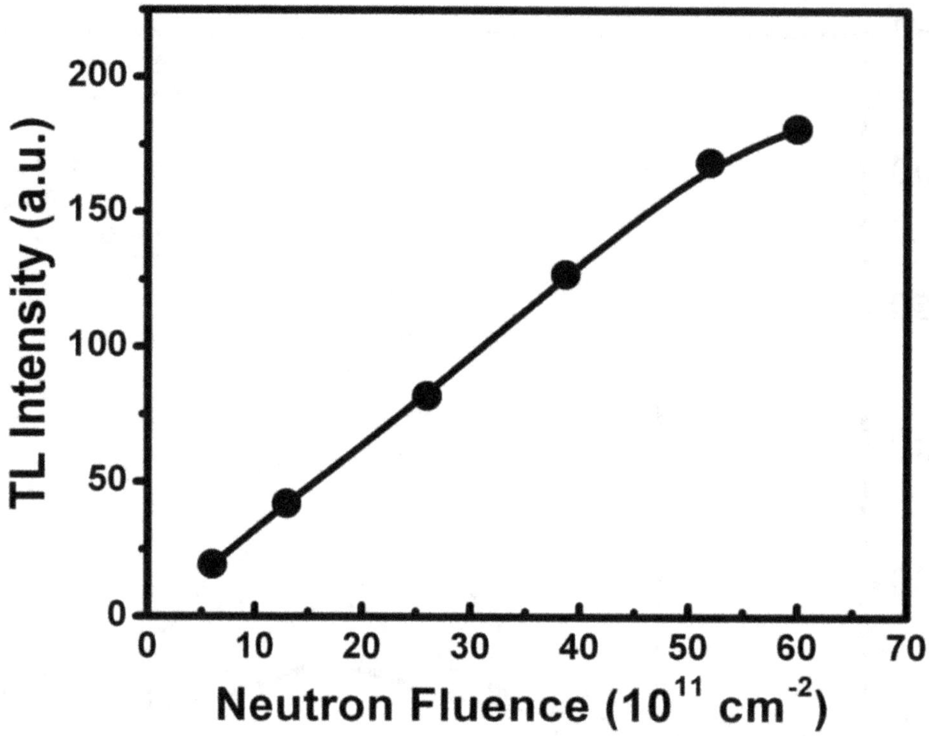

A highly Sensitive CaF_2: Dy Nanophosphor as an Efficient Low Energy Ion Dosimetry

Structural and Morphological Characterizations

The structural characterization has been done with the help of XRD studies. XRD spectra shown in Figure 10 for pre and post irradiated CaF_2: Dy nanophosphor different ion species at fluences ranging from 1×10^{13} to 5×10^{14} ion/cm².

Pristine CaF_2: Dy nanophosphor having average crystalline size around 45-50 nm showed peaks at 28.29, 32.766, 47.07, 55.85, 68.63, 75.81 and 87.41 respectively which corresponds to (111), (200), (220), (311), (400), (331) and (420) planes derived from JCPDS data (code 77-2245). The obtained d-spacing for (111) plane is ~3.1513 Å. After irradiation, the intensity of all these peaks decreased with increase in fluence indicating the production of defects in nanophosphor. It is well known that ion continuously loses its energy due to elastic and inelastic collisions with atom in the target materials. But, in case of low energy ion beams, the elastic collision with the nuclei of the atoms in which energy is transferred to the target atom as entire. Hence, it produces nuclear collision in CaF_2: Dy phosphor which is referred as nuclear energy loss of that atom from elastic collision (Bhadane et al., 2016). Therefore, nuclear collisions creates lattice disorder by the displacement of atoms from their positions and results in linearly decrease of the peak intensity of 100 keV H, 200 keV Ar and 350 keV N irradiated CaF_2: Dy nanophosphor.

Figure 10. XRD pattern of the Pristine and LEIB irradiated CaF_2:Dy phosphor at different fluences from 1 x 10^{13} to 5 x 10^{14} ion/cm^2 with (a) Argon (E= 200 keV), (b) Hydrogen (E= 100 keV) and (c) Nitrogen (E= 350 keV) ion species

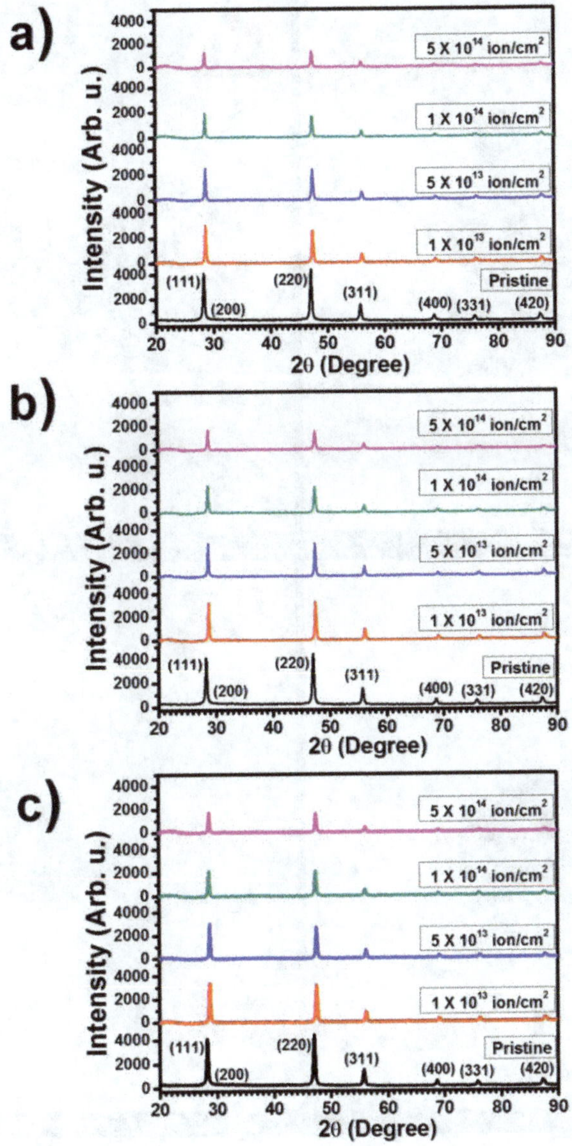

Figure 11 shows the morphological TEM images for pristine and LEIB irradiated CaF_2: Dy nanophosphor. As can be seen from Figure 11(a) CaF_2: Dy nanoparticles are asymmetrical cuboids shape having an average particle size ~68 nm. Figure 11 (b-d) shows TEM images for (b) Ar, (c) H and (d) N irradiated CaF_2: Dy nanophosphor at fluence of 5×10^{14} ions/cm^2. It is noticed that the particle size is unaltered after LEIB irradiation.

A representative HRTEM image at the edge of an individual CaF_2 fringes is shown in Figure 14, which further supports the claim of crystallinity for CaF_2 nanophosphors. The lattice fringes are clearly

Figure 11. TEM images of (a) pristine and different ion beam species (b) Ar, (c) H, (d) N irradiated CaF_2:Dy nanophosphor at a fluence of 5 x 10^{14} ion/cm^2 route

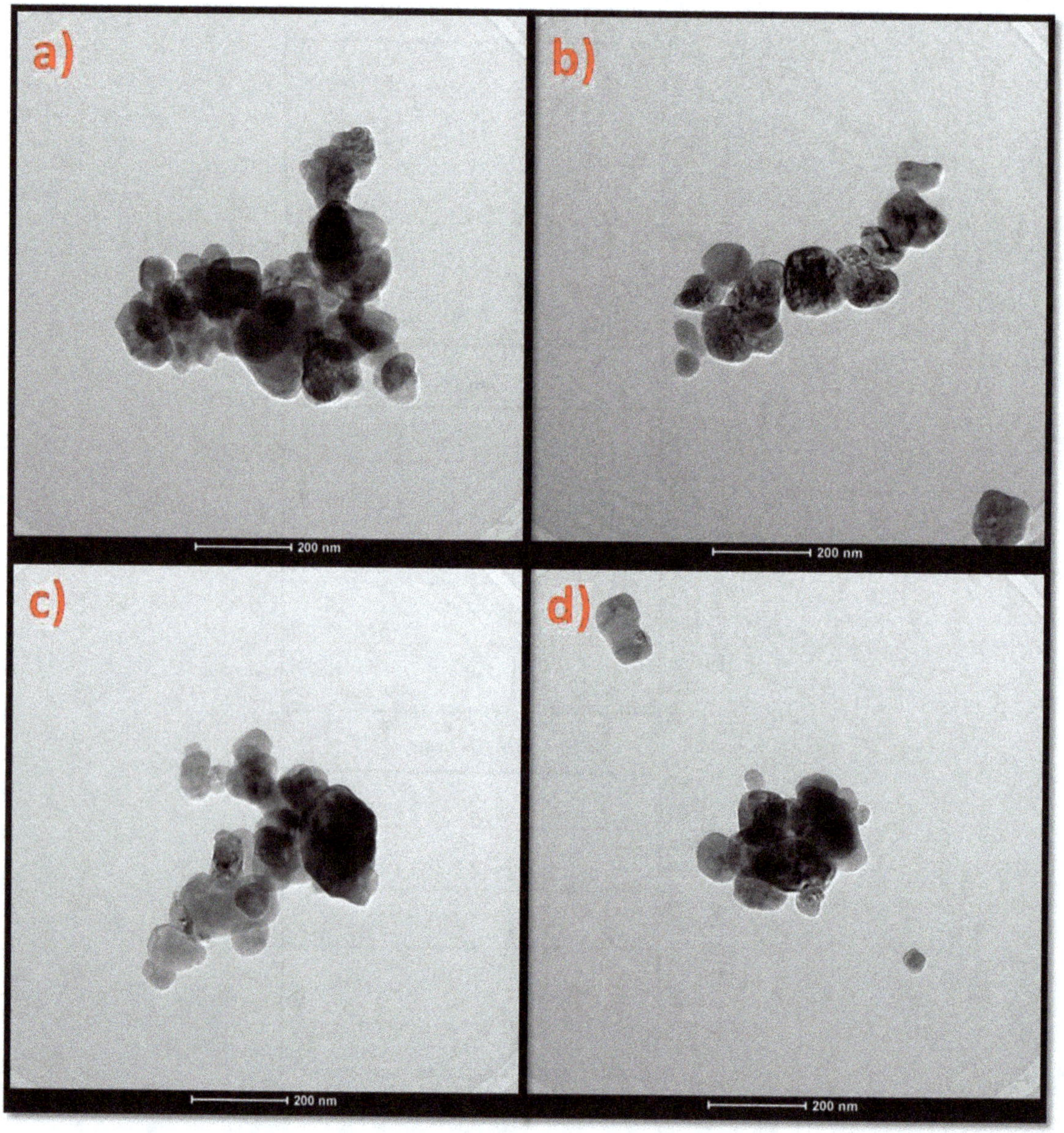

visible with a spacing of 3.09 Å, which agrees well with the lattice spacing of (111) of cubic CaF_2 as can be observed from XRD result. Selective area energy dispersive (SAED) image (inset in Figure 12) reveals the highly single-crystalline nature of the CaF_2 nanophosphors. This SAED pattern exhibits a set of sharp concentric rings which can be well indexed hkl planes.

Figure 12. A HRTEM image of the CaF_2: Dy nanophosphor. Inset shows the SAED pattern

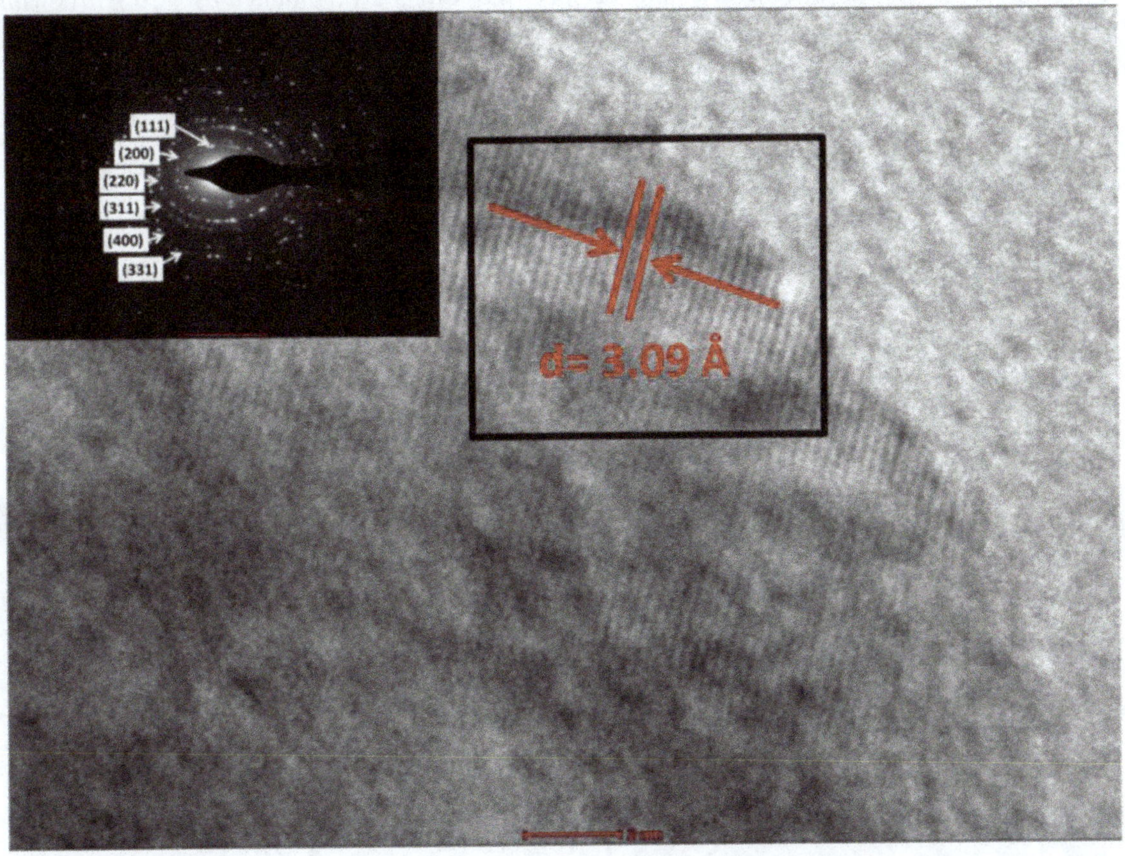

TL Studies

TL spectra of CaF_2: Dy nanophosphor for LEIB irradiated with Ar, H and N ion at different fluences are shown in Figure 13 (a-c). 200 keV Ar ion irradiated CaF_2: Dy showed a single peak around 207 °C at a fluence of 1×10^{13} ion/cm^2 can easily seen from Figure 13 (a). With increase in fluence, the intensity of this peak increased indicating the increase in trapping populations which results in enhancing the intensity (Bos, 2001). Figure 13 (b) shows the TL spectra of 100 keV H ion irradiated CaF_2: Dy phosphor at different fluences. In H irradiated phosphor, the spectra showed a peak around 203° C. In this case, the peak intensity also increases with increase in the fluence which may due to increase in trapping centers. As can be seen from Figure 13 (c), the 350 keV N ion irradiated CaF_2: Dy showed a strong peak around 216° C and a weak peak around 270° C due to their two trapping levels.

For comparision with LEIB with the irradiated phosphor, the gamma irradiated CaF_2:Dy nanophosphor showed a peak around 184° C with remarkable and significant changes in the TL spectral intensity as well as peak position shown in Figure 13 (d). The variation of spectral structure and peak position changes with change in ion species (Lakshmanan, 2001; Salah et al., 2006; Shinsho et al., 2008; Bhadane et al., 2016). However, the intensity of TL spectra for low energy ion beam irradiated CaF_2: Dy nanophosphor is more, indicating the effectual sensitive and more capable as analogous to TLD-200 (^{60}Co γ-rays irradiated at a dose of 10 Gy).

Figure 13. TL spectra of LEIB irradiated CaF_2: Dy for (a) Ar, (b) H ion, (c) N ion with the fluence from 1×10^{13} to 5×10^{14} ions/cm², (d) comparison of TL spectra of CaF_2: Dy irradiated by ^{60}Co γ-rays with a dose 10 Gy along at a LEIB with fluence of 5×10^{14} ion/cm²

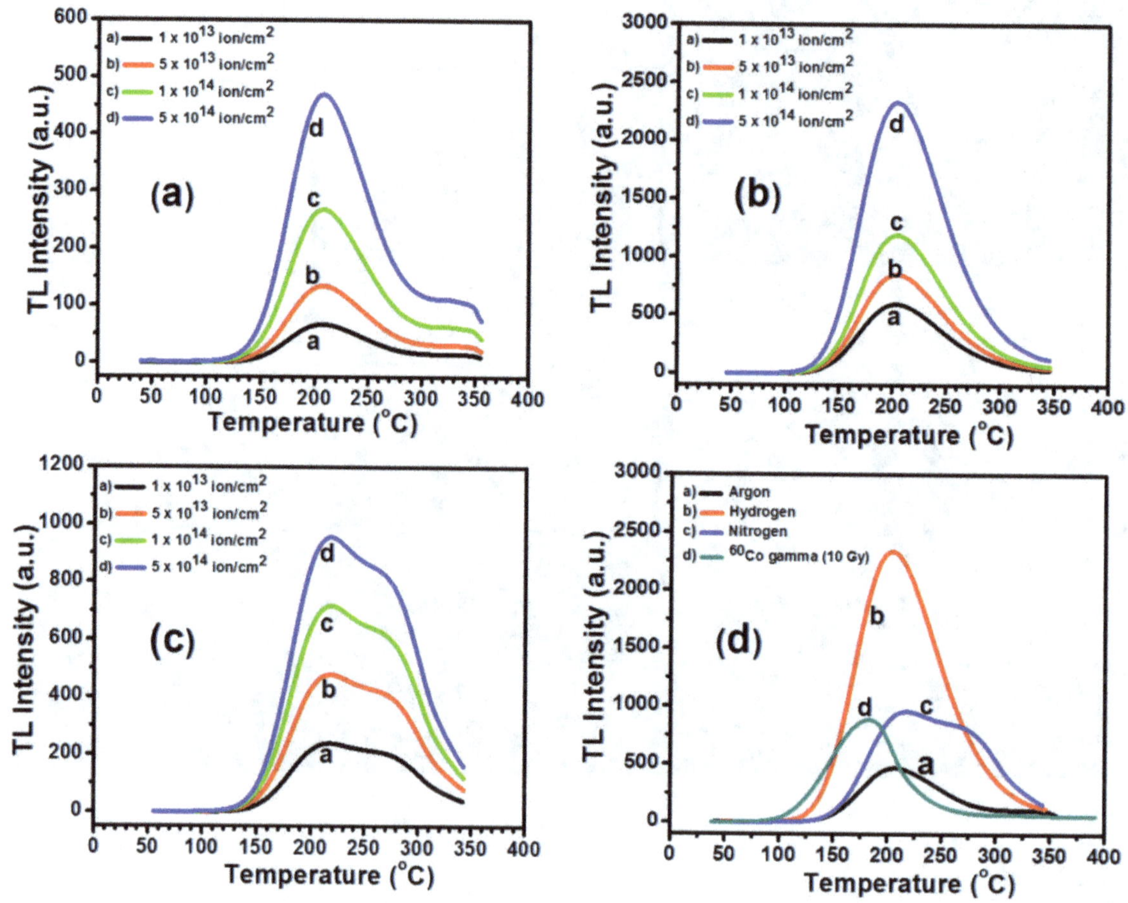

TL Response

Figure 16 indicates the maximum intensity of TL spectra versus fluences for CaF_2: Dy irradiated with Ar, H and N ions. As per the earlier discussion, H ion irradiated sample showed a higher intensity, but not sharp linear response than Ar and N ion irradiated phosphor. The phosphors irradiated by Ar and N ions showed superior linear response with slope ~1 in the fluences range from 1×10^{13} to 5×10^{14} ion/cm². Finally, it shows that the 200 keV Ar and 350 keV N irradiated CaF_2: Dy nanophosphors are good materials for the measurement ion dose and can be effectively used for the ion dosimetry.

Figure 14. TL response of LEIB irradiated with (a) Ar ion, (b) H ion and (c) N ion at different fluences from 1×10^{13} to 5×10^{14} ion/cm²

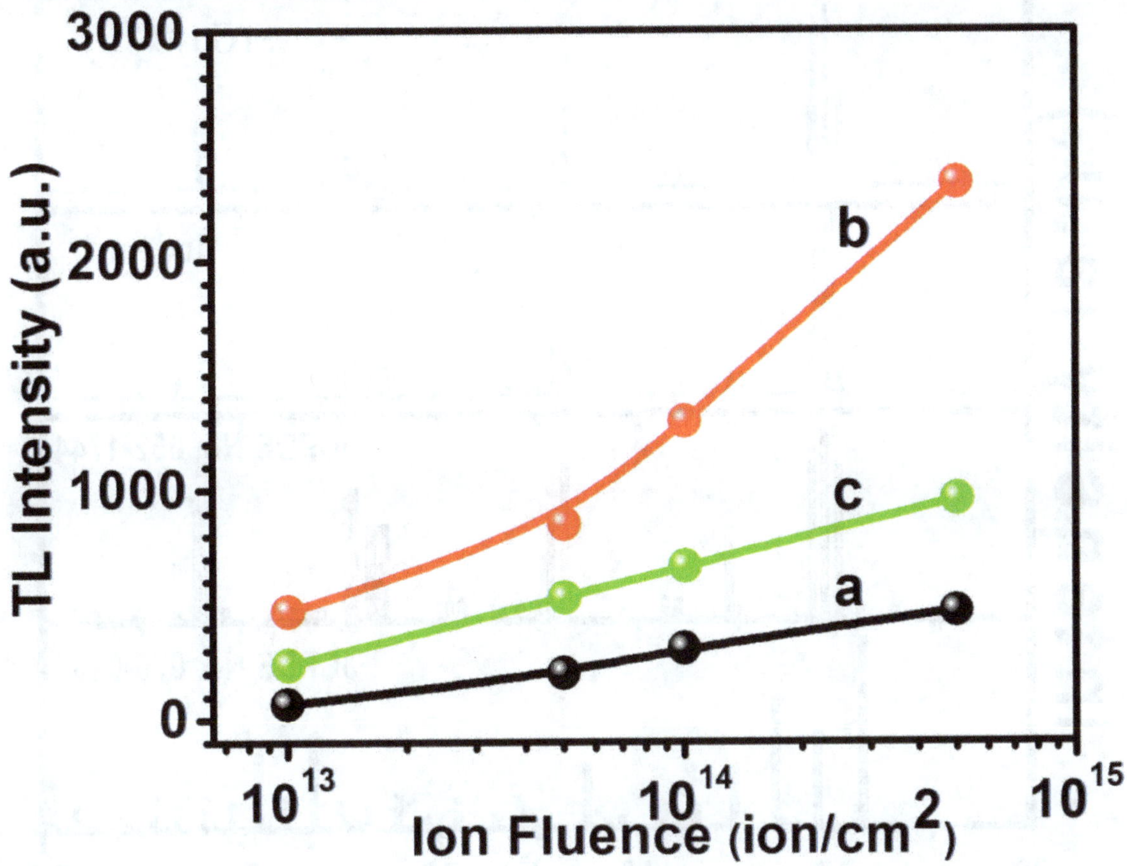

A Highly Sensitive $SrGd_2O_4$: Dy Phosphor as Gamma Ray Dosimetry

XRD Studies and Morphology

Figure 15 represents the XRD patterns of virgin and Dy^{3+} doped $SrGd_2O_4$ phosphors. The XRD patterns of as-synthesized samples exhibit a set of sharp and well-defined diffraction peaks which are in good agreement with the pure orthorhombic phase of $SrGd_2O_4$ (JCPDS 052-1744). However, a less intense peak is observed at around $2\theta=61°$ attributed to the second phase of cubic Gd_2O_3 (JCPDS card no.076-0155). No impurity peaks were detected due to the doping which indicates that the Dy^{3+} ions are successfully incorporated into the sites of Gd^{3+} ions of $SrGd_2O_4$ matrices.

The microstructure of as-synthesized $SrGd_2O_4$: Dy^{3+} (3 mol%) phosphors are studied and shown in Figure 15 a-b. The FESEM images show single morphology with different particle size. The particles can be categorized into two ways: one that shows particle sizes in the range 0.01-0.08 μm and the other shows comparatively bigger particle sizes in the range 0.09-0.2 μm, respectively.

Figure 15. XRD patterns of undoped and $SrGd_2O_4$:Dy^{3+} samples

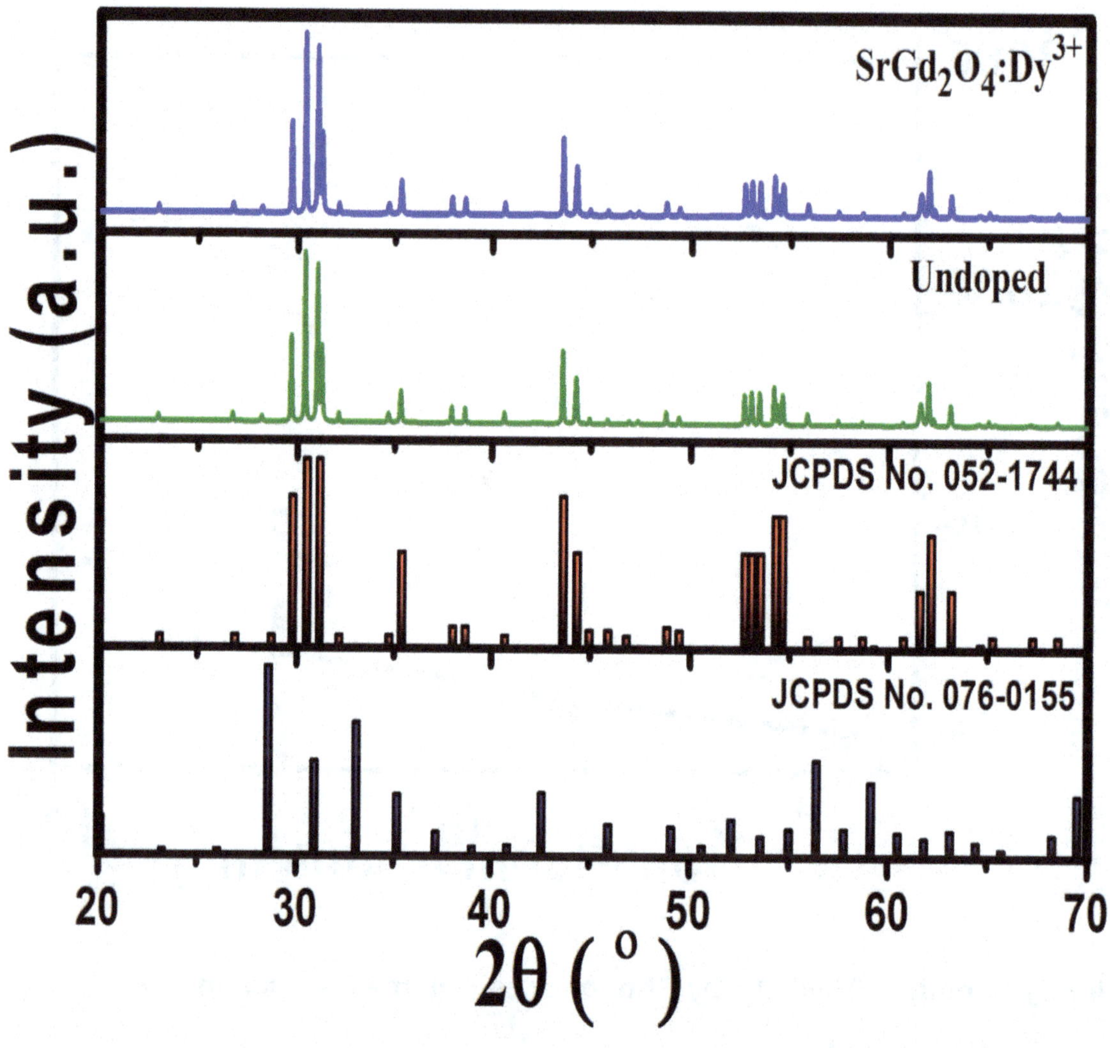

The FESEM micrographs display granular morphology having the particle sizes in the range 0.01-0.2 µm. The non-uniform distribution of temperature and mass flow during the synthesis is the reason behind the non-uniformity in the particle sizes.

To examine the microstructural analysis of as-synthesized samples more precisely, transmission electron microscope (TEM) of $SrGd_2O_4$: Dy^{3+} (3 mol%) samples were also carried out (Figure 16c). TEM images reveal granular morphology having average sizes ~ 0.01-0.2 µm which have a good consistency with SEM data.

Thermoluminescence Studies

The TL response of $SrGd_2O_4$: Dy^{3+} (3 mol%) samples after γ-irradiation was recorded in order to study the TL process. Figure 19 represents the TL glow curves of Dy^{3+} doped $SrGd_2O_4$ phosphors at a dose of 2 kGy. On the basis of previously reported results of rare earth doped phosphors (Manam & Das, 2008;

Figure 16. FESEM images at (a) lower (b) higher magnification and (c) corresponding TEM images of $SrGd_2O_4$: 3 mol% Dy^{3+} samples

Figure 17. TL glow curve of 2 kGy γ-irradiated $SrGd_2O_4$: Dy^{3+} samples

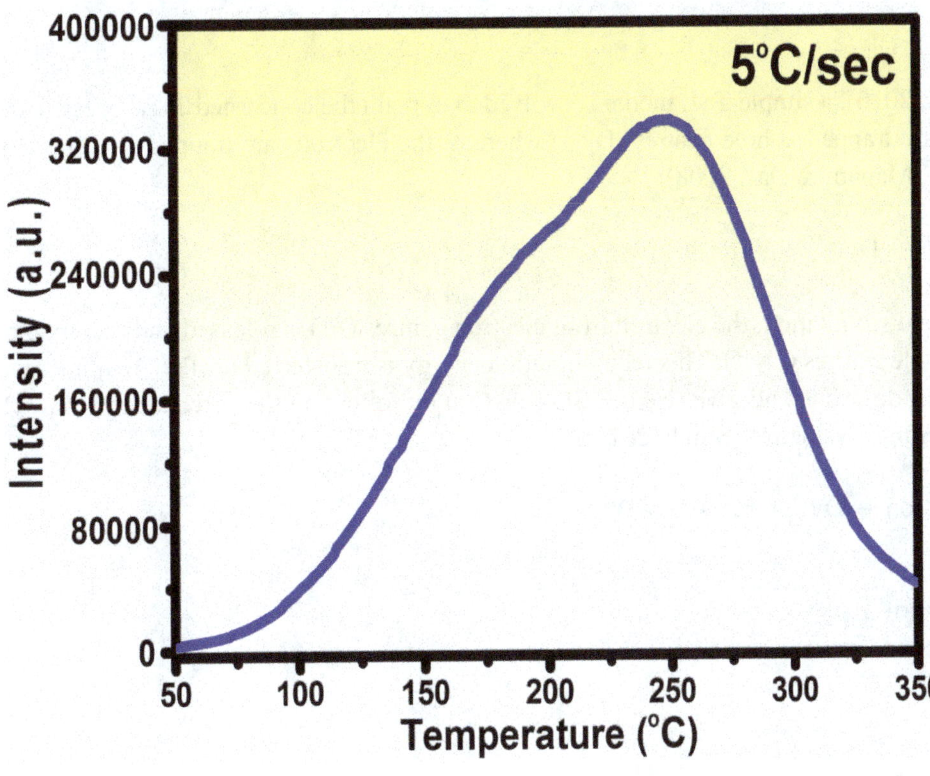

Figure 18. TL glow curves of γ-irradiated $SrGd_2O_4$: Dy^{3+} samples for different dose (inset: TL response with dose)

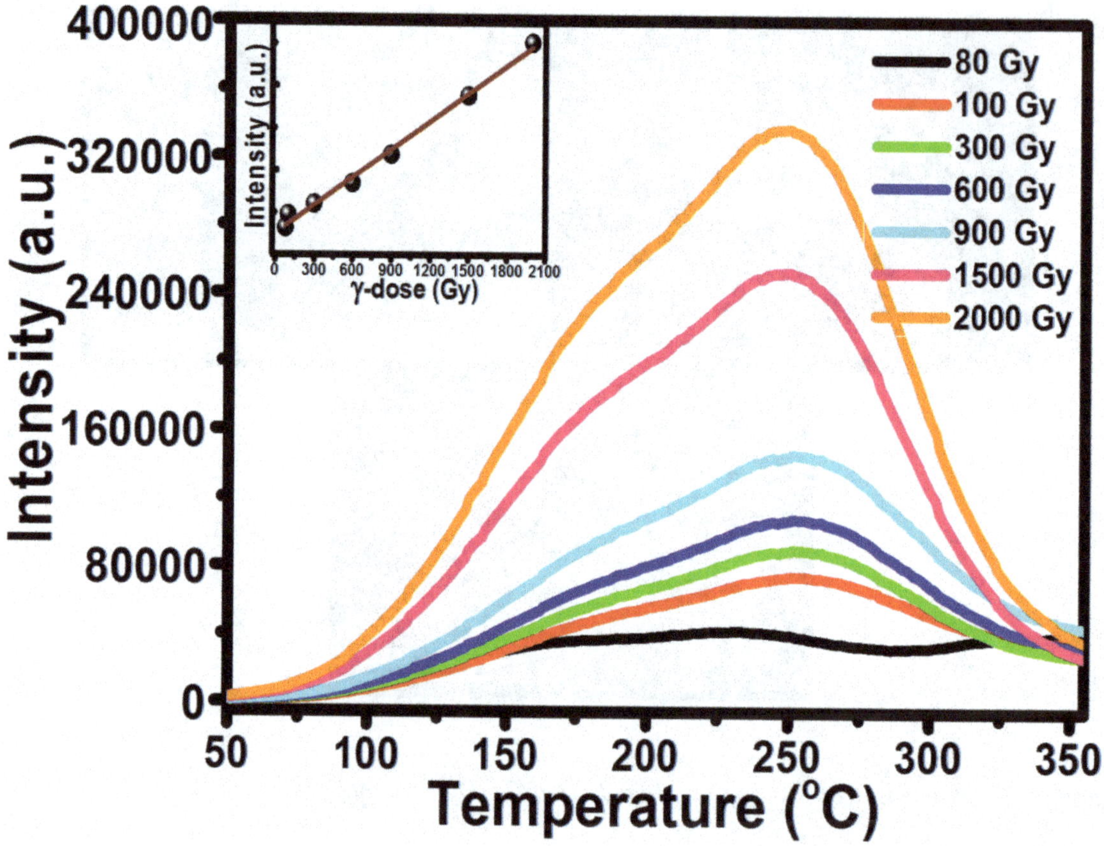

Dubey et al., 2016), a simple TSL model is evolved to explain the concerned TL process. Upon irradiation, holes are trapped at hole centres (Dy^{4+}) whereas, the electrons are trapped in respective electron centres (F^+) (Manam & Das, 2008):

$$Dy^{3+} + \text{Hole} \rightarrow Dy^{4+} \quad (1)$$

On thermal stimulation, the electron from electron centres (F^+) is released and recombines with the hole at the hole centres (Dy^{4+}). This recombination energy is transferred to (Dy^{3+}) luminescent centres, which further de-excite and give rise to TSL emission by following the mechanism below (Manam & Das, 2008; Singh, Manam & Singh, 2017):

$$Dy^{4+} + \text{electron} \rightarrow Dy^{3+} + h\nu \rightarrow (Dy^{3+})^* \quad (2)$$

$$(Dy^{3+})^* \rightarrow Dy^{3+} + h\nu \quad (3)$$

Figure 19. CGCD fitted glow curves of $CaSO_4$: Dy microparticles irradiated with thermal neutron at a fluence of 26×10^{11} n/cm²

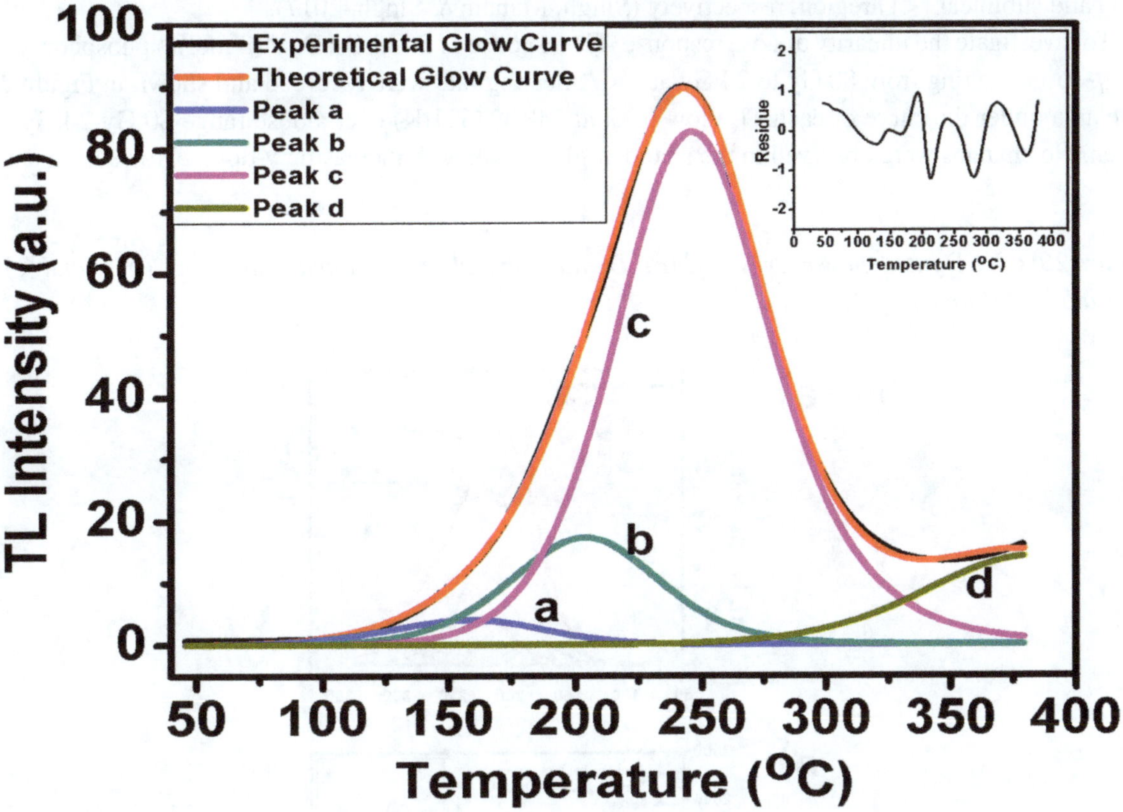

The TL glow peak of any TL materials effectively depends on the heating rate at which the material is heated. Therefore, thermoluminescence is also known as thermally stimulated emission (Sun et al., 2014).

TL Response

The linear behaviour of the TL dose response is the essence of TL materials utilized for radiation dosimeters. Furthermore, the amount of absorbed ionizing radiation is directly related to TL response of any TL material. To find the calibration of dose easily, a linear dose response starts from zero to a finite value of doses is very helpful. The normalized dose response function can be expressed as (Singh, Manam & Singh, 2017):

$$F(D) = \frac{f(D)}{D} \bigg/ \frac{f(D_0)}{D_0} \tag{4}$$

where f(D) denotes the TL dose response at a particular dose (D) whereas $f(D_0)$ represents the TL dose response at the low dose (D_0). The function F(D) has different values assigned for linear (-1), supralinear (>1) and sublinear (<1) region, respectively (Singh, Manam & Singh, 2017).

To investigate the linearity of dose response, TL response of $SrGd_2O_4$: Dy^{3+} (3 mol%) phosphors over the γ-doses ranging from 80 Gy to 2 kGy at 5°C/s heating rate were recorded and shown in Figure 20. The area under the curve of each TL glow peak at 248°C (521 K) over a dose range 80 Gy-2 kGy are taken. No changes were observed in the shape of glow peak with increasing γ-doses. However, a small

Figure 20. CGCD fitted glow curves of CaF_2: Dy nanophosphor irradiated with (a) Ar, (b) H and (c) N ions at5 X 10^{14} ion/cm^2

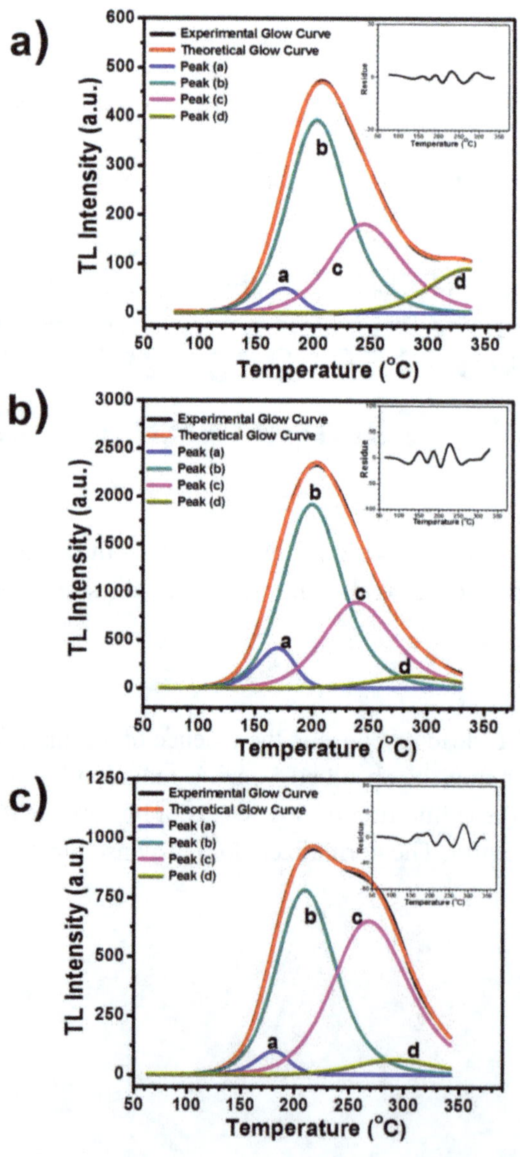

shift was observed of the maximum peak temperature towards the lower temperature region suggesting filling of traps on elevating γ-dose value (Manam & Das, 2008). A linear increment in the TL intensity of $SrGd_2O_4$: Dy^{3+} samples with the dose of γ-rays were observed in the inset of Figure 20.

Calculation of Trapping Parameters by CGCD Fitting and Glow Curves Analysis by Kitis Equations

The Computerized Glow Curve Deconvolution (CGCD) fitting in the case of all the phosphors such as $CaSO_4$: Dy microcrystalline, CaF_2: Dy and $SrGd_2O_4$: Dy materials have been carried out using glow curve deconvolution (GCD) functions, suggested by Kitis & Gomez-Ros, 1998.

The curve fitting was carried out using the first order, the second order and the general order equations as given below:

For first order:

$$I(T) = I_m \exp\left[1 + \frac{E}{kT}\left(\frac{T - T_m}{T_m}\right) - \frac{T^2}{T_m^2}\exp\left(\frac{E}{kT}\left(\frac{T - T_m}{T_m}\right)\right)\left(1 - \frac{2kT}{E}\right) - \frac{2kT_m}{E}\right] \tag{5}$$

For second order:

$$I(T) = 4I_m \exp\left(\frac{E}{kT}\left(\frac{T - T_m}{T_m}\right)\right)\left[\frac{T^2}{T_m^2}\exp\left(\frac{E}{kT}\left(\frac{T - T_m}{T_m}\right)\right)\left(1 - \frac{2kT_m}{E}\right) + 1 + \frac{2kT_m}{E}\right]^{-2} \tag{6}$$

The general order:

$$I(T) = I_m b^{\left(\frac{b}{b-1}\right)} \exp\left(\frac{E}{kT}\left(\frac{T - T_m}{T_m}\right)\right)\left[(b-1)\frac{T^2}{T_m^2}\left(1 - \frac{2kT}{E}\right)\exp\left(\frac{E}{kT}\left(\frac{T - T_m}{T_m}\right)\right) + 1 + (b-1)\frac{2kT_m}{E}\right]^{-\frac{b}{b-1}} \tag{7}$$

I (T) is the TL intensity at temperature T (K), I_m, the maximum peak intensity, T_m, is the temperature corresponding to maximum peak intensity I_m, E, is the trap depth or the thermal activation energy (eV) needed to free the trapped electrons, b, is the order of kinetics, and k, is the Boltzmann's constant (8.6 $\times 10^{-5}$ eVK^{-1}).

Similarly, the frequency factor S was calculated from following the first order, the second order and the general order equations:

The first order:

$$S = \frac{\beta E}{kT_m^2}\exp\left(\frac{E}{kT_m}\right) \tag{8}$$

The second and the general order:

$$S = \frac{\beta E}{kT_m^2\left(1+(b-1)\dfrac{2kT_m}{E}\right)}\exp\left(\frac{E}{kT_m}\right) \quad (9)$$

where, β is the linear heating rate (K/sec) and b is the Order of kinetics.

There are two very important dosimetric applications of CGCD fitted spectra i.e. one is an improvement in precision and minimum measurable dose, and other one is improvements in dose re-estimation capability.

The following equation (9) shows goodness of the fitting, i.e. figure of merit (FOM) is judged objectively and in percent defined by,

$$FOM = \sum_{j_f}^{j_i}\frac{\left|Y_j - Y(X_j)\right|}{A}\times 100 \quad (10)$$

Where J*i*= initial temperature in the fit region, J*f*= final or ending temperature in the fit region, Y*j*= PMT tube current at temperature *j*, Y(X*j*)= Value of the function at channel *j* and A= area under the peak, i.e., integral of the fit function between J*i* and J*f*.

Figure 19, Figure 20 a-c and Figure 21 show the curve fitting technique which is capable of evaluating all the glow peaks in a multi-peak spectra of actual $CaSO_4$: Dy (exposed to 26 X 10^{11} n/cm^2, CaF_2: Dy (Ar, H, and N ion beam irradiated at fluence of 5 X 10^{14} ion/cm^2) and $SrGd_2O_4$: Dy (2 kGy gamma-irradiated) phosphors, respectively.

The figures also show the experimental and theoretically deconvoluted TL curve as well as multi-curve glow peak of three trapping parameters fit such as order of kinetics (b), activation energy (E) and frequency factor (S) respectively and determined parameters mentioned in the Table 1 for $CaSO_4$:Dy microphosphor, Table 2 for CaF_2:Dy and Table 3 for $SrGd_2O_4$: Dy, respectively.

The FOM values is found to be around 1.52% ($CaSO_4$: Dy), 0.685% for Ar, 0.994% for H, 1.43% for N (CaF_2: Dy) and 1.26% ($SrGd_2O_4$: Dy), respectively. These results show that the experimental and theoretical glow curves are in good agreement and very much overlapping on each side. The residue curve of the theoretically fitted TL spectra is also shown in the inset of Figure 20 and Figure 21 respectively. The calculated trapping parameters for thermal neutron irradiated microcrystalline $CaSO_4$: Dy (fluence of 26 X 10^{11} n/cm^2), different ion beam irradiated CaF_2: Dy (fluence of 5 X 10^{14} ion/cm^2) and γ-ray irradiated $SrGd_2O_4$: Dy samples are summarized in Table 1, 2 and 3.

CONCLUSION

The phosphors, $CaSO_4$: Dy, CaF_2: Dy and $SrGd_2O_4$: Dy was successfully synthesized via acid recrystallization method, chemical co-precipitation method and homogenous precipitation cum auto-combustion method, respectively. The structural and morphological characterization revealed the confirmation of phase

Figure 21. Deconvolution of TL glow curve of 2 kGy γ-irradiated $SrGd_2O_4$: Dy^{3+} samples

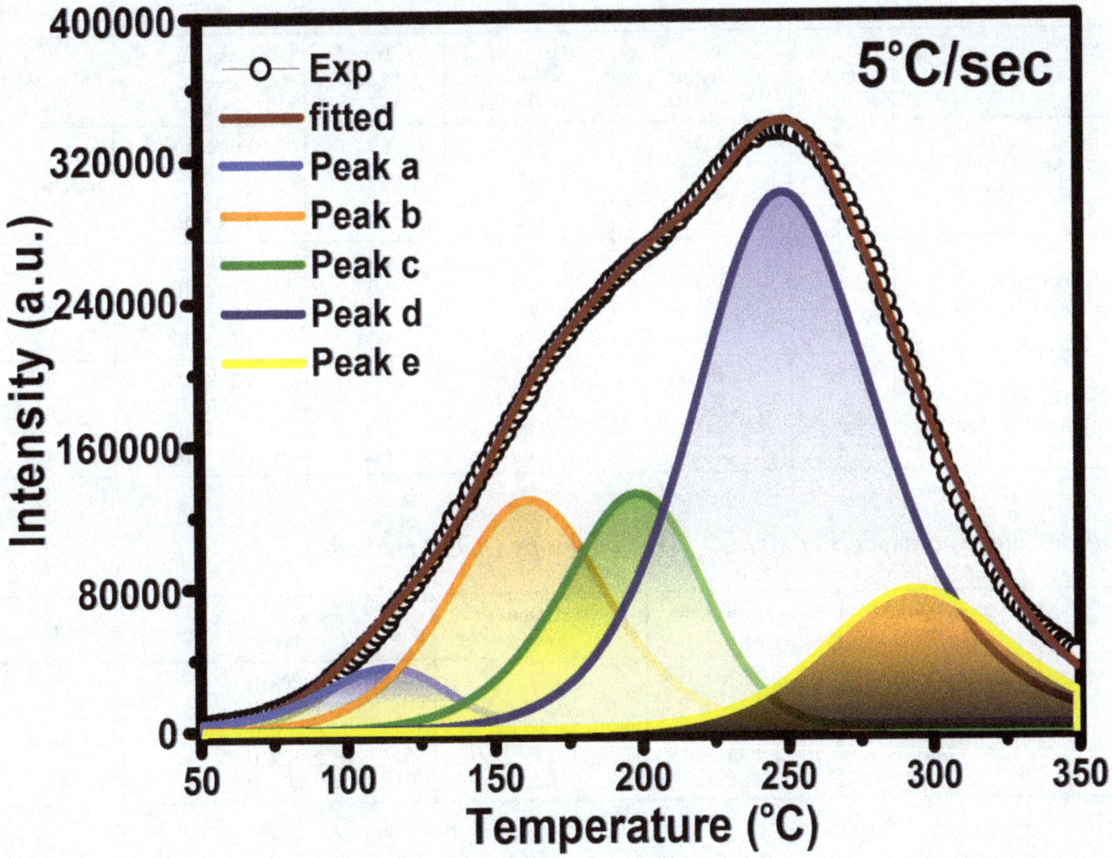

Table 1. Summary of obtained values of a $CaSO_4$: Dy microphosphor irradiated at thermal neutron fluence of 26×10^{11} n/cm²

Sample Name $CaSO_4$: Dy	Peak	Peak Temp. Tm (°C)	Order of Kinetics (b)	Trap Depth E (eV)	Frequency Factor (s-1) S	FOM
26×10^{11} n/cm²	a	158	1.32	0.57	4.54 X 10^8	
	b	202	1.62	0.825	6.64 X 10^9	1.52
	c	240	1.90	1.00	7.95 X 10^{10}	
	d	382	1.80	1.09	1.92 X 10^9	

formation, crystal structure and surface morphology of all the phosphors. Thermoluminescence results showed that the Dy dopant has played very important role and have produce a more trapping centers, excellent efficiency in $CaSO_4$, CaF_2 and $SrGd_2O_4$ materials. Linear profiles of TL dose response for all the phosphor systems reveal the possibility of using $CaSO_4$: Dy, CaF_2: Dy and $SrGd_2O_4$: Dy system as a

Table 2. Summary of obtained values from CGCD for different peaks of LEIB irradiated CaF2: Dy at a fluence of 5×10^{14} ion/cm^2

LEIB	Peak	Peak Temp. T_m (°C)	Order of Kinetics (b)	Trap Depth E(eV)	Frequency Factor S(s^{-1})	FOM
Argon (Ar) 5 × 10^{14} ion/cm^2	a b c d	175.2 202.5 244.0 334.7	1.30 2.00 2.00 1.80	1.32 1.05 1.02 1.20	2.62 X 10^{14} 3.36 X 10^{10} 1.79 X 10^{9} 1.58 X 10^{9}	0.685%
Hydrogen (H) 5 × 10^{14} ion/cm^2	a b c d	170.2 199.5 239.0 290.0	1.30 2.00 2.00 1.80	1.20 1.06 1.03 1.00	1.54X 10^{13} 8.20 X 10^{9} 2.80 X 10^{9} 1.52 X 10^{8}	0.994%
Nitrogen (N) 5 × 10^{14} ion/cm^2	a b c d	181.0 209.5 267.9 291.6	1.45 1.98 2.00 2.00	1.40 1.05 1.00 1.00	1.31X 10^{15} 2.25 X 10^{10} 4.12 X 10^{8} 1.40 X 10^{8}	1.430%

Table 3. Kinetic parameters of $SrGd_2O_4$: Dy samples by CGCD methods

Peak	PeakTemp. T_m (°C)	Order of Kinetics (b)	Trap Depth E(eV)	Frequency Factor S(s^{-1})	FOM
a b c d e	114 161 198 248 293	1.12 1.88 1.45 1.76 1.96	0.58 0.79 1.00 0.85 1.01	3.05 X 10^{6} 1.04 X 10^{8} 1.00 X 10^{10} 2.49 X 10^{7} 8.48 X 10^{7}	1.26%

TLD material in various radiation dosimetries. The Computerized Glow Curve Deconvolution (CGCD) fitting using glow curve deconvolution (GCD) functions, suggested by Kitis was used to calculate trapping parameters. A close agreement was observed in the calculated kinetic parameters.

REFERENCES

Avasthi, D. K., & Mehta, G. K. (2011). *Swift heavy ions for materials engineering and nanostructuring* (Vol. 145). Springer Science & Business Media. doi:10.1007/978-94-007-1229-4

Ayyangar, K., Lakshmanan, A. R., Chandra, B., & Ramadas, K. (1974). A comparison of thermal neutron and gamma ray sensitivities of common TLD materials. *Physics in Medicine and Biology, 19*(5), 665–676. doi:10.1088/0031-9155/19/5/007 PMID:4374709

Azorin, J., Furetta, C., & Scacco, A. (1993) Preparation and properties of thermoluminescent materials. *Physica Status Solidi (A), 138*, 9-46.

Bhadane, M. S., Akhtar, S., Dahiwale, S. S., Hareesh, K., Asokan, K., Kanjilal, D., & Dhole, S. D. (2017). Evaluation of thermoluminescence of 200 keV carbon ion irradiated $CaSO_4$: Dy nanophosphors for medical dosimetry. *Journal of Luminescence, 192*, 695–700. doi:10.1016/j.jlumin.2017.07.042

Bhadane, M. S., Hareesh, K., Dahiwale, S. S., Sature, K. R., Patil, B. J., Asokan, K., ... Dhole, S. D. (2016). A highly sensitive CaF_2: Dy nanophosphor as an efficient low energy ion dosimetry. *Nuclear Instruments and Methods in Physics Research Section B, 386*, 61–69. doi:10.1016/j.nimb.2016.09.017

Bhadane, M. S., Mandlik, N. T., Patil, B. J., Dahiwale, S. S., Sature, K. R., Bhoraskar, V. N., & Dhole, S. D. (2016). $CaSO_4$:Dy microphosphor for thermal neutron dosimetry. *Journal of Luminescence, 170*, 226–230. doi:10.1016/j.jlumin.2015.10.025

Bos, A., & Dielhof, J. (1991). The analysis of Thermoluminescencent Glow Peaks in CaF_2:Tm (TLD-300). *Radiation Protection Dosimetry, 37*, 231–239.

Bos, A. J. J. (2001). High sensitivity thermoluminescence dosimetry. *Nuclear Instruments & Methods in Physics Research. Section B, Beam Interactions with Materials and Atoms, 184*(1-2), 3–28. doi:10.1016/S0168-583X(01)00717-0

Bull, R. K. (1986). Thermoluminescence and its applications: An introduction. *Nuclear Tracks and Radiation Measurements, 11*(1-2), 105–113. doi:10.1016/1359-0189(86)90029-4

Campos, L. L., & Lima, M. F. (1987). Thermoluminescent $CaSO_4$:Dy Teflon Pellets for Beta Radiation Detection. *Radiation Protection Dosimetry, 18*(2), 95–97. doi:10.1093/oxfordjournals.rpd.a079889

Deilhof, J. B., Bos, A. J. J., Zoetelief, J., & Broerse, J. J. (1988). Sensitivity of CaF_2 thermoluminescent materials to fast neutrons. *Radiation Protection Dosimetry, 23*(1-4), 405–408. doi:10.1093/oxfordjournals.rpd.a080207

Dubey, V., Kaur, J., Parganiha, Y., Suryanarayana, N. S., & Murthy, K. V. R. (2016). Study of formation of deep trapping mechanism by UV, beta and gamma irradiated Eu^{3+} activated SrY_2O_4 and $Y_4A_{12}O_9$. *Applied Radiation and Isotopes, 110*, 16–27. doi:10.1016/j.apradiso.2015.12.047 PMID:26748019

Fain, J., Montret, M., & Sahraoui, L. (1980). Thermoluminescent response of CaF_2: Dy and LiF: Mg, Ti under heavy ion bombardment. *Nuclear Instruments and Methods, 175*(1), 37–39. doi:10.1016/0029-554X(80)90245-1

Gnaser, H. (1999). *Introduction.* Springer Berlin Heidelberg.

Gupta, K. K., Kadam, R. M., Dhoble, N. S., Lochab, S. P., Singh, V., & Dhoble, S. J. (2016). Photoluminescence, thermoluminescence and evaluation of some parameters of Dy^{3+} activated $Sr_5(PO_4)_3F$ phosphor synthesized by sol-gel method. *Journal of Alloys and Compounds, 688*, 982–993. doi:10.1016/j.jallcom.2016.07.114

Jain, I. P., & Agarwal, G. (2011). Ion beam induced surface and interface engineering. *Surface Science Reports, 66*(3-4), 77–172. doi:10.1016/j.surfrep.2010.11.001

Jayasudha, S., Madhukumar, K., Nair, C. M. K., Nair, R. G., Rajesh, S., Elias, T. S., ... Gopakumar, N. (2017). TL dosimetric characterization of gamma irradiated $SrSO_4$:Eu phosphors. *Journal of Luminescence, 183*, 259–265. doi:10.1016/j.jlumin.2016.11.044

Kitis, G., Gomez-Ros, J. M., & Tuyn, J. W. N. (1998). Thermoluminescence glow-curve deconvolution functions for first, second and general orders of kinetics. *Journal of Physics. D, Applied Physics, 31*(19), 2636–2641. doi:10.1088/0022-3727/31/19/037

Kumar, P., Rodrigues, G., Lakshmy, P. S., Kanjilal, D., & Kumar, R. (2008). Charge-state distributions of metallic electron cyclotron resonance plasmas. *Journal of Vacuum Science & Technology. A, Vacuum, Surfaces, and Films*, *26*(1), 97–102. doi:10.1116/1.2823486

Kumar, S., Gathania, A. K., Vij, A., & Kumar, R. (2016). Gamma induced thermoluminescence and color centers study of Dy doped LiF micro-cubes. *Ceramics International*, *42*(13), 14511–14517. doi:10.1016/j.ceramint.2016.06.062

Lakshmanan, A. R. (2001) A new high sensitive $CaSO_4$: Dy thermo stimulated luminescence phosphor. *Physica Status Solidi (A)*, *186*, 153–166.

Lakshmanan, A. R., Jose, M. T., & Annalakshmi, O. (2008). High-sensitive $CaSO_4$: Dy thermoluminescent phosphor synthesis by co-precipitation technique. *Radiation Protection Dosimetry*, *132*(1), 42–50. doi:10.1093/rpd/ncn215 PMID:18716066

Madhusoodanan, U., Jose, M. T., Tomita, A., Hoffmann, W., & Lakshmanan, A. R. (1999). A new thermostimulated luminescence phosphor based on CaSO4:Ag,Tm for applications in radiation dosimetry. *Journal of Luminescence*, *82*(3), 221–232. doi:10.1016/S0022-2313(99)00044-7

Manam, J., & Das, S. (2008). Thermally stimulated luminescence studies of undoped, Cu- and Mn-doped $CaSO_4$ compounds. *Radiation Effects and Defects in Solids*, *163*(12), 955–965. doi:10.1080/10420150802163869

Manam, J., & Das, S. (2010). Characterization and TSL dosimetric properties of Mn doped $BaSO_4$ phosphor prepared by recrystallisation method. *Journal of Alloys and Compounds*, *489*(1), 84–90. doi:10.1016/j.jallcom.2009.09.018

Marrale, M., Longo, A., Bartolotta, A., Basile, S., D'Oca, M. C., Tomarchio, E., ... Di. (2012). Thermoluminescence response of sodalime glass irradiated with proton and neutron beams. *Nuclear Instruments and Methods in Physics Research Section B*, *292*, 55–58. doi:10.1016/j.nimb.2012.10.003

Mondragon, G., Mendoza-Anaya, D., Nicho-Diaz, M. E., Garcia, R., & Reyes-Gasga, J. (2008). Thermoluminescence response of previously heated NaCl:Eu crystals to UV radiation, *Journal of Applied Physics D. Applied Physics (Berlin)*, *41*, 8.

Mondragón-Galicia, G., Mendoza-Anaya, D., Nicho-Diaz, M. E., García, R., & Reyes-Gasga, J. (2008). Thermoluminescence response of previously heated NaCl: Eu crystals to UV radiation. *Journal of Physics. D, Applied Physics*, *41*(4), 045103. doi:10.1088/0022-3727/41/4/045103

Pagonis, V., Kitis, G., & Furetta, C. (2006). *Numerical and practical exercises in thermoluminescence*. Springer Science & Business Media.

Patra, et al. (2015). Development of Thermal Neutron Detector based on Lithium Tetraborate (LTB) Single Crystal. *BARC News Letter*, 9-13.

Rajurkar, N., Jha, A., & Bhadane, R. (1994). Instrumental neutron activation analysis of some ancient punchmark coins from India. *Journal of Radioanalytical and Nuclear Chemistry*, *186*(5), 393–400. doi:10.1007/BF02162623

Reddy, S. S., Nagabhushana, K. R., & Singh, F. (2016). Thermoluminescence studies of γ-irradiated Al_2O_3:Ce^{3+} phosphor. *Nuclear Instruments and Methods in Physics Research Section B*, *379*, 146–151. doi:10.1016/j.nimb.2016.03.051

Salah, N. (2008). Carbon ions irradiation on nano- and microcrystalline $CaSO_4$: Dy *Journal of Applied Physics D. Applied Physics (Berlin)*, *41*, 155302–155308.

Salah, N., Sahare, P. D., Lochab, S. P., & Kumar, P. (2006). TL and PL studies on $CaSO_4$: Dy nanoparticles. *Radiation Measurements*, *41*(1), 40–47. doi:10.1016/j.radmeas.2005.07.026

Salah, N., Sahare, P. D., & Rupasov, A. A. (2007). Thermoluminescence of nanocrystalline LiF: Mg, Cu, P. *Journal of Luminescence*, *124*(2), 357–364. doi:10.1016/j.jlumin.2006.04.004

Shani, G. (2000). *Radiation Dosimetry-Instrumentation and Methods* (2nd ed.). CRC Press.

Shinsho, K., Harada, K., Yamamoto, Y., & Urushiyama, A. (2008). Differences in glow curves structure of nano- and microcrystals of $CaSO_4$:Dy measured at a low heating rate. *Radiation Measurements*, *43*(2-6), 236–240. doi:10.1016/j.radmeas.2007.11.037

Singh, J., Manam, J., & Singh, F. (2017). Synthesis and thermoluminescence studies of γ-irradiated Dy^{3+} doped $SrGd_2O_4$ phosphor. *Materials Research Bulletin*, *94*, 113–121. doi:10.1016/j.materresbull.2017.05.052

Sun, X.-Y., Liu, Y., Liu, X.-L., Cao, R.-P., Li, Y.-N., & Lin, L.-W. (2014). Substitution of Y^{3+} for Gd^{3+} on the luminescent properties of $BaGd_2O_4$:Eu^{3+} scintillating phosphors. *Optical Materials*, *36*(9), 1478–1483. doi:10.1016/j.optmat.2014.03.044

Sunitha, D. V., Manjunatha, C., Shilpa, C. J., Nagabhushana, H., Sharma, S. C., Nagabhushana, B. M., & Chakradhar, R. P. S. (2012). $CdSiO_3$: Pr^{3+} nanophosphor: Synthesis, characterization and thermoluminescence studies. *Spectrochimica Acta. Part A: Molecular and Biomolecular Spectroscopy*, *99*, 279–287. doi:10.1016/j.saa.2012.08.057 PMID:23085285

Triolo, A., Marrale, M., & Brai, M. (2007). Neutron–gamma mixed field measurements by means of MCP–TLD600 dosimeter pair. *Nuclear Instruments and Methods in Physics Research Section B*, *264*(1), 183–188. doi:10.1016/j.nimb.2007.08.009

Wagiran, H., Hossain, I., Bradley, D., Yaakob, A. N. H., & Ramli, T. (2012). Thermoluminescence Responses of Photon and Electron Irradiated Ge- and Al-Doped SiO_2 Optical Fibres. *Chinese Physics Letters*, *29*(2), 027802. doi:10.1088/0256-307X/29/2/027802

Wang, X-J., Jia, Z.Y., Strek, W., Pazik, R., Hreniak, D., & Yen, W. M. (2005). Crystal size dependence of the persistent phosphorescence in $Sr_2ZnSi_2O_7$: Eu^{2+}, Dy^{3+}. *Microelectronics Journal*, *36*, 546–548.

Wulf, H. C., & Gniadecka, M. (1996). CaF_2:Dy and CaF_2 crystal-based UV dosimeters. *Skin Research and Technology*, *2*(3), 108–113. doi:10.1111/j.1600-0846.1996.tb00071.x PMID:27327385

Yamashita, T., Nada, N., Onishi, H., & Kitamura, S. (1971). Calcium sulfate activated by thulium or dysprosium for thermoluminescence dosimetry. *Health Physics*, *21*(2), 295–300. doi:10.1097/00004032-197108000-00016 PMID:5094199

Yazici, A. N., Chen, R., Solak, S., & Yegingil, Z. (2002). The analysis of thermoluminescent glow peaks of CaF_2: Dy (TLD-200) after β-irradiation. *Journal of Applied Physics D Applied Physics (Berlin), 35*, 2526–2535.

Zahedifar, M., Sadeghi, E., & Mohebbi, Z. (2012). Synthesis and thermoluminescence characteristics of Mn doped CaF_2 nanoparticles. *Nuclear Instruments and Methods in Physics Research Section B, 274*, 162–166. doi:10.1016/j.nimb.2011.12.015

Section 2

Emerging Synthesis Techniques of Up-Conversion Luminescent Materials and Its Applications

Various synthesis techniques are reported for up-conversion luminescent materials and its application in various fields.

Chapter 4
Photon Upconversion in Lanthanide-Activated Inorganic Luminescent Materials

Anurag Pandey
Rhodes University, South Africa

Surya Prakash Tiwari
Indian Institute of Technology (ISM) Dhanbad, India

Viresh Dutta
Indian Institute of Technology Delhi, India

Vinod Kumar
Indian Institute of Technology Delhi, India

ABSTRACT

The growing demand for energy-efficient materials needs advancement in existing approaches. Photon upconversion is an effective way to overcome this issue. Novel upconverting materials show their potential in the advancement of energy conservation and lighting technology. In this contest, lanthanide-activated inorganic luminescent materials play a key role. The unique features of these materials like non-toxic nature, narrow multicolour emissions, long luminescence lifetime, high signal/noise ratio, and good chemical durability make them more promising. In this chapter, an overview of photon upconversion process in various lanthanide doped/codoped materials is given. Introductory idea about upconversion mechanism, criteria of material selection, novel synthesis roots, reports on upconversion-emitting materials, and emergent applications of these materials are presented.

INTRODUCTION

There is a great deal of research interest in development of the visible light emitting materials for different photonics, sensing and biomedical applications [Mader et al., 2010; Xu et al., 2010; Wade et al., 2003; Ye et al., 2010; Le et al., 2006; Kim et al., 2000]. Inorganic phosphors are very efficient materials for

DOI: 10.4018/978-1-5225-5170-6.ch004

such task and used heavily by researchers working in the field of photonics. These materials can exhibit the desired emission by controlling several parameter e.g. materials ingredients, methods of preparation and process of pumping. Two main processes are famous for photon conversion in phosphor materials, namely, upconversion (UC) and downconversion (DC). Photon UC is the process of generating higher energy photon via pumping with a lower one. It is also termed as anti-stoke type emission whereas DC is the stoke type process, observed on excitation of material by high energy than the emitted radiations [Wang et al., 2009]. The term UC was proposed first time in 1959 by Bloembergen and then experimentally verified in 1966 by Auzel [Gao et al., 2004]. It is more effective way of generating visible emission by absorption of two or more low energy photons via different process such as, ground state absorption (GSA), excited state absorption (ESA), energy transfer (ET) etc. [Auzel, 2004]. Lanthanides (Ln^{3+}) are very suitable candidates for the photon UC owing to their large number of energy levels, narrow emission spectra, long lifetime of the excited states, good chemical durability and they can be easily populated by the near infrared (NIR) radiations [Prasad, 2004; Dieke et al., 1968].

Actually Ln^{3+} serves as activator and/or sensitizer and gives intense multicolour light emissions throughout visible region of electromagnetic spectrum, when embedded into a suitable host lattice. Mostly, the NIR diode lasers have been used as excitation source in UC process, which are less expensive with compare to the ultra violet (UV) lasers. Also, NIR radiation leads low autofluorescence, less scattering and absorption, and deep penetration to the target [Wang et al., 2009; Auzel, 2004]. A schematic representation of thetwo process of photon conversion is shown in Figure 1. It is clearly shown in the figure that NIR photon is converting in to the visible/UV by UC process. On the other hand, a UV or blue photon is converting into other visible or NIR photons by DC process. A three level energy scheme is also shown for both processes. In UC process, the atoms/ions (also known as luminescent-centers/active-centers) in ground state (E_1) are jump to excited state (E_2) via GSA process (by absorbing energy $h\nu_1$) and then to excited state (E_3) through ESA process (by absorbing a second photon of energy $h\nu_2$; $h\nu_1$ may equal to $h\nu_2$). A radiative transition from E_3 to E_1 gives the UC emission. In DC process, atoms in ground state (E_1) reaches directly to excited state (E_3) by absorbing a photon of energy $h\nu_1$ and relax back to ground state by releasing slightly lower photon $h\nu_2$ ($h\nu_1 > h\nu_2$) involving some non radiative relaxation between intermediate level (E_2). A more clear strategy of UC processes will present in the next section of the chapter.

It is mandatory to understand the basics of photon UC process and their related aspects, for the newcomers in the field. As we know that, research articles are representing a particular research output, not the detail scenario of related phenomena's. So, the demand of background study in the form of reviews and book chapters are still maintain. Various reviews and book chapters were reported by different researcher to get insight UC process [DaCosta et al., 2014; Liu et al., 2014; Dou et al., 2014; Zhou et al., 2014; Wang et al., 2011a; Zhou et al., 2012; Chen et al., 2012; Tiwari et al., 2016; Rai, 2012], which need to update regularly by adding new outcomes. Herein, the authors tried to present a simplest approach to understand the photon UC in Ln^{3+}activated luminescent materials. This chapter deals with Ln^{3+} doped/codoped upconverting materials (mainly focus to inorganic phosphors). A way of phosphor's preparation via various famous/novel synthesis techniques is presented along with detail theoretical aspects of mechanism involved in UC process. Results of UC emissions from different Ln^{3+} doped/codoped phosphors are discussed with their possible application areas. Overall, this chapter will develop a good understanding about synthesis of upconverting phosphor materials, their possible characterizations and the most important UC mechanism involved to achieve desired optical emissions.

Figure 1. Schematic representation of photon conversion process

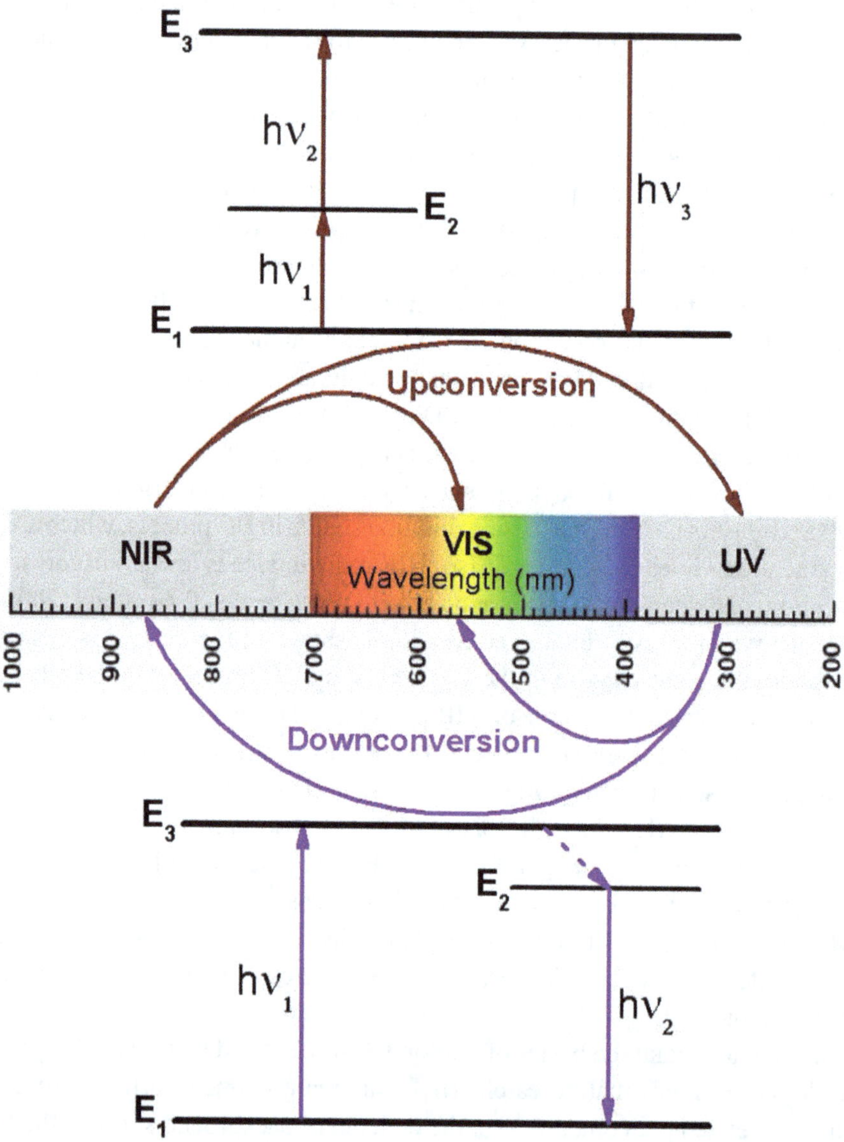

BACKGROUND

The non-linear process under which conversion of low photon energy into high photon energy takes place is known as UC process. Since long decades of time span (1960) the UC mechanism is being studied. Researchers have tried to extend the application of such materials in various fields after couture the shape and size of particles since last two decades. In this case different light ions along with rare earth (RE) ions have also been doped/co-doped to enhance the quantum efficiency of developed UC materials. Prior to insertion of light elements like zinc, lithium, manganese and many more into host materials the UCET mechanism plays an important role to be understood. Actually, the luminescence of materials is basically caused through the process that is recognized as Stoke's law. In this process the higher energy of incident photons emit low energy of photons. On the basis of this mechanism, it may be presumed that the efficiency of luminescent material always lies below one. But in the process of UC, the emissive energy of photons is always more than 10 to 100 fold ofthermal energy. Moreover, the UC emission takes place when f-ions of the RE are doped in certain host lattice. However, the clear feature of UC mechanism was too illusive around 1960s. The classical absorption system of photons known as -ESA was defined explicitly. This described process is well understood in which an ion transfers its energy to other ions [Hehlen et al., 2013; Auzel, 1990; Auzel, 1992]. Moreover, the ET process, co-operative energy transfer and photon avalanche (PA) effect also play a crucial role for UC emission.

Energy Transfer Plan

The energy transfer phenomenon occurs between two different RE doped ions, when absorption and emission process do not take place. The amount of ET depends on the concentration of dopants. Because the immigration of energy between two centres increases with concentration of dopants of two ions. This is remarkable to note that this phenomenon of ET occurs without charge transformation. In this situation multi-ion process have been found by single f and d ions that results the radiative ET process. Thus two different cases of radiative and non-radiative energy transformation along with phonon assisted energy migration takes place due to the resonant case. The consecutive ET phenomenon occurs between two unlike sites. In this process the energy migration between two ions due to mutual connections is based on the optimized concentration of doped/codoped ions in host matrices. In the ET mechanism the sensitizer is excited first and consequently the energy is migrated to other ion called activator. Sometimes the sensitizer is said to be donor whereas activator as acceptor. The ET phenomena based on its evolved mechanism may be considered as (a) radiative, (b) non-radiative and (c) phonon assisted (Figure 2).

The radiative ET occurs when the energy liberated from sensitizer is absorbed by acceptor ions. The excess amount of energy depends on the cross section of sensitizer i.e. large cross sectional area of ion

Figure 2. Consecutive ET mechanism of (a) radiative, (b) non-radiative, and (c) phonon assisted while 'x' shows the energy miss-matching factor

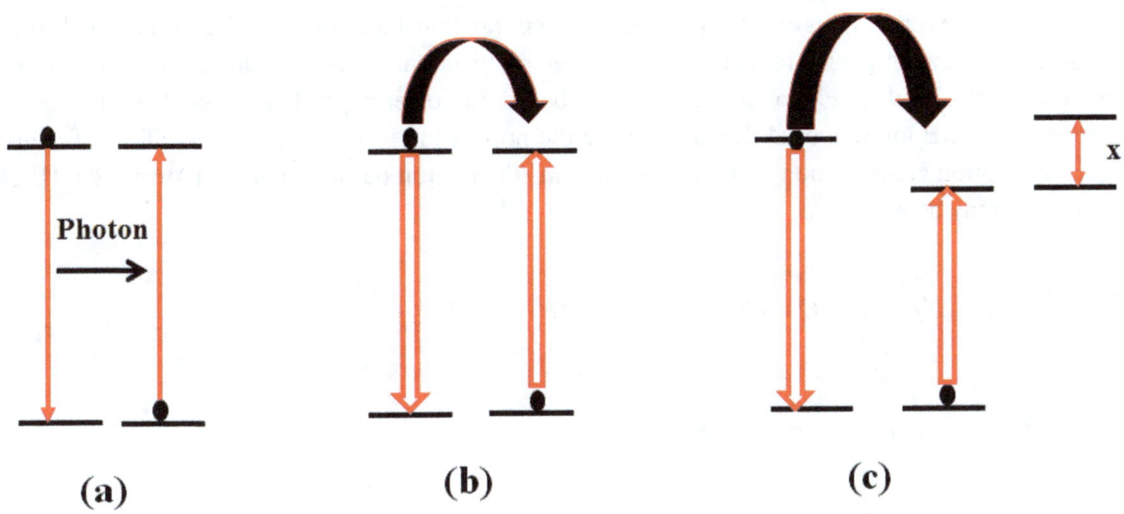

is able to transfer the huge amount of photons to the activators. Moreover, the ET phenomenon also depends on the concentration of activators. This means the emission spectra can be characterized with the concentration of activator ions. So, treating the re-absorption effects, the radiative and non-radiative ET phenomena may be differentiated. Thus the probability of ET (G_{SA}) may be related to a function R by the equation:

$$G_{SA} = \frac{\sigma}{4\pi R^2 \tau_s} \int g_S(x) g_A(x) dx$$

where, τ_s and σ are lifetime of sensitizer and absorption cross section, respectively. $1/R^2$ is the distance between two activator ions. The resonant radiative transfer phenomenon occurs due to long range energy allocation connecting to the similar ion. The photon trapping effect occurs in the gases due to this resonant radiative energy transfer. Thus the life time observed in this situation enhances for REh ions doping. Here, we assume two ions system [Figure 2(b)] to discuss the energy transfer phenomena. The non-radiative ET takes place between two ions having almost equal energy due to the interactions of ions presented in different states. Among all the electronic levels due to suitable interactions of ions, the non-radiative ET gives the channel to other level from there the luminescence centre starts. Now, the probability of ET (G_{SA}) may be given as:

$$G_{SA} = \frac{R_0}{R}^{n} / \tau_S$$

where, R_0 is transfer distance, τ_S is lifetime and 'n' is interaction (for dipole - dipole interaction n=6, dipole- quadrupole interaction n=8 and quadrupole-quadrupole interaction n=10). These remarkable interactions are computed by theoretical model known as Judd-ofelt's analysis. This theoretical analysis may be found elsewhere [Di-Bartolo et al., 2006]. Next step of ET depends over phonon-assisted that happen between two ions and dissimilar energy transportation. It is remarkable to note from above equation that the phonon-assisted energy transfer phenomena would occur if energy integral parts overlap and supposed to be zero. In this case the possibility for energy transfer through the ions is nominal. When the migrated energy is higher at room temperature then multi-phonon process enhances rapidly. The rate equation has been used for ET to the UC emission based on power dependence. For this purpose four excited levels of RE ion was used. Let us suppose the population inversion is $G_{i,}$ the ET coefficient is $Q_{i,}$, α is absorption cross section of sensitizer ion and ∂ is photon number density [Auzel, 2004]. The rate equation may be written as:

$$\frac{dG_1}{dt} = \partial_P \alpha_0 G_0 - 2Q_1 G_1 G_1 - Q_2 G_1 G_2 - Q_3 G_1 G_3 - A_1 G_1 + \beta_2 A_2 G_{2,}$$

$$\frac{dG_2}{dt} = Q_1 G_1 G_1 - Q_2 G_1 G_2 - A_2 G_2 + \beta_3 A_3 G_{3,}$$

$$\frac{dG_3}{dt} = Q_2G_1G_2 - Q_3G_1G_3 - A_3G_3 + \beta_4A_4G_{4,}$$

$$\frac{dG_4}{dt} = Q_3G_1G_3 - A_4G_4.$$

whereas,

$G_0 \cong$ should be treated as invariant

The above rate equations are solved to get the maximum and minimum UC, such as:

(A) For maximum UC emission:

The solutions; when $\beta_i = 0$

$$G_i = \frac{1}{2}Q_1^{1/2}Q_i^{-1}\left(\partial_p\alpha_0G_0\right)^{1/2} \quad where,\ i = 1,2,3\ldots.g\ G_r = \frac{1}{4}\frac{1}{A_n}\partial_r\alpha_0G_0$$

(B) For minimum UC emission:

The solution is as;

$$G_i = \frac{1}{A_1^i}\prod_{j=2,3,..i}\left(R_{j-1}A_j^{-1}\right)\left(\partial_r\alpha_0G_0\right)^i \quad where\,i = 1,2,3\ldots..g,$$

Upconversion Mechanism

To explain the energy transfer phenomena the classical law was proposed prior to the 1966s (Figure 2). Beyond this classical assumption it was Auzel who first time explained quantum mechanically the ET mechanism by assuming the acceptor ion previously presented in excitation state (Figure 3). Since, activator ion has many energy states so the photon is added on this state. Laser action of Er^{3+} was explained, explicitly, by considering this mechanism. In this system, the Yb^{3+} ions was used as a sensitizer [Snitzer and Woodcock, 1965]. The confirmation of this effect was efficiently achieved for green emission of Er^{3+} [Auzel, 1966].

A comparative study between (a) ET UC effect, (b) two steps absorption system, (c) co-operative sensitization, (d) co-operative emission procedure, (e) second-harmonic generation (SHG) and (f) photonic absorption system are shown in Figure 3. The active ion in this situation transfers energy by dipole-dipole interaction and emits photons with shorter life span. In ET UC process the dipole-dipole energy transfer occurs, which is eventually accepted for lasing actions. The detailed description about ET UC process may be found elsewhere [Chen el al.,2012]. From the excited state the RE ion decays to ground state either by emitting photons (radiative) or phonons (non-radiative). Due to non-radiative decay process the ET UC is said to be as superfluous effect that results heating in the host matrix system. Moreover, the concentration quenching is also result of ET UC because of collection of ions in the host matrices.

Figure 3. The upconversion mechanism (a) ET upconversion effect, (b) two steps absorption system, (c) co-operative sensitization, (d) co-operative emission procedure, (e) second harmonic generation and (f) two photon absorption system

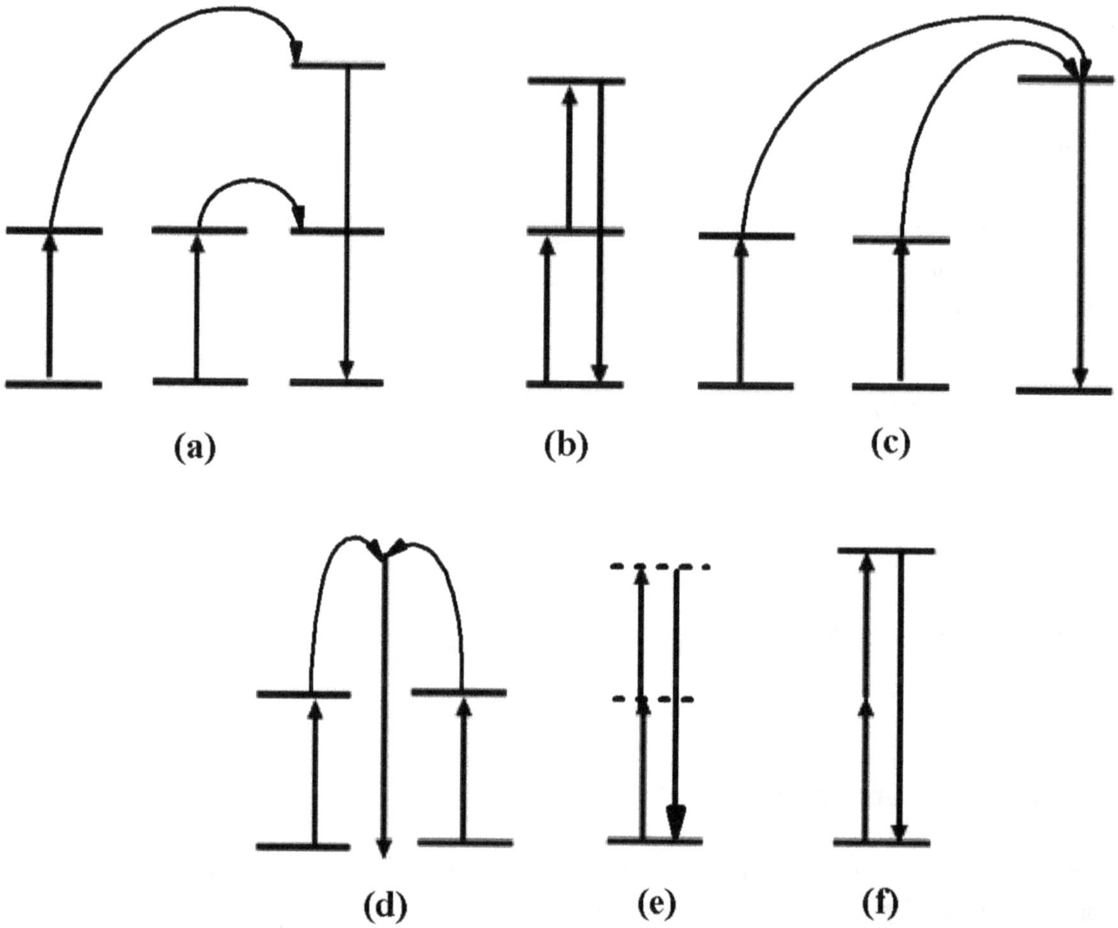

The excitation of ions from ground to excited state for similar energy state the two photon absorption process takes place. This photon absorption process is considered as a third order process because of lower linear absorption at weak emission intensity. Linear absorption process is quite different from two photon absorption process. Hence, the non-linear process is very strong process to get the high UC emission intensity. In the Yb^{3+}-Tb^{3+} co-doped phosphor system the co-operative UC mechanism has been demonstrated explicitly, and may be found elsewhere [Nakazawa et al., 1970]. Lee *et al.* have shown the co-operative luminescence process for the Pr^{3+} activated inorganic host system [Lee et al. 1984]. Auzel in 1973 has explained the second harmonic generation and absorption excitation process for RE ions (Eu^{2+}) doped CaF_2 host matrix [Auzel, 1973].

Selection Criteria of Materials

For high efficient UC phosphor material the selection of suitable host material is very crucial. To this criterion the host should be transparent, high refractive index, chemically stable and optically damage threshold. As far as concern to choose the appropriate host is important because it affects the emission intensity by the local crystal field and phonon frequency. The non-radiative channels starts in the material consequently due to enhanced phonon induced process and it is decreased the efficiency of UC. Here it is worth to note that the luminescence intensity is rather responsive at phonon frequency. Therefore, the low cut off phonon frequency of host material is needed to reduce the non-radiative decay channels and hence enhance the emission efficiency. Some hosts like Y_2O_3 (cut off phonon frequency 550 cm^{-1}) [Martinez et al., 2010], $NaYF_4$ (cut off phonon frequency 350 cm^{-1}) [Ivaturi et al., 2013], YVO_4 (cut off phonon frequency 890 cm^{-1}) [Mialon et al., 2010], Y_2O_2S (cut off phonon frequency 520 cm^{-1}) [Kumar et al., 2012a], ZrO_2 (cut off phonon frequency 500 cm^{-1}) [Patra et al., 2003] are the appropriate hosts investigated by many researchers for enhanced emission efficiency. The local crystal field of hosts also affects the emission of emitters. Because the luminescent center of emitters are surrounded by the local crystal field symmetry. Therefore, for the selection of suitable hosts, low phonon frequency and low crystal field symmetry is significant to achieve the high UC efficiency.

PHOSPHOR'S SYNTHESIS ROUTE

The way of phosphor synthesis plays an important role to achieve good quality phosphor material. Several approaches have been used to synthesize UC phosphors. A short detail on few of important preparation methods is presented in this section.

Solid State Reaction

Solid state reaction method is a high temperature synthesis route of preparing UC phosphor material. The solids are not reacted together at room temperature, so it is necessary to heat them at higher temperatures, often 1000 to 1500 °C. The factors on which the feasibility and rate of a solid state reaction depend include, reaction conditions, structural properties of the reactants, surface area of the solids, their reactivity and thermodynamic free energy [West, 2007]. In a typical synthesis procedure, the starting materials uses in oxide precursors. The reaction starts spontaneously during the mixing process, accompanied by the release of heat and water vapor. The required amount of the chemicals are weighed and ground to a fine powder using agate mortar and pestle. Sometimes, acetone is used as a mixing medium to get homogeneous mixture. The ground sample is placed in alumina/platinum crucibles and heated at high temperature for a particular time, followed by dry grinding. In some cases two step annealing process is used to prepare the UC phosphors [Pandey et al., 2015].

Continuous Spray Pyrolysis

A better way to prepare well controlled nano sized phosphors is theContinuous spray pyrolysis (CoSP). The experimental CoSP set-up was composed of three zones, i.e. spray zone, evaporation zone and collection zone. Each zone is important for controlling the shape and size of the nanoparticle. Each zone has a separate temperature controller giving the flexibility in control of the temperature distribution in the range of 100-1000°C. The first zone (spray zone)- having an atomizer that converted the starting solution into droplets. Each droplet acted as an individual phase-separated reactor, which is transported through the heating zone. The second zone is the heating zone in which the liquid droplet gets pyrolyzed. The last or third zone is meant for the trapping of produced nanoparticles. An arrangement is made in the third zone for keeping the substrates (horizontal/vertical) for film formation. The outlet is connected to a round bottom flask with a glass tube for collection of the produced nanoparticles. The CoSP technique has the advantage of simultaneously fabricating both nanoparticles and films. For synthesis of UC nanoparticles, acetate precursor of starting materials was used for preparation of solution whereas acetic acid and toluene were added as catalysts. The aqueous precursor solution was sprayed into the reactor and the nanoparticles were collected in a collection chamber. Oxygen or nitrogen gas was used as the carrier gas. The gas pressure was maintained at 2.0 kgf/cm^2 and the solution flow rate was 2 ml/min. Atomization can be affected by applying an electric field, a DC voltage of 1 kV was applied between the spray nozzle and a circular metal electrode placed 2 mm in front of the nozzle for creating the electric pressure on the spray droplets [Dwivedi and Dutta, 2013].

Solution Combustion Method

Solution combustion is very famous method of phosphor preparation. In this method, the starting materials are used in nitrate form and urea as organic fuel. The precursor materials and urea were dissolved in distilled water to get transparent solutions. The solution is concentrated by heating to 60 °C on a magnetic stirrer until free water evaporated and spontaneous ignition occurred. In the beginning, the solution underwent dehydration with the liberation of large amount of gaseous products. At the spontaneous combustion point, the solution started burning and released the heat. The solution vaporized instantly with carbon and nitrogen oxide gas products liberation and became a burning solid when the combustion was finished. When being cooled to room temperature, the resultant fluffy mass like products crushed properly by mortar and pestle into fine powder form. The prepared powder phosphors are sometimes annealed at different temperatures for better crystalline material [Tamrakar and Bisen, 2015]. A pictorial representation of combustion process is shown in Figure 4 to visualize the process in better way. This method is very useful for phosphor synthesis because of its low processing temperature, homogeneous mixing and smaller particle formation [Pandey and Rai, 2014a].

Sol-Gel Method

The sol-gel process is a typical wet-chemical technique for synthesizing UC phosphors. It can be generally divided into three types: (i) sol-gel route based on the hydrolysis and condensation of molecular precursors (ii) gelation route based on condensation of the aqueous solutions containing metal-chelates (iii) polymerizable complex route [Lin et al, 2007]. In sol-gel process, a RE nitrate salt or metal alkoxide is generally used as the starting reactants. The reaction is started by mixing the reactants in liquid phase through hydrolysis and condensation reactions, followed by annealing at high temperature for a certain period of time. Prasad et al. [Patra et al., 2002] have reported a sol emulsion-gel method to produce Er^{3+} doped ZrO_2 nano-phosphor. The annealing procedure (temperature and time) is the key step in the preparation process of the sol-gel method, which can seriously determine the quality of the samples. A general flow chart of phosphor preparation by sol-gel method is given in Figure 5. The sol-gel synthesis

Figure 4. Pictorial representation of the sample preparation by solution combustion process

method can be used for large-scale production and the products usually offer high luminescence intensity due to the high crystallinity formed at high annealing temperature. The sol-gel derived nanocrystals generally have broad particle size distribution, irregular morphology, and are insoluble in water, which compose the shortcomings of this method.

Hydrothermal Method

Hydrothermal is a chemical process that can react between negative ions and positive ions that usually precipitate from the solvent under high/medium temperature and high pressure, to generate nanoscale materials after proper processing. Wang et al. (2013) reported the hydrothermal synthesis and tunable multicolor UC emission of Y_2O_3 nanoparticles. For a typical protocol, 1 m mol (total amounts) of the all nitrate precursor of starting materials was added into 20 ml deionized water and then obtained solution was rapidly adjusted to pH value of 14 by adding NH_4OH solution. The obtained mixture was then transferred into a 50 ml stainless Teflon lined autoclave and reacted at 120°C for 12 h. After reaction, the system was naturally cooled to room temperature. The resulting products were collected, washed several times with de-ionized water to remove residual organic compounds, and then dried at 80°C for 3 h. This method is very useful in the formation of a well-shaped material [Zhang et al., 2004].

Figure 5. Flow chart of phosphor preparation by sol-gel method

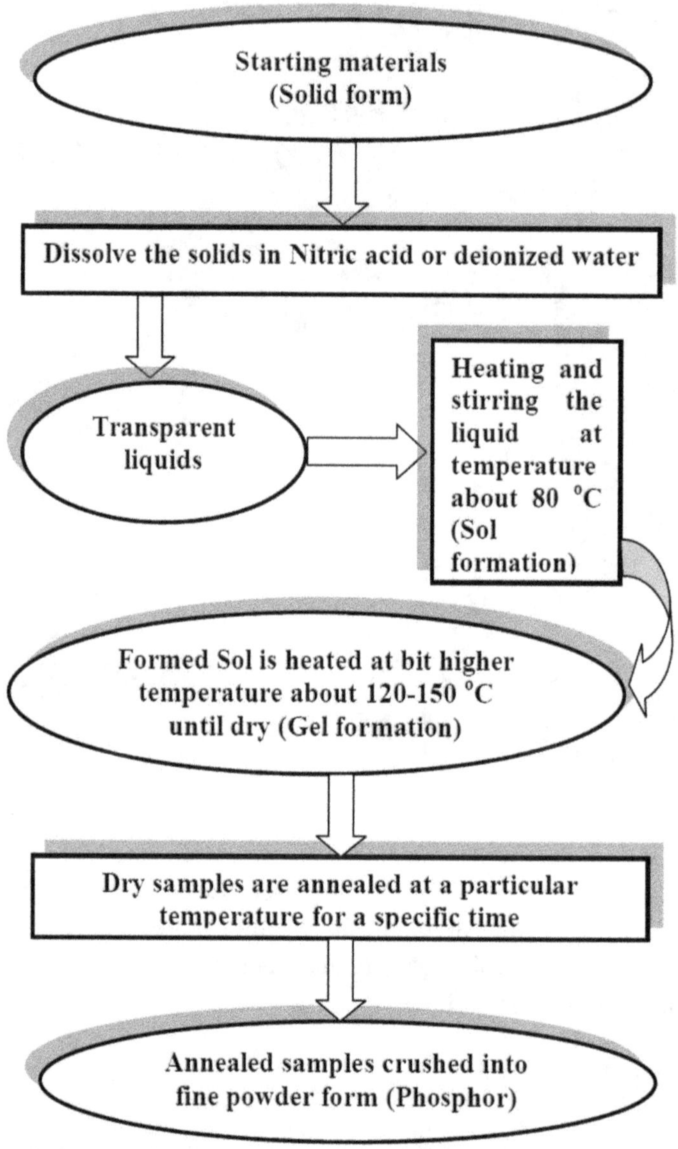

Thermal Decomposition Method

The thermal decomposition process is a traditional method for preparation the inorganic nanocrystals comprising of dissolution of organic/ inorganic precursors in an organic solvent with high-boiling point. Zhang et al. (2005) reported a novel synthesis of LaF_3 triangular nanoplates via the thermal decomposition of lanthanum trifluoroacetates. This method has become a common route for synthesizing high quality Ln^{3+} doped $NaYF_4$ UCNPs since its discovery. Mai et al. (2006) have reported a general synthesis of high-quality $NaREF_4$ (RE = Pr to Lu, Y) NPs using $Na(CF_3COO)$ and $RE(CF_3COO)_3$ as precursors. The combined solvents were composed of both the coordinating solvent and non-coordinating solvent. 1-octadecene (ODE) with a high boiling point (315 °C) was used as the non-coordinating solvent to provide a high temperature environment. Oleic acid (OA) and oleyamine (OM), having good coordinate abilities, acted as the coordinating solvent which aids in capping the surface of NPs to prevent agglomeration. This method is allowed the successful synthesis of high quality UC phosphor materials.

Co-Precipitation Method

In co-precipitation technique, the oxide form of starting materials is dissolved in deionized water. To form the nitrate precursors, concentrated HNO_3 was then added into the solutions. The solutions were heated for 10-15 minutes at a constant temperature 60°C. To start precipitation, NH_4OH solution is added drop wise into nitrate precursors. To achieve complete precipitate, the solution is left for about 20 hours. After that the precipitate is filtered using filter paper and washed twice with deionized water and twice with ethanol to remove the impurities. The precipitate is dried and then annealed at high temperatures for certain time in order to completely remove the water and nitrates remain in the sample. A typical synthesis of ZrO_2 based phosphor by co-precipitation method is reported by Pandey et al. [Pandey et al., 2014a].

UPCONVERSION STUDY OF VARIOUS LANTHANIDE DOPED PHOSPHORS

In previous sections, authors discussed about constituents of UC phosphor materials, important UC mechanisms and some famous synthesis process in detail. Now, authors will explore the UC emission feature of different Ln^{3+} activated UC phosphors one by one. As authors discussed earlier, phosphors are the combination of host and doped materials. Several materials such as oxide, halide, sulfide, nitride, silicate, phosphate, etc. have been used as host materials but those having lower cut-off phonon frequency are more attracted for luminescence purpose. The small amount of metal impurities (Ln^{3+} and/or transition metals) is used as dopants. Abundant energy structure of lanthanide ions make them very useful as dopant in development of phosphor materials. Ln^{3+} are mostly stable in their triply ionized state but few have 2+ (Sm, Eu and Yb) and 4+ (Ce, Pr and Tb) stable states as well. In UC studies, generally NIR excitation at 808, 976/980, 1342 and 1530 nm wavelengths have been used to excite the materials. Though, each Ln^{3+} having no such energy level structure that they can easily excite by a commonly used 980 nm laser, but the sensitizer like Yb make this possible. Figure 6 presented a typical energy level structure of Ln^3 ions with possible UC emission transitions. The levels are drawn using energy values of a defined level of Ln^{3+} ions given by Dieke (1968). The figure confirms that most of Ln^3 can give UC emissions except cerium (Ce) and gadolinium (Gd) (not presented here). These sharp spectral emissions are spread over full visible region, indicating multicolour emission characteristic of Ln^3 activated materials. The Er^{3+}, Ho^{3+} and Tm^{3+} activated UC phosphors have been widely investigated by researchers whereas others like Pr^{3+}, Dy^{3+}, Tb^{3+}, Eu^{3+}, Nd^{3+} and Sm^{3+} were less studied because of their comparatively low UC emission efficiency.

Start from Er^{3+} doped phosphor, which displays very good green UC emission around 523 and 548 nm through the ${}^2H_{11/2}\rightarrow{}^4I_{15/2}$ and ${}^4S_{3/2}\rightarrow{}^4I_{15/2}$ transition, respectively [Kumar et al., 2012b]. Also the other emissions in UV-blue, red and NIR regions peaking at 335, 366, 380, 408, 475, 661 and 800 nm arising from the ${}^4G_{9/2}\rightarrow{}^4I_{13/2}$, ${}^2G_{9/2}\rightarrow{}^4I_{15/2}$, ${}^4G_{11/2}\rightarrow{}^4I_{15/2}$, ${}^2H_{9/2}\rightarrow{}^4I_{15/2}$, ${}^4F_{3/2}\rightarrow{}^4I_{15/2}$, ${}^4F_{9/2}\rightarrow{}^4I_{15/2}$ and ${}^4I_{9/2}\rightarrow{}^4I_{15/2}$ transitions, respectively [Joshi et al., 2009; Singh et al., 2009]. The emissions from singly Er^{3+} doped phosphors are less intense, so codoping with suitable sensitizer (mostly Yb^{3+}) boost up the emission efficiency. As

Figure 6. Partial energy level diagram of Ln^{3+} ions with possible UC emission transitions

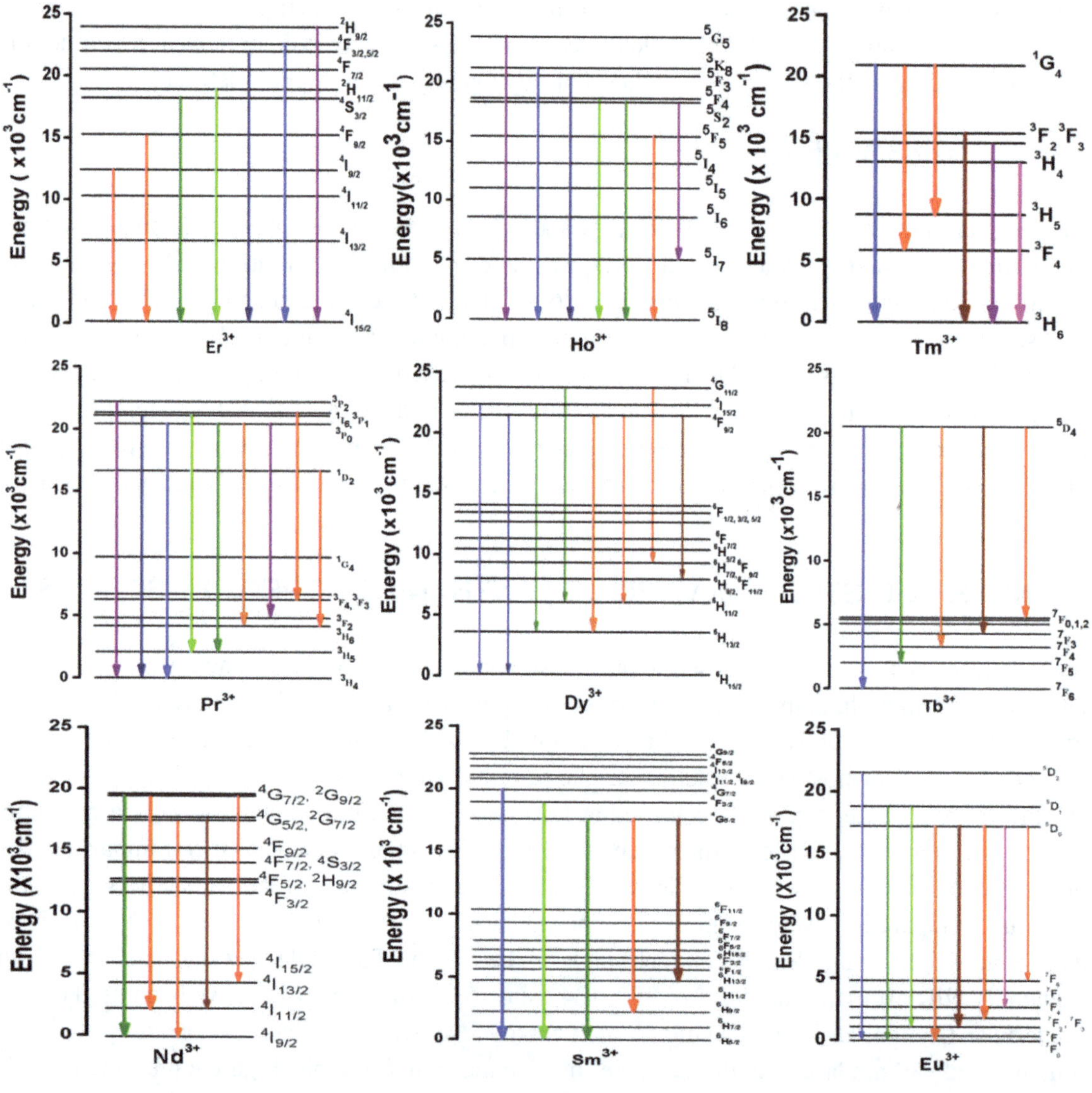

980 nm diode lasers often used as excitation source and Yb^{3+} ion has broad absorption cross-session (870-1050 nm) in this range, so absorb most of pump energy in the codoped system. Also, the excited energy level $^2F_{5/2}$ of Yb^{3+} ion is in resonance with $^4I_{11/2}$level of the Er^{3+} ion, lead energy transfer from Yb^{3+} to Er^{3+} significantly [Rai et al., 2013b]. The emissions observed from green to NIR range are due to two photon UC process, whereas three-four NIR photons are responsible for emissions in UV-blue region [Singh et al., 2009]. To predict about number of pump photons involved in UC process, a power dependence UC study need to do. As well known that UC emission intensity is non-linearly depend upon power of excitation source and related to the following equation-

$$Intensity \propto (Power)^n$$

where, *n* represent the number of pump photons. Generally, the value of "n" is found 2 or more than that, but sometimes, it comes in between 1 to 2 due to involvement of more non-radiative relaxation channel and effect of saturation [Pandey et al., 2013b]. Attempt has also made to improve the emission efficiency via codoping of other impurities into Yb^{3+}- Er^{3+} codoped phosphors [Cao et al., 2011; Tiwari et al., 2015a]. Multicolour emissions have reported in Er^{3+} activated tri-doped phosphors for white light generation [Giri et al., 2011; Pang et al., 2011; Hu et al., 2012; Rai et al., 2013a].

The UC emission spectra of Ho^{3+} doped phosphors also shows intense green emission around 537 and 550 nm through $^5F_4 \rightarrow ^5I_8$ and $^5S_2 \rightarrow ^5I_8$ transitions, respectively [Rai et al., 2012; Verma et al., 2012]. The other UC emission peaks centred at 362, 393, 426, 465, 489, 658, and 757 nm have reported from the $^3H_5 \rightarrow ^5I_8$, $^5G_4 \rightarrow ^5I_8$, $^5G_5 \rightarrow ^5I_8$, $^3K_8 \rightarrow ^5I_8$, $^5F_3 \rightarrow ^5I_8$, $^5F_5 \rightarrow ^5I_8$ and $^5S_2 \rightarrow ^5I_7$ transitions, respectively [Pandey et al., 2013b]. These emissions were also observed through three photon UC process. The influence of Yb^{3+} codoping enhances the emission intensity of all UC bands very largely via efficient energy transfer process [Dwivedi et al., 2011; Rai et al., 2012; Verma et al., 2012; Kumar et al., 2016]. Furthermore, codoping of non-lanthanides like zinc and magnesium also improved the UC emission from Ho^{3+}-Yb^{3+}codoped phosphors [Pandey and Rai, 2013; Dey et al., 2015]. Pandey et al. (2017) obtained a dominant red UC emission rather than green in the $CaMoO_4$:Ho^{3+} phosphor. It's happened in few cases due to occurrence of more non-radiative relaxations from the $^5F_4/^5S_2$ to $^5F_5$ states of the Ho^{3+} ion and cross-relaxation $^3K_8$; $^5I_8 \rightarrow ^5F_5$; $^5I_7$ transitions. Moreover, multicolour tunable emissions [Pandey and Rai, 2012] white light generations [Chung et al., 2012] from tri-doped Ho^{3+} activated phosphors were also reported.

UC emissions from singly Tm^{3+} doped materials are hardly found owing to unavailability of energy level corresponding to famous 980 nm excitation. Therefore, reports were available in Tm^{3+}-Yb^{3+}codoped phosphors mainly. The Tm^{3+} activated phosphors are generally give strong blue, red and NIR emission around 478, 650 and 800 nm through the $^1G_4 \rightarrow ^3H_6$,$^1G_4 \rightarrow ^3F_4$ and $^1G_4 \rightarrow ^3H_5$ (or $^3H_4 \rightarrow ^3H_6$) transitions, respectively [Mishra et al., 2011; Kumari et al., 2016]. The UV-Blue emission peaks were also identified by Mishra et al. (2011) in Tm^{3+}-Yb^{3+} codoped Y_2O_3 phosphor. They observed UV-blue emissions at 300, 346, 364, 394 and 467 nm corresponding to the $^1I_6 \rightarrow ^3H_6$, $^1I_6 \rightarrow ^3F_4$, $^1D_2 \rightarrow ^3H_6$, $^1I_6 \rightarrow ^3H_5$ and $^1D_2 \rightarrow ^3F_4$ transitions, respectively. Also a weak green peak was detected in 530-555 nm range due to the $^1D_2 \rightarrow ^3H_5$ transition [Kumari et al., 2016]. Additionally, weak NIR emission around 700 nm is observed and assigned for the $^3F_4 \rightarrow ^3H_6$ transitions [Mishra et al., 2011]. Enhancement in UC emissions from Tm^{3+}-Yb^{3+}codoped phosphors were also reported via incorporation of Li^+, Zn^{2+} and Mg^{2+} ions [Kumari et al., 2014a; Kumari et al., 2017].

The Dy^{3+} is also very popular Ln^{3+} for optical investigation but UC study of Dy^{3+} activated phosphor is very rare. Cheng-Ren et al. (2012) studied UC behaviour of Dy^{3+}-Yb^{3+} codoped Al_2O_3 phosphor prepared by sol-gel method. They found several UC emission bands centred at 378, 408, 543 and 663 nm corresponding to the $^4G_{9/2} \rightarrow ^6H_{13/2}$, $^4G_{9/2} \rightarrow ^6H_{11/2}$, $^4I_{15/22} \rightarrow ^6H_{13/2}$ and $^4F_{9/2} \rightarrow ^6H_{11/2}$ transitions, respectively. The power dependence study confirmed 3 photon UC process responsible for these UC emissions. Peng et al. (2016) have reported UC emissions of Dy^{3+}-Yb^{3+} codoped Gd_2O_3 phosphor by optimizing the concentration of dopants. Biju et al. (2016) is also used Dy^{3+} as sensitizer to improve the UC luminescence of $NaGdF_4$:Yb^{3+}- Er^{3+} nanomaterials.

To obtain UC emission, it is necessary to excite the $^5D_4$ state of Tb^{3+} ions, which is not possible by using a 980 nm diode laser, in singly Tb^{3+} doped system. As we mentioned earlier, Yb^{3+} is ideal sensitizer and can excite the $^5D_4$ state of Tb^{3+} ions via cooperative energy transfer process [Pandey et al., 2014b]. The research on UC luminescence from Tb^{3+} activated phosphors performed by using the Yb^{3+} as sensitizer. The UC emission bands assigned due to the Tb^{3+} ions in Tb^{3+}-Yb^{3+} codoped phosphors are centred at 379 ($^5D_3 \rightarrow {}^7F_6$), 414 ($^5D_3 \rightarrow {}^7F_5$), 437 ($^5D_3 \rightarrow {}^7F_4$), 488 ($^5D_4 \rightarrow {}^7F_6$), 547 ($^5D_4 \rightarrow {}^7F_5$), 590 ($^5D_4 \rightarrow {}^7F_4$), 623 ($^5D_4 \rightarrow {}^7F_3$) and 660 nm ($^5D_4 \rightarrow {}^7F_{0,1,2}$) wavelengths [Pandey et al., 2014b; Kore et al., 2017]. The room temperature UC emission spectra of Y_2O_3: Tb^{3+}-Yb^{3+} and Y_2O_3: Tb^{3+}-Yb^{3+}-Li^+ phosphors upon a 980 nm excitation is shown in Figure 7(a). The figure concludes that Tb^{3+} exhibits several UC emission peaks throughout visible region and improved via codoping of Li^+ ion. The concentration of Li^+ ion was optimized to achieve maximum luminescence intensity as shown in right-inset of Figure 7(a). Interestingly,

Figure 7. (a) UC emission spectra of Y_2O_3: Tb^{3+}-Yb^{3+} and Y_2O_3: Tb^{3+}-Yb^{3+}-Li^+ phosphors at 980 nm excitation with optimizing Li^+ concentration (inset-right) and optical photograph of light emitted from sample (inset-left), and (b) Energy level diagram of Tb^{3+} and Yb^{3+} ions to understand the energy transfer characteristics. The image is reproduced with permission from reference [Pandey et al., 2014b]. Copyright @ 2014 Elsevier

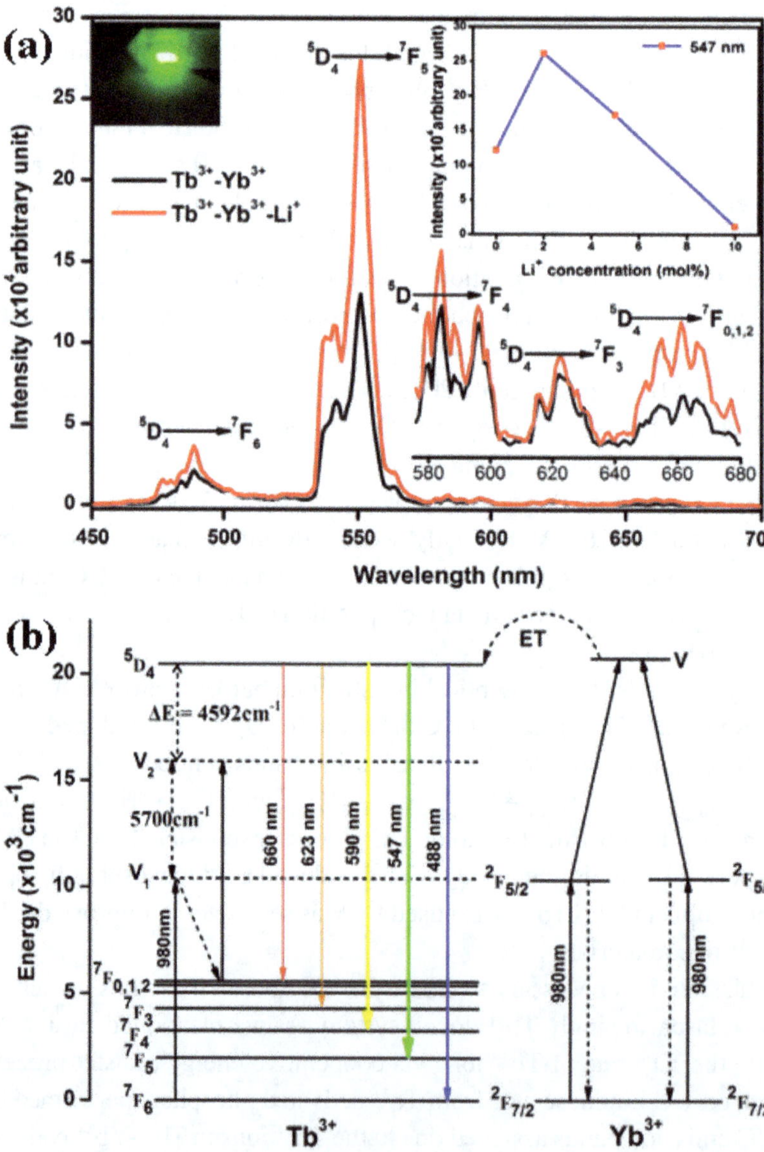

a bright green colour light was observed as given in left-inset of Figure 7(a). To show these emissions transitions and involved UC process, a schematic energy level diagram is presented as given in Figure 7(b). Cooperative energy transfer from Yb^{3+} to Tb^{3+} ion was found responsible for UC emissions obtained [Pandey et al., 2014b]. The effect of non-lanthanides [Kore et al., 2017], lanthanides [Yang et al., 2007] and multimodal emission [Molina et al., 2011] in Tb^{3+}- Yb^{3+} codoped phosphors were also studied.

The UC emission study of Pr^{3+} activated phosphors given several emission peaks around 475, 489, 522, 546, 619, 631, 660, 800 and 850 nm wavelengths corresponding to the ${}^3P_1\rightarrow{}^3H_4$, ${}^3P_0\rightarrow{}^3H_4$, ${}^3P_1\rightarrow{}^3H_5$, ${}^3P_0\rightarrow{}^3H_5$, ${}^3P_0\rightarrow{}^3H_6$, ${}^3P_0\rightarrow{}^3F_2$, ${}^3P_1\rightarrow{}^3F_3$, ${}^1D_2\rightarrow{}^3H_6$ and ${}^1I_6\rightarrow{}^1G_4$ electronic transitions upon a 980 nm excitation [Pandey and Rai, 2014a; Yadav et al., 2015]. The codoping with Yb^{3+} ions enhanced the intensity of all UC emission bands due to efficient energy transfer from the Yb^{3+} to Pr^{3+} ions [Maciel et al., 2009]. A three photon UC and two photon co-operative energy transfer process were reported for the UV-Vis UC emissions from the Pr^{3+}- Yb^{3+} codoped phosphors and suggested an efficient material for lighting technology.

The Nd^{3+} ions has been vastly used in development of glass-laser but the report on UC emission from Nd^{3+} activated phosphors are very rare. Dey et al. (2012) observed intense green UC emission from Nd^{3+}-Yb^{3+} codoped Y_2O_3 phosphor upon a 980 nm excitation and further enhancement via Zn^{2+} codoping. Only two UC emission bands were detected around 550 and 664 nm arising from ${}^4G_{7/2}\rightarrow{}^4I_{9/2}$ and ${}^4G_{7/2}\rightarrow{}^4I_{13/2}$ electronic transitions of Nd^{3+} ion. As the ${}^4F_{3/2}$ state of Nd^{3+} can hardly populate by 980 nm pumping, so the cooperative sensitization from Yb^{3+} ion is main cause of these UC emissions. On the other hand, 808 nm pumping can easily excite Nd^{3+} atoms through excited state absorption process and resulted in to strong red UC luminescence [Sukul et al., 2017]. The other emission transitions ${}^4G_{7/2}\rightarrow{}^4I_{11/2}$ and ${}^4F_{9/2}\rightarrow{}^4I_{11/2}$ in yellow and NIR regions are also reported [Tian et al., 2014].

Samarium (Sm) is one of the admired Ln^{3+} reported for its good fluorescence characteristics. Like the others, it has rich energy structure which display several emission transition upon suitable excitation. UC property of the Sm^{3+}-Yb^{3+} codoped Y_2O_3 phosphor upon 980 nm excitation is reported by Pandey S. et al. (2013). They optimized the concentration of Sm^{3+} and Yb^{3+} as 0.2 and 1 mol%, respectively. The UC emission peaks were detected around 489, 525, 551, 661 and 800 nm, corresponding to ${}^4G_{7/2}\rightarrow{}^6H_{5/2}$, ${}^4F_{3/2}\rightarrow{}^6H_{5/2}$, ${}^4G_{5/2}\rightarrow{}^6H_{5/2}$, ${}^4G_{5/2}\rightarrow{}^6H_{9/2}$ and ${}^4G_{5/2}\rightarrow{}^6H_{13/2}$ transitions of Sm^{3+} ion. A dominant green emission obtained and improved via post annealing and codoping. Strong red UC luminescence was also reported upon 976 nm excitation [Liu et al., 2017]. Dosev et al. (2006) studied UC fluorescence of Sm doped Gd_2O_3 nanopowder by changing excitation wavelength.

The most popular Ln^{3+} ion for red emission is Eu^{3+}, whose UC studies were performing since decades. Due to lack of energy level corresponding to the 980 nm excitation, its UC property investigated by using Yb^{3+} as sensitizer into variety of phosphor hosts. The 970 and 980 nm diode laser excited green emitting Eu^{3+}-Yb^{3+} codoped Y_2O_3 phosphors were reported via cooperative energy transfer from Yb^{3+} to Eu^{3+} ions [Wang et al., 2008; Pandey et al., 2013a]. Bright red UC emitted phosphors were also investigated in Eu^{3+}-Yb^{3+} codoped systems [Kaur et al., 2010; Dwivedi and Zilio, 2013]. Several emission bands are observed at 352, 405, 417, 427, 452, 467, 511, 537, 550, 581, 592, 611, 652 and 707 nm through the ${}^5D_4\rightarrow{}^7F_0$, ${}^5D_3\rightarrow{}^7F_{0,\,1,\,2\,\&\,3}$, ${}^5D_2\rightarrow{}^7F_0$, ${}^5D_1\rightarrow{}^7F_{0,\,1,\,2}$ and ${}^5D_0\rightarrow{}^7F_{0,\,1,\,2,\,3\&\,4}$ transitions of Eu^{3+} ion, respectively [Kaur et al., 2010]. A NIR emission around 796 nm for the ${}^5D_0\rightarrow{}^7F_6$ transition has also been reported in yellow-orange emitting Eu^{3+}-Yb^{3+} codoped $BaTiO_3$ phosphor [Kumari et al., 2014b]. Dong et al. (2016) investigated Eu^{3+}-Yb^{3+} activated core/shell nanoparticles with dual band excitation.

The Yb^{3+} has been established as best sensitizer for UC emissions from different Ln^{3+} activated phosphors. Beyond this, it has ability of blue UC emission through cooperative emission process when pumped with a 980 nm excitation. Several papers have been reported on the cooperative blue UC emission from Yb^{3+} ions in different phosphor materials [Dey et al., 2012; Rai and Pandey, 2016; Mokoena et al., 2017]. Mokoena et al. (2017) studied the combustion synthesized Al_2O_3:Yb^{3+}phosphor powder and observed broad blue emission band when excited with a 980 nm laser. It was reported that two Yb^{3+} ions cooperatively emits energy double that of excitation. Broad range and different sharp peaks are supposed to be due to transition between different stark sublevels of excited and ground level of Yb^{3+} ion.

EMERGENT APPLICATIONS OF UPCONVERTING PHOSPHORS

The demand of Ln^{3+}activated UC phosphors are increasing day by day owing to their potential application in various fields. So, it is worth to discuss some emergent technological application area of UC materials. In this step, the authors will survey the utilization of UC phosphors for solar cells, sensors, security and biomedical application.

Application in Solar Cell

Global energy consumption is increasing regularly and it is expected to double by the end of 2050. Most of our energy supply comes from fossil fuels; however, their finite nature and the environmental concerns associated with fossil fuels have triggered the search for alternative renewable energy sources. There are existing different potential renewable energy source of technologies such as biomass, geothermal, hydropower, hydrogen fuel cells, wind, ocean, and solar energy. Sunlight is a freely available abundant source of solar energy. It is irradiated the Earth's surface with 120 000 TW of solar power. It has been reported that the Earth's surface receives nearly 10 000 times more sunlight than what we consume. Harnessing solar energy through photovoltaic (PV) technology has the potential to meet our energy demands in a cleaner way. In the literature, different approaches have been suggested to enhance the efficiency of solar cells above the theoretical limit of 30% calculated by Shockley and Queisser (1961) for the conventional solar cells.

A typical solar cell have different kind of losses caused by reflection losses, recombination losses, resistive losses and thermal losses as shown in Figure 8 [Verma et al., 2012]. The most common loss is due to the absorption of photons, having energy greater than the band gap of the materials and it is leading to the thermalization of the charge carriers. The thermalization of charge carriers can be reduced using the layer of DC materials in devices structure. On the other hand, the transmission loss is due to the photon having energy less than the band gap of the materials. The transmission losses can be reduced by using UC materials. Trupke et al. (2002) have theoretically evaluated the maximum efficiency for Si solar cell in non-concentrated sunlight to about 47.6% via UC process. The Er^{3+} doped UC material suggested an increase in spectral response of the 1489-1580 nm wavelength for a bi-facial solar cell. Chander et al. (2015) reported the improvement in the short wavelength spectral response of silicon solar cells by spray deposition of YVO_4: Eu^{3+} DS nanophosphor. The optimized quantity of phosphor NPs is provided a photocurrent enhancement of ~2.1%. Xu et al. (2011) have shown significant improvement in the efficiency of dye sensitized solar cells (DSSCs) by adding an organic fluorescent dye into the liquid electrolyte where the unabsorbed photons have gone through the DC process. Gibart et al. (1996) reported first time the application of UC phosphors for enhancing the efficiency of solar cell in 1996. They combined the GaAs solar cell with a vitro ceramic material doped with Yb^{3+} and Er^{3+} and

Figure 8. Schematic diagram of a solar cell to visualize different kind of losses

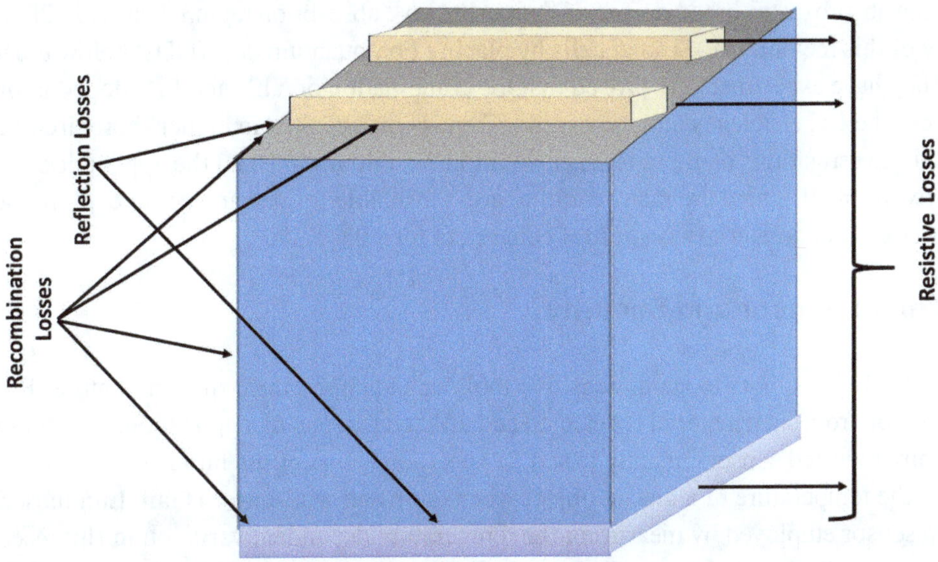

obtained an efficiency of 2.5% under high excitation densities. Since then, many research groups have been involved in the development of UC based solar cells. UC materials with singly doped Er^{3+} or Ho^{3+} are highly suitable for narrow band gap c-Si solar cells (Eg~1.12 eV; 1100 nm). They can be converted a long wavelength (~1500 nm) NIR light into short-wavelength (980, 800, 660, 550 nm) NIR as well as in the visible emissions. For other kind of solar cells, such as amorphous silicon (a-Si), DSSC, perovskite and organic solar cells, the co-doping of Yb^{3+} with Er^{3+} or Ho^{3+} UC materials is very convenient due to efficient ETU from the Yb^{3+} sensitizer to the Er^{3+}/Ho^{3+} activators. In 2010, Shan et al. (2010) reported first time, the application of UC nanoparticles for enhancing NIR sunlight harvesting in DSSCs. After this, several research groups have successfully modified this strategy to improve the performance of DSSCs [Chang et al., 2013; Wu et al., 2013; Green et al., 2017].

In 2011, Wang et al. (2011b) demonstrated the feasibility of UC in P3HT: PCBM organic solar cells. A commercial Er^{3+}, Yb^{3+} codoped YF_3: UC phosphor material with a quantum efficiency of 0.19% (excitation density: 250 mW/cm^2) was used as UC source and obtained a photocurrent density of 16.5 mA/cm^2. Later in 2012, they reported the application of dual-functional upconverting Yb^{3+}-Er^{3+} codoped MoO_3 film in P3HT: PCBM solar cells. The MoO_3 matrix is used as a hole extraction layer, while the Yb^{3+} as a sensitizer and Er^{3+} as activator to convert the NIR photons into visible light. The contribution of UC in short-circuit current was less than 1% under AM 1.5G illumination. Adikaari et al. (2012) demonstrated the application of Y_2BaZnO_5: Yb^{3+}-Ho^{3+} UC nanoparticles to PCDTBT: PCBM organic solar cells in order to utilize the NIR solar spectrum. Guo et al. (2014) reported the $NaYF_4$: Yb^{3+}-Er^{3+} nanocomposites into a PCDTBT: PCBM blend as a bifunctional additive to improve the performance of inverted polymer solar cells. The doped UC nanocomposites exhibited both the NIR light-harvesting as well a slight-scattering properties, which enhanced the device performance. The PCE is obtained 6.87% for optimized devices, which is 19.1% higher than the device without UC layer.

In UC process, only a very small fraction of sub-band-gap sunlight can be converted due to the narrow absorption cross-section of Ln^{3+} ions. The absorption spectrum broadening is one another area needs to work out by researchers. The broadening of the absorption spectrum for Ln^{3+} upconverters can be achieved by using external sensitizers such as quantum dots or organic dyes [Zou et al., 2012]. These sensitizers can absorb over a broad range and emit in the Ln^{3+} absorption region. Pan et al. (2010) reported the viability of this method in c-Si solar cells by placing PbS quantum dots (QDs) below the upconverting layer. They have experimentally proved that the combination of Yb^{3+} and Er^{3+} doped phosphor with QDs enhances the UC efficiency and increased UC emissions resulted in higher short circuit currents in the solar cell. Investigations on these issues are still in the initial stage and the application to solar cells is comparatively small. There are many exciting areas available for researchers to explore the practical application of UC nanomaterials as spectral converters for solar cells.

Application in Sensor and Security

The ability of UC phosphors to act as sensitive tool depends upon their spectral feature. Basically the spectral variation from such material is detected and utilized for sensing application. Temperature sensor is one of many reported sensors, mainly based on spectral change of the luminescence from materials. It can detect the temperature of a distant object where conventional sensor is fail. Luminescence based temperature sensor employed by measuring the temperature dependent variation in fluorescence emissions of two thermally coupled levels (TCL) of Ln^{3+} ions [Wade et al., 2003]. The energy levels having separation in the range 200-2000 cm^{-1} are declared as thermally coupled [Wang et al., 2015]. Several lanthanides such as Er, Ho, Tm, Nd, Eu, Pr, Dy, and Sm have available TCL while the stark sublevels of Yb, Tm, Er have also used for optical thermometry. A list of transitions from TCL of Ln^{3+} ions reported by different researchers is presented in Table 1. Fluorescence lifetime dependent temperature monitoring

Table 1. List of transitions from thermally coupled levels of different Ln^{3+} ions

S. No.	Lanthanide Ions	Thermally Coupled Transitions
1	Er^{3+}	$^2D_{7/2}\rightarrow{}^4I_{15/2}$ and $^4G_{9/2}\rightarrow{}^4I_{15/2}$ $^2H_{11/2}\rightarrow{}^4I_{15/2}$ and $^4S_{3/2}\rightarrow{}^4I_{15/2}$ $^2H_{11/2}\rightarrow{}^4I_{13/2}$ and $^4S_{3/2}\rightarrow{}^4I_{13/2}$ $^4S_{3/2}\rightarrow{}^4I_{15/2}$ (stark sublevels) $^4I_{13/2}\rightarrow{}^4I_{15/2}$ (stark sublevels)
2	Ho^{3+}	$^3K_8\rightarrow{}^5I_8$ and $^5F_3\rightarrow{}^5I_8$ $^5F_4\rightarrow{}^5I_8$ and $^5S_2\rightarrow{}^5I_8$ $^5F_{2,3}/{}^3K_8\rightarrow{}^5I_8$ and $^5F_1/{}^5G_6\rightarrow{}^5I_8$
3	Tm^{3+}	$^1D_2\rightarrow{}^3F_4$ and $^1G_4\rightarrow{}^3H_6$ $^3F_{2,3}\rightarrow{}^3H_6$ and $^3H_4\rightarrow{}^3H_6$ $^1G_4\rightarrow{}^3H_6$ (stark sublevels)
4	Nd^{3+}	$^4F_{7/2}\rightarrow{}^4I_{9/2}$ and $^4F_{5/2}\rightarrow{}^4I_{9/2}$ $^4F_{5/2}/{}^2H_{9/2}\rightarrow{}^4I_{9/2}$ and $^4F_{3/2}\rightarrow{}^4I_{9/2}$ $^4F_{7/2}/{}^4S_{3/2}\rightarrow{}^4I_{9/2}$ and $^4F_{3/2}\rightarrow{}^4I_{9/2}$
5	Eu^{3+}	$^5D_1\rightarrow{}^7F_1$ and $^5D_0\rightarrow{}^7F_1$ $^5D_0\rightarrow{}^7F_1$ and $^5D_0\rightarrow{}^7F_2$ $^5D_0\rightarrow{}^7F_2$ and $^5D_0\rightarrow{}^7F_4$
6	Pr^{3+}	$^3P_1\rightarrow{}^1G_4$ and $^3P_0\rightarrow{}^1G_4$ $^3P_1\rightarrow{}^3H_5$ and $^3P_0\rightarrow{}^3H_5$
7	Dy^{3+}	$^4I_{15/2}\rightarrow{}^6H_{15/2}$ and $^4F_{9/2}\rightarrow{}^6H_{15/2}$ $^4I_{15/2}\rightarrow{}^6H_{13/2}$ and $^4F_{9/2}\rightarrow{}^6H_{13/2}$
8	Sm^{3+}	$^4F_{3/2}\rightarrow{}^6H_{5/2}$ and $^4G_{5/2}\rightarrow{}^6H_{5/2}$
9	Yb^{3+}	$^2F_{5/2}\rightarrow{}^2F_{7/2}$ (stark sublevels)

is another popular technique but fluorescence intensity ratio (FIR) is more functional. A detail theoretical background of FIR based temperature sensor is given by Pandey and Rai (2014b).

Mostly the UC based temperature sensing studies were performed using $^2H_{11/2}$ and $^4S_{3/2}$ levels of Er^{3+} ion, because of high temperature detection with good sensitivity. Temperature dependent UC spectra of $SrWO_4$:Er^{3+}-Yb^{3+} phosphor is given in Figure 9. The variation in FIR of the $^2H_{11/2}\rightarrow{}^4I_{15/2}$ and $^4S_{3/2}\rightarrow{}^4I_{15/2}$ transitions (inset of Figure 9) confirms the emissions of these transitions are influenced by absolute temperature. To evaluate the ability of such materials as temperature sensor, a parameter named sensor sensitivity is evaluated [Pandey et al., 2015]. A good temperature sensor should have very high temperature range with high sensor sensitivity. Apart from optical thermometry, concept of laser induced optical heating was also established on FIR technique. It is very useful in development of phosphor based nano-heater and can be utilized in hyperthermia based cancer treatment [Tikhomirov et al., 2009]. Suo et al. (2017) proposed all-in-one thermometer-heater upconverting platform via FIR based optical thermometry of YF_3:Tm^{3+}/Yb^{3+} micro-crystals.

The Ln^{3+} activated UC phosphors have a potential application for security purpose. There are two main security applications, security ink and latent finger mark detection as demonstrated by researchers. The UC phosphors have ability to disperse in common solvents, can be excite by invisible NIR radiation

Figure 9. Green UC emission spectra of $SrWO_4$:Er^{3+}-Yb^{3+}phosphor on increasing temperature and the variation of FIR as function of absolute temperature (inset). The image is reproduced with permission from reference [Pandey et al., 2015]. Copyright @ 2015 Elsevier

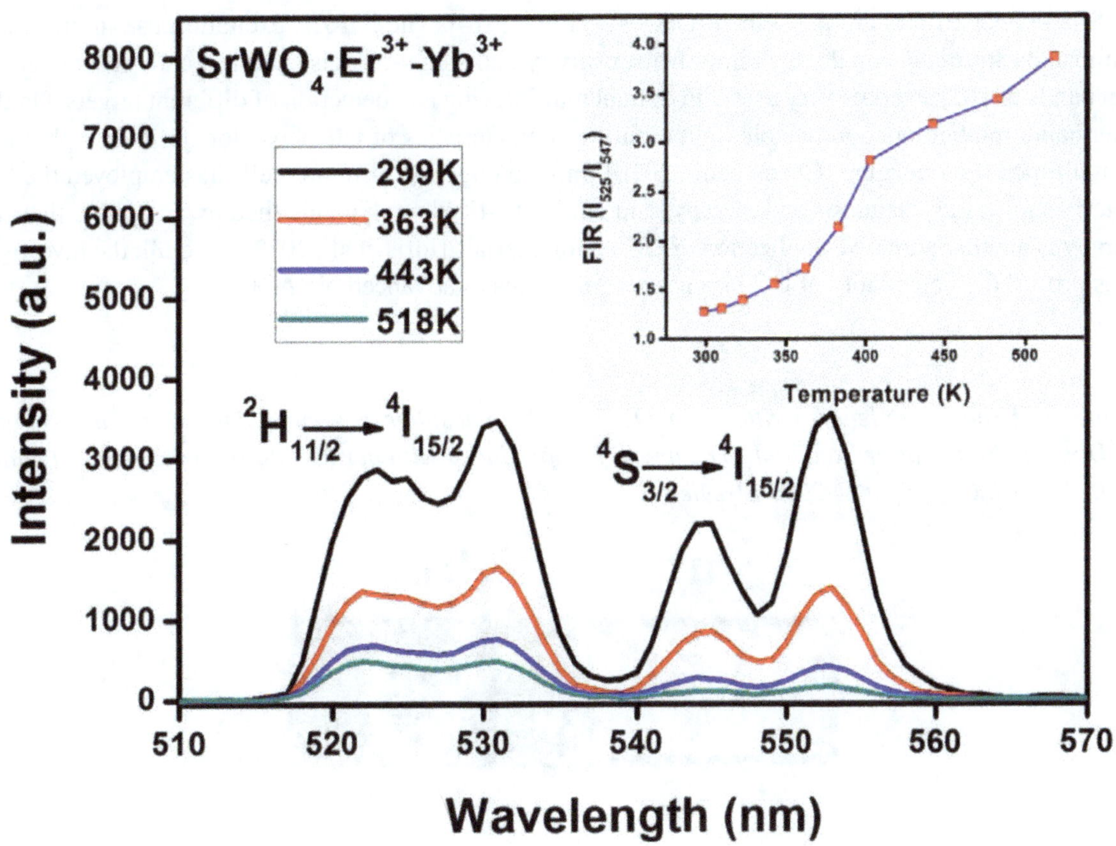

and produce bright visible emissions. So, it's worth to use as security ink as well as in latent finger mark detection. Pandey et al. (2013b) have demonstrated the use of $Y_2O_3:Ho^{3+}-Yb^{3+}$ phosphor as security ink. The written letters were exposed by NIR radiation is shown in Figure 10. Figure 10 (a) represents the picture of letters taken in day light, while Figure 10(b) represent the same irradiated by 980 nm diode laser in darken condition. It is clear that hidden letters can be visualized by a particular radiation, so can be use in writing a secure letter. Latent finger mark on various difficult surfaces was presented by Kumar et al. (2016) using $Gd_2O_3:Ho^{3+}-Yb^{3+}$ nanomaterials. Also, Tiwari et al. (2015b) developed the technique of observing latent finger marks by using $La_2O_3:Er^{3+}-Yb^{3+}$ phosphor powder. The UC based security materials has no background interference as observed in down conversion materials. Several reports have been found on security application of UC phosphors [Singh et al., 2009; Ma et al., 2011].

Application in Biomedical

Modern research on inorganic luminescent materials is focused on its application in bioassay and bio-imaging because of excellent photostability and non-toxic nature of these materials [DaCosta et al., 2014; Liu et al., 2014; Dou et al., 2014]. Low energy coherent excitation also minimized the damage of biological materials and overcome to the difficulties of UV excited materials as chemical instability, auto-fluorescence and inherent cytotoxicity, low penetration depth, etc. [Mader et al., 2010]. Many of research and review articles have been focused on the application of Ln^{3+} activated UC materials for bioconjugation, bioimaging, biolabeling, biosensing, drug delivery and photodynamic therapy. For targeting material, conjugation of biomolecules such as proteins, nucleic acids, antibodies to the surface modified nanoparticles are performed [DaCosta et al., 2014]. The NIR/IR excitation use in bioassays diminish background signal which improve sensitivity and signal-to-noise. Additionally narrow emission bands of UC materials very useful in optical multiplexing and detection of different targets. On the other hand, multicolour and tunable light emitting characteristics of UC phosphors provide a platform for multimodal bioimaging [Cheng et al., 2010]. In vivo targeting of tumor cells has employed the UC materials for tumor diagnosis and therapy [Liu et al., 2014]. Photodynamic therapy and photo thermal therapy is another attractive application of UC nanomaterials [Idris et al., 2012]. Overall, the investigations proved the potentiality of UC luminescent materials as advanced bio probe.

Figure 10. Exposure of letters written by $Y_2O_3:Ho^{3+}-Yb^{3+}$ phosphor as security ink (a) In Sun light and (b) Upon a 980 nm laser excitation. The image is reproduced with permission from reference [Pandey et al., 2013b]. Copyright @ 2013 Elsevier

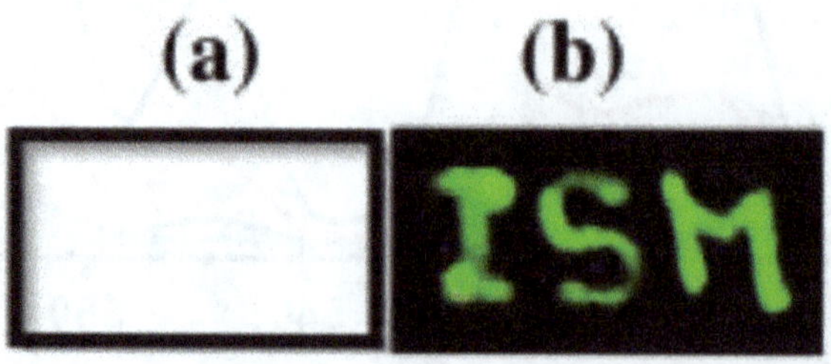

CONCLUSION

This chapter summarizes the recent progress on UC based inorganic phosphors, mainly focused to synthesis, properties and applications. The particular interest in Ln^{3+} activated phosphors is due to their excellent optical properties, employing them in development of novel optical devices, sensor and biomedical tools. There are several factors such as the luminescence efficiency, chemical stability, multicolour emissions, biological compatibility, nanosize particles, etc. attracted to the researchers for working in the field of luminescent materials. Herein, the UC features of Ln^{3+} activated phosphors are explored from basic to standard level. Ingredient of inorganic phosphors, their selection criteria, synthesis route and possible UC mechanisms are detailed. A study on Ln^{3+} activated phosphors, their emission characteristics and the factors influenced them are presented. It was found that the luminescence intensity is not only depend upon electronic structure and concentration of doping Ln^{3+} ions but also influenced by the structure, shape, size, and synthetic method used. The results suggested that UC based inorganic phosphor materials having potential application in the field of solar cells, temperature sensors, security, bioimaging and therapy. Overall, authors tried to give best information to understand the background of UC mechanism and optical features of lanthanide doped/codoped phosphors with possible application areas, which will definitely assist to the readers for future prospects.

ACKNOWLEDGMENT

Anurag Pandey would like to acknowledge Claude Leon Foundation, South Africa for financial support in the form of Postdoctoral fellowship. The research supports from the National Research Foundation of South Africa and Rhodes University Grahamstown are also acknowledged. One of the authors, Vinod Kumar is acknowledge to DST New Delhi, India for the award of Inspire Faculty (DST/INSPIRE/04/2015/001497).

REFERENCES

Adikaari, A. D., Etchart, I., Guéring, P. H., Bérard, M., Silva, S. R. P., Cheetham, A. K., & Curry, R. J. (2012). Near infrared up-conversion in organic photovoltaic devices using an efficient Yb^{3+}:Ho^{3+} Co-doped Ln_2BaZnO_5 (Ln= Y, Gd) phosphor. *Journal of Applied Physics, 111*(9), 094502. doi:10.1063/1.4704687

Auzel, F. (1966). Compteur quantique par transfert d'energie entre deux ions de terres rares dans un tungstate mixte et dans un verre. *CR Acad. Sci. Paris, 262*, 1016–1019.

Auzel, F. (1973). Materials and devices using double-pumped-phosphors with energy transfer. *Proceedings of the IEEE, 61*(6), 758–786. doi:10.1109/PROC.1973.9155

Auzel, F. (1990). Upconversion processes in coupled ion systems. *Journal of Luminescence, 45*(1-6), 341–345. doi:10.1016/0022-2313(90)90189-I

Auzel, F. (1992). Upconversion process in coupled ions systems. In R. P. Rao (Ed.), *Luminescence: phenomena, materials and devices* (pp. 33–37). New York: Nova Science Publisher.

Auzel, F. (2004). Upconversion and anti-stokes processes with f and d ions in solids. *Chemical Reviews*, *104*(1), 139–174. doi:10.1021/cr020357g PMID:14719973

Biju, S., Harris, M., Vander Elst, L., Wolberg, M., Kirschhock, C., & Parac-Vogt, T. N. (2016). Multifunctional β-$NaGdF_4$: Ln^{3+} (Ln= Yb, Er, Dy) nanoparticles with NIR to visible upconversion and high transverse relaxivity: a potential bimodal contrast agent for high-field MRI and optical imaging. *RSC Advances*, *6*(66), 61443–61448. doi:10.1039/C6RA09450C

Cao, B. S., He, Y. Y., Feng, Z. Q., Song, M., & Dong, B. (2011). Crystalline-structure-dependent green and red upconversion emissions of Er^{3+}–Yb^{3+}–Li^{+} codoped TiO_2. *Optics Communications*, *284*(13), 3311–3314. doi:10.1016/j.optcom.2011.03.005

Chander, N., Sardana, S. K., Parashar, P. K., Khan, A. F., Chawla, S., & Komarala, V. K. (2015). Improving the short-wavelength spectral response of silicon solar cells by spray deposition of YVO_4: Eu^{3+} downshifting phosphor nanoparticles. *IEEE Journal of Photovoltaics*, *5*(5), 1373–1379. doi:10.1109/JPHOTOV.2015.2438633

Chang, J., Ning, Y., Wu, S., Niu, W., & Zhang, S. (2013). Effectively Utilizing NIR Light Using Direct Electron Injection from Up-Conversion Nanoparticles to the TiO_2 Photoanode in Dye-Sensitized Solar Cells. *Advanced Functional Materials*, *23*(47), 5910–5915. doi:10.1002/adfm.201301158

Chen, J., & Zhao, J. X. (2012). Upconversion nanomaterials: Synthesis, mechanism, and applications in sensing. *Sensors (Basel)*, *12*(3), 2414–2435. doi:10.3390120302414 PMID:22736958

Cheng, L., Yang, K., Zhang, S., Shao, M., Lee, S., & Liu, Z. (2010). Highly-sensitive multiplexed in vivo imaging using PEGylated upconversion nanoparticles. *Nano Research*, *3*(10), 722–732. doi:10.100712274-010-0036-2

Cheng-Ren, L., Shu-Feng, L., Bin, D., Jing-Chang, S., Xiao-Feng, B., & Xu-Nan, F. (2012). Intense up-conversion emissions of Yb^{3+}/Dy^{3+} co-doped Al_2O_3 nanopowders prepared by non-aqueous sol-gel method. *Chinese Physics B*, *21*(9), 097803. doi:10.1088/1674-1056/21/9/097803

Chung, J. H., Lee, S. Y., Shim, K. B., & Ryu, J. H. (2012). White lighting upconversion in $Tm^{3+}/Ho^{3+}/Yb^{3+}$ co-doped $CaWO_4$. *Applied Physics Express*, *5*(5), 052602. doi:10.1143/APEX.5.052602

DaCosta, M. V., Doughan, S., Han, Y., & Krull, U. J. (2014). Lanthanide upconversion nanoparticles and applications in bioassays and bioimaging: A review. *Analytica Chimica Acta*, *832*, 1–33. doi:10.1016/j.aca.2014.04.030 PMID:24890691

Dey, R., Kumari, A., Soni, A. K., & Rai, V. K. (2015). $CaMoO_4$: Ho^{3+}–Yb^{3+}–Mg^{2+} upconverting phosphor for application in lighting devices and optical temperature sensing. *Sensors and Actuators. B, Chemical*, *210*, 581–588. doi:10.1016/j.snb.2015.01.007

Dey, R., Rai, V. K., & Pandey, A. (2012). Green upconversion emission in Nd^{3+}–Yb^{3+}–Zn^{2+}: Y_2O_3 phosphor. *Spectrochimica Acta. Part A: Molecular and Biomolecular Spectroscopy*, *99*, 288–291. doi:10.1016/j.saa.2012.09.001 PMID:23085286

Di Bartolo, B., & Forte, O. (Eds.). (2006). *Advances in spectroscopy for lasers and sensing* (Vol. 231). Springer Science & Business Media. doi:10.1007/1-4020-4789-4

Dieke, G. H., Crosswhite, H. M., & Crosswhite, H. M. (1968). *Spectra and energy levels of rare earth ions in crystals* (Vol. 5). New York: Interscience Publishers.

Dong, H., Sun, L. D., Wang, Y. F., Xiao, J. W., Tu, D., Chen, X., & Yan, C. H. (2016). Photon upconversion in Yb^{3+}–Tb^{3+} and Yb^{3+}–Eu^{3+} activated core/shell nanoparticles with dual-band excitation. *Journal of Materials Chemistry. C, Materials for Optical and Electronic Devices*, *4*(19), 4186–4192. doi:10.1039/C6TC00413J

Dosev, D., Kennedy, I. M., Godlewski, M., Gryczynski, I., Tomsia, K., & Goldys, E. M. (2006). Fluorescence upconversion in Sm-doped Gd_2O_3. *Applied Physics Letters*, *88*(1), 011906. doi:10.1063/1.2161400

Dou, Q. Q., Guo, H. C., & Ye, E. (2014). Near-infrared upconversion nanoparticles for bio-applications. *Materials Science and Engineering C*, *45*, 635–643. doi:10.1016/j.msec.2014.03.056 PMID:25491873

Dwivedi, C., & Dutta, V. (2013). Effect of seed layer on the self assembly of spray pyrolyzed Al-doped ZnO nanoparticles. *AIP Advances*, *3*(3), 032127. doi:10.1063/1.4795762

Dwivedi, Y., Bahadur, A., & Rai, S. B. (2011). Optical avalanche in Ho-Yb: Gd_2O_3 nanocrystals. *Journal of Applied Physics*, *110*(4), 043103. doi:10.1063/1.3622624

Dwivedi, Y., & Zilio, S. C. (2013). Infrared cascade and cooperative multicolor upconversion emissions in $Y_8V_2O_{17}$:Eu-Yb nanophosphors. *Optics Express*, *21*(4), 4717–4727. doi:10.1364/OE.21.004717 PMID:23482004

Gao, G., Turshatov, A., Howard, I. A., Busko, D., Joseph, R., Hudry, D., & Richards, B. S. (2017). Up-Conversion Fluorescent Labels for Plastic Recycling: A Review. *Advanced Sustainable Systems*.

Gibart, P., Auzel, F., Guillaume, J. C., & Zahraman, K. (1996). Below band-gap IR response of substrate-free GaAs solar cells using two-photon up-conversion. *Japanese Journal of Applied Physics*, *35*(8R), 4401–4402. doi:10.1143/JJAP.35.4401

Giri, N. K., Mishra, K., & Rai, S. B. (2011). Upconversion based tunable white-light generation in Ln: Y_2O_3 nanocrystalline phosphor (Ln= Tm/Er/Yb). *Journal of Fluorescence*, *21*(5), 1951–1958. doi:10.100710895-011-0894-3 PMID:21614491

Green, M. A., & Bremner, S. P. (2017). Energy conversion approaches and materials for high-efficiency photovoltaics. *Nature Materials*, *16*(1), 23–34. doi:10.1038/nmat4676 PMID:27994249

Guo, W., Zheng, K., Xie, W., Sun, L., Shen, L., Liu, C., ... Zhang, Z. (2014). Efficiency enhancement of inverted polymer solar cells by doping $NaYF_4$: Yb^{3+}, Er^{3+} nanocomposites in PCDTBT: PCBM active layer. *Solar Energy Materials and Solar Cells*, *124*, 126–132. doi:10.1016/j.solmat.2014.01.038

Hehlen, M. P., Brik, M. G., & Krämer, K. W. (2013). 50th anniversary of the Judd–Ofelt theory: An experimentalist's view of the formalism and its application. *Journal of Luminescence*, *136*, 221–239. doi:10.1016/j.jlumin.2012.10.035

Hu, H., & Bai, Y. (2012). Upconversion white luminescence of TeO_2:Tm^{3+}/Er^{3+}/Yb^{3+} nanoparticles. *Journal of Alloys and Compounds*, *527*, 25–29. doi:10.1016/j.jallcom.2012.02.129

Idris, N. M., Gnanasammandhan, M. K., Zhang, J., Ho, P. C., Mahendran, R., & Zhang, Y. (2012). In vivo photodynamic therapy using upconversion nanoparticles as remote-controlled nanotransducers. *Nature Medicine*, *18*(10), 1580–1585. doi:10.1038/nm.2933 PMID:22983397

Ivaturi, A., MacDougall, S. K., Martín-Rodríguez, R., Quintanilla, M., Marques-Hueso, J., Krämer, K. W., ... Richards, B. S. (2013). Optimizing infrared to near infrared upconversion quantum yield of β-$NaYF_4$: Er^{3+} in fluoropolymer matrix for photovoltaic devices. *Journal of Applied Physics*, *114*(1), 013505. doi:10.1063/1.4812578

Joshi, C., Kumar, K., & Rai, S. B. (2009). Upconversion and anomalous power dependence in $Ca_{12}Al_{14}O_{33}$: Er^{3+}/Yb^{3+} single phase nanophosphor. *Journal of Applied Physics*, *105*(12), 123103. doi:10.1063/1.3148294

Kaur, G., Singh, S. K., & Rai, S. B. (2010). Eu^{3+} and Yb^{3+} codoped Gd_2O_3 single phase nanophosphor: An enhanced monochromatic red emission through cooperative upconversion and downconversion. *Journal of Applied Physics*, *107*(7), 073514. doi:10.1063/1.3373628

Kim, C. H., Kwon, I. E., Park, C. H., Hwang, Y. J., Bae, H. S., Yu, B. Y., ... Hong, G. Y. (2000). Phosphors for plasma display panels. *Journal of Alloys and Compounds*, *311*(1), 33–39. doi:10.1016/S0925-8388(00)00856-2

Kore, B. P., Kumar, A., Pandey, A., Kroon, R. E., Terblans, J. J., Dhoble, S. J., & Swart, H. C. (2017). Spectroscopic Investigation of Up-Conversion Properties in Green Emitting $BaMgF_4$: Yb^{3+}, Tb^{3+} Phosphor. *Inorganic Chemistry*, *56*(9), 4996–5005. doi:10.1021/acs.inorgchem.7b00044 PMID:28406623

Kumar, A., Tiwari, S. P., Singh, A. K., & Kumar, K. (2016). Synthesis of Gd_2O_3: Ho^{3+}/ Yb^{3+} upconversion nanoparticles for latent fingermark detection. *Applied Physics. B, Lasers and Optics*, *122*(7), 1–10. doi:10.100700340-016-6468-y

Kumar, G. A., Pokhrel, M., Martinez, A., Dennis, R. C., Villegas, I. L., & Sardar, D. K. (2012a). Synthesis and spectroscopy of color tunable Y_2O_2S:Yb^{3+},Er^{3+} phosphors with intense emission. *Journal of Alloys and Compounds*, *513*, 559–565. doi:10.1016/j.jallcom.2011.11.006

Kumar, G. A., Pokhrel, M., & Sardar, D. K. (2012b). Intense visible and near infrared upconversion in M_2O_2S: Er (M= Y, Gd, La) phosphor under 1550nm excitation. *Materials Letters*, *68*, 395–398. doi:10.1016/j.matlet.2011.10.087

Kumari, A., Pandey, A., Dey, R., & Rai, V. K. (2014a). Simultaneous influence of Zn^{2+}/Mg^{2+} on the luminescent behaviour of La_2O_3: Tm^{3+}–Yb^{3+} phosphors. *RSC Advances*, *4*(42), 21844–21851. doi:10.1039/C4RA01400F

Kumari, A., Rai, V. K., & Kumar, K. (2014b). Yellow–orange upconversion emission in Eu^{3+}/Yb^{3+} codoped $BaTiO_3$ phosphor. *Spectrochimica Acta. Part A: Molecular and Biomolecular Spectroscopy*, *127*, 98–101. doi:10.1016/j.saa.2014.02.023 PMID:24632162

Kumari, A., Soni, A. K., & Rai, V. K. (2016). Rare Earth Doped Upconverting Phosphors for Display and Temperature Sensing Applications. *Materials Focus*, *5*(3), 187–194. doi:10.1166/mat.2016.1328

Kumari, A., Soni, A. K., & Rai, V. K. (2017). Near infrared to blue upconverting $Tm^{3+}/Yb^{3+}/Li^{+}$: $Gd_2(MoO_4)_3$ phosphors for light emitting display devices. *Infrared Physics & Technology*, *81*, 313–319. doi:10.1016/j.infrared.2017.01.024

Le Toquin, R., & Cheetham, A. K. (2006). Red-emitting cerium-based phosphor materials for solid-state lighting applications. *Chemical Physics Letters*, *423*(4), 352–356. doi:10.1016/j.cplett.2006.03.056

Lee, L. S., Rand, S. C., & Schawlow, A. L. (1984). Cooperative energy transfer among Pr^{3+} ions in LaF_3. *Physical Review B: Condensed Matter and Materials Physics*, *29*(12), 6901–6906. doi:10.1103/PhysRevB.29.6901

Lin, J., Yu, M., Lin, C., & Liu, X. (2007). Multiform oxide optical materials via the versatile pechini-type sol– gel process: Synthesis and characteristics. *The Journal of Physical Chemistry C*, *111*(16), 5835–5845. doi:10.1021/jp070062c

Liu, Q., Feng, W., & Li, F. (2014). Water-soluble lanthanide upconversion nanophosphors: Synthesis and bioimaging applications in vivo. *Coordination Chemistry Reviews*, *273*, 100–110. doi:10.1016/j.ccr.2014.01.004

Liu, X., Aidilibike, T., Guo, J., Li, Y., Di, W., & Qin, W. (2017). Upconversion luminescence of Sm^{2+} ions. *RSC Advances*, *7*(23), 14010–14014. doi:10.1039/C7RA00071E

Ma, R., Bullock, E., Maynard, P., Reedy, B., Shimmon, R., Lennard, C., ... McDonagh, A. (2011). Fingermark detection on non-porous and semi-porous surfaces using $NaYF_4$: Er, Yb up-converter particles. *Forensic Science International*, *207*(1), 145–149. doi:10.1016/j.forsciint.2010.09.020 PMID:20980110

Maciel, G. S., Guimaraes, R. B., Barreto, P. G., Carvalho, I. C. S., & Rakov, N. (2009). The influence of Yb 3+ doping on the upconversion luminescence of Pr^{3+} in aluminum oxide based powders prepared by combustion synthesis. *Optical Materials*, *31*(11), 1735–1740. doi:10.1016/j.optmat.2009.05.007

Mader, H. S., Kele, P., Saleh, S. M., & Wolfbeis, O. S. (2010). Upconverting luminescent nanoparticles for use in bioconjugation and bioimaging. *Current Opinion in Chemical Biology*, *14*(5), 582–596. doi:10.1016/j.cbpa.2010.08.014 PMID:20829098

Mai, H. X., Zhang, Y. W., Si, R., Yan, Z. G., Sun, L. D., You, L. P., & Yan, C. H. (2006). High-quality sodium rare-earth fluoride nanocrystals: Controlled synthesis and optical properties. *Journal of the American Chemical Society*, *128*(19), 6426–6436. doi:10.1021/ja060212h PMID:16683808

Martinez, A., Morales, J., Diaz-Torres, L. A., Salas, P., De la Rosa, E., Oliva, J., & Desirena, H. (2010). Green and red upconverted emission of hydrothermal synthesized Y_2O_3: Er^{3+}–Yb^{3+} nanophosphors using different solvent ratio conditions. *Materials Science and Engineering B*, *174*(1), 164–168. doi:10.1016/j.mseb.2010.04.028

Mialon, G., Türkcan, S., Dantelle, G., Collins, D. P., Hadjipanayi, M., Taylor, R. A., ... Boilot, J. P. (2010). High up-conversion efficiency of YVO_4: Yb, Er nanoparticles in water down to the single-particle level. *The Journal of Physical Chemistry C*, *114*(51), 22449–22454. doi:10.1021/jp107900z

Mishra, K., Giri, N. K., & Rai, S. B. (2011). Preparation and characterization of upconversion luminescent Tm^{3+}/Yb^{3+} co-doped Y_2O_3 nanophosphor. *Applied Physics. B, Lasers and Optics*, *103*(4), 863–875. doi:10.100700340-011-4379-5

Mokoena, T. P., Linganiso, E. C., Swart, H. C., Kumar, V., & Ntwaeaborwa, O. M. (2017). Cooperative luminescence from low temperature synthesized α-Al_2O_3: Yb^{3+} phosphor by using solution combustion. *Ceramics International*, *43*(1), 174–181. doi:10.1016/j.ceramint.2016.09.130

Molina, P., Vasyliev, V., Villora, E. G., & Shimamura, K. (2011). Tb^{3+}-Yb^{3+} cooperative down and up conversion processes in $Tb_{0.81}Ca_{0.19}F_{2.81}$: Yb^{3+} single crystals. *Journal of Applied Physics*, *110*(12), 123527. doi:10.1063/1.3671406

Nakazawa, E., & Shionoya, S. (1970). Cooperative Luminescence in $YbPO_4$. *Physical Review Letters*, *25*(25), 1710–1712. doi:10.1103/PhysRevLett.25.1710

Pan, A. C., Del Cañizo, C., Cánovas, E., Santos, N. M., Leitão, J. P., & Luque, A. (2010). Enhancement of up-conversion efficiency by combining rare earth-doped phosphors with PbS quantum dots. *Solar Energy Materials and Solar Cells*, *94*(11), 1923–1926. doi:10.1016/j.solmat.2010.06.028

Pandey, A., Dey, R., Pandey, S., & Rai, V. K. (2014a). Visible Light-Emitting Er^{3+}-Tm^{3+}–Yb^{3+} Codoped ZrO_2 Phosphor for Sensing and Photonics Applications. *Journal of Display Technology*, *10*(6), 494–499. doi:10.1109/JDT.2014.2305773

Pandey, A., Dey, R., & Rai, V. K. (2013a). Sensitization Effect of Yb^{3+} in Upconversion Luminescence of Eu^{3+} Codoped Y_2O_3 Phosphor. *Journal of Physical Chemistry & Biophysics*, *3*(5), 1. doi:10.4172/2161-0398.1000129

Pandey, A., Kumar, V., Som, S., Yousif, A., Kroon, R. E., Coetsee, E., & Swart, H. C. (2017). Photon and electron beam pumped luminescence of Ho^{3+} activated $CaMoO_4$ phosphor. *Applied Surface Science*, *423*, 1169–1175. doi:10.1016/j.apsusc.2017.06.326

Pandey, A., & Rai, V. K. (2012). Colour emission tunability in Ho^{3+}–Tm^{3+}–Yb^{3+} co-doped Y_2O_3 upconverted phosphor. *Applied Physics. B, Lasers and Optics*, *109*(4), 611–616. doi:10.100700340-012-5224-1

Pandey, A., & Rai, V. K. (2013). Improved luminescence and temperature sensing performance of Ho^{3+}– Yb^{3+}–Zn^{2+}: Y_2O_3 phosphor. *Dalton Transactions (Cambridge, England)*, *42*(30), 11005–11011. doi:10.1039/c3dt50592h PMID:23797536

Pandey, A., & Rai, V. K. (2014a). Pr^{3+}–Yb^{3+} codoped Y_2O_3 phosphor for display devices. *Materials Research Bulletin*, *57*, 156–161. doi:10.1016/j.materresbull.2014.04.071

Pandey, A., & Rai, V. K. (2014b). Rare earth doped materials for temperature sensors. In Y. Dwivedi, S. B. Rai, & J. P. Singh (Eds.), *Spectroscopic techniques for security forensic and environmental applications* (pp. 279–292). Nova Publisher.

Pandey, A., Rai, V. K., Dey, R., & Kumar, K. (2013b). Enriched green upconversion emission in combustion synthesized Y_2O_3: Ho^{3+}– Yb^{3+} phosphor. *Materials Chemistry and Physics*, *139*(2), 483–488. doi:10.1016/j.matchemphys.2013.01.043

Pandey, A., Rai, V. K., & Kumar, K. (2014b). Influence of Li^{+} codoping on visible emission of Y_2O_3:Tb^{3+},Yb^{3+} phosphor. *Spectrochimica Acta. Part A: Molecular and Biomolecular Spectroscopy*, *118*, 619–623. doi:10.1016/j.saa.2013.08.109 PMID:24095771

Pandey, A., Rai, V. K., Kumar, V., Kumar, V., & Swart, H. C. (2015). Upconversion based temperature sensing ability of Er^{3+}–Yb^{3+} codoped $SrWO_4$: An optical heating phosphor. *Sensors and Actuators. B, Chemical*, *209*, 352–358. doi:10.1016/j.snb.2014.11.126

Pandey, S., Pandey, A., & Rai, V. K. (2013). Synthesis and Characterization of Sm^{3+}–Yb^{3+} Codoped Y_2O_3 Phosphor. *Journal of Display Technology*, *9*(12), 989–994. doi:10.1109/JDT.2013.2275964

Pang, X. L., Jia, C. H., Li, G. Q., & Zhang, W. F. (2011). Bright white upconversion luminescence from Er^{3+}–Tm^{3+}–Yb^{3+} doped $CaSnO_3$ powders. *Optical Materials*, *34*(1), 234–238. doi:10.1016/j.optmat.2011.08.019

Patra, A., Friend, C. S., Kapoor, R., & Prasad, P. N. (2002). Upconversion in Er^{3+}:ZrO_2 nanocrystals. *The Journal of Physical Chemistry B*, *106*(8), 1909–1912. doi:10.1021/jp013576z

Patra, A., Friend, C. S., Kapoor, R., & Prasad, P. N. (2003). Effect of crystal nature on upconversion luminescence in Er^{3+}:ZrO_2 nanocrystals. *Applied Physics Letters*, *83*(2), 284–286. doi:10.1063/1.1592891

Peng, L., Huang, M., Cao, S., Liu, B., Han, T., & Zhao, C. (2016). Enhanced upconversion in Dy^{3+}, Yb^{3+} co-doped Gd_2O_3 monodisperse nanocrystals. *Journal of Sol-Gel Science and Technology*, *78*(2), 307–312. doi:10.100710971-015-3952-5

Prasad, P. N. (2004). *Nanophotonics*. John Wiley & Sons. doi:10.1002/0471670251

Rai, M., Mishra, K., Singh, S. K., Verma, R. K., & Rai, S. B. (2012). Infrared to visible upconversion in Ho^{3+}/Yb^{3+} co-doped Y_2O_3 phosphor: Effect of laser input power and external temperature. *Spectrochimica Acta. Part A: Molecular and Biomolecular Spectroscopy*, *97*, 825–829. doi:10.1016/j.saa.2012.07.071 PMID:22902580

Rai, V. K. (2012). Frequency Upconversion in Rare earth ions. In A. H. Al-Khursan (Ed.), *Solid State Laser* (pp. 209–224). Intech Publisher.

Rai, V. K., Dey, R., & Kumar, K. (2013a). White upconversion emission in Y_2O_3: Er^{3+}–Tm^{3+}–Yb^{3+} phosphor. *Materials Research Bulletin*, *48*(6), 2232–2236. doi:10.1016/j.materresbull.2013.02.064

Rai, V. K., & Pandey, A. (2016). Efficient color tunable $ZnWO_4$: Er^{3+}-Yb^{3+} phosphor for high temperature sensing. *Journal of Display Technology*, *12*(11), 1472–1477. doi:10.1109/JDT.2016.2602503

Rai, V. K., Pandey, A., & Dey, R. (2013b). Photoluminescence study of Y_2O_3: Er^{3+}–Eu^{3+}–Yb^{3+} phosphor for lighting and sensing applications. *Journal of Applied Physics*, *113*(8), 083104. doi:10.1063/1.4793265

Shan, G. B., & Demopoulos, G. P. (2010). Near-Infrared Sunlight Harvesting in Dye-Sensitized Solar Cells Via the Insertion of an Upconverter-TiO_2 Nanocomposite Layer. *Advanced Materials*, *22*(39), 4373–4377. doi:10.1002/adma.201001816 PMID:20809511

Shockley, W., & Queisser, H. J. (1961). Detailed balance limit of efficiency of p-n junction solar cells. *Journal of Applied Physics*, *32*(3), 510–519. doi:10.1063/1.1736034

Singh, S. K., Kumar, K., & Rai, S. B. (2009). Multifunctional Er^{3+}–Yb^3codoped Gd_2O_3 nanocrystalline phosphor synthesized through optimized combustion route. *Applied Physics. B, Lasers and Optics*, *94*(1), 165–173. doi:10.100700340-008-3261-6

Snitzer, E., & Woodcock, R. (1965). Yb^{5+}-Er^{3+}-Ybw glass laser. *Applied Physics Letters*, *6*(3), 5. doi:10.1063/1.1754157

Sukul, P. P., Mahata, M. K., & Kumar, K. (2017). NIR optimized dual mode photoluminescence in Nd doped Y_2O_3 ceramic phosphor. *Journal of Luminescence*, *185*, 92–98. doi:10.1016/j.jlumin.2017.01.008

Suo, H., Hu, F., Zhao, X., Zhang, Z., Li, T., Duan, C., ... Guo, C. (2017). All-in-one thermometer-heater up-converting platform YF_3:Yb^{3+},Tm^{3+} operating in the first biological window. *Journal of Materials Chemistry. C, Materials for Optical and Electronic Devices*, *5*(6), 1501–1507. doi:10.1039/C6TC05449H

Tamrakar, R. K., & Bisen, D. P. (2015). Thermoluminescence studies of ultraviolet and gamma irradiated erbium (III)-and ytterbium (III)-doped gadolinium oxide phosphors. *Materials Science in Semiconductor Processing*, *33*, 169–188. doi:10.1016/j.mssp.2015.01.044

Tian, L., Xu, Z., Zhao, S., Cui, Y., Liang, Z., Zhang, J., & Xu, X. (2014). The Upconversion Luminescence of Er^{3+}/Yb^{3+}/Nd^{3+} Triply-Doped β-$NaYF_4$ Nanocrystals under 808-nm Excitation. *Materials (Basel)*, *7*(11), 7289–7303. doi:10.3390/ma7117289 PMID:28788246

Tikhomirov, V. K., Driesen, K., Rodriguez, V. D., Gredin, P., Mortier, M., & Moshchalkov, V. V. (2009). Optical nanoheater based on the Yb^{3+}-Er^{3+} co-doped nanoparticles. *Optics Express*, *17*(14), 11794–11798. doi:10.1364/OE.17.011794 PMID:19582094

Tiwari, S. P., Kumar, A., &Kumar,K. (2016). Upconversion phosphor materials for beginners: synthesis and applications. *Research Frontiers in Sciences,* 23-42.

Tiwari, S. P., Kumar, K., & Rai, V. K. (2015a). Plasmonic enhancement in upconversion emission of La_2O_3: Er^{3+}/Yb^{3+} phosphor via introducing silver metal nanoparticles. *Applied Physics. B, Lasers and Optics*, *121*(2), 221–228. doi:10.100700340-015-6223-9

Tiwari, S. P., Kumar, K., & Rai, V. K. (2015b). Latent fingermarks detection for La_2O_3: Er^{3+}/Yb^{3+} phosphor material in upconversion emission mode: A comparative study. *Journal of Applied Physics*, *118*(18), 183109. doi:10.1063/1.4935279

Trupke, T., Green, M. A., & Würfel, P. (2002). Improving solar cell efficiencies by up-conversion of sub-band-gap light. *Journal of Applied Physics*, *92*(7), 4117–4122. doi:10.1063/1.1505677

Verma, D., Saetre, T. O., & Midtgård, O. M. (2012, June). Review on up/down conversion materials for solar cell application. In *Photovoltaic Specialists Conference (PVSC), 2012 38th IEEE* (pp. 002608-002613). IEEE. 10.1109/PVSC.2012.6318129

Verma, R. K., Singh, S. K., & Rai, S. B. (2012). Upconversion, avalanche effect and controlled optical switching in Yb^{3+}, Ho^{3+} co-doped $Ca_{12}Al_{14}O_{33}$ phosphor. *Current Applied Physics*, *12*(6), 1481–1484. doi:10.1016/j.cap.2012.04.018

Wade, S. A., Collins, S. F., & Baxter, G. W. (2003). Fluorescence intensity ratio technique for optical fiber point temperature sensing. *Journal of Applied Physics*, *94*(8), 4743–4756. doi:10.1063/1.1606526

Wang, F., & Liu, X. (2009). Recent advances in the chemistry of lanthanide-doped upconversion nanocrystals. *Chemical Society Reviews, 38*(4), 976–989. doi:10.1039/b809132n PMID:19421576

Wang, G., Peng, Q., & Li, Y. (2011a). Lanthanide-doped nanocrystals: Synthesis, optical-magnetic properties, and applications. *Accounts of Chemical Research, 44*(5), 322–332. doi:10.1021/ar100129p PMID:21395256

Wang, H., Duan, C. K., & Tanner, P. A. (2008). Visible upconversion luminescence from Y_2O_3: Eu^{3+}, Yb^{3+}. *The Journal of Physical Chemistry C, 112*(42), 16651–16654. doi:10.1021/jp8046505

Wang, H., Qian, C., Yi, Z., Rao, L., Liu, H., & Zeng, S. (2013). Hydrothermal Synthesis and Tunable Multicolor Upconversion Emission of Cubic Phase Y_2O_3 Nanoparticles. *Advances in Condensed Matter Physics*.

Wang, H. Q., Batentschuk, M., Osvet, A., Pinna, L., & Brabec, C. J. (2011b). Rare-Earth Ion Doped Up-Conversion Materials for Photovoltaic Applications. *Advanced Materials, 23*(22-23), 2675–2680. doi:10.1002/adma.201100511 PMID:21823249

Wang, H. Q., Stubhan, T., Osvet, A., Litzov, I., & Brabec, C. J. (2012). Up-conversion semiconducting MoO 3: Yb/Er nanocomposites as buffer layer in organic solar cells. *Solar Energy Materials and Solar Cells, 105*, 196–201. doi:10.1016/j.solmat.2012.06.005

Wang, X., Liu, Q., Bu, Y., Liu, C. S., Liu, T., & Yan, X. (2015). Optical temperature sensing of rare-earth ion doped phosphors. *RSC Advances, 5*(105), 86219–86236. doi:10.1039/C5RA16986K

West, A. R. (2007). *Solid state chemistry and its applications*. John Wiley & Sons.

Wu, X., & Wang, L. (2013). Dual-Functional Upconverter-Doped TiO_2 Hollow Shells for Light Scattering and Near-Infrared Sunlight Harvesting in Dye-Sensitized Solar Cells. *Advanced Energy Materials, 3*(6), 704–707. doi:10.1002/aenm.201200933

Xu, J., Zhang, H., Xiong, Q., Liang, G., Wang, L., Shen, X., ... Xu, W. (2011). Large Enhancement in short-circuit current densities of dye-sensitized solar cells via photon conversion. *Journal of Physics: Conference Series, 276*(1), 012195. doi:10.1088/1742-6596/276/1/012195

Xu, X. G., Chen, J., Deng, S. Z., Xu, N. S., & Lin, J. (2010). Cathodoluminescent properties of nanocrystalline Lu 3 Ga 5 O 12: Tb 3+ phosphor for field emission display application a. *Journal of Vacuum Science & Technology B. Nanotechnology and Microelectronics: Materials, Processing, Measurement, and Phenomena, 28*(3), 490–494.

Yadav, R. S., Verma, R. K., Bahadur, A., & Rai, S. B. (2015). Infrared to infrared upconversion emission in Pr 3+/Yb 3+ co-doped La 2 O 3 and La (OH) 3 nano-phosphors: A comparative study. *Spectrochimica Acta. Part A: Molecular and Biomolecular Spectroscopy, 142*, 324–330. doi:10.1016/j.saa.2015.01.095 PMID:25706603

Yang, C. H., Pan, Y. X., Zhang, Q. Y., & Jiang, Z. H. (2007). Cooperative energy transfer and frequency upconversion in Yb^{3+}–Tb^{3+} and Nd^{3+}–Yb^{3+}–Tb^{3+} Codoped $GdAl_3(BO_3)_4$ Phosphors. *Journal of Fluorescence, 17*(5), 500–504. doi:10.100710895-007-0200-6 PMID:17609865

Ye, S., Xiao, F., Pan, Y. X., Ma, Y. Y., & Zhang, Q. Y. (2010). Phosphors in phosphor-converted white light-emitting diodes: Recent advances in materials, techniques and properties. *Materials Science and Engineering R Reports*, *71*(1), 1–34. doi:10.1016/j.mser.2010.07.001

Zhang, H., Yang, D., Ji, Y., Ma, X., Xu, J., & Que, D. (2004). Low temperature synthesis of flowerlike ZnO nanostructures by cetyltrimethylammonium bromide-assisted hydrothermal process. *The Journal of Physical Chemistry B*, *108*(13), 3955–3958. doi:10.1021/jp036826f

Zhang, Y. W., Sun, X., Si, R., You, L. P., & Yan, C. H. (2005). Single-crystalline and monodisperse LaF_3 triangular nanoplates from a single-source precursor. *Journal of the American Chemical Society*, *127*(10), 3260–3261. doi:10.1021/ja042801y PMID:15755126

Zhou, J., Liu, Q., Feng, W., Sun, Y., & Li, F. (2014). Upconversion luminescent materials: Advances and applications. *Chemical Reviews*, *115*(1), 395–465. doi:10.1021/cr400478f PMID:25492128

Zhou, J., Liu, Z., & Li, F. (2012). Upconversion nanophosphors for small-animal imaging. *Chemical Society Reviews*, *41*(3), 1323–1349. doi:10.1039/C1CS15187H PMID:22008740

Zou, W., Visser, C., Maduro, J. A., Pshenichnikov, M. S., & Hummelen, J. C. (2012). Broadband dye-sensitized upconversion of near-infrared light. *Nature Photonics*, *6*(8), 560–564. doi:10.1038/nphoton.2012.158

KEY TERMS AND DEFINITIONS

Emission: An ejection of energy when atoms relax from excited to ground state.
Energy Level Diagram: The pictorial representation of energy states of an atom.
Energy Transfer: It takes place between two different dopant ions.
Excitation: A process in which atoms/ions are jumped from ground to higher states.
Host Material: The suitable compound provides the suitable environment for doping elements.
Phosphor: The material which can emit the light when excited by electromagnetic radiation.
Rare Earth: The lanthanum series of periodic table including Yttrium and Scandium.
Synthesis Process: The method by which the desired product is obtained using its ingredients.

Chapter 5
Upconversion Luminescence Behaviour of Er^{3+}/Yb^{3+} Doped MY_2O_4 (M=Ba, Ca, Sr) Phosphors:
Upconversion Study of MY_2O_4:Er^{3+}/Yb^{3+}

Vikas Dubey
Bhilai Institute of Technology, Raipur, India

Neha Dubey
ICFAI University, Kumhari, India

Ravi Shrivastava
ICFAI University, Kumhari, India

Jagjeet KaurSaluja
Govt. V.Y.T.PG. Auto. College Durg, C.G., India

Sudipta Som
Indian School of Mines, Dhanbad, India

Somrita Dutta
Indian School of Mines, Dhanbad, India

Rituraj Chandrakar
CSIT, Kolihapur, Durg, India

ABSTRACT

The upconversion (UC) of the rare earth doped MY2O4 (M=Ba, Sr, Ca) has been extensively investigated due to their potential applications in many fields, such as color display, high density memories, optical data storage, sensor and energy solar cell, etc. Many series of them, especially the Er3+, Yb3+ doped MY2O4 were studied in this chapter, due to the thermal and mechanical toughness, high optical trans-

DOI: 10.4018/978-1-5225-5170-6.ch005

mittance from the ultraviolet to the infrared regions, and a low nonlinear refractive index compared to the other commercial laser glasses. The energy transfer (ET) mechanism of rare earth doped phosphor plays an important role in the upconversion process.

INTRODUCTION

It is widely known that some rare earth ions can convert the wavelength of incident light to a shorter one via a multiphonon absorption mechanism. This phenomenon is of great interest because it can be used to build up a new laser system, in particular, a light amplification in the visible to ultraviolet region. It is an anti-Stokes process defined as upconversion. In recent decades, the upconversion (UC) materials have been extensively investigated due to their potential applications in many fields, such as color display, optical data storage, sensor energy solar cell, etc.

The whole field of the UC in ion-doped systems can be traced back to an idea of Bloembergen in 1959. Many reports described the UC emission phenomenon in different matrix environments like phosphors, glasses, crystals, nanoparticles, photonic crystal, and so on Zhou et al. (2015), Zhou et al. (2009), Xing et al. (2014), Cao et al. (2015), Yang et al. (2011) and Ye et al. (2014) . Among various optical systems, rare-earth doped halide glasses, such as fluoride Kawamoto et al (1993), Takahashi at al. (1994) and Zhang et al. (1993), Hirao et al. (1993) chloride Shojiya et al. (1997) and Shojiya et al. (1994), bromide Kawamoto et al. (1995), iodide Shojiya et al. (1995), and chalcogenide glasses Kadono et al. (1997)are considered to possess remarkably greater potentiality than optical devices. Some of the reasons are described as follows:

(1) Doped rare-earth ions are confined in crystalline environments of low phonon energies.
(2) Consequently excited state lifetimes and optical absorption cross sections of the doped rare earth ions become large compared with those in vitreous environments QIU et al. (2016).

Upconversion luminescent materials can convert infrared radiation into visible light. In the last few decades, much attention has been paid to study of upconversion phosphors since near-infrared (NIR) pump lasers with high power are easily obtainable. Now a days upconversion phosphors play a significant role due to their excellent luminescent properties in many areas such as all-solid compact lasers, NIR quantum counting devices, two-photon confocal microscope, optical data storage and fluorescent labels for sensitive detection of bimolecular Downing et al. (1996), Maciel et al. (1997), Hebert et al. (1990), Bhargava et al (1994), Guo et al. (2004), Chen et al. (2007), Chen et al. (2007), Shalav et al. (2005), Zhang et al. (2010), Zhou et al. (2010), Chen at al. (2011), Chen et al. (2010), Dutta et al. (2011) . Capobianco et al. (2002) reported the enhancement of red emission in Er-doped nanocrystalline yttria. Among all the rare earth ions, the Er^{3+}ion can efficiently emit photons in the green and red regions of the spectrum and has the ability to convert infrared light to visible one. Er^{3+} ion provides intermediate electronic energy levels ($^4I_{11/2}$, $^4I_{13/2}$) with long lifetimes, which can be conveniently pumped by low-cost commercial diodes. As an upconversion active ion, Er^{3+} ion is widely used in upconversion phosphors Lin et al. (2003) and Riseberg et al. (1968).Upconversion efficiency is principally governed by nonradiation process of host materials Vetrone et al. (2003). The upconversion performance could be enhanced significantly by suitable selection of host matrix. Numerous rare earth oxides such as yttrium oxide, gadolinium oxide, and cerium oxide have been selected as a host material for upconversion studies due to their excellent properties Vetrone et al. (2004), Guo et al. (2007) and Laversenne et al. (2011).

Among the candidates of upconversion materials, tripositive erbium (Er^{3+}) has excellent upconversion properties. In the near-infrared region, Er^{3+}has a favorable energy level structure with two transitions $^4I_{15/2}$-$^4I_{11/2}$ (at 978 nm) and $^4I_{15/2}$ - $^4I_{9/2}$ (at 800 nm) that can be excited with a semiconductor laser, yielding blue, green and red emissions. However, the upconversion luminescence cannot be increased effectively by increasing the Er^{3+} concentration in crystals because cross relaxation between Er^{3+} ions quenches the green emissions at a higher dopant concentration Vetrone et al. (2003). The Er^{3+} ion has a relatively low absorption cross section in the near-infrared region. However, the ytterbium ion (Yb^{3+}) exhibits a much larger absorption cross-section in this region. Moreover, the energy level gap between $^2F_{5/2}$ and $^2F_{7/2}$ of Yb^{3+} is matched to the gap between $^4I_{15/2}$and $^4I_{11/2}$, the gap between $^4I_{11/2}$ and $^4F_{7/2}$, the gap between $^4I_{13/2}$and$^4F_{9/2}$ of Er^{3+}, which results in an efficient energy transfer from Yb^{3+} to Er^{3+}. Thus, the Er^{3+}/Yb^{3+} codoped material is a good choice for increasing upconversion luminescence. In the year 2000 Capobianco et al. reported the near infrared to visible UCL in erbium-doped nanocrystals yttria for the first time Capobianco et al. (2000). Since this report, infrared to visible UCL in nanocrystals has attracted considerable attention due to its potential application in infrared detectors, molecule recognition and three-dimensional displays Heer et al. (2003). Hongwei Song et al. reported an upconversion luminescence mechanism for the green levels in Er^{3+}/Yb^{3+}codoped cubic nanocrystalline yttria Song et al. (2004). Now our study is related to Er^{3+}/Yb^{3+}doped MY_2O_4 (M=Ba, Ca, Sr) phosphors for suitable optical applications.

Figure 1 shows the energy level diagrams of Er^{3+}/Yb^{3+} and the UCL process under 978-nm laser excitation. First, the ground state ytterbium ions Yb^{3+} ($^2F_{7/2}$) are excited to $^2F_{5/2}$ energy level through absorbing 978 nm photons. As the $^4I_{11/2}$ energy level of Er^{3+} and the $^2F_{5/2}$ energy level of $Yb^{3+}$are very close, Yb^{3+} ($^2F_{5/2}$)ions transfer their energy to Er^{3+} ($^4I_{15/2}$) ions very easily leading themselves come back to $^2F_{7/2}$energy level and the Er^{3+} ($^4I_{15/2}$) ions reach $^4I_{11/2}$energy level simultaneously, which is called resonance energy transfer. Then some Er^{3+}ions on the $^4I_{11/2}$ energy level are excited to the $^4F_{7/2}$energy level through resonance energy transfer. Because the space between $^4F_{7/2}$ and $^2H_{11/2}$ or $^4S_{3/2}$ is small, the Er^{3+} on the $^4F_{7/2}$ energy level can relax into the $^2H_{11/2}$ or $^4S_{3/2}$energy level rapidly without radiation and the energy differences are adjusted by phonons of the host. We call this process nonradiative relaxation. When the transitions of $^2H_{11/2}$/ $^4S_{3/2}$ to $^4I_{15/2}$ occur, the green emission starts to generate. In addition, populating $^4F_{7/2}$ level of Er^{3+}can also be realized by excited state absorption upconversion. Er^{3+} ($^4I_{15/2}$) ions reach $^4I_{11/2}$ energy level by absorbing 978-nm photons, then reach $^4F_{7/2}$ level by absorbing 978-nm photons again. The excited state absorption upconversion does happen in the nanowires, but it is not dominant, which can be proved by a much stronger emission intensity in Er^{3+}/Yb^{3+}codoped nanowires in comparison with that in Er^{3+} doped ones and strong ground state absorption at 978-nm for Yb^{3+}. The other Er^{3+} ions on the $^4I_{11/2}$energy level relax into the $^4I_{13/2}$ energy level by nonradiative relaxation and then are excited to $^4F_{9/2}$ through resonance energy transfer or excited state absorption, generating the red emissions when they come back to ground state $^4I_{15/2}$ energy level Yang et al. (2009).

Besides the requirement of high optical conversion efficiency, the thermal and chemical stability of the host material is very important for solar cell applications Ivanoval et al. (2009), Maciel et al. (2006) and Ivanova et al. (2008), and a stable oxide host material is usually the optimal option. SrY_2O_4 presents an ordered $CaFe_2O_4$ structure composed of a double octahedral $(R_2O_4)^{2-}$ (R=rare earth metal) framework with rare-earth ions residing within this framework. In this structure, $Ca^{2+}$is in the octahedral position, whereas Sr^{2+} is in the triangular prism position. Rare-earth ions are located at

Figure 1. The energy level diagrams of Er^{3+} and $Yb^{3+}$and the UCL process under 978 nm excitation

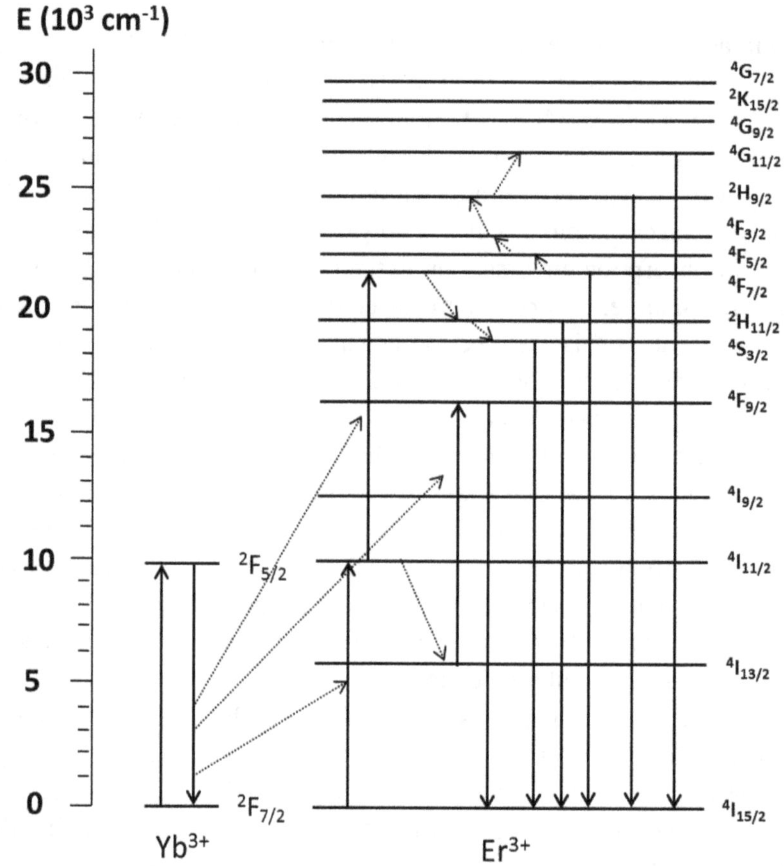

the crystallization and hexahedral structural positions in these materials Park et al. (1999), Xu et al. (2001) and Pavitra et al. (2005). Because of its thermochemical stability, SrY_2O_4 is used as a thermal barrier coating material Kurosaki et al. (2005) and Kurosaki et al. (2005). Er^{3+}/Yb^{3+} co-doped SrY_2O_4 phosphors present green emission under 980nm excitation Yang et al. (2009) and Qin et al. (2002).

Lanthanides and their Optical Properties

Lanthanides having unique luminescent properties are considered to be non-toxic. They emit radiations in a wide range of wavelengths, covering the ultraviolet (UV), visible and near-infrared (NIR) regions, which makes them ideal candidates for many applications, including lasers, lighting and optical fibres Szabadvary et al. (1988), Haley et al. (1979) and Hatakoshi (2007). The lanthanides are the group of fifteen elements in the periodic table, from atomic numbers 57 to 71 (La, Ce, Pr, Nd, Pm, Sm, Eu, Gd, Tb, Dy, Ho, Er, Tm, Yb, Lu). It is worth noting that Yttrium (Z = 39) and Scandium (Z = 21) are often associated with the 15 lanthanides within the group of rare-earth elements. The neutral lanthanide atoms have the ground state electronic configuration of a Xe core ($1s^2 2s^2 2p^6 3s^2 3p^6 4s^2 3d^{10} 4p^6 5s^2 4d^{10} 5p^6$), with two or three outer electrons ($6s^2$ or $6s^2 5d^1$), and a number of $4f$ electrons varying from 0 to 14.

Movement along the lanthanide group corresponds to the successive filling of the 4*f* electronic shell, from La ($4f^0$) to Lu ($4f^{14}$).

Lanthanides are particularly common in their ionized trivalent state, having lost their 5*d* and 6*s* electrons. The electrons of the partially filled 4*f* shell are shielded from interactions with external forces by the overlying $5s^2$ and $5p^6$ shells. As a result, lanthanide energy levels are only slightly affected by the environment surrounding the ions, and remain practically invariant, for a given ion, in various hosts. The transitions involved in the upconversion processes presented in this chapter are 4*f*-4*f* transitions. The absorption and emission cross-sections associated with these transitions are very small. This results in relatively long lanthanide luminescence lifetimes (up to a few ms).

4*F* ENERGY LEVELS OF LANTHANIDE ELEMENTS

4*f* Energy Level Splitting

In this section, *L* and *S* are the quantum numbers associated with the total orbital angular momentum and the total spin of the lanthanide element under consideration, respectively. *L* and *S* are the sums of the quantum numbers *l* and *s* associated with the orbital angular momenta and the spin of each of the 4*f* electrons of the lanthanide ion considered, respectively. *J* is the quantum number associated with the total angular momentum, defined as $\vec{J} = \vec{L} + \vec{S}$ (the values of *J* are restricted to $|L - S| \leq J \leq |L + S|$). Table 1 lists the values of *S*, *L* and *J* for trivalent lanthanide ions in their ground state [56]. When the lanthanide ions are doped into a solid, the degeneracy of the free-ion levels is removed by coulombic, spin-orbit and crystal-field interactions (Figure 2). Coulombic (or electrostatic) interactions arise from the existence of repulsions between the 4*f* electrons of the lanthanide ion considered. This results in a splitting of the free-ion levels into ${}^{2S+1}L$energy levels, which are each $(2L + 1)$ $(2S + 1)$-fold degenerate. The electrostatic energy splitting of the ${}^{2S+1}L$ terms is typically around 10^4 cm^{-1}. Spin-orbit interactions (arising from the electromagnetic interaction between the electron spin and the magnetic field created by the electron's motion) split each ${}^{2S+1}L$ term into $(2J + 1)$ states denoted ${}^{2S+1}L_J$. The typical splitting of ${}^{(2S+1)}L_J$ terms is on the order of 10^3 cm^{-1}. Finally, the interaction of the 4*f* electrons with the crystal field of the surrounding environment results in a Stark splitting of the ${}^{(2S+1)}L_J$ terms into $(2J + 1)$ terms (if the number n of 4*f* electrons is even) or $(J + 1/2)$ terms (if n is odd) denoted ${}^{(2S+1)}L_{J,}$. In the case where n is odd, Kramer's doublets are observed. The Stark splitting is typically on the order of 10^2 cm^{-1}.

It is worth noting that for lanthanides, the crystal field interaction ($\sim 10^2 cm^{-1}$) is smaller than the spin-orbit interaction ($\sim 10^3$ cm^{-1}). This arises from the shielded character of the electrons in the 4*f* shell. This is not the case for transition-metal ions.

The Dieke Diagram

The 4*f* energy levels of lanthanide ions are characteristics of each ion and are not affected much by the environment of the ion, as mentioned earlier. The 4*f*-4*f* transition energies are therefore relatively invariant for a given ion in different host lattices. The characteristic ${}^{2S+1}L_J$ energy levels of lanthanide ions doped into $LaCl_3$ Dieke (1968) and Dieke (1963) were measured by Dieke et al. in the1960's.

Table 1. Electronic configurations of trivalent lanthanide ions in their ground state

Ion	Z	Number of 4f electrons (n)							S = ∑s	L = ∑l	J = L - S (if n<7) J = L + S (if n≥7)
La^{3+}	57								0	0	0
Ce^{3+}	58	↑							1/2	3	5/2
Pr^{3+}	59	↑	↑						1	5	4
Nd^{3+}	60	↑	↑	↑					3/2	6	9/2
Pm^{3+}	61	↑	↑	↑	↑				2	6	4
Sm^{3+}	62	↑	↑	↑	↑	↑			5/2	5	5/2
Eu^{3+}	63	↑	↑	↑	↑	↑	↑		3	3	0
Gd^{3+}	64	↑	↑	↑	↑	↑	↑	↑	7/2	0	7/2
Tb^{3+}	65	↑↓	↑	↑	↑	↑	↑	↑	3	3	6
Dy^{3+}	66	↑↓	↑↓	↑	↑	↑	↑	↑	5/2	5	15/2
Ho^{3+}	67	↑↓	↑↓	↑↓	↑	↑	↑	↑	2	6	8
Er^{3+}	68	↑↓	↑↓	↑↓	↑↓	↑	↑	↑	3/2	6	15/2
Tm^{3+}	69	↑↓	↑↓	↑↓	↑↓	↑↓	↑	↑	1	5	6
Yb^{3+}	70	↑↓	↑↓	↑↓	↑↓	↑↓	↑↓	↑	1/2	3	7/2
Lu^{3+}	71	↑↓	↑↓	↑↓	↑↓	↑↓	↑↓	↑↓	0	0	0

This work resulted in the elaboration of the so-called *Dieke diagram*, which has since then become a reference used to approximate the energy levels of the $^{2S+1}L_J$ multiplet manifolds of trivalent lanthanide ions embedded in any host lattice. The Dieke diagram was extended to higher energies by Ogasawara et al.(2004). The width of each state in Figure 3 indicates the magnitude of the crystal-field splitting, and the center of each multiplet approximates the location of the free ion $^{2S+1}L_J$ energy level.

Lanthanide Excitation Processes

In this section, the processes involved upon excitation of a lanthanide ion by an electromagnetic radiation are described.

Figure 2. Interactions responsible for the loss of the lanthanide 4f free-ion degeneracy

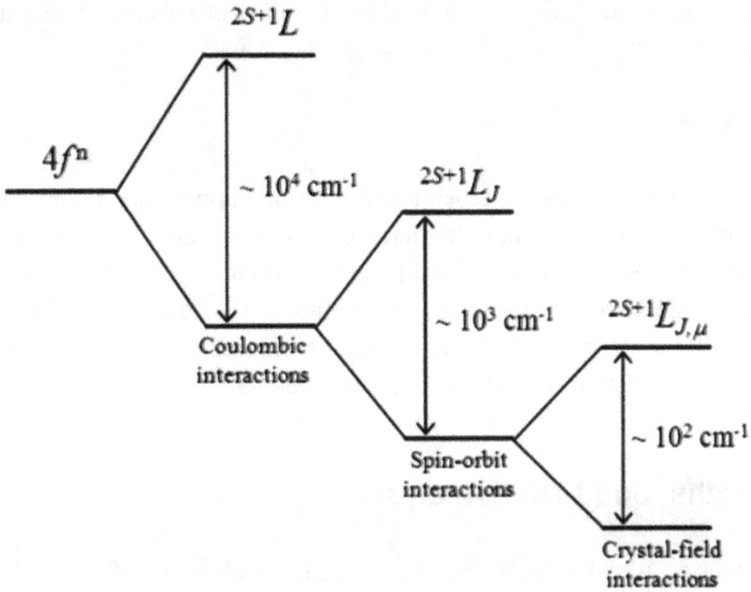

Three Types of Energy Absorption Transition

Lanthanide ions can exhibit three types of excitation transitions: *4f*-*4f* transitions, *4f*-*5d* transitions, and charge-transfer state transitions (Figure 3) Hatakoshi (2007).

4f-4f Transitions

4f-*4f* transitions are very important in spectroscopy and they involve movements of electrons between the different energy levels of the *4f* orbitals of the same lanthanide ion. *4f*-*4f* electric-dipole transitions are in theory forbidden by the parity rule., however, they are in practice generally observed, but the corresponding emission bands are usually weak and narrow.

4f-5d Transitions

In *4f*-*5d* transitions, one of the *4f* electrons is excited to a *5d* orbital of higher energy Hatakoshi (2007). This type of excitation absorption is usually denoted $4f^n$-$4f^{n-1}5d$ and is typically observed in Ce^{3+} ions ($4f^1$ configuration). Since empty, half-filled or completely filled electron shell configurations are the most stable ones, the excess *4f* electron of $Ce^{3+}$is easily transferred to the *5d* orbital. Unlike *4f*-*4f* transitions, *4f*-*5d* transitions are allowed, which results in strong and broad absorption cross-sections.

Charge-Transfer State (CTS) Transitions

The last possible excitation mechanism is the charge transfer state (CTS) transition (figure 4), in which *2p* electrons from the neighboring anions (e.g. O^{2-}in oxides) are transferred to a *4f* orbital Hatakoshi

(2007).This is typically observed in Eu^{3+} ($4f^6$) which needs one additional electron to reach the half -filled configuration. These transitions are allowed and result in broad and intense absorptions Etchart (2010).

Judd-Ofelt Analysis

In 1962, Judd [61] and Ofelt, [62] two independent researchers, published their work on the theory of *4f-4f* transition line intensities for trivalent lanthanide ions in crystals. It is a powerful tool to predict the optical properties of lanthanide ions in a crystal environment (e.g. spontaneous radiative emission rates, branching ratios, radiative lifetimes, etc). According to the Judd- Ofelt theory Judd (1962) and Ofelt (1962), the oscillator strength for the absorption band corresponding to the electronic transition from an initial state $^{2S+1}L_J$ to a final state $^{2S'+1}L'_{J'}$ can be estimated as GarcíaSolé et al. (2005) and Carnall et al. (1968)

Oscillator Strengths: Judd-Ofelt Analysis

Optically, the J-O theory allowed analyzing the possible mechanism of relaxation of the trivalent lanthanum (Ln^{3+}) ions affected by the emission of the host. It has been widely used to calculate the oscillator strength for an induced electric dipole transition from the excited state to a ground energy level from the emission spectra according to important equation Judd (1962) and Ofelt (1962)

$$f = \frac{8\pi^2 mcv}{3h\left(2J+1\right)} \frac{(n^2+2)^2}{9n} \sum_{J=2,4,6}^{n} \Omega_\lambda (\Psi_J \parallel U^J \parallel \Psi'_J)^2 \quad (1)$$

The factor $\frac{(n^2+2)^2}{9n}$ takes into account that the Ln^{3+} ion was not in vacuum but in a dielectric medium where 'n' was refractive index of the medium. J was the total angular momentum of ground state degenerated (2J+1) stark level 'v' was the wave number of transition. Ω_J (J=2, 4 and 6) were Judd-Ofelt intensity parameter and $\|U^J\|^2$(J=2, 4 and 6) were doubly reduced matrix element evaluated in the intermediate coupling approximation for a transition Ψ_J - Ψ'_J. They were usually considered to be host invariant. The transition intensity f depends on $\|U^J\|^2$ values between Ψ_J and Ψ'_J manifold. Due to selection rules and the unique nature of transition intensities for Ln^{3+} ion, any one of the $\|U^J\|^2$parameters decides the intensities of the transitions since the remaining two were zero GarcíaSolé et al. (2005)

The three phenomenological intensity parameter Ω_J (J=2, 4 and 6) describing the intensity of induced electric dipole transition according to Judd-Ofelt theory Ebendor-Heidepriem et al. (1996), Peng and Izumitani (1994) can be determined by Eq. (2). For that, the reduced matrix element was used for the emission spectra Peacock (1975) and Reisfeld (2009). Thus, Ω_J (J=2, 4 and 6) parameters can be evaluated separately from the emission transition of $^5D_0$- $^7F_2$, $^5D_0$- $^7F_4$ and $^5D_0$- $^7F_6$ respectively Antic et al. (2009).The intensity of the $^5D_0$- $^7F_1$ fluorescent transition was independent of host matrix whereas that of the $^5D_0$- ^7F_J (J=2, 4 and 6) transition depends solely on the Ω_λ parameters. This characteristic trend helps one to determine the Ω_λ parameters by the ratio of intensities of the $^5D_0$- $^7F_{2,4,6}$ transitions to the intensity of $^5D_0$- $^7F_1$ transition as follows (in the case of Eu^{3+} ion) Liu et al. (2008).

$$\frac{\int I_J dv}{\int I_1 dv} = \frac{A_J}{A_1} = \frac{e^2 v_J^3 n(n^2+2)^2}{S_{md1} v_1^3 9n^3} \Sigma \Omega_J \parallel U^J \parallel^2 \tag{2}$$

Radiative Properties

The electric (S_{ed}) and magnetic (S_{md}) dipole lines strengths were calculated according to the relation:

$$S_{ed} = e^2 \sum_{\lambda=2,4,6} (x\Omega_J (U_\lambda \parallel \Psi^1 J^1)^2 \tag{3}$$

$$S_{md} = \frac{e^2 h^2}{16\pi^2 m^2 c^2} (\Psi J \parallel L + 2S \parallel \Psi^1 J^1)^2 \tag{4}$$

The calculated dipole lines strength was used to evaluate electric (A_{ed}) and magnetic (A_{md}) dipole radiative transition probabilities according the following expressions:

$$A_{ed} = \frac{64\pi^4 v^3}{3h\left(2J+1\right)} \frac{n(n^2+2)^2}{9} S_{ed} \tag{5}$$

$$A_{md} = \frac{64\pi^4 v^3}{3h\left(2J+1\right)} n^3 S_{md} \tag{6}$$

The radiative transition probability (A) for a transition $A = \Psi_J - \Psi'_J$ was obtained by the sum of electric (A_{ed}) and magnetic (A_{md}):

$$A\left(\Psi J,\ \Psi^1 J^1\right) = A_{ed} + A_{md} \tag{7}$$

The total radiative transition probabilities (A_T) for an excited state was given as the sum of the A ($\Psi_J - \Psi'_J$) terms calculated overall the terminal states

$$A_T\left(\Psi J\right) = \sum_{\Psi^1 J^1} A\left(\Psi_J,\ \Psi_J^1\right) \tag{8}$$

Once all emission probabilities that depopulate as initial level $^{2S+1}L_J$ have been calculated, they can be used to determine how fast that level was depopulated. This rate of depopulation of an excited state was given by radiative lifetime $\tau_R(\Psi J)$:

$$\tau_R(\Psi J) = \frac{1}{A_T(\Psi J)} \tag{9}$$

Herein, for the first time, we reported the UCL properties of the Er^{3+}/Yb^{3+}co-doped MY_2O_4 (M=Ba, Sr, Ca) powders under 980 nm excitation. The Er^{3+}/Yb^{3+} co-doped MY_2O_4powders are prepared using a solid-state reaction method, and the UCL properties of the phosphors were studied under excitation wavelengths 980 nm. Results showed that the sample achieved highly pure green - red emissions under 980nm excitation. Because the rare-earth powder is an important tricolor phosphor used in full displays, the preparation of green phosphors often presents a significant challenge. The work presented in this study provided a new route for synthesizing high-quality green - red phosphors.

MATERIALS AND METHODS

To prepare MY_2O_4 with various concentrations of erbium ion (1 mol% to 3 mol%) with constant mol% of Yb^{3+} ion (0.5mol%), stoichiometric amounts of reactant mixture is taken in an alumina crucible and is fired in air at 1000°C for 2 hours in a muffle furnace. Every heating is followed by intermediate grinding using agate mortar and pestle. The Er^{3+}/ Yb^{3+}activated MY_2O_4phosphor were prepared via high temperature modified solid state diffusion. The starting materials $SrCO_3$, $CaCO_3$, $BaCO_3$, Y_2O_3, Eu_2O_3 and H_3BO_3 (as a flux) in a molar ratio (1% to 3% of Er^{3+} and Yb^{3+} 1 mol%) were used to prepare the phosphor. The mixture of reagents was grounded together for 45minute to obtain a homogeneous powder. Powder was transferred to alumina crucible and then heated in a muffle furnace at 1350°C for 3 hr Dubey et al. (2015a) and Dubey et al. (2015b). The phosphor materials were cooled to room temperature naturally (shown in the equations).

$$SrCO_3 + Y_2O_3 \rightarrow SrY_2O_4 + CO_2$$

$$BaCO_3 + Y_2O_3 \rightarrow BaY_2O_4 + CO_2$$

$$CaCO_3 + Y_2O_3 \rightarrow CaY_2O_4 + CO_2$$

The samples were characterized by using Photoluminescence (PL), XRD, FEGSEM (field emission gun scanning electron microscopy) and HRTEM (high-resolution transmission electron microscopy). The XRD measurements were carried out using Bruker D8 Advance X-ray diffractometer. The X-rays were produced using a sealed tube and the wavelength of X-ray was 0.154 nm (Cu K-alpha). The X-rays were detected using a fast counting detector based on Silicon strip technology (Bruker LynxEye detector). Particle morphology was investigated by FEGSEM (field emission gun scanning electron microscope) (JEOL JSM-6360). The photoluminescence (PL) emission and excitation spectra were recorded at room temperature Dubey et al. (2013), Dubey et al. (2010), Dubey et al. (2015), . UCL spectra (500–900 nm) were measured using a Hitachi F-4500 fluorescence spectrophotometer equipped with a power-tunable 980 nm laser diode (LD). The dependence of the upconverted emission intensities of different samples on the pumping power was determined by varying the excitation powers at room temperature. For direct comparison of luminescence intensity, all measurement conditions were maintained for different samples Mai et al. (2007) and Nyk et al. (2008).

RESULTS AND DISCUSSION

Phase Identification and Structural Refinement

The XRD pattern of the sample (MY_2O_4, M=Ba, Ca, Sr) is shown in figure 3. The width of the peak increases as the crystallite size decreases. The crystallite size has been computed from the full width half maximum (FWHM) of the intense peak using Scherer's formula. Formula used for calculation is

$$D = \frac{0.9\lambda}{\beta \cos\theta} \tag{10}$$

Here D is crystallite sizeβ is FWHM (full width half maximum)λ is wavelength of X-ray sourceθ is angle of diffraction

From XRD pattern corresponding miller indices values are calculated which matches with JCPDS card no. 74-0264(black line) as shown in Figure 3 (SrY_2O_4). Sample show orthorhombic structure. The crystallite size calculated using Scherer's formula is 66nm. For XRD pattern corresponding miller indices values are calculated which matches with JCPDS (blue line) card no. 75-0306 (CaY_2O_4), sample show cubic structure. The crystallite size calculated using Scherer's formula is 56.41nm.The diffraction peaks of CaY_2O_4 phosphor matched with the cubic structure of NaCl because more information about cubic CaY_2O_4 lattice construction is not available up to now [38]. Similarly, BaY_2O_4phosphor (red line) shows a cubic structure which is more similar to the CaY_2O_4 phosphor. Structural parameters and crystallite size are shown in Table 2.

Morphology of MY_2O_4: Er^{3+}/Yb^{3+} Phosphors

FEGSEM micrographs recorded for MY_2O_4:Er^{3+}/Yb^{3+} phosphors showninFigure4 (a-c). From figures indicates clearly that the particle is of uniform size and having good connectivity with grain. The average particles have shown nano-grass with the particle size 56-75 nanometres for CaY_2O_4: Er^{3+}/Yb^{3+}(Figure 4a), similarly for BaY_2O_4: Er^{3+}/Yb^{3+} phosphor shows spherical size structure of prepared phosphor having particle size distribution. SrY_2O_4: Er^{3+}/Yb^{3+} shows good morphology and formation of nano rods which is confirmed by HRTEM image of SrY_2O_4: Er^{3+}/Yb^{3+}(3mol%) phosphor.

The images revealed that the Er^{3+}/Yb^{3+} doped SrY_2O_4 phosphor (figure 5c) was a collection of nanorods 55-58 nm long and 2-6 nm thick. The synthesized sample showed the nanorod shapes, with no critical difference in morphology, this pattern is quite similar recorded by FEGSEM image. The

Figure 3. XRD pattern of MY_2O_4: Er^{3+}/Yb^{3+} phosphors

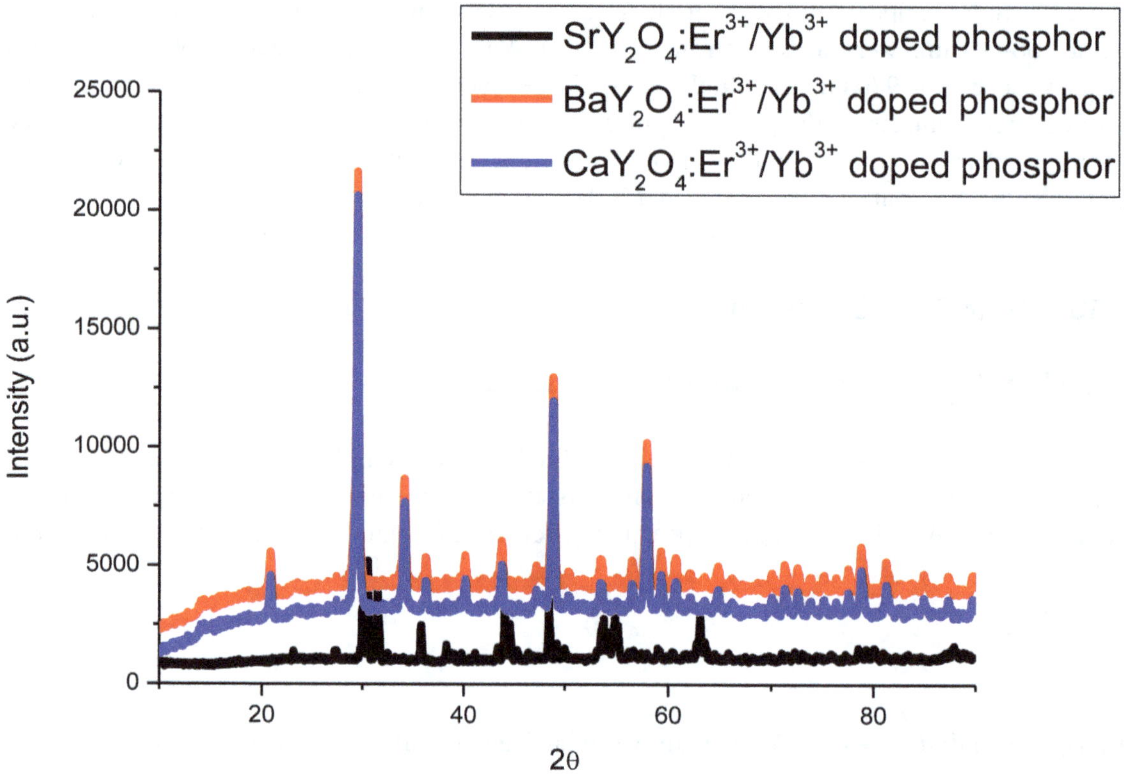

Table 2. Structural parameter of phosphor

Phosphor	Structure	Crystallite Size	Matched With
CaY_2O_4:Er^{3+}/Yb^{3+}	Cubic	56.41nm	NaCl (75-0306)
BaY_2O_4:Er^{3+}/Yb^{3+}	Cubic	86nm	NaCl (75-0306)
SrY_2O_4: Er^{3+}/Yb^{3+}	Orthorhombic	66nm	JCPDS 74-0264

nano-rod formation is responsible for the very good PL spectra from the prepared phosphor. It is found spherical size particle for CaY_2O_4 phosphor doped with Er^{3+}/Yb^{3+} which shows good morphology and very useful for luminescence efficiency (figure 5a). BaY_2O_4: Er^{3+}/Yb^{3+} agglomerated with a nano-rod phosphor which was prepared by solid state reaction method (figure 5b).

Photoluminescence Study and Up-Conversion Mechanism for SrY_2O_4: Er^{3+}/Yb^{3+}

Excited by laser excitation and corresponding room temperature upconversion emission spectra of SrY_2O_4: Er^{3+}: Yb^{3+} doped phosphor (figure 6). All the spectra show the typical emission peaks of Er^{3+} ions behave like activator and Yb^{3+} ions behave like sensitizers. According to the energy levels of Er^{3+}, the observed peaks in the green region 527–536 nm and 555–567 nm are assigned to the transitions $^2H_{11/2}\rightarrow{}^4I_{15/2}$ and $^4S_{3/2}\rightarrow{}^4I_{15/2}$ of Er^{3+}, respectively.

Figure 4. (a-c)FEGSEM image of (a) CaY_2O_4:Er^{3+}/Yb^{3+} (b) BaY_2O_4:Er^{3+}/Yb^{3+} (c) SrY_2O_4:Er^{3+}/Yb^{3+} phosphor

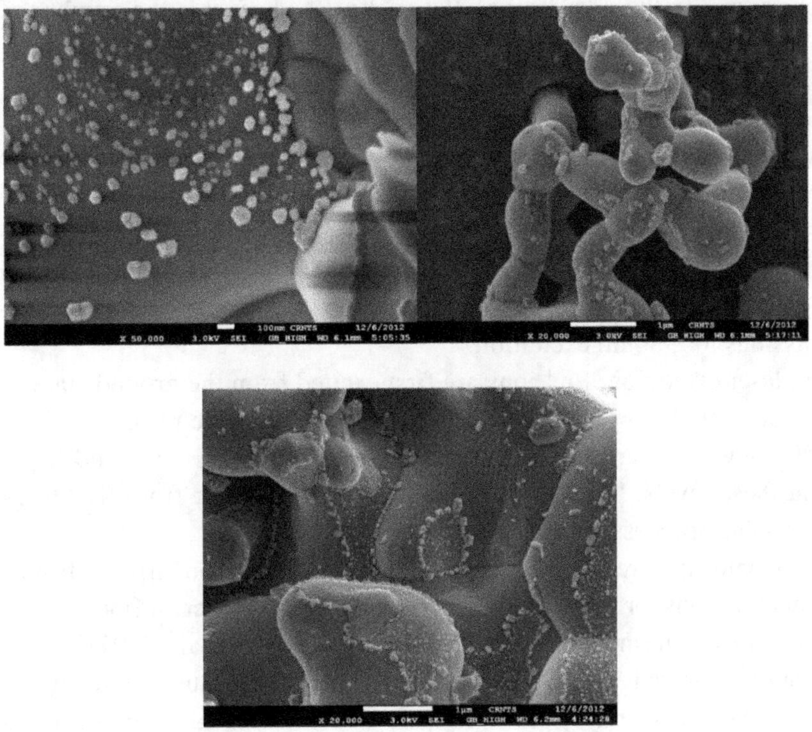

Figure 5. (a-c)HRTEM image of (a) CaY_2O_4: Er^{3+}/Yb^{3+} (b) BaY_2O_4: Er^{3+}/Yb^{3+} (c) SrY_2O_4: Er^{3+}/Yb^{3+} phosphor

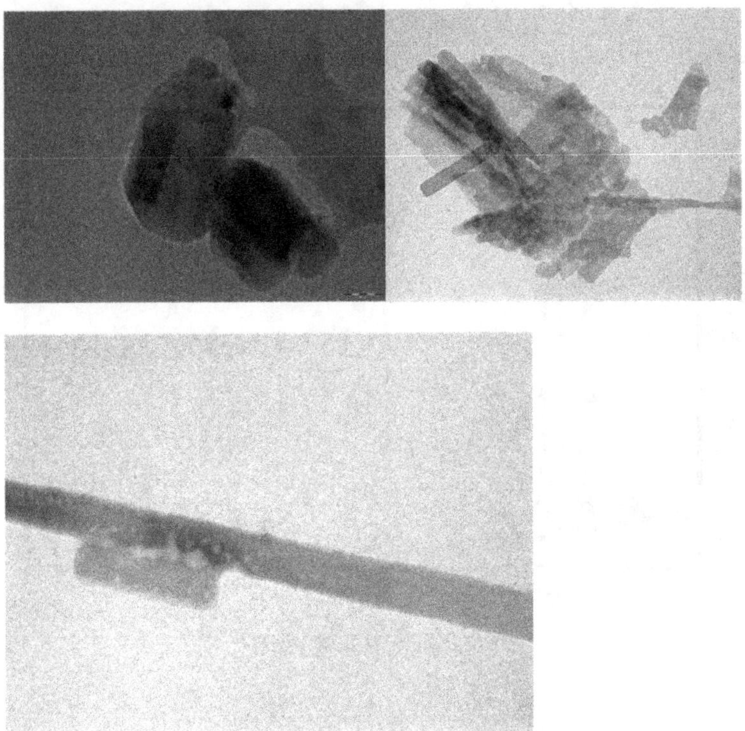

In order to understand the upconversion mechanisms of the upconverted emission bands, the upconversion intensity *I* of the transition was measured as a function of pump power *P*. In upconversion process, *I* is proportional to the nth power of *P*, i.e., $I \alpha P^n$ where *n* is the number of pump photons absorbed per upconverted photon emitted. A plot of lg*I* versus lg*P* yields a straight line with slope *n*. Under 980 nm excitation, the slope value *n* obtained for 545 nm ($^2H_{11/2} \rightarrow ^4I_{15/2}$) emissions, respectively. These results indicate that two-photon process is responsible for green upconversion (figure 7b).

The excited states for upconversion can be populated with several well-known mechanisms: (1) excited state absorption (ESA), (2) energy transfer (ET) and (3) photon avalanche Bhargava et al. (1994), Guo et al. (2004) and Chen et al. (2007). Photon avalanche was ruled out as a possible mechanism for upconversion because no inflection point was observed in the power study. Figure7 (a) presents energy level diagram of Er^{3+}/Yb^{3+} ions as well as the probable upconversion mechanisms accounting for the green emissions for 980nm excitation.

Under 980 nm laser excitation, Er^{3+} ions are first excited from the ground state to $^4I_{15/2}$ level, and then excited by another 980 nm photon, the ions in the $^4I_{15/2}$ level are raised to $^4F_{7/2}$ levels (in Figure 6). The ions in $^4F_{7/2}$ levels undergo multi-phonon relaxation to $^2H_{11/2}$, $^4S_{3/2}$ and $^4F_{9/2}$ levels, and give green emission at these levels. Herein Yb^{3+} ions charge the transfer from $^2F_{5/2}$ to $Er^{3+} \rightarrow ^4F_{7/2}$ in upconversion luminescence process.

To date, the vast majority of research and application has focused on infrared-to-visible conversion. There are three main reasons for this: (i) the advent of compact, powerful fiber diode lasers has made IR, particularly at ~980 nm, the most practical excitation source Qin et al. (2002); (ii) Yb^{3+} conveniently absorbs strongly at this wavelength and provides efficient ETU when used with an erbium activator as mentioned above; and (iii) such phosphors work effectively in biological applications, as they can be synthesized precisely on the nanoscale and IR radiation easily penetrates biological tissues Nyk et al. (2008) and Hilderbrand et al. (2009). Thus, research is active in the application of UC nanoparticles to bioimaging, detection, and photodynamic therapy, wherein nanoparticles are surface-functionalized

Figure 6. Up-conversion luminescence emission spectra under 980nm laser excitation

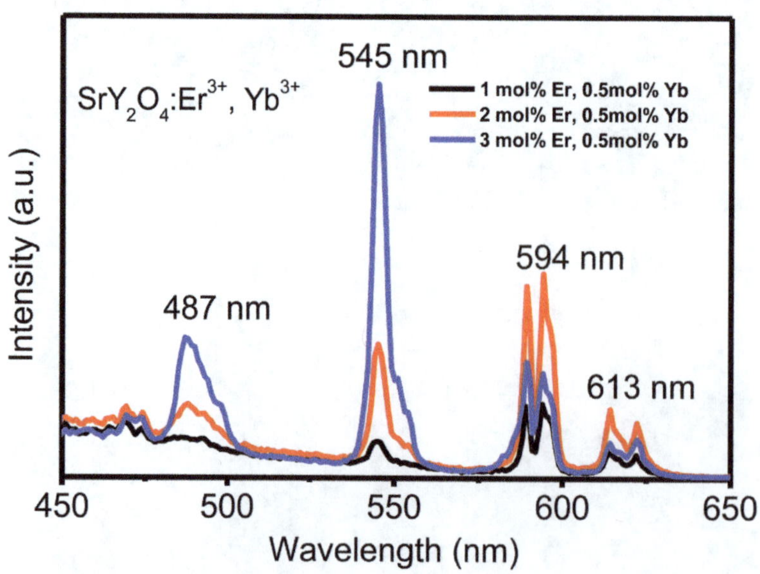

Figure 7. (a) Energy transfer mechanism for up-conversion SrY_2O_4: Er^{3+}/Yb^{3+} phosphor Bhargava et al. (1994), Guo et al. (2004) and Chen et al. (2007)

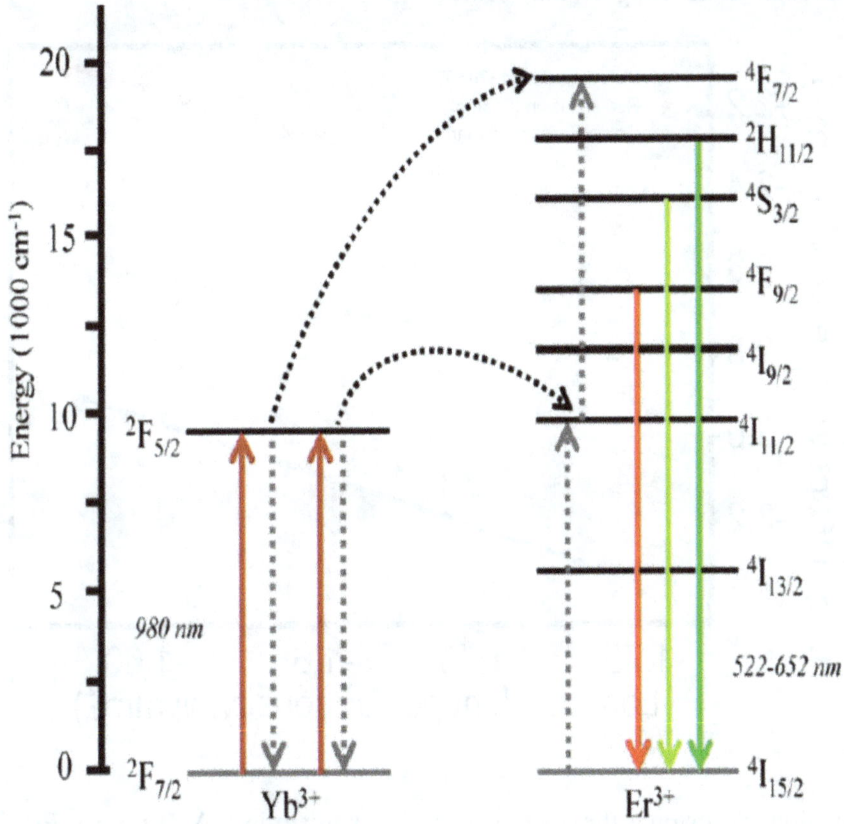

to interact with specific cells or tissues Kumar et al. (2009) and Idris et al. (2012). An IR laser source may then excite the particles through the tissue to induce visible light emission that results in imaging or photosensitization. Other applications include the use of UC for spectral conversion in solar cells and photocatalysis systems – described in detail in a later section – and finally, UC phosphors are already being employed for anti-counterfeiting in UC luminescent inks Wang et al. (2006). Figure 7 (a) shows the luminescence mechanism and emission spectra of prepared phosphor including CIE coordinate which indicates the lightening application for prepared phosphor such as solar and security encoding materials.

Pump Power Dependence of Upconversion Emission

Pump power density dependences of the blue $Er^3 \rightarrow {}^4F_{7/2} \rightarrow {}^4I_{15/2,}$ green Er^{3+} $({}^4S_{3/2}, {}^2H_{11/2})$ $-{}^4I_{15/2}$and red $Er^3 \rightarrow {}^4F_{9/2} \rightarrow {}^4I_{15/2}$ emission powers were investigated at room temperature under 980nm excitation. Here blue, green and red emission powers show a quadratic dependence on the excitation power in the low power regime, which confirms that two-photon steps are involved in the upconversion process (Figure 7b). It is worth mentioning that for ~ 25.10^{-3} W/mm² excitation power densities, about half of the incident light is absorbed by the sample, i.e. $P_{abs}^{IR} \sim \frac{1}{2} P_{inc}^{IR}$.

Figure 8. Pump power density dependence of the blue $Er^{3+4}F_{7/2} - {}^4I_{15/2}$green Er^{3+} (${}^4S_{3/2}$, ${}^2H_{11/2}$) - ${}^4I_{15/2}$ and red $Er^{3+4}F_{9/2}{}^4I_{15/2}$ emission powers at room temperature under 980nm excitation. The results are presented in a decadic double-logarithmic scale

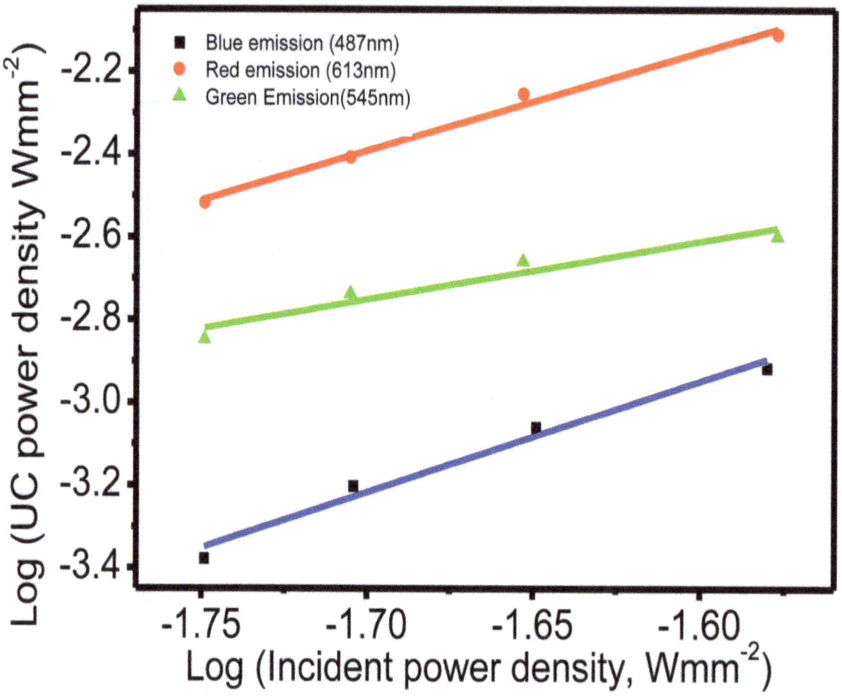

The rate equations describing the upconversion mechanism in SrY_2O_4: Yb^{3+}, Er^{3+}simplified and solved in the steady-state regime and at low incident power. A quadratic dependence was found between the blue $Er^3 \rightarrow {}^4F_{7/2} \rightarrow {}^4I_{15/2,}$ green Er^{3+} (${}^4S_{3/2}$, ${}^2H_{11/2}$) -${}^4I_{15/2}$and red $Er^3 \rightarrow {}^4F_{9/2} \rightarrow {}^4I_{15/2}$ emitting level populations and the incident pump power. Since upconversion emission powers are proportional to the population densities of the emitting levels, the experimental power dependence results are in good agreement with the theoretical predictions.

Colour Characteristics for SrY_2O_4:Er^{3+}/Yb^{3+} Phosphor

Figure 9 (a-c) represents the CIE coordinate of the present phosphor as a function of various concentrations of dopants. As seen from Figure 8 (a-c), with the change in Er^{3+} concentration with fixed concentration of Yb^{3+} (0.5mol%), the color coordinates shift from green region to towards white light boundary. The CIE coordinates for the optimized (2mol% Er^{3+}/Yb^{3+} doped SrY_2O_4) phosphor are estimated to be (0.33, 0.36) which is very nearer to the commercial white light phosphor Y_2O_2S:Eu^{3+} (0.33, 0.33) (Figure 8a-c). Our previous studies support this result which is calculated for SrY_2O_4: Eu^{3+} phosphor Dubey et al. (2015). Coordinates are listed in Table 3.

The color purity of the emitted color was calculated using Dubey et al. (2015, 2016)

Figure 9. (a-c) CIE coordinate for Er^{3+}/Yb^{3+} doped SrY_2O_4 (1-3mol%) phosphor

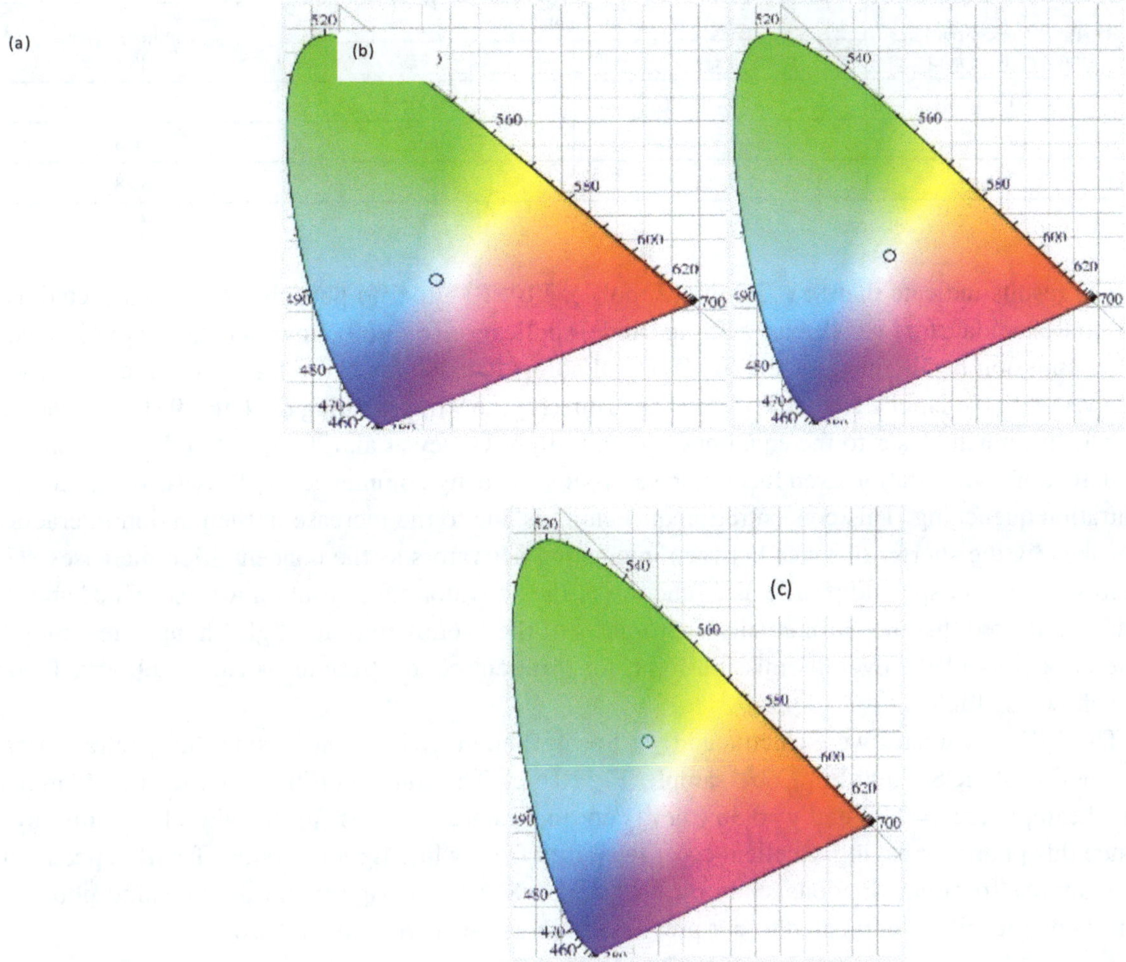

$$ColourPurity = \frac{\sqrt{(x_s - x_i)^2 + (y_s - y_i)^2}}{\sqrt{(x_d - x_i)^2 + (y_d - y_i)^2}} \times 100\% \tag{11}$$

Where (x_s, y_s) are the coordinates of a sample point, (x_d, y_d) are the coordinates of the dominant wavelength and (x_i,y_i) are the coordinates of the illuminant point.According to National Television Standard Committee (NTSC) system, the ideal white light chromaticity is (0.33, 0.33)[85]. It is seen from PL emission spectra (figure 6) the relative PL intensity especially for green emission increases with increasing the concentration of Er^{3+} ion shows the highest for 3mol% of Er^{3+} (shown in Figure 9c).

For 2 mol% of Er^{3+} ion, the blue, green and red emission intensity are nearly same and it gives intense white light emission (Figure 9b) which is very closer to white light x=0.33 and y= 0.36 (standard reported x=0.33 and y=0.33).

Table 3. CIE coordinates for Er^{3+}/Yb^{3+} doped SrY_2O_4(1-3mol%) phosphor

RE(Er^{3+}) Concentration (mol%) Yb^{3+} Fixed	Colour Coordinate		Colour Purity (%)
	x	y	
1	0.28	0.31	89.9
2	0.33	0.36	96.3
3	0.29	0.43	88.8

The results indicate that SrY_2O_4: Er^{3+}(2 mol%), Yb^{3+} (0.5 mol%) phosphors can be selected as a potential candidate for LED (Light Emitting Diode) application (Ex.980).However, the relative intensity of the emission bands provides the fundamental colors balance for white-light emission which was achieved with the 2mol% sample with the spectrum (Figure 9 b) providing the CIE 1931 chromaticity coordinates much closer to the equal-energy white-light Dubey et al. (2015, 2016). If one increases the activator concentration even further, the emission intensity commences to decrease owing to concentration quenching. This concentration quenching is due to the increase in the ion–ion interaction provoked by the shorter distance between interacting activators as the concentration increases. The fluorescence light spectral profile as a function of the activator concentration was examined and the results indicated that the chromaticity coordinates of the overall emission light changed resulting in different colors of the overall emission light, for different concentrations as can be observed from data shown in Table 3.

The CIE coordinates were calculated by a Spectrophotometric method using the spectral energy distribution of the SrY_2O_4: Er^{3+}, Yb^{3+} sample (Fig 10 a-c). The color coordinates for the Er^{3+} (2 mol%) doped sample are x=0.33 and y=0.36 (these coordinates are very near to the white light emission). Hence this phosphor having excellent color tenability from white light emission. The PL spectra are quite similar from our previous study on Eu^{3+} doped SrY_2O_4 phosphor and the prepared phosphor shows good results as compared to our previous work Dubey et al. (2015, 2016).

PHOTOLUMINESCENCE STUDY AND UP-CONVERSION MECHANISM FOR BAY_2O_4: ER^{3+}/YB^{3+}

The UCL spectra of BaY_2O_4: Er^{3+}/Yb^{3+}under 980 nm laser excitation shown in Figure 9. Under the 980nm laser excitation, the visible emission peaks in the regions of 545–585 nm (Green) and 612–621 nm (Red), which can be attributed to the transitions of $^4S_{3/2}\rightarrow{}^4I_{15/2}$ (Green) and $^4F_{9/2}\rightarrow{}^4I_{15/2}$ (Red) of Er^{3+} ions, are observed. In the case of 980 nm laser excitation, the red emission intensity of the samples is weaker and the green emission one is stronger. Here the upconversion luminescence intensity increases with increasing the concentration of Er^{3+} ions which stabilized by Yb^{3+} (0.5 mol%) ion. It seems that the codoped Yb^{3+} concentration of 0.5mol% in this system does reach the quenching solubility. It is different from other material system prepared by Su et al. (2015). All the phenomena are in agreement with those in UCL spectra. So the color of the proposed phosphors can be tuned by tailoring the Yb^{3+} doping concentration.

Figure 10. Emission spectra of prepared phosphor under 980nm (LD) excitation

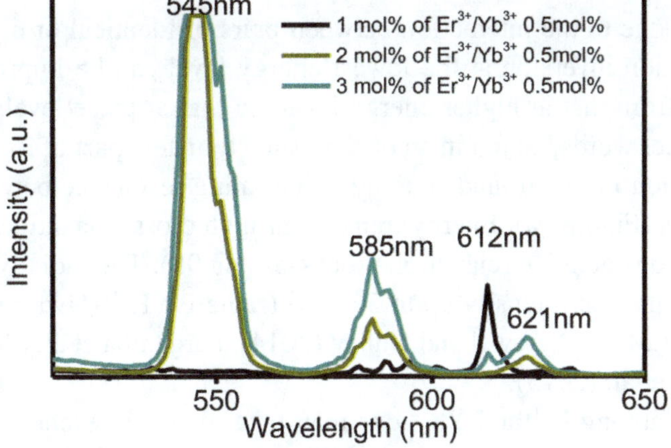

Figure 11. Transition level diagram for various transitions of Er^{3+} and Yb^{3+}

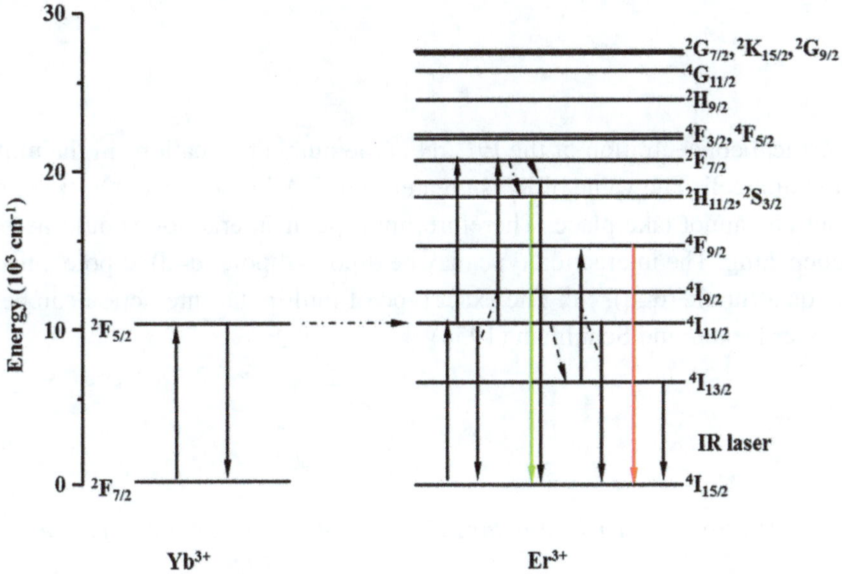

Changes in the intensity ratio of red to green emissions (I_{Red}/I_{Green}) of BaY_2O_4: Er^{3+}/Yb^{3+} as functions of the LD (laser diode) pumping current and Yb^{3+}co doping concentrations under 980 nm excitation are presented in Figure 9. As shown in Figure9, the BaY_2O_4 sample of Er^{3+}/Yb^{3+}(BaY_2O_4:Er^{3+}/Yb^{3+}, Exi = 980 nm) shows increasing intensity of red emissions with increasing Er^{3+}ion concentration and excitation power under 980 nm pumping, i.e., the I_{Red}/I_{Green} increases from roughly 1 to 4 as the Er^{3+} ion concentration increases from 2 mol.% to 3 mol%. The I_{Red}/I_{Green} of BaY_2O_4: Er^{3+}/Yb^{3+}decreases with increasing excitation power, i.e., the intensity of green emission improved. For example, the I_{Red}/I_{Green} decreases from roughly 2.86 to 1.96 as the excitation power increases from 0.2 to 0.65 A.

Cross – Relaxation and Concentration Quenching

Cross-relaxation arises due to the interaction between pairs of identical or different rare earth ions. In this process, population inversion at the lowest energy level can be improved by increasing the relaxation energy rate from that at higher energy levels to that at these levels by means of induced ET between ions. In other words, an ion in an excited state transfers part of its excitation energy non-radiatively to a nearby ion in the ground state by maintaining resonance between the two transition energy states in the ions (Figure 9a). Energy transfer through cross-relaxation has been explained in terms of radiative and non-radiative relaxation processes (fig 9b). The radiative relaxation process is involved in increased emission intensity up to 3 mol% (radiative ET) Tiwari et al. (2016), Dubey et al. (2016), Dubey et al. (2014), Agrawal and Dubey (2014), Parganiha et al. (2015), Shrivastava et al. (2014) and Shrivastava et al. (2015).

The energy transfer among $Er^{3+}$and Yb^{3+}ions may take place via exchange interaction or multi-pole–multipole interaction. The type of interaction can be identified via the calculation of the critical distance (R_c) between Er^{3+}/Yb^{3+} in BaY_2O_4host as per the formula of Blasse Blasse (1986):

$$R_c \approx 2\left[\frac{3V}{4\pi CN}\right]^{1/3} \tag{12}$$

where C is the critical concentration of the Er^{3+}, N is the number of cations in the unit cell and V is the volume of the unit cell. The value of R_c is greater than 5 Å in the present host indicating that the exchange interaction cannot take place. Therefore, multipolar interaction should be responsible for concentration quenching. The interaction type may be dipole–dipole (d–d), dipole–quadrupole (d–q), and quadrupole–quadrupole (q–q), etc. The exact type of multipolar interaction can be evaluated via the theory of Dexter Dexter and Schulman (1954):

$$\frac{I}{C} = \frac{k_1}{\beta C^{s/3}} \tag{13}$$

Figure 12. (b) concentration vs intensity plot for Er^{3+} (1-3mol%) with fixed Yb^{3+} concentration 0.5mol%

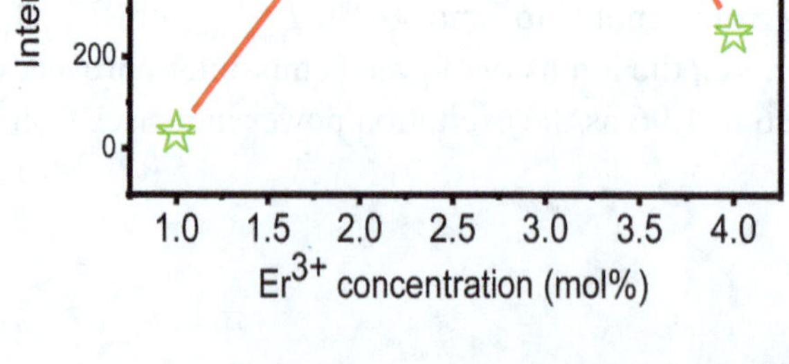

Figure 13. (a-c) CIE coordinate for Er^{3+}/Yb^{3+} doped BaY_2O_4(1-3mol%) phosphor

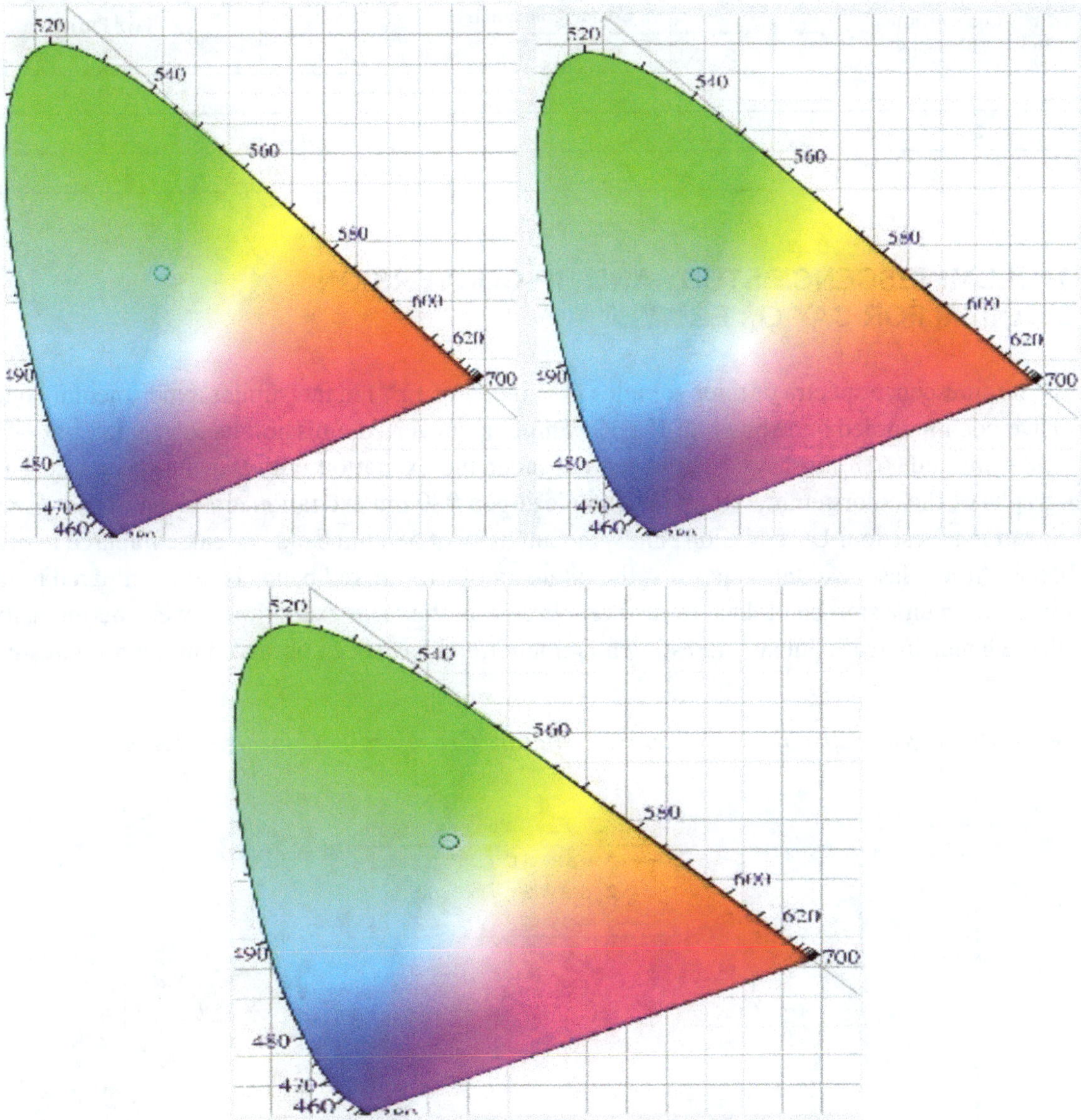

where I is the emission intensity, C *is* the activator concentration involved in self-concentration quenching, k_1 and β are constants for each interaction in the same excitation conditions for a given host lattice.

Figure 13 (a-c) represents the CIE coordinate of the present phosphor as a function of the variable concentration of Er^{3+} ions. As seen from Figure 13(a-c), the coordinates does not vary much with the change in the variable concentration. However, with the change in Er^{3+} concentration, the color coordinates minute shifts from green region to deep green region (see table 4).The color coordinates are shown in table 4 and it give the information about the emission spectrum.

Table 4. CIE coordinates for Er^{3+}/Yb^{3+} doped BaY_2O_4(1-3mol%) phosphor

Re(Er^{3+}) Concentration (Mol%) Yb^{3+} Fixed	Colour Coordinate		Colour Purity (%)
	X	y	
1	0.24	0.45	83.9
2	0.25	0.45	86.3
3	0.26	0.47	89.8

PHOTOLUMINESCENCE STUDY AND UP-CONVERSION MECHANISM FOR CAY_2O_4: ER^{3+}/YB^{3+}

The UC luminescence spectra of three samples (1-3 mol% of Er^{3+}) in the similar experimental conditions have been recorded and shown in Fig 12. Three upconverted emission bands peaking at ~545 nm, ~584 nm and~613 nm have been observed under the excitation with 980 nm NIR laser. The photographs of the colors emitted from the samples upon 980 nm excitation are given in the inset of Fig 16. Among these four UC emission bands, the intensity of 584 nm band is weak compared to other two bands. At low laser excitation powers, the intensity of 613 nm red bands is less compared to that of 545nm green emission band. Due to further increase in the laser excitation power, the intensities of both the bands increase but an interesting increment trend have been observed in the present study.

Figure 14. PL emission spectra under 980nm excitation CIE coordinate for its emission

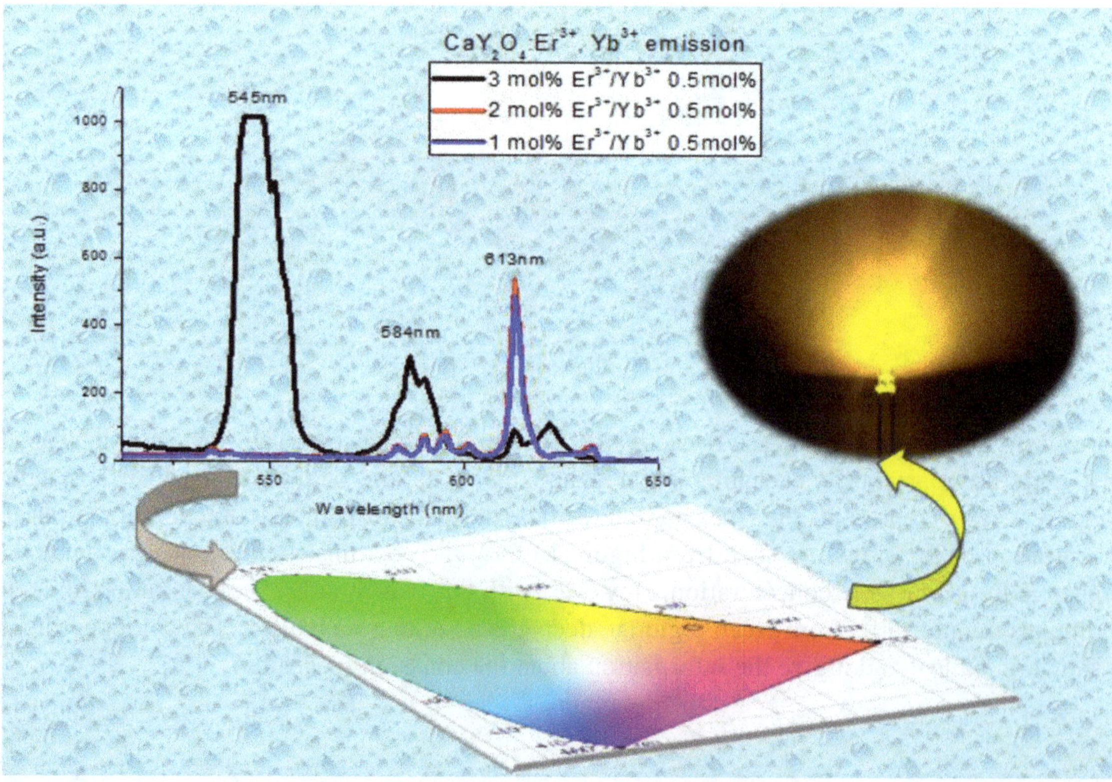

Figure 15. (a-c) CIE coordinate for Er^{3+}/Yb^{3+} doped CaY_2O_4 (1-3mol%) phosphor

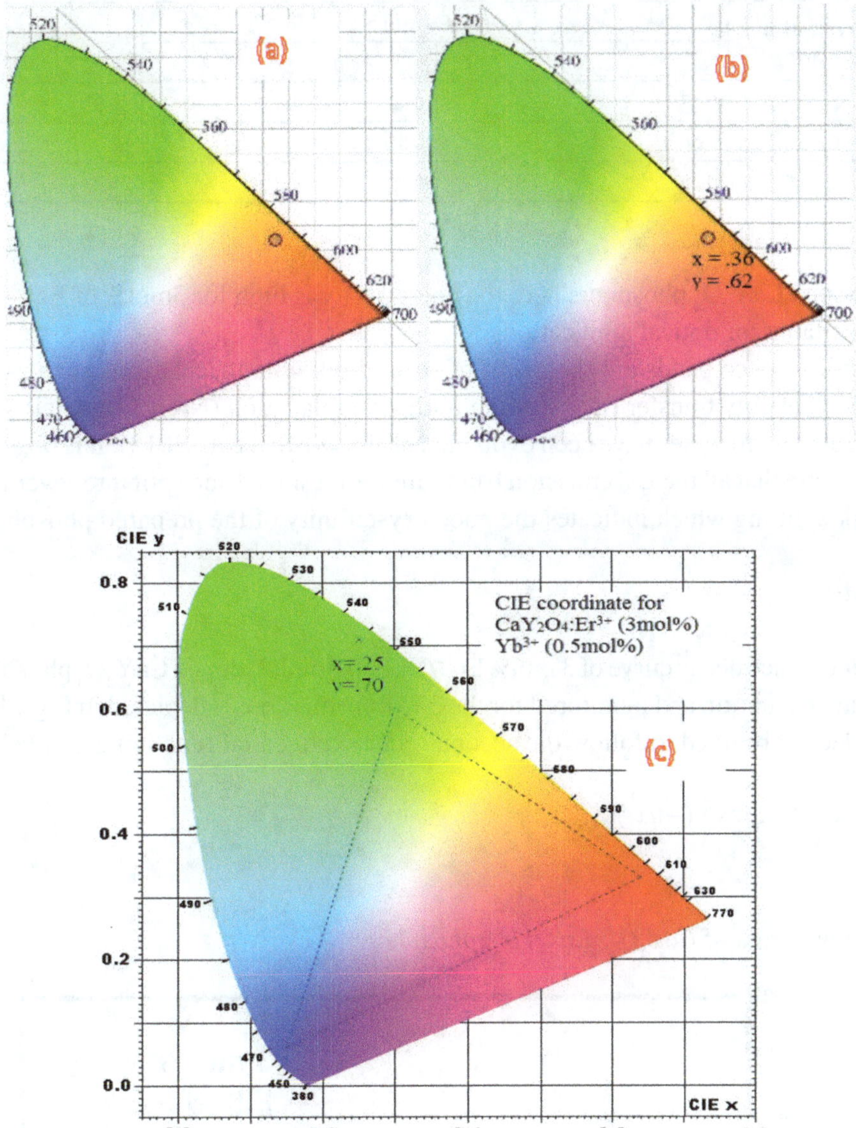

Although both the UC emission peaks show enhancement in UC emission intensity with laser excitation power, the rate of increase of intensity for ~613 nm UC emission band is greater in comparison to that of ~545 nm band (for 2mol% of Er^{3+}). The intensity ratio of green to red UC emission bands for the three samples. This fact confirms that the prepared phosphor powders can be used as intense green upconverting phosphors. The intensity ratio of green/red UC emission bands may be tuned with several factors Su et al. (2015). Herein for 1mol% of Er^{3+} ion and 0.5mol% Yb^{3+} ion strong yellow emission was found (Figure 12). Further, the increasing concentration of Er^{3+} wavelength shifted towards the green light emission and highest for 3mol% of Er^{3+}ion (figure 15a). CIE coordinates for the variable concentration of Er^{3+} ions represented in Figure 13(a-c).Table 5 shows the x and y coordinates

Table 5. CIE coordinates for Er^{3+}/Yb^{3+} doped CaY_2O_4 (1-3mol%) phosphor

RE (Er^{3+}) Concentration (mol%) Yb^{3+} Fixed	Colour Coordinate		Colour Purity (%)
	X	y	
1	0.57	0.42	80.9
2	0.56	0.42	82.5
3	0.25	0.70	85.2

for Er^{3+}/Yb^{3+} doped CaY_2O_4 phosphors here the color purity is high for 3mol% of Er^{3+} ion more than 85% which is suitable for optical applications.

The UC luminescence bands are the result of several possible mechanisms such as excited state absorption (ESA), energy transfer (ET), and nonradiative relaxation (NR). The details about the UC emissions and related processes have been explained elsewhere Su et al. (2015). It is observed from the UC emission spectra that all the UC emission bands are well resolved and split into several components due to the Stark splitting which indicates the good crystallinity of the prepared phosphor samples.

Decay Kinetics

The photoluminescence decay curve of 3 mol% Er^{3+}/Yb^{3+}(0.5mol%) doped CaY_2O_4 phosphor, recorded under the 980 nm excitation and monitored for the 545 nm emission, is depicted in Fig 14. The respective emission decay observed to follow the second order exponential function given below:

$$I = A_1 \exp(-t/\tau_1) + A_2 \exp(-t/\tau_2) \quad (14)$$

Figure 16. Decay curves of CaY_2O_4: Er^{3+}/Yb^{3+} phosphor

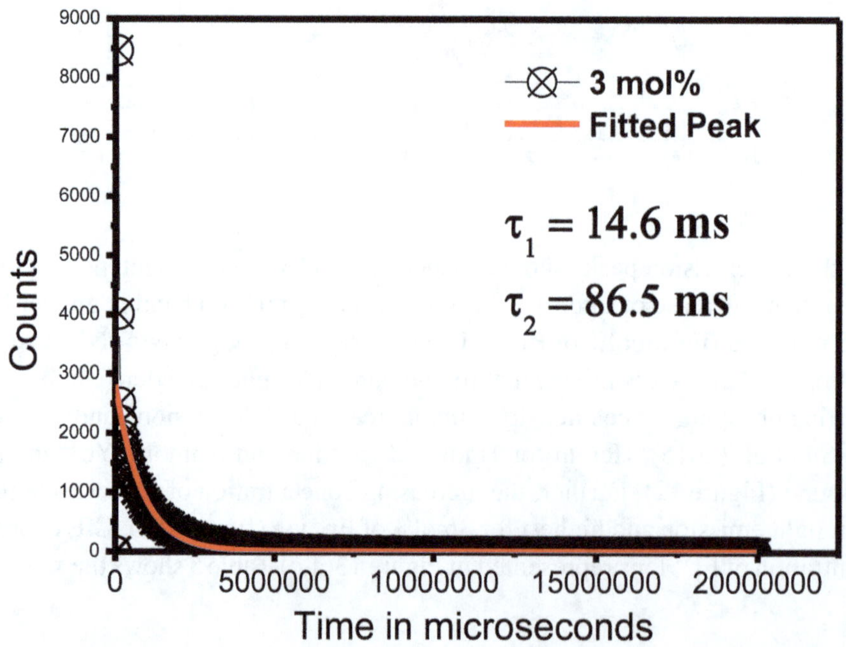

where I is the fluorescence intensity, A_1, A_2 are constants, t is time, τ_1 and τ_2 are decay times for the exponential components.

CONCLUSION

In this work MY_2O_4: Er^{3+}/Yb^{3+} (M=Ba, Ca, Sr) phosphor synthesized via solid state reaction method is suitable for large-scale production and industrial applications. Structural and morphological studies for all prepared phosphors discussed briefly inside the manuscript. The Up-conversion behavior of prepared phosphor shows good PL emission spectra under 980nm laser excitation and the possible applications are a medical diagnosis, solar cell, optical fibers, Li-Fi technology etc. Here in the possible mechanism of Up-conversion behavior and cross relaxation explained in details. Laser pump power excitation of prepared phosphor discussed briefly. It is found that the color purity increases with increasing concentration of Er^{3+} ions and found maximum for 3mol% sometimes the shifting found to red to green emission. This observation concludes that the prepared phosphor is good emitting phosphor for the IR light excitations and useful for commercial applications. The emission spectra with variable concentration of Er^{3+} shows strong peaks at 545nm and 594nm, the dominant peak at 545nm due to the ($^2H_{11/2}$,$^4S_{3/2}$) transition and weaker transition at 594nm with peak associated to 613nm. For Er^{3+}/Yb^{3+} doped SrY_2O_4 phosphor where Er^{3+} (2mol%) shows near white light emission (x=0.33, y=0.36) which is applicable for white application of optical displays. The upconversion (UC) of the rare earth doped MY_2O_4 (M=Ba, Sr, Ca) has been extensively investigated due to their potential applications in many fields, such as color display, high-density memories, optical data storage, sensor, and energy solar cell, etc.

ACKNOWLEDGMENT

Author (Dr.Vikas Dubey) is very thankful to Chhattisgarh Council of Science and Technology (CCOST) for funding through mini research project. Very thankful to Director General of CCOST Prof. M.M. Hambarde for the kind corporation.I am also thankful to Dr. D. Harnath NPL New Delhi for Up-conversion analysis of the prepared sample. Thankful to SAIF IIT, Bombay for characterization of our samples.

REFERENCES

Agrawal, S., & Dubey, V. (2014). Article. *Jour. Rad. Res. Appl. Sci.*, *7*(4), 601–606.

Antic, Z., Krsmanovic, R., Dordevic, V., Dramicanin, T., & Dramicanin, M. D. (2009). Optical Properties of Y_2O_3:Eu^{3+} Red Emitting Phosphor Obtained via Spray Pyrolysis. *Acta Physica Polonica A*, *4*(116), 622–624. doi:10.12693/APhysPolA.116.622

Bhargava, R. N., Gallagher, D., Hong, X., & Nurmikko, A. (1994). Optical properties of manganese-doped nanocrystals of ZnS. *Physical Review Letters*, *72*(3), 416–419. doi:10.1103/PhysRevLett.72.416 PMID:10056425

Blasse, G. (1986). Energy transfer between inequivalent Eu2+ ions. *Journal of Solid State Chemistry, 62*(2), 207–211. doi:10.1016/0022-4596(86)90233-1

Bloembergen, N. (1959). Solid state infrared quantum counters. *Physical Review Letters, 2*(3), 84–85. doi:10.1103/PhysRevLett.2.84

Cao, Z. M., Chen, Y., Zhou, S., Duan, C., & Yin, M. (2015). Optical properties of Er^{3+} doped KYb_xF_{3x+1} (x=2, 3) upconverting nanoparticles. *Journal of Rare Earths, 33*(9), 911–914. doi:10.1016/S1002-0721(14)60504-4

Capobianco, J. A., Vetrone, F., Alesio, T. D., Tessari, G., Speghini, A., & Bettinelli, M. (2000). Optical spectroscopy of nanocrystalline cubic Y2O3:Er3+ obtained by combustion synthesis. *Physical Chemistry Chemical Physics, 2*(14), 3203–3207. doi:10.1039/b003031g

Capobianco, J. A., Vetrone, F., Boyer, J. C., Speghini, A., & Bettinelli, M. (2002). Visible upconversion of Er^{3+} doped nanocrystalline and bulk Lu_2O_3. *Optical Materials, 19*(2), 259–268. doi:10.1016/S0925-3467(01)00188-4

Capobianco, J. A., Vetrone, F., Boyer, J. C., Speghini, A., & Bettinelli, M. (2002). Enhancement of red emission ($^4F_{9/2}\rightarrow{}^4I_{15/2}$) via upconversion in bulk and nanocrystalline cubic Y_2O_3:Er^{3+}. *The Journal of Physical Chemistry B, 106*(6), 1181–1187. doi:10.1021/jp0129582

Carnall, & Field, & Rajnak. (1968). Article. *J. Chem. Phys. 49*, 104253.

Carnall, W. T. (1978). Optical Spectroscopy of f-Element Compounds. Report Argonne National Laboratory, Chemistry Division.

Chen, D., Huang, P., Yu, Y., Huang, F., Yang, A., & Wang, Y. (2011). Dopant-induced phase transition: A new strategy of synthesizing hexagonal upconversion $NaYF_4$ at low temperature. *Chemical Communications (Cambridge), 47*(20), 5801–5803. doi:10.1039/c0cc05722c PMID:21491047

Chen, D., Yu, Y., Huang, F., Huang, P., Yang, A., & Wang, Y. (2010). Modifying the size and shape of monodisperse bifunctional alkaline-earth fluoride nanocrystals through lanthanide doping. *Journal of the American Chemical Society, 132*(29), 9976–9978. doi:10.1021/ja1036429 PMID:20593808

Chen, G., Somesfalean, G., Liu, Y., Zhang, Z., Sun, Q., & Wang, F. (2007). Upconversion mechanism for two-color emission in rare-earth-ion-doped ZrO_2nanocrystals. *Physical Review B: Condensed Matter and Materials Physics, 75*(19), 195204. doi:10.1103/PhysRevB.75.195204

Chen, G. Y., Liu, Y., Zhang, Y. G., Somesfalean, G., Zhang, Z. G., Sun, Q., & Wang, F. P. (2007). Bright white upconversion luminescence in rare-earth-ion-doped Y_2O_3nanocrystals. *Applied Physics Letters, 91*(13), 133103. doi:10.1063/1.2787893

Dexter, D. L., & Schulman, J. H. (1954). Theory of Concentration Quenching in Inorganic Phosphors. *The Journal of Chemical Physics, 22*(6), 1063–1070. doi:10.1063/1.1740265

Dieke, G. H., & Crosswhite, H. M. (1963). The Spectra of the Doubly and Triply Ionized Rare Earths. *Applied Optics, 2*(7), 675. doi:10.1364/AO.2.000675

Downing, E., Hesselink, L., Ralston, J., & Macfarlane, R. (1996). A three-color, solid-state, three-dimensional display. *Science*, *273*(5279), 1185–1189. doi:10.1126cience.273.5279.1185

Dubey, Tiwari, Tamrakar, Markande, Rathore, & Pradhan. (2015). Article. *Indian Journal of Materials Science*. 10.1155/2015/745052

Dubey, Kaur, Parganiha, Suryanarayana, & Murthy. (2016). Article. *Appl. Rad. Isotop.s*, *110*, 16-27.

Dubey, & Suryanarayana, & Kaur. (2010). Kinetics of TL Glow Peak of Limestone from Patharia of CG Basin (India). *Journal of Minerals & Materials Characterization & Engineering*, *9*(12), 1101–1111. doi:10.4236/jmmce.2010.912080

Dubey, V., & Jagjeet Kaur, N. S. (2013). Thermoluminescence and chemical characterization of natural calcite collected from Kodwa mines. *Research on Chemical Intermediates*, *39*(8), 3689–3697. doi:10.100711164-012-0872-7

Dubey, V., Kaur, J., & Agrawal, S. (2015). Effect of europium doping levels on photoluminescence and thermoluminescence of strontium yttrium oxide phosphor. *Materials Science in Semiconductor Processing*, *31*, 27–37. doi:10.1016/j.mssp.2014.10.052

Dubey, V., Kaur, J., & Agrawal, S. (2015). Effect of europium concentration on photoluminescence and thermoluminescence behavior of Y2O3:Eu3+ phosphor. *Research on Chemical Intermediates*, *41*(7), 4727–4739. doi:10.100711164-014-1563-3

Dubey, V., Kaur, J., Agrawal, S., Suryanarayana, N. S., & Murthy, K. V. R. (2013). Optik – Int. *J.Light Electron Opt.*, *124*(22), 5585–5587. doi:10.1016/j.ijleo.2013.03.153

Dubey, V., Kaur, J., Agrawal, S., Suryanarayana, N. S., & Murthy, K. V. R. (2014). Effect of Eu3+ concentration on photoluminescence and thermoluminescence behavior of YBO3:Eu3+ phosphor. *Superlattices and Microstructures*, *67*, 156–171. doi:10.1016/j.spmi.2013.12.026

Dubey, V., Tiwari, R., Tamrakar, R. K., Kaur, J., Dutta, S., Das, S., ... Som, S. (2016). Estimation of spectroscopic parameters and colour purity of the red-light-emitting YBa3B9O18 phosphor: Judd–Ofelt approach. *Journal of Luminescence*, *180*, 169–176. doi:10.1016/j.jlumin.2016.08.029

Dubey, V., Tiwari, R., Tamrakar, R. K., Rathore, G. S., Sharma, C., & Tiwari, N. (2014). Infrared spectroscopy and upconversion luminescence behaviour of erbium doped yttrium (III) oxide phosphor. *Infrared Physics & Technology*, *67*, 537–541. doi:10.1016/j.infrared.2014.09.014

Dutta, P., & Rai, S. (2011). Optical transitions and frequency upconversions of Ho^{3+} and Ho^{3+}/Yb^{3+} ions in $Al(NO_3)_3$–SiO_2 sol–gel glasses. *Optik (Stuttgart)*, *122*(10), 858–863. doi:10.1016/j.ijleo.2010.05.029

Ebendor-Heidepriem, H., & Ehrt, D. (1996). Article. *Journal of Non-Crystalline Solids*, *208*, 205. doi:10.1016/S0022-3093(96)00524-8

GarcíaSolé, Bausa, & Jaque. (2005). *An Introduction to the Optical Spectroscopy of Inorganic Solids*. John Wiley and Sons Ltd.

Guo, H. (2007). Green and red upconversion luminescence in CeO_2:Er^{3+}powders produced by 785 nm laser. *Journal of Solid State Chemistry*, *180*(1), 127–131. doi:10.1016/j.jssc.2006.10.003

Guo, H., Dong, N., Yin, M., Zhang, W., Lou, L., & Xia, S. (2004). Visible upconversion in rare earth ion-doped Gd_2O_3 nanocrystals. *The Journal of Physical Chemistry B*, *108*(50), 19205–19209. doi:10.1021/jp048072q

Hebert, T., Wannemacher, R., Lenth, W., & Macfarlane, R. M. (1990). Blue and green cw upconversion lasing in Er:$YLiF_4$. *Applied Physics Letters*, *57*(17), 1727–1729. doi:10.1063/1.104048

Hilderbrand, S. A., Shao, F., Salthouse, C., Mahmood, U., & Weissleder, R. (2009). Upconverting luminescent nanomaterials: Application to in vivo bioimaging. *Chemical Communications*, (28): 4188–4190. doi:10.1039/b905927j PMID:19585016

Hirao, K., Tamai, K., Tanabe, S., & Soga, N. (1993). Frequency upconversion its new mechanism in Tm3+-doped fluoro aluminate glasses. *Journal of Non-Crystalline Solids*, *160*(3), 261–267. doi:10.1016/0022-3093(93)91270-D

Idris, N. M., Gnanasammandhan, M. K., Zhang, J., Ho, P. C., Mahendran, R., & Zhang, Y. (2012). In vivo photodynamic therapy using upconversion nanoparticles as remote-controlled nano transducers. *Nature Medicine*, *18*(10), 1580–1585. doi:10.1038/nm.2933 PMID:22983397

Ivanova, S., Pellé, F., Tkachuk, A., Joubert, M. F., Guyot, Y., & Gapontzev, V. P. (2008). Upconversion luminescence dynamics ofEr-doped fluoride crystals for optical converters. *Journal of Luminescence*, *128*(5-6), 914–917. doi:10.1016/j.jlumin.2007.11.031

Ivanova, S., & Strong, P. F. (2009). 153 μm to NIR-VIS-UV upconversion in Er-doped fluoride glass for high-efficiency solar cells. *J. Opt. Soc. B*, *29*, 1930.

Jianbei, Q. I. U., Qing, J. I. A. O., Dacheng, Z. H. O. U., & Zhengwen, Y. A. N. G. (2016).. . *Journal of Rare Earths*, *34*(4), 341. doi:10.1016/S1002-0721(16)60034-0

Judd, B. R. (1962). Optical Absorption Intensities of Rare-Earth Ions. *Physical Review*, *127*(3), 750–761. doi:10.1103/PhysRev.127.750

Kadono, K., Takahashi, M., Higuchi, H., Shojiya, M., Ohtsuki, T., Kawamoto, Y., & Peyghambarian, N. (1997). Frequency upconversion of Er3+ in Ga2S3-GeS2-La2S3 glasses. *SPIE Proc.*, *2996*, 25.

Kawamoto, Y., Kanno, R., Shojiaya, M., & Kadono, K. (1995). Structural study of ion-conducting MX-K (or Rb) X-CsX-BaX_2halideglasses (M=Li, Cu or Ag) with single X species (X=Cl, Br or I). *Journal of Non-Crystalline Solids*, *184*, 155–159. doi:10.1016/0022-3093(94)00608-3

Kawamoto, Y., Kanno, R., Yokota, R., Takahashi, M., Tanabe, S., & Hirao, K. (1993). Compositional dependence of upconversion luminescence of Er^{3+} in ZrF_4 based glasses III. anion substitution effect. *Journal of Solid State Chemistry*, *103*(2), 334–340. doi:10.1006/jssc.1993.1108

Kumar, R., Nyk, M., Ohulchanskyy, T. Y., Flask, C. A., & Prasad, P. N. (2009). Combined Optical and MR Bioimaging Using Rare Earth Ion-Doped NaYF4 Nanocrystals. *Advanced Functional Materials*, *19*(6), 853–859. doi:10.1002/adfm.200800765

Kurosaki, K., Tanaka, T., Maekawa, T., & Yamanaka, S. (2005). Thermophysical properties of SrY_2O_4. *Journal of Alloys and Compounds*, *398*(1-2), 304–308. doi:10.1016/j.jallcom.2005.02.034

Kurosaki, K., Tanaka, T., Maekawa, T., & Yamanaka, S. (2005). Thermal properties of polycrystalline sintered SrY_2O_4. *Journal of Alloys and Compounds*, *395*(1-2), 318–321. doi:10.1016/j.jallcom.2004.11.041

Laversenne, L., Guyot, Y., Goutaudier, C., Cohen-Adad, M. T., & Boulon, G. (2001). Optimization of spectroscopic properties of Yb^{3+}-doped refractory sesquioxides: Cubic Y_2O_3, Lu_2O_3 and monoclinic Gd_2O_3. *Optical Materials*, *16*(4), 475–483. doi:10.1016/S0925-3467(00)00095-1

Lin, H., Meredith, G., Jiang, S., Peng, X., Luo, T., Peyghambarian, N., & Pun, E. Y. B. (2003). Optical transitions and visible upconversion in Er^{3+} doped niobic tellurite glass. *Journal of Applied Physics*, *93*(1), 186–191. doi:10.1063/1.1527209

Maciel, G. S., de Araújo, C. B., Messaddeq, Y., & Aegerter, M. A. (1997). Frequency upconversion in Er 3 + -doped fluoroindate glasses pumped at 1.48 μm. *Physical Review B: Condensed Matter and Materials Physics*, *55*(10), 6335–6342. doi:10.1103/PhysRevB.55.6335

Maciel, G. S., Rakov, N., Fokine, M., Carvalho, I. C. S., & Pinheiro, C. B. (2006). Strong upconversion from $Er_3Al_5O_{12}$ceramic powders prepared by low temperature direct combustion synthesis. *Applied Physics Letters*, *89*(8), 109. doi:10.1063/1.2338558

Mai, H.-X., Zhang, Y.-W., Sun, L.-D., & Yan, C.-H. (2007). Highly Efficient Multicolor Up-Conversion Emissions and Their Mechanisms of Monodisperse NaYF4:Yb,Er Core and Core/Shell-Structured Nanocrystals. *The Journal of Physical Chemistry C*, *111*(37), 13721–13729. doi:10.1021/jp073920d

Neha, T., & Dubey, V. (2016). Luminescence studies and infrared emission of erbium-doped calcium zirconate phosphor. *Luminescence*, *31*(3), 837–842. doi:10.1002/bio.3039 PMID:26456020

Nyk, M., Kumar, R., Ohulchanskyy, T. Y., Bergey, E. J., & Prasad, P. N. (2008). High Contrast in Vitro and in Vivo Photoluminescence Bioimaging Using Near Infrared to Near Infrared Up-Conversion in Tm3+ and Yb^{3+} Doped Fluoride Nanophosphors. *Nano Letters*, *8*(11), 3834–3838. doi:10.1021/nl802223f PMID:18928324

Ofelt, G. S. (1962). Intensities of Crystal Spectra of Rare-Earth Ions. *The Journal of Chemical Physics*, *37*(3), 511–520. doi:10.1063/1.1701366

Ogasawara, K., Watanabe, S., Sakai, Y., Toyoshima, H., Ishii, T., Brik, M. G., & Tanaka, I. (2004). Calculations of Complete $4f^n$ and $4f^{n-1}5d^1$ Energy Level Schemes of Free Trivalent Rare-Earth Ions. *Japanese Journal of Applied Physics*, *43*(5A), L611–L613. doi:10.1143/JJAP.43.L611

Parganiha, Y., Kaur, J., Dubey, V., & Chandrakar, D. (2015). Synthesis, characterization, thermoluminescence and optical studies of Eu 3+ doped Y 2 SiO 5 phosphor. *Superlattices and Microstructures*, *77*, 152–161. doi:10.1016/j.spmi.2014.11.010

Park, S. G., Park, C. H., Yu, B. Y., Bae, H. S., Kim, C. H., & Pyun, C. H. (1999). Structure and luminescence of SrY_2O_4:Eu. *Journal of the Electrochemical Society*, *146*(10), 3903. doi:10.1149/1.1392570

Pavitra, E., Raju, G. S. R., Park, W., & Yu, J. S. (2005). Concentration and penetration depth dependent tunable emissions from Eu3+ co-doped SrY_2O_4:Dy^{3+}nanocrystalline phosphor. *Journal of Chemistry*, *38*, 163.

Peacock, R. D. (1975). The intensities of lanthanide f ↔ f transitions. *Structure and Bonding*, *22*, 83–122. doi:10.1007/BFb0116556

Peng, B., & Izumitani, T. (1994). The Fluorescence Properties of Eu3+ in Various Glasses and the Energy Transfer Between Eu3+ and Sm3+ in Borosilico-phosphate Glass. *Rev. Laser Eng.*, *22*(1), 16–27. doi:10.2184/lsj.22.16

Qin, G. S., Qin, W. P., Huang, S. H., Wu, C. F., Zhao, D., Chen, B. J., ... Shulin, E. (2002). Infrared-to-violet upconversion from Yb^{3+} and Er^{3+}codoped amorphous fluoride film prepared by pulsed laser deposition. *Journal of Applied Physics*, *92*(11), 6936–6938. doi:10.1063/1.1521801

Riseberg, L. A., & Moos, H. W. (1968). Multiphonon orbit-lattice relaxation of excited states of rare-earth ions in crystals. *Physical Review*, *174*(2), 429–438. doi:10.1103/PhysRev.174.429

Shalav, A., Richards, B. S., Trupke, T., Kramer, K. W., & Güdel, H. U. (2005). Application of NaYF4:Er^{3+} up-converting phosphors for enhanced near-infrared silicon solar cell response. *Applied Physics Letters*, *86*(1), 013505. doi:10.1063/1.1844592

Shojiya, M., Takahashi, M., Kanno, R., Kawamoto, Y., & Kadono, K. (1994). Upconversion luminescence of Er^{3+}in chloride glasses based on $ZnCl_2$ or $CdCl_2$. *Applied Physics Letters*, *65*(15), 1874–1876. doi:10.1063/1.112874

Shojiya, M., Takahashi, M., Kanno, R., Kawamoto, Y., & Kadono, K. (1995). Upconversion luminescence of Er^{3+} in CdX_2system glasses (X=Cl, Br, I). *Applied Physics Letters*, *67*(17), 2453–2455. doi:10.1063/1.114605

Shojiya, M., Takahashi, M., Kanno, R., Kawamoto, Y., & Kadono, K. (1997). Optical transitions of Er^{3+} ions in $ZnCl_2$-based glass. *Journal of Applied Physics*, *82*(12), 6259–6266. doi:10.1063/1.366545

Shrivastava, R., Kaur, J., Dubey, V., & Jaykumar, B. (2014). Photoluminescence, trap states and thermoluminescence decay process study of Ca2MgSi2O7: Eu2+, Dy3+ phosphor. *Bulletin of Materials Science*, *37*(4), 925–929. doi:10.100712034-014-0027-0

Shrivastava, R., Kaur, J., Dubey, V., Jaykumar, B., & Loreti, S. (2015). Photoluminescence and Thermoluminescence Investigation of Europium- and Dysprosium-Doped Dibarium Magnesium Silicate Phosphor. *Spectroscopy Letters*, *48*(3), 179–183. doi:10.1080/00387010.2013.872666

Song, H., Sun, B., Wang, T., Lu, S., Yang, L., Chen, B., ... Kong, X. (2004). Three-photon upconversion luminescence phenomenon for the green levels in Er3+/Yb3+ codoped cubic nanocrystalline yttria. *Solid State Communications*, *132*(6), 409–413. doi:10.1016/j.ssc.2004.07.044

Su, M. Y., Zhou, Y. F., Wang, K., Huang, D. C., Xu, W. T., & Cao, Y. G. (2015). Effect of Yb^{3+} concentration on upconversion luminescence of AlON:Er^{3+} phosphors. *Journal of Rare Earths*, *33*(3), 227–230. doi:10.1016/S1002-0721(14)60407-5

Takahashi, M., Kanno, R., Kawamoto, Y., Tanabe, S., & Hirao, K. (1994). Compositional dependence of Er^{3+}upconversion luminescence in MFLiFZrF_4 glasses (M: Alkali metals). *Journal of Non-Crystalline Solids*, *168*(1), 137–143. doi:10.1016/0022-3093(94)90129-5

Vetrone, F., Boyer, J. C., Capobianco, J. A., Speghini, A., & Bettinelli, M. (2003). Concentration-dependent near-infrared to visible upconversion in nanocrystalline and bulk Y_2O_3: Er^{3+}. *Chemistry of Materials, 15*(14), 2737–2743. doi:10.1021/cm0301930

Vetrone, F., Boyer, J. C., Capobianco, J. A., Speghini, A., & Bettinelli, M. (2003). Effect of Yb^{3+} Codoping on the Upconversion Emission in Nanocrystalline Y_2O_3:Er^{3+}. *The Journal of Physical Chemistry B, 107*(5), 1107–1112. doi:10.1021/jp0218692

Vetrone, F., Boyer, J. C., Capobianco, J. A., Speghini, A., & Bettinelli, M. (2004). Significance of Yb^{3+} concentration on the upconversion mechanisms in codoped Y_2O_3:Er^{3+}, Yb^{3+} nanocrystals. *Journal of Applied Physics, 96*(1), 661–667. doi:10.1063/1.1739523

Wang, L., & Li, Y. (2006). Green upconversion nanocrystals for DNA detection. *Chemical Communications*, (24): 2557–2559. doi:10.1039/b604871d PMID:16779476

Xing, M. M., Ma, Y. B., Luo, X. X., Fu, Y., Jiang, T., Wang, H., & Duan, X. L. (2014). Design and achieving of multicolor upconversion emission based on rare-earth-doped tellurite. *Journal of Rare Earths, 32*(5), 394–398. doi:10.1016/S1002-0721(14)60083-1

Xu, W., Jia, W. Y., Revira, I., Monge, K., & Liu, H. M. (2001). Optical properties of multiple sites of Eu^{3+} in SrY_2O_4 single-crystal fibers. *Journal of the Electrochemical Society, 148*(12), 179. doi:10.1149/1.1415724

Yang, J. K., Xiao, S. G., Ding, J. W., Yang, X. L., & Wang, X. F. (2009). Preparation and photoluminescence properties of SrY_2O_4:Yb^{3+},Er^{3+} powders. *Journal of Alloys and Compounds, 474*(1-2), 424–427. doi:10.1016/j.jallcom.2008.06.158

Yang, Z. W., Zhu, K., Song, Z. G., Zhou, D. C., Yin, Z. Y., Yan, L., & Qiu, J. B. (2011). Significant reduction of upconversion emission in $CaTiO_3$:Yb,Er inverse opals. *Thin Solid Films, 519*(16), 5696–5699. doi:10.1016/j.tsf.2011.03.027

Ye, Y. S., Jiang, Z. H., Wang, Q. Z., Zhu, Z. S., Wang, X., Sui, Z. L., ... Ding, Z. J. (2014). Upconversion luminescence of NaYF4:Yb,Er nanocrystals with high uniformity. *Journal of Rare Earths, 32*(9), 802–805. doi:10.1016/S1002-0721(14)60144-7

Zhang, C., Sun, L., Zhang, Y., & Yan, C. (2010). Rare earth upconversion nanophosphors: Synthesis, functionalization and application as biolabels and energy transfer donors. *Journal of Rare Earths, 28*(6), 807–819. doi:10.1016/S1002-0721(09)60206-4

Zhang, X. X., Hong, P., Bass, M., & Chai, B. H. T. (1993). Ho^{3+} to Yb^{3+}back transfer and thermal quenching of upconversion green emission in fluoride crystals. *Applied Physics Letters, 63*(19), 2606–2608. doi:10.1063/1.110445

Zhou, D. C., Song, Z. G., Chi, G. W., & Qiu, J. B. (2009). NIR broadband luminescence and energy transfer in Er^{3+}-Tm^{3+}-co-doped tellurite glasses. *Journal of Alloys and Compounds, 481*(1), 881–884. doi:10.1016/j.jallcom.2009.03.141

Zhou, D. L., Hu, Y. B., Qiu, J. B., Yang, Z. W., & Song, Z. G. (2010). Characteristics and mechanism of upconversion luminescence in $Er^{3+}/Yb^{3+}/Tb^{3+}$co-doped oxyfluorogermanateglasses. *Journal of Inorganic Materials*, *25*, 551–556.

Zhou, T. M., Zhang, Y. Q., Wu, Z. L., & Chen, B. J. (2015). Concentration effect and temperature quenching of upconversion luminescence in $BaGd_2ZnO_5$:Er^{3+}/Yb^{3+} phosphor. *Journal of Rare Earths*, *33*(7), 686–692. doi:10.1016/S1002-0721(14)60471-3

Section 3
Emerging Synthesis Techniques of Nano-Materials for Solid State Lighting

In this section, emerging syntheses are reported for solid state lighting phosphors, and their applications for white light emitting diodes and various synthesis techniques are reported.

Chapter 6
White Light Emitting Phosphors for Solid State Lighting

Sivakumar V.
National Institute of Technology, Rourkela, India

Marikumar R.
National Institute of Technology, Rourkela, India

Kasturi S.
National Institute of Technology, Rourkela, India

Satish Kumar S.
National Institute of Technology, Rourkela, India

ABSTRACT

Phosphor materials are an integral and important part in the white light generation in LEDs and have participated in the global warming reduction significantly. In this chapter, the authors focus on different types of phosphor materials belonging to a family of apatites, fluoroapatites, silicates, oxides, phosphates, and borates with different crystal structures. The detailed investigations on their crystal chemical structure, synthesis, and photoluminescence properties are briefly discussed. The improved optical properties make the phosphor potential candidate for smart panel displays and white light emitting diodes (including solid state lighting). The crystal structure has a great influence on the chemical and luminescence properties of any phosphor; hence, a great change of activator ion (Eu^{2+}, Ce^{3+}, Mn^{2+}, and Tb^{3+}) concentrations can be achieved in the phosphor performances. The chapter correlates the structure-compositions-property of the phosphor materials with special emphasis on white LEDs.

INTRODUCTION

In past two decades, research on inorganic luminescent materials have gained significant attention, due to emerging technology with the lighting industry as well as their wide spectral applications, such as cathode ray tubes (CRTs), light emitting diodes (LEDs), field-emission displays (FEDs), vacuum fluo-

DOI: 10.4018/978-1-5225-5170-6.ch006

rescent displays (VFDs), and plasma display panels (PDPs)(Wang, 2013; Xie, 2004; Liu, 2013; Hoppe, 2009; Dai, 2011; G. Li, 2014; K. Li, 2016). Especially significant revolution has occurred in the field of lighting industry (starting from the incandescent bulb to the current era of energy efficient white light emitting diodes (WLEDs). Increasing global energy demand and carbon emission pushes the use of renewable energy or energy saving smart devices. The energy efficient devices or smart displays play vital role in reducing the cost of the energy by half. The discovery of blue LED or solid state lighting based white LEDs considered as a triumph in the field of materials chemistry and subsequently awarded Nobel prize in physics 2011 (Sivakumar & Kasturi, 2017).

The cool white LEDs have successfully replaced the existing florescent lamps as due to their improved performances and lifetime. Further, as time goes the compact fluorescent lamps (CFLs) have been surpassed by the warm white LEDs. In the solid state lighting (SSL) devices (including LEDs), violet, blue, and green LEDs are based upon InGaN semiconductors, while the red and amber LEDs are based upon AlInGaP semiconductors (Phillips, 2007; Krames, 2007; Schubert, 2005). To overcome the disadvantages generated from InGaN and AlInGaN LEDs the use of down converted phosphor or inorganic luminescent materials came into scenario to be used for generating the yellow and green light for highly efficient white LEDs. There has been wide research and development for phosphors used in CFLs, cathode-ray tubes (CRTs), and X-ray films but the most of the traditional phosphors do not meet the need for the LED applications. Since, it fails to absorb either in the near UV or blue region, where the LED emission occurs. There are several ways; one can produce white light from the LED. The selective methods based on phosphor + LEDs are depicted in Figure 1.

The understanding of energy transfer behavior between the sensitizer (D) and activator (A) is important task, as it involves in the overall improvement of the phosphor quantum efficiency as well as the LED performance.

Figure 1. Schematic diagram of LED fabrication with (a) RGB and (b) single host white phosphor

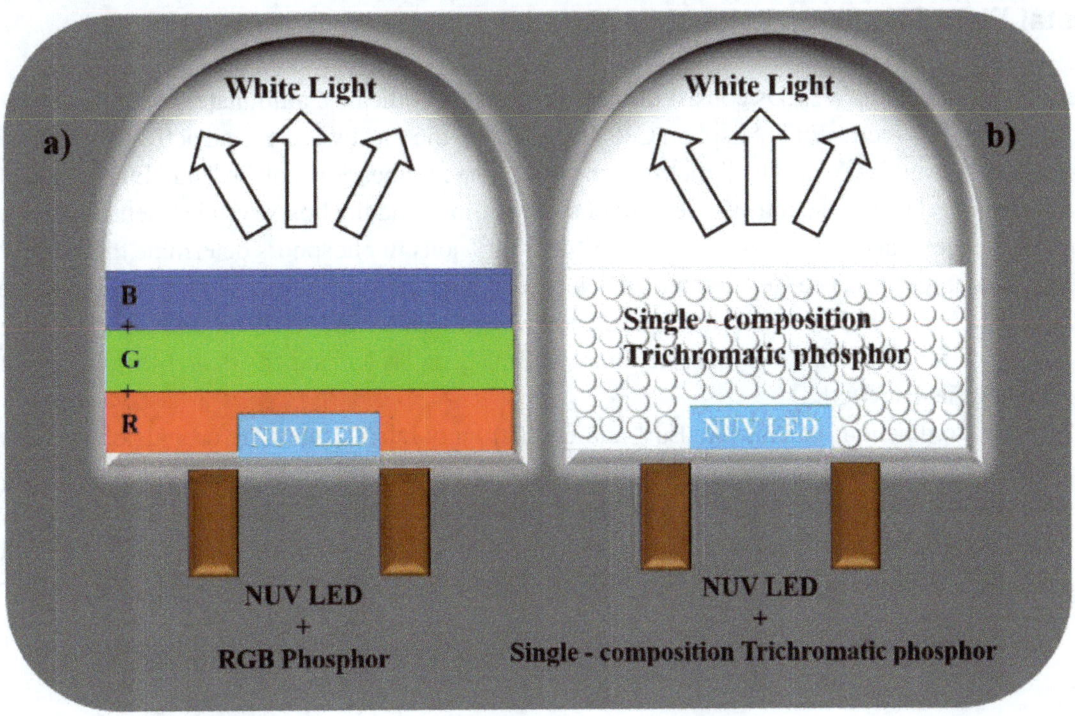

ENERGY TRANSFER

The migration of an excitation energy from one ion to another ion in known as energy transfer process. Energy transfer is one of the important processes in luminescentmaterials; mainly the energy transfer depends upon the sensitizer(D) and activator(A) ion present in the host lattices. The sensitizer species will transfer the excitation energy to the activator species by the following process

1. **Resonant Energy Transfer**: The excitation energy transfer from one ion to another ion is in the same ground state.
2. **Sensitization of Luminescence**: This energy transfer process is frequently used in practical phosphors to enrich the emission efficiency. The sensitizers (D) absorb the excitation energy and very efficiently transfer to the activator(A). The resulting emission is significantly increased. For example, the emission of Tb^{3+} is sensitized by Ce^{3+} ions in many green-emitting phosphors. The schematic energy level diagram is shown in Figure 58.
3. **Concentration Quenching of Luminescent Materials**: An undesirable, and very common, feature of energy transfer is decrease in emission intensities when the emitting ions lie too close to each other.

The concentration quenching depends on the following criteria.

a. Because of the cross-relaxation between the acceptor ions, the excitation energy is lost from the emitting excited state.
b. In order to make the energy reach remote killers or the crystal surface act as quenching centers, excitation migration due to the resonance between the acceptor ions increases with their concentration.
c. The acceptor ions are paired, forming a quenching center.

Crystal-Field Effects on Phosphor Materials

Generally, phosphors in WLEDs consist of inactive hosts like silicates, aluminates, etc. and activators, defined as the "host-activator" phosphor. Ce^{3+} and Eu^{2+} are the most widely used activators whose sole purpose is to convert NUV or blue light to visible light. The luminescence of Ce^{3+} and Eu^{2+} doped phosphors is clearly related to the intrinsic feature of activator ions and the host crystal structure. According to energetics, the 4f and 5d energy levels of Ce^{3+} and Eu^{2+} ions in phosphors determine their excitation and emission properties. The Ce^{3+} ion with $4f^1$electronic configuration shows broadband luminescence due to parity allowed 4f-5d electric dipole transition whereas Eu^{2+} with $4f^7$ configuration has a emission which is partly forbidden due to spin selection rule. Hence, its broadband emission is solely dependent on the crystal field strength.

The 5d wave functions of ions like Eu^{2+} and Ce^{3+} strongly interact with the orbitals of surrounding ligands and ions. Hence, the splitting is generally larger than the spin-orbit coupling. Therefore, the crystal field at an emission center ion in Ce^{3+}- and Eu^{2+}-doped phosphors has much more drastic influence on luminescence properties of the phosphors in comparison to other trivalent rare earth-doped phosphors relying on 4f–4f transitions. Crystal-field splitting depends on several factors, some of which are as follows:

Figure 2. Schematic diagram of 5d levels and emissions of rare earth ions such as Eu^{2+} or Ce^{3+} lying on a center with cubic Oh symmetry

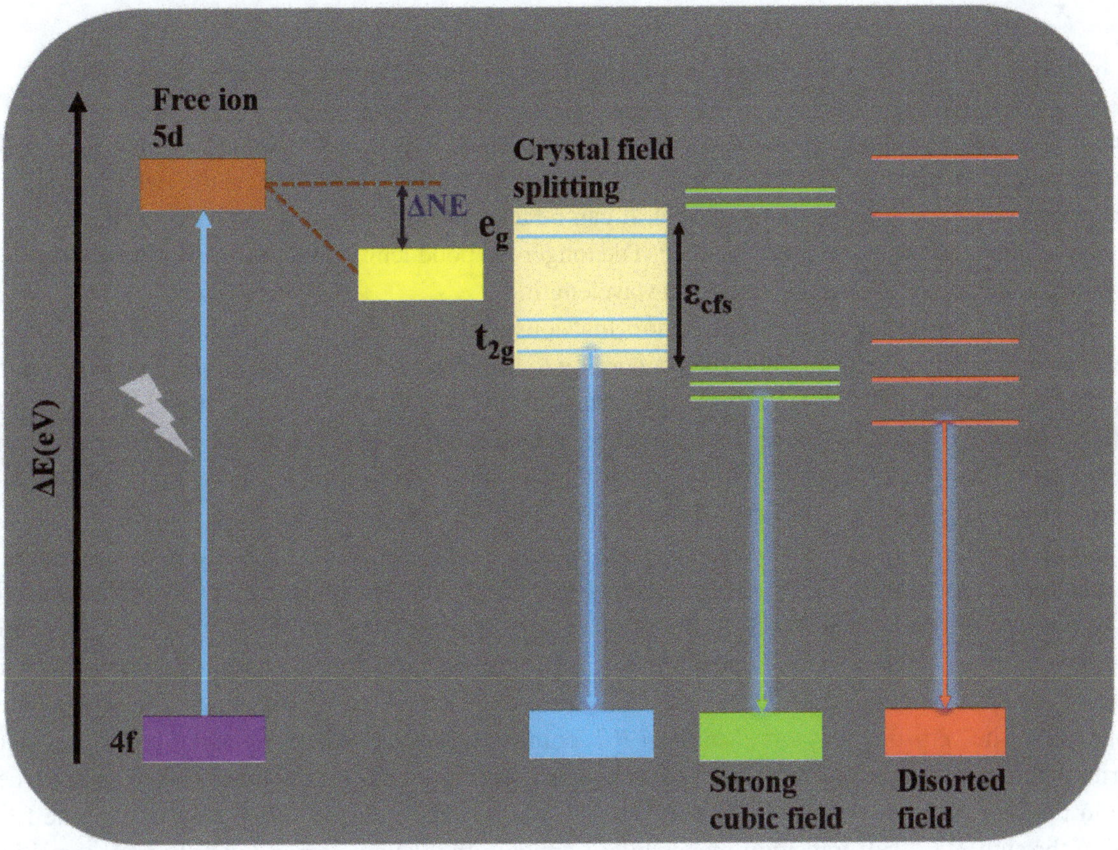

1. Number of electrons in the d orbitals.
2. Oxidation state of the ion (a high oxidation state will lead to a larger splitting).
3. Arrangement of the ligands around the crystal (symmetry).
4. Nature of the ligands.

A schematic diagram of 5d levels and emissions of rare earth ions such as Eu^{2+} or Ce^{3+} lying on a center with cubic Oh symmetry is illustrated in Figure 2. Firstly, the nephelauxetic effect, which is caused by the electron cloud expansion of the ligand towards the metal ion, lowers the energy of the 5d orbitals, which is also known as "centroid shift". Secondly, the five degenerate d orbitals of the octahedral crystal field are divided into the higher energy eg orbitals and the lower energy t_{2g} orbitals. In luminescence phenomena of Eu^{2+}- or Ce^{3+}- activated phosphors, the excitation/emission processes correspond to 4f↔5d electronic transitions. The last section of the figure corresponds to the effect of substitution of a less electronegative atom. For example, nitrogen is less electronegative but more polarizable than oxygen, which increases bond covalency. Inter-electronic repulsion decreases and the nephelauxetic effect increases. As a result, the crystal-filed splitting becomes larger as a consequence of higher charge of nitride. These effects induce important changes in properties.

Generally, the crystal field splitting (D_q) can be determined by the following equation.

$$D_q = \frac{x}{y} Ze^2 \frac{e^4}{R^5}$$

Where, D_q is the magnitude of the 5d energy level separation, Z is the anion charge or valence, e is the electron charge, r is the radius of the d wave function, and R is the bond length. Although this equation was derived by using a point charge model, it can be used as an approximation for describing crystal field splitting trends with the bond distance. The longer the bond length and the weaker the crystal field splitting, and then the shorter the emission wavelength.

Many of the existing commercial phosphors that consist of Eu^{3+}, Mn^{2+} or Tb^{3+} ions as an activator ion, the transitions of these ions are forbidden according to the selection rules causing phosphor quenching due to saturation from the high LED radiation flux on the phosphor. Use of Ce^{3+} and Eu^{2+} phosphors with $5d^1 \rightarrow 4f^1$ Ce^{3+} or $4f^65d^1 \rightarrow 4f^7$ Eu^{2+} electronic transitions have decay times of <100 ns and <3 ms, respectively, it will certainly helpful to overcome the phosphor luminescence quenching. Another major factor is sustaining the emission intensity against the temperature quenching effect. An ideal phosphor should retain it luminescence intensity and chemical stability beyond 150 °C and is the device operating temperature for most of the LEDs. Over past ten years' efforts have been made by many of the researchers to improve the performances and utility of phosphors. During this time, the field of LED phosphors has moved from a single family of phosphor compositions, the Ce^{3+}-doped aluminate garnets to a wide range of silicate, aluminate, nitride, oxynitride, sulfide, and fluoride compositions, leading to commercial LEDs. The very firstly reported in the 1967, commercial garnet based phosphor a Ce^{3+} activated yttrium aluminum garnet ($Y_3Al_5O_{12}$:Ce^{3+}) was widely used in the cathode ray tubes (CRT) prior to their use in the phosphor converted LEDs (pcLEDs). It works in the pcLEDs by absorbing the blue LED region through $4f^1 \rightarrow 5d^1$ transition meanwhile it emits in the yellow region via $5d^1 \rightarrow 4f^1$ transition. The emission from the lowest excited $5d^1$ level is to the spin–orbit split $4f^1$ ground states, leading to an extremely broad emission band with a FWHM > 100 nm. In the YAG:Ce phosphor the blue and yellow emission recombine to give a white light emission with color correlation temperature (CCT) close to that of day light > 4000 K and a color rendering index (CRI) value between 70-80. Hence, this can be used for the backlighting purpose in many of the display devises. The major advantages of the YAG:Ce phosphor are due to their parity allowed broad transitions they can yield strong absorption in the blue region which prevent the phosphor quenching along with a high value of quantum yield (QY) that is > 85% under blue LED excitation at 200°C (Bachmann, Ronda, & Meijerink, 2009). The phosphor is also decay resistant and least affected by the moisture attack. The synthesis of the presently used phosphor is clean and straight forward which used highly purified commercial precursors like Al_2O_3, Y_2O_3 and Ce_2O_3. The only drawback associated with this phosphor material is they fail to mimic the CCT and CRI value of the natural light due to lack of an appropriate red component in their white light emission spectrum. It is possible to tune the $4f^1 \rightarrow 5d^1$ emission towards red shift by substituting the existing Y^{3+} by Gd^{3+} ion or substitution of Al^{3+} by either Mg^{2+} or Si^{4+} (Robertson, Van Tol, Smits, & Heynen, 1981). The restrictions for red emitting Ce^{3+}-activated aluminate garnets directed to further investigation of alternate garnet compositions. In the meanwhile, the silicon based garnets came into scenario. However, any orange–red garnet still has stronger high temperature quenching related to YAG:Ce. As this additional quenching is moderately independent of composition, it may be intrinsic and points to a limitation for Ce^{3+}-doped garnets for orange-red phosphors.

As one of the important constituents in LEDs different color emitting phosphors have been widely explored and investigated to be used in backlighting or displays. It is well acknowledged that Eu^{2+}, Ce^{3+} and Bi^{3+} can emit broad bands with different colors in various hosts based on their $4f^6\,5d^1 - 4f^7$, $5d^1 - 4f^1$ and $^3P_1 - ^1S_0$ allowed transitions (Yuan, 2012; Yu, 2014; Li, 2011).The emission color tuning can be executed in Eu^{2+}, Ce^{3+} and Bi^{3+} singly doped phosphors by ion substitution methods, which have been well discussed in a recent review (Li, Tian, Zhao, & Lin, 2015). However, Eu^{2+}, Ce^{3+} and Bi^{3+} can also act as proficient sensitizers to transfer their energy to activators like Mn^{2+}, Tb^{3+}, Dy^{3+}, Eu^{3+} and Sm^{3+} to produce rich and tunable emission colors including white light through adjusting activator ion concentrations in the host lattice. Hence, the energy transfer process is mostly utilized for color tunability between activator ions in the lattice (Kang, 2014; Guo, 2010; Xie, 2015; Park, 2014; Lee, 2013; Gedam, 2013; Maggay, 2014; Ju, 2012).

The Eu^{2+} ions are choice of activators for most of phosphor components due to their broad excitation and emission band which are originated from the allowed 4f–5d electronic dipole transitions. Several Eu^{2+} activated host materials have been studied such as $Ca_2Al_3O_6F{:}Eu^{2+}$, $K_2Al_2B_2O_7{:}Eu^{2+}$, $Sr_9Mg_{1.5}(PO_4)_7{:}Eu^{2+}$, $Sr_3Ce(PO_4)_3{:}Eu^{2+}$, $BaLa_2Si_2S_8{:}Eu^{2+}$, etc. Eu^{2+} and Mn^{2+} co-doped phosphor is considered as a promising competitor in the use of white LEDs.Eu^{2+} and Ce^{3+} are two essential activators that have been widely explored for LED/SSL applications. The 5d states of Ce^{3+} and Eu^{2+} are outer orbitals, and the coordination environments collectively impact the energies of these ions. In this manner, emissions rising from 5d–4f transitions are wavelength tunable because they are precise to crystal-field splitting and nephelauxetic effects. These qualities of Ce^{3+} and Eu^{2+} particles empower them to have the capacity of encouraging white-light-emission and proven to be potential candidate for the LED applications.

Ce^{3+} activated phosphor for LEDs

$Sr_{2.975-x}Ba_xCe_{0.025}AlO_4F$

The oxyfluoride phosphor Sr_3AlO_4F (SAF) was reported by Kennedy and co-workers (Vogt, 1999; Prodjosantoso, 2003; Park, 2009). SBAF was prepared by conventional high-temperature solid state method. Structurally SBAF belongs to the Cs_3-$CoCl_5$-family (tetragonal, space group *I4/mcm*), (Drofenik & Golic, 1979) and is also closely related to $Ba_3M^{4+}O_5$-type compounds (tetragonal, space group *P4/ncc*) (Glasser & Glasser, (1965). The lattice is crystallize in tetragonal structure with space group *I4/mcm* (S.G. #140), and the cell parameters are $a{=}b{=}6.7720(1)$ Å $c{=}$ 11.1485(2) Å and the unit cell representation is shown in Figure 3. The 8h and 4a sites in this structure are fully occupied by Sr and Ba, respectively, and the 4b site is occupied by Al. F is believed to occupy the 4c site and the 16*l* site is occupied by O. Sr and Ba sites corresponding to polyhedra with 8-coordination (Sr) and 10-coordination (Ba) are both substituted by Ce ions. Both SrO_6F_2 and BaO_8F_2 polyhedra comprise two F atoms and six or eight (respectively) O atoms(Im, Brinkley, Hu, Mikhailovsky, DenBaars, & Seshadri, 2010)

PL spectra of SBAF:Ce^{3+} (x=0) as a function of the Ce^{3+} concentration y with 400 nm excitation are displayed in Figure 4(a). The emission spectrum is broad, with a maximum at about 460 nm. Figure 4b shows the PL excitation and emission spectra of SBAF:Ce^{3+} (x=1.0) at room temperature. The results shown here suggests that the PL intensity (λ_{ex}=400 nm) is about 150%that of commercialYAG:Ce^{3+} (λ_{ex}=460nm). The quantum efficiency of SBAF:Ce^{3+} (x=1) was measured and found to be around 95%

Figure 3. Crystal structure of Sr_2BaAlO_4F (SBAF). Redrawn from ref (Im, Brinkley, Hu, Mikhailovsky, DenBaars, & Seshadri, 2010)

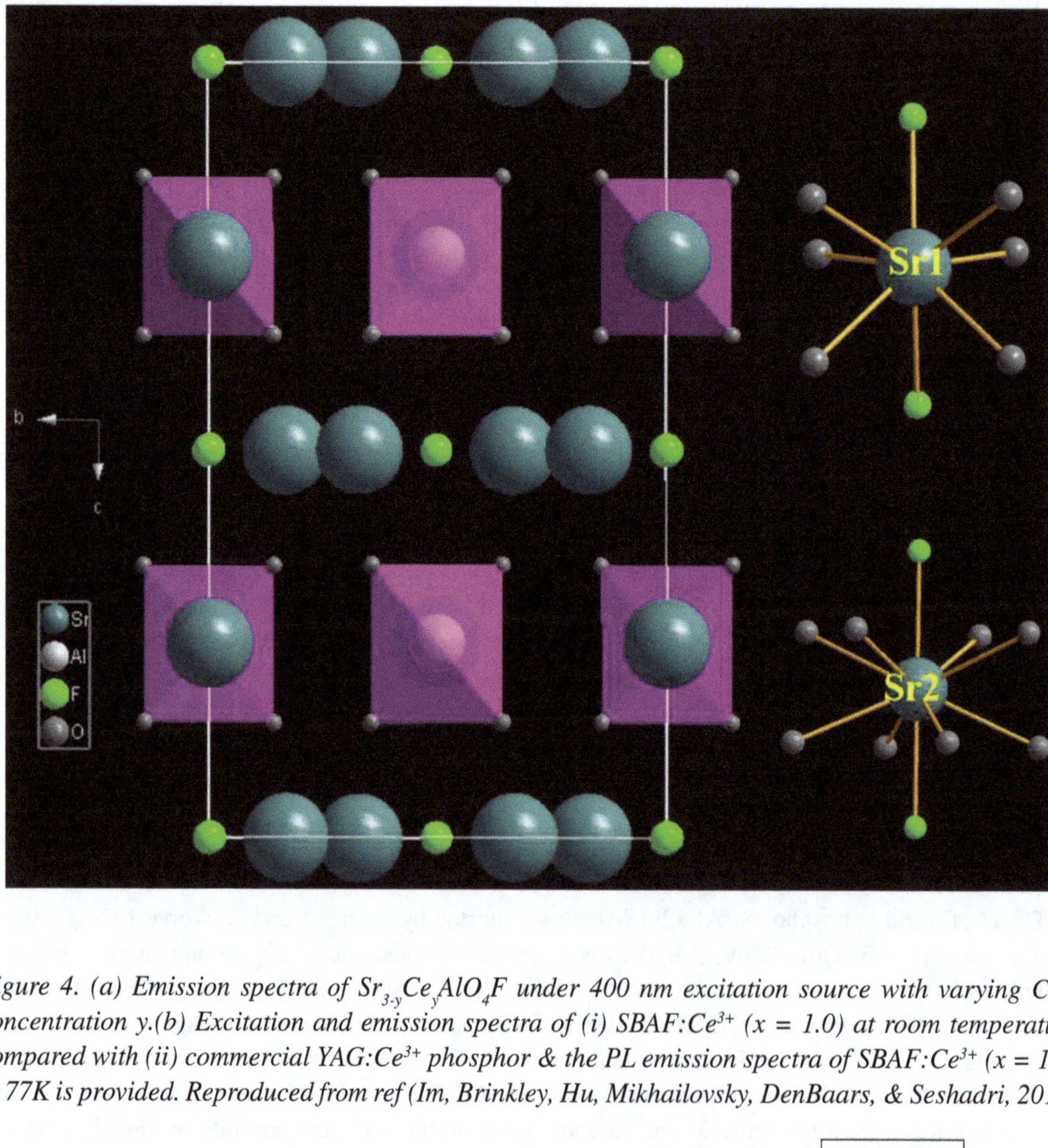

Figure 4. (a) Emission spectra of $Sr_{3-y}Ce_yAlO_4F$ under 400 nm excitation source with varying Ce^{3+} concentration y.(b) Excitation and emission spectra of (i) SBAF:Ce^{3+} (x = 1.0) at room temperature compared with (ii) commercial YAG:Ce^{3+} phosphor & the PL emission spectra of SBAF:Ce^{3+} (x = 1.0) at 77K is provided. Reproduced from ref (Im, Brinkley, Hu, Mikhailovsky, DenBaars, & Seshadri, 2010)

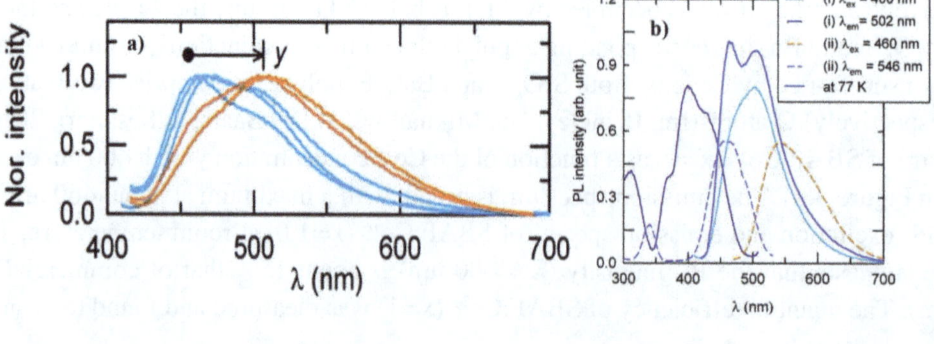

Figure 5. Crystal structure of $Ca_{5.45}Li_{3.55}(SiO_4)_3O_{0.45}F_{1.55}$ and coordination spheres and bond lengths (Å) of the two different Ca^{2+} sites in $Ca_{5.45}Li_{3.55}(SiO_4)_3O_{0.45}F_{1.55}$. Redrawn from ref (Zhou, Xia, Yang, & Shen, 2012)

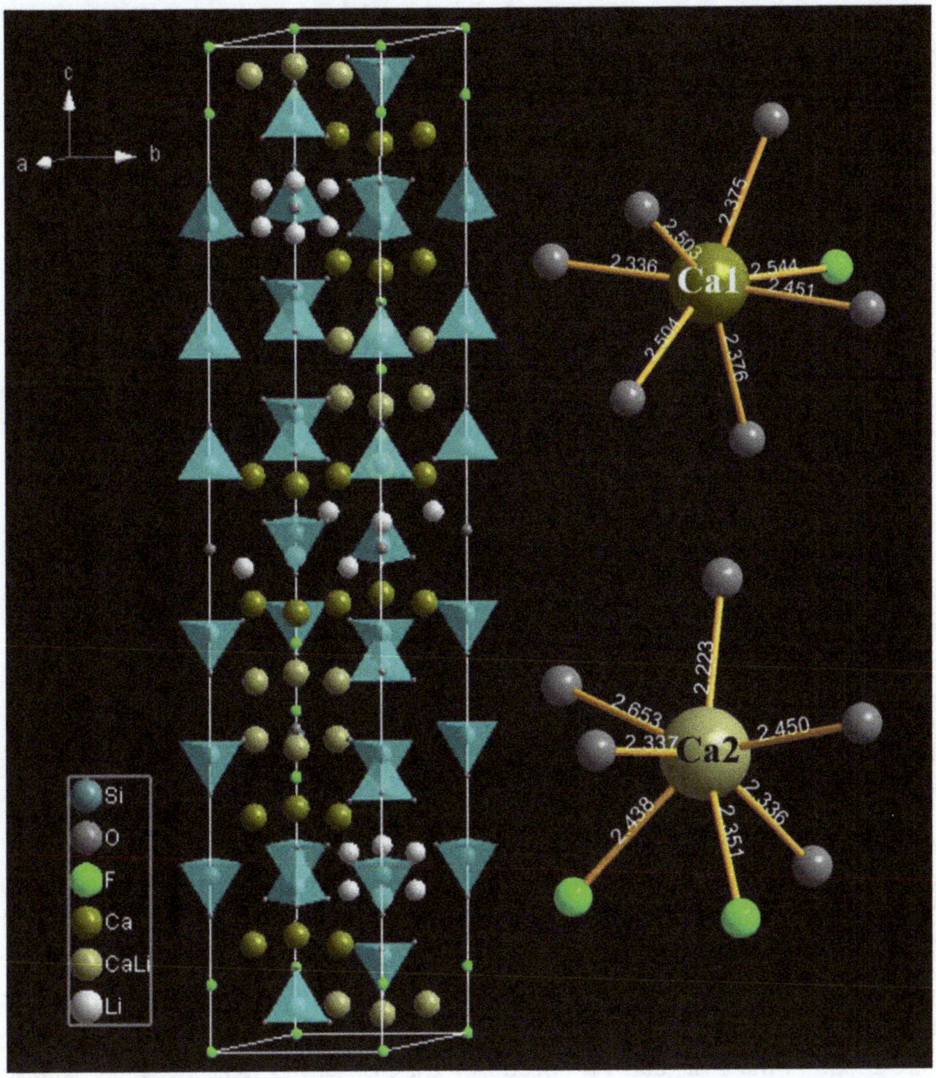

at RT, which is twice in comparison to LSA:Ce^{3+}(Saradhi & Varadaraju, 2006). Upon fabrication of an InGaN LED with the SBAF:Ce^{3+} (x=1.0) phosphor, the observed CIE chromaticity coordinates are (0.28, 0.42) with a CCT of 6900 K at 20mA and R_a of about 62.

$Ca_{5.45}Li_{3.55}(SiO_4)_3O_{0.45}F_{1.55}$:$Ce^{3+}$

In 2010, Krüger first synthesized the new F-containing compound, $Ca_{5.42}Li_{3.55}(SiO_4)_3O_{0.45}F_{1.55}$, as a single crystal and revealed its chemical composition and detailed crystal structure (Krüger, 2010). $Ca_{5.45}Li_{3.55}(SiO_4)_3O_{0.45}F_{1.55}$ phosphor crystallizes in the trigonal crystal structure with space group $R\bar{3}m$ and the lattice parameters are a = 7.1365Å and c = 41.459Å . Figure 5 shows the crystal structure of $Ca_{5.45}Li_{3.55}(SiO_4)_3O_{0.45}F_{1.55}$ and it contains three independent [SiO_4] tetrahedra, and three coordinating

Figure 6. PLE and PL spectra of $Ca_{5.45}Li_{3.55}(SiO_4)_3O_{0.45}F_{1.55}$:$_x$$Ce^{3+}$ phosphors. Reproduced from ref (Zhou, Xia, Yang, & Shen, 2012)

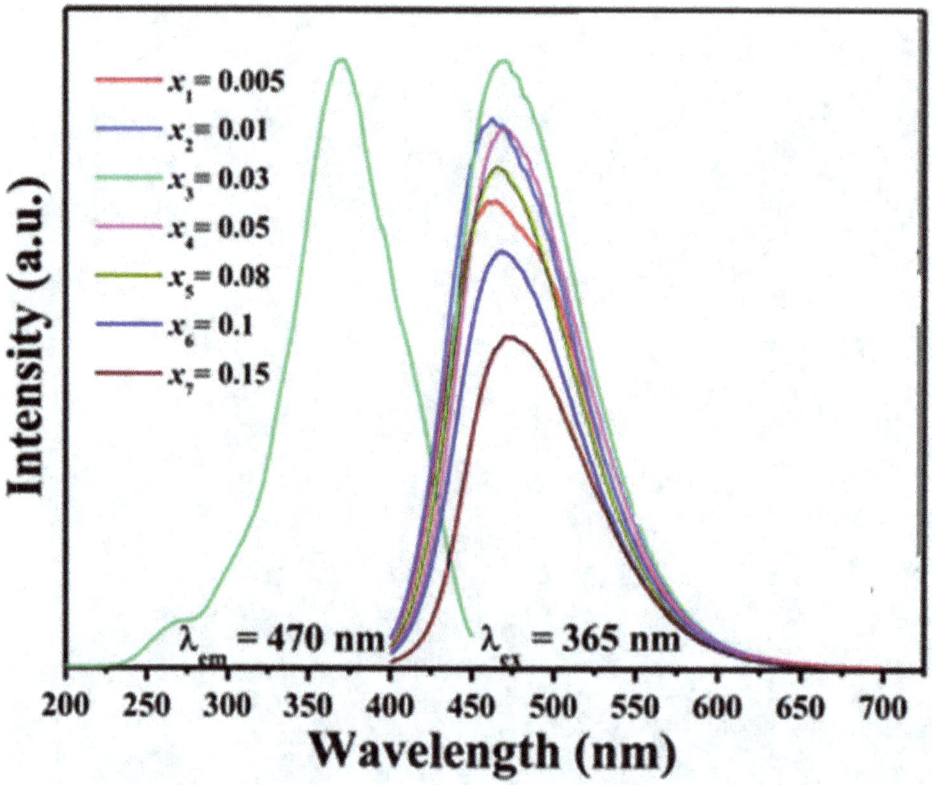

oxygen atoms of each tetrahedron are equivalent by symmetry and lie within a plane perpendicular to the *c* axis. There are two calcium atoms showing the same seven-fold coordination. Ca1 has six oxygen neighbours (2.34–2.50 Å), which all belong to silicate tetrahedra, and the seventh neighbour is a fluorine atom at 2.54 Å. The Ca2 site is coordinated by five oxygen atoms of silicate tetrahedra(Zhou, Xia, Yang, & Shen, 2012).

The Figure 6 shows the PLE and PL spectra of $Ca_{5.45}Li_{3.55}(SiO_4)_3O_{0.45}F_{1.55}$:$_x$$Ce^{3+}$ phosphors. The phosphors show strong broad absorption bands in the UV range, and bright blue emissions are obtained upon 365 nm near UV excitation. The strong broad absorption appeared in the range from 200–420 nm n-UV range, which is attributed to the 4f–5d electronic transition absorption of the Ce^{3+}. The absorption intensities increase progressively with increasing Ce^{3+} concentration. It has been verified that the Ce^{3+} luminescence centre at 470 nm originates from the seven-coordinate Ca^{2+} sites in the lattice. The optimal doping concentration of Ce^{3+} is found to be 3 mol%, and the concentration quenching mechanism was determined to be dipole–dipole interaction. Upon excitation at 365 nm, the composition-optimized $Ca_{5.42}Li_{3.55}(SiO_4)_3O_{0.45}F_{1.55}$:$0.03Ce^{3+}$ exhibited a high QY of 82%, and its luminescence intensity at 150 °C decreases to 71% of the initial value at RT. Hence, $Ca_{5.45}Li_{3.55}(SiO_4)_3O_{0.45}F_{1.55}$:$Ce^{3+}$ is a promising candidate as a blue-emitting n-UV convertible phosphor for application in white LEDs (w-LEDs).

$Ca_2LaZr_2Ga_3O_{12}$:Ce^{3+}

Figure 7. (a) Crystal structure of $Ca_2LaZr_2Ga_3O_{12}$ and (b) the coordination environment of cations. Redrawn from ref (Zhongn et al., 2015)

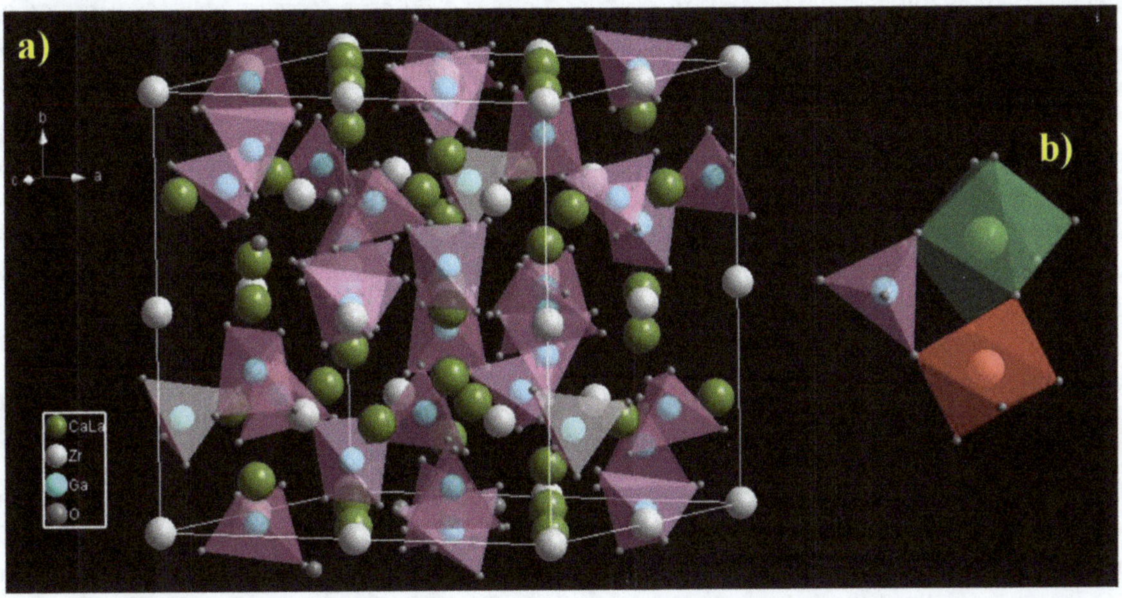

Figure 8. (a) PLE and PL spectra of $Ca_2La_{0.94}Zr_2Ga_3O_{12}$:$0.06Ce^{3+}$ phosphor.(b) CIE coordinates of $Ca_2La_{1-x}Zr_2Ga_3O_{12}$:$xCe^{3+}$ (x = 0.02– 0.2). Reproduced from ref (Zhongn et al., 2015)

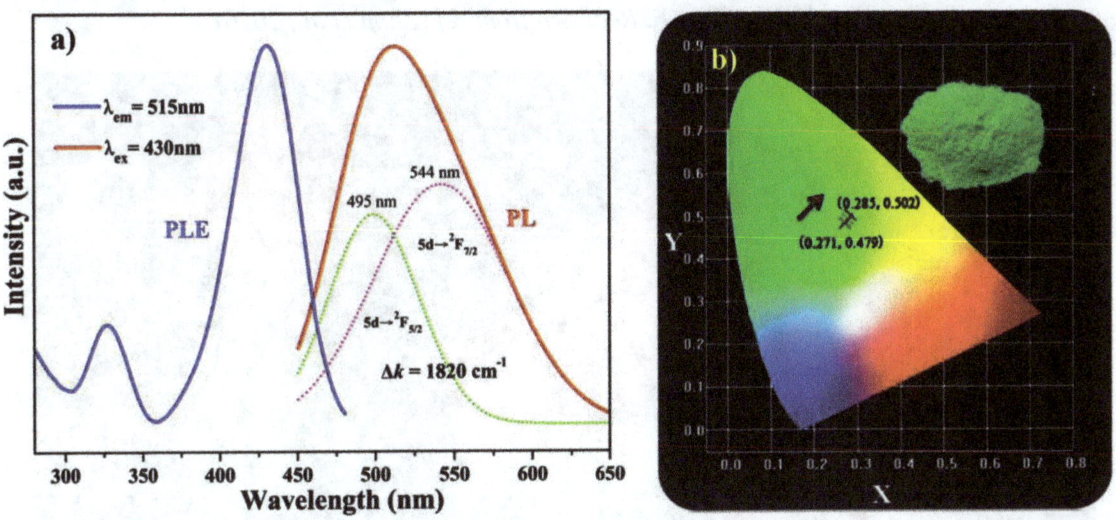

$Ca_2LaZr_2Ga_3O_{12}$ was synthesized by conventional solid state method. $Ca_2LaZr_2Ga_3O_{12}$ crystallizes in cubic symmetry in the space group $Ia\overline{3}d$ with lattice parameters of $a = b = c = 12.7539$Å and $\beta = 90°$. Figure 7 represents the crystal structure of $Ca_2LaZr_2Ga_3O_{12}$. Zr^{4+} and Ga^{3+} are octahedral and tetrahedral coordinated, respectively. The $[ZrO_6]$ octahedrons and $[GaO_4]$ tetrahedrons are connected by O^{2-} points, and they share edges or corners with [Ca/LnO$_8$] dodecahedra to form a three dimensional network, which can be observed clearly from Figure 7b(Zhongn et al., 2015).

The excitation (PLE) and emission (PL) spectra of a selected $Ca_2La_{1-x}Zr_2Ga_3O_{12}$:x$Ce^{3+}$ (x = 0.06) phosphor is shown in Figure 8a. The photoluminescence results revealed that these phosphors have a broad excitation band in the blue region and a broad green emission band centered at about 515 nm, and both of the excitation and emission spectra show red shift with Ce^{3+} content increasing, mainly due to large crystal field splitting. Temperature- dependent PL studies reveal that the emission intensity decreases rapidly as temperature increases, which can be ascribed to smaller band gap and higher energy levels of excited 5d-states. The CIE chromaticity coordinates for $Ca_2La_{1-x}Zr_2Ga_3O_{12}$:x$Ce^{3+}$ phosphors excited at 430 nm were determined and the CIE coordinates are calculated to be ranged from (0.271, 0.479) to (0.285, 0.502) (depicted in Figure 8b). Also the internal quantum efficiency (QE) of $Ca_2La_{0.94}Zr_2Ga_3O_{12}$:0.06$Ce^{3+}$ phosphor is determined to be 35.2% under 430 nm excitation.

$LiY_9(SiO_4)_6O_2$:Ce^{3+}

As mentioned earlier, the tuneable 4f–5d emissions characteristics of Ce^{3+} ion in different host lattices have gained applications in phosphors for smart lighting and displays (George, 2013; Xia, 2015). Due to the great impact of coordination environment on the outer 5d state energies, different emissions of Ce^{3+} can be realized through appropriate selection of suitable host compounds. Hence tuning of spectral emissions can be achieved with ideal coordination environments of Ce^{3+} via crystal chemical substitution with in the host lattice [16]. $LiY_9(SiO_4)_6O_2$ system belongs to the class of apatite compounds

Figure 9. Crystal structure of $LiY_9(SiO_4)_6O_2$. Redrawn from ref (Zhou et al., 2016)

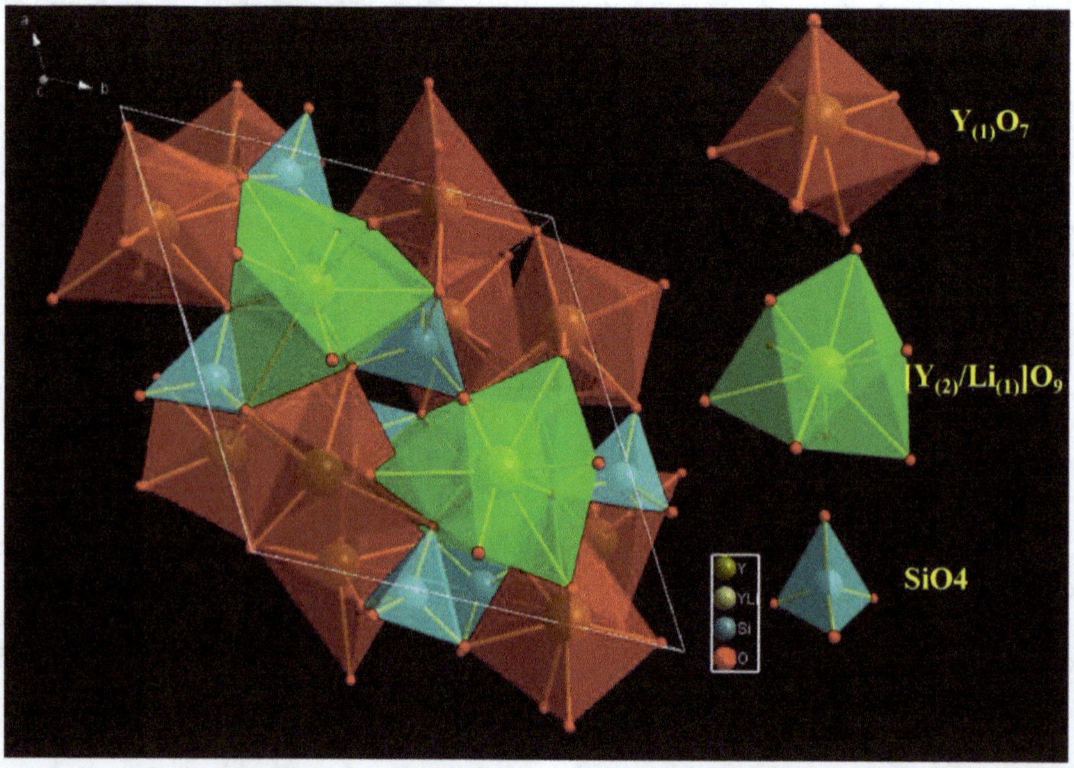

$M_{10}(AO_4)_6B_2$ (M = Ca, Sr, Ba, La, Y, ...; A = P, V, Si, ...; B = F, Cl, Br, OH, ...), which has two types of Y^{3+} sites (Wyckoff 4f and 6h) (Redhammer & Roth, (2003). $LiY_9(SiO_4)_6O_2$ has a hexagonal structure with space group $P6_3/m$ and the lattice parameters are a = b = 9.3287(1)Å, c = 6.7134(1)Å, V = 505.97(1) Å^3. Due to possible diverse emission wavelengths of Ce^{3+} at these two sites, and due to the collective influences of site occupancies, energy transfer, and thermal quenching of Ce^{3+} at two sites, the tuneable emission is expected to be achieved by either varying the doping concentrations of Ce^{3+} or changing the temperature of materials.

Figure 9 shows the crystal structure of $LiY_9(SiO_4)_6O_2$. All polyhedra around Y^{3+} ions share their edges or corners with the $[SiO_4]^{4-}$ tetrahedra, creating a structural framework. In the $Y(1)O_7$ polyhedron of the 6h site there is the so-called free oxygen ion $O(1)^{2-}$ which fails to participate in the formation of an $[SiO_4]^{4-}$ tetrahedron. The bond length (2.193 Å) between this free oxygen ion $O(1)^{2-}$ and $Y(1)^{3+}$ is relatively shorter than those of other Y(1)–O bonds, which indicates more covalency of this Y(1)–O(1) (free) bond. This nature may have a significant influence on the luminescence of Ce^{3+} at the $Y(1)^{3+}$ sites. The percentage of Ce^{3+} ions in 6h sites increases but that in 4f sites decreases with an increase in doping concentration, implying preferential occupancies of Ce^{3+} ions in 6h sites at high-doping levels(Zhou et al., 2016).

The PL excitation and emission of $LiY_{8.995}Ce_{0.005}(SiO_4)_6O_2$ shown in Figure 10. In the photoluminescence studies, with an increase in wavelength of excitation, the corresponding emission spectra show different shapes and their maximum positions gradually shift toward higher wavelengths. The typical emission spectra under 280, 300 and 340 nm excitations imply that two types of Ce^{3+} luminescence exist. The energies of the two band maxima are at ~2.65 eV (~468 nm) and ~2.92 eV (~425 nm), which can be assigned to the doublet emission bands from the relaxed lowest 5d state to the $^2F_{5/2}$ and $^2F_{7/2}$ 4f levels, respectively (Figure 10). With an increase in Ce^{3+} concentration, the joint influences of site occupancies, energy transfer, and emission band broadening push the total emission of Ce^{3+} toward the longer wavelength side, result in the concentration-dependent emission properties of phosphors. Moreover, the phosphors also possess tuneable emission from cyan to blue with an increase in temperature. Through temperature-dependent luminescence decay measurements for these two Ce^{3+} ions, the emission from $Ce(1)^{3+}$ is found to be more prone to thermal quenching in comparison to $Ce(2)^{3+}$.Because of the tuneable

Figure 10. Excitation and emission of $LiY_{8.995}Ce_{0.005}(SiO_4)_6O_2$ phosphor. Reproduced from ref (Zhou et al., 2016)

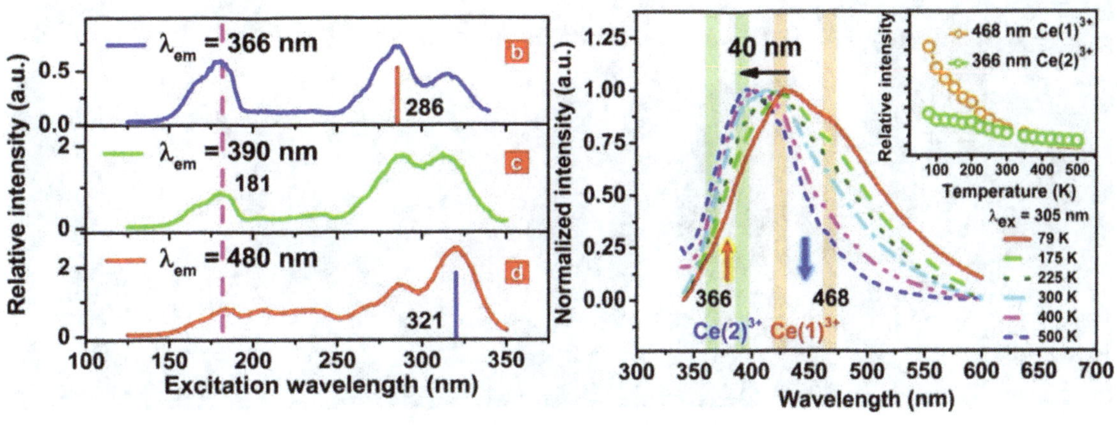

emissions with increasing temperature, the phosphors $LiY_9(SiO_4)_6O_2:Ce^{3+}$ are endowed with possible thermometric applications.

$Ca_{10}(PO_4)_6O:Ce^{3+}$

Apatite compounds are one of the most attractive host materials due to their facile synthesis, low cost, and high physical and chemical stability (Zhai, 2014; Tao, 2013). This structure commonly contains two kinds of cation sites, the nine-coordinated 4f sites with C3 point symmetry and the seven- or eight-coordinated 6h sites with Cs point symmetry, and both sites are easily occupied by a variety of rare earth ions and transition metal ions (Li, 2011; Lv, 2013; Qi, 2014). Moreover, the coordination environments of the 4f and 6h sites could be affected by changes to the composition of the compound, generating a color-tunable emission.

$Ca_8Ce_2(PO_4)_6O$ crystallizes in a hexagonal phase with space group $P6_{3/m}$ (173), $a = b = 9.4367$Å, $c = 6.9114$Å, $\alpha = \beta = 90°$, $\gamma = 120°$, $V = 533.02$Å^3 and $Z = 1$ and the crystal structure is depicted in Figure 11. Ca1 and Ca2 atoms were found to be located at 4f sites and each coordinate with nine oxygen atoms to form tetrakaidecahedrons. Although Ca1 and Ca2 each coordinated nine oxygen atoms,

Figure 11. Crystal structure of $Ca_{10}(PO_4)_6O$ along the c-axis. Redrawn from ref (Li et al., 2016)

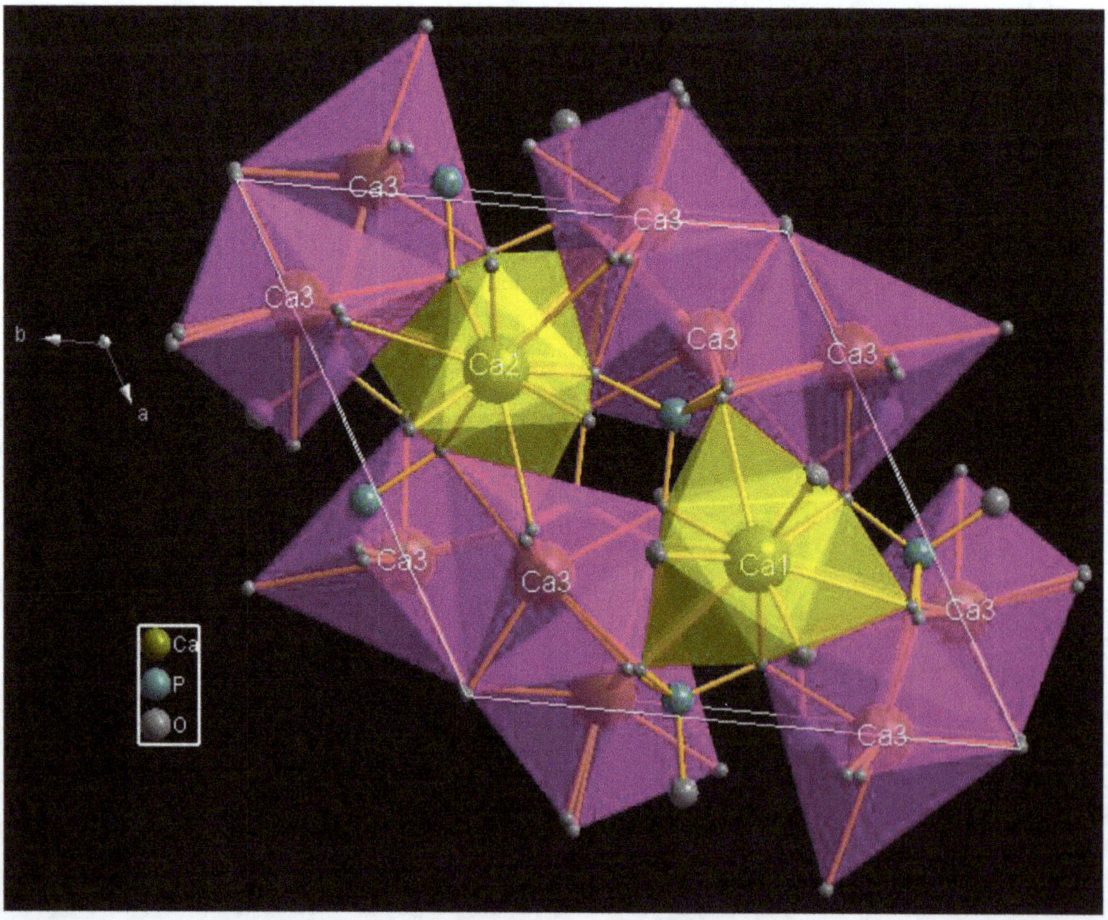

their different observed Ca–O bond lengths lead to different crystal field strengths. Along the c axis, any two tetrakaidecahedrons were observed to be connected to each other by sharing a triangular plane, constructing a 4f site tunnel. While the Ca3 atom located at the 6h site formed a decahedron with its surrounding seven oxygen atoms, and such decahedrons in the lattice were connected to each other by sharing vertices. Finally, tetrakaidecahedrons were observed to be connected to neighboring decahedrons through tetrahedral PO_4 groups and sharing edges(Li et al., 2016).

Figure 12. (a) The normalized photoluminescence excitation (PLE, λ_{em} = 410–510 nm). (b) photoluminescence (PL, λ_{ex} = 330–355 nm) spectra of $Ca_{10(1-x)}Ce_{10x}(PO_4)_6O$ (x = 0–0.30) samples. (c) Photographs of luminescent $Ca_{10(1-x)}Ce_{10x}(PO_4)_6O$ (x = 0–0.30) under 320 nm-wavelength UV light. Reproduced from ref (Li et al., 2016)

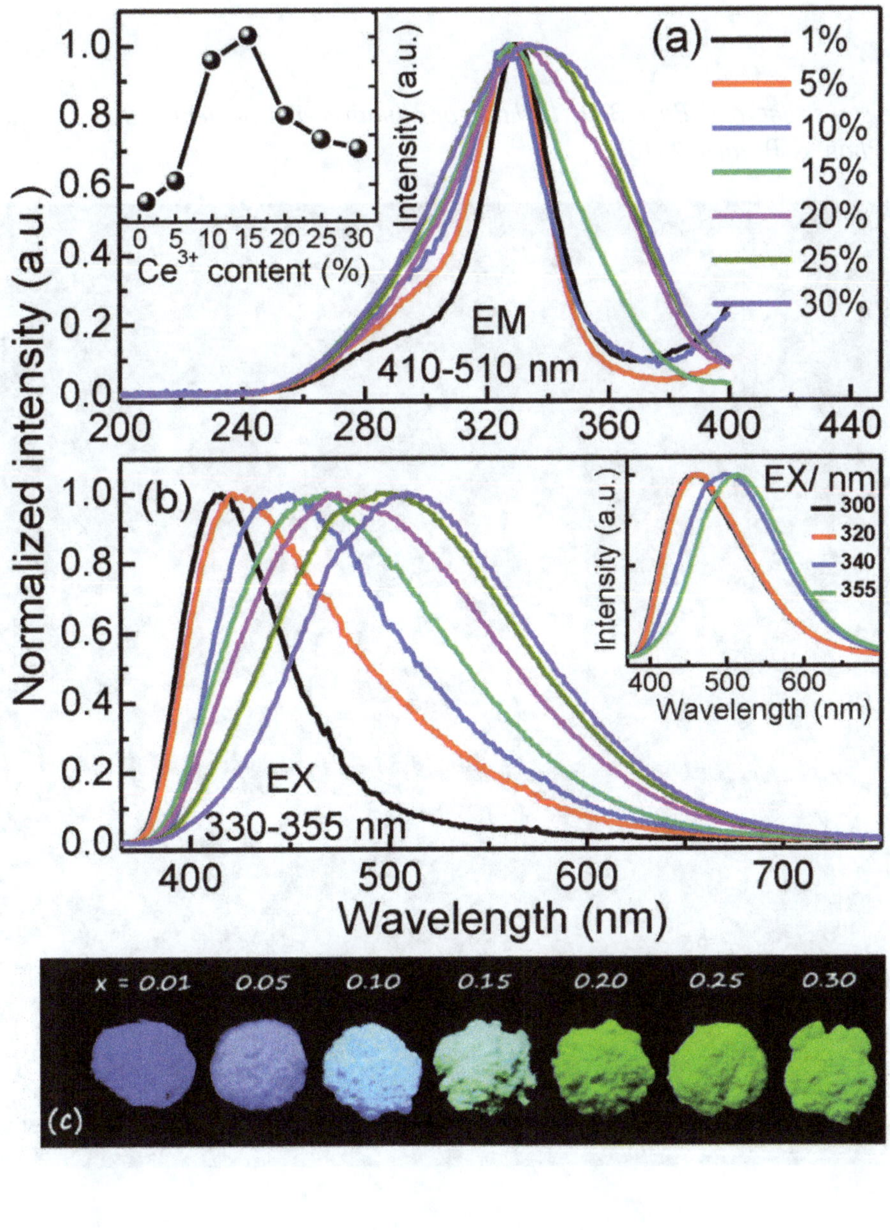

Under near-UV (n-UV) excitation, the colors emitted by the luminescent $Ca_{10}(PO_4)_6O:Ce^{3+}$ phosphors could be continuously tuned from blue light (410 nm) to yellowish-green light (510 nm) with increasing levels of doped Ce^{3+}. This effect may have been due to a changing preference of the crystallographic site that was occupied by doped Ce^{3+} for the different concentrations of Ce^{3+}. The PLE spectrum of CPO:0.01Ce and its absorption spectrum showed one broad band of absorption from a wavelength of 270 nm to 360 nm centered at 330 nm, which can be attributed to the electronic transitions from the 4f ground state of Ce^{3+} ($^2F_{5/2}$ and $^2F_{7/2}$) to its 5d excited state. Under 330 nm wavelength UV light, CPO:0.01Ce exhibited an asymmetric emission band (blue color) ranging from 370 to 500 nm and peaking at 410 nm giving a highly pure blue-light emission with CIE color coordinate (0.165, 0.056) as depicted in Figure 12. The thermal stability of CPO:xCe gradually worsened with the increasing Ce^{3+} concentration. However, the thermal stability at 423 K was still 76% of the room-temperature emission intensity for x = 0.15 and 51% for x = 0.20.

$Ba_2Y_5B_5O_{17}:Ce^{3+}$

Figure 13. Crystal structure of $Ba_2Y_5B_5O_{17}$ and the coordination environment of Ba and Y. Redrawn from ref (Hermus, Phan, & Brgoch, 2016)

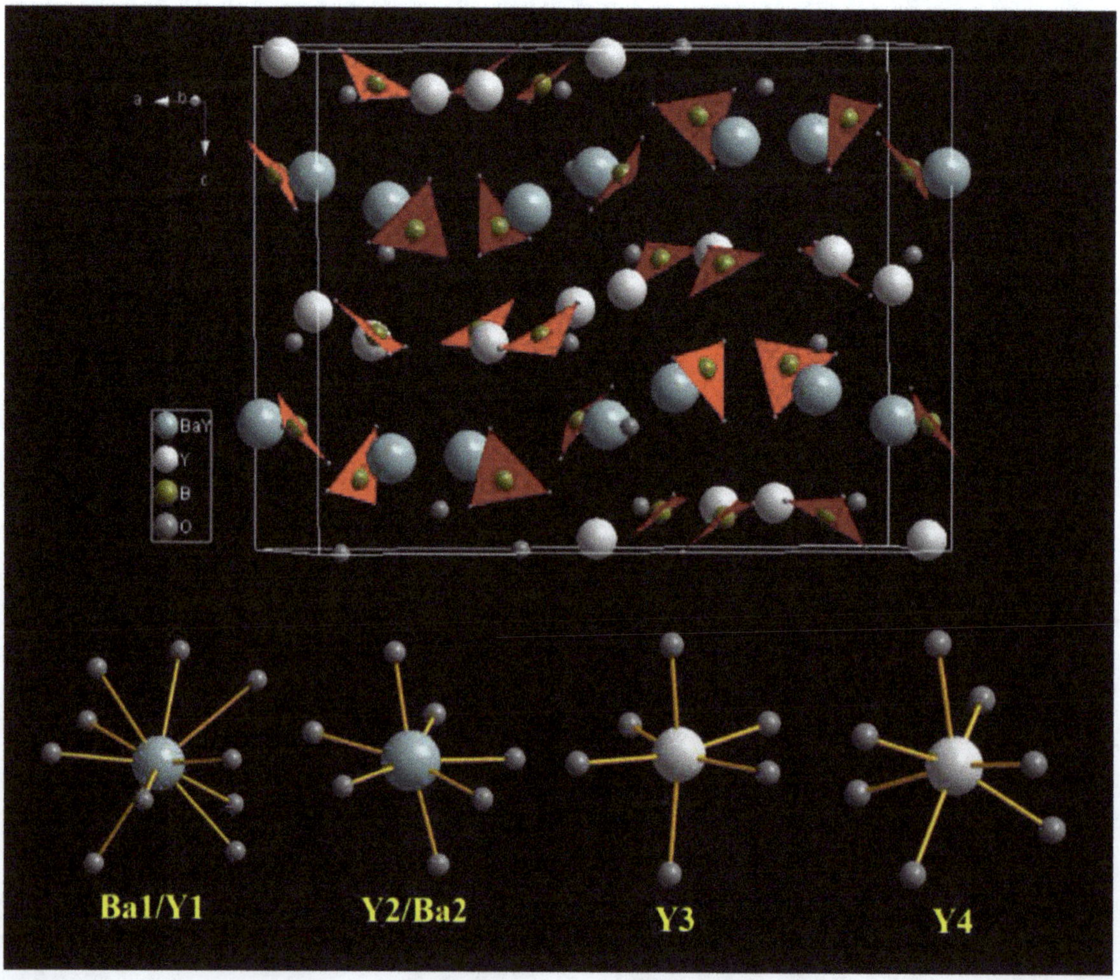

$Ba_2Y_5B_5O_{17}$ was synthesized by conventional solid state method under a reducing atmosphere. $Ba_2Y_5B_5O_{17}$ crystallizes in an orthorhombic unit cell with space group *Pbcn* and lattice parameters *a* = 17.3946(5)Å, *b* = 6.6586(2)Å, and *c* = 13.0486(4)Å. The crystal structure obtained is shown in Figure 13. The structure consists of edge- and corner-sharing Y- and Ba-centered polyhedra, which build a three-dimensional net. There are two crystallographically independent, fully occupied Y^{3+} sites and a third Y^{3+} site that is shared with Ba^{2+}. These coordination polyhedra form a distorted octahedron around atom Y3, a distorted capped trigonal prism around atom Y4, and a distorted pentagonal bipyramid around the mixed Y2/Ba2 site, as depicted in Figure 13(Hermus, Phan, & Brgoch, 2016).

The excitation spectrum contains three distinctive peaks (λ_{ex} = 265, 308, and 350 nm) as shown in Figure 14a. The emission spectra collected at all three excitation, the emission maxima are found to be the same, indicating these peaks likely arise from the excitations between the Ce^{3+} 4f orbitals to multiple Ce^{3+} excited states (5d orbitals). Under 308 nm excitation, the emission spectrum shows a rather broad emission band with a peak maximacentered at 443 nm and the corresponding CIE shown in Figure 14b. The photoluminescence is highly efficient (PLQY = 70%; λ_{ex} = 365 nm) at RT, while the temperature-dependent luminescence measurements show this new borate is capable of operating at temperatures approaching 400K with a minimal impact on the performance of the materials.

β-$NaSrBO_3$:Ce^{3+}

As mentioned earlier that the most important prerequisite for a LED phosphor is that it should be thermally stable and its luminescence property should not get much affected by temperature. But most phosphors undergo phase transformation or have a thermal quenching at a high temperature of ~150°C. Hence, for practical application of phosphors, stabilization of host structure and temperature-dependent photoluminescence property is required.The alkali borate α-$NaSrBO_3$reported by Yi et al., (2016) underwent phase transition at a temperature >100°C. As per requirement, β-$NaSrBO_3$ was synthesized using traditional high-temperature solid state reaction method. β-$NaSrBO_3$ crystallized in a monoclinic unit cell in the space group $P2_1/c$ with lattice parameters *a* = 6.06214(8)Å, *b* = 5.41005(7)Å, *c* = 9.1468(1) Å, and β = 102.116(1)°. As depicted in Figure 15, the fundamental building units of β-$NaSrBO_3$ are the parallel distributed planar $[BO_3]^{3-}$ anionic groups. Both Na2 and Sr2 (collectively denoted M2) and

Figure 14. (a) Excitation and emission (λ_{ex} = 308 nm) spectra at room temperature of $Ba_2Y_5B_5O_{17}$:Ce^{3+}. (b) The blue emission of $Ba_2Y_5B_5O_{17}$:Ce^{3+} shows CIE coordinates of (0.159, 0.155). Reproduced from ref (Hermus, Phan, & Brgoch, 2016)

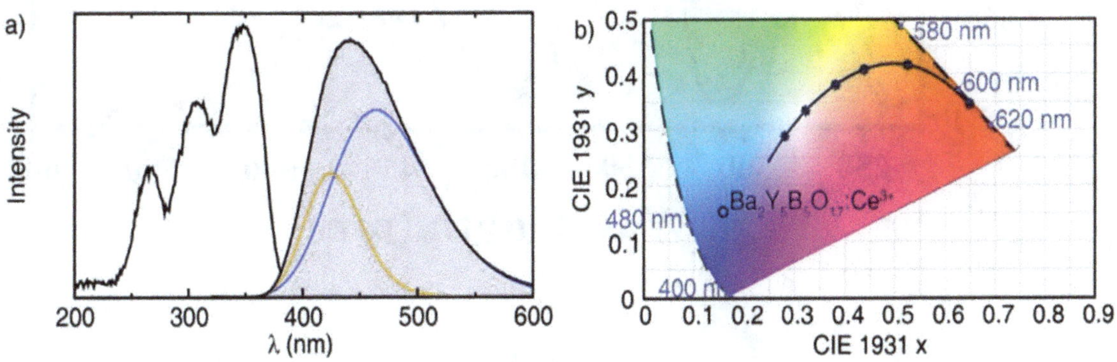

Figure 15. Crystal structure projection of β-$NaSrBO_3$ along different directions. Redrawn from ref (Yi et al., 2016)

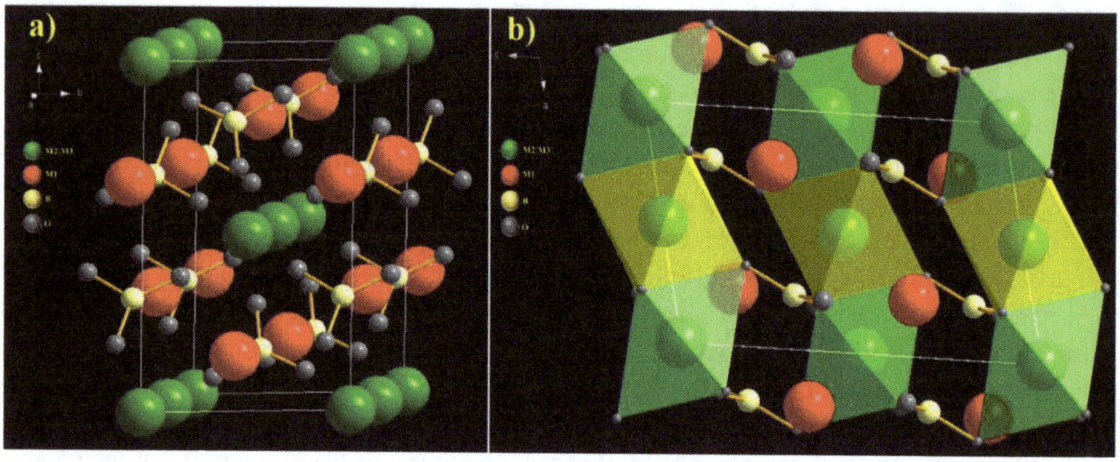

Figure 16. Excitation (λ_{em} = 424 nm) and emission (λ_{ex} = 358 nm) spectra of α $NaSrBO_3$:0.01Ce^{3+} (dashed lines) and β-$NaSrBO_3$:0.05Ce^{3+} (solid lines). Reproduced from ref (Yi et al., 2016)

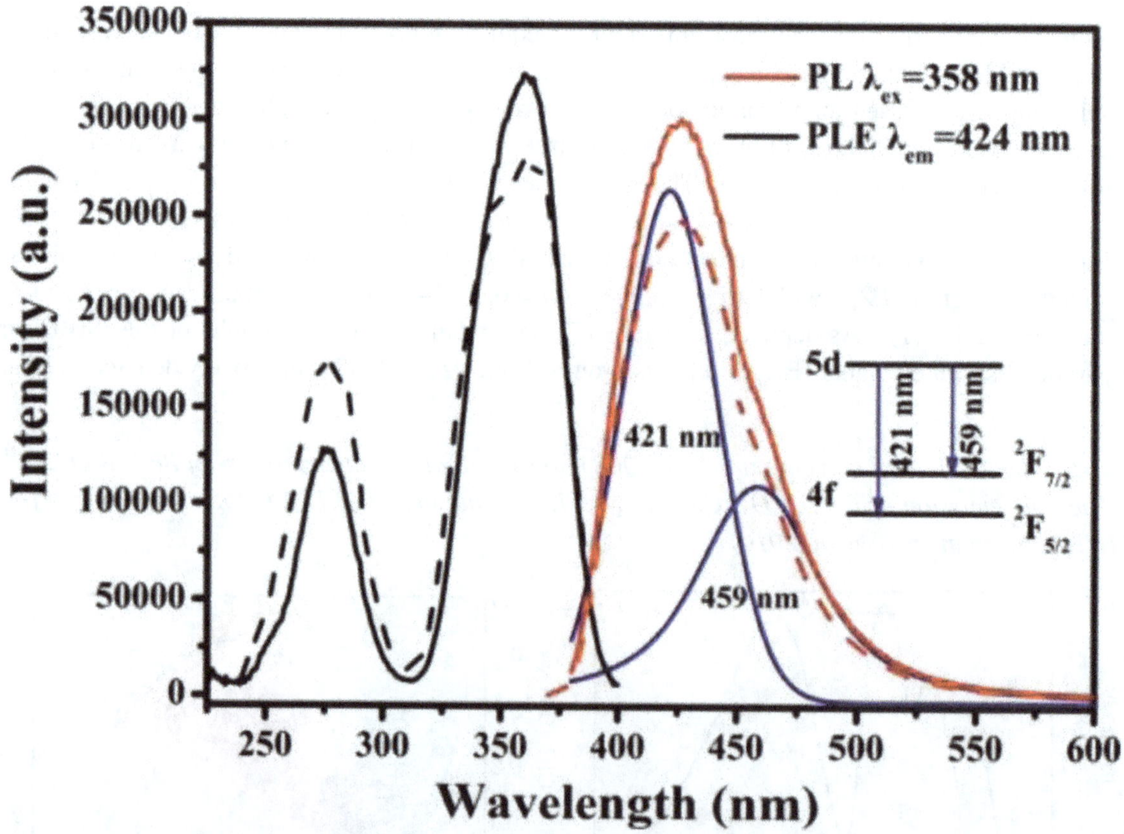

Na3 and Sr3 (collectively denoted M3) atoms are coordinated with six oxygen atoms, forming distorted octahedra. The $M2O_6$ and $M3O_6$ octahedra share faces and alternately align along the a-axis, forming an infinite chain. The neighboring BO_3 planar triangles share corners with the chains. There are nine O atoms around the Sr1 and Na1 (collectively denoted M1) atoms, forming tricapped trigonal prisms. Sr and B atoms are arranged alternately along the a-axis. The $M2O_6$ and $M3O_6$octahedra and $M1O_9$polyhedra share edges with each other, forming an infinite three-dimensional network(Yi et al., 2016).

The PLE and PL spectra of β-$NaSrBO_3$:$0.05Ce^{3+}$ is shown in Figure 16. Two main broad excitation bands peaking at 275 and 358 nm are attributed to the crystal field splitting of Ce^{3+}. PLE spectra verify that β-$NaSrBO_3$:Ce^{3+} has an absorption range that closely matches the emission of NUV-LEDs. The PL spectrum consisting of a broad asymmetric emission band peaking at 424 nm is associated with the spin-allowed d−f transitions of Ce^{3+}. The QE of β- $NaSrBO_3$:$0.05Ce^{3+}$ is determined to be 43.9%. The temperature- dependent PL intensities at 75 and 150 °C remain 75.8 and 51.2% of that at RT, respectively. α- $NaSrBO_3$ will transform into β-$NaSrBO_3$ when the temperature is >117.8 °C, which corresponds to the LED working temperature. Hence, β-$NaSrBO_3$:$0.05Ce^{3+}$ shows thermal stability better than that of α-$NaSrBO_3$:$0.01Ce^{3+}$. Upon illumination at 250mA, the fabricated w-LED of β- $NaSrBO_3$:$0.05Ce^{3+}$ has optical parameters of white light generation: CIE (x = 0.3821, y = 0.3430), correlated color temperature (CCT = 3654 K), and average color-rendering index (R_a = 92.8).

$Ba_2(Y_{1-x}Lu_x)_5B_5O_{17}$:$Ce^{3+}$

One of the more intriguing borates reported, $Ba_2Y_5B_5O_{17}$:Ce^{3+}, produces a bright blue emission ($\lambda_{em, max}$ = 443 nm) under UV radiation excitation (λ = 365 nm) with PLQY of 70% (Hermus et al., 2016). This phosphor can also be co-substituted with Ce^{3+} and Tb^{3+} to yield an efficient green emitting material (Xiao et al., 2017). The high efficiency arises from the densely packed Y−O polyhedral units and high-energy B−O phonon modes, which give rise to superior structural rigidity. Unfortunately, the Ce^{3+} substituted phosphor does not perform well at elevated temperatures with the emission decreasing to 50% of its low temperature (80 K) integrated intensity by only 408 K (Hermus et al., 2016). Therefore, to consider incorporating this borate phosphor in a device, it is necessary to improve the PLQY as well as increase the thermal quenching temperature to >423 K, which is considered the normal operating temperature for most LED packages. $Ba_2Lu_5B_5O_{17}$:Ce^{3+} is a potential substitute for the garnet phosphor. The partial substitution of the rare-earth ions by Ce^{3+} yields an efficient, thermally stable blue emitting phosphor when excited by UV light. Varying Lu^{3+} and Y^{3+} content causes a slight shift in the emission and negligible changes to the excitation bands. More importantly, atomic substitution leads to an improvement in the PLQY as well as the thermal stability, with a 49 K increase in the quenching temperature across the solid solution.

$Ba_2Lu_5B_5O_{17}$:Ce^{3+} crystallizes isostructurally to the parent compound, $Ba_2Y_5B_5O_{17}$, with space group *Pbcn* (No. 60) (Hermus et al., 2016). $Ba_2Y_5B_5O_{17}$ also shows atomic mixing on the Ba-site (Wyck. *8d*), with that compound containing 13% Y^{3+} on the 8d site. The crystal structure, illustrated in Figure 17, contains lutetium and barium centered polyhedra as well as boron centered trigonal planar units. There are four crystallographically independent metal atom sites; one Ba site forms a 10-coordinate, highly distorted polyhedra (Ba1), and Lu2 sits on Wyckoff site 4c and is shared with Ba2 at the center of a distorted pentagonal bipyramid. Lu also occupies a distorted octahedron (Lu3) and a distorted capped trigonal prism (Lu4)(Hermus et al., 2017).

Figure 17. Crystal structure of $Ba_2Lu_5B_5O_{17}$:Ce^{3+} with different polyhedral units highlighted. Redrawn form ref (Hermus et al., 2017)

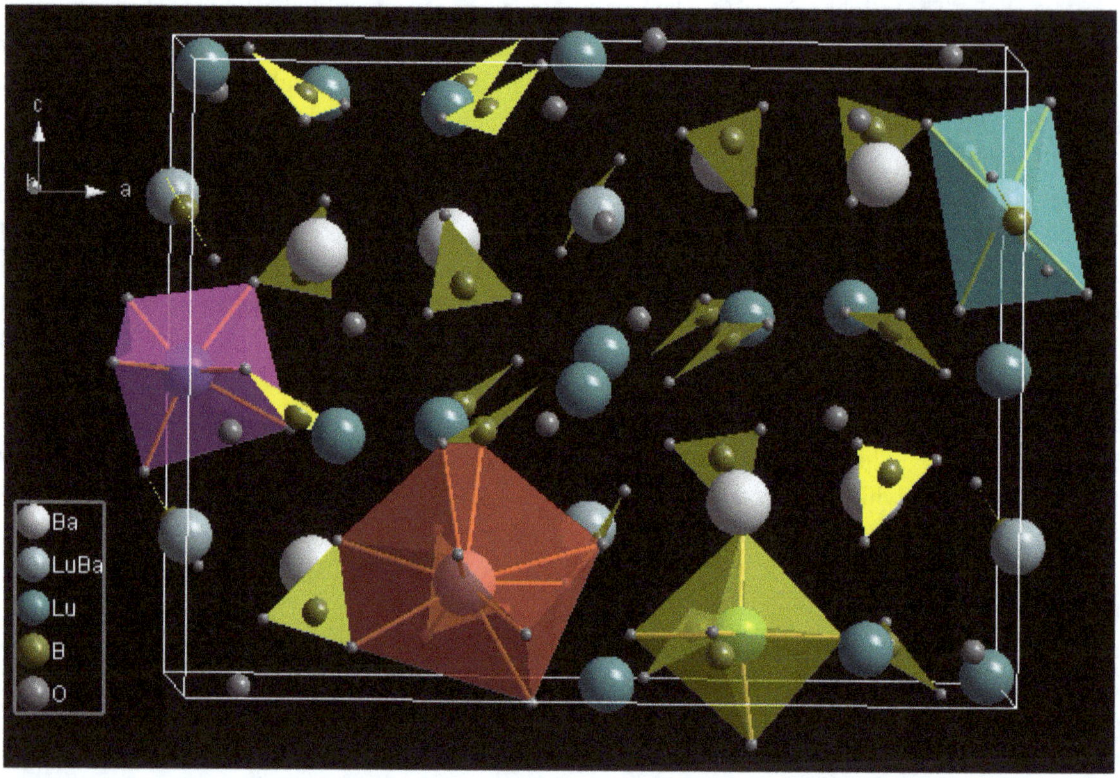

Figure 18. (a) Photoluminescent excitation (λ_{em} = 447 nm) and emission spectra (λ_{ex} = 340 nm) of $Ba_2Lu_{4.75}Ce_{0.25}B_5O_{17}$.(b) The CIE coordinates. Reproduced from ref (Hermus et al., 2017)

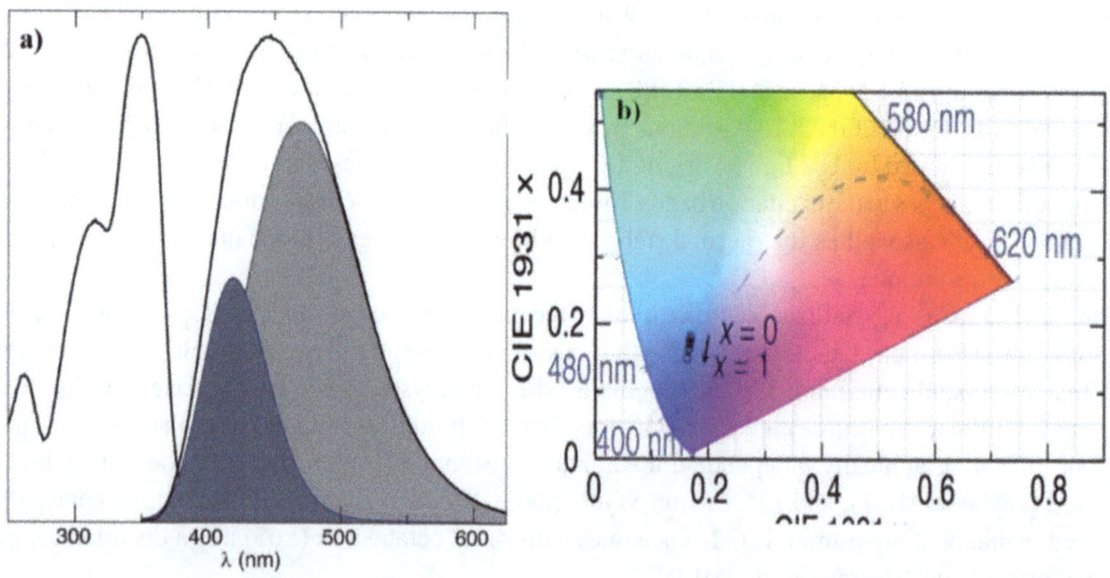

The photoluminescence excitation spectrum (Figure 18) of $Ba_2Lu_{4.75}Ce_{0.25}B_5O_{17}$ shows broad band excitation spanning from 250 nm to ≈390 nm with three excitation peaks (258 nm, 315 nm, and 349 nm) arising from multiple 4f to 5d electronic transitions in Ce^{3+} as well as the likely presence of more than one independent luminescent center, as observed in $Ba_2Y_{4.75}Ce_{0.25}B_5O_{17}$(Hermus et al., 2016). Exciting $Ba_2Lu_{4.75}Ce_{0.25}B_5O_{17}$ at each of these three peaks produces a bright blue emission spanning from 360 to 640 nm with $\lambda_{em,max}$ = 447 nm (Figure 18). Increase in x as seen in CIE co-ordinates (Figure 18b) clearly shows that the phosphors are not only in the blue spectral region but there is a red shift with decreasing Lu^{3+} content.

Eu^{2+} activated phosphor for LEDs:

$Li_2SrSiO_4:Eu^{2+}$

Li_2SrSiO_4 was first synthesized by Haferkorn & Meyer, (1998) by using sol-gel method.Saradhi et al., (2006) reported Eu^{2+} doped Li_2SrSiO_4 phosphor by sol-gel combustion method. The molar ratio of the reactants taken is 2:0.995:0.005: 1:2.5 $LiNO_3$:$Sr(NO_3)_2$:$Eu(NO_3)_3$:SiO_2:CH_6N_4O. The mixture was taken in a glass beaker and stirred until slurry resulted. The beaker was kept in a preheated furnace at 550 °C. Within a few minutes, the slurry boiled and was ignited to produce a self-propagating flame. The product obtained was heated in a reducing atmosphere (5% H_2 + 95% N_2) at various temperatures, viz. 600°C /3 h, 800°C /3 h and 800°C/12 h. Li_2SrSiO_4 crystallizes in a hexagonal crystal system with space group $P3_121$ with a = 5.023(5) Å and c = 12.457(1) Å. Sr^{2+}, Li^+ and Si^{4+} is coordinated to eight, four and four oxygens, respectively (Saradhi et al., 2006).

Figure 19a and b. shows the PLE and PL spectra of $Li_2SrSiO_4:_{0.005}Eu^{2+}$ phosphor respectively. The PL excitation spectra of Li_2SrSiO_4:0.005Eu^{2+} clearly indicates an absorption band arising at 308 nm and a broad absorption from 400 to 470 nm with the excitation bands rising due to $4f^7 \rightarrow 4f^65d^1$ transitions. The phosphor illustrates a strong orange-yellow emission band with a maximum at 562 nm extending to

Figure 19. (a) Photoluminescence excitation spectrum of $Li_2SrSiO_4:_{0.005}Eu^{2+}$.(b) Photoluminescence emission spectra of $Li_2Sr_{0.995}SiO_4:_{0.005}Eu^{2+}$ synthesized by solid-state reaction at various temperatures. Reproduced from ref (Saradhi et al., 2006)

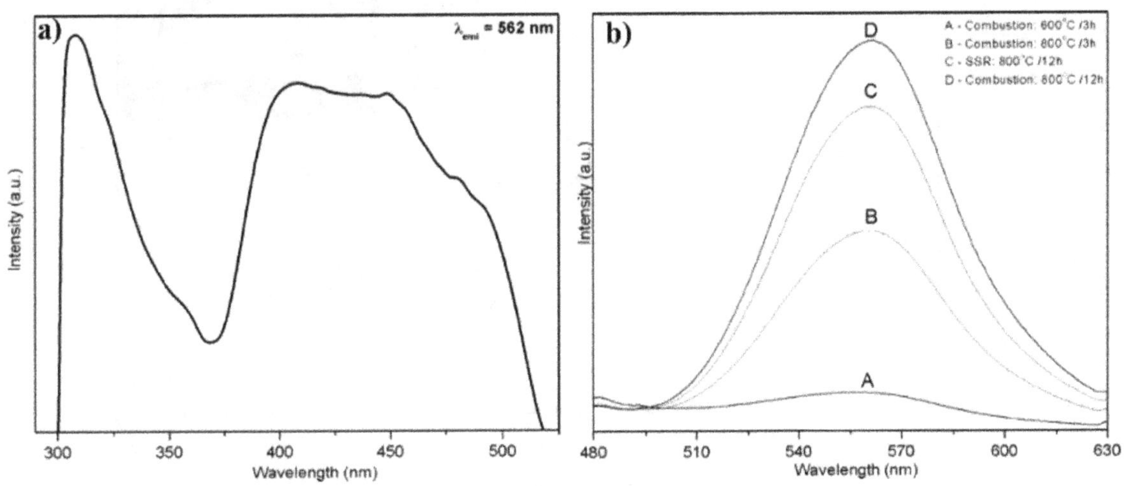

Figure 20. PLE and PL spectra of the $Ca_{3.95}(PO_4)_2O{:}_{0.05}Eu^{2+}$.Reproduced from ref (Deng et al., 2013)

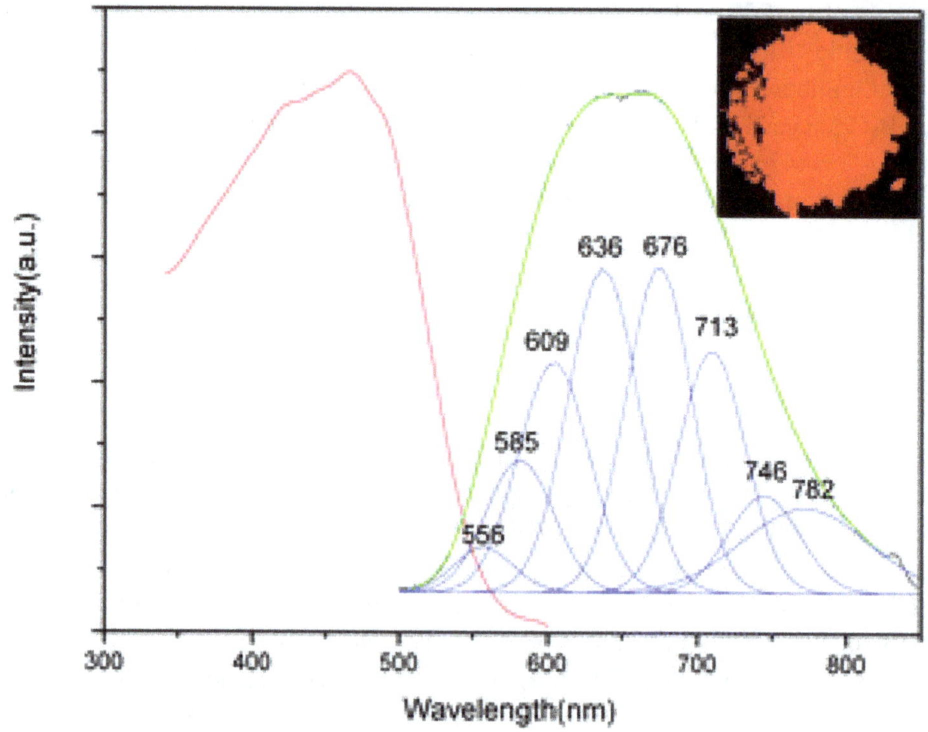

Figure 21. Electroluminescence spectrum of white LED using $Sr_2SiO_4{:}Eu^{2+}$ + $Ca_{3.93}(PO_4)_2O{:}_{0.07}Eu^{2+}$ under various applied currents.Reproduced from ref (Deng et al., 2013)

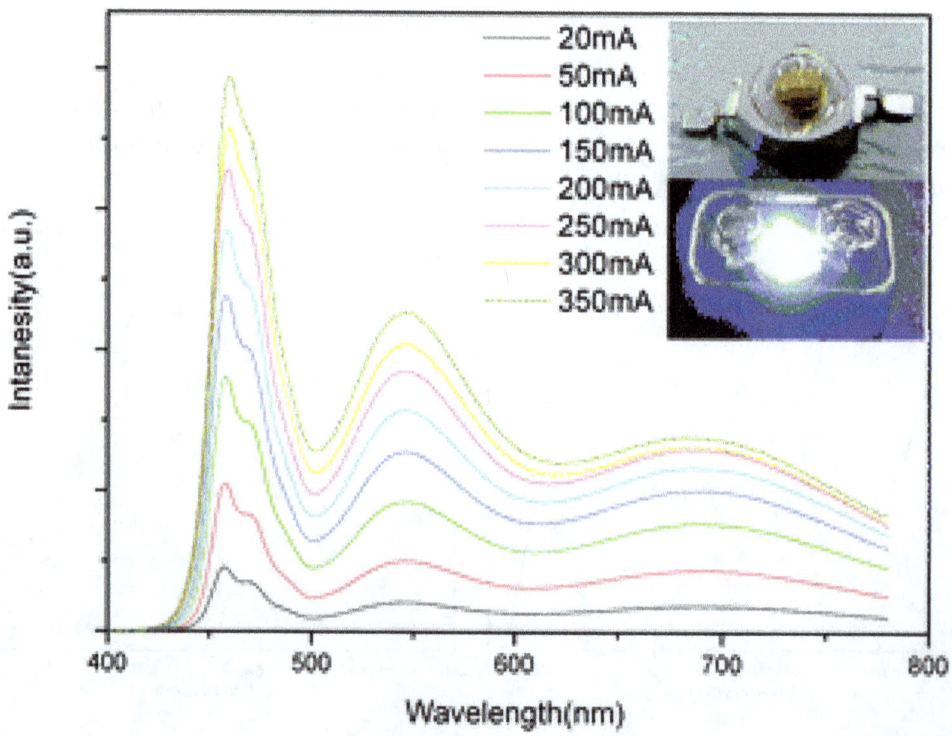

650 nm. The critical concentration is found to 0.005 mol of Eu^{2+} (R_c = 34 Å) per formula unit afterwards with increasing Eu^{2+} concentration, emission intensity decreases. The dipole-dipole interaction is behind the mechanism of concentration quenching of Eu^{2+} in Li_2SrSiO_4. The combination of a blue-emitting LED (420 nm) and orange-yellow phosphor results white light and the CIE values (0.3346, 0.3401) are better than those of the YAG: Ce phosphor coated LED (455 nm) (0.3069, 0.3592). Li_2SrSiO_4:Eu^{2+} is a potential phosphor for SSL lighting using InGaN (420 nm) to produce white light.

$Ca_4(PO_4)_2O$:Eu^{2+}

Eu^{2+}- activated $LiSrPO_4$ shows a higher absorption efficiency along with high value of conversion efficiency than the commercially available phosphor for example $BaMgAl_{10}O_{17}$:Eu^{2+}(Liu, 2008; Lin, 2010). Eu^{2+}-doped $KSrPO_4$ phosphor showed highly thermal stable luminescence properties than $Y_3Al_5O_{12}$:Ce^{3+} phosphor (Tang, Hu, Lin, Bagkar, & Liu, 2007). Similarly, Eu^{2+}-doped $Sr_8MgLn(PO_4)_7$:Eu^{2+} (Ln = Sc, Y, La) exhibits strong yellow emissions with improved thermal stability than $(Ba,Sr)_2SiO_4$ commercial phosphors. The warm white-emitting near-UV LEDs with assorted phosphors of $BaMgAl_{10}O_{17}$:Eu^{2+} and $Sr_8MgLn(PO_4)_7$:Eu^{2+} exhibit excellent CRI and lower CCT compared to those of conventional white LEDs based on YAG:Ce^{3+} pumped with blue LED chips (Huang, Chen, Y. C., Chen, T. M., Chan, & Sheu, 2011).

$Ca_4(PO_4)_2O$ has adopted a monoclinic unit cell with space group *P21* (crystal structure of Ca4(PO4)2O is shown in Figure 82) and the lattice constants are a = 7.023 (11) Å, b = 11.986 (13)Å, c = 9.473 (9) Å, β = 90.9(2)°, and cell volume = 797.3167(15)$Å^3$. For crystallographic information please refer the page no92see ref 85).

Figure 20 shows the PLE and PL spectra of the $Ca_{3.95}(PO_4)_2O$:$_{0.05}Eu^{2+}$, the excitation spectrum has a maxima at the 460 nm due to the $4f^7$– 4f ^{6}5d^1 transition of the Eu^{2+} ions. In the emission spectrum the broad emission was experimental at 665 nm with FWHM 154 nm which can be ascribed to the typical $4f^65d^1$– $4f^7$ transition of Eu^{2+} ions. Moreover the stoke shift of the $Ca_4(PO_4)_2O$:Eu^{2+} has the 6354.5 cm^{-1} which is larger than that of other Eu^{2+} doped oxide red phosphors. The emission spectrum was decomposed into eight Gaussian profiles by Gaussian deconvolution and these eight peaks are designated to the eight different emission sites, which could be occupied by Eu^{2+} ions in place of Ca^{2+} site of different coordination environment. The $Ca_4(PO_4)_2O$ structure has eight crystallographically different Ca^{2+} sites with average Ca–O band distance as follows: Ca(1)–O 2.443 Å; Ca(2)–O 2.468 Å; Ca(3)–O 2.526 Å; Ca(4)–O 2.442 Å ; Ca(5)–O 2.603 Å; Ca(6)–O 2.556 Å; Ca(7)–O 2.414 Å and Ca(8)–O 2.467 Å (Deng et al., 2013).

Figure 21 shows the electroluminescence (EL) spectra of the white LED, was fabricated with highly efficient blue-InGaN chips. The EL spectra clearly indicates blue bands at around 460 nm, green emitting bands corresponding to the Sr_2SiO_4:Eu^{2+} phosphor at ~ 540 nm, and red-emitting bands due to $Ca_{3.93}(PO_4)_2O$:$_{0.07}Eu^{2+}$ at around 665 nm. The emission intensity of the mixed phosphor (Sr_2SiO_4:Eu^{2+} + $Ca_{3.93}(PO_4)_2O$:$_{0.07}Eu^{2+)}$ was almost not saturated at a high forward current of 350 mA.

$K_2Ba_7Si_{16}O_{40}$:Eu^{2+}

The luminescent property is strongly dependent on the structure of host lattice, researchers have devoted a lots of effort in developing garnet based phosphors such as $Lu_2CaMg_2(Si,Ge)_3O_{12}$:$Ce^{3+}$ (yellow) or $Ca_3Sc_2(SiO_4)_3$:Ce^{3+} (green) subsequently, the most commonly used yellow phosphor $Y_3Al_5O_{12}$:Ce^{3+} is

Figure 22. (a) The crystal structure of $K_2Ba_7Si_{16}O_{40}$. (b) The infinite two-dimensional layers of corner-sharing [SiO4] tetrahedral seen from (201). (c) Coordination geometry of Ba^{2+} ions. Reproduced from ref (Lv et al., 2014)

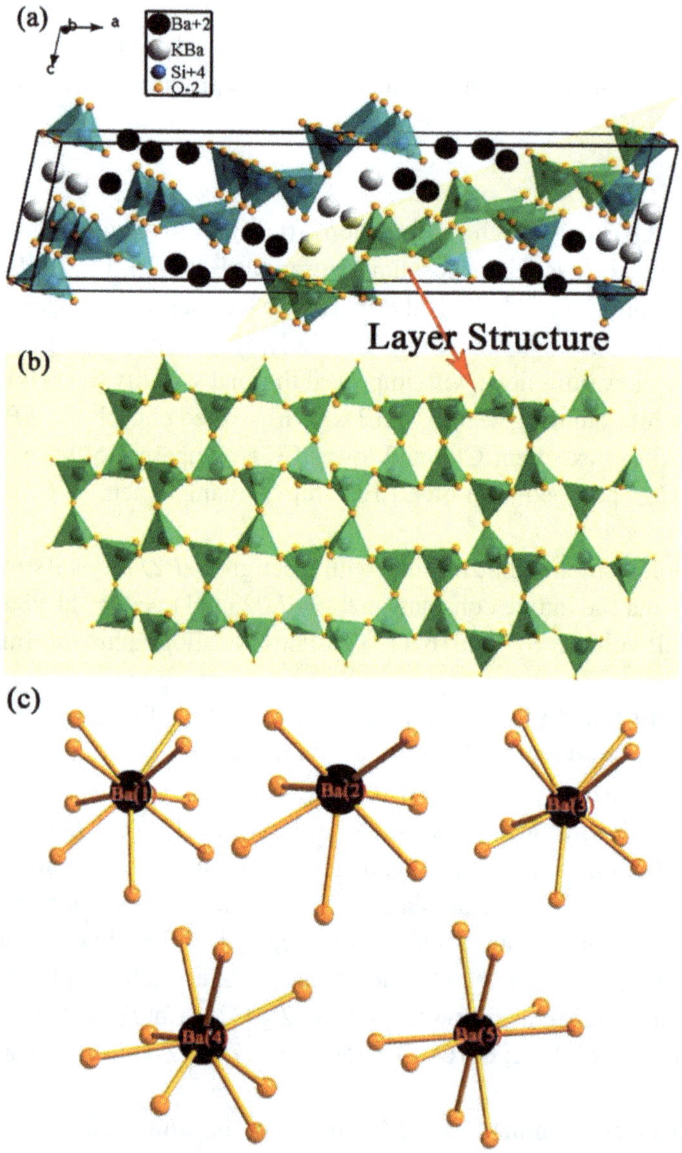

developed from a garnet system (Setlur, 2006; Liu, 2011). $K_2Ba_7Si_{16}O_{40}$:Eu^{2+} is also a garnet based phosphor widely used in white LEDs. $K_2Ba_7Si_{16}O_{40}$:Eu^{2+} crystallises a monoclinic structure with space group *C2/m* and the corresponding lattice parameters are a/Å = 32.046(7); b/Å = 7.705(6); c/Å = 8.224(6), V/ $Å^3$ = 1994.825. The crystal structure of $K_2Ba_7Si_{16}O_{40}$is displayed in Figure 22. The K ions and Ba ions are arranged at the interstitial positions between the infinite layers as seen in the Figure 22. It is clearly manifest from the crystal structure that the two-dimensional layer is composed of corner-sharing [SiO_4] tetrahedral which is parallel to (201) and extends infinitely along the b-direction. Specially, there are five

Figure 23. PLE and PL spectra of $K_2Ba_{6.72}Si_{16}O_{40}$:$_{0.28}Eu^{2+}$ phosphor Reproduced from ref(Lv et al., 2014)

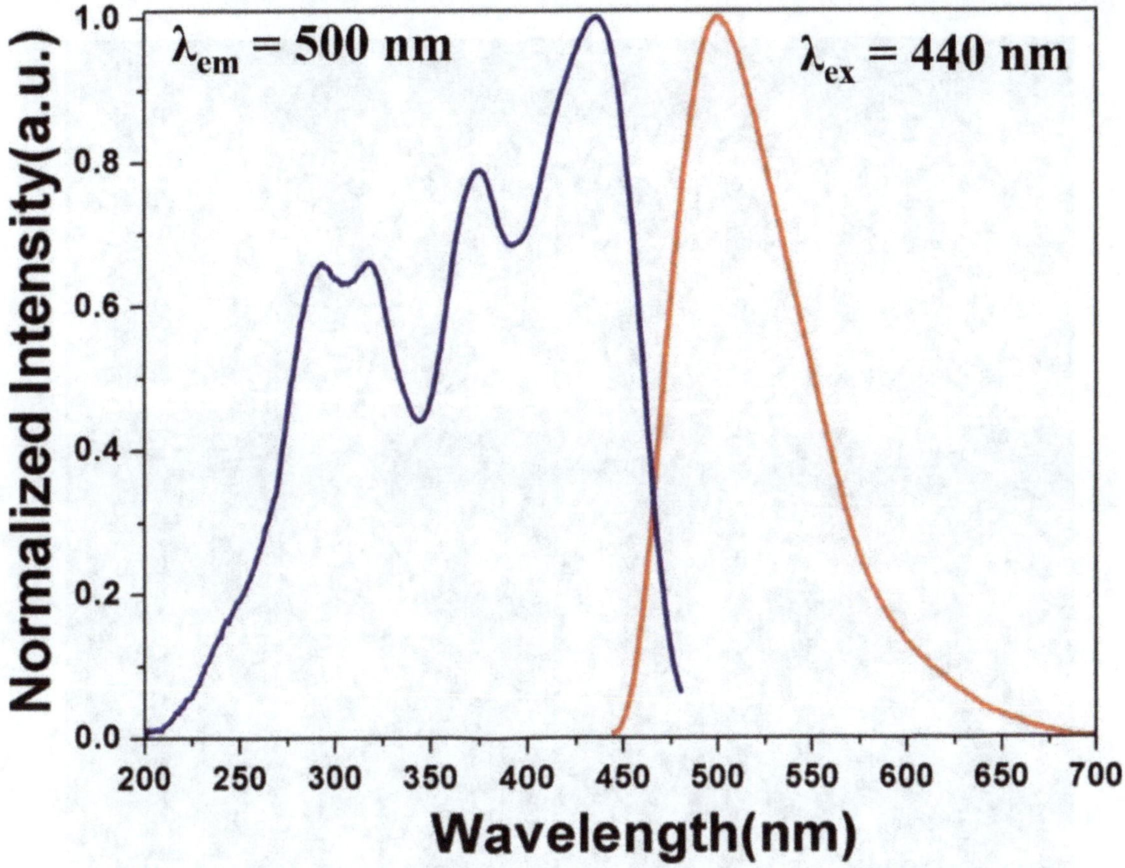

distinct crystallographic sites of Ba^{2+} ions with coordination numbers varying from 7 to 9. The Ba(1) and Ba(3) have a higher coordination number of 9 and a analogous geometrical arrangement. The Ba(2) has a coordination number of 7 with O atoms. The coordination of Ba(4) and Ba(5) consist of eight O atoms. On account of the ionic radii and valence, the Eu^{2+} ions are expected to randomly occupy the Ba^{2+} ions (Lv et al., 2014).

PLE and PL spectra of $K_2Ba_{6.72}Si_{16}O_{40}$:$_{0.28}Eu^{2+}$ phosphor is shown in Figure 23. The PL spectrum of $K_2Ba_{6.72}Si_{16}O_{40}$:$0.28Eu^{2+}$ consists of one broad non-symmetrical band along with a peak at about 500 nm and a tail at the longer wavelength side, which is expected to arise due to the electric dipole allowed $4f^65d^1$–$4f^7$ transition of the Eu^{2+} ions; hence giving blue light (CIE co-ordinates x=0.223, y=0.507). A minor red shift and a slight broadening of the emission band can be observed upon raising the Eu^{2+} concentration. The absolute quantum yield of $K_2Ba_{6.93}Si_{16}O_{40}$:$0.07Eu^{2+}$ under the excitation of 440 nm is found to be 61.9% and it shows a reduced amount of thermal quenching at 150°C as compared to Sr_2SiO_4:Eu^{2+}(Lv et al., 2014). Moreover, a reduction of 35% of the initial intensity at 150°C for the $K_2Ba_{6.93}Si_{16}O_{40}$:$0.07Eu^{2+}$ is inferior than that of 45% of the commercial YAG:Ce^{3+}, signifying its suitability to be used in WLEDs.

$Sr_{1.75}Ca_{1.25}(PO_4)_2$:$Eu^{2+}$

Figure 24. (a) Crystal structure of the sample $Sr_{1.75}Ca_{1.15}Eu_{0.10}(PO_4)_2$, and (b) coordination of the cation ions with oxygen atoms. Redrawn from ref (Ji et al., 2014)

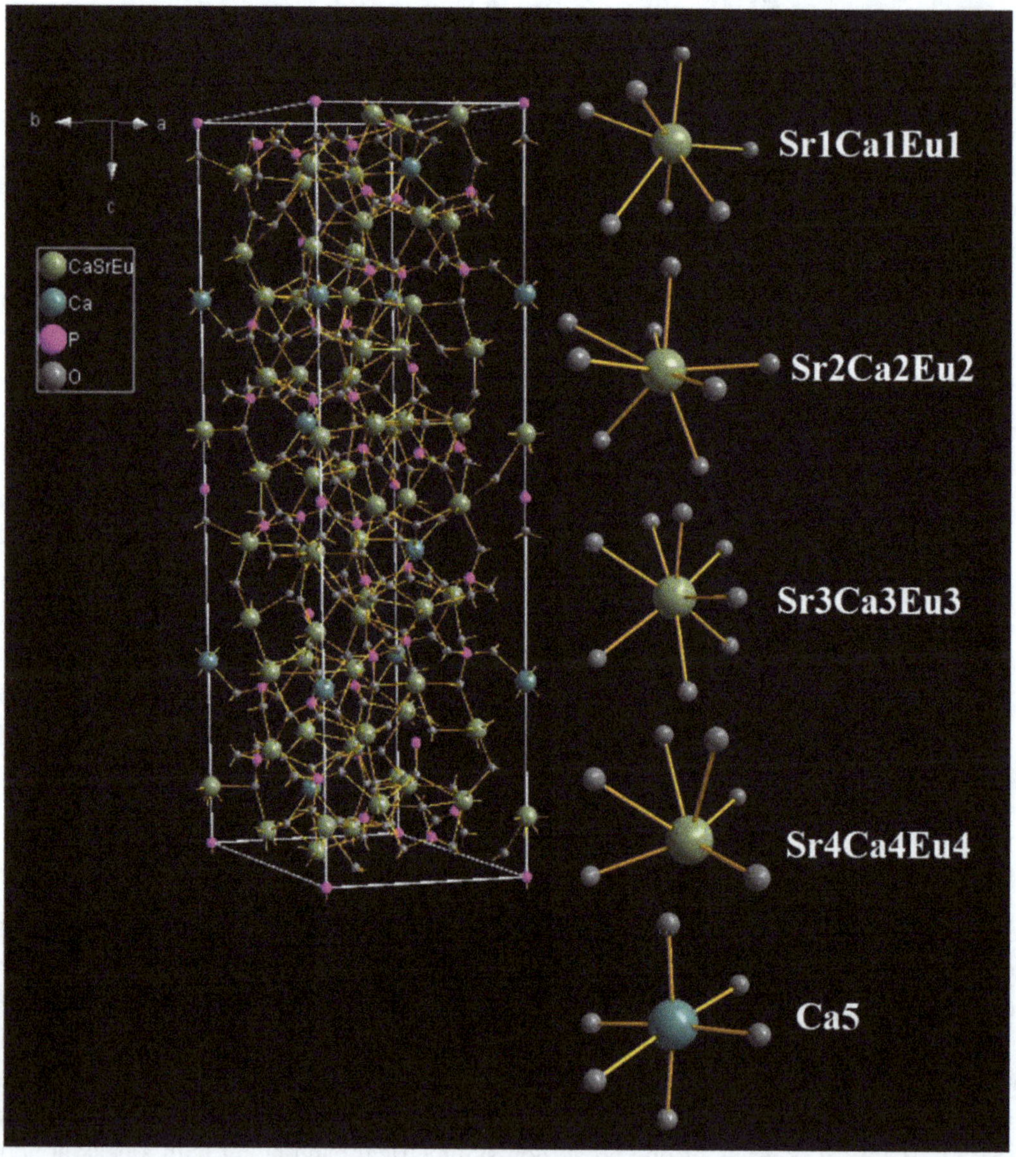

Rare earth ions activated orthophosphate is being widely investigated as RGB phosphors as because of their low sintering temperature, great chemical stability, and satisfactory absorption in the ultraviolent (UV) – near UV region. For example, the end members and some solid solution compositions in the $Sr_3(PO_4)_2$–$Ca_3(PO_4)_2$ lattice have been reported to be host for Eu^{2+}ion(Poort, Van Krevel, Stomphorst, Vink, & Blasse, 1996). The $Sr_{1.75}Ca_{(1.25-x)}(PO_4)_2{:}_xEu_{2+}$phosphor synthesized by conventional solid state method. The $Sr_{1.75}Ca_{1.15}Eu_{0.1}(PO_4)_2$ solidified in a rhombohedral unit cell with space group *R3c* and the lattice constants are a = 10.6498(3)Å and c = 38.787(1)Å, and cell volume V = 3809.8(2)$Å^3$. Figure 24 displayed the crystal structure of $Sr_{1.75}Ca_{1.15}Eu_{0.10}(PO_4)_2$, it has five different crystallographic sites

Figure 25. Excitation (λem = 514 nm) and emission (λex = 279 and 365 nm) spectra of $Sr_{1.75}Ca_{1.24}(PO_4)_2:_{0.01}Eu^{2+}$.Reproduced from ref (Ji et al., 2014)

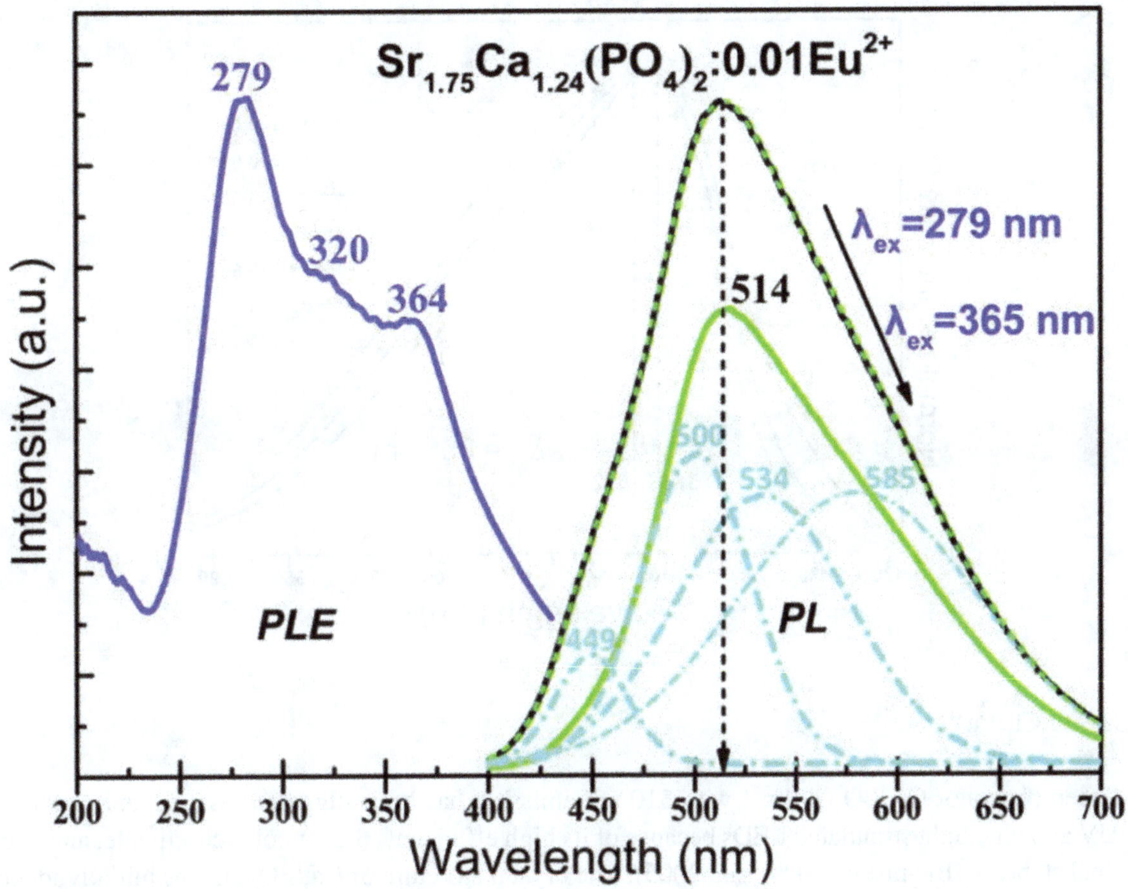

they are Sr_1/Ca_1, Sr_2/Ca_2, Sr_3/Ca_3, and Sr_4/Ca_4 positions are coordinated with seven, eight, eight, and six oxygen atoms, and the Ca_5 position is coordinated with six oxygen atoms(Ji et al., 2014).

PL excitation and emission spectrum of $Sr_{1.75}Ca_{1.24}(PO_4)_2:_{0.01}Eu^{2+}$ shown in Figure 25.The excitation spectrum observed in the green region with peaking at 514 nm. It has three major absorption peaks observed at 279, 320, and 364 nm in the range of 200 to 430 nm. The broad excitation band is a resultant of the $^4f^7$ ($^8S_{7/2}$) ground state to $^4f^6\,^5d^1$ excited state transition of Eu^{2+} ions. The wide asymmetric emission spectrum peaking at 514 nm demonstrates that Eu^{2+} has more than one emission center in $Sr_{1.75}Ca_{1.25}(PO_4)_2$ which is attributed to 5d–4f transitions of Eu^{2+} ions. The emission spectrum concludes that the Eu^{2+} ions occupied the four Ca^{2+} sites.

Figure 26 shows the normalized emission spectra of $Sr_{1.75}Ca_{(1.25-x)}(PO_4)_2:_xEu^{2+}$ (x = 0.005–0.15) phosphors under 365 nm UV light excitation. It clearly indicate that the emission peak of the phosphors has shifted to the longer wavelength region (from 515 to 523 nm) with increasing the Eu^{2+} ion concentration in the lattice. This is due CFs around the coordination environment around Eu^{2+} ion in the lattice. The WLED was fabricated $Sr_{1.75}Ca_{1.24}(PO_4)_2:_{0.01}Eu^{2+}$ phosphor mixing with commercial blue emitting $BaMgAl_{10}O_{17}:Eu^{2+}$.

Figure 26. Normalized emission spectra of $Sr_{1.75}Ca_{(1.25-x)}(PO_4)_2:_xEu^{2+}$ (x = 0.005–0.15) phosphors under 365 nm UV light excitation. Reproduced from ref (Ji et al., 2014)

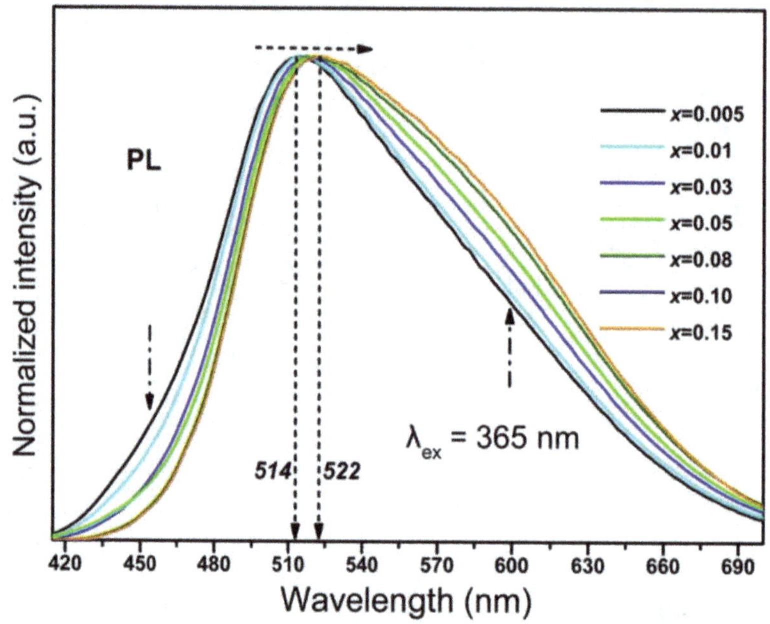

β-$Ca_3SiO_4Cl_2$:Eu^{2+}

Green phosphor $Ca_3SiO_4Cl_2$:Eu^{2+} with 510 nm emission has been suggested as a decent candidate for UV and blue light stimulated LEDs because of its high efficiency, proper color coordinate, and good thermal stability (Baginskiy, 2009; Liu, 2005). The crystal structure of $Ca_3SiO_4Cl_2$ was initially identified by Czaya & Bissert, (1971). $Ca_3SiO_4Cl_2$ crystallizes in the monoclinic system, which consists of alternating layers of Ca_2SiO_4Cl and Ca_2Cl_2 with a 2:1 ratio. β-$Ca_3SiO_4Cl_2$ has a monoclinic system with space group $P2_1/c$ (No. 14) (lattice parameters, a = 5.91234(1)Å, b = 10.20128(1)Å, c = 10.98866(1)Å, β = 90.3423(1)°). The crystal structure of β-$Ca_3SiO_4Cl_2$ isshown in Figure 27. Ca atoms are coordinated with O and Cl atoms to form irregular polyhedra, $Ca(1)O_4Cl_3$, $Ca(2)O_5Cl_2$, and $Ca(3)O_3Cl_4$ (Figure 27). The environment of Ca^{2+} ions is a crucial factor to determine the luminescent property of substituted Eu^{2+} ions(Kim, T. G., Kim, T., Kim, J., Kim, S. J., & Im, 2014).

β-$Ca_3SiO_4Cl_2$: $0.05Eu^{2+}$ emits yellowish orange light with a 635 nm peak and a 150 nm bandwidth for 450 nm excitation, arising from $4f^6d^1 \leftrightarrow 4f^7$ transition (Figure 28). The red-shifted and broad-band emission excited at 450 nm is a suitable feature for the warm white LED illumination with a high color rendering index. The quantum efficiency (QE) of β-$Ca_3SiO_4Cl_2$:$0.05Eu^{2+}$ is about 70%. With increase in excitation wavelength, a red shift (bathochromic) is observed in the emission peaks. The luminescence variation at different excitations and emissions was caused by the multi-luminescence centres with different electronic environments in a single phosphor (Lee & Sohn, 2010). Three humps in excitation spectra around 350, 430, and 480 nm would correspond to three types of Ca^{2+} sites in β-$Ca_3SiO_4Cl_2$. The red shift of the emission from green to yellowish orange over 600 nm with increasing the Eu^{2+}-doping concentrations is attributed to the active energy transfer between Eu^{2+} ions in the different crystallographic

Figure 27. Crystal structure of β-$Ca_3SiO_4Cl_2$. Redrawn from ref (Kim, T. G., Kim, T., Kim, J., Kim, S. J., & Im, 2014)

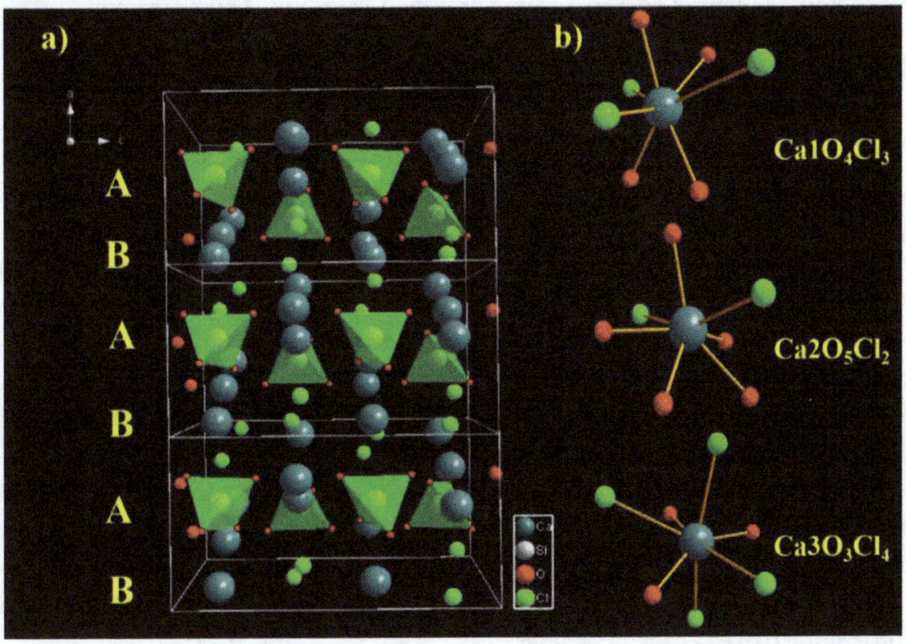

Figure 28. (a) Excitation spectrum for the emission at 650 nm, and the emission spectra excited at 300, 350, 400, 450, and 500 nm. (b) Excitation spectra for the emissions at 500, 550, 600, 650, 700, and 750 nm, and the emission spectra excited at 400 nm. Reproduced from ref Kim, T. G., Kim, T., Kim, J., Kim, S. J., & Im, 2014)

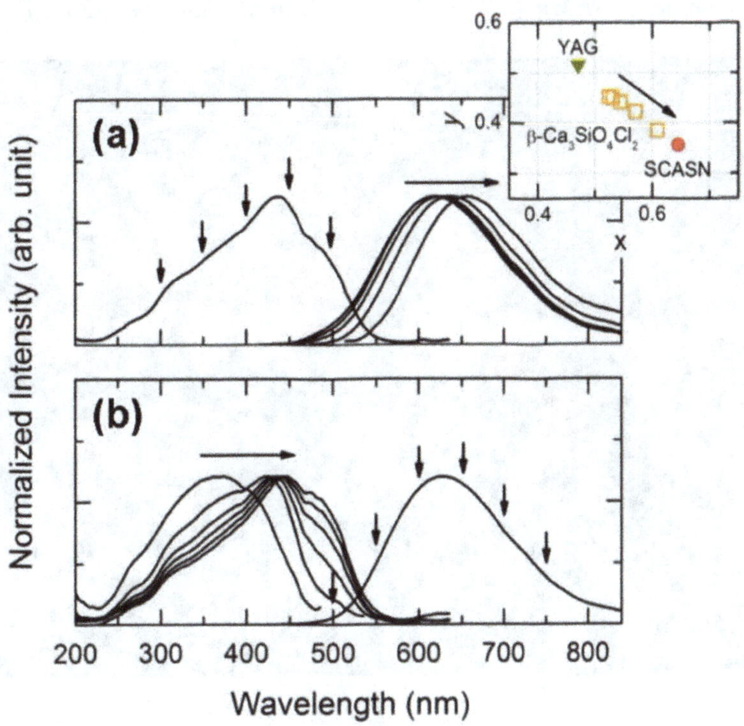

sites as Ca1 → Ca2, Ca1 → Ca3, and Ca2 → Ca3. Because of the active energy transfer between the different Eu^{2+} sites at the concentration of 0.05, the temperature stability of the phosphor is dominated by luminescence with the lowest energy, i.e., the emission from Eu^{2+} at the Ca3 site.

$Sr_9Mg_{1.5}(PO_4)_7:Eu^{2+}$

The alkaline-earth phosphate phosphors have been widely investigated and used the same in white LEDs owing to their excellent chemical and thermal stability, high luminescence efficiency, and low cost. Among them, $Sr_9Mg_{1.5}(PO_4)_7:Eu^{2+}$is pulling in consideration as a promising candidate to substitute the viable yellow-emitting $YAG:Ce^{3+}$. The $Sr_9Mg_{1.5}(PO_4)_7:Eu^{2+}$ phosphor synthesized by conventional high temperature CO reduced atmosphere and it crystallize in trigonal structure with space group $R\bar{3}m$, lattice parameters = b = 10.582Å, c = 19.723Å, Z = 3, and cell volume of V = 1912.727Å^3. Figure 29 shows the crystal structure of $Sr_9Mg_{1.5}(PO_4)_7$ and it has three Sr (Sr1, Sr31, and Sr32) crystallographic sites, two Mg sites (Mg4 and Mg5) and two P sites (P1 and P2).The highly disordered $Sr_9Mg_{1.5}(PO_4)_7$ unit cell has three Sr crystallographic sites (Sr1, Sr31 and Sr32), two Mg sites (Mg4 and Mg5) and two P sites (P1 and P2). The Sr3 sites i.e. Sr31 and Sr32 sites are disordered over two 18h sites near the symmetry center; $P1O_4$ tetrahedra are disordered in orientation; the Mg4 and Mg5 sites are 25% and 100% occupied, respectively. The coordination number (CN) of Sr1 site is fixed to 8, while those of Sr31 and Sr32 sites vary with the orientation of $P1O_4$ tetrahedra: 8 or 9 for Sr31 site, and 6 or 7 for Sr32 site(Belik, 2001;Sun, 2015).

Figure 30 shows the PLE and PL spectra of SMPO:$0.06Eu^{2+}$. The excitation spectrum of 523 and 620 nm wavelengths, both consist of broad bands ranging from 220 nm–550 nm with maxima at ~460 nm which are allotted to the $4f^7$–$4f^65d_1$ transition of Eu^{2+} ions. The emission spectra were observed two

Figure 29. Crystal structure of $Sr_9Mg_{1.5}(PO_4)_7$Redrawn from ref (Sun et al., 2015)

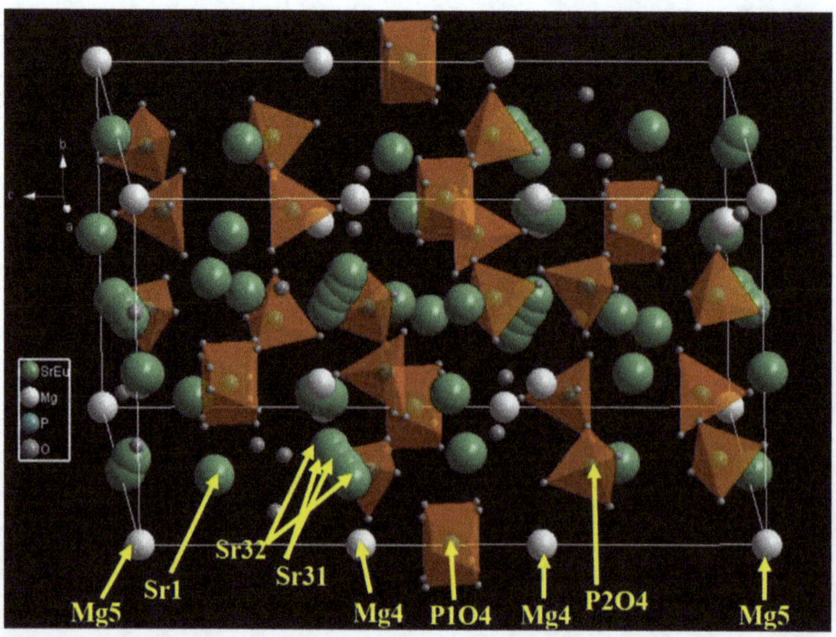

Figure 30. PLE and PL spectra of SMPO:$_{0.06}$Eu^{2+} phosphor. Reproduced from ref (Sun et al., 2015)

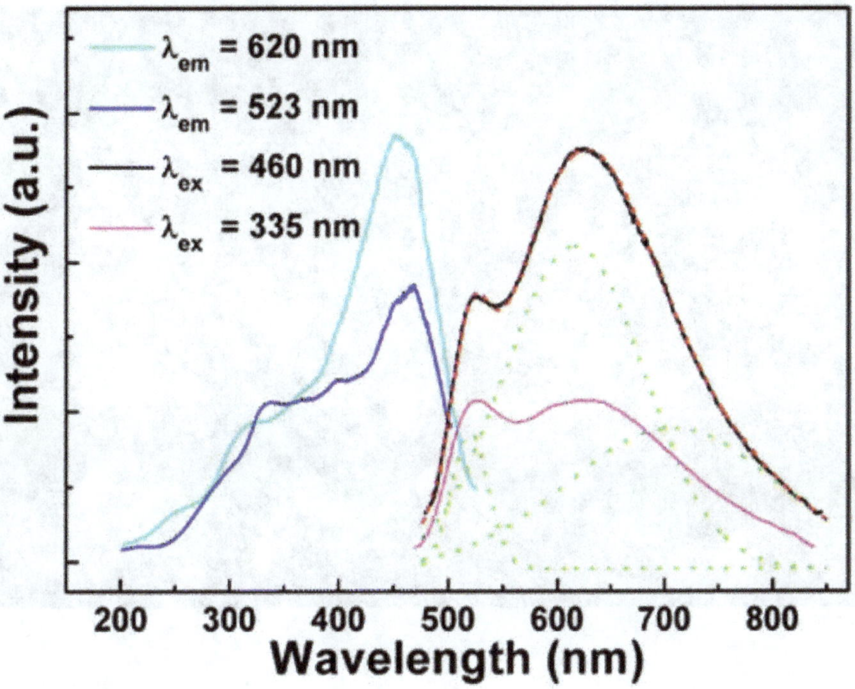

different wavelengths 523 and 620 nm as Eu^{2+} ion has two different emitting centers(He, 2014; Miao, 2014).The SMPO:0.06Eu^{2+} phosphor has intense red emission as compared to that of yellow-emitting YAG:Ce^{3+} commercial phosphor. The emission intensity of SMPO:0.06Eu^{2+} decreases with the rising temperature from 300 to 448 K. When compared with the initial intensity, the emission intensity of phosphor at 373 K it is upto 44.32%. LED device was fabricated by using 460 nm blue InGaN chip and SMPO:0.06Eu^{2+} phosphor.

$BaSc_2Si_3O_{10}$:Eu^{2+}

In silicate matrices, scandium based silicates display excellent photoluminescence properties on account of their stable physical and chemical properties and rich crystal chemistry environment for the doped rare earth ions,(Maria, Uwe, & Ekkehart, 2010) such as $Ca_3Sc_2Si_3O_{12}$:Ce^{3+}, (Liu et al., 2011) and $NaBaScSi_2O_7$:Eu^{2+}(Zhu, Shi, Mikami, Shimomura, & Wang, 2014). A family of trisilicates $BaRe_2Si_3O_{10}$ (Re = Gd, Er, Yb, Sc) was first reported by Wierzbicka-Wieczorek *et al.*in 2010 (Wierzbicka-Wieczorek,Kolitsch, & Tillmanns, 2010). $BaSc_2Si_3O_{10}$ has a monoclinic structure with the space group *$P2_1/m$*, and the lattice parameters are a = 5.27162Å, b = 11.91488Å, c = 6.58826Å, the crystal structure of $BaSc_2Si_3O_{10}$as shown in Figure 31.Their topology is characterized by horseshoe-shaped trisilicate (Si_3O_{10}) groups and zigzag chains of edge-sharing distorted ReO_6 octahedra (Re = Gd, Er, Yb, Sc). There are two coordination environments of cations: Ba^{2+} having 8-fold coordination and Sc^{3+} having 6-fold coordination in $BaSc_2Si_3O_{10}$. There is only one suitable cationic site (Ba^{2+}) available for doping of Eu^{2+} ion in the $BaSc_2Si_3O_{10}$ host lattice(Wang et al., 2015).

Figure 31. Crystal structure of $BaSc_2Si_3O_{10}$ .Redrawn from ref(Wang et al., 2015)

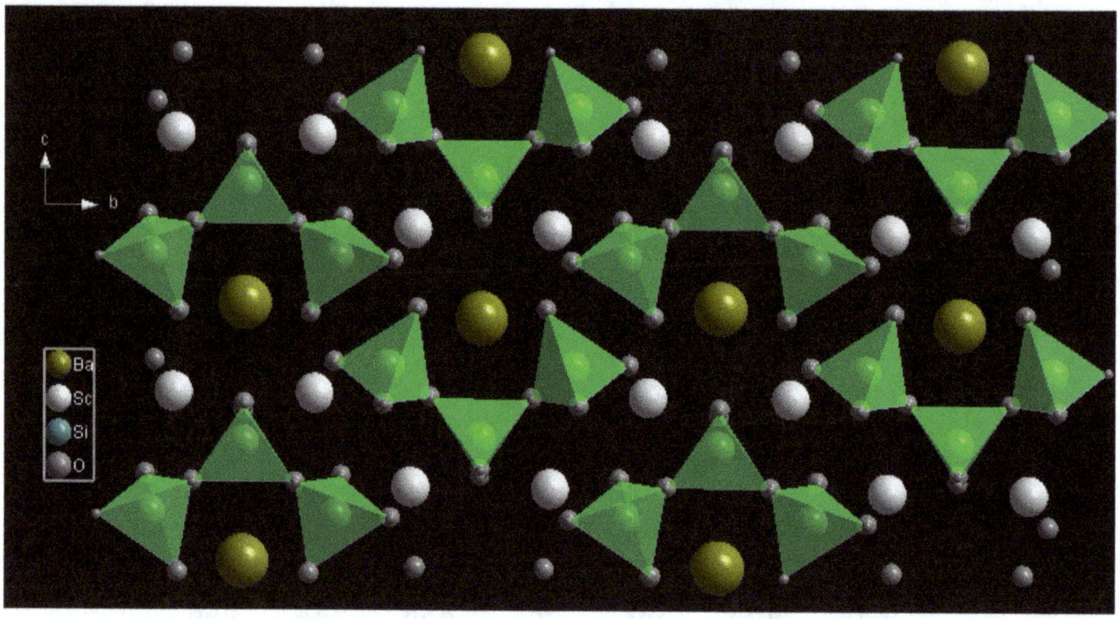

The PLE and PL spectra of $BaSc_2Si_3O_{10}$:$0.03Eu^{2+}$ are shown in Figure32a. $BaSc_2Si_3O_{10}$:Eu^{2+} phosphor shows a broad absorption band ranging from 250 to 420 nm (well covered in the UV to near UV spectral region), which can match well with the UV-LEDs. The $BaSc_2Si_3O_{10}$:Eu^{2+} exhibits blue broad band emission peaking at 446 nm with a tail on the longer wavelength side and the CIE coordinates are (0.15, 0.05). As the Eu^{2+} concentration increases, the emission spectrum shifts to a longer wavelength (red shift) and then to a shorter wavelength (blue shift). Eu^{2+} likely occupy the Sc^{3+} site at a low concentration of Eu^{2+}, and then occupy the Ba^{2+} site with the increase of Eu^{2+} content. The concentration quenching mechanism has been investigated to be a quadrupole–quadrupole interaction. The emission intensities of $BaSc_2Si_3O_{10}$:Eu^{2+} dependence on different temperatures show a similar thermal property compared to BAM as seen in Figure 32b. Upon excitation at 330 nm, the external quantum efficiency of this sample is found to be 42.65%.

$Ca_{2+x}La_{8-x}(SiO_4)_{6-x}(PO_4)_xO_2$:$Eu^{2+}$

Generally the apatites have the general chemical formula of $A_{10}(XO_4)_6Z_2$ (A = Ca^{2+},Ba^{2+},Ce^{3+},La^{3+},Y^{3+}, etc., X = P^{5+},As^{5+},Si^{4+}, etc., and Z = O^{2-},F^-,Cl^-,OH^-, etc.)(Xia, Molokeev, Im, Unithrattil, & Liu, 2015). The phosphor $Ca_{2+x}La_{8-x}(SiO_4)_{6-x}(PO_4)_xO_2$ derived from the general formula, and it was synthesized by the solid state method. The phosphor crystallizes in the hexagonal crystal structure with space group *P63/m*(176) and the cell parameters are a = b = 9.651Å and c = 7.151Å. Figure 33a exhibited the crystal structure of the $Ca_2La_8(SiO_4)_2(PO_4)_4O_2$. The crystal structure contains two kinds of cation sites M(I) and M(II) and their local symmetry's are C_3 and Cs, respectively(Ryan, Warren, Hopkins, & Murphy, 1978). Figure 33b displayed the M(I) site was contained by 9 oxygen atoms to form a mono-caped square antiprism, which was linked with tetrahedral PO_4/SiO_4 groups.The M(II) site framed a pentagonal bipyramid with the encompassing 7 oxygen iotas and these bipyramids were associated with each other through

Figure 32. (a) Excitation and emission spectra of $BaSc_2Si_3O_{10}:_xEu^{2+}$, the inset shows the normalized emission spectra. (b) Temperature-dependent PL spectra of $BaSc_2Si_3O_{10}:_{0.03}Eu^{2+}$ (the inset shows the thermal stability comparison of $BaSc_2Si_3O_{10}:_{0.03}Eu^{2+}$ and the commercial BAM). Reproduced from ref (Wang et al., 2015)

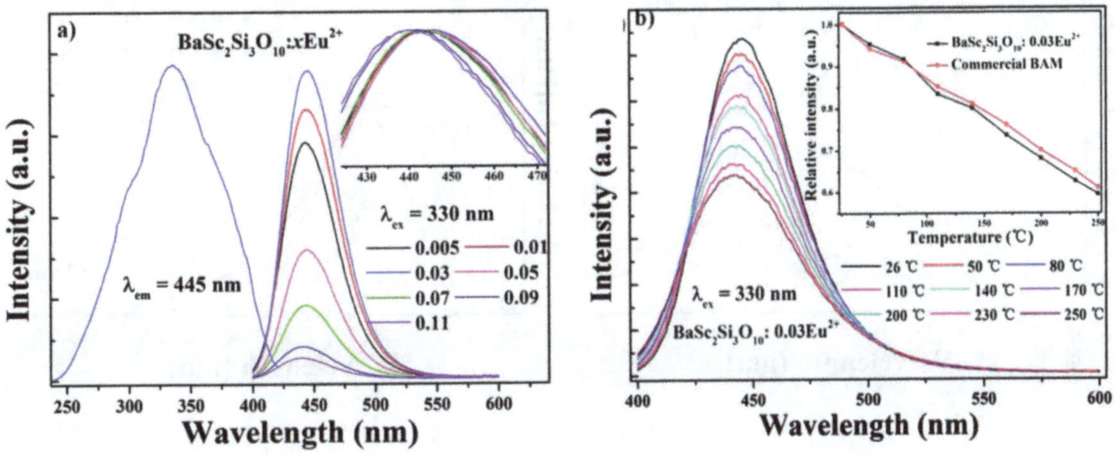

Figure 33. Crystal structure of the 2 × 2 × 2 unit cells of $Ca_2La_8(SiO_4)_2(PO_4)_4O_2:_{0.02}Eu^{2+}$.Redrawn from ref (Xia et al., 2016)

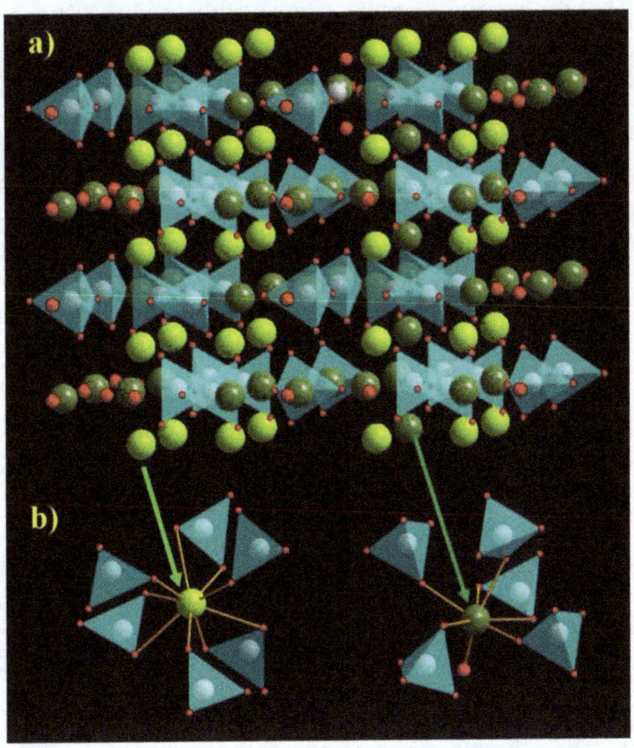

Figure 34. The excitation and emission spectrum of the $Ca_{2+x}La_{8-x}(SiO_4)_{6-x}(PO_4)_xO_2{:}_{0.02}Eu^{2+}$ (x = 0,2,4,6) phosphors.Reproduced from ref (Xia et al., 2016)

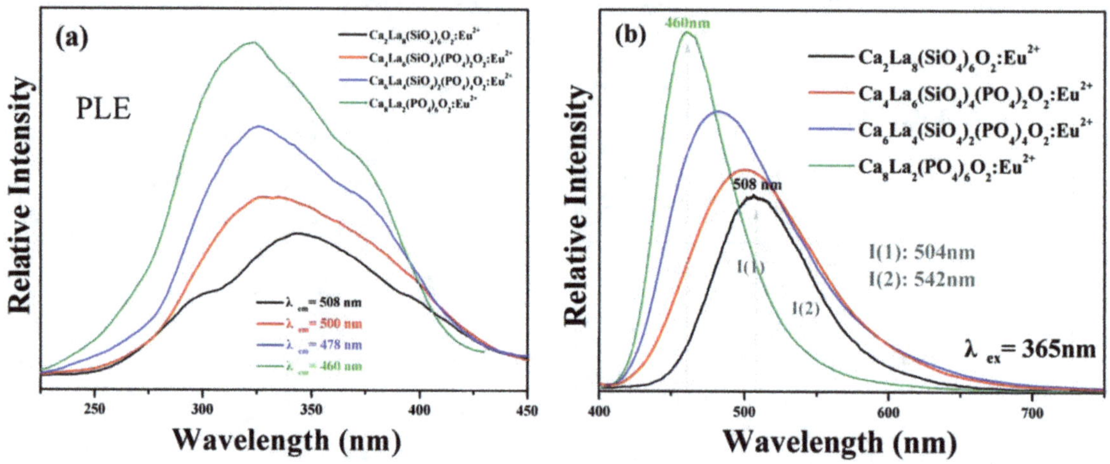

a vertex. The host matrix has four kinds of emitting blocks Ca(I), Ca(II), La(I) and La(II), initially the Eu^{2+} ion can substitute the La^{3+} and Ca^{2+} sites in the $Ca_{2+x}La_{8-x}(SiO_4)_{6-x}(PO_4)_xO_2$ (x = 0,2,4,6) crystal structure. When the rate of Ca^{2+} ions substitution was increased, the occupation rate of La(II) is reduced faster than that of La(I) and the M(II) site is totally occupied by Ca^{2+} ions. So finally at the sage of $Ca_8La_2(PO_4)_6O_2$ (x = 6), the Eu^{2+} ion is occupying the M(I) site(Xia et al., 2016).

Figure 34 a and b shows the excitation and emission spectrum of the $Ca_{2+x}La_{8-x}(SiO_4)_{6-x}(PO_4)_xO_2{:}_{0.02}Eu^{2+}$ (x = 0,2,4,6) phosphors. The excitation spectra are monitored within the range of 460 nm–508 nm, broad excitation bands between 250 nm and 450 nm, which were attributed to the $4f^7 \rightarrow 4f^6 5d^1$ transition of the Eu^{2+} ion. The emission spectra (under 365 nm excitation) exhibit a series of wide emission bands from 400 nm to 650 nm, which are attributed to the 5d → 4f transition of the Eu^{2+} ion.

Table 1 shows the, FWHM of the emission spectra, stoke shift and Crystal field splitting was decreases with increases the concentration x. Reproduced from ref (Xia et al., 2016).

$(Sr,Ba)_2SiO_4{:}Eu^{2+}$

In 1986, L. Stenberg *et.al.* reported the α'-Sr_2SiO_4 structure(Stenberg, Sellar, & Hyde, 1986).*β*-$Sr_2SiO_4{:}Eu^{2+}$ and *α*-$(Ba,Sr)_2SiO_4{:}Eu^{2+}$ are well-known as green-yellow phosphors with high

Table 1.

X Values	λ_{ex} Range (nm)	FWHM of λ_{ex} (nm)	λ_{em} (nm)	Stokes Shift (cm^{-1})	Crystal Field Splitting (cm^{-1})
0	250–450	142	508	6135	22230
2	250–450	127	500	6082	20833
4	250–450	118	478	6039	19231
6	250–400	113	460	5985	18190

Figure 35. 1 × 1 × 2 unit cell of Sr_2SiO_4 (upper) and coordinate environments at Sr1 and Sr2 sites (bottom). Redrawn from ref (Wen et al., 2016)

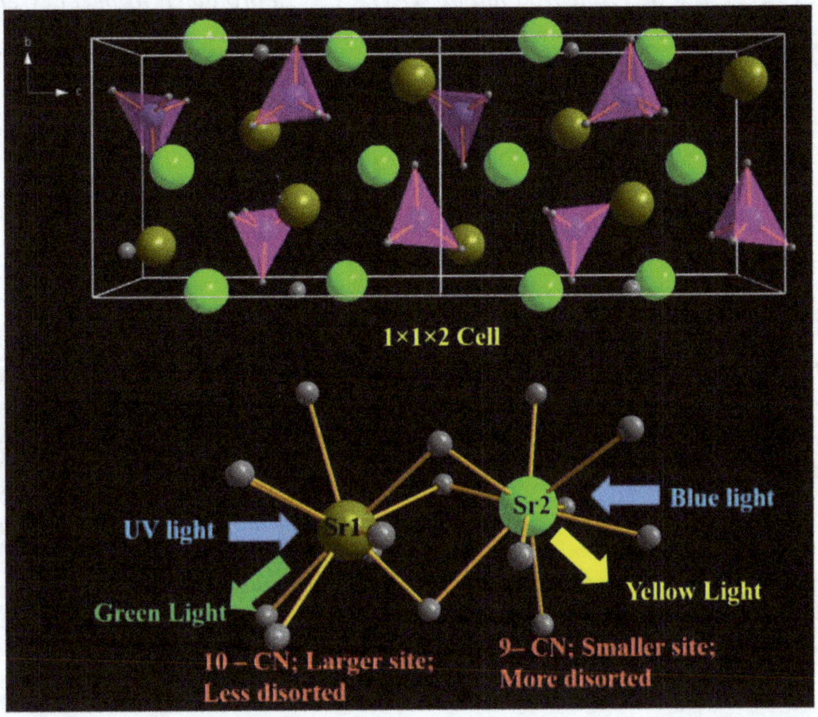

efficiency. For β-Sr_2SiO_4, Eu^{2+} gives green emission when excited by ultraviolet (UV) light. As Ba^{2+} is doped or the concentration of Eu^{2+} increases, the β-phase transforms into the α-phase (Lakshminarasimhan, & Varadaraju, 2005).The α-Sr_2SiO_4 crystallizes in the orthorhombic system with *Pnmb* space group and the cell parameters are a = 5.674Å, b = 7.100Å and c = 9.748Å, the crystal structure of Sr_2SiO_4 shown in Figure 35. In α-Sr_2SiO_4, there are two distinct crystallographic sites for Sr ions, Sr1 and Sr2 (Catti, & Gazzoni, 1983).The coordination number of Sr1 and Sr2 are 10 and 9, with average bond lengths to oxygen of 2.85 and 2.70 Å as well as the distortion indexes of 0.04 and 0.07, respectively (Baur, 1974, pp. 1195–1215) implying that Sr1 sites are larger and less distorted. Eu^{2+} ions in Sr1 site can be excited by UV light and give green emission whereas those in Sr2 site can be excited by UV and blue light and give yellow emission (Rack, & Holloway, 1998). By increasing the ratio of Sr/Ba, the emission of α-$(Ba,Sr)_2SiO_4$:Eu^{2+} can be tuned from green to yellow(Wen et al ., 2016).

The PL spectra measured under excitation of 450 nm shows that current emission is arising from from Eu^{2+} not at Sr1 site (Eu1) but at Sr2 site (Eu2) because lack of sensitivity of Eu1 for the incident beam at 450 nm. From the PL studies, it can be concluded that as the *x* value increased the threshold of the PLE spectra and whole of PL spectra shifted to longer wavelength region. The samples with x = 0.25 and 0.4 have more sufficient red emission component than commercial YAG:Ce^{3+} (P46) as presented in Figure 36. The internal quantum efficiencies (IQE)/external quantum efficiencies (EQE) of $(Sr,Ba)_{1.5}Eu_{0.5}SiO_4$ and $(Sr,Ba)_{1.2}Eu_{0.8}SiO_4$ were 75% /65% and 64% /56% (λ_{Ex} = 450 nm), higher than the IQE of 53% in the previous work due to the homogeneous reaction in the synthesis process (Sato et al., 2015). The Eu^{2+} ion selectively occupies in the Sr1 site caused preventing from the concentration quenching for the Eu2

Figure 36. PLE (a), PL (b), spectra of of the $(Sr,Ba)_{2(1-x)}Eu_{2x}SiO_4$.PLE spectra were monitored at 585, 589, 603 and 616 nm for x = 0.05, 0.1, 0.25 and 0.4, respectively. PL spectra were taken under excitation at 450 nm for $Sr_{2(1-x)}Eu_{2x}SiO_4$ and 460 nm for YAG:Ce^{3+}. Reproduced from ref (Wen et al., 2016)

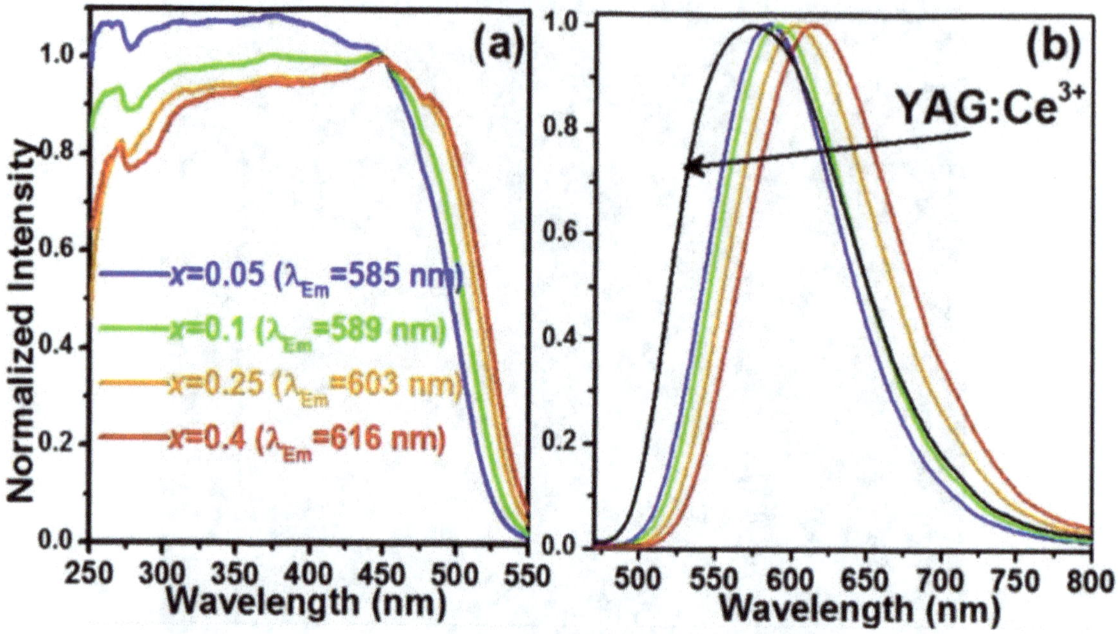

emission. The local environmental model, in which Eu1 and Eu2 are adjacent to each other, indicates that the lowering energy difference between f and d orbitals occurs through 5d orbital interaction between of Eu1 and Eu2 in the situation of high Eu^{2+} concentration. According to these results, it is concluding that interesting red shifts were induced by combination of surviving concentration quenching even with large Eu substitution and the orbital interaction (Kuwahara et al., 2016).

$Ba_4Gd_3Na_3(PO_4)_6F_2$:Eu^{2+}

As mentioned earlier that the apatite-type structure exist in the hexagonal system, the general formula of the apatite is $A_{10}(XO_4)_6Z_2$, where A is cation (Ca^{2+}, Sr^{2+}, Ba^{2+}, La^{3+}, Y^{3+}, etc.), X is cation (P^{5+}, Si^{4+}, etc.), and Z is anion (F^-, Cl^-, etc.) (Que, 2013; Ci, 2014; Zhang, 2012; Liu, 2014; Ding, 2015). In recent years lot of apatite phosphors have been investigated such as$La_{9.33}(SiO_4)_6O_2$:Ce^{3+},Eu^{2+},Mn^{2+}(Lü, Jia, Lv, Zhao, & You, 2013), $Ca_6Y_2Na_2(PO_4)_6F_2$:Eu^{2+},Mn^{2+}(Guo, You, Jia, Ouyang, & Wu, 2014), $Mg_2Y_8(SiO_4)_6O_2$:Ce^{3+},Mn^{2+},Tb^{3+}(Li et al., 2011), $Sr_3GdNa(PO_4)_3F$:Eu^{2+},Mn^{2+}(Jiao et al., 2014) and etc.

The $Ba_{4-x}Gd_3Na_3(PO_4)_6F_2$:$_x Eu^{2+}$ (BGNPF) phosphors have been synthesized by a typical solid state reaction and all the compositions crystallized in trigonal structure with space group $P\bar{3}$ with the cell parameters of a = b = 9.738Å, c = 7.210Å, V = 592.144Å^3. The crystal structure of the $Ba_4Gd_3Na_3(PO_4)_6F_2$ host and crystallographic sites are shown in Figure 37. The M cation, which contains a mixture of 2/3Ba, 1/6Gd, and 1/6Na, is coordinated by nine atoms (seven oxygen atoms and two fluoride), and the Gd site is coordinated with only nine oxygen atom. The Eu^{2+} occupied both Ba^{2+} site and as well as the Gd^{3+} site, this is clearly observed in PL spectrum(Fu, Lü, Jiao, & You, 2016).

Figure 37. Crystal structure of the $Ba_4Gd_3Na_3(PO_4)_6F_2$ host and crystallographic sites. Redrawn from ref (Fu, Lü, Jiao, & You, 2016)

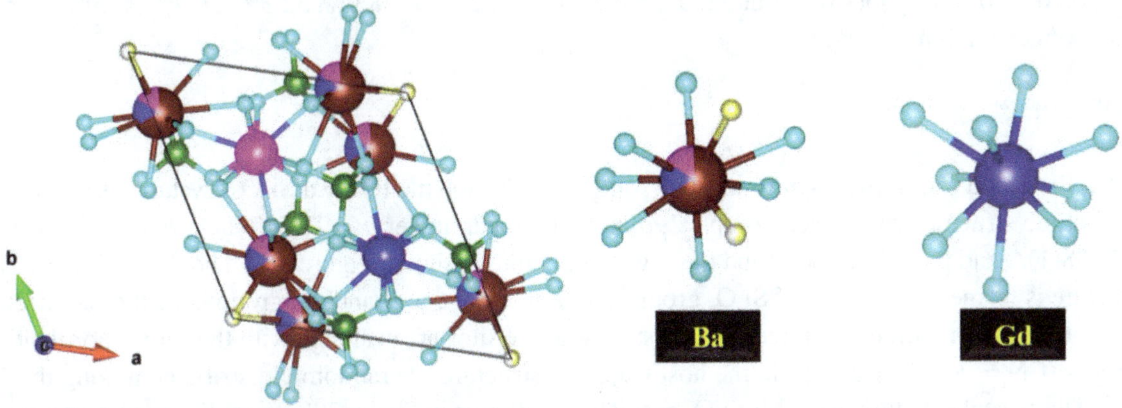

Figure 38. PL and PLE spectra of the BGNPF:0.04Eu²⁺ phosphor. Reproduced from ref (Fu, Lü, Jiao, & You, 2016)

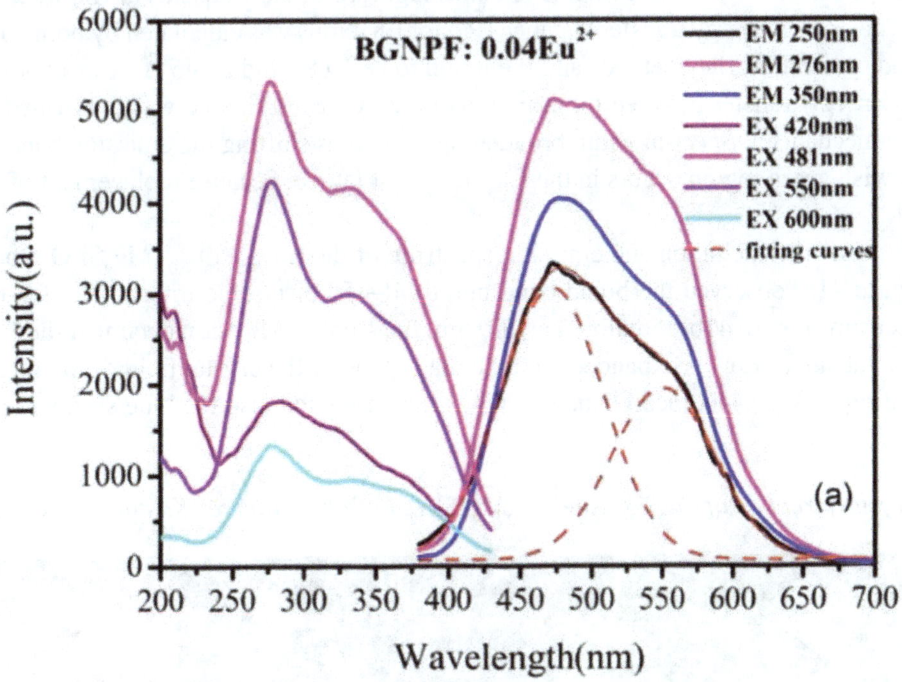

PL and PLE spectrum of the BGNPF:0.04Eu^{2+} phosphor is shown in Figure 38. The excitation wavelengths are stable at 4], 550, and 600 nm, separately. It can be detected that the PLE displays a wide band from 230 to 430 nm due to the $4f^7$ $-4f^6$ $5d^1$ transition of Eu^{2+} ions.Upon excitation at 350 nm, the emission spectrum of BGNPF:$_{0.04}Eu^{2+}$ displays a broad band from 400 to 650 nm, with its peak at 481 nm with shoulder around 550 nm, and this is assigned to the electric dipole permitted $4f^6$ $5d^1$ $-4f^7$

move of Eu^{2+} ions.The BGNPF:0.04Eu^{2+} phosphor retains 50% of its initial emission intensity, when the temperature rises to 150 °C. The absolute QY of BGNPF:xEu^{2+} (x = 0.005, 0.01, 0.02, 0.04, 0.08, 0.12, 0.16,0.20) phosphors was found to be 5.6.%, 15.1%, 27.5%, 48.9%, 52.3%, 56.7%, 41.9%, 32.5% at 365 nm excitation.

$Ca_3Si_2O_7$:Eu^{2+}

$Ca_3Si_2O_7$ and $Ca_2MgSi_2O_7$synthesized by solid state reaction method, $Ca_3Si_2O_7$ crystallizes in a single monoclinic structure with space group $P2_1/a$ and lattice parameter a = 10.557(1)Å, b = 8.885(3)Å, c = 7.858(1)Å and β = 119.586(6)° and the crystal structure is shown Figure 39a. The three dimensional structure is formed as an array of Si_2O_7 groups parallel to the c axis and the group is further connected to Ca atoms. In this structure, three Ca^{2+} crystallographic sites are available with the same coordination number (CN = 7). The Eu^{2+} ion in the host $Ca_3Si_2O_7$ structure are randomly distributed among the Ca sites. The crystal structure of $Ca_2MgSi_2O_7$ host lattice is shown in Figure 39b, crystallize in the tetragonal phasewith space group *P421m*(Kasturi & Sivakumar, 2016).

$Ca_3Si_2O_7$ has a broad excitation band ranging from 200 to 500 nm, which matches with the NUV or blue LED chip emission range. The emission spectra of all compositions depicted in Figure 40 show yellowish-orange color and x = 0.015 composition show highest emission intensity having a peak maximum at ~622 nm. The critical energy transfer distance (R_c) for Eu^{2+} ions was calculated by both concentration quenching and spectral overlap methods and the found to be 19.85 and 21.45 Å, respectively. Based on the R_c values energy transfer between the Eu^{2+} ions in the $Ca_3Si_2O_7$ lattice was confirmed to be via a dipole–dipole mechanism. Spectral width broadening as well as shifting the emission band (blue shift) towards yellowish green region occurs in the $Ca_3Si_2O_7$ host lattice, because replacement of Ca^{2+} by the smaller Mg^{2+}.

Figure 41 shows the excitation and emission spectrum of the $Ca_{2.985-y}Eu_{0.015}Mg_ySi_2O_7$, the excitation spectrum(Figure 41a) observed the broad band due to 4f→5d electronic transitions of Eu^{2+} ion. The emission spectrum (Figure 41b) monitored at 400 nm, the smaller Mg^{2+} ion occupied the Ca^{2+} site due to that the crystal site of Eu^{2+} is expanded, and the magnitude of the crystal field decreases. Increasing the concentration of Mg^{2+} ion, the5d band of Eu^{2+} is increased and also the blue shift was observed in

Figure 39. Crystal structure of $Ca_3Si_2O_7$ and $Ca_2MgSi_2O_7$. Redrawn from ref(Kasturi & Sivakumar, 2016)

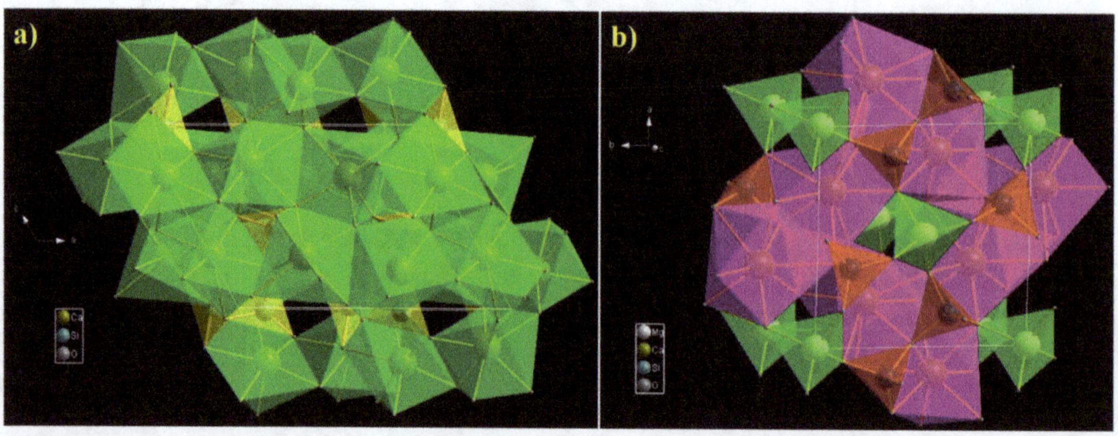

Figure 40. Photoluminescence excitation and emission spectrum of $Ca_{2.985}Eu_{0.015}Si_2O_7$ Reproduced from ref(Kasturi & Sivakumar, 2016)

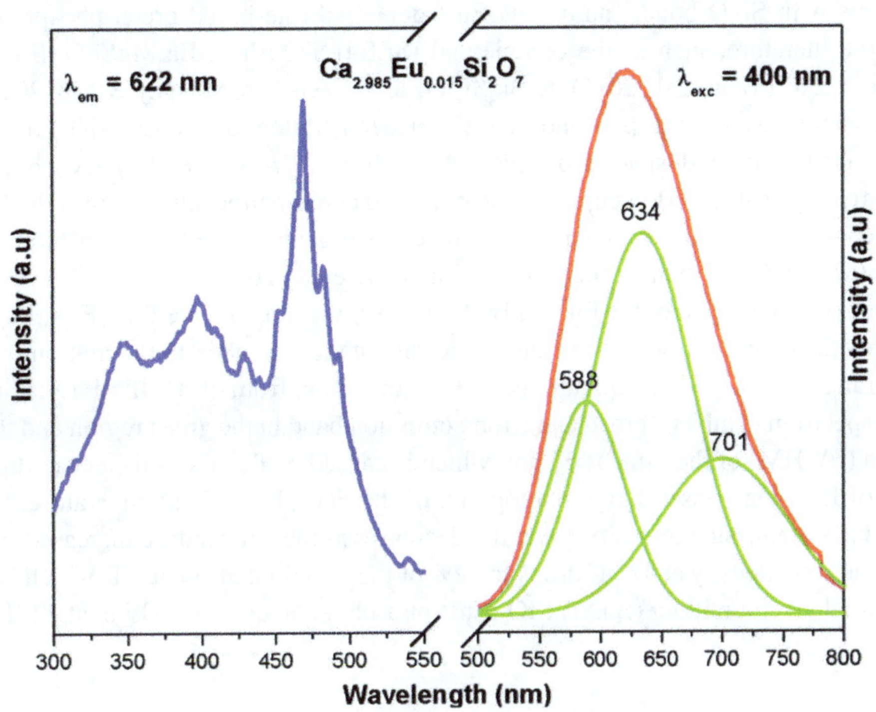

Figure 41. The PL excitation and emission spectrum of $Ca_{2.985-y}Eu_{0.015}Mg_ySi_2O_7$ [(y = 0–1, insteps of 0.25)].Reproduced from ref(Kasturi & Sivakumar, 2016)

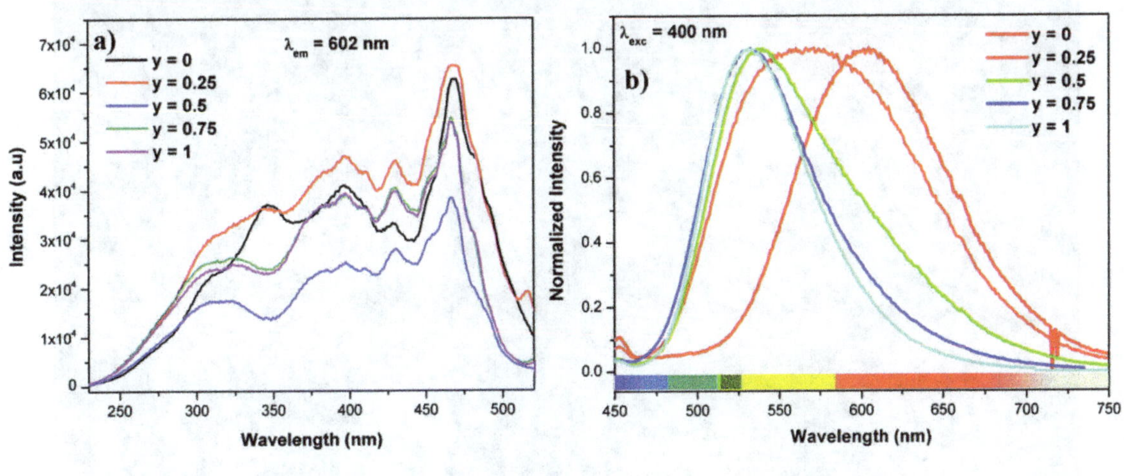

the emission spectrum.The quantum efficiencies were found to be 28, 26.5, 30, 32.5, and 27.4% for y = 0, 0.25, 0.5, 0.75, and 1, respectively.

$Ba_3La_6(SiO_4)_6:Eu^{2+}$

Silicate materials are known to be efficient host materials as because of their good chemical stability and lavish crystal structure (Liu, Zhang, Hao, Wang, & Zhang, (2011). Due to their energetic and reduced systems with Si–O bonds, numerous Eu^{2+}-doped silicate-based green phosphors have been described in the literature, such as the commercial $(Sr,Ba)_2SiO_4:Eu^{2+}$, $Ba_2MgSi_2O_7:Eu^{2+}$(Tam et al., 2014), $K_2ZrSi_2O_7:Eu^{2+}$(Tang et al., 2016), $K_2Ba_5Si_{12}O_{30}:Eu^{2+}$(Yeh, Lin, Maggay, & Liu, 2016) and so on. Especially the apatite based silicate phosphors have gained much attention. $Ba_3La_6(SiO_4)_6:Eu^{2+}$ crystalizes in the hexagonal structure with space group *P63/m*, a = b = 9.77Å, c = 7.31, V (Å^3) = 604.31. In the crystal structure (Figure 42), Ba1 occupies in the nine-fold coordination site, whereas the Ba2 occupies the sevenfold coordination which are connected with tetrahedral unit of SiO_4. The Eu^{2+} ion preferentially occupied the Ba2 site due to their similar ionic radii(Zhong et al., 2017).

The optimum concentration of the Eu^{2+} in $Ba_3La_6(SiO_4)_6$ was found to be $Ba_{2.95}Eu_{0.05}La_6(SiO_4)_6$. The Figure 43 shows the PL excitation and emission spectrum of the phosphor, the excitation spectra consist of two bands ranging from 250 to 450 nm due to Eu^{2+} transition from $4f^7$ to $4f^65d^1$ (Xiao et al., 2015). The emission spectrum exhibits a broad and strong emission band in the green region and the full-width half maximum (FWHM) of the same is 67 nm, which is caused by the allowed electric-dipole $4f^65d \rightarrow 4f^7$ transition of Eu^{2+} ions (inset digital photograph of the phosphor). The temperature dependent PL studies show that the emission intensity gradually decreases as the temperature increased from 303 K to 503 K. The emission intensity at 423 K drops to 52% of the initial intensity at RT, which is comparable to the commercial green-emitting $(Ba,Sr)_2SiO_4:Eu^{2+}$ phosphors (about 49%) (Li et al., 2015).

$Ba_{10}(PO_4)_6F_2$: Eu^{2+}

Figure 42. Crystal structure $Ba_3La_6(SiO_4)_6$. Redrawn from ref (Zhong et al., 2017)

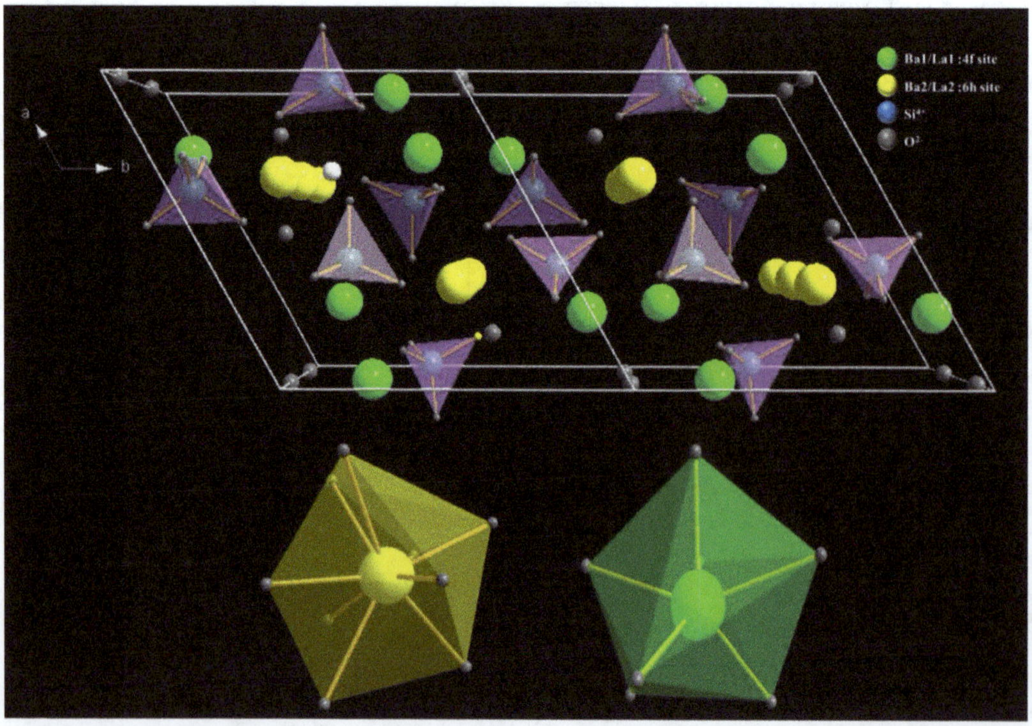

Figure 43. Excitation and emission spectrum of $Ba_3La_6(SiO_4)_6$:$_{0.05}Eu^{2+}$.Reproduced from ref (Zhong et al., 2017)

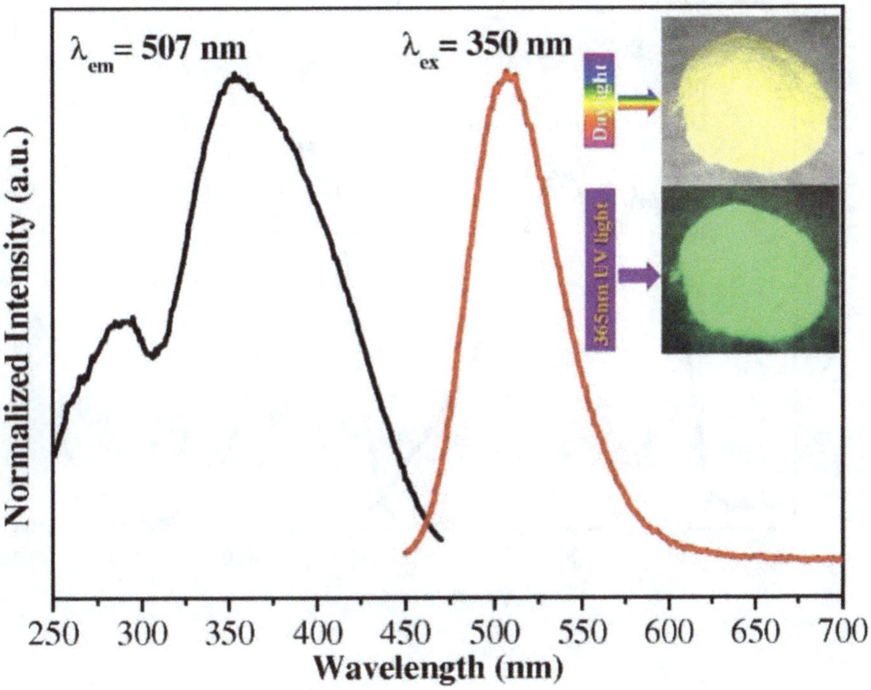

Figure 44. Crystal structure of $Ba_{10}(PO_4)_6F_2$ .Redrawn from ref (Huang, Zhang, & Pan, 2017)

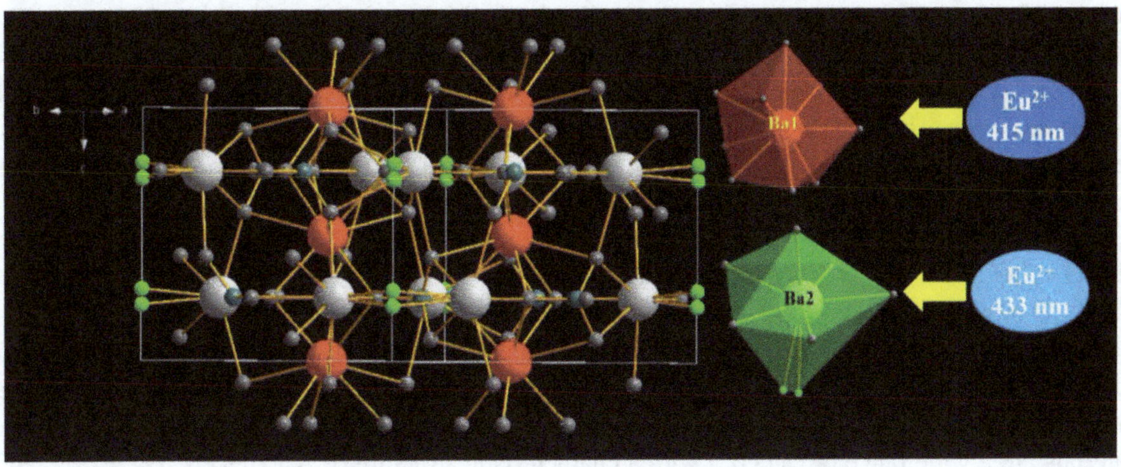

$Ba_{10}(PO_4)_6F_2$ (BPF) phosphors were synthesized by high temperature solid-state reaction method. $Ba_{10}(PO_4)_6F_2$ apatite has crystallized in a hexagonal structure with space group P63/m, a = b = 10.153Å, c = 7.733Å, α = β = 90°, γ = 120°, Z = 2, and V = 690.35Å^3. The crystal structure of $Ba_{10}(PO_4)_6F_2$ shown in the Figure 44. In this structure the Ba has the two types of sites to accommodate incoming Eu^{2+} ion. Ba^{2+}(1) is coordinated with nine oxygen atoms and the Ba^{2+}(2) is coordinated with six oxygen atoms

Figure 45. PL and PLE spectra of $Ba_{10}(PO_4)_6F_2$ .Reproduced from ref (Huang, Zhang, & Pan, 2017)

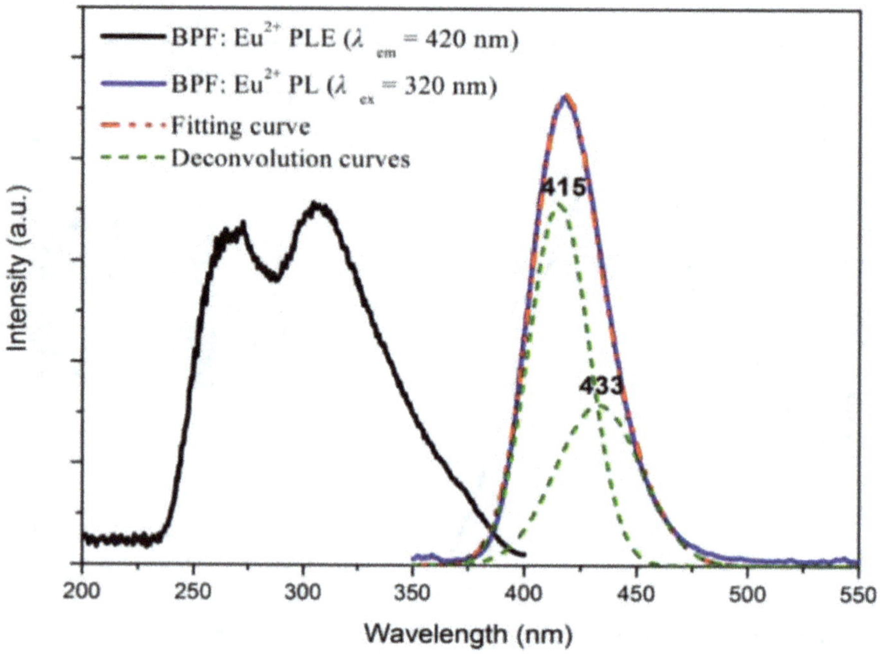

and one fluorine atom. The Eu^{2+} occupied both Ba(1) and Ba(2) sites based on their identical ionic radius(Huang, Zhang, & Pan, 2017).

The PL and PLE spectra of $Ba_{10}(PO_4)_6F_2$:$_{0.05}Eu^{2+}$ are exhibited in Figure 45. The excitation spectra show broad absorption band ranging from 240 to 400 nm and further it has two bands at 270 nm and 320 nm which are attributed to the $4f^7 \rightarrow 4f^65d^1$ transition of the Eu^{2+} ions. The PL emission spectrum The PL spectrum exhibits an emission peak around 420 nm, which are attributed due to the Eu^{2+} $4f^65d^1 \rightarrow 4f^7$ allowed transition. The FWHM of PL spectrum is ~40 nm. The phosphor intends good thermal stability, the relative intensity of the emission was decreased with increasing temperature. It retains 85% of initial intensity value at 150°C and then 81.5% at 180°C.

$Ba_5(PO_4)_2SiO_4$:Eu^{2+}

In recent years, apatite-like compounds have also been widely investigated as phosphor hosts for white light emitting diodes (w-LEDs) because of their variable structure and decent photoluminescence properties; $Ba_5(PO_4)_3F$ being one of them (Sato, 1994; Huang, 2011; Xin, 2013). Apatite-based materials are a wide range of category with a characteristic structure of $A_{10}(BO_4)_6X_2$ (A = Ca^{2+}, Sr^{2+}, Ba^{2+}; BO_4 = $[PO_4]^{3-}$, $[VO_4]^{3-}$, $[AsO4]^{3-}$; X = F^-, Cl^-, OH^-) (Nagpure, 2009; Ju, 2012; Chen, 2014). In the $Ba_5(PO_4)_2SiO_4$:1%Eu^{2+} phosphor, one $[PO_4]^{3-}$ and one F^- ions were substituted with one $[SiO_4]^{4-}$ ion. The $Ba_5(PO_4)_2SiO_4$ phosphor crystallises in a hexagonal crystal system with space group *P63/m* (176) and the cell parameters are a = 10.1762(1)Å, c = 7.7557(1)Å, V = 695.54(1)Å.The crystal structure of $Ba_5(PO_4)_2SiO_4$ is depicted in Figure 46. As two kinds of Ba^{2+} ions are present in the lattice, Ba(1) is belonging to 7-coordination and Ba(2) prefers to be in 6-coordination. There is only one kind of P site available in the tetrahedral structure. Si ion has the same site as P, signifying that these two kinds of

Figure 46. Crystal structure of $Ba_5(PO_4)_2SiO_4$. Redrawn from on ref (Ding & Wang, 2017)

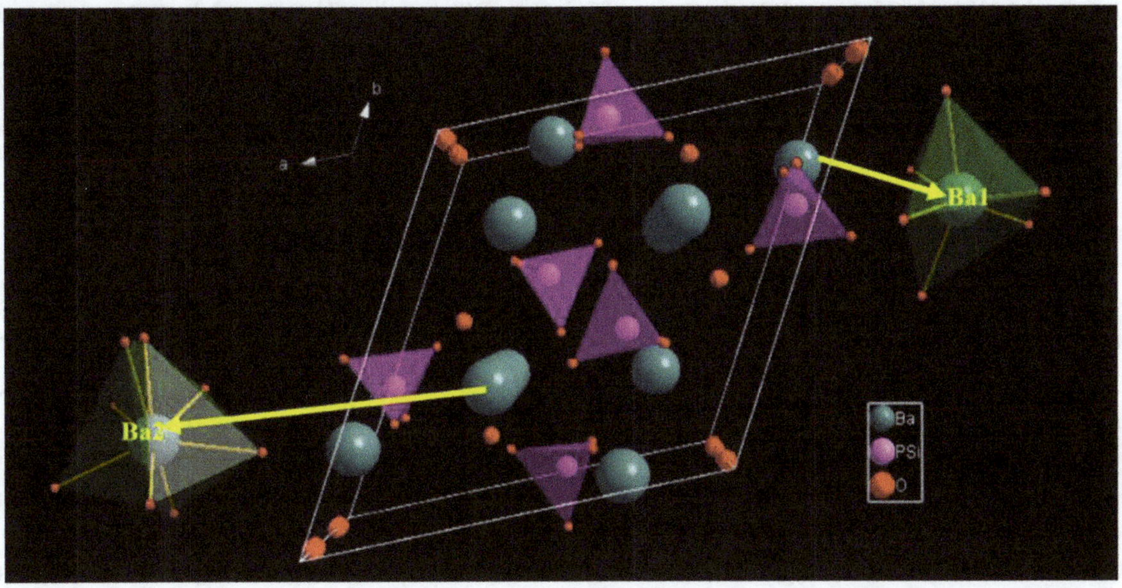

Figure 47. (a) PLE and (b) PL spectra of series $Ba_5(PO_4)_2SiO_4{:}_xEu^{2+}$ phosphor. Reproduced from ref (Ding & Wang, 2017)

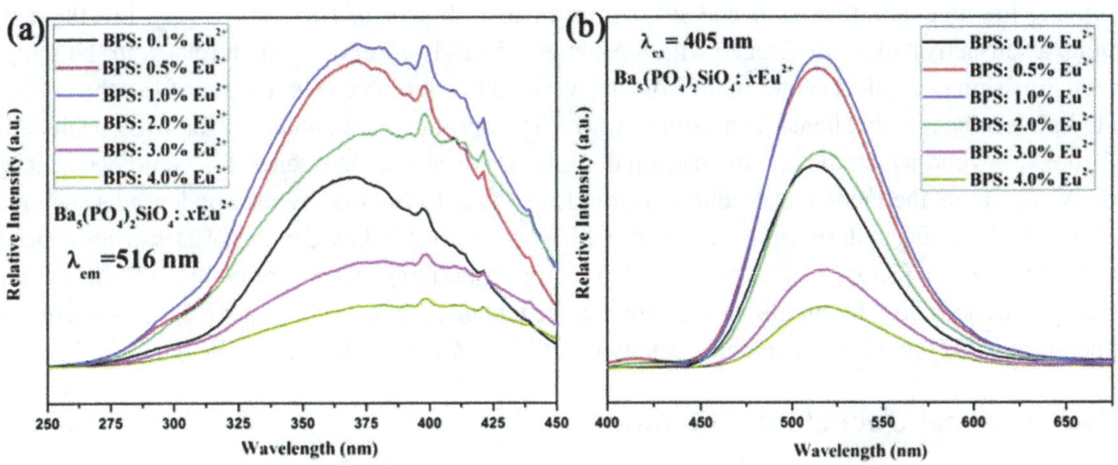

anions occupy the sites randomly and the corresponding occupation rates are 1/3 and 2/3 for Si and P, respectively(Ding & Wang, 2017).

The photoluminescence excitation (PLE) and photoluminescence (PL) spectra are shown in Figure 47. The photoluminescence studies of $Ba_5(PO_4)_2SiO_4{:}Eu^{2+}$ suggest that the phosphors own a very broad excitation band which is ranging from 300 to 450 nm, almost covers the UV and near UV (NUV) region, where the InGaN LED chip emits. Also, with increasing Eu^{2+} content, the intensity of the PLE spectra initially increases up to 1%Eu^{2+}, and decreases beyond 1% Eu^{2+}, is attributed to concentration quenching effect of Eu^{2+} ion. Under excitation of 405 nm NUV light, the phosphor emits green light at around

516 nm arising from the $4f^65d \rightarrow 4f^7$ transition of Eu^{2+}. The full-width at half-maximum (FWHM) value for the above is 75 nm and the quantum efficiency (QE) worth is 31.89% for the brightest sample ($Ba_5(PO_4)_2SiO_4$:1%Eu^{2+}).

By mixing RGB phosphors: $CaAlSiN_3:Eu^{2+}$, $Ba_5(PO_4)_2SiO_4$:1%Eu^{2+}, and BAM:Eu^{2+}, the white-LED was fabricated using a 405 nm GaN NUV chip driven by 30 mA current can get warm-white with co-ordinates at (0.355, 0.342) and CCT 4287 K.

$Ba_2Y_3(SiO_4)_3F:Eu^{2+}$

$Ba_2Y_3(SiO_4)_3F$:Eu belongs to the family of apatite's with a general molecular formula of $M_{10}(ZO_4)_6X_2$ (M is mono-, di- or trivalent cation, ZO_4 is a trivalent anion group, and X is a monovalent anion) (Yu, 2014; Kim, 2016; Huang, 2011). $Ba_2Y_3(SiO_4)_3F$ crystallizes in the hexagonal space group *P63/m* with cell parameters of $a = b = 9.7379$Å, and $c = 7.0182$Å. As revealed by the crystal structure of $Ba_2Y_3(SiO_4)_3F$ in Figure 48, there are two types of cationic sites labelled M1 (Wyckoff position 4f) and M2 (Wyckoff position 6h) for Eu doping. The M1 site possesses C_3 symmetry with nine surrounding oxygen atoms coordinated to form a mono-capped square anti-prism, while the M2 site is seven-fold coordinated with six O atoms and one F atom to formulate a pentagonal bi-pyramid with C_s symmetry. Y and Ba statistically occupy mutually the M1 and M2 sites with a Ba/Y molar ratio of 2/3(Wang et al., 2017).

Previously, Qiet al (2014). reported the luminescence properties of $Eu^{2+,3+}$ activated $M_2Y_3[SiO_4]_3F$ (M = Sr, Ba). Upon 350 nm UV light excitation (Figure 49a, b), the phosphor exhibits two main broad bands peaking at 470 nm and 600 nm, which are ascribed to the Eu^{2+}: $4f_6$ $5d_1$-$4f_7$ transitions of Eu^{2+} occupying the nine-coordinated 4f and seven-coordinated 6h sites of Ba, respectively. For the x = 0 sample, a self-activated luminescence with a peak at 415 nm is observed, which relates to the trapped excited emission of a defect center in the host. By varying the doping concentration of Eu, the obtained white light can be readily altered from cool to warm. The optimized Eu concentration was concluded to be 1 mol%, beyond which the d–d interaction dominated concentration quenching would occur. It was acknowledged that the incomplete reduction from Eu^{3+} to Eu^{2+} is caused by the condensed structure of $Ba_2Y_3(SiO_4)_3F$. Owing to the even spectral power distribution, the UV based white LED was fabricated by integrating the $Ba_2Y_3(SiO_4)_3F$:Eu phosphor/silicone composite along with a commercial 370 nm-emitting UV chip, produces bright white light with good color rendering index (CRI) of 83.5, color correlation temperature (CCT) of 3530 K and chromaticity coordinate (CIE) of (0.402, 0.371).

A Zero-Thermal-Quenching Phosphor

One of the most significant task of pc-WLEDs is the thermal quenching, in which the phosphor loses the emission intensity with increasing temperature during high-power LED operation. Blue-emitting $Na_{3-2x}Sc_2(PO_4)_3:xEu^{2+}$ phosphor was first reported by Wanget al., (2016), detailed fundamental studies of this phosphor investigated by Kim et al., (2017). $Na_{3-2x}Sc_2(PO_4)_3:xEu^{2+}$ phosphor was synthesized by conventional solid state method. The $Na_{3-2x}Sc_2(PO_4)_3:xEu^{2+}$ phosphor has adopted monoclinic *C2/c* and lattice parameters are a = 16.0937Å, b = 8.9275Å, c = 9.1070Å, β = 127.021° and V = 1044.700Å^3. The crystal structure of NSPO shown in Figure 50 and it has three different crystallographic Na sites namely Na1, Na2 and Na3 coordinated with six, seven and eight oxygens, respectively. The Eu^{2+} ions could occupy any of the three crystallographic Na sites, because of their similar cationic size. The Na^+

Figure 48. Crystal structure of $Ba_2Y_3(SiO_4)_3F$, and the coordination environment of the M(I) 4f site with the local symmetry of C_3 and the M(II) 6h site with the local symmetry of C_S. Redrawn from ref (Wang et al., 2017)

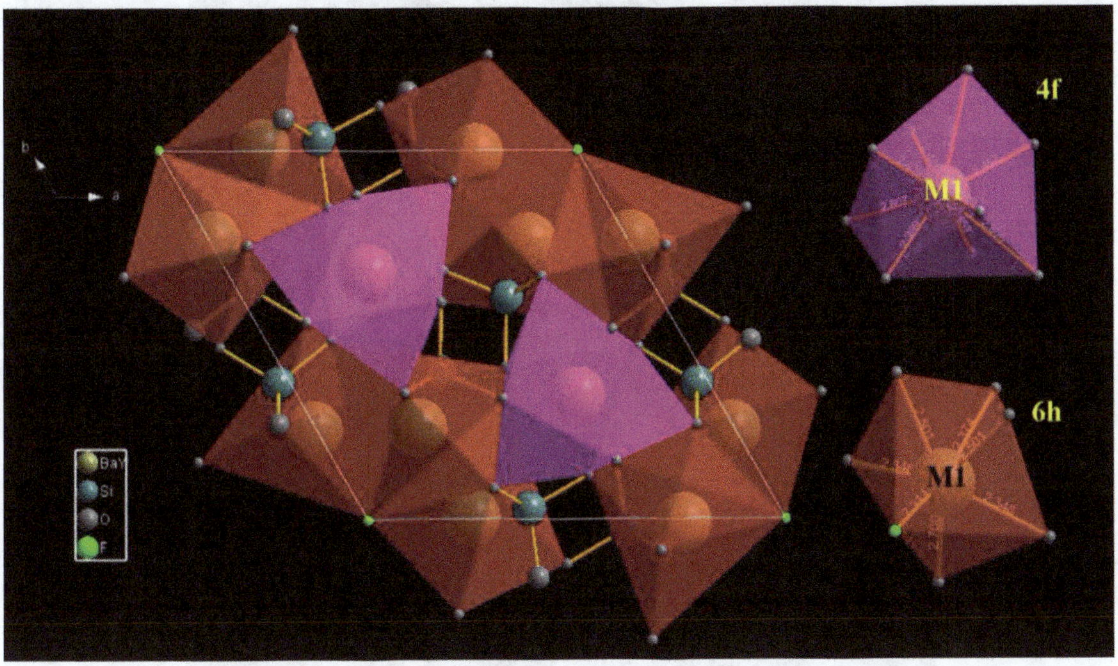

Figure 49. (a) PL spectrum of $Ba_2Y_3(SiO_4)_3F:_xEu$ upon excitation at 350 nm. (b) Corresponding CIE chromaticity diagram of the $Ba_2Y_3(SiO_4)_3F:_xEu$ phosphors. Reproduced from ref (Wang et al., 2017)

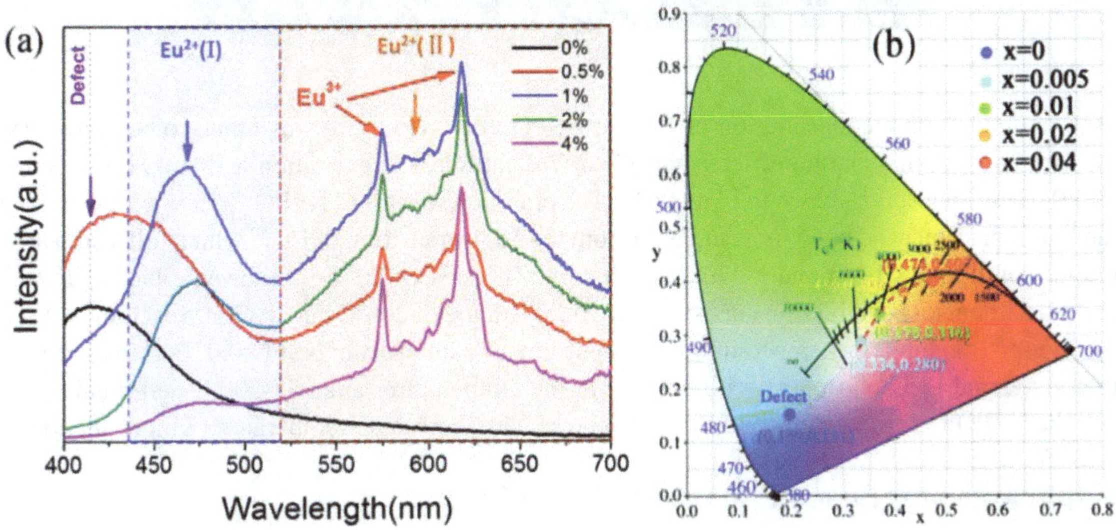

disorder in the host rises with temperature, and may thus disturb the optical properties of the Eu^{2+} activator (Wickleder, 2002, pp. 2011-2088).

Figure 50. Crystal structure of NSPO and crystallographic sites of Na. Redrawn from ref by (Kim et al., 2017)

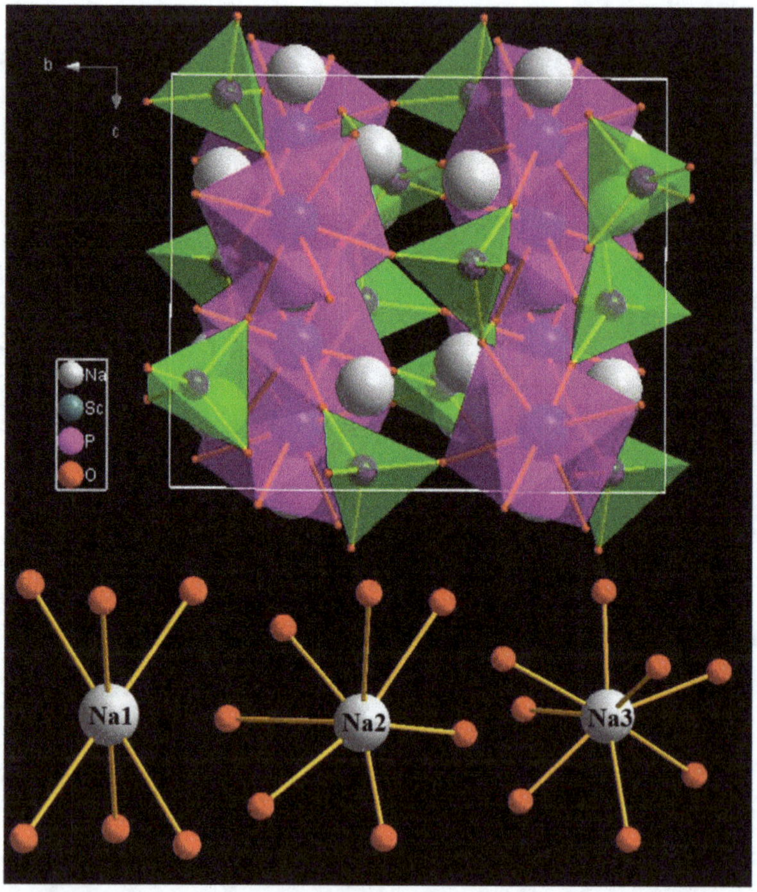

The optimal doping concentration of Eu^{2+} in NSPO:xEu^{2+} (Fig. 2b) was found to be 3mol% (x = 0.03) with an internal quantum efficiency (QE) of 74% at RT. The excitation and emission spectra of NSPO:0.03Eu^{2+} phosphor shown in Figure 51a. The emission spectra of NSPO:$_{0.03}$$Eu^{2+}$ exposes a single broad blue-emission band at 453nm emerging from the 5d-4f transition of Eu^{2+}, with a full-width at half maximum of around 44 nm under 370nm excitation. The NSPO:0.07Eu^{2+} showed a thermally stable emission, without loss of emission intensity up to a temperature of 200°C under 370nm excitation (Figure 51b). Figure 51c exhibits the temperature-dependent emission intensity of NSPO:0.07Eu^{2+} in comparison with commercial LED phosphors with low TQ in the temperature range 25-200°C and this is clearly shows the NSPO:$_{0.03}$$Eu^{2+}$ phosphor zero emission loss with increasing temperature(Kim et al., 2017)

Multiple Activator Based Phosphors for LEDs

The advantage of having multiple activator in the phosphor host lattices is that it can be very much useful in generating tunable emission as well as white light emission. Using the single component white

Figure 51. (a) The excitation and emission spectra of NSPO:$_{0.03}$Eu^{2+} phosphor. (b) Temperature-dependent emission spectra of NSPO:$_{0.07}$Eu^{2+} phosphor under 370nm excitation in the temperature range 25–200°C. (c) Temperature-dependent normalized emission spectra of NSPO:$_{0.07}$Eu^{2+} under 370nm excitation in comparison with commercial LED phosphors upon heating to 200°C. Reproduced from refby (Kim et al., 2017)

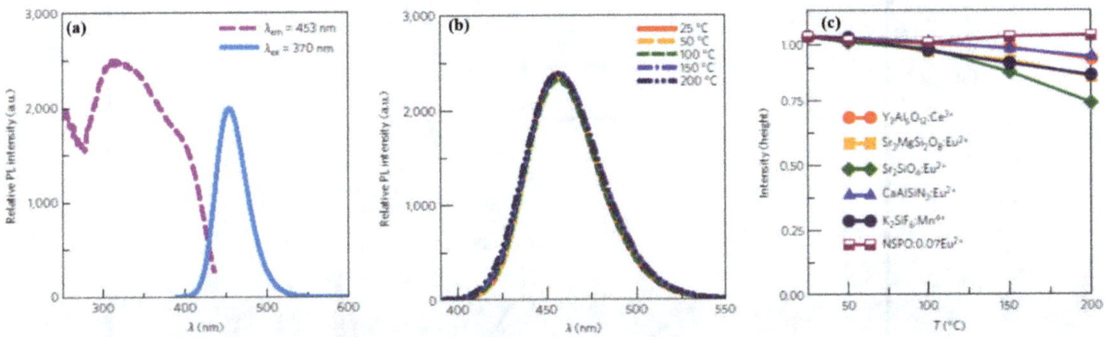

light emitting phosphor can utilize the whole spectral range to mimic sunlight spectrum and overcome the reabsorption of multiple phosphor.

$Na_6Ca_3Si_6O_{18}$:Ce^{3+}, Tb^{3+}

As a vast phosphor family, silicates were one of the first hosts fabricated in to single-phased full-color emitting phosphors. They can parade efficient luminescent properties due to their high chemical and physical stability, as well as their water-resistant properties (Park, 2003; Kim, 2004, 2005). The calcining temperatures of the rare-earth doped alkaline earth silicates are lower in comparison with alkaline earth silicates. Moreover, it has been well-known that a terbium (Tb^{3+}) ion is well known as a green emitting activator owing to its chief $^5D_4$–$^7F_5$ transition peak a ~ 550 nm [(Li, 2012; Fu, 2013)]. However, the intensities of Tb^{3+} absorption peaks in the ultraviolet (UV) region are very feeble. In order to pacify the absorption strength of Tb^{3+} in the UV region, one of the approaches is to use Ce^{3+} as a sensitizer because it has a strong excitation band originating from allowed 4f→5d transitions, which could efficiently absorb the UV light and transfer the excitation energy to Tb^{3+}ion.

The $Na_6Ca_3Si_6O_{18}$phosphor crystallizes in the rhombohedral crystal structure with space group *R* $\bar{3}m$ (166), and the cell parameters are a = b = 10.50Å and c = 13.184Å. The excitation spectrum of $Na_6Ca_3Si_6O_{18}$:$0.04Ce^{3+}$ (Figure 52)monitored at 440 nm exhibits a broad band with a maximum at about 450 nm under the excitation of UV light (330 nm). The PLE spectrum of $Na_6Ca_3Si_6O_{18}$:$0.1Tb^{3+}$ monitored at 550 nm exhibits a broad band ascribed to the 4f→5d transitions of the Tb^{3+} ions and other several peaks due to the intra-4f^8 transitions of Tb^{3+} ions. Moreover, a noticeable spectral overlap amid the emission band of Ce^{3+} and the excitation band of Tb^{3+} is obtained in Figure 52, indicating an efficient possibility of energy transfer from Ce^{3+} to Tb^{3+} ions (Xu et al., 2017).

The Tb^{3+} emission affectedly increased with the doping of the Ce^{3+} ions compared to the emission spectrum of $Na_6Ca_3Si_6O_{18}$: $0.04Tb^{3+}$ in Figure 53a. This phenomenon indicated that Ce^{3+} could absorb energy efficiently and then transferred the excess energy to Tb^{3+}, acting as a sensitizer. Hence, under the excitation of 330 nm (Figure 53b), the emission spectra appear not only as a strong blue band of the

Figure 52. PL and PLE spectra of $Na_6Ca_3Si_6O_{18}$:$_{0.04}Ce^{3+}$ and $Na_6Ca_3Si_6O_{18}$:$_{0.1}Tb^{3+}$ phosphor. Reproduced from ref (Xu et al., 2017)

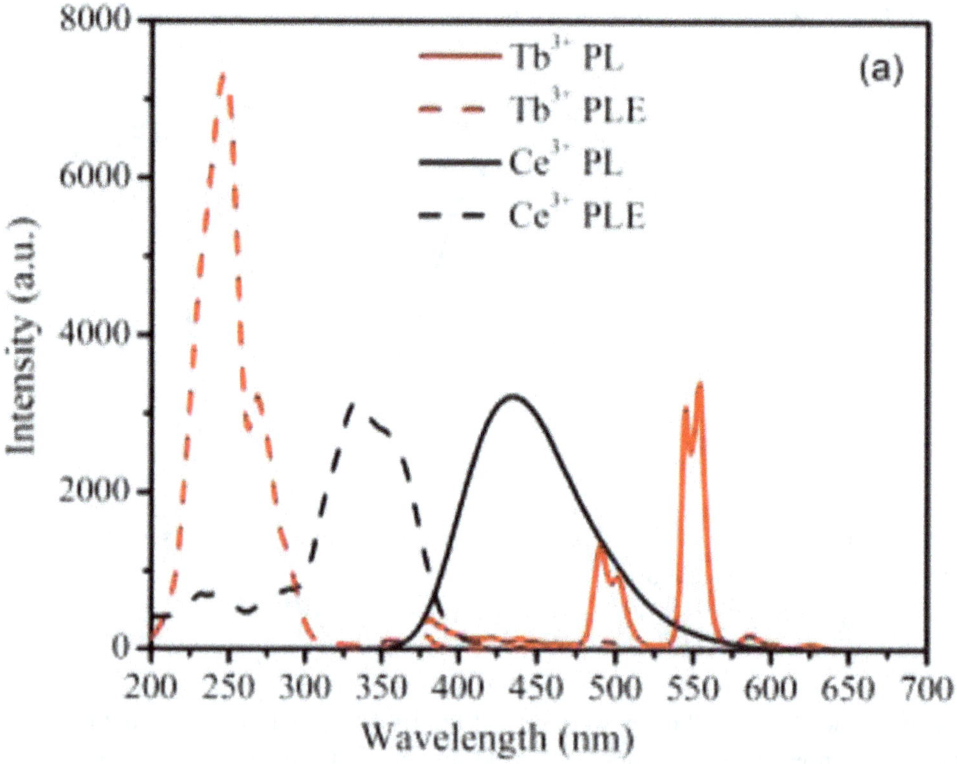

Figure 53. (a) PL spectra of $Na_6Ca_3Si_6O_{18}$: $_xCe^{3+}$, $_{0.04}Tb^{3+}$ phosphors with various Ce^{3+} contents (λ_{ex}=350 nm). (b) PL spectra of $Na_6Ca_3Si_6O_{18}$:$_{0.04}Ce^{3+}$, $_yTb^{3+}$. Reproduced from ref (Xu et al., 2017)

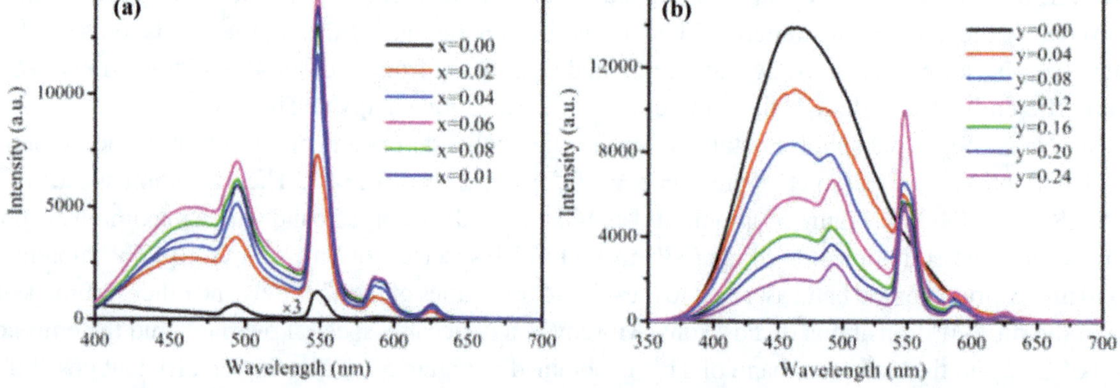

Ce^{3+} ions, but also as a strong green band of the Tb^{3+} ions. The absolute PL QE can achieve about 43% under UV radiation, indicating that the phosphor may serve as a potential green-emitting UV-convertible phosphor.

$NaBaScSi_2O_7$:Ce^{3+},Tb^{3+}

As a promising sensitizer for Tb^{3+} ions, Ce^{3+} ions have been widely used in many hosts (Huang, 2011; Xia, 2012; Hsu, 2011). Ce^{3+} ion is also an excellent activator of blue emitting luminescent materials. By co-doping Ce^{3+} and Tb^{3+} ions in suitable hosts, high efficiency blue-green emission for UV-LEDs can be simultaneously realized in single-phase phosphors. The phosphorescence properties of Eu^{2+} doped $NaBaScSi_2O_7$ were studied in detail by Li et al.(2015). Its photoluminescence properties also have been studied by Liu et al. (2013). The silicate $NaBaScSi_2O_7$ compound has a monoclinic system with space-group symmetry of $P2_1/m$ and the cell parameters are a = 6.86117Å, b = 5.61656Å, c = 8.81239Å, which has been considered to be an excellent host for luminescence (Li et al., 2016).

The NBSSO:Ce^{3+} phosphors can emit intense blue light. By comparison from figure 54, it is found that the PLE spectrum of NBSSO:0.04Ce^{3+},0.04Tb^{3+} monitored at 425 nm (Ce^{3+} emission) is similar to that monitored at 542 nm (Tb^{3+} emission) except for the difference in the relative intensity, and the two PLE spectra consist of the characteristic excitation of both Ce^{3+} and Tb^{3+} ions, simultaneously. The presence of the broad-band transition from Ce^{3+} ions in the PLE spectrum monitored at the $^5D_4$ - $^7F_5$ transition of Tb^{3+} ions proves the occurrence of ET from Ce^{3+} to Tb^{3+} ions. For the Ce^{3+}/Tb^{3+} co-doped samples, ET from Ce^{3+} to Tb^{3+} ions has been demonstrated to be of the resonant type via the dipole–dipole interaction. Tunable colors from blue to green can be realized by solely varying the doping concentration of Tb^{3+} ions. The phosphors obtained have a broad excitation band ranging from 250 to 400 nm which matches well with the commercial UV-LED chips.

$Ba_2Y_5B_5O_{17}$:Ce^{3+}, Tb^{3+}

Figure 54. (a) PLE spectra of NBSSO:0.04Ce^{3+}, NBSSO:0.04Tb^{3+} and NBSSO:0.04Ce^{3+},0.04Tb^{3+}. (b) PL spectra of NBSSO:0.04Ce^{3+}, NBSSO:0.04Tb^{3+} and NBSSO:0.04Ce^{3+},yTb^{3+} ($0.01 \le y \le 0.07$). Reproduced from ref (Li et al., 2016)

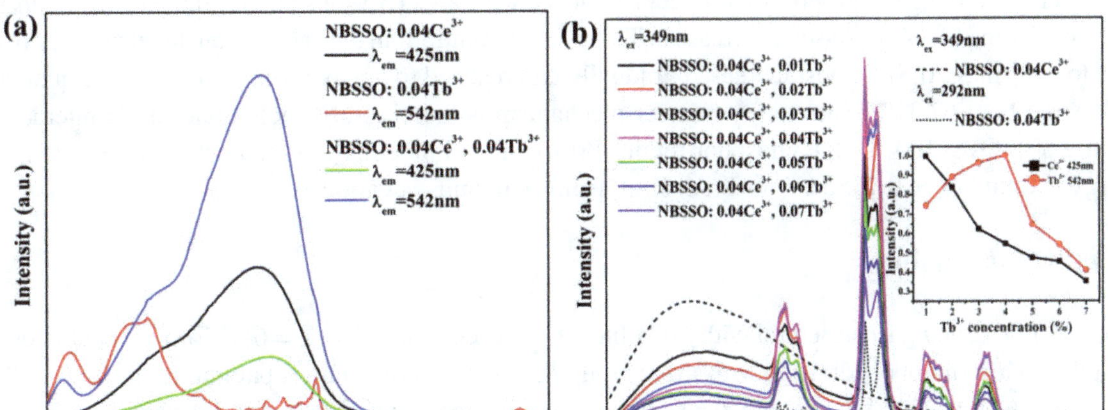

Hermus (2016) reported a novel efficient blue-emitting $Ba_2Y_5B_5O_{17}$:Ce^{3+} (BYBO:Ce^{3+}) phosphor for near-UV excitation . The crystal structure is shown in Figure 13, crystallographic information given in page 20-21 see ref (Hermus, Phan, & Brgoch, 2016).

The Tb^{3+} singly doped BYBO: 2% Tb^{3+} presents the well-known green fluorescence with the main emission line at 543 nm (Figure 55a top). The PLE spectrum of BYBO: 1% Ce^{3+}, 2% Tb^{3+} was monitored Tb^{3+} emission at 543 nm exhibits only the characteristic PLE bands of Ce^{3+} in the near-UV region (Figure 55a down). Emission color tuning from blue to green was achieved by continuously increasing Tb^{3+} concentration to enhance energy transfer. The green emission of Tb^{3+} dominates the luminescence for Tb^{3+} concentration of at least 10%. The strongest Tb^{3+}emission was observed at 20% Tb^{3+} with the external quantum efficiency of 55%. The blue emission of Ce^{3+} appears the similar thermal stability to Ce^{3+} singly doped sample. In Figure 55b, the green emissions of all the samples at 423 K keep an average of 92% of their intensities at room temperature, including the best one, which shows no intensity decreasing from room temperature to 483 K for 30% Tb^{3+}. The high thermal stability of the green emission upon Ce^{3+} excitation is ascribed to suppression of Ce^{3+} thermal de-excitation by fast energy transfer to thermally stable Tb^{3+} emitters. The relatively low thermal stability of Ce^{3+} emission in the co-doped BYBO is attributed to a slow transfer rate of the emitting Ce^{3+} ions that hardly compete with the thermal de-excitation (Xiao et al., 2017).

$SrAl_2Si_2O_8$: Ce^{3+}/Tb^{3+}

Aluminosilicate compounds are potential candidates as host materials due to numerous advantages, such as excellent physical and chemical properties, high efficiency and thermal stability (Lemanski, 2016; Lv, 2014). The host $SrAl_2Si_2O_8$ was first reported by A.F. Reid and co-workers (Reid & Ringwood, 1969). $SrAl_2Si_2O_8$ has a monoclinic structure (Figure 56) with a space group of *C2/m* (a = 8.39Å and c = 7.12Å). There is only one Sr^{2+} crystallographic sites in the monoclinic unit cell, which is bonded to eight oxygen atoms forming an irregular hexahedron. The Al^{3+} and Si^{4+} connect four O atoms to form three-dimensional $(Al/Si)O_4$ frameworks, which creates an infinite net structure by corner-sharing (Ma et al., 2017).

The $SrAl_2Si_2O_8$:0.04Ce^{3+} and $SrAl_2Si_2O_8$:0.04 Tb^{3+} phosphors can emit blue and green light, respectively. Under 365 nm excitation, the emission color of $SrAl_2Si_2O_8$:Ce^{3+}, Tb^{3+} can be adjusted from blue to green by changing the proportion (concentration) of Ce^{3+} and Tb^{3+} ions based on the energy transfer (Figure 57 a and b). With constant increase in Tb^{3+} ions concentration, the energy transfer efficiency from Ce^{3+} to Tb^{3+} in $SrAl_2Si_2O_8$ host increased gradually and reached to a maximum at 80.92%, the quantum yield was about 67.37%. The energy transfer mechanism is found to be dipole-quadrupole interaction (Figure 58). Over 100 °C, the phosphor retains 90% of its initial emission intensity and the activation energy is found to be 0.2352 eV for $SrAl_2Si_2O_8$ from Arrhenius equation.

$Ca_9Y(PO_4)_7$:Eu^{2+},Mn^{2+}

$Ca_9Y(PO_4)_7$, being isostructural with β-$Ca_9In(PO_4)_7$ (space group $R3c$, $Z = 6$) (Morozov et al., 2002), crystallizes in a rhombohedral cell with space group $R3c$ (No. 161) and lattice parameters a = 10.4442Å, c = 37.324Å, V = 3525.89 nm^3, and Z = 6. It has seven independent cation crystallographic sites available the unit cell: among them, three Ca^{2+} site (Ca^{I}, Ca^{II}, Ca^{III} with C.N. = 8, 8 and 9 respectively), one Y^{3+} site with C.N. = 6 and three P^{5+} sites with C.N. = 4. The Mn^{2+} and Eu^{2+} ions occupy the Ca^{2+} sites

Figure 55. (a) PLE and PL spectra of BYBO: 2% Tb^{3+} and BYBO: 1% Ce^{3+}, 2% Tb^{3+}.(b) Temperature dependence of (a) Ce^{3+} emission intensity and (b) Tb^{3+} emission intensity in BYBO: 1% Ce^{3+}, x% Tb^{3+} under 345 nm excitation. Reproduced from ref (Xiao et al., 2017)

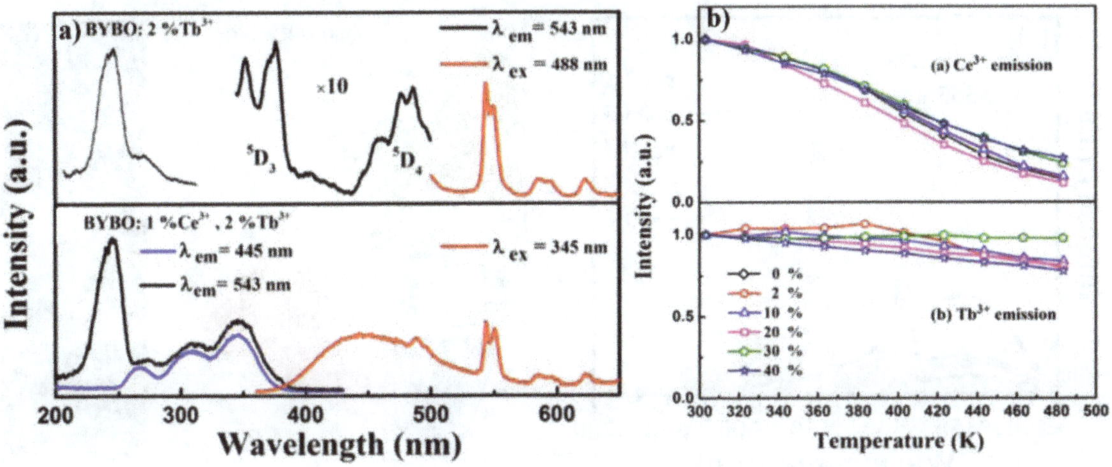

Figure 56. Crystal structure of $SrAl_2Si_2O_8$. Redrawn from ref (Ma et al., 2017)

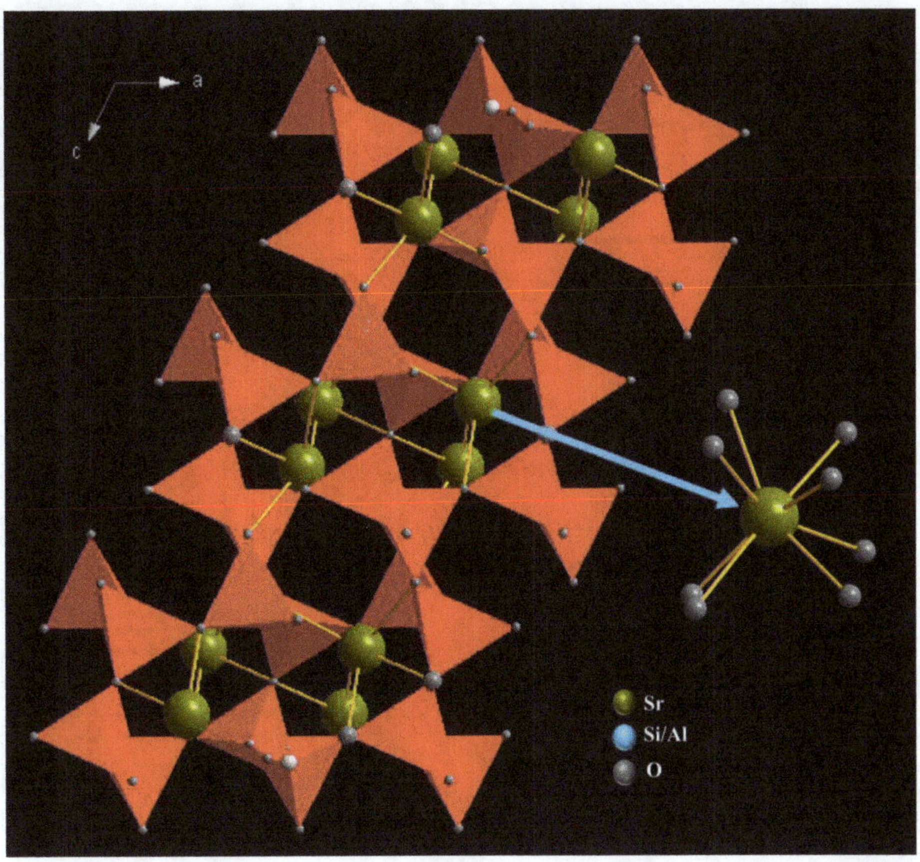

Figure 57. (a) PL spectra for SASO: $0.04Ce^{3+}$,zTb^{3+} (z = 0 - 0.06) under 365 nm excitation. (b) CIE coordinates of SASO:$0.04Ce^{3+}$, zTb^{3+} (z = 0 - 0.06) under and the corresponding photographs under 365 nm UV lamp excitation. Reproduced from ref (Ma et al., 2017)

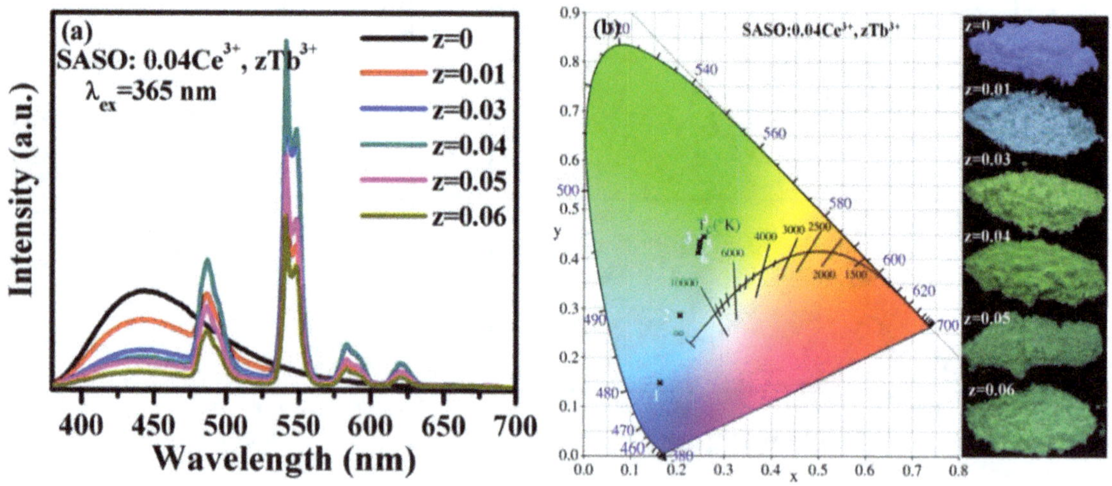

Figure 58. Schematic energy-level diagram showing the excitation and emission mechanism of SASO: Ce^{3+}, Tb^{3+} phosphors. Redrawn from ref (Ma et al., 2017)

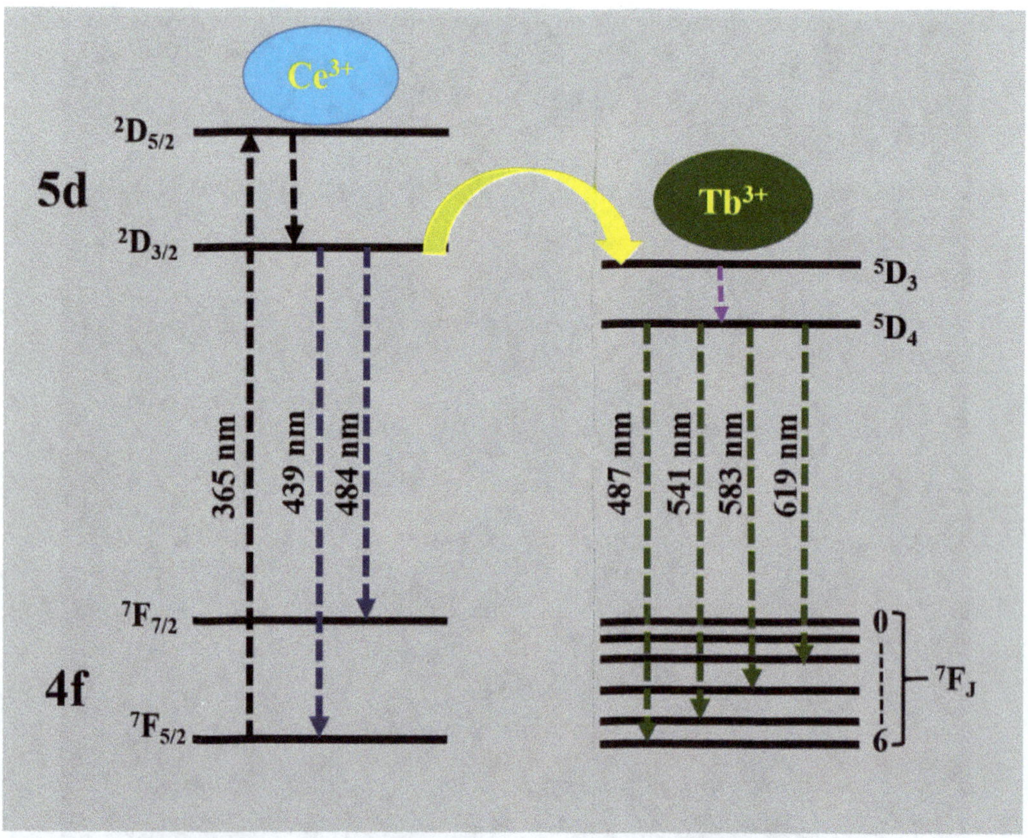

Figure 59. (a) Reflectance spectra of CYP. (b) Reflectance and PL/PLEs pectra of CYP:$_{0.01}$Eu^{2+} and (c) CYP:0.01Eu^{2+}, 0.015Mn^{2+}. Reproduced from ref (Huang et al., 2010)

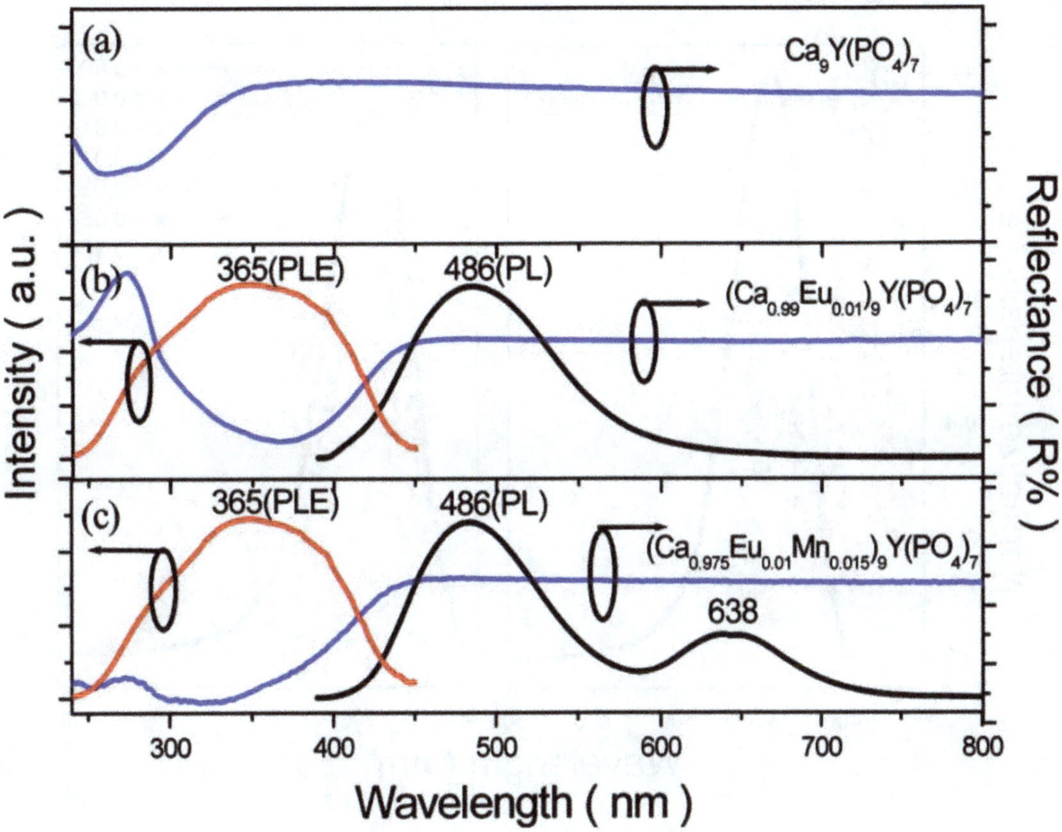

randomly because the ionic radii for eight- and nine-coordinated Ca^{2+} are 1.12 and 1.18 Å, respectively whereas for eight- and nine-coordinated Eu^{2+} and eight-coordinated Mn^{2+} are 1.25, 1.30 and 0.95 Å, respectively (Huang et al., 2010).

The PLE spectra at λ_{em} = 486 nm, shows a broad hump between 250 to 440 nm, which is cons the reflection spectrum (Figure 59a-c). The $Ca_9Y(PO_4)_7$: Eu^{2+},Mn^{2+} exhibits two broad blue-greenish and red emission bands raising at 486 and at 638 nm ($^4T_1$ (^4G) → $^6A_1$ (^6S) transition of Mn^{2+}), respectively. For $Ca_9Y(PO_4)_7$: 0.01Eu^{2+}, $x$$Mn^{2+}$ in Figure 60, the PL emission intensity at 486 nm of Eu^{2+} decreases with the increasing Mn^{2+} content x, and the emission intensity of Mn^{2+} at 638 nm gradually increases with increasing Mn^{2+} content x until the concentration quenching occurs in Mn^{2+} content of 0.05. This is indirect evidence for energy transfer process from Eu^{2+} to Mn^{2+}. The energy transfers from Eu^{2+} to Mn^{2+} in $Ca_9Y(PO_4)_7$ host matrix occurs through a dipole-quadrupole mechanism with critical distance of ~ 11Å.

When a white LED was fabricated by integration of a 365 nm UV chip with a single-phased white-light phosphor $(Ca_{0.975}Eu_{0.01}Mn_{0.015})_9Y(PO_4)_7$, the corresponding phosphor shows a cool white light with color temperature of 7200 K, *Ra* of 76, and color coordinates of (0.30, 0.31) (Figure 61). These results specify that the performance of a white pc-LED with $(Ca_{0.975}Eu_{0.01}Mn_{0.015})_9Y(PO_4)_7$ is superior than that of white YAG:Ce^{3+} pc-LED pumped with a blue LED chip in all aspects.

$Ca_9Gd(PO_4)_7$:Eu^{2+},Mn^{2+}

Figure 60. (a). Spectral overlap between PL spectrum of CYP:Eu^{2+} (solid line) and PLE spectrum of CYP:Mn^{2+} (dash line). (b) PL spectra of CYP:0.01Eu^{2+}, xMn^{2+} phosphors excited at 365 nm. Reproduced from ref (Huang et al., 2010)

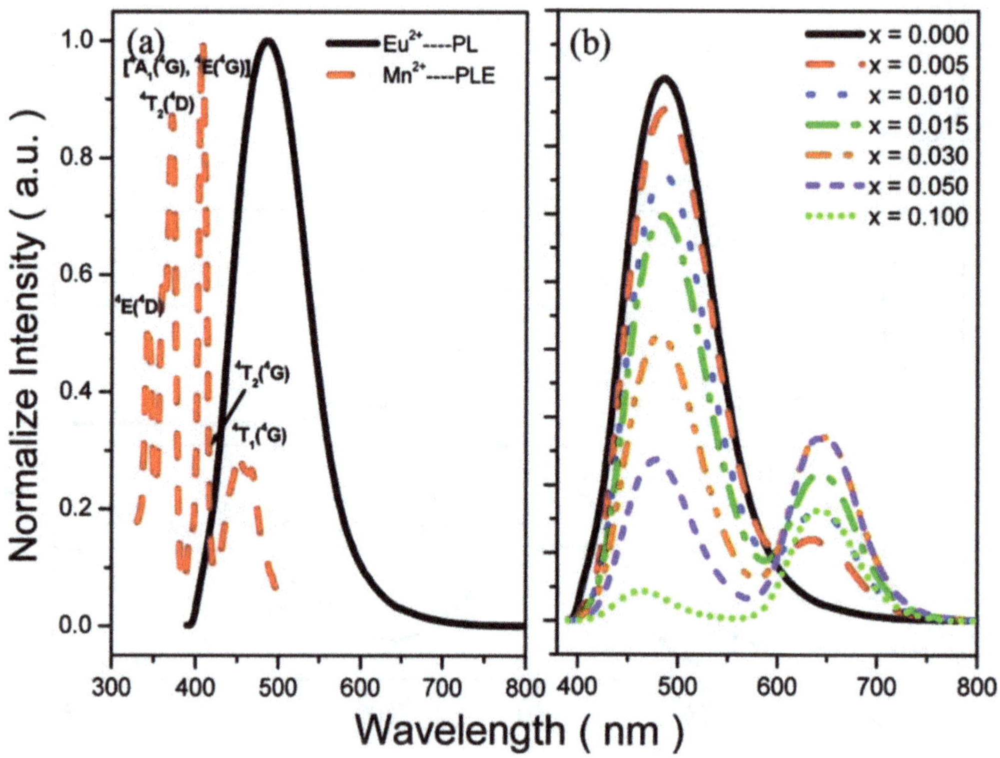

The $Ca_9Gd(PO_4)_7$ is isostructural with the $Ca_9Y(PO_4)_7$and crystallizes in a rhombohedral cell with space group *R3c* (No. 161) and lattice parameters a = 10.4442Å, c = 37.324Å, V = 3525.89 nm^3, and Z = 6 (Huang, Liu, & Chen, 2010). The PL spectrum of $Ca_9Gd(PO_4)_7$:Eu^{2+} illustrations a broad band ranging between 400 and 750 nm, which is especially because the electric dipole allowed $4f^65d^1 \rightarrow 4f^7$ transitions of Eu^{2+}. The excitation peaks of $Ca_9Gd(PO_4)_7$:Mn^{2+} contain a number of bands centred at 251, 340, 357, 407, and 502 nm (Figure 62). Varying the Mn^{2+} concentration from 0.01 to 0.1 mol of in $Ca_9Gd(PO_4)_7$:0.007Eu^{2+}, $x$$Mn^{2+}$ phosphors yield blue-greenish and red emission bands centering at 494 nm ($4f^65d^1 \rightarrow 4f^7$ transition of Eu^{2+}) and 652 nm ($^4T_1(4G) \rightarrow {}^6A_1(^6S)$ transition of Mn^{2+}). The intensity of Eu^{2+} blue-green emission at 494 nm decreases with increasing Mn^{2+} content up to x. The intensity of red emission at 652 nm increases with increasing concentration of Mn^{2+} to x = 0.05, afterwards with further increasing the concentration quenching befalls.

The energy transfer process exhibiting from sensitizer Eu^{2+} to activator Mn^{2+} in $Ca_9Gd(PO_4)_7$ host through a nonradiative dipole-quadrupole mechanism. A white-light near-UV LED was fabricated by integrating a near-UV 380 nm chip pumped by a single-phase white-light $Ca_9Gd(PO_4)_7$:0.007Eu^{2+},0.02Mn^{2+} phosphor operated by a 350 mA current, producing a white light with a correlated CCT of 6569 K and the CIE of (0.312, 0.327).

$NaScSi_2O_6$:Eu^{2+},Mn^{2+}

Figure 61. EL spectrum of a pc-LED lamp fabricated with a 365 nm LED chip and a white-emitting phosphor CYP:0.01Eu²⁺,0.015Mn²⁺, driven by 350 mA current. Reproduced from ref (Huang et al., 2010)

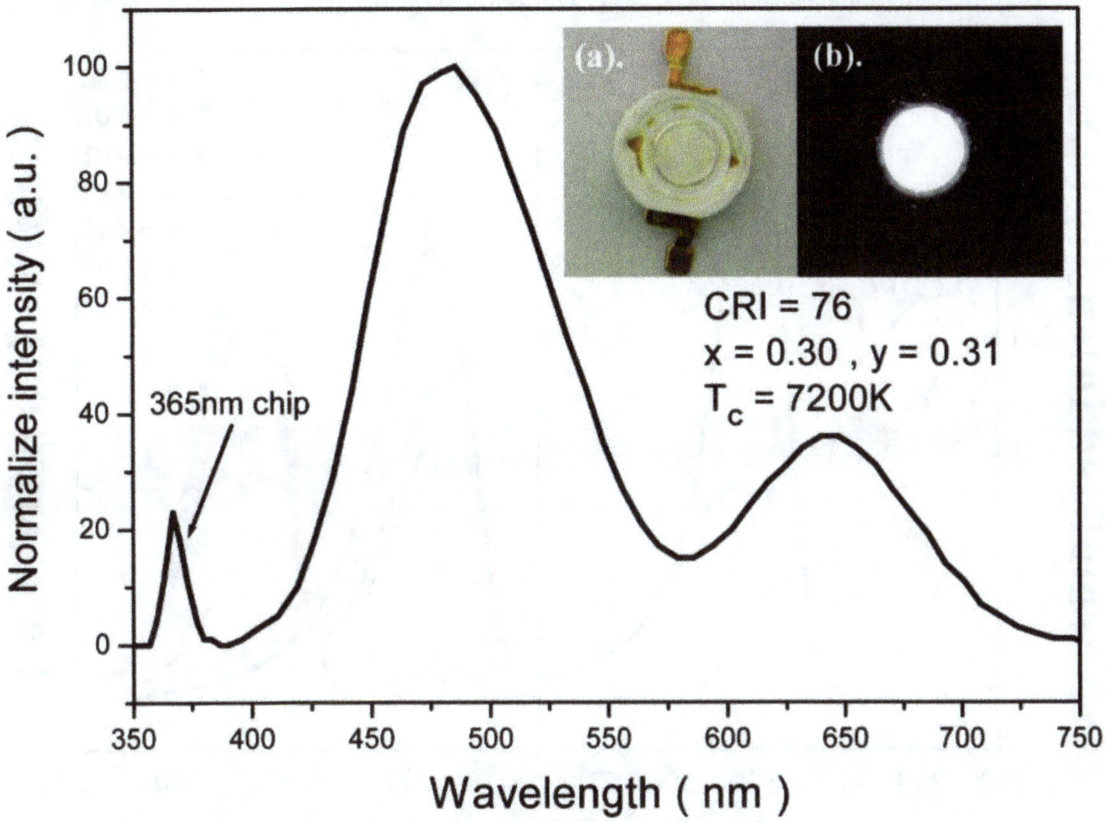

In recent days, the white LEDs are fabricated using near ultraviolet (n-UV) light (350–420 nm) chips combined with the multicolor-emitting phosphors have been widely reported (Huang, 2011; Liu, 2011; Lu, 2013). Among them, the progression of new yellow emitting phosphors is remarkable.As per literature reports, number of novel yellow phosphors for the n-UV white LEDs have been reported, for example, $Sr_8ZnSc(PO_4)_7$:Eu^{2+}, $Ba_2Gd(BO_3)_2Cl$:Eu^{2+}, $Y_2(CN_2)_3$:Ce^{3+}, $Na_3K(Si_{1-x}Al_x)_8O_{16}$:$Eu^{2+}$, etc.

$NaScSi_2O_6$ phosphors were synthesized by sol-gel technique. First the $EuCO_3$ and $MnCO_3$ were dissolved in concentrated HNO_3 acid to get the dried $Eu(NO_3)_3$ and $Mn(NO_3)_2$ powders. The stoichiometric amounts of $NaNO_3$ and $Sc(NO_3)_3$ were dissolved in ethanol with constant stirring. Then the stoichiometric amounts of $NaNO_3$ and $Sc(NO_3)_3$ were dissolved in ethanol under stirring. After this, the designed amounts of $Eu(NO_3)_3$, $Mn(NO_3)_2$, and $Si(OC_2H_5)_4$ were added successively in the solution. The resultant mixtures were stirred at 80°C for 30 min until the formation of homogeneous gels. After drying at 120°C for 12 h in an oven, the gels were ground and preferred at 800°C for 20 h in the air and then fully ground and sintered at 1200°C in the CO reducing atmosphere for 5h. After that, the samples were furnace-cooled to RT and ground again into fine powders. Thus, the Eu^{2+} and Eu^{2+}/Mn^{2+} doped $NaScSi_2O_6$ phosphors were finally obtained for the following investigation. The $NaScSi_2O_6$ the crystalized in the monoclinic structure with space group*C2/c* and the unit cell parameters are a = 9.831Å, b = 9.068Å, c = 5.341Å, β = 107.01° and V = 456.01(3). The crystal structure of $NaScSi_2O_6$ is shown in Figure 63. On the basis

Figure 62. (a) Spectral overlap between the Eu^{2+} PL spectrum (solid line) and the Mn^{2+} PLE spectrum (dash line). (b) the emission spectra of $(Ca_{0.993-x})_9Gd(PO_4)_7$:$0.007Eu^{2+}$,$xMn^{2+}$ phosphors under near-UV 380 nm excitation. Reproduced from ref (Huang, Liu, & Chen, 2010)

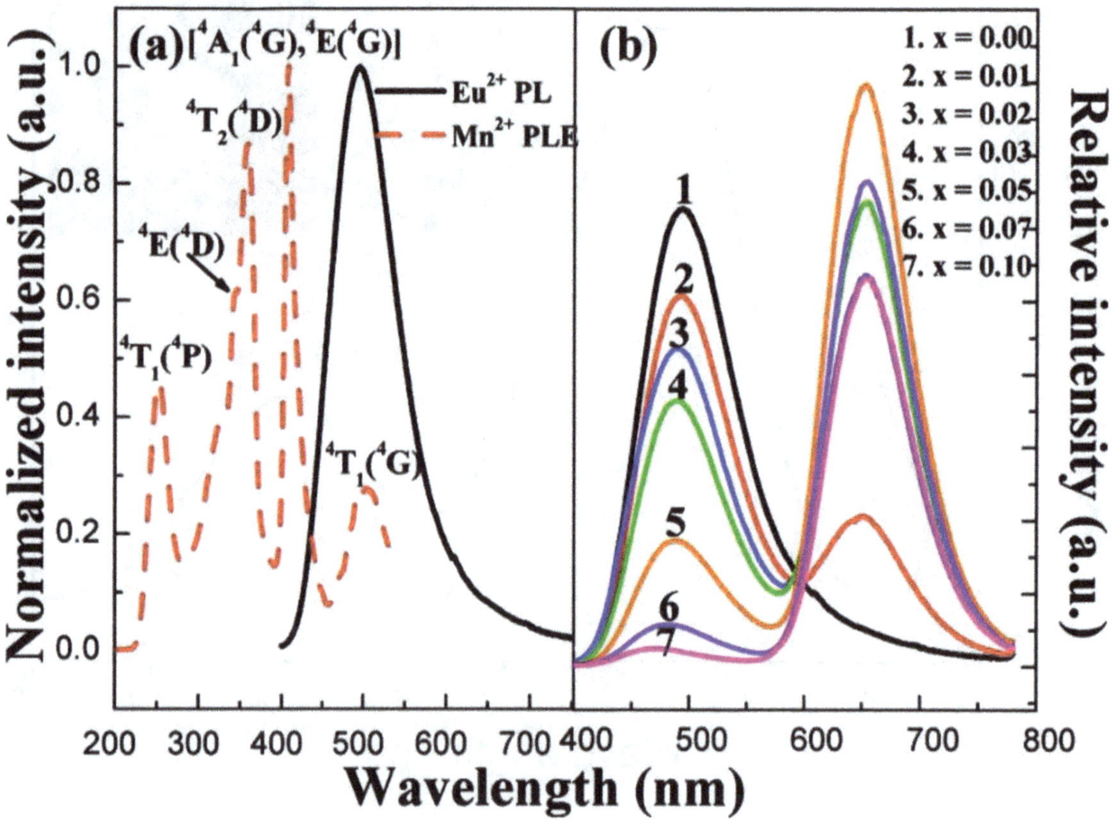

Figure 63. Crystal structure of $NaScSi_2O$. Redrawn from ref (Xia, Zhang, Molokeev, & Atuchin, 2013)

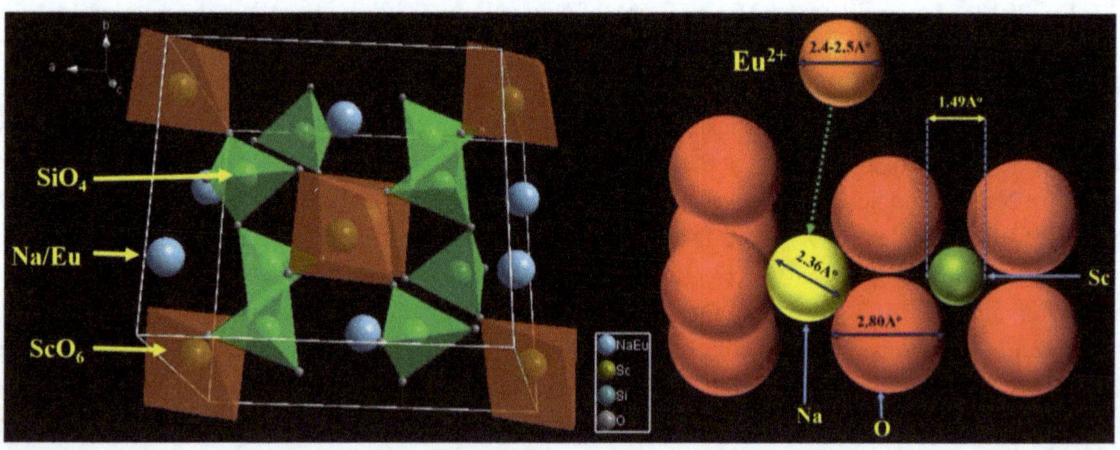

Figure 64. PLE and PL spectra of $NaScSi_2O_6$:$_{0.05}Eu^{2+}$, $NaScSi_2O_6$:$_{0.10}Mn^{2+}$, and $NaScSi_2O_6$:$_{0.05}Eu^{2+}$,$_{0.10}Mn^{2+}$ phosphors. Reproduced from ref (Xia, Zhang, Molokeev, & Atuchin, 2013)

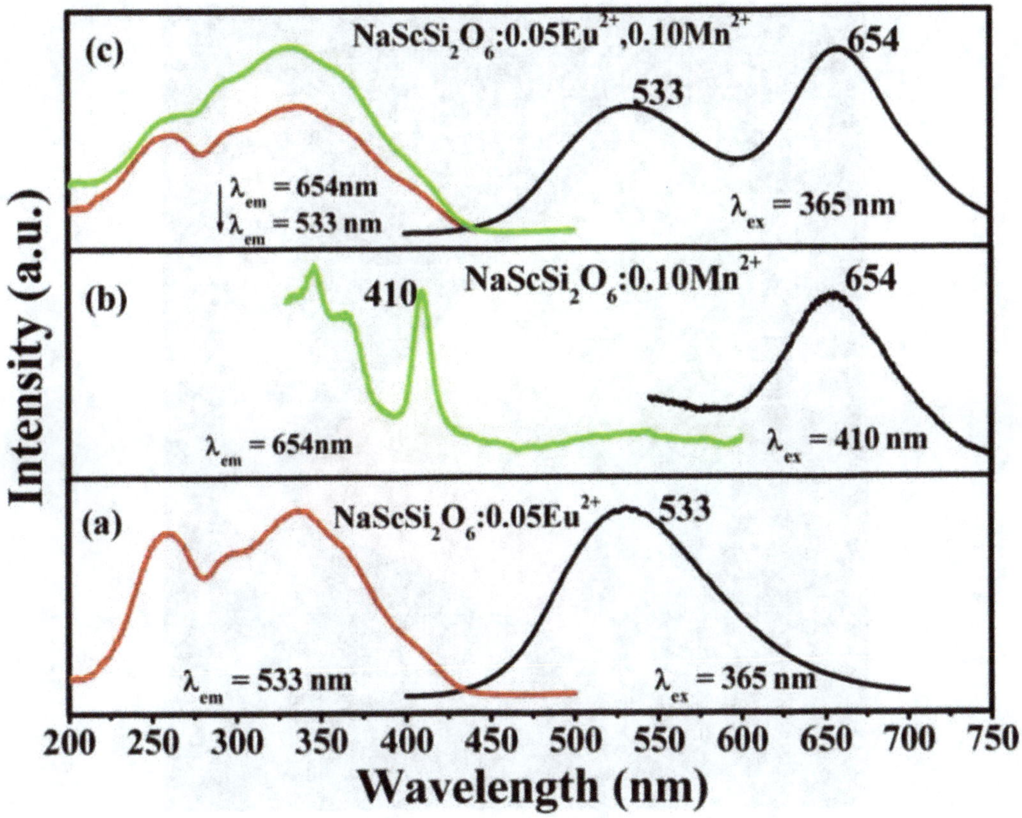

Figure 65. (a) The excitation and emission spectrum $Ca_{8.82}Mg(PO_4)_6F_2$:$_{0.18}Eu^{2+}$. (b) Excitation and emission spectra of $Ca_{8.82}Mg(PO_4)_6F_2$:$_{0.18}Eu^{2+}$ (i), $Ca_9Mg_{0.82}(PO_4)_6F_2$:$_{0.18}Mn^{2+}$(ii), and $Ca_{8.82}Mg_{0.82}(PO_4)_6F_2$:$_{0.18}$ Eu^{2+},$_{0.18}Mn^{2+}$ (iii) phosphors. Reproduced from ref (Li et al., 2014)

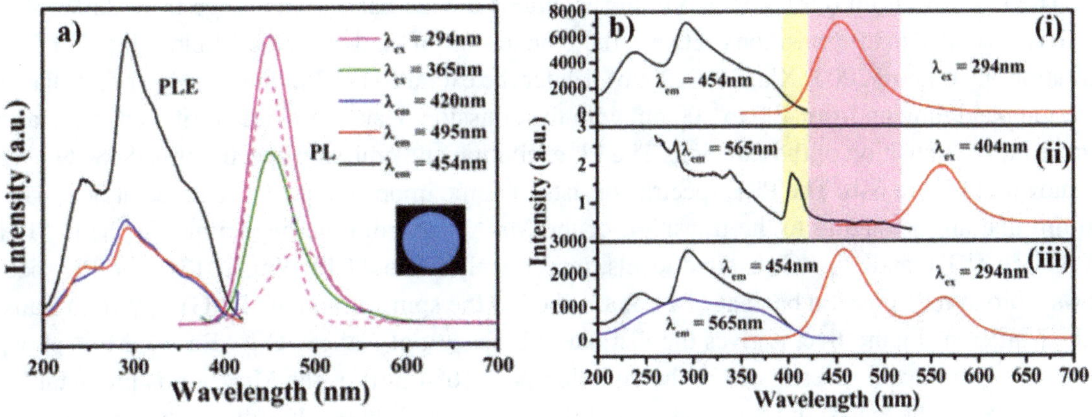

Figure 66. CIE chromaticity coordination of CMPF:$_{0.18}Eu^{2+}$,$_{y}Mn^{2+}$ samples (A–H corresponds to y = 0, 0.02, 0.06, 0.10, 0.18, 0.26, 0.34, 0.38, respectively.). Reproduced from ref (Li et al., 2014)

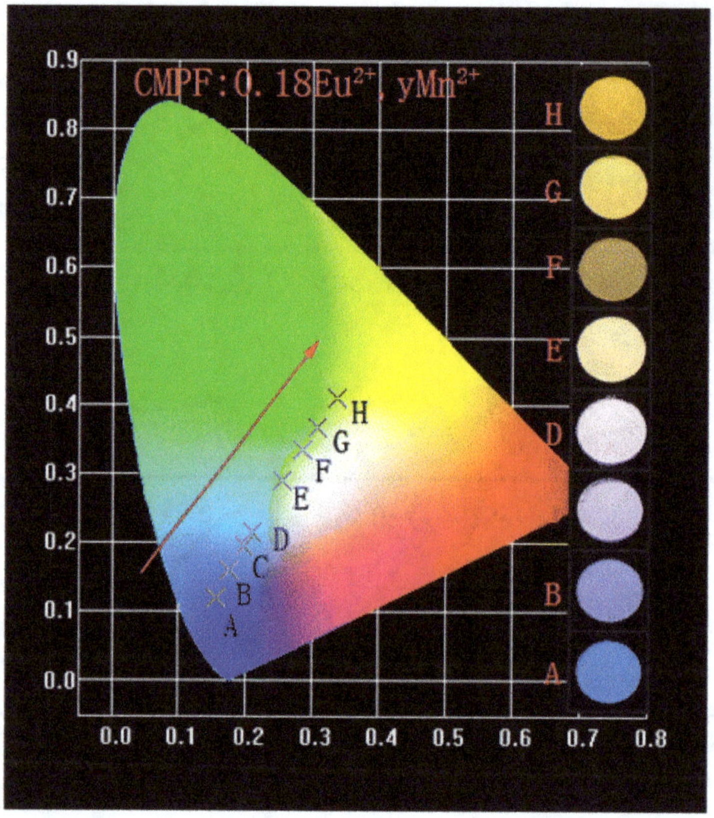

of ionic radii in the crystal structure the Eu^{2+} ion occupied in the Na^{+} site and the Mn^{2+} ion occupied in the Sc^{3+} site (Xia, Zhang, Molokeev, & Atuchin, 2013).

The PLE (λ_{em} = 533 nm) and PL (λ_{ex} = 365 nm) spectra of the $NaScSi_2O_6$:$_{0.05}Eu^{2+}$ is shown in Figure 64a. The PLE spectrum (emission at 533 nm) contains a wide band in the range from 220 to 450 nm which is attributed to the transitions between the ground state $4f^7$ and the crystal-field splitted $4f^65d^1$ configuration (Dorenbos, 2003; Xia, 2007, 2006). Under the excitation at 365 nm, the PL spectra has broad emission band ranging from 400 to 700 nm with the emission peak at 533 nm attributed to the allowed electric dipole transition of the Eu^{2+} ions.The PL excitation and emission spectra of $NaScSi_2O_6$:$_{0.10}Mn^{2+}$ are shown in Figure 64b. The PLE spectra consists of some important peaks centered at 362, 410, and 536 nm that are belonging to the transition of the Mn^{2+} ions from the beginning $^6A_1(^6S)$ to $^4T_2$ (^4D), [$^4A_1{}^4G$), ^4E (^4G)], and $^4T_1$ (^4G) excited levels, respectively (Guo, 2013; Wu, 2013). The PL spectrum shows a broad red emission band at 654 nm ascribed to the spin-forbidden $^4T_1(^4G) \rightarrow {}^6A_1(^6S)$ transition of the Mn^{2+} ions.Figure 64c, it gives the PLE and PL spectra of $NaScSi_2O_6$:$_{0.05}Eu^{2+}$,$_{0.10}Mn^{2+}$ phosphor, and the PLE spectrum investigated at the emission peak (654 nm) of the Mn^{2+} ion is predictable with that observed at the emission peak (533 nm) of the Eu^{2+} ions. As a result of this behavior the emission spectrum of $NaScSi_2O_6$:0.05Eu^{2+}, 0.10Mn^{2+} has combining yellow emission of Eu^{2+} and orange red emission of Mn^{2+}. With increasing the Mn^{2+} concentration, the emission intensity of Eu^{2+} ions decrease

Figure 67. Crystal structure of $Ca_6Y_2Na_2(PO_4)_6F_2$ (a), and the coordination environment of M1 (b) and M2 (c). Redrawn from ref (Guo, You, Jia, Ouyang, & Wu, 2014)

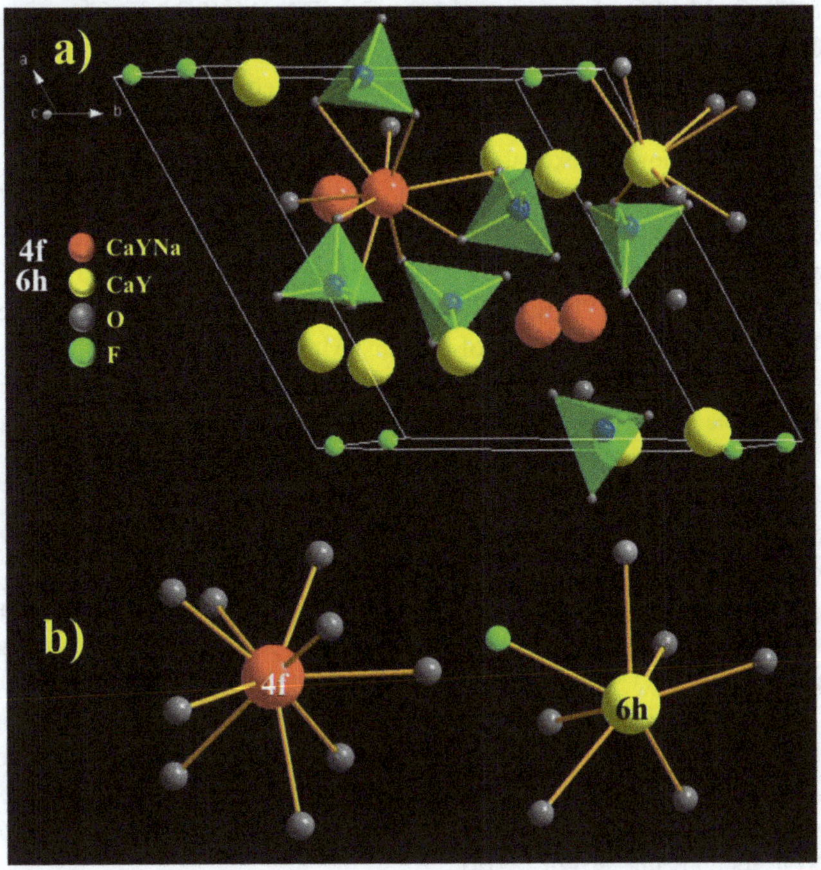

continuously because of the energy transfer from Eu^{2+} to Mn^{2+} in the $NaScSi_2O_6$ host. The white LED device was fabricated with $NaScSi_2O_6$:$0.05Eu^{2+}$, $0.10Mn^{2+}$ and commercial blue phosphors exhibits the CIE color coordinates of (0.358, 0.378), CCT of 4666 K, and Ra value of 92.2.

$Ca_9Mg(PO_4)_6F_2$: Eu^{2+}, Mn^{2+}

The single phase host white phosphor materials are fascinating in the display technology applications. The successful method to procure white light is to use the energy transfer from a sensitizer (Eu^{2+}, Ce^{3+}) to an activator (Mn^{2+}, Tb^{3+}, Dy^{3+} and so on.), which can be well suited in various hosts inferable from its great color reproducibility and low expenses. Numerous phosphors are reported $Ca_8MgY(PO_4)_7$:Eu^{2+},Mn^{2+} (Wen, Dong, Shi, Gong, & Wu, 2013) $Ca_8MgGd(PO_4)_7$:Ln^{3+}, Mn^{2+} ($Ln^{3+} = Ce^{3+}$, Tb^{3+}, Dy^{3+}) (J. Zhang, Wang, F. Zhang, & Huang, 2011).

The $Ca_9Mg(PO_4)_6F_2$ is solidified in the hexagonal system with space group *P63/m* and lattice parameters a = 9.3507Å, b = 9.3507Å, c = 6.8451Å, V = 518.327Å^3. In the $Ca_9Mg(PO_4)_6F_2$ structure, the Mg^{2+} ion substituted in the Ca^{2+}site . The Ca^{2+} has the two different crystallographic lattices involving 9-fold-coordinated with C_3 point symmetry [represented as the Ca^{2+}(1)] and 7- fold-coordinated 6h sites with Cs point symmetry [represented as $Ca^{2+}(2)$] (Li et al., 2014). The analogous substitution of Eu^{2+} to Ca^{2+} and Mn^{2+} to Mg^{2+} is caused due to their similar ionic radii and valence states. The concentration quenching of Eu^{2+} in $Ca_9Mg(PO_4)_6F_2$ was found to be $Ca_{8.82}Eu_{0.18}Mg(PO_4)_6F_2$. The excitation and emission spectrum CMPF:$_{0.18}Eu^{2+}$ have been depicted in Figure 65a. In the excitation spectrum two bands are observed at 294 and 365 nm, the emission band was observed from 380 to 580 nm centered at 454 nm, ascribed to the $4f^65d^1 \rightarrow 4f^7$ allowed transition of Eu^{2+}.

The Figure 65b (i – iii) show the excitation and emission spectra of CMPF:$_{0.18}Eu^{2+}$, CMPF:$_{0.18}Mn^{2+}$ and CMPF:$_{0.18}Eu^{2+}$,$_{0.18}Mn^{2+}$samples, respectively. The emission spectrum of CMPF:$_{0.18}Mn^{2+}$ has broad band at 565 nm, which is attributed to the $^4T_{1g}(G)–^6A_{1g}(S)$ spin-forbidden transition of Mn^{2+} ions. The excitation spectrum has several peaks at 280, 335, 404, 495 nm, corresponding to the typical transitions of Mn^{2+} ion from $^6A_{1g}(S)$ to $^4E_g(D)$, $^4T_{2g}(D)$, [$^4A_{1g}$, $^4E_{1g}(G)$], and $^4T_{1g}(G)$ excited levels, respectively. The emission spectrum has the both the emission of Eu^{2+} and Mn^{2+} ion, this clearly indicates the possibility of resonance type energy transfer from Eu^{2+} to Mn^{2+} ions in this host. CIE color coordinate of the phosphor indicates the tunable emission when Mn^{2+} is doped keeping Eu^{2+} concentration fixed (Figure 66).

$Ca_6Y_2Na_2(PO_4)_6F_2$: Eu^{2+} and Mn^{2+}

The $Ca_6Y_2Na_2(PO_4)_6F_2$ adopts the similar crystal structure as that of $Ca_6Eu_2Na_2(PO_4)_6F_2$. Subsequently, $Ca_6Y_2Na_2(PO_4)_6F_2$ phosphor crystallizes in the hexagonal crystal system with space group *P63/m* lattice parameters are a = b = 9.385Å and c = 6.893Å (Mayer, 1974, 1983). Figure 67 demonstrates a schematic crystal structure of the $Ca_6Y_2Na_2(PO_4)_6F_2$ and necessary information regarding coordination environment of the cations. The $Ca_6Y_2Na_2(PO_4)_6F_2$ host there available two kind cationic sites for substitution, the M1 has nine coordination with oxygen atom and the point group is C_3, the M1 has seven coordination with six oxygen atom and one fluoride atom the point group is C_s. On the premise of the ionic radii and valence state, we can sensibly expect that the Eu^{2+}/Mn^{2+} ions would possess the sites of the Ca^{2+} (M1 and M2) (Guo, You, Jia, Ouyang, & Wu, 2014).

Figure 68 shows the PL spectrum of the Eu^{2+} and/or Mn^{2+} in the CYNP host lattice. Figure 68a displays the PLE and PL spectra of CYNP: $_{0.01}Eu^{2+}$, the excitation (under 458 nm) spectrum has the broad band ranging from 225 to 410 nm, the emission spectrum (under 350nm) has the strong band at 458 nm and a weak band at around 523 nm, which are ascribed to the electric dipole allowed $4f^65d^1 \rightarrow 4f^7$ transition of the Eu^{2+} ions. Fig. 68b shows the PLE and PL spectra of CYNP:$_{0.05}Mn^{2+}$. The PLE spectra demonstrates a few broad bands along with some remarkable peaks in the range of 325–525 nm, which can be attributed to the transition of Mn^{2+} ions from ground level $^6A_1(^6S)$ to $^4T_2(^4D)$, [$^4A_1(^4G)$, $^4E(^4G)$], $^4T_2(^4G)$, and $^4T_1(^4G)$ excited levels, respectively (Shi, Huang, & Seo, 2010).Under excitation at 404 nm, the PL spectrum shows a broad emission band focused at 570 nm credited to the spin – forbidden $^4T_1(^4G)–^6A_1(^6S)$ move of the Mn^{2+} ions. Figure 68c displays more confirmation for the energy exchange from Eu^{2+} to Mn^{2+} in CYNP: Eu^{2+}, Mn^{2+} samples, as the PLE spectra which checked the emission of the Mn^{2+} ions are predictable with that which observed the emission of the Eu^{2+} ions. These components demonstrate that there is clear energy transfer from Eu^{2+} to Mn^{2+} in the CYNP: Eu^{2+},Mn^{2+} sample.

Figure 69 represents the PL spectra of CYNP:$_{0.01}Eu^{2+}$,$_nMn^{2+}$ samples, with increment of Mn^{2+} ions the emission intensity of Eu^{2+} was found to decreased while Mn^{2+} emission intensity increased. The observation clearly indicates the energy transfer is occurring from Eu^{2+} to Mn^{2+} ion. With increasing temperature up to 420 K, the emission intensity of the Eu^{2+} and Mn^{2+} ions decreased up to 65.3% and 72.6% of the initial value, respectively.

Figure 68. PLE and PL spectra of (a) the CYNP:$_{0.01}$Eu^{2+} phosphor, (b) the CYNP:$_{0.05}$Mn^{2+} phosphor, and (c) the CYNP:$_{0.01}$Eu^{2+},$_{0.025}$Mn^{2+} phosphor. Reproduced from ref (Guo, You, Jia, Ouyang, & Wu, 2014)

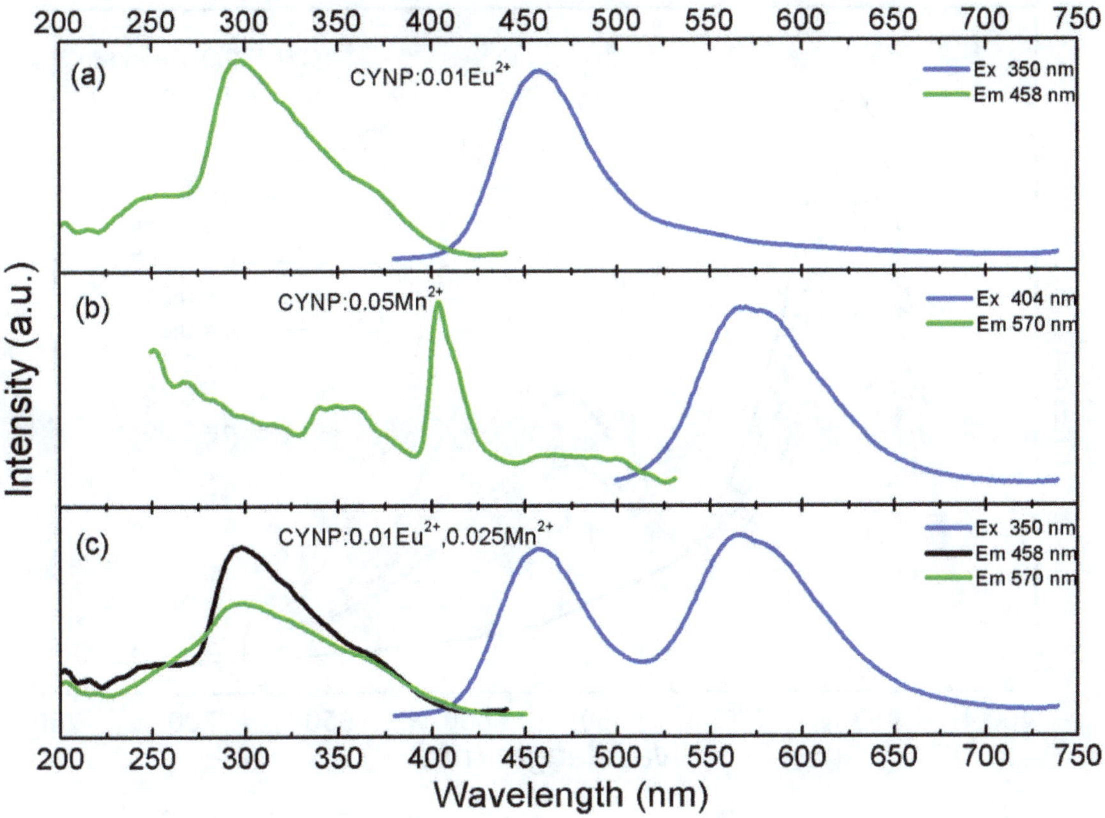

$Na_2BaCa(PO_4)_2$:Eu^{2+},Mn^{2+}

Xia et al. (2016) reported the discovery and structural design of color conversion phosphors for LEDs, including the β-K_2SO_4 type phosphors. The β-K2SO4 structure-type $NaBaPO_4$ (JCPDS: No. 33-1210, space group: $P\bar{3}m1$) (Zhang & Huang, 2011) and NaCaPO4 (JCPDS: No. 29-1193, space group: $Pn2_1a$) (Amara, Vlasse, Flem, & Hagenmuller, 1983) have been both reported with the formula of $ABPO_4$. Kim *et al.* (2015) reported $NaBa_{1-x}Ca_xPO_4$ synthesized by a polymerizable complex method (PC method), and new phases were observed at x = 1/3 and 0.5, *i.e.*, $Na_3Ba_2Ca(PO_4)_3$ and $Na_2BaCa(PO_4)_2$, respectively. According to their work, NBCP has the same space group ($P\bar{3}m1$) and similar cell parameters as $NaBaPO_4$.

The crystalline structure of the phosphor is based on the space group $P\bar{3}m1$ (164), belonging to the trigonal system and the unit cell parameters are $a = b = 5.42796(9)$ Å, $c = 7.34839(15)$ Å and $V = 187.498(7)$ Å^3. Figure 70 shows the crystal structure of NBCP. It is a three-dimensional framework with a specific arrangement of its various atoms. Ba^{2+} and Ca^{2+} ions are arranged layer by layer as a basic framework with the Na–P polyhedral (called a double coordination pyramid) layers located between these cations. The polyhedra of Ca^{2+} and Ba^{2+} is coplanar, sharing the edge O^{2-} ions (O2), as shown in Figure 70b. When the substitution occurs, Eu2+ prefers to occupy the site of Ba^{2+} and Ca^{2+} in consideration

Figure 69. PL spectra for CYNP:$_{0.01}Eu^{2+}$,$_{n}Mn^{2+}$ phosphors with varying Mn^{2+} doping content (n). Reproduced from ref (Guo, You, Jia, Ouyang, & Wu, 2014)

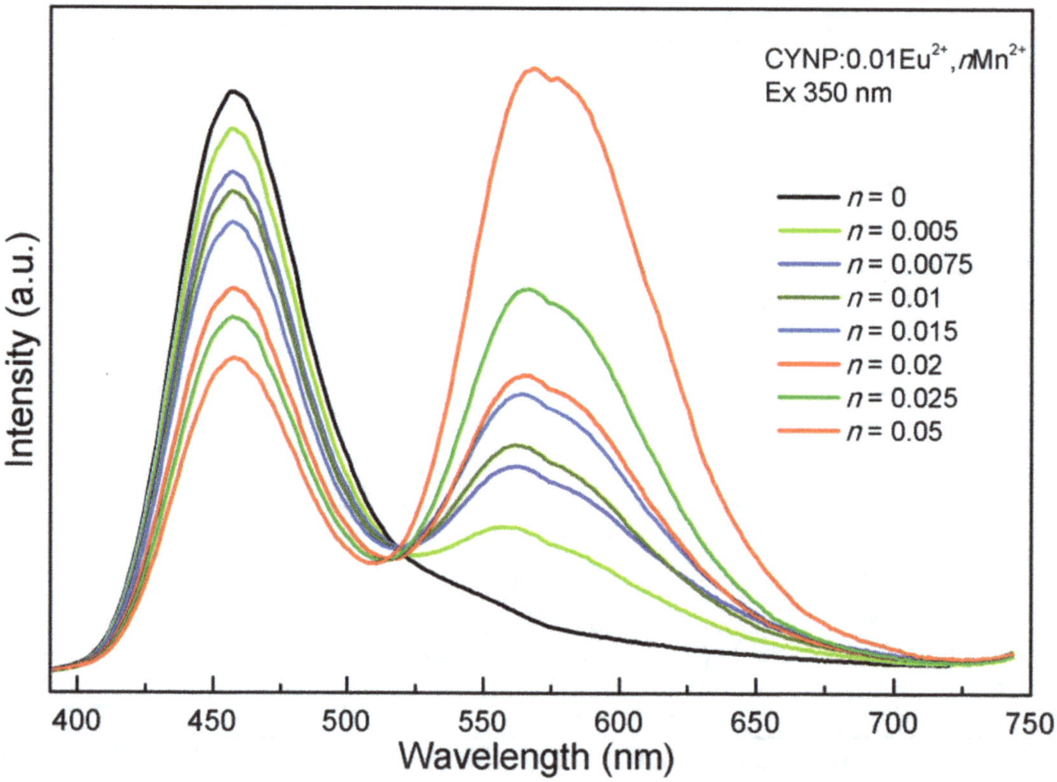

Figure 70. Crystal structure of $Na_2BaCa(PO_4)_2$. Redrawn from ref (Cao, Ding, X. Wang, & Y. Wang, 2017)

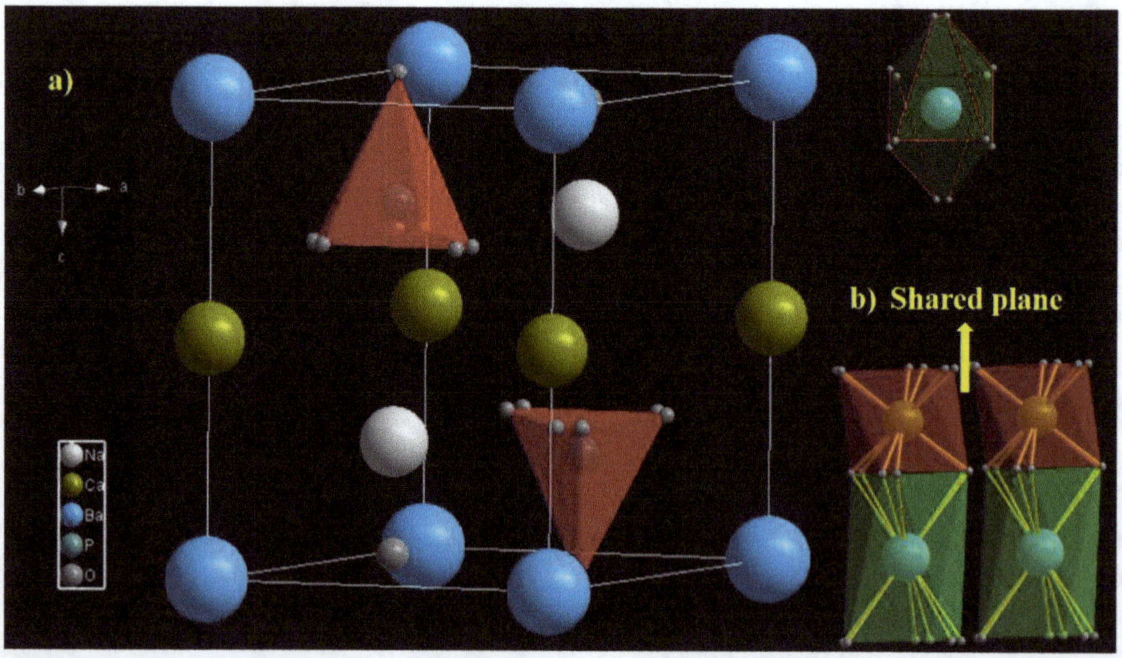

Figure 71. (a) PLE and PL spectra of NBCP:0.01Eu2+; (b) PLE and PL spectra of NBCP:$_{0.01}$Mn^{2+} (c) PLE (monitored at 451 and 576 nm, respectively) and PL spectra of NBCP:$_{0.01}$Eu,$_{0.05}$Mn; (d) PL spectra of NBCP:$_{0.01}$Eu,$_{0.09}$Mn and NBCP:$_{0.01}$Eu^{3+} while the inset shows the PL spectrumof NBCP:$_{0.01}$Eu^{2+} and the PLE spectrum of NBCP:$_{0.01}$Eu^{3+}.Reproduced from ref (Cao, Ding, X. Wang, & Y. Wang, 2017)

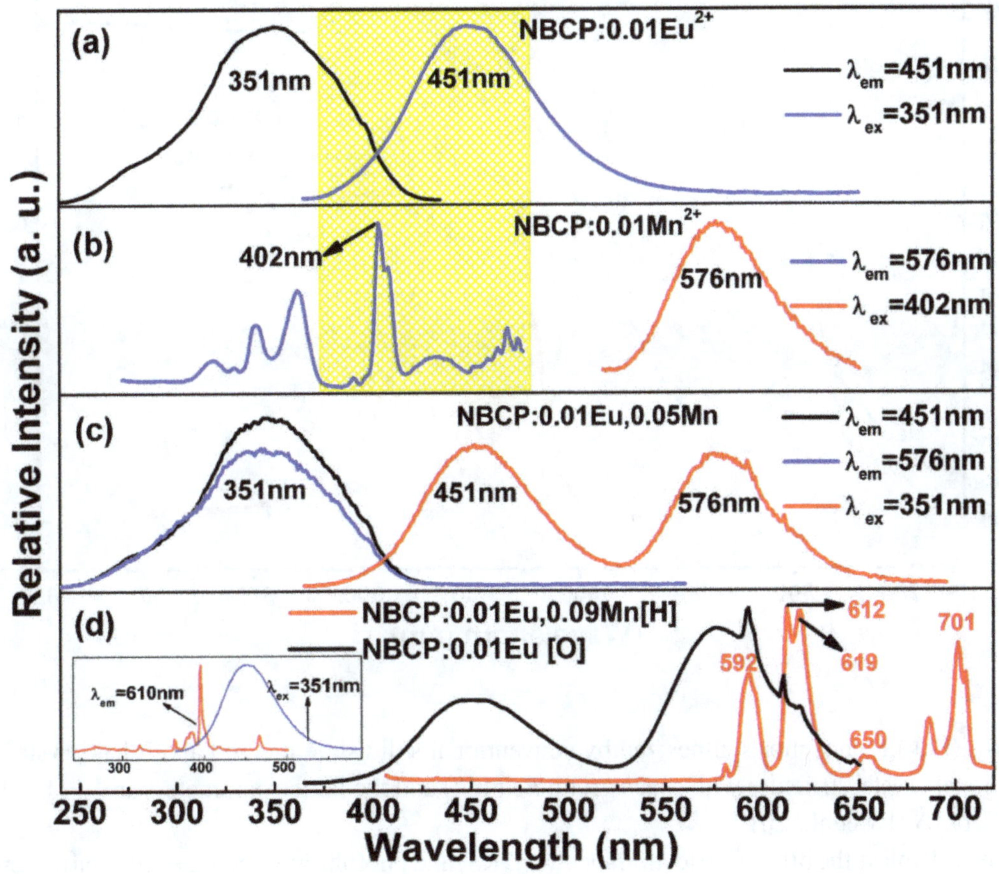

of the radii, valence and coordination, while Mn^{2+} is likely to substitute Ca^{2+} (Cao, Ding, X. Wang, & Y. Wang, 2017).

The PLE and PL spectra of NBCP:0.01Eu^{2+}, NBCP:0.01Mn^{2+}, NBCP:0.01Eu, 0.05Mn are shown in Figure 71. In NBCP, the emission of Mn^{2+} presents a band from 525–650 nm while peaking at 576 nm. The PL spectrum of NBCP:0.01Eu^{2+} presents broad band (375–550 nm, attributed to the $4f^65d^1$–$4f^7$ transition) peaks at 451 nm, while the PLE spectrum of NBCP:0.01Mn^{2+} consists of sharp lines in the region of 300-550 nm. The excitation spectra were measured and monitored at 451 nm for the 4f–5d transition of Eu^{2+} and 576 nm for the $^4T_1$–$^6A_1$ transition of Mn^{2+}. CL intensity attributed to the Eu^{2+} gradual decrease with the increasing Mn^{2+} content. The intensity assigned to Mn^{2+} increases first and then falls, with a maximum when Mn^{2+} reaches 0.03 mol.

$Sr_3GdLi(PO_4)_3F$:Eu^{2+}, Mn^{2+}

Figure 72. Photoluminescence excitation and emission spectra of the SGLP:$_{0.01}$Eu^{2+} (solid lines) together with diffuse reflection spectra of SGLP and SGLP:$_{0.01}$Eu^{2+} (dotted lines). Reproduced from ref (Lv et al., 2017)

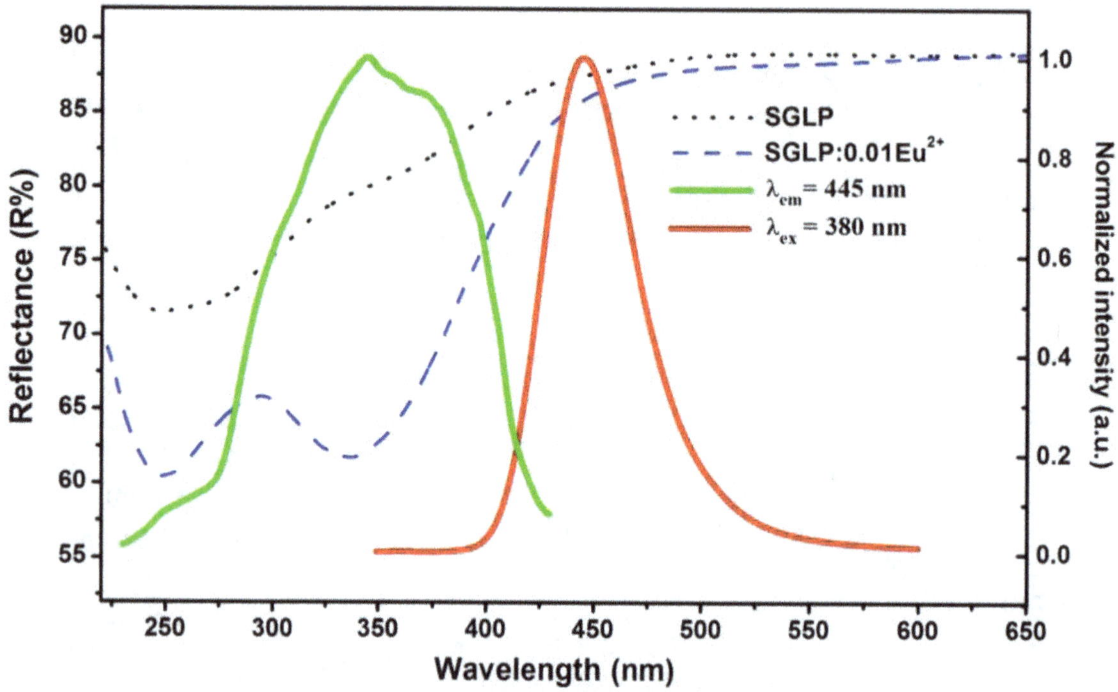

$Sr_3GdLi(PO_4)_3F$ phosphor synthesized by conventional solid state method and it has crystallized in the hexagonal crystal structure with space group $P\bar{3}$ (147) and the lattice parameters are a = b = 9.666Å and c = 7.177Å (Lv et al., 2017).

Figure 72 displays the diffuse reflection spectra (DRS) and photoluminescence excitation and emission spectra of the SGLP:$_{0.01}$Eu^{2+}. In the DRS, the host component drop around 250 nm, which is attributed to the host absorption. When the Eu^{2+} is incorporated in the host lattice an intense decrease of reflectivity between 250 and 430 nm is observed, which is caused by the Eu^{2+}absorption. The excitation spectrum was monitored by the emission at 445 nm, the PLE spectrum exhibits a broad band with a range from 230 to 430 nm, which is attributed to the electron transition from $4f^7$ to excitation state of Eu^{2+} ions. The emission spectra were monitored by the excitation at 380 nm, the PL range demonstrates an expansive blue emission band going from 400 to 550 nm focused at 445 nm which is caused by $4f^65d^1$–$4f^7$ transition of the Eu^{2+} ions.

Figure 73 exhibit the excitation and emission spectra of the SGLP:$_{0.10}$Mn^{2+}. The PLE range of SGLP:$_{0.10}$Mn^{2+} demonstrates a few excitation peaks at 340, 357, 404, 438 and 455 nm which are induced by the movement of electron from the $^6A_1(^6S)$ ground state to $^4E(^4D)$, $^4T_2(^4D)$, [$^4A_1(^4G)$, $^4E(^4G)$], $^4T_2(^4G)$, and $^4T_1(^4G)$ excited states, respectively (Chan, 2008; Huang, 2010). As per the existing selection rules, the d–d transitions of Mn^{2+} are spin and parity forbidden, hence emission from SGLP:$_{0.10}$Mn^{2+} is moderately weak. The Mn^{2+} emission intensity increased and as well as the Eu^{2+} intensity decreased with increment of Mn^{2+} ions in SGLP:$_{0.01}$Eu^{2+}, $_y$Mn^{2+}. Against increasing temperature at around 135 °C SGLP:$_{0.01}$Eu^{2+}, $_{0.12}$Mn^{2+} emission intensity decreased half.

Figure 73. Photoluminescence excitation and emission spectra of the SGLP:$_{0.10}$Mn^{2+}. Reproduced from ref (Lv et al., 2017)

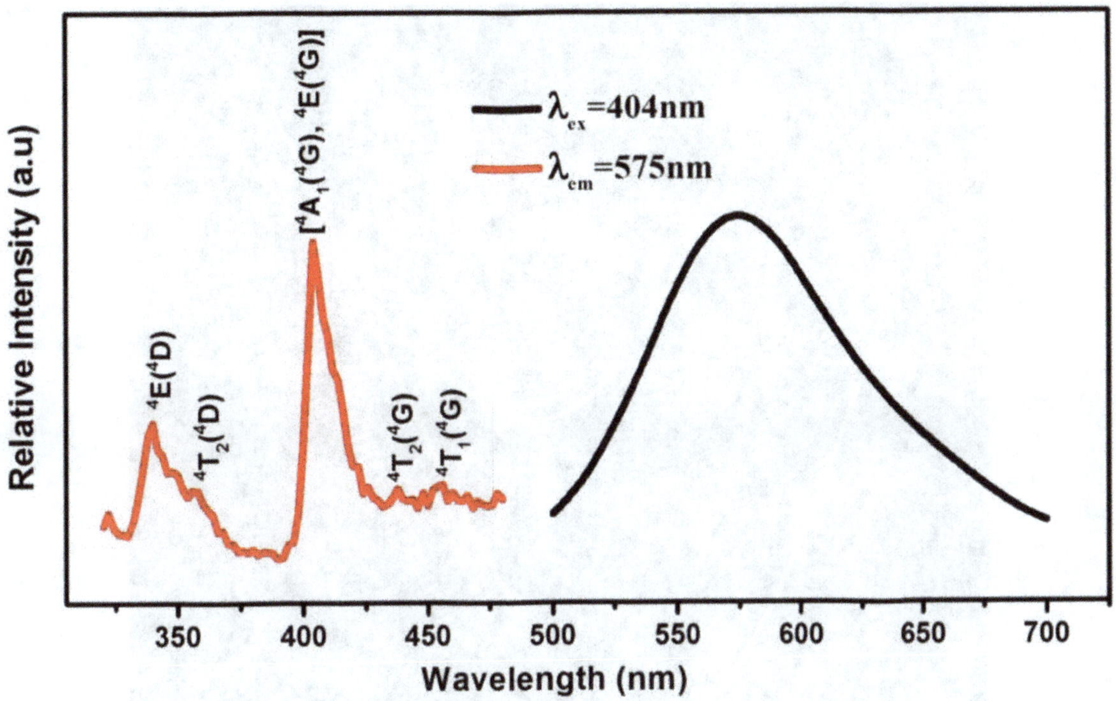

Schematic configuration coordinate diagram shown in Figure 74, Upon excitation by UV light, the electrons are transferred from the ground state to excited state, then the electrons move to lower 5d state through nonradioactive way and finally jump to the ground state emitting blue light emission.With increase of temperature, the electrons at the 5d excited states have more affinity producing extra electron-phonon interaction and moving to the intersection of the excited and ground states,(Guo et al., 2010) and then come back to the ground state through a nonradioactive process. Consequently, the reduction of emission intensity is detected.

$BaMgSiO_4$:Ce^{3+},Eu^{2+}

The lattice of $BaMgSiO_4$ belongs to the large structural family of stuffed tridymite (SiO_2) derivatives (Liu & Barbier, 1993). In 2007 the detailed luminescent characteristics of Ce_{3+} and Eu^{2+} doped $BaMgSiO_4$ lattice was studied by Sivakumar and Varadaraju (2007). The $BaMgSiO_4$ host lattice synthesized by solid state reaction method and it crystallized in the hexagonal crystal system with space group *P63* and the cell parameters are a = 9.1226(7)Å and c = 8.7496(15)Å. The $BaMgSiO_4$ crystal structure has the two different tetrahedral groups such as MgO_4 and SiO_4, these tetrahedral structures forms the three-dimensional network by sharing the O (4) atoms (Sivakumar and Varadaraju, 2007).

The Figure75 shows the PL excitation and emission spectrum of the $Ba_{0.9}Ce_{0.05}Li_{0.05}MgSiO_4$ and $Ba_{0.925}Eu_{0.075}MgSiO_4$ phosphors. The excitation spectrum shows a broad band from 220 to 400 nm with peaking maximum at 310 nm (Figure 75 a). The excitation spectrum is arise due to $4f^1 - 4f^05d^1$ transi-

Figure 74. The schematic configuration coordinate diagram of SGLP:Eu^{2+}, Mn^{2+}. Redrawn from ref (Lv et al., 2017)

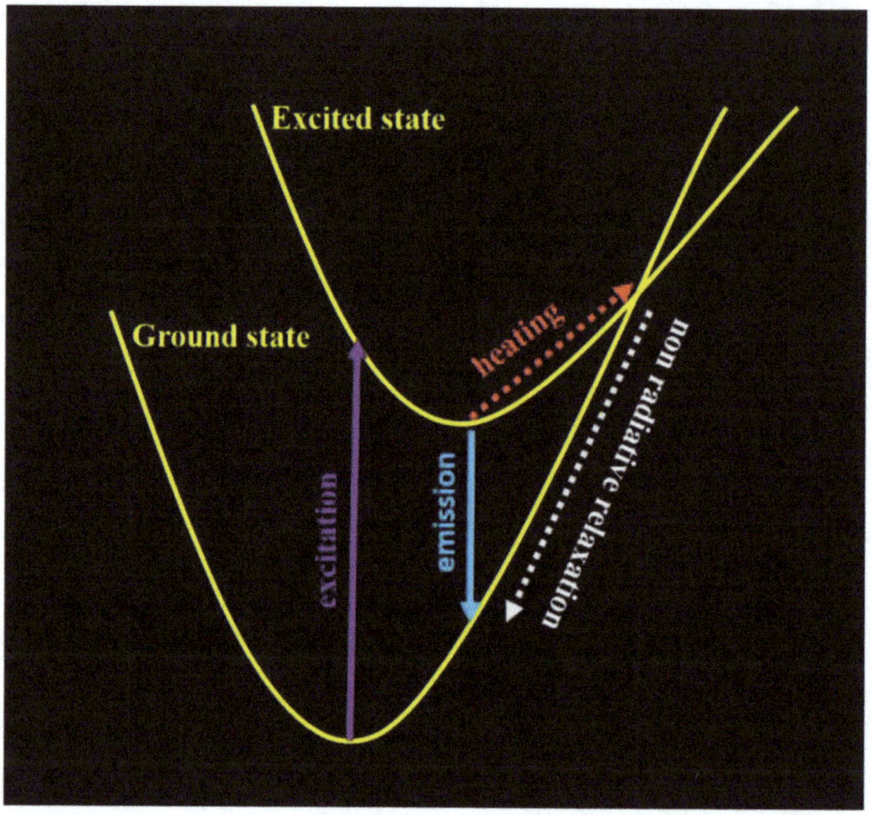

Figure 75. Photoluminescence excitation and emission spectra of $Ba_{0.9}Ce_{0.05}Li_{0.05}MgSiO_4$ and $Ba_{0.925}Eu_{0.075}MgSiO_4$. Reproduced from ref (Sivakumar and Varadaraju, 2007)

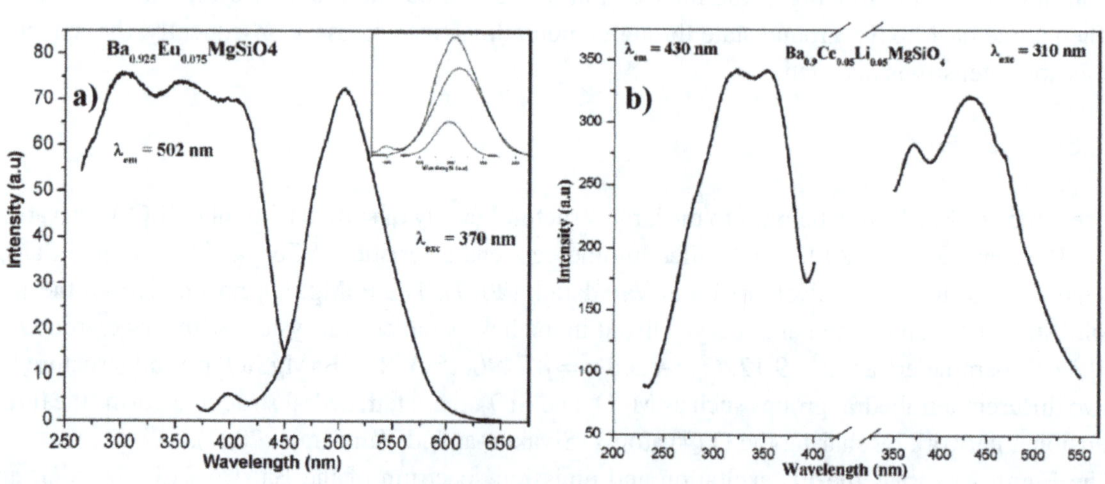

tion. The excitation spectrum of Eu^{2+} doped $BaMgSiO_4$ host lattice shows a broad absorption from 270 to 430 nm with two maxima at ~ 300 and ~ 370 nm (Figure 75 b). The $BaMgSiO_4$ structure has the two different Ba^{2+} crystallographic sites, in these sites the Eu^{2+} ions are equally distributed in the host lattice.

The Figure 76 shows the excitation and emission spectra of $Ba_{0.99-2x}Eu_{0.01}Ce_xLi_xMgSiO_4$ (x = 0.01−0.1). In the excitation spectrum the absorption band of Eu^{2+} is enlarged in the near-UV region when increasing the concentration of Ce^{3+} ion in the host lattice, so there is spectral overlap between Ce^{3+} (S) and the Eu^{2+} (A) ions. Green emission was observed in the emission spectrum at 510 nm due to the Eu^{2+} ion codoped in the lattice, the highest green emission intensity have been observed for $Ba_{0.97}Eu_{0.01}Ce_{0.01}Li_{0.01}MgSiO_4$ phosphor (Figure 76b). These results reveal that the Ce^{3+} acts as a sensitizer in $BaMgSiO_4$:Eu^{2+}.

Figure 76. Photoluminescence excitation (a) and emission (b) spectra of $Ba_{0.99-2x}Eu_{0.01}Ce_xLi_xMgSiO_4$ (x = 0.01−0.1). Reproduced from ref (Sivakumar and Varadaraju, 2007)

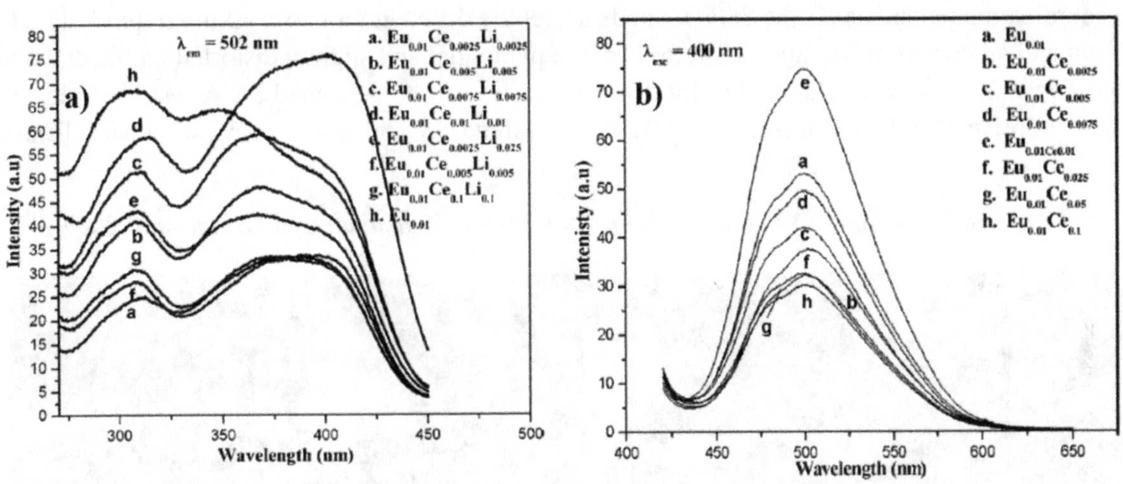

Figure 77. Photoluminescence excitation (a) and emission (b) spectra of $Ba_{0.8-x}Eu_xCe_{0.1}Li_{0.1}MgSiO_4$ (x = 0.025−0.015). Reproduced from ref (Sivakumar and Varadaraju, 2007)

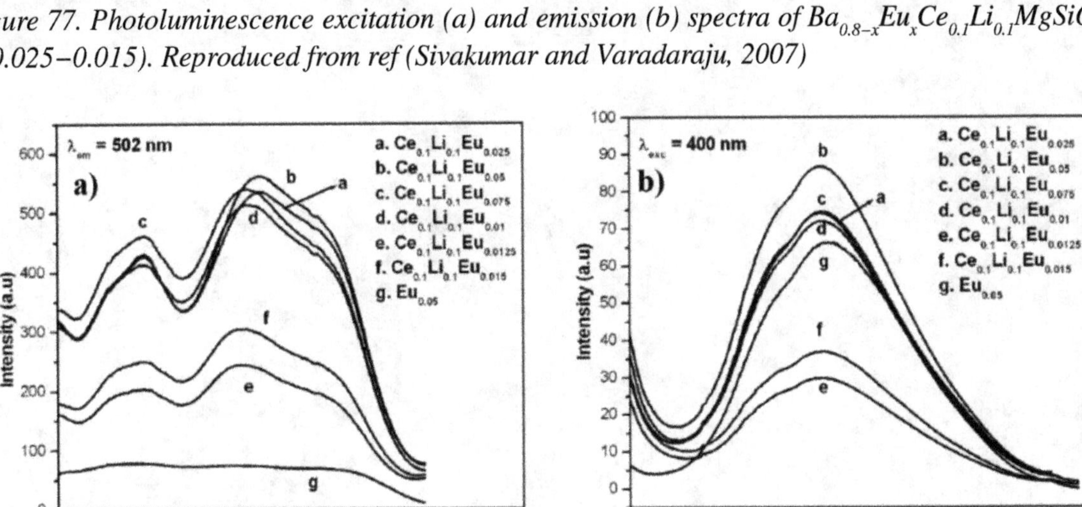

In order to find the critical concentration of the Eu^{2+} ion the Ce^{3+} ion concentration was fixed in the host lattice and varied the Eu^{2+} ion concentration. The excitation and emission spectrum of $Ba_{0.8-x}Eu_xCe_{0.1}Li_{0.1}MgSiO_4$ is shown in Figure 77. In excitation spectrum (Figure 77a) broad absorption spectrum was observed ranging from 250 to 450 nm, and the maximum intensity was found in x =0.5. The emission spectrum (Figure 77b) shows the green emission under UV and near UV.

Sr_2LiSiO_4F:Ce^{3+},Eu^{2+}

Sr_2LiSiO_4F has been synthesized and its structural properties have been reported by Akella and Keszler (1995). The compound Sr_2LiSiO_4F crystallizes with monoclinic structure with space group $P2_1/m$ (no. 11): The lattice parameters are $a = 6.5825(9)$Å, $b = 5.4158(8)$Å, $c = 6.9266(6)$Å and $\beta = 112.528°$. The crystal structure of Sr_2LiSiO_4F has been depicted in Figure 78. The structure has two crystallographically distinct Sr sites. They are 10-coordinated, with eight oxygen atoms and two fluorine atoms. Even though the coordination number is 10, the Sr–O bond distances are different for two crystallographic sites. The lithium and silicon atoms occupy a distorted square pyramidal and slightly distorted tetrahedral sites, respectively. The two crystallographically distinct Sr sites can be described as: A distorted bicapped square antiprism [Sr(I)] is formed by one O1, four O2, three O3, and two F atoms, whereas a distorted

Figure 78.Crystal structure of Sr_2LiSiO_4F. Redrawn from ref (Sivakumar and Varadaraju, 2009)

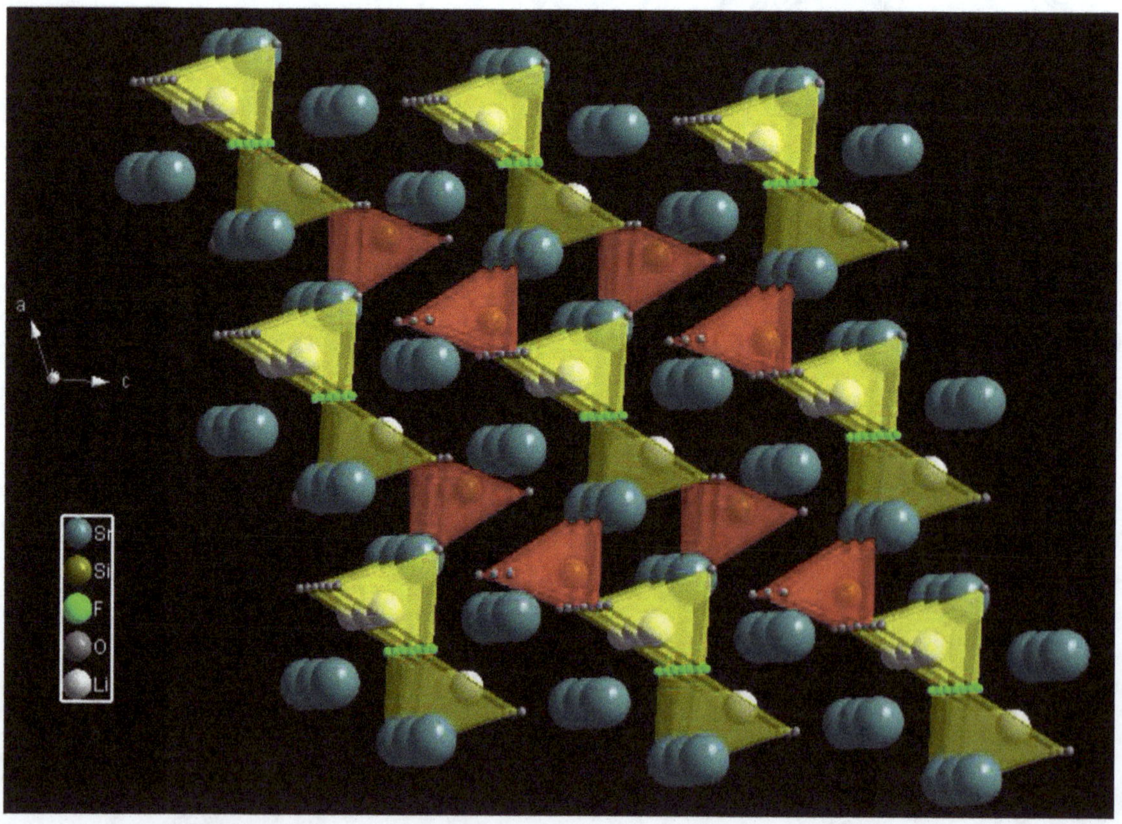

Figure 79. The excitation and emission spectra of (a) $Sr_{1.9925}Eu_{0.0075}LiSiO_4F$ and (b) $Sr_{1.98}Ce_{0.01}Li_{0.01}LiSiO_4F$ respectively. Reproduced from ref (Sivakumar and Varadaraju, 2009)

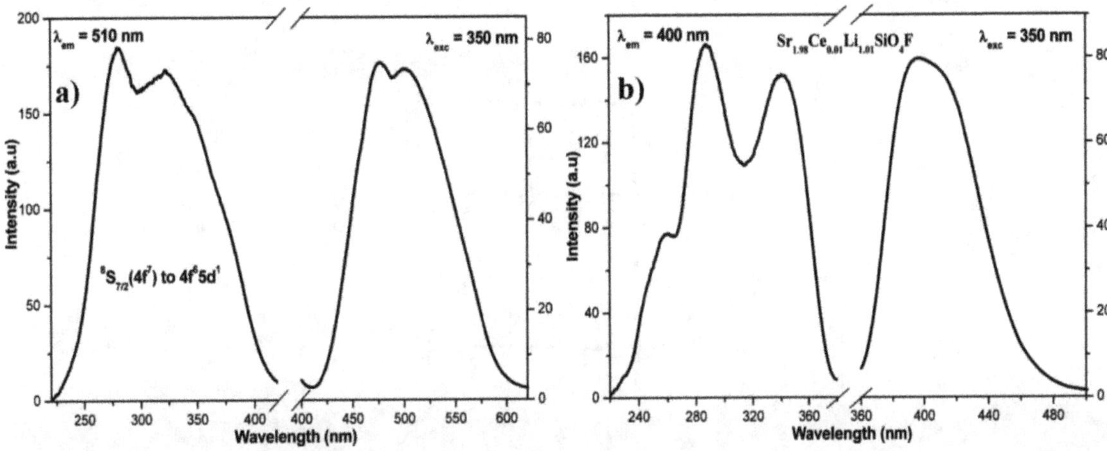

bicapped dodecahedron [Sr(II)] is formed by three O1, four O2, one O3, and two F atoms (Sivakumar and Varadaraju, 2009).

PLE and PL spectra of $Sr_{2-x}Eu_xLiSiO_4F$ (x = 0.0075) are given in Figure 79a. The spectrum shows a broad absorption band in the UV to near-UV region peaking at 280 and 330 nm. The emission spectrum shows a broad emission band starting from 420 nm and extending to the 620 nm region under a 350 nm excitation. Whatever the excitation (250–380 nm), a green emission band has been observed because the absorption band is very broad. In Figure 79b, Ce^{3+}-doped samples show an intense blue emission with a peak maximum at around 400 nm. Ce^{3+} co-doped $Sr_2LiSiO_4F:Eu^{2+}$ shows a broad emission covering the wavelength range from 360 to 620 nm when excited with Ce^{3+} excitation wavelength (350 nm); the co-doped compositions show emission bands corresponding to both Ce^{3+} and Eu^{2+}.

The presence of multiple absorption maxima indicates the lowering of the local site symmetry of Ce^{3+} from cubic to C_{2h}, which causes the splitting of the 5d orbital into A_g, B_g, A_u, and B_u, respectively (Figure 80). The emission peak at 400 nm indicates the presence of Ce^{3+} at the Sr(I) site and the emission at 430 nm corresponds to Ce^{3+} at the Sr(II) site in Sr_2LiSiO_4F. The Ce^{3+} emission band shows a doublet structure due to spin-orbit splitting of the groundstate ($^2F_{7/2}$ and $^2F_{5/2}$). Blue emission was observed at ~ 400 nm with a shoulder at ~ 430 nm clearly indicates that the spin-orbit splitting plays a role in the presently studied compound. When the energy difference between the ground state and the excited state of the two centers *i.e.* Eu and Ce is equal and there is a possible interaction between them, an energy transfer is likely to occur between them. This is satisfied by the spectral overlap between sensitizer (*S*) and the activator (A) absorption band.Figure 81 shows the photoluminescence excitation spectrum of Eu^{2+} (A) and the emission spectrum of Ce^{3+} (S). There is a spectral overlap between Ce^{3+} emission and Eu^{2+} absorption which clearly shows that they are in resonance. Because the optical transitions involved are dipole-allowed 4f–5d electronic transitions, the possible interaction between Ce^{3+}(S) and Eu^{2+} (A) is electric multipolar interaction. These results clearly indicate that the partial energy transfer occurs from Ce^{3+} to Eu^{2+} ion.

$Ca_4(PO_4)_2O$: Ce^{3+},Eu^{2+}

Figure 80. Schematic illustration of the energy-level diagram for $Sr_{1.98}Ce_{0.01}Li_{0.01}LiSiO_4F$. The solid and dotted arrows indicate the radiative and non-radiative processes, respectively. Redrawn from ref (Siva-kumar and Varadaraju, 2009)

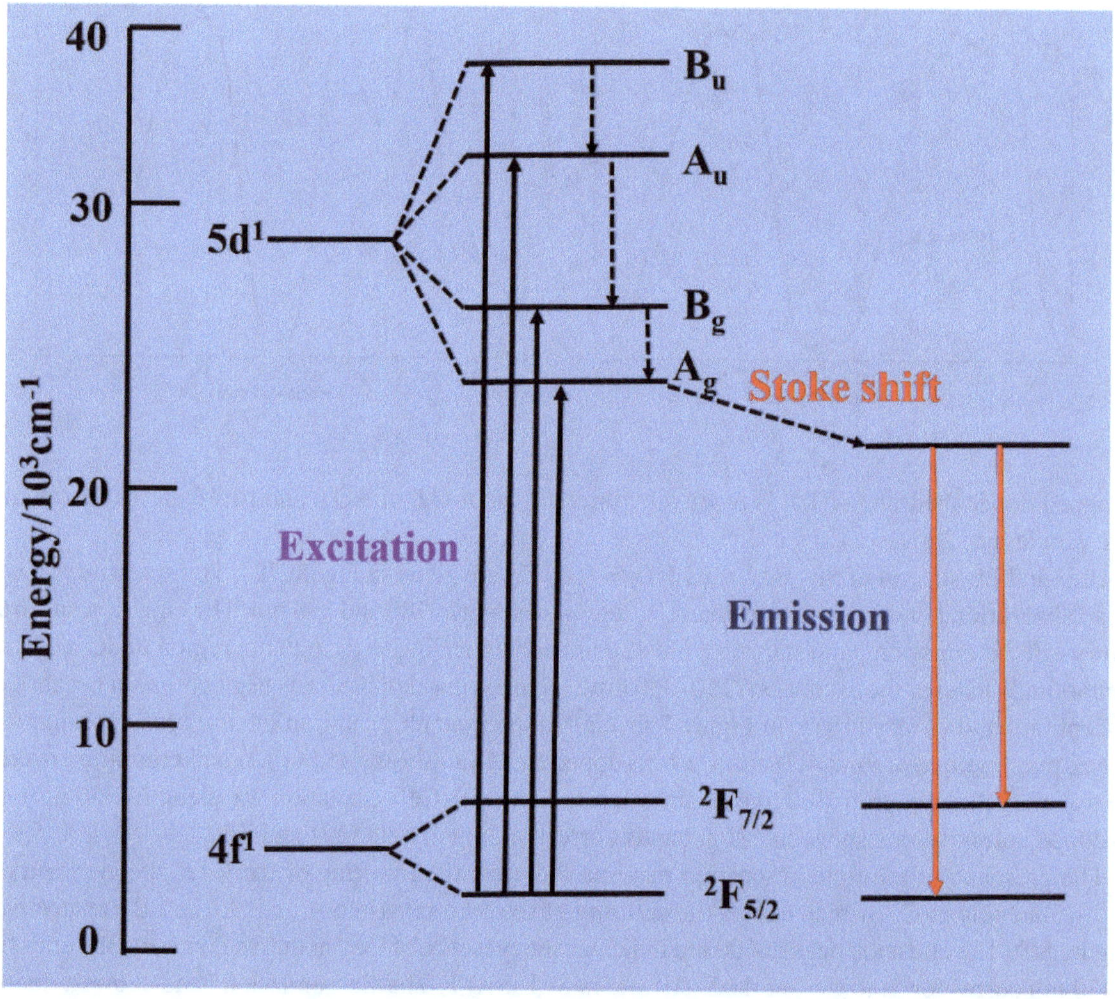

$Ca_4(PO_4)_2O$ crystallizes in the monoclinic clinic structure with space group *P21*, the parameters included a = 7.014Å, b = 11.978Å, c = 9.465Å, β = 90.87° and V = 795.16. Figure 82 displays the crystal structure of $Ca_4(PO_4)_2O$, in this crystal structure has eight crystallographic sites. The coordination for the Ca(6) site is eight, though that for the other calcium sites are seven. The coordination geometry for Ca (1), Ca(3), Ca(4), Ca(5), and Ca(8) sites is around a pentagonal bipyramid, though the coordination polyhedron for Ca(2), Ca(6), and Ca(7) is uneven (Jia et al., 2015).

Figure 83a exhibits the PLE and PL spectra of the $Ca_4(PO_4)_2O:_{0.02}Ce^{3+}$ phosphor.The PLE spectrum displays a wide absorption ranging from 260 nm to 430 nm that arises from $4f^1 \rightarrow 5d^1$ transitions of Ce^{3+} ions; this broad absorption coordinates well with the InGaN-based LED.Under the excitation of 380 nm, the $Ca_4(PO_4)_2O:_{0.02}Ce^{3+}$ phosphor shows deep blue emission produced by the 5d→4f electronic dipole-permitted transition of Ce^{3+};this emission contains of a solid asymmetric broad band (400–600 nm) with a maximal at 453 nm.

Figure 81. The spectra overlap between Ce^{3+} emission and Eu^{2+} excitation. Reproduced from ref (Sivakumar and Varadaraju, 2009)

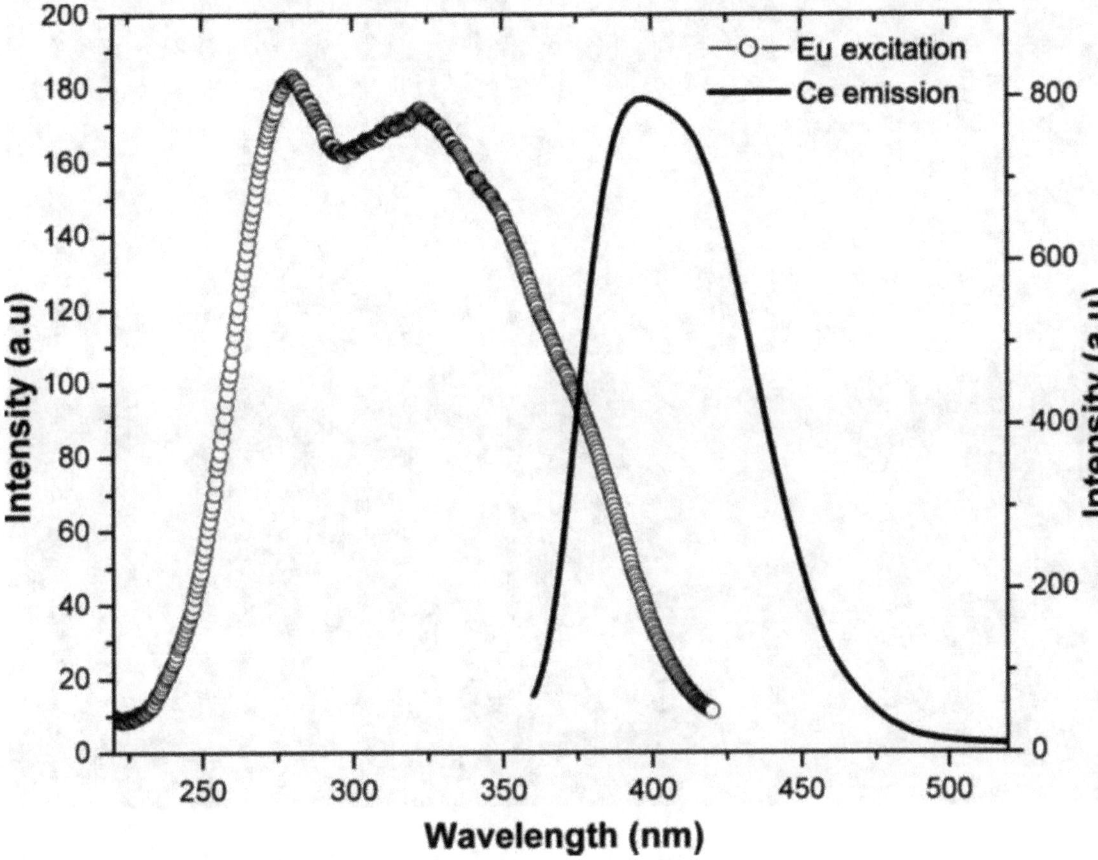

The PL and PLE spectra of $Ca_4(PO_4)_2O{:}_{0.07}Eu^{2+}$ is displayed in Figure 83b.The PLE range exhibits a broad absorption that covers the entire UV and blue region from 300–550 nm with a most extreme at 460 nm. $Ca_4(PO_4)_2O{:}_{0.07}Eu^{2+}$ exhibits a broad red emission from 550 nm to 800 nm intensifying at 650 nm under 460 nm excitation; this band can be ascribed to the $4f^65d^1 \rightarrow 4f^7$ allowed transition of Eu^{2+}ions.

Figure 83c exhibits the emission spectra of $Ca_4(PO_4)_2O{:}_{0.02}Ce^{3+}, {}_yEu^{2+}$ (y = 0–0.026). Under 380 nm excitation, the broad emission band is observed ranging from 400 nm to 550 nm, which is attributed to the allowed 4f–5d transition of the Ce^{3+}, whereas the emission band is observed maximum at 650 nm is ascribed to the f–d transition of the Eu^{2+} ion. With increasing the Eu^{2+}concentration, the emission intensity of Ce^{3+} is found to decreased while the intensity of Eu^{2+} increased. These results indicate that the existence of efficient energy transfer from Ce^{3+} to Eu^{2+} ions.The energy transfer from Ce^{3+} to Eu^{2+} ions is mainly attributed to the dipole–dipole interaction, the schematic energy structure of Ce 4f and 5d levels in $Ca_4(PO_4)_2O{:}Ce^{3+},Eu^{2+}$ is proposed by the basis of the spectra and energy transfer. The schematic energy level diagram of Ce 4f and 5d levels is shown in Figure 84. A WLED device was created through the mix of a NUV 380 nm chip and the $Ca_4(PO_4)_2O{:}_{0.02}Ce^{3+},{}_{0.012}Eu^{2+}$ phosphor. This device produces an extreme warm white light with a low CCT value (4124 K).

Figure 82. Crystal structure of $Ca_4(PO_4)_2O$. Redrawn from ref (Jia et al., 2015)

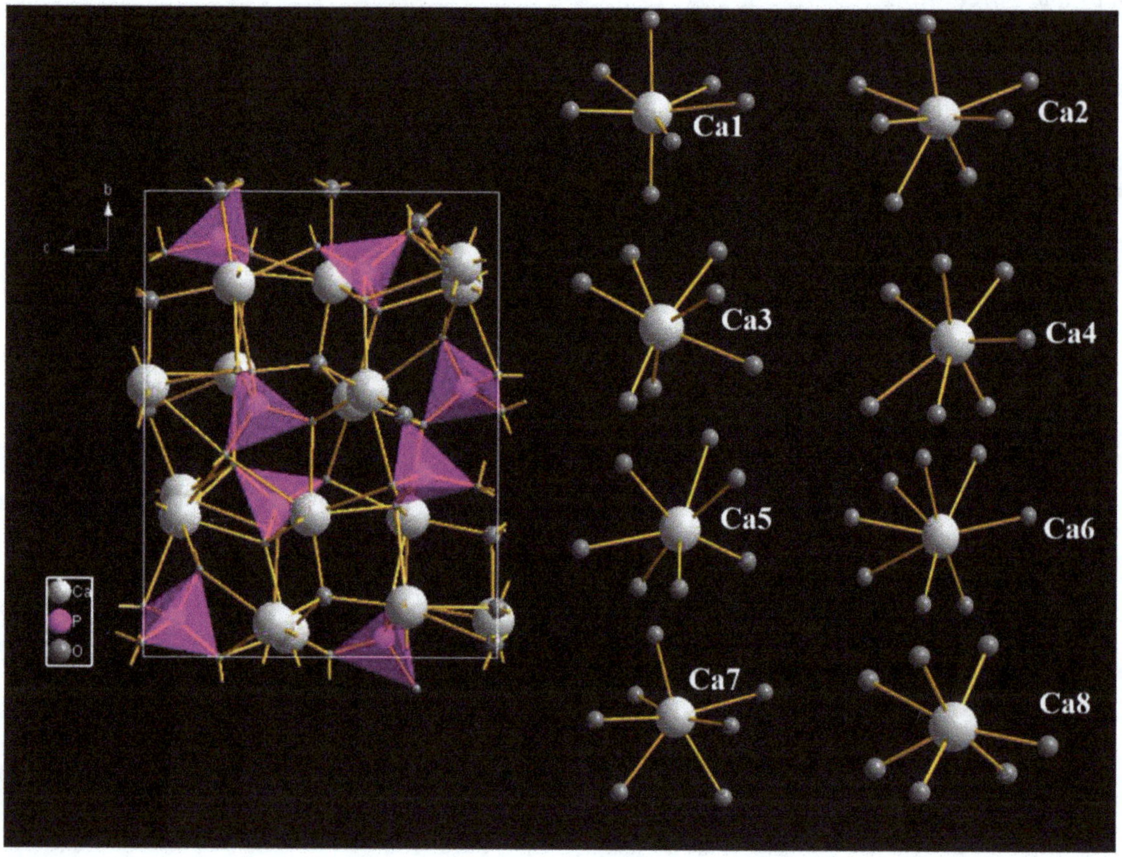

$Sr_3Ce(PO_4)_3$:Eu^{2+}

In recent days, an incredible number of eulytite-sort $M^{II}_3Ln^{III}(PO_4)_3$ compounds (where M^{II} = divalent cation and Ln^{III} = trivalent cation) acting as host materials for phosphors have drawn increasing attention due to their chemical and mechanical stability (Guo, 2011, 2012a, 2012b, 2013a, 2013b). Recently, numerous investigations have been studied on Eu^{2+}doped $B_3M(PO_4)_3$ compounds (B = Ca, Sr, Ba, M = Bi, Gd, Lu and Y) to display novel luminescence properties (Wang, 2014; Liang, 2004).

$Sr_3Ce(PO_4)_3$ phosphor synthesized by conventional solid state reaction and it is crystallized in cubic structure with space group *I43d* and the cell parameters are a = 10.169(7)Å and V = 1051.791(1)Å^3, Figure 85 displays the crystal structure of $Sr_3Ce(PO_4)_3$ and the Sr and P sites are coordinated with seven and four oxygen atoms respectively (Dai et al., 2016).

Figure 86 (a). exhibited the PL and PLE spectrum of the SCPO host under excitation at 325 nm, a wide emission band centered at 402 nm, which is ascribed due to the transitions of Ce^{3+} from the 5d excited state to the $^2F_{5/2}$ and $^2F_{7/2}$ ground states (Kalaji, Mikami, & Cheetham, 2014). The excitation spectrum (monitored at λ_{em} = 402 nm) extended from 250 to 360 nm and is ascribed to the 5d–4f move of Ce^{3+}. Figure 86b displayed the PL spectra of the SCPO:$_x$$Eu^{2+}$ phosphors (0.01 $\geq$ 0.2) under excitation

Figure 83. PL and PLE spectra of (a) $Ca_4(PO_4)_2O{:}_{0.02}Ce^{3+}$, (b) $Ca_4(PO_4)_2O{:}_{0.07}Eu^{2+}$, (c) $Ca_4(PO_4)_2O{:}_{0.02}Ce^{3+},_{0.012}Eu^{2+}$. Reproduced from ref (Jia et al., 2015)

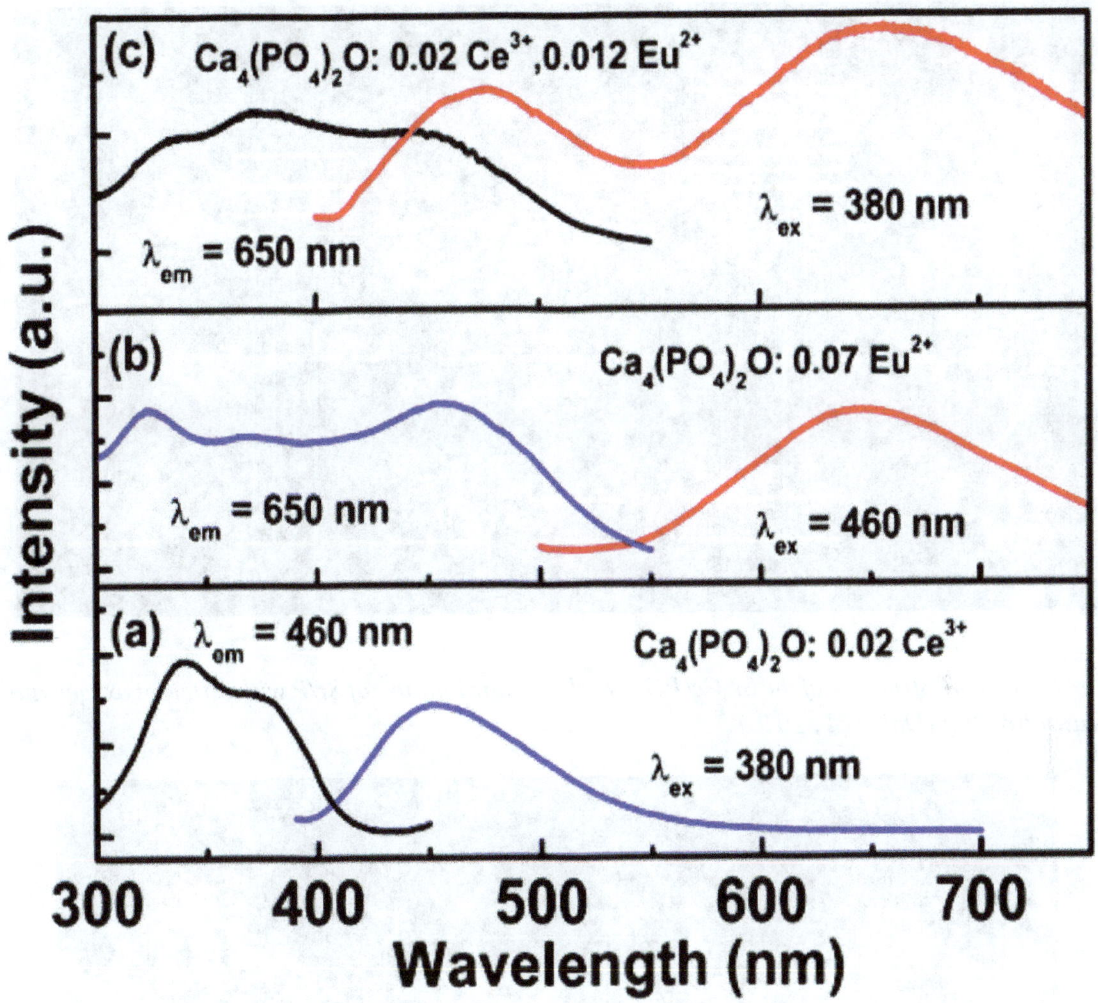

at 375 nm, when the Eu^{2+} ion doped in SCPO host the emission spectrum shows an extreme broad-band yellow emission from 400 to 800 nm under excitation at 375 nm. The FWHM and Stokes shift of the $SCPO{:}_{0.01}Eu^{2+}$ sample is found to be around 200 nm and 8380 cm^{-1}, respectively.

The PL spectrum of SCPO overlaps well with the PLE spectrum of SCPO:0.07Eu^{2+} in the range of 365–450 nm, indicating that the energy transfers from Ce^{3+} to Eu^{2+} can occur when the sample is excited with a shorter wavelength (<360 nm). Therefore, the obtained broadband yellow emission (λ_{ex} = 375 nm) totally originates from the 5d - 4f transition of Eu^{2+} without the contribution from the Ce^{3+} ions. Moreover, a WLED, fabricated by integrating a 405 nm UV LED chip with phosphor blends of yellow-emitting $Sr_3Ce(PO_4)_3$: 0.07Eu^{2+} and the blue-emitting $BaMgAl_{10}O_{17}{:}Eu^{2+}$, exhibits bright white-light with a CCT of 5996 K, a Ra of 86.5, and CIEs of (0.32,0.38) (Figure 87).

Figure 84. Schematic energy structure of Ce 4f and 5d levels in $Ca_4(PO_4)_2O$:Ce^{3+},Eu^{2+}. Redrawn from ref (Jia et al., 2015)

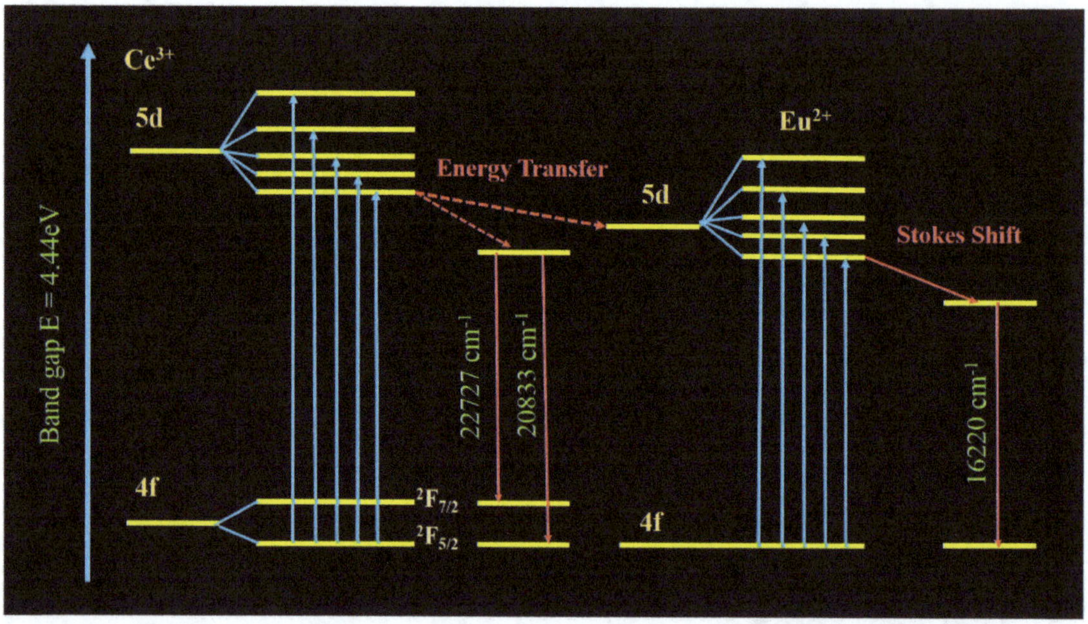

Figure 85. Crystal structure of the $Sr_3Ce(PO_4)_3$ and the coordination of Sr/P with different oxygen atoms. Redrawn from ref (Dai et al., 2016)

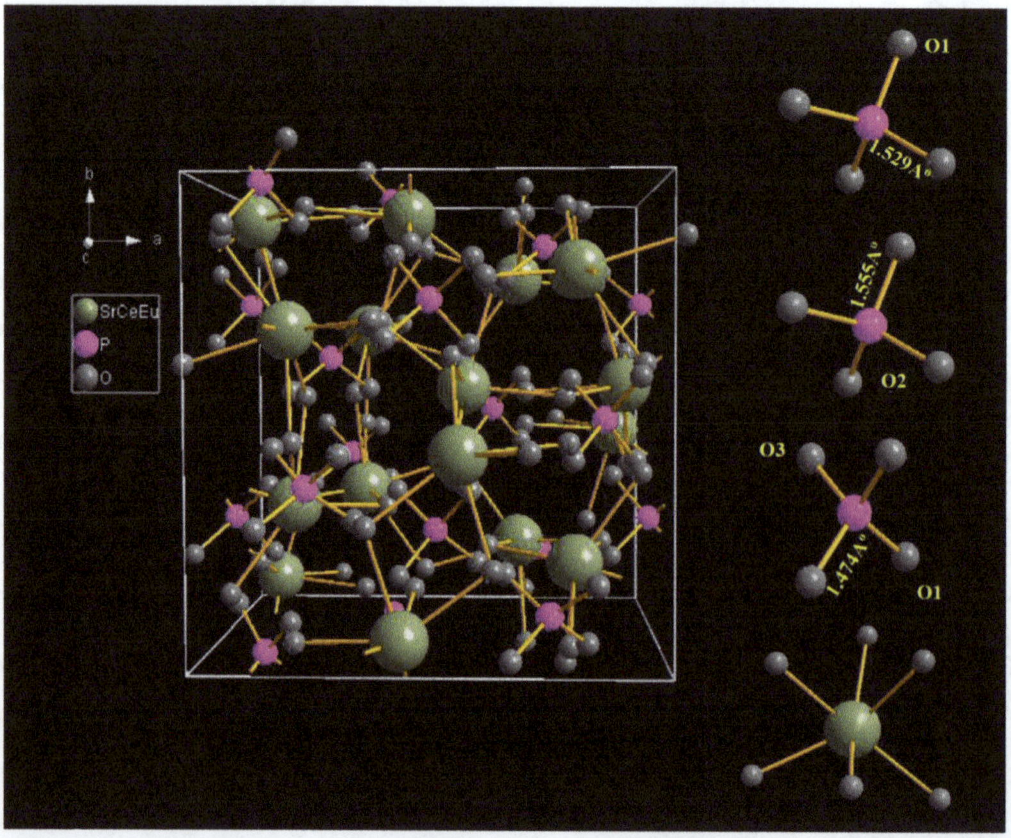

Figure 86. The normalized PL (lex = 325 nm) and PLE (lem = 402 nm) spectrum of the host SCPO. b) The PL spectra of the SCPO:xEu^{2+} phosphors (0.01 r x r 0.2) under excitation at 375 nm.Reproduced from ref (Dai et al., 2016)

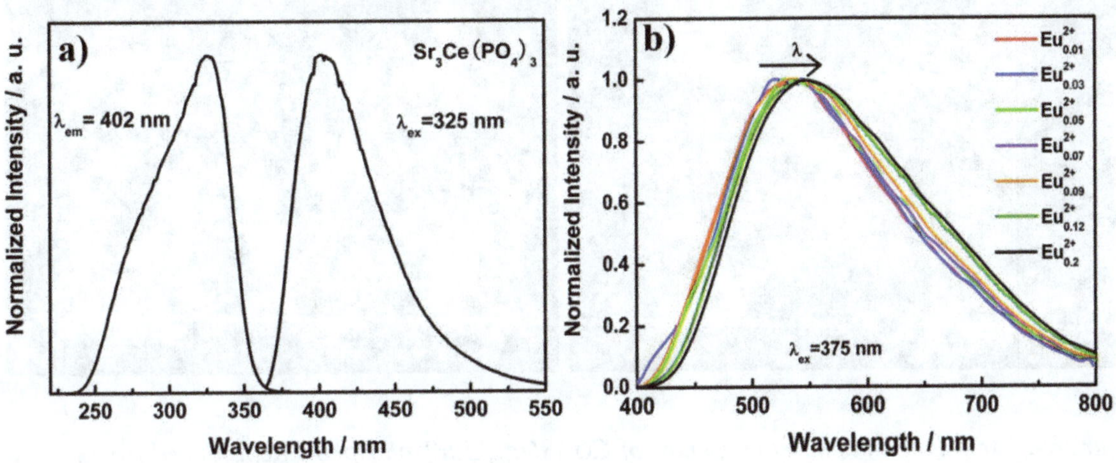

Figure 87. The EL spectrum of WLEDs fabricated by using a 405 nm NUVLED chip. The phosphor is combining a blue-emitting phosphor (BAM:Eu^{2+}) and the yellow-emitting phosphor (SCPO:Eu^{2+}) under the forward-bias current of 350 mA. The inset show the photographs of the as-prepared WLED driven by 350 mA current.Reproduced from ref (Dai et al., 2016)

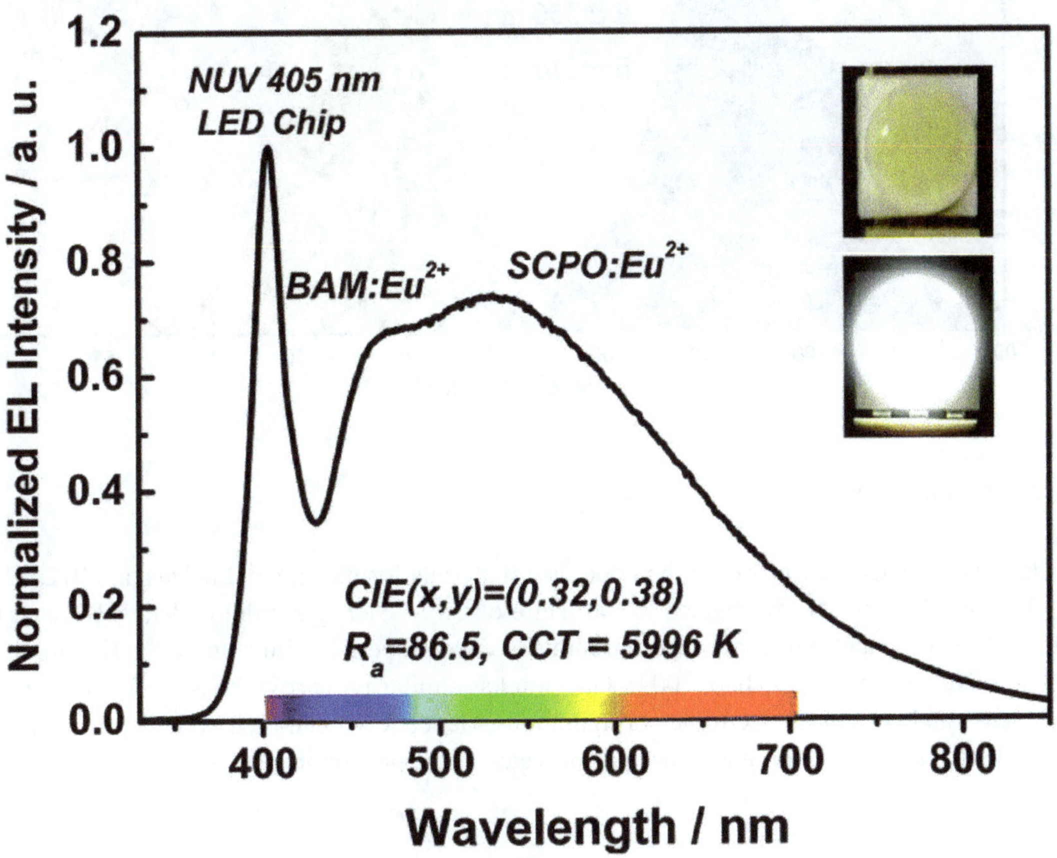

Figure 88. Cell structure of CGASO viewing along c and b direction. Redrawn from ref (Jiao, Xu, Yang, & You, 2017)

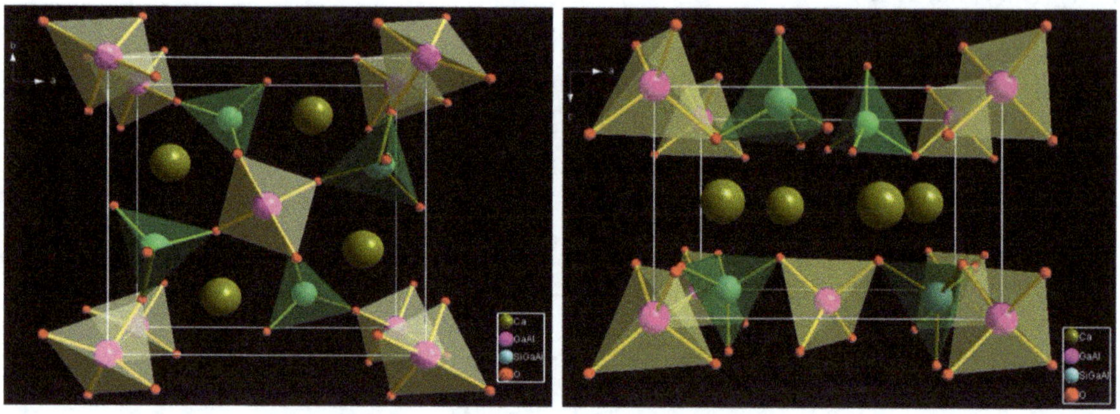

Figure 89. Excitation and emission spectra of CGASO:$_{0.012}$Eu^{2+} phosphor. Reproduced from ref (Jiao, Xu, Yang, & You, 2017)

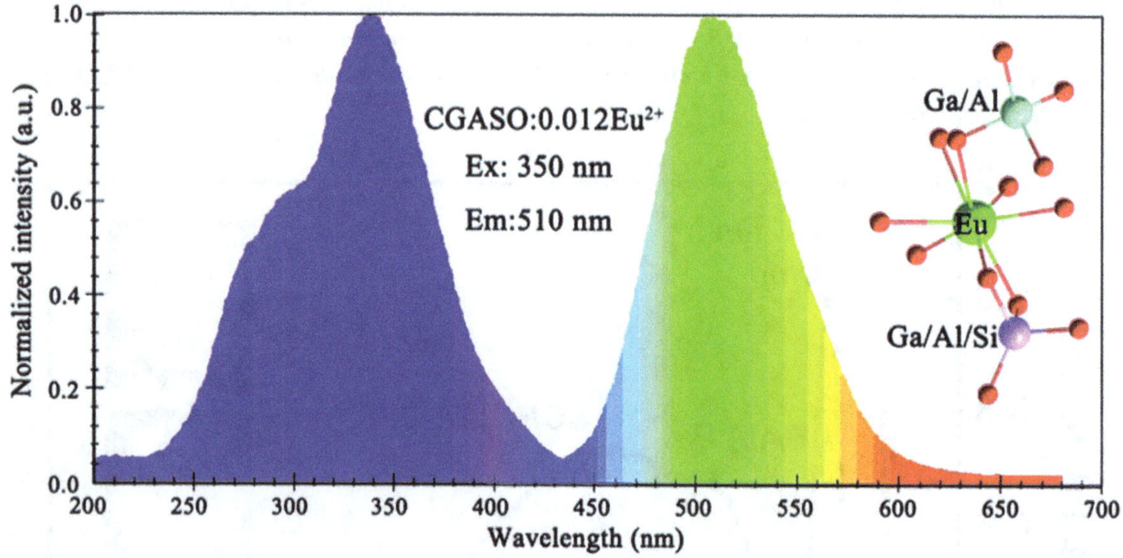

$Ca_2GaAlSiO_7$:Eu^{2+}/Ce^{3+}

Gallate based oxides are proven to be good host materials for phosphor (Puchalska, 2012; Zhang, 2016). In 1982, M. Kimata investigated the structure of $Ca_2Al_2SiO_7$ gehlenite in detail (Kimata, Ii, & Ibaraki, 1982). Yang and Jiao et al., reported luminescence property of Eu^{2+} and Ce^{3+}/Tb^{3+} ion in this host, respectively (Yang, 2012; Jiao, 2009). Ga^{3+} ion has similar property with Al^{3+} due to their same valence states and outermost electronic configurations. Hence, Ga^{3+} can be used to tune the emission wavelength and improve the luminescence performance of the phosphor.

Figure 90. (a) CIE coordinates and digital photographs of CGASO:$_{0.012}Eu^{2+}$ and CGASO:$_{0.02}Ce^{3+}$,$_{0.02}Li^{+}$ samples. (b) EL spectrum of a white LED device fabricated by "CGASO:$_{0.012}Eu^{2+}$ + CGASO:$_{0.02}Ce^{3+}$,$_{0.02}Li^{+}$ + red phosphor + 365 nm LED chip". The inset shows the corresponding LED device photograph. Reproduced from ref (Jiao, Xu, Yang, & You, 2017)

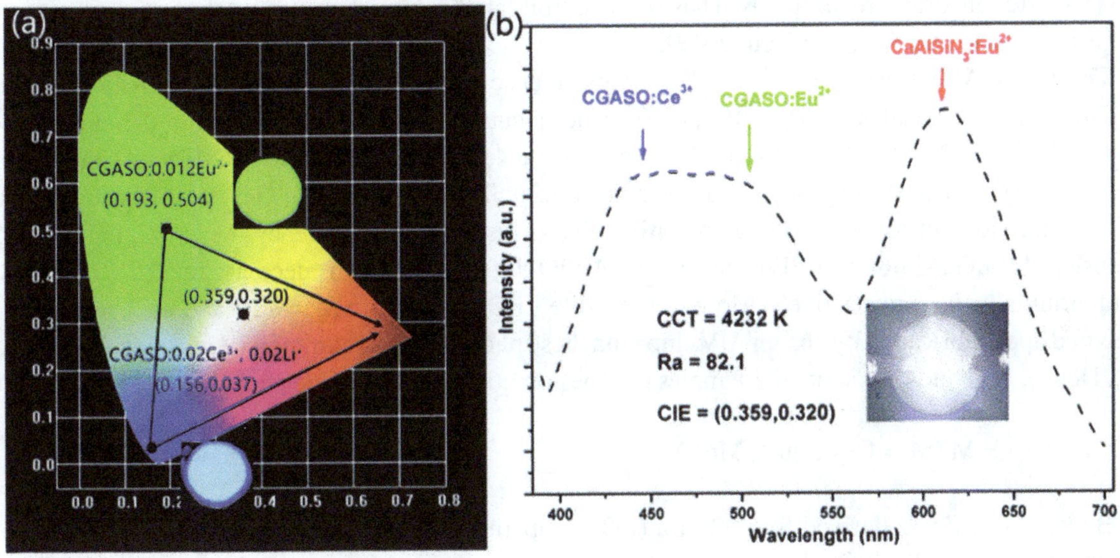

Figure 91. Schematic diagram of $Ba_3Ca_2(PO_4)_3F$ structure and coordination environment of Ca^{2+} and Ba^{2+} cations. Redrawn from ref (Meng et al., 2017)

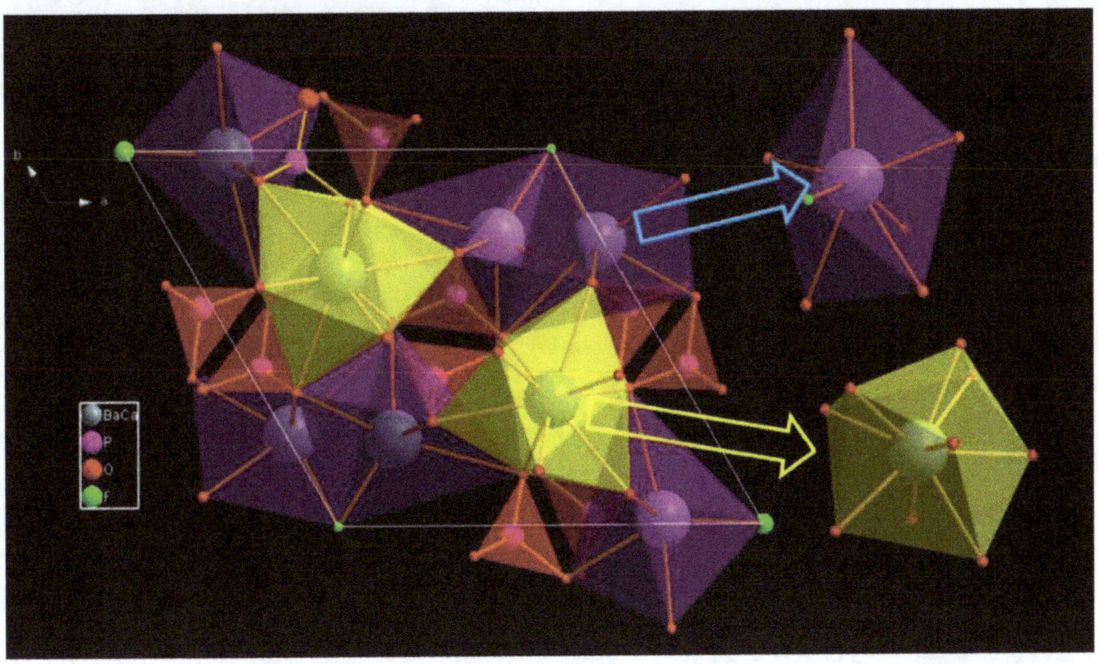

The $Ca_2GaAlSiO_7$ host crystallises in tetragonal crystal system and $P\bar{4}2_1m$ (113) space group with lattice parameters $a = b = 7.7441(2)$Å, $c = 5.1039(2)$Å and $V = 306.08(1)$ Å^3. In the lattice, the Ca^{2+} site is coordinated by eight oxygen ions forming polyhedron, which can be substituted by Eu^{2+}/Ce^{3+} ion forming emission centre. The Ga^{3+}, Al^{3+} and Si^{4+} cations are 4-fold coordinated and form [GaO_4], [AlO_4] and [SiO_4] tetrahedrons, respectively. These tetrahedrons are further connected by sharing oxygen atoms (Figure 88) (Jiao, Xu, Yang, & You, 2017).

The $Ca_2GaAlSiO_7$:0.012Eu^{2+} phosphor emits intense green light at 350 nm irradiation (Figure 89) with CIE coordinates being (0.193, 0.504) and quantum efficiency 65.0%, while the emission of $Ca_2GaAlSiO_7$:0.02Ce^{3+}, 0.02Li^+ locates in blue region at 410 nm irradiation with CIE coordinates being (0.156, 0.037) and quantum efficiency value being 65.2%. The substitution of Ga^{3+} for Al^{3+} causes a blue shift of emission due to the decrement of cation–ligand distance and covalency. At 150°C, the emission intensity of $Ca_2GaAlSiO_7$:0.012Eu^{2+} and $Ca_2GaAlSiO_7$:0.02Ce^{3+}, 0.02Li^+ decrease to 50.2% and 71.8% of their initial values, respectively. Moreover, a white LED device with $R_a = 82.1$ and CCT = 4232 K is prepared upon fabrication by 365 nm UV chip with these phosphors and commercial red phosphor (Figure 90). Hence, Eu^{2+} and Ce^{3+} activated samples may be potential green and blue phosphors for UV WLEDs.

$Ba_3Ca_2(PO_4)_3F$:M (M = Ce^{3+}, Eu^{2+}, Mn^{2+})

$Ba_3Ca_2(PO_4)_3F$ was derived from the $Ba_5(PO_4)_3F$ apatite formula. The $Ba_3Ca_2(PO_4)_3F$ has the same crystal structure as $Ba_5(PO_4)_3F$. It is crystallized in the hexagonal structure with space group *P63/m*

Figure 92. The excitation and emission spectrum of the BCPF: $_{0.1}Ce^{3+}$ phosphor. Reproduced from ref (Meng et al., 2017)

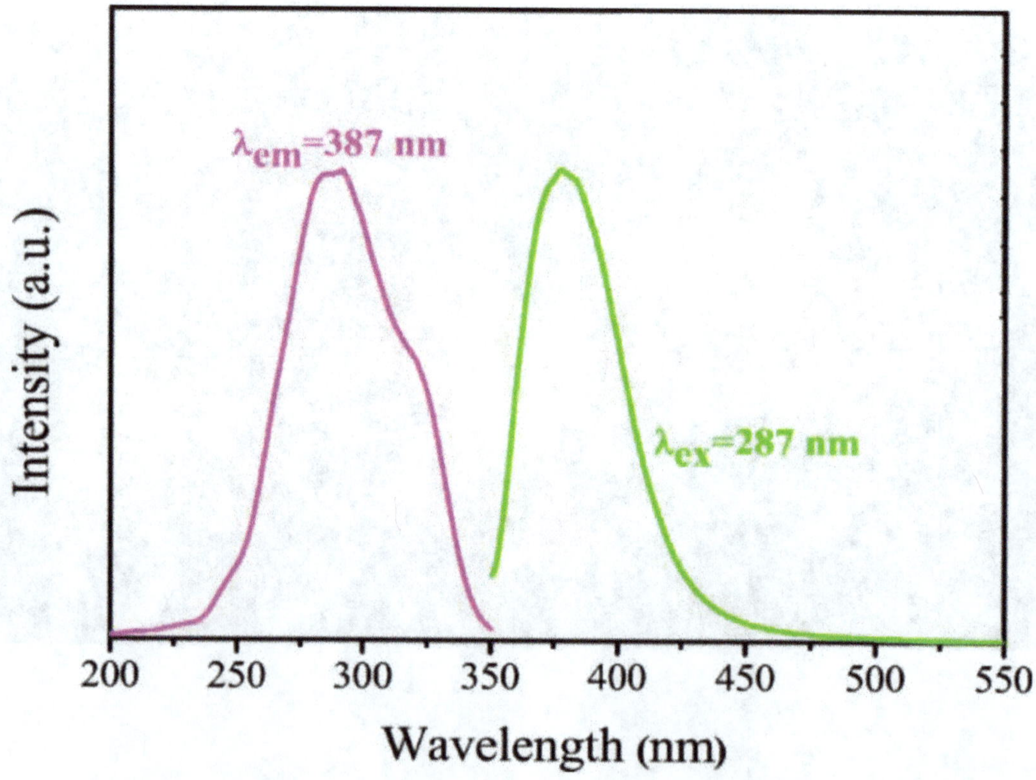

Figure 93. PL and PLE spectra of $Ba_3Ca_2(PO_4)_3F:_{0.03}Eu^{2+}$. The insert describes the emission intensity as a function of Eu^{2+} concentration in $Ba_3Ca_2(PO_4)_3F$. Reproduced from ref (Meng et al., 2017)

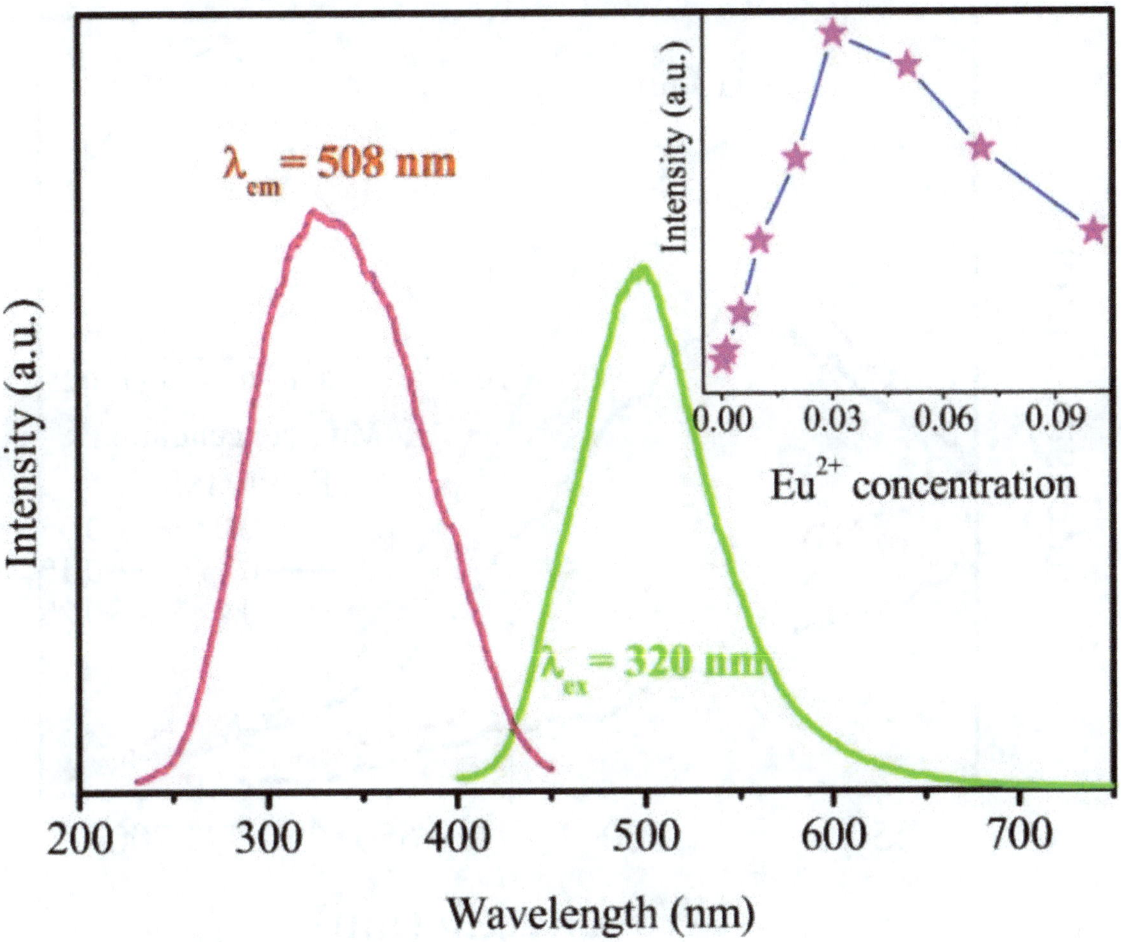

(No.176), a = 9.8450Å, c = 7.3590Å, and V = 617.71Å³. The crystal structure shown in Figure 91, it has two different type cationic sites M1(4f) and M2(6h). The M1 sites are nine-fold coordination surrounded by 9 O^- atom with C_3 point group symmetry. The M1 sites are seven-fold coordinated surrounded by six oxygen atom and one fluoride atom with C_3 point group symmetry (Meng et al., 2017).

Figure 92 represents the excitation and emission spectrum of the BCPF: $_{0.1}Ce^{3+}$ phosphor. The excitation spectra (under 387nm emission) contain the stronger band at 287 nm and a weaker band at 324 nm. The emission spectrum (under 287 nm excitation) of BCPF: $_{0.1}Ce^{3+}$ has a near UV band around 387 nm, which originates from the 5d to 4f ground state transition. The Ce^{3+} ion occupied the two cationic sites.

The excitation and emission spectrum of BCPF: $_{0.03}Eu^{2+}$ are shown in Figure 93. The excitation spectrum was observed at around 508 nm, shows an extensive band from 250 nm to 450 nm and it covers the entire UV region with a peak at 320 nm. Under 320 nm excitation BCPF: $_{0.03}Eu^{2+}$ phosphor has a broad green emission band centered at 508 nm is observed which ascribed to the move from the $4f^6$ $5d^1$ to the $4f^7$ levels of Eu^{2+} ions.

Figure 94. PL spectra of $Ba_3Ca_2(PO_4)_3F$:tMn^{2+}. The inset shows the emission intensity as a function of Mn^{2+} concentration in $Ba_3Ca_2(PO_4)_3F$. Reproduced from ref (Meng et al., 2017)

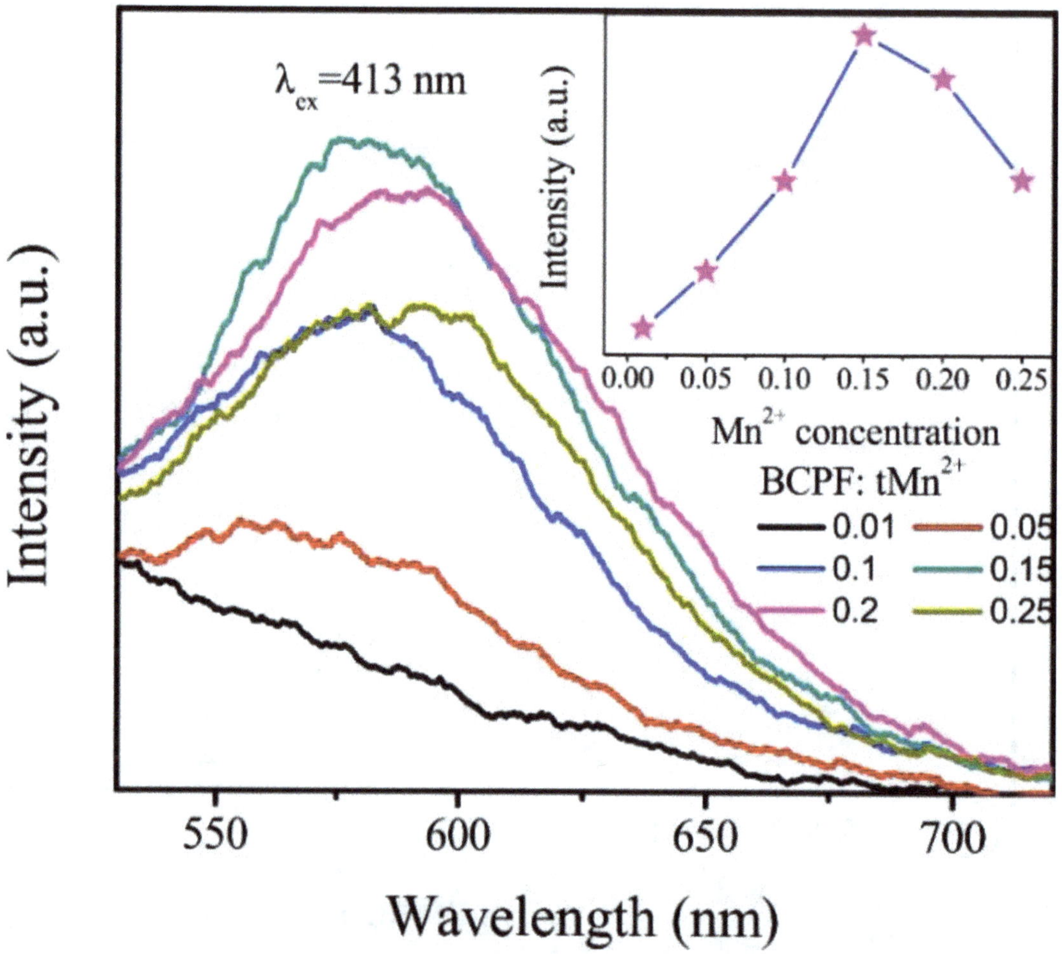

The PL spectra of Mn^{2+} doped BCPF acquired under 413 nm excited are appeared in Figure 94. It shows a broad band at 580 nm, ascribed to the transition from the $^4T_1$ to the $^6A_1$ levels of Mn^{2+} ions.

Figure 95 represents the PL spectra of BCPF: $_{0.1}Ce^{3+}$, $_xEu^{2+}$ phosphors with different Eu^{2+} ion concentration. Under the excitation of 287 nm, the PL spectra display is violet emission of Ce^{3+} ions and green emission of Eu^{2+} ions. The obtained emission spectrum suggests that there might be an efficient energy transfer from the Ce^{3+} ion to the Eu^{2+}, in which the Ce^{3+} ion act as sensitizer and the Eu^{2+} ion act as activator.

The PL spectra of BCPF: $_{0.1}Ce^{3+}$, $_yMn^{2+}$ under 324 nm excitation wavelength were shown in Figure 96. The inset demonstrates the emission intensity variation of Ce^{3+} and Mn^{2+}, it can be seen that the emission intensity of Ce^{3+} reduced with raise in Mn^{2+} content. It clearly indicates that Ce^{3+} may act as the sensitizer and Mn^{2+} act as the activator. Figure 96 shows the PLE spectra of BCPF: $_{0.1}Ce^{3+}$, $_yMn^{2+}$

Figure 95. PL spectra of $Ba_3Ca_2(PO_4)_3F$:$_{0.1}Ce^{3+}$, $_xEu^{2+}$. The inset shows the emission intensity as a function of Eu^{2+} concentration in $Ba_3Ca_2(PO_4)_3F$. Reproduced from ref (Meng et al., 2017)

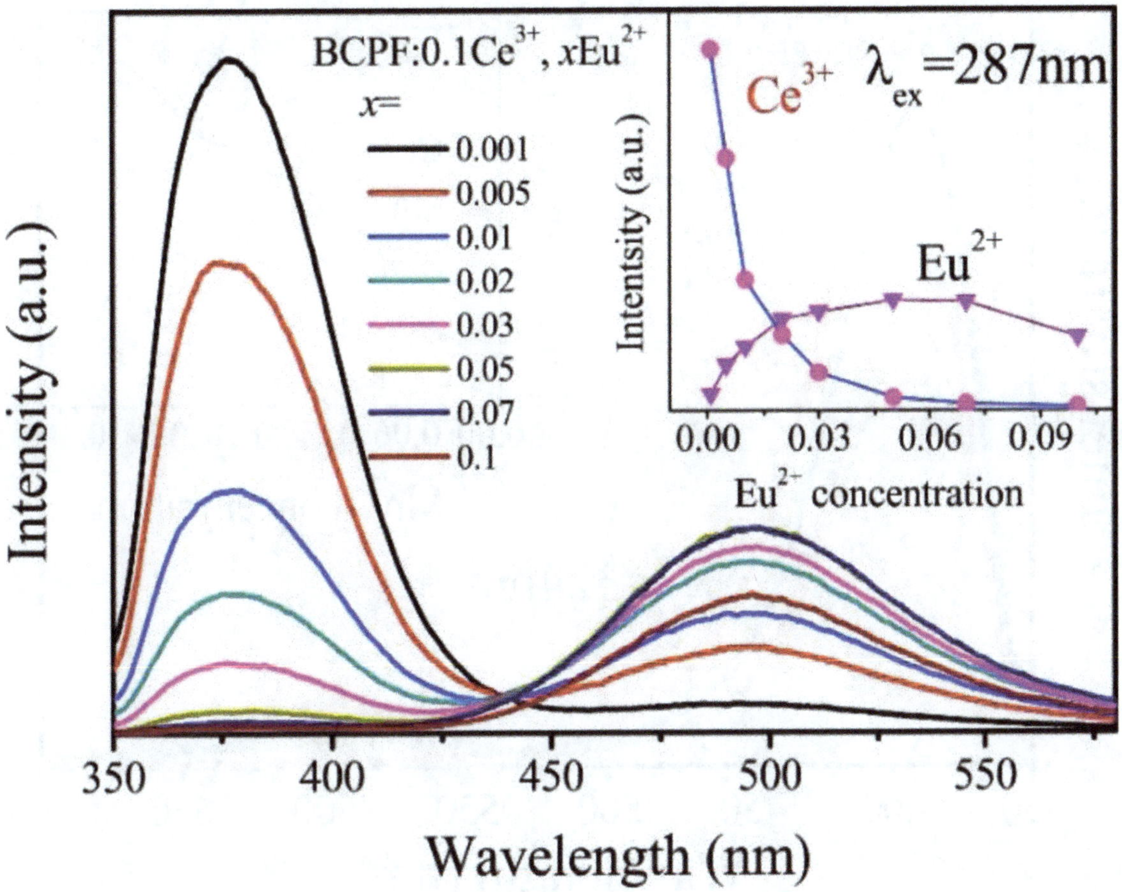

monitored at 580 nm, comprise of not only a weak band with a peak at 413 nm of Mn^{2+}, but also a stronger excitation band at 324 nm of Ce^{3+} which confirms the energy transfer from Ce^{3+} to Mn^{2+} in BCPF:$_{0.1}Ce^{3+}$, $_yMn^{2+}$phosphor.

Figure 97(a) demonstrate the PL spectra of BCPF: $_{0.3}Eu^{2+}$, $_zMn^{2+}$, which consist of a strong peak at 508 nm from Eu^{2+} and a weak peak at around 580 nm from Mn^{2+}. As displayed in Figure 97 (b), the excitation spectra of Mn^{2+} is similar that of Eu^{2+}, which can be a strong evidence for the energy transfer process from Eu^{2+} to Mn^{2+}. As shown in Figure 97(c), the emission intensity of Eu^{2+} steadily diminished with raising the Mn^{2+} concentration, which also state that there is an adequate energy transfer from Eu^{2+} to Mn^{2+}(Huang, 2010; Yang, 2015; Zhou, 2014). Here around 623 nm some weak peaks are observed, which is due to the some unstable Eu^{2+} will be oxidized to Eu^{3+} ion in the BCPF host (Zheng & Tang, 2016).

Figure 98 shows the PL spectrum of BCPF: $_{0.1}Ce^{3+}$, $_ZEu^{2+}$, $_{0.25}Mn^{2+}$. With increment of Eu^{2+} concentration, the emission intensity of Ce^{3+} (387 nm) continuously decrease and the emission intensity of Mn^{2+} (580 nm) increases, which proves that the energy transfer mechanism of $Ce^{3+} \rightarrow Eu^{2+} \rightarrow Mn^{2+}$ in the phosphor.

Figure 96. Emission spectra of $Ba_3Ca_2(PO_4)_3F{:}_{0.1}Ce^{3+}$, $_yMn^{2+}$. The inset describes emission intensity as a function of Mn^{2+} concentration. Reproduced from ref (Meng et al., 2017)

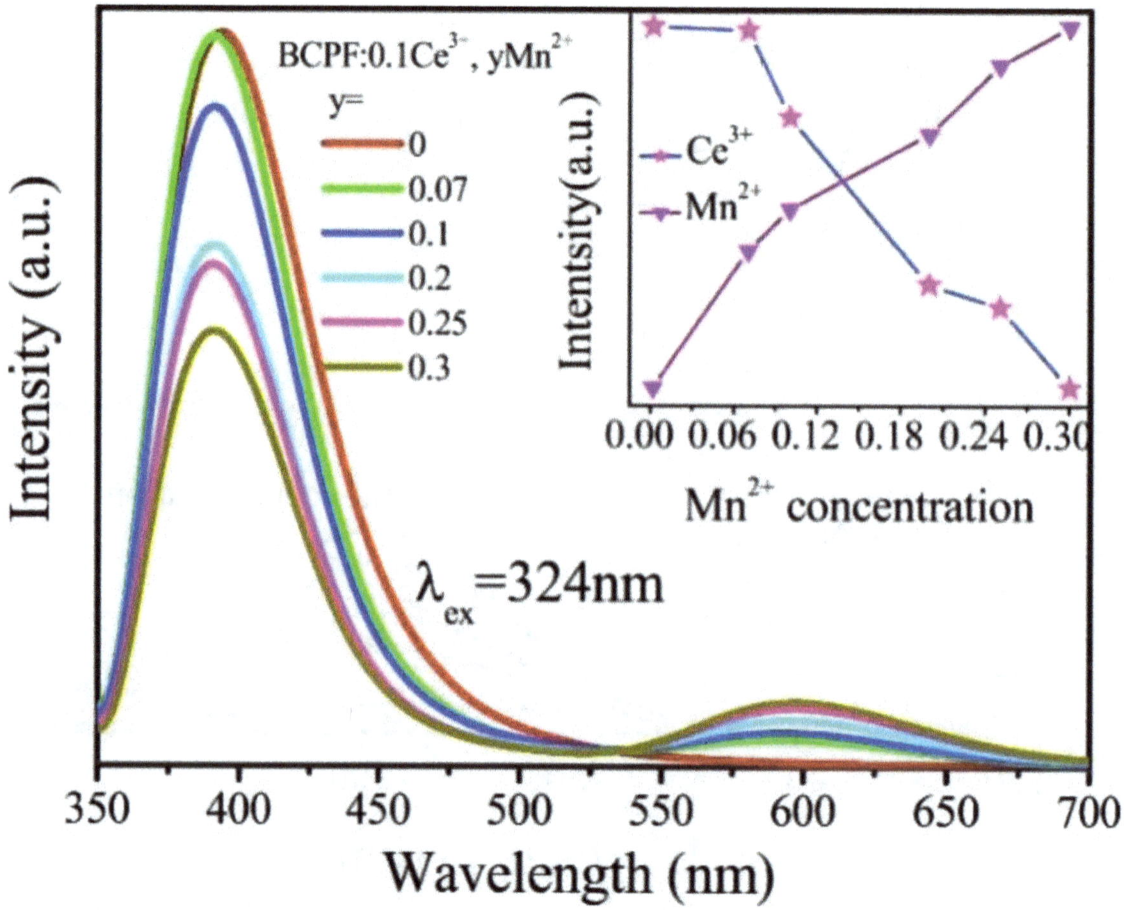

The schematic energy level chart for BCPF: Ce^{3+}, Eu^{2+}, Mn^{2+} is shown in Figure 99, which represents that the ET process may occurs in the lattice. The Ce^{3+} consumes UV light, and afterward exchanges (transfers) the energy to Eu^{2+} levels. Furthermore, probably there are two different possibilities for the Eu^{2+} ion to come back to ground state: one perspective, Eu^{2+} ions at the $4f^65d^1$state may transfer partially from $4f^65d^1$ to $4f^7$ emitting a band peaking at 508 nm. Then again, Eu^{2+} ions at the excited state may exchange the consumed energy to 4A1 (4G) state of Mn^{2+} through resonance energy transfer mechanism. Hence, Mn^{2+} ions will ultimately relax to the ground state $^6A_1$ (^6S) and radiate a red emission band focused at 580 nm (Zheng & Tang, 2016).

All the phosphor compositions which already been discussed in the book chapter and their related properties summarized in Table 2.

Figure 97. (a) PL spectra of $Ba_3Ca_2(PO_4)_3F$:$_{0.03}Eu^{2+}$, $_zMn^{2+}$. (b): excitation spectra of Eu^{2+} (508 nm) and Mn^{2+} (580 nm). (c): emission intensity as a function of Mn^{2+} concentration. Reproduced from ref (Meng et al., 2017)

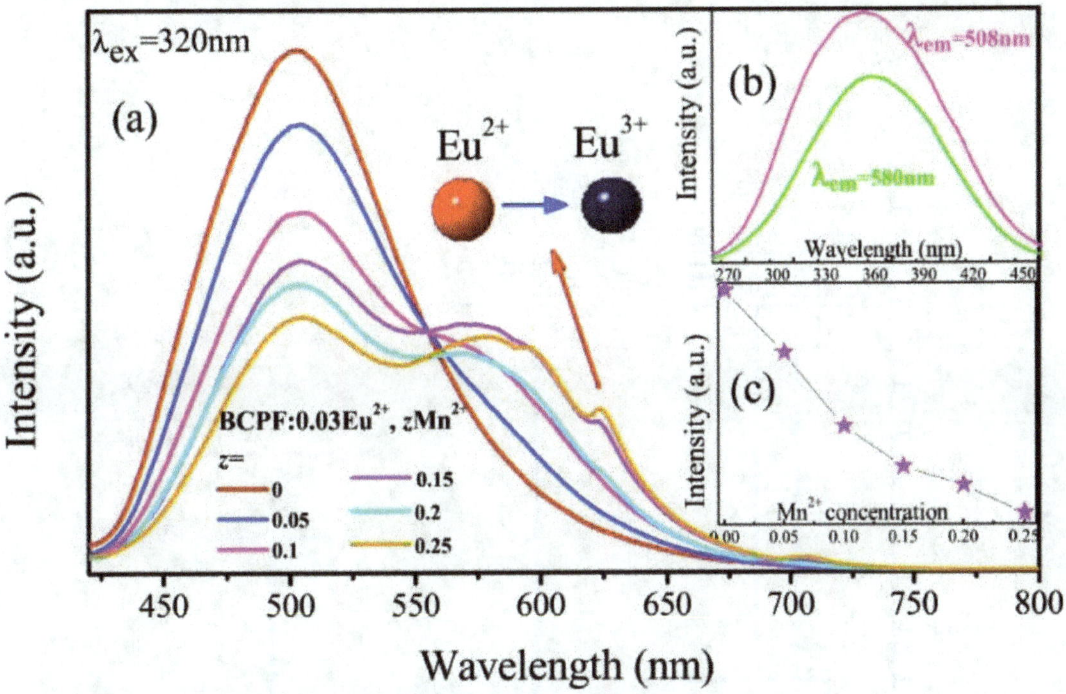

Figure 98. PL spectra of $Ba_3Ca_2(PO_4)_3F$:$_{0.1}Ce^{3+}$, $_ZEu^{2+}$, $_{0.25}Mn^{2+}$. The inset shows the emission intensity as a function of Eu^{2+} concentration. Reproduced from ref (Meng et al., 2017)

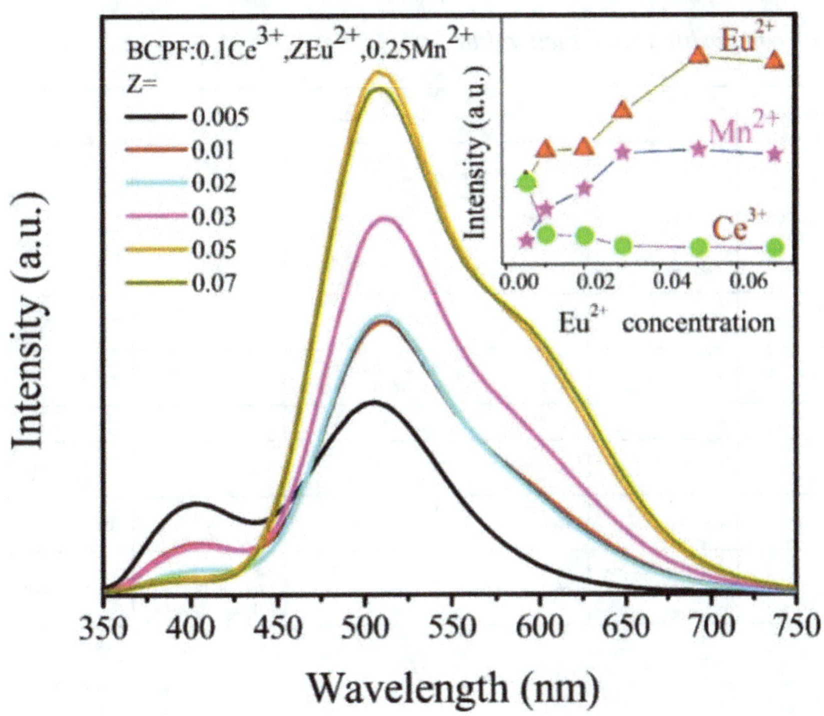

Figure 99. Energy levels schemes of $Ce^{3+}/Eu^{2+}/Mn^{2+}$ and possible energy transfer mode in the BCPF. Redrawn from ref (Meng et al., 2017)

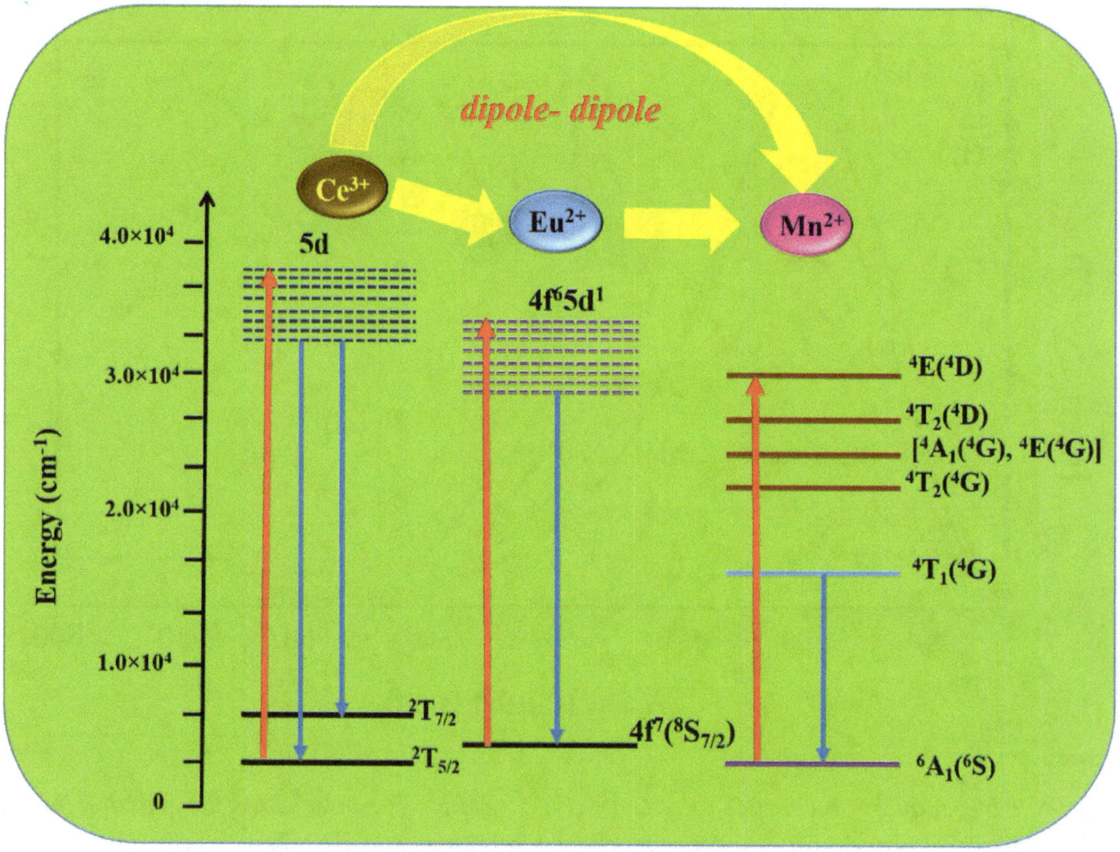

Table 2. Phosphor compositions and their related properties

S.No	Phosphor Compositions	λ_{em} (nm)	λ_{exc} (nm)	PL CIE	LED	QY	Ref
1	$Sr_{2.975-x}Ba_xCe_{0.025}AlO_4F:Ce^{3+}$	502, 546	400, 460	-	Ra-87	-	[12] Im, Brinkley, Hu, Mikhailovsky, DenBaars, & Seshadri, (2010); [14] Zhou, Xia, Yang, & Shen, (2012);
2	$Ca_{5.45}Li_{3.55}(SiO_4)_3O_{0.45}F_{1.55}$ $:Ce^{3+}$	470	365	0.154,0.188	-	82%	Zhou, Xia, Yang, & Shen, (2012);
3	$Ca_2LaZr_2Ga_3O_{12}:Ce^{3+}$	515	430	0.271, 0.479 0.285, 0.502	-	IQE-35.2%	Zhongn et al., (2015);
4	$LiY_9(SiO_4)_6O_2:Ce^{3+}$	366	305	-	-	-	Zhou et al., (2016);
5	$Ca_{10}(PO_4)_6O:Ce^{3+}$	410 - 510	330-355	0.165, 0.056 0.298, 0.434	-	-	Li et al., (2016);
6	$Ba_2Y_5B_5O_{17}:Ce^{3+}$	413	308	0.159, 0.155	-	70%	Hermus, Phan, & Brgoch, (2016);
7	β-$NaSrBO_3:Ce^{3+}$	424	358	-	Ra-92.8 CCT- 3654K CIE-0.3821, 0.3430	43.9% and 49.0%	Yi et al., (2016);

continued on following page

Table 2. Continued

S.No	Phosphor Compositions	λ_{em} (nm)	λ_{exc} (nm)	PL CIE	LED	QY	Ref
8	$Ba_2(Y_{1-x}Lu_x)_5B_5O_{17}:Ce^{3+}$	447	340	-	-	90%	Hermus et al., (2017);
9	$Li_2SrSiO_4:Eu^{2+}$	562	420	0.3677, 0.5148	CIE - 0.3346,0.3401	-	Saradhi et al., (2006);
10	$Ca_4(PO_4)_2O:Eu^{2+}$	665	460	-	R_a-90.5 CCT-6446K CIE-0.3135, 0.3316	37.5%	Deng et al., (2013);
11	$K_2Ba_7Si_{16}O_{40}:Eu^{2+}$	500	440	0.233, 0.507	-	61.9%	Lv et al., (2014);
12	$Sr_{1.75}Ca_{1.25}(PO_4)_2:Eu^{2+}$	514	279, 365	0.374, 0.588	Ra-85.1. CCT-4529K CIE-0.368, 0.414	-	Ji et al., (2014);
13	β-$Ca_3SiO_4Cl_2:Eu^{2+}$	635	450	-	-	70%	Kim, T. G., Kim, T., Kim, J., Kim, S. J., & Im, (2014);
14	$Sr_9Mg_{1.5}(PO_4)_7:Eu^{2+}$	523, 620	335, 460	0.492, 0.478	Ra-86.07 CCT-3437K CIE-0.393, 0.352	IQE-46.2 EQE-26.9	Sun et al., (2015);
15	$BaSc_2Si_3O_{10}:Eu^{2+}$	421 445	360 330	0.15, 0.05	-	EQE-42.65%.	Wang et al., (2015);
16	$Ca_{2+x}La_{8-x}(SiO_4)_{6-x}(PO_4)_xO_2:Eu^{2+}$	508, 500, 478, 460	365	0.1846, 0.1851	-	IQE46.5%	Xia et al., (2016);
17	$(Sr,Ba)_2SiO_4:Eu^{2+}$	585,589, 603,616	450	-	-	$(Sr,Ba)_{1.5}Eu_{0.5}SiO_4$ (IQE)/(EQE) - 75%/65% $(Sr,Ba)_{1.2}Eu_{0.8}SiO_4$ (IQE)/(EQE) - 64%/56%	Wen et al., (2016);
18	$Ba_4Gd_3Na_3(PO_4)_6F_2:Eu^{2+}$	420, 481, 550, 600	250, 276, 350	0.266, 0.401	-	48.9%	Fu, Lü, Jiao, & You, (2016);
19	$Ca_3Si_2O_7:Eu^{2+}$	622 602	400	0.5681, 0.4143 to 0.3616, 0.5876	-	32.5%	Kasturi & Sivakumar, (2016);
20	$Ba_3La_6(SiO_4)_6:Eu^{2+}$	507	350	-	CRI - 82.9 CCT -4818K CIE -0.3320, 0.3418	-	Zhong et al., (2017);
21	$Ba_{10}(PO_4)_6F_2: Eu^{2+}$	420	320	0.163, 0.024	-	-	Huang, Zhang, & Pan, (2017);
22	$Ba_5(PO_4)_2SiO_4:Eu^{2+}$	516	405	-	CIE - 0.355, 0.342 CCT - 4287K	-	Ding & Wang, (2017);
23	$Ba_2Y_3(SiO_4)_3F:Eu^{2+}$	470, 560, 618	350	0.370, 0.336	CRI - 83.5 CCT - 3530K CIE - 0.402, 0.371	η_{int} - 11.2% η_{ext} - 4.9%	Wang et al., (2017);
24	$Na_{3-2x}Sc_2(PO_4)_3: xEu^{2+}$	453	370	-	CRI-93 CIE- 0.2989 0.3502	74%	Kim et al., (2017);
25	$Na_6Ca_3Si_6O_{18}:Ce^{3+}, Tb^{3+}$	440 550	330 350	Ce -0.162, 0.214 Ce,Tb - 0.212, 0.336	-	43%	Xu et al., (2017);
26	$NaBaScSi_2O_7:Ce^{3+},Tb^{3+}$	425 542	349 292,	0.1738, 0.1823 to 0.3384, 0.5335	-	-	Li et al., (2016);
27	$Ba_2Y_5B_5O_{17}:Ce^{3+}, Tb^{3+}$	543, 445	448, 345	0.1519, 0.1067 to 0.2820, 0.5028	-	-	Xiao et al., (2017);
28	$SrAl_2Si_2O_8: Ce^{3+}/Tb^{3+}$	541	365	-	-	80.92%	Ma et al., (2017);
29	$Ca_9Y(PO_4)_7:Eu^{2+},Mn^{2+}$	486	365	0.310, 0.330	R_a- 76 CCT-7200K CIE-0.30, 0.31	-	Huang et al., (2010);
30	$Ca_9Gd(PO_4)_7:Eu^{2+},Mn^{2+}$	-	380	0.326, 0.328	CCT-6569K CIE-0.312, 0.327	-	Huang, Liu, & Chen, (2010);

continued on following page

Table 2. Continued

S.No	Phosphor Compositions	λ_{em} (nm)	λ_{exc} (nm)	PL CIE	LED	QY	Ref
31	$NaScSi_2O_6:Eu^{2+},Mn^{2+}$	Mn-654 Eu -533 Eu,Mn–654, 533	410 365 365	0.313, 0.529 0.350, 0.505	Ra - 92.2 CCT – 4666 K CIE - 0.358, 0.378	Eu-71.7% Eu, Mn-40.0%	Xia, Zhang, Molokeev, & Atuchin, (2013);
32	$Ca_9Mg(PO_4)_6F_2$: Eu^{2+}, Mn^{2+}	495, 565	294	Eu- 0.155, 0.121 Eu,Mn-0.255, 0.290	-	Eu -18% Eu,Mn-22%	Li et al., (2014);
33	$Ca_6Y_2Na_2(PO_4)_6F_2:Eu^{2+}$ Mn^{2+}	486	365	0.310, 0.330	R_a- 76 CCT-7200K CIE-0.30, 0.31	-	Guo, You, Jia, Ouyang, & Wu, (2014);
34	$Na_2BaCa(PO_4)_2:Eu^{2+},Mn^{2+}$	451, 576	351, 402	0.15, 0.10 to 0.38 0.34	-	19.92% and 23.62%,	Guo, You, Jia, Ouyang, & Wu, (2014);
35	$Sr_3GdLi(PO_4)_3F:Eu^{2+}$, Mn^{2+}	445, 575	380,404	Eu-0.479 0.509 Eu,Mn-0.308, 0.270	-	-	Lv et al., (2017);
36	$BaMgSiO_4:Ce^{3+},Eu^{2+}$	502, 430, 500	370, 310 400	0.23, 0.44	-	-	Sivakumar and Varadaraju, (2007);
37	Sr_2LiSiO_4F	510 400	350	-	-	-	Sivakumar and Varadaraju, (2009);
38	$Ca_4(PO_4)_2O$: Ce^{3+},Eu^{2+}	380, 460,380	460, 650, 650	-	Ra - 84 CCT-4124K CIE-0.359, 0.310	-	Jia et al., (2015);
39	$Sr_3Ce(PO_4)_3:Eu^{2+}$	402	325, 375	0.368 0.463	Ra- 86.5 CCT- 5996K CIE-0.32, 0.38	-	Dai et al., (2016);
40	$Ca_2GaAlSiO_7:Eu^{2+}/Ce^{3+}$	510	350	Ce – 0.156, 0.037 Eu – 0.193, 0.504	Ra - 82.1 CCT – 4232K CIE – 0.359, 0.320	-	Jiao, Xu, Yang, & You, (2017);
41	$Ba_3Ca_2(PO_4)_3F:M$ (M = Ce^{3+}, Eu^{2+}, Mn^{2+})	387; 508; 580	287; 320; 413	Ce-0.1943, 0.1372 Eu-0.2408, 0.3707 Ce,Eu,Mn-0.3572, 0.40	-	-	Meng et al., (2017).

CONCLUSION

A number of host materials activated with mainly Ce^{3+} and Eu^{2+} are investigated in detail in the current chapter. The Ce^{3+} absorbs UV light, and transfer the same energy to excited Eu^{2+} levels.The process of energy transfer is well explained. The structural properties and influence of the same on the energy transfer process and luminescence properties are given widely.The concentration quenching phenomena with increasing activator ion concentration in different hosts have also been discussed. The phosphors like apatites, phosphates, fluorophosphates and borates have given special preferences. The thermal stability and performances of the phosphor in white LEDs have also been discussed in detailed. Overall the structure-compositions-property correlation was made for the phosphor materials. Designing of new phosphor composition or modifying the existing phosphors with high absorption in the near UV to blue region will certainly push the solid state lighting into general lightings.

ACKNOWLEDGMENT

The authors thank Naval Research Board (NRB), DRDO for funding.

REFERENCES

Akella, A., & Keszler, D. (1995). Sr_2LiSiO_4F: Synthesis, Structure, and Eu^{2+} Luminescence. *Chemistry of Materials*, *7*(7), 1299–1302. doi:10.1021/cm00055a005

Amara, M. B., Vlasse, M., Flem, G. L., & Hagenmuller, P. (1983). Structure of the low-temperature variety of calcium sodium orthophosphate, $NaCaPO_4$. *Acta Crystallographica. Section C, Crystal Structure Communications*, *39*(11), 1483–1485. doi:10.1107/S0108270183008963

Bachmann, V., Ronda, C., & Meijerink, A. (2009). Temperature Quenching of Yellow Ce^{3+} Luminescence in YAG:Ce. *Chemistry of Materials*, *21*(10), 2077–2084. doi:10.1021/cm8030768

Baginskiy, I., & Liu, R. S. (2009). Significant Improved Luminescence Intensity of Eu^{2+}-Doped $Ca_3SiO_4Cl_2$ Green Phosphor for White LEDs Synthesized Through Two-Stage Method. *Journal of the Electrochemical Society*, *156*(5), G29–G32. doi:10.1149/1.3089366

Baur, W. (1974). The geometry of polyhedral distortions. Predictive relationships for the phosphate group. *Acta Crystallographica. Section B, Structural Crystallography and Crystal Chemistry*, *30*(5), 1195–1215. doi:10.1107/S0567740874004560

Belik, A. A., Lazoryak, B. I., Pokholok, K. V., Terekhina, T. P., Leonidov, I. A., Mitberg, E. B., ... Kellerman, D. G. (2001). Synthesis and Characterization of New Strontium Iron(II) Phosphates, $SrFe_2(PO_4)_2$ and $Sr_9Fe_{1.5}(PO_4)_7$. *Journal of Solid State Chemistry*, *162*(1), 113–121. doi:10.1006/jssc.2001.9363

Cao, Y., Ding, J., Wang, X., & Wang, Y. (2017). Tunable white light of multi-cation-site $Na_2BaCa(PO_4)_2$:Eu,Mn phosphor: synthesis, structure and PL/CL properties. *Journal of Materials Chemistry. C, Materials for Optical and Electronic Devices*, *5*(5), 1184–1194. doi:10.1039/C6TC04827G

Catti, M., & Gazzoni, G. (1983). The $\beta \leftrightarrow \alpha'$ phase transition of Sr_2SiO_4. II. X-ray and optical study, and ferroelasticity of the β form. *Acta Crystallographica. Section B, Structural Crystallography and Crystal Chemistry*, *39*(6), 679–684. doi:10.1107/S0108768183003225

Chan, T. S., Liu, R. S., & Baginskiy, I. (2008). Synthesis, Crystal Structure, and Luminescence Properties of a Novel Green-Yellow Emitting Phosphor $LiZn_{1-x}PO_4$:xMn for Light Emitting Diodes. *Chemistry of Materials*, *20*(4), 1215–1217. doi:10.1021/cm7028867

Chen, X., Dai, P. P., Zhang, X. T., Li, C., Lu, S., Wang, X. L., ... Liu, Y. C. (2014). A Highly Efficient White Light (Sr3,Ca,Ba)$(PO_4)_3$Cl:Eu^{2+}, Tb^{3+}, Mn^{2+} Phosphor via Dual Energy Transfers for White Light-Emitting Diodes. *Inorganic Chemistry*, *53*(7), 3441–3448. doi:10.1021/ic402870p PMID:24649898

Ci, Z. P., Sun, Q. S., Sun, M. X., Jiang, X. J., Qin, S. C., & Wang, Y. H. (2014). Structure, photoluminescence and thermal properties of Ce^{3+}, Mn^{2+} co-doped phosphosilicate $Sr_7La_3[(PO_4)_{2.5}(SiO_4)_3(BO_4)_{0.5}](BO_2)$ emission-tunable phosphor. *Journal of Materials Chemistry. C, Materials for Optical and Electronic Devices*, *2*(29), 5850–5856. doi:10.1039/C4TC00217B

Czaya, R., & Bissert, G. (1971). Die Kristallstruktur von Tricalciummonosilikatdichlorid (Ca_2SiO_4. $CaCl_2$). *Acta Crystallographica. Section B, Structural Crystallography and Crystal Chemistry*, *27*(4), 747–752. doi:10.1107/S0567740871002875

Dai, P., Lee, S. P., Chan, T. S., Huang, C. H., Chiang, Y. W., & Chen, T. M. (2016). $Sr_3Ce(PO_4)_3$:Eu^{2+}: a broadband yellow-emitting phosphor for near ultraviolet-pumped white light-emitting devices. *The Journal of Physical Chemistry C, 4*, 1170–1177.

Dai, Q., Foley, M. E., Breshike, C. J., Lita, A., & Strouse, G. F. (2011). Ligand-Passivated Eu:Y_2O_3 Nanocrystals as a Phosphor for White Light Emitting Diodes. *Journal of the American Chemical Society, 133*(39), 15475–15486. doi:10.1021/ja2039419 PMID:21863840

Deng, D., Yu, H., Li, Y., Hua, Y., Jia, G., Zhao, S., ... Xu, S. (2013). $Ca_4(PO_4)_2O$:Eu^{2+} red-emitting phosphor for solid-state lighting: structure, luminescent properties and white light emitting diode application. *Journal of Materials Chemistry. C, Materials for Optical and Electronic Devices, 1*(19), 3194–3199. doi:10.1039/c3tc30148f

Ding, J. Y., Wu, Q. S., Li, Y. Y., Long, Q., Wang, C., & Wang, Y. H. (2015). $Sr_{7.3}Ca_{2.7}(PO_4)_6F_2$: Eu^{2+}, Mn^{2+}: a novel single-phased white light-emitting phosphor for NUV-LEDs. *Dalton Transactions (Cambridge, England), 44*(20), 9630–9636. doi:10.1039/C5DT00907C PMID:25924106

Ding, X., & Wang, Y. (2017). Structure and photoluminescence properties of a novel apatite green phosphor $Ba_5(PO_4)_2SiO_4$:Eu^{2+} excited by NUV light. *Physical Chemistry Chemical Physics, 19*(3), 2449–2458. doi:10.1039/C6CP08300E PMID:28058425

Dorenbos, P. (2003). Energy of the first $4f^7 \rightarrow 4f^6 5d$ transition of Eu^{2+} in inorganic compounds. *Journal of Luminescence, 104*(4), 239–260. doi:10.1016/S0022-2313(03)00078-4

Drofenik, M., & Golic, L. (1979). Refinement of the Sr_2EuFeO_5 and Sr_3EuAlO_5 structures. *Acta Crystallographica. Section B, Structural Crystallography and Crystal Chemistry, B35*(5), 1059–1062. doi:10.1107/S0567740879005574

Fu, X., Lü, W., Jiao, M., & You, H. (2016). Broadband Yellowish-Green Emitting $Ba_4Gd_3Na_3(PO_4)_6F_2$:Eu^{2+} Phosphor: Structure Refinement, Energy Transfer, and Thermal Stability. *Inorganic Chemistry, 55*(12), 6107–6113. doi:10.1021/acs.inorgchem.6b00648 PMID:27249557

Fu, X. Y., Fang, L. J., Niu, S. Y., & Zhang, H. W. (2013). Luminescence properties and energy transfer investigations of $SrMgSi_2O_6$:Ce, Tb phosphors. *Journal of Luminescence, 142*, 163–166. doi:10.1016/j.jlumin.2013.03.055

Gedam, S. C., Dhoble, S. J., & Park, K. (2013). Resonant energy transfer from $Ce^{3+} \rightarrow Dy^{3+}$ and Mn^{2+} in $NaZnSO_4Cl$ chlorosulphate phosphor. *Journal of Luminescence, 139*, 47–51. doi:10.1016/j.jlumin.2013.02.010

George, N. C., Pell, A. J., Dantelle, G., Page, K., Llobet, A., Balasubramanian, M., ... Seshadri, R. (2013). Local Environments of Dilute Activator Ions in the Solid-State Lighting Phosphor $Y_{3-x}Ce_xAl_5O_{12.}$ *Chemistry of Materials, 25*(20), 3979–3995. doi:10.1021/cm401598n

Glasser, L. S. D., & Glasser, F. P. (1965). Silicates M_3SiO_5. I. Sr_3SiO_5. *Acta Crystallographica, 18*(3), 453–454. doi:10.1107/S0365110X65000993

Guo, C., Ding, X., Luan, L., & Xu, Y. (2010). Two-color emitting of Eu^{2+} and Mn^{2+} co-doped $Sr_2Mg_3P_4O_{15}$ for UV LEDs. *Sensors and Actuators. B, Chemical, 143*(2), 712–715. doi:10.1016/j.snb.2009.10.023

Guo, N., Huang, Y. J., Jia, Y. C., Lv, W. Z., Zhao, Q., Lu, W., ... You, H. P. (2013). A novel orange-yellow-emitting $Ba_3Lu(PO_4)_3$:Eu^{2+}, Mn^{2+} phosphor with energy transfer for UV-excited white LEDs. *Dalton Transactions (Cambridge, England)*, *42*(4), 941–947. doi:10.1039/C2DT31657A PMID:23108262

Guo, N., Huang, Y. J., Yang, M., Song, Y. H., Zheng, Y. H., & You, H. P. (2011). A tunable single-component warm white-light $Sr_3Y(PO_4)_3$: Eu^{2+}, Mn^{2+} phosphor for white-light emitting diodes. *Physical Chemistry Chemical Physics*, *13*(33), 15077–15082. doi:10.1039/c1cp20635d PMID:21773615

Guo, N., Jia, Y. C., Lü, W., Lv, W. Z., Zhao, Q., Jiao, M. M., ... You, H. P. (2013). Direct warm-white-emitting $Sr_3Sc(PO_4)_3$:Eu^{2+}, Mn^{2+} phosphor with tunable photoluminescence via efficient energy transfer. *Dalton Transactions (Cambridge, England)*, *28*(16), 5649–5654. doi:10.1039/c3dt32946a PMID:23429499

Guo, N., Lu, W., Jia, Y. C., Lv, W. Z., Zhao, Q., & You, H. P. (2013). Eu^{2+}& Mn^{2+}-Coactivated $Ba_3Gd(PO_4)_3$ Orange-Yellow-Emitting Phosphor with Tunable Color Tone for UV-Excited White LEDs. *ChemPhysChem*, *14*(1), 192–197. doi:10.1002/cphc.201200836 PMID:23161526

Guo, N., You, H., Song, Y., Yang, M., Liu, K., Zheng, Y., ... Zhang, H. (2010). White-light emission from a single-emitting-component $Ca_9Gd(PO_4)_7$:Eu^{2+},Mn^{2+} phosphor with tunable luminescent properties for near-UV light-emitting diodes. *Journal of Materials Chemistry*, *20*(41), 9061–9067. doi:10.1039/c0jm01860k

Guo, N., You, H. P., Jia, C. Z., Ouyang, R. Z., & Wu, D. H. (2014). A Eu $^{2+}$ and Mn $^{2+}$-co-activated fluoro-apatite-structure $Ca_6Y_2Na_2(PO_4)_6F_2$ as a standard white-emitting phosphor via energy transfer. *Dalton Transactions (Cambridge, England)*, *43*(32), 12373–12379. doi:10.1039/C4DT01021C PMID:24993422

Guo, N., Zhang, X. Y., Jia, Y. C., Qiao, H., & You, H. P. (2012). Warm-white-emitting from Eu^{2+}/Mn^{2+}-codoped $Sr_3Lu(PO_4)_3$ phosphor with tunable color tone and correlated color temperature. *The Journal of Physical Chemistry C*, *116*(1), 1329–1334. doi:10.1021/jp209891b

Guo, N., Zheng, Y. H., Jia, Y. C., Qiao, H., & You, H. P. (2012). A tunable warm-white-light $Sr_3Gd(PO_4)_3$:Eu^{2+}, Mn^{2+} phosphor system for LED-based solid-state lighting. *New Journal of Chemistry*, *36*(1), 168–172. doi:10.1039/C1NJ20532C

Haferkorn, B., & Meyer, G. (1998). Li_2EuSiO_4, einEuropium(II)-dilithosilicat:Eu[$(Li_2Si)O_4$]. *Zeitschrift fur Anorganische und Allgemeine Chemie*, *624*(7), 1079–1081. doi:

He, Y., Zhang, J., Zhou, W., Han, J., Qiu, Z., Yu, L., ... Lian, S. (2014). Multicolor Emission in a Single-Phase Phosphor $Ca_3Al_2O_6$:Ce^{3+}, Li^+: Luminescence and Site Occupancy. *Journal of the American Ceramic Society*, *97*(5), 1517–1522. doi:10.1111/jace.12801

Hermus, M., Phan, P., & Brgoch, J. (2016). Ab Initio Structure Determination and Photoluminescent Properties of an Efficient, Thermally Stable Blue Phosphor, $Ba_2Y_5B_5O_{17}$:Ce^{3+}. *Chemistry of Materials*, *28*(4), 1121–1127. doi:10.1021/acs.chemmater.5b04542

Hermus, M., Phan, P., Duke, A. C., & Brgoch, J. (2017). Tunable Optical Properties and Increased Thermal Quenching in the Blue-Emitting Phosphor Series: $Ba_2(Y_{1-x}Lu_x)_5B_5O_{17}$:$Ce^{3+}$ (x = 0–1). *Chemistry of Materials*, *29*(12), 5267–5275. doi:10.1021/acs.chemmater.7b01416

Hoppe, H. A. (2009). Recent Developments in the Field of Inorganic Phosphors. *Angewandte Chemie*, *48*(20), 3572–3582. doi:10.1002/anie.200804005 PMID:19350600

Hsu, C. H., & Lu, C. H. (2011). Microwave-hydrothermally synthesized $(Sr_{1-x-y}Ce_xTb_y)Si_2O_{2-\delta}N_{2+\mu}$ phosphors: Efficient energy transfer, structural refinement and photoluminescence properties. *Journal of Materials Chemistry*, *21*(9), 2932–2939. doi:10.1039/c0jm02501a

Huang, C. H., & Chen, T. M. (2010). $Ca_9La(PO_4)_7$:Eu^{2+},Mn^{2+}: an emission-tunable phosphor through efficient energy transfer for white light-emitting diodes. *Optics Express*, *18*(5), 5089–5099. doi:10.1364/OE.18.005089 PMID:20389521

Huang, C. H., & Chen, T. M. (2011). Novel yellow-emitting $Sr_8MgLn(PO_4)_7$:Eu^{2+} (Ln= Y, La) phosphors for applications in white LEDs with excellent color rendering index. *Inorganic Chemistry*, *50*(12), 5725–5730. doi:10.1021/ic200515w PMID:21591766

Huang, C. H., Chen, T. M., Liu, W. R., Chiu, Y. C., Yeh, Y. T., & Jang, S. M. (2010). A Single-Phased Emission-Tunable Phosphor $Ca_9Y(PO_4)_7$:Eu^{2+},Mn^{2+} with Efficient Energy Transfer for White-Light-Emitting Diodes. *ACS Applied Materials & Interfaces*, *2*(1), 259–264. doi:10.1021/am900668r

Huang, C. H., Chen, Y. C., Chen, T. M., Chan, T. S., & Sheu, H. S. (2011). Near UV-pumped yellow-emitting $Sr_8MgSc(PO_4)_7$:Eu^{2+} phosphor for white-light LEDs with excellent color rendering index. *Journal of Materials Chemistry*, *21*(15), 5645–5649. doi:10.1039/c0jm04524a

Huang, C. H., Kuo, T. W., & Chen, T. M. (2010). Novel Red-Emitting Phosphor $Ca_9Y(PO_4)_7$:Ce^{3+},Mn^{2+} with Energy Transfer for Fluorescent Lamp Application. *ACS Applied Materials & Interfaces*, *2*(5), 1395–1399. doi:10.1021/am100043q PMID:20429527

Huang, C. H., Liu, W. R., & Chen, T.-M. (2010). Single-phased white-light phosphors $Ca_9Gd(PO_4)_7$: Eu^{2+}, Mn^{2+} under near-ultraviolet excitation. *The Journal of Physical Chemistry C*, *114*(43), 18698–18701. doi:10.1021/jp106693z

Huang, C. H., Luo, L. Y., & Chen, T. M. (2011). An investigation on the luminescence and $Ce^{3+} \rightarrow Eu^{2+}$ energy transfer in $Ca_9Y(PO_4)_7$: Ce^{3+}, Eu^{2+} phosphor. *Journal of the Electrochemical Society*, *158*(11), J341–J344. doi:10.1149/2.029111jes

Huang, Y., Zhang, X., & Pan, J. (2017). Synthesis and photoluminescence of violet-blue phosphor $Ba_{10}(PO_4)_6F_2$: Eu^{2+}. *Solid State Sciences*, *63*, 9–15. doi:10.1016/j.solidstatesciences.2016.11.007

Huang, Y. L., Gan, J. H., & Seo, H. J. (2011). Luminescence Investigation of Eu-Activated $Sr_5(PO_4)_2SiO_4$ Phosphor by Combustion Synthesis. *Journal of the American Ceramic Society*, *94*(4), 1143–1148. doi:10.1111/j.1551-2916.2010.04190.x

Huang, Y. L., Gan, J. H., Zhu, R., Wang, X. G., & Seo, H. J. (2011). Structural Phase Formation and Tunable Luminescence of Eu^{2+}-Activated Apatite-Type $(Ca,Sr,Ba)_5(PO_4)_2(SiO_4)$. *Journal of the Electrochemical Society*, *158*(11), J334–J340. doi:10.1149/2.026111jes

Im, W. B., Brinkley, S., Hu, J., Mikhailovsky, A., DenBaars, S. P., & Seshadri, R. (2010). $Sr_{2.975-x}Ba_xCe_{0.025}AlO_4F$: A Highly Efficient Green-Emitting Oxyfluoride Phosphor for Solid State White Lighting. *Chemistry of Materials*, *22*(9), 2842–2849. doi:10.1021/cm100010z

Ji, H., Huang, Z., Xia, Z., Molokeev, M. S., Atuchin, V. V., Fang, M., & Huang, S. (2014). New Yellow-Emitting Whitlockite-type Structure $Sr_{1.75}Ca_{1.25}(PO_4)_2:Eu^{2+}$ Phosphor for Near-UV Pumped White Light-Emitting Devices. *Inorganic Chemistry*, *53*(10), 5129–5135. doi:10.1021/ic500230v PMID:24773050

Jia, Y., Pang, R., Li, H., Sun, W., Fu, J., Jiang, L., ... Liu, R. S. (2015). Single-phased white-light-emitting $Ca_4(PO_4)_2O:Ce^{3+},Eu^{2+}$ phosphors based on energy transfer. *Dalton Transactions (Cambridge, England)*, *44*(25), 11399–11407. doi:10.1039/C5DT01018G PMID:26017804

Jiao, H., & Wang, Y. (2009). $Ca_2Al_2SiO_7:Ce^{3+},Tb^{3+}$: A White-Light Phosphor Suitable for White-Light-Emitting Diodes. *Journal of the Electrochemical Society*, *5*(5), J117–J120. doi:10.1149/1.3097191

Jiao, M., Xu, Q., Yang, C., & You, H. (2017). Cation substitution induced novel gehlenite $Ca_2GaAlSiO_7:Eu^{2+}/Ce^{3+}$ phosphor with green/blue emission for UV-WLEDs. *RSC Advances*, *7*(46), 28647–28654. doi:10.1039/C7RA04105E

Jiao, M. M., Jia, Y. C., Lü, W., Lv, W. Z., Zhao, Q., Shao, B. Q., & You, H. P. (2014). $Sr_3GdNa(PO_4)_3F:Eu^{2+}$, Mn^{2+}: a potential color tunable phosphor for white LEDs. *Journal of Materials Chemistry. C, Materials for Optical and Electronic Devices*, *2*(1), 90–97. doi:10.1039/C3TC31837K

Ju, G., Hu, Y., Chen, L., Wang, X., Mu, Z., Wu, H., & Kang, F. (2012). Luminescence properties of $Y_2O_3:Bi^{3+}$, Ln^{3+} (Ln=Sm, Eu, Dy, Er, Ho) and the sensitization of Ln^{3+} by Bi^{3+}. *Journal of Luminescence*, *132*(8), 1853–1859. doi:10.1016/j.jlumin.2012.03.020

Ju, G. F., Hu, Y. H., Chen, L., & Wang, X. J. (2012). Persistent luminescence and its mechanism of $Ba_5(PO_4)_3Cl:Ce^{3+},Eu^{2+}$. *Journal of Applied Physics*, *111*(11), 113508. doi:10.1063/1.4723646

Kalaji, A., Mikami, M., & Cheetham, A. K. (2014). Ce^{3+}-Activated γ- Ca_2SiO_4 and Other Olivine-Type $ABXO_4$ Phosphors for Solid-State Lighting. *Chemistry of Materials*, *26*(13), 3966–3975. doi:10.1021/cm501516m

Kang, F., Peng, M., Yang, X., Dong, G., Nie, G., Liang, W., ... Qiu, J. (2014). Broadly tuning Bi^{3+} emission via crystal field modulation in solid solution compounds $(Y,Lu,Sc)VO_4:Bi$ for ultraviolet converted white LEDs. *Journal of Materials Chemistry. C, Materials for Optical and Electronic Devices*, *2*(30), 6068–6076. doi:10.1039/C4TC00238E

Kasturi, S., & Sivakumar, V. (2016). Eu^{2+} luminescence in $Ca_3Si_2O_7$ and spectral widening and tuning of Eu^{2+} emission color (orangish-red to green) by crystal chemical substitution. *RSC Advances*, *6*(101), 98652–98662. doi:10.1039/C6RA24258H

Kim, D., Kim, S. C., Bae, J. S., Kim, S., Kim, S. J., & Park, J. C. (2016). Eu^{2+}-Activated Alkaline-Earth Halophosphates, $M_5(PO_4)_3X:Eu^{2+}$ (M = Ca, Sr, Ba; X = F, Cl, Br) for NUV-LEDs: Site-Selective Crystal Field Effect. *Inorganic Chemistry*, *55*(17), 8359–8370. doi:10.1021/acs.inorgchem.6b00637 PMID:27494550

Kim, J. S., Jeon, P. E., Choi, J. C., & Park, H. L. (2005). Emission color variation of $M_2SiO_4:Eu^{2+}$ (M=Ba, Sr, Ca) phosphors for light-emitting diode. *Solid State Communications*, *133*(3), 187–190. doi:10.1016/j.ssc.2004.10.017

Kim, J. S., Jeon, P. E., Choi, J. C., Park, H. L., Mho, S. I., & Kim, G. C. (2004). Warm-white-light emitting diode utilizing a single-phase full-color $Ba_3MgSi_2O_8$:Eu^{2+},$Ba_3MgSi_2O_8$:Eu^{2+}, Mn^{2+} phosphor. *Applied Physics Letters*, *84*(15), 2931–2933. doi:10.1063/1.1695441

Kim, T. G., Kim, T., Kim, J., Kim, S. J., & Im, S. J. (2014). Interplay between Crystal Structure and Photoluminescence Properties of β-$Ca_3SiO_4Cl_2$:Eu^{2+}. *The Journal of Physical Chemistry C, 118*(23), 12428–12435. doi:10.1021/jp5002379

Kim, Y. H., Arunkumar, P., Kim, B. Y., Unithrattil, S., Kim, E., Moon, S. H., . . . Lee, J.-S. Im, … W. B. (2017). A zero-thermal-quenching phosphor. Nature Materials, 16, 543–550.

Kimata, M., Ii, N. I., & Ibaraki, N. (1982). The structural property of synthetic gehlenite, $Ca_2Al_2SiO_7$. *Neues Jahrbuch für Mineralogie. Abhandlungen*, *3*, 254–267.

Krames, M. R., Shchekin, O. B., Mueller-Mach, R., Mueller, G. O., Zhou, L., Harbers, G., & Craford, M. G. (2007). Status and Future of High-Power Light-Emitting Diodes for Solid-State Lighting. *Journal of Display Technology*, *3*(2), 160–175. doi:10.1109/JDT.2007.895339

Krüger, H. (2010). $Ca_{5.45}Li_{3.55}[SiO_4]_3O_{0.45}F_{1.55}$ and $Ca_7K[SiO_4]_3F_3$: single-crystal synthesis and structures of two trigonal oxyfluorides. *Zeitschrift für Kristallographie – Crystalline Materials, 225,* 418-424.

Kuwahara, H., Sato, Y., Kato, H., Kobayashi, M., Masaki, T., & Kakihana, M. (2016). Effect of Site Occupancies on Deep-red Emission from Eu^{2+}-activated Ca_2SiO_4 Phosphor. *Chemistry Letters*, *45*(3), 321–323. doi:10.1246/cl.151145

Lakshminarasimhan, N., & Varadaraju, U. V. (2005). White-Light Generation in Sr_2SiO_4:Eu^{2+},Ce^{3+} under Near-UV Excitation A Novel Phosphor for Solid-State Lighting. *Journal of the Electrochemical Society*, *152*(9), H152–H156. doi:10.1149/1.1997169

Lee, S., & Park, S. (2013). Preparation and luminescent properties of Tb^{3+} and Tb^{3+}–Ce^{3+} doped $Ba_9Y_2Si_6O_{24}$ phosphors. *Journal of Luminescence*, *143*, 215–218. doi:10.1016/j.jlumin.2013.05.008

Lee, S., & Sohn, K. S. (2010). Effect of inhomogeneous broadening on time-resolved photoluminescence in $CaAlSiN_3$:Eu^{2+}. *Optics Letters*, *35*(7), 1004–1006. doi:10.1364/OL.35.001004 PMID:20364198

Lemanski, K., Walerczyk, W., & Deren, P. J. (2016). Luminescent properties of europium ions in $CaAl_2SiO_6$. *Journal of Alloys and Compounds*, *672*, 595–599. doi:10.1016/j.jallcom.2016.02.208

Li, G., & Lin, J. (2014). Recent progress in low-voltage cathodoluminescent materials: Synthesis, improvement and emission properties. *Chemical Society Reviews*, *43*(20), 7099–7131. doi:10.1039/C4CS00109E PMID:24960634

Li, G., Tian, Y., Zhao, Y., & Lin, J. (2015). Recent progress in luminescence tuning of Ce^{3+} and Eu^{2+}-activated phosphors for pc-WLEDs. *Chemical Society Reviews*, *44*(23), 8688–8713. doi:10.1039/C4CS00446A PMID:26421319

Li, G., Wang, Y., Zeng, W., Chen, W., Han, S., Guo, H., & Li, Y. (2016). Photo-/cathodoluminescence and energy transfer properties of novel Ce^{3+} singly doped and Ce^{3+}/Tb^{3+} codoped $NaBaScSi_2O_7$ phosphors. *Journal of Materials Chemistry. C, Materials for Optical and Electronic Devices*, *4*(15), 3304–3312. doi:10.1039/C5TC04396D

Li, G., Wang, Y. H., Zeng, W., Chen, W. B., Han, S. C., Guo, H. J., & Wang, X. C. (2015). Luminescence properties of a new green afterglow phosphor $NaBaScSi_2O_7:Eu^{2+}$. *Dalton Transactions (Cambridge, England), 44*(40), 17572–17578. doi:10.1039/C5DT02748A PMID:26391314

Li, G., Zhao, Y., Wei, Y., Tian, Y., Quan, Z., & Lin, J. (2016). Novel yellowish-green light-emitting $Ca_{10}(PO_4)_6O:Ce^{3+}$ phosphor: structural refinement, preferential site occupancy and color tuning. *Chem-Comm, 52*, 3376–3379. PMID:26880115

Li, G. G., Geng, D. L., Shang, M. M., Peng, C., Cheng, Z. Y., & Lin, J. (2011). Tunable luminescence of Ce^{3+}/Mn^{2+}-coactivated $Ca_2Gd_8(SiO_4)_6O_2$ through energy transfer and modulation of excitation: Potential single-phase white/yellow-emitting phosphors. *Journal of Materials Chemistry, 21*(35), 13334–13344. doi:10.1039/c1jm11650a

Li, G. G., Geng, D. L., Shang, M. M., Zhang, Y., Peng, C., Cheng, Z. Y., & Lin, J. (2011). Color Tuning Luminescence of $Ce^{3+}/Mn^{2+}/Tb^{3+}$-triactivated $Mg_2Y_8(SiO_4)_6O_2$ via Energy Transfer: Potential Single-Phase White-Light-Emitting Phosphors. *The Journal of Physical Chemistry C, 115*(44), 21882–21897. doi:10.1021/jp204824d

Li, G. H., Lan, S., Li, L. L., Li, M. M., Bao, W. W., Zou, H. F., ... Gan, S. C. (2012). Tunable luminescence properties of $NaLa(MoO_4)_2:Ce^{3+},Tb^{3+}$ phosphors for near UV-excited white light-emitting-diodes. *Journal of Alloys and Compounds, 513*, 145–149. doi:10.1016/j.jallcom.2011.10.008

Li, J., Liu, J., & Yu, X. (2011). Synthesis and luminescence properties of Bi^{3+}-doped YVO_4 phosphors. *Journal of Alloys and Compounds, 509*(41), 9897–9900. doi:10.1016/j.jallcom.2011.07.079

Li, K., Geng, D., Shang, M., Zhang, Y., Lian, H., & Lin, J. (2014). Color-Tunable Luminescence and Energy Transfer Properties of $Ca_9Mg(PO_4)_6F_2:Eu^{2+}$, Mn^{2+} Phosphors for UV-LEDs. *The Journal of Physical Chemistry C, 118*(20), 11026–11034. doi:10.1021/jp501949m

Li, K., Shang, M., Lian, H., & Lin, J. (2016). Recent development in phosphors with different emitting colors via energy transfer. *Journal of Materials Chemistry. C, Materials for Optical and Electronic Devices, 4*(24), 5507–5530. doi:10.1039/C6TC00436A

Li, K., Xu, J., Cai, X. C., Fan, J., Zhang, Y., Shang, M. M., ... Lin, J. (2015). An efficient green-emitting α-$Ca_{1.65}Sr_{0.35}SiO_4:Eu^{2+}$ phosphor for UV/n-UV w-LEDs: synthesis, luminescence and thermal properties. *Journal of Materials Chemistry. C, Materials for Optical and Electronic Devices, 3*(24), 6341–6349. doi:10.1039/C5TC00796H

Liang, H. B., Tao, Y., Xu, J. H., He, H., Wu, H., Chen, W. X., ... Su, Q. (2004). Photoluminescence of Ce^{3+},Pr^{3+} and Tb^{3+} activated $Sr_3Ln(PO_4)_3$ under VUV-UV excitation. *Journal of Solid State Chemistry, 177*(3), 901–908. doi:10.1016/j.jssc.2003.09.037

Lin, C. C., Xiao, Z. R., Guo, G. Y., Chan, T. S., & Liu, R. S. (2010). Versatile phosphate phosphors $ABPO_4$ in white light-emitting diodes: Collocated characteristic analysis and theoretical calculations. *Journal of the American Chemical Society, 132*(9), 3020–3028. doi:10.1021/ja9092456 PMID:20155924

Liu, B., & Barbier, J. (1993). Structures of the Stuffed Tridymite Derivatives, $BaMSiO_4$ (M = Co, Zn, Mg). *Journal of Solid State Chemistry, 102*(1), 115–125. doi:10.1006/jssc.1993.1013

Liu, C. Y., Xia, Z. G., Lian, Z. P., Zhou, J., & Yan, Q. F. (2013). Structure and luminescence properties of green-emitting $NaBaScSi_2O_7:Eu^{2+}$ phosphors for near-UV-pumped light emitting diodes. *Journal of Materials Chemistry. C, Materials for Optical and Electronic Devices*, *1*(43), 7139–7147. doi:10.1039/C3TC31423E

Liu, H. K., Luo, Y., Mao, Z. Y., Liao, L. B., & Xia, Z. G. (2014). A novel single-composition trichromatic white-emitting $Sr_{3.5}Y_{6.5}O_2(PO_4)_{1.5}(SiO_4)_{4.5}$[thin space(1/6-em)]:[thinspace(1/6-em)]$Ce^{3+}/Tb^{3+}/Mn^{2+}$ phosphor: synthesis, luminescent properties and applications for white LEDs. *Journal of Materials Chemistry. C, Materials for Optical and Electronic Devices*, *2*, 1619–1627. doi:10.1039/c3tc32003k

Liu, J., Lian, H., Sun, J., & Shi, C. (2005). Characterization and Properties of Green Emitting $Ca_3SiO_4Cl_2:Eu^{2+}$ Powder Phosphor for White Light-emitting Diodes. *Chemistry Letters*, *34*(10), 1340–1341. doi:10.1246/cl.2005.1340

Liu, J., Wu, Z., & Gong, M. (2008). Thermally stable luminescence of blue $LiSrPO_4$: Eu^{2+} phosphor for near-UV light-emitting diodes. *Applied Physics. B, Lasers and Optics*, *93*(2-3), 583–587. doi:10.100700340-008-3169-1

Liu, Y., Tu, D., Zhu, H., & Chen, X. (2013). Lanthanide-doped luminescent nanoprobes: Controlled synthesis, optical spectroscopy, and bioapplications. *Chemical Society Reviews*, *42*(16), 6924–6958. doi:10.1039/c3cs60060b PMID:23775339

Liu, Y., Zhang, X., Hao, Z., Wang, X., & Zhang, J. (2011). Generation of broadband emission by incorporating N^{3-} into $Ca_3Sc_2Si_3O_{12}:Ce^{3+}$ garnet for high rendering white LEDs. *Journal of Materials Chemistry*, *21*(17), 6354–6358. doi:10.1039/c0jm04404k

Liu, Y., Zhang, X., Hao, Z., Wang, X., & Zhang, J. (2011). Tunable full-color-emitting $Ca_3Sc_2Si_3O_{12}:Ce^{3+}$, Mn^{2+} phosphor via charge compensation and energy transfer. *ChemComm*, *47*, 10677–10679. PMID:21892443

Liu, Y. F., Zhang, X., Hao, Z. D., Luo, Y. S., Wang, X. J., & Zhang, J. H. (2011). Generating yellow and red emissions by co-doping Mn^{2+} to substitute for Ca^{2+} and Sc^{3+} sites in $Ca_3Sc_2Si_3O_{12}:Ce^{3+}$ green emitting phosphor for white LED applications. *Journal of Materials Chemistry*, *21*(41), 16379–16384. doi:10.1039/c1jm11601k

Liu, Y. F., Zhang, X. D., Hao, Z., Wang, X. J., & Zhang, J. H. (2011). Tunable full-color-emitting $Ca_3Sc_2Si_3O_{12}:Ce^{3+}$, Mn^{2+} phosphor via charge compensation and energy transfer. *ChemComm*, *47*, 10677–10679. PMID:21892443

Lu, W., Guo, N., Jia, Y. C., Zhao, Q., Lv, W. Z., Jiao, M. M., ... You, H. P. (2013). Tunable color of $Ce^{3+}/Tb^{3+}/Mn^{2+}$-coactivated $CaScAlSiO_6$ via energy transfer: A single-component red/white-emitting phosphor. *Inorganic Chemistry*, *52*(6), 3007–3012. doi:10.1021/ic3024666 PMID:23445353

Lü, W., Jia, Y. C., Lv, W. Z., Zhao, Q., & You, H. P. (2013). An orange-emitting phosphor via the efficient Ce^{3+}–Mn^{2+} and Eu^{2+}–Mn^{2+} energy transfers in $La_{9.33}(SiO_4)_6O_2$ for UV or near-UV LEDs. *New Journal of Chemistry*, *37*(11), 3701–3705. doi:10.1039/c3nj00758h

Lv, W., Jia, Y., Zhao, Q., Jiao, M., Shao, B., Lü, W., & You, H. (2014). Crystal Structure and Luminescence Properties of $Ca_8Mg_3Al_2Si_7O_{28}$:Eu^{2+} for WLEDs. *Advanced Optical Materials*, *2*(2), 183–188. doi:10.1002/adom.201300448

Lv, W., Jia, Y., Zhao, Q., Lü, W., Jiao, M., Shao, B., & You, H. P. (2014). Synthesis, Structure, and Luminescence Properties of $K_2Ba_7Si_{16}O_{40}$:Eu^{2+} for White Light Emitting Diodes. *The Journal of Physical Chemistry C*, *118*(9), 4649–4655. doi:10.1021/jp500662a

Lv, W., Jia, Y. C., Lv, W. Z., Zhao, Q., & You, H. P. (2013). An orange-emitting phosphor via the efficient Ce^{3+}–Mn^{2+} and Eu^{2+}–Mn^{2+} energy transfers in $La_{9.33}(SiO_4)_6O_2$ for UV or near-UV LEDs. *New Journal of Chemistry*, *37*(11), 3701–37050. doi:10.1039/c3nj00758h

Lv, Y., Jin, Y., Zhang, S., Wang, C., Ju, G., Xue, F., ... Hu, Y. (2017). $Sr_3GdLi(PO_4)_3F$:Eu^{2+}, Mn^{2+}: A tunable blue-white color emitting phosphor via energy transfer for near-UV white LEDs. *Ceramics International*, *43*(12), 8824–8830. doi:10.1016/j.ceramint.2017.04.015

Ma, P., Yuan, B., Sheng, Y., Zheng, K., Wang, Y., Xu, C., ... Song, Y. (2017). Tunable emission, thermal stability and energy-transfer properties of $SrAl_2Si_2O_8$: Ce^{3+}/Tb^{3+} phosphors for w-LEDs. *Journal of Alloys and Compounds*, *714*, 627–635. doi:10.1016/j.jallcom.2017.04.296

Maggay, I. V. B., Lin, P. C., & Liu, W. R. (2014). Enhanced luminescence intensity of novel red-emitting phosphor -$Sr_3Lu_2(BO_3)_4$:Bi^{3+},Eu^{3+} via energy transfer. *Journal of Solid State Lighting*, *1*(1), 13. doi:10.118640539-014-0013-6

Maria, W. W., Uwe, K., & Ekkehart, T. (2010). The Crystal Structures of three new complex silicates of Scandium. *Canadian Mineralogist*, *48*(1), 51–68. doi:10.3749/canmin.48.1.51

Mayer, I., & Cohen, S. (1983). The crystal structure of $Ca_6Eu_2Na_2(PO_4)_6F_2$. *Journal of Solid State Chemistry*, *48*(1), 17–20. doi:10.1016/0022-4596(83)90054-3

Mayer, I., Roth, R. S., & Brown, W. E. (1974). Rare earth substituted fluoride-phosphate apatites. *Journal of Solid State Chemistry*, *11*(1), 33–37. doi:10.1016/0022-4596(74)90143-1

Meng, X., Qiu, K., Tian, Z., Shi, X., You, J., Wang, Z., ... Yang, Z. (2017). Tunable-emission single-phase phosphors $Ba_3Ca_2(PO_4)_3F$:M (M=Ce^{3+},Eu^{2+},Mn^{2+}): Crystal structure, luminescence and energy transfer. *Journal of Alloys and Compounds*, *719*, 322–330. doi:10.1016/j.jallcom.2017.05.197

Miao, S., Xia, Z., Zhang, J., & Liu, Q. (2014). Increased Eu^{2+} Content and Codoping Mn^{2+} Induced Tunable Full-Color Emitting Phosphor $Ba_{1.55}Ca_{0.45}SiO_4$:Eu^{2+}, Mn^{2+}. *Inorganic Chemistry*, *53*(19), 10386–10393. doi:10.1021/ic501450f PMID:25254295

Minsung, K., Kobayashi, M., Kato, H., Yamane, H., Sato, Y., & Kakihana, M. (2015). Crystal structures and luminescence properties of Eu2+-activated new $NaBa_{0.5}Ca_{0.5}PO_4$ and $Na_3Ba_2Ca(PO_4)_3$. *Dalton Transactions (Cambridge, England)*, *44*(4), 1900–1904. doi:10.1039/C4DT03024A

Morozov, V. A., Belik, A. A., Stefanovich, S. Y., Grebenev, V. V., Lebedev, O. I., Van Tendeloo, G., & Lazoryak, B. I. (2002). High-Temperature Phase Transition in the Whitlockite-Type Phosphate $Ca_9In(PO_4)_7$. *Journal of Solid State Chemistry*, *165*(2), 278–288. doi:10.1006/jssc.2001.9521

Nagpure, I. M., Shinde, K. N., Dhoble, S. J., & Kumar, A. (2009). Photoluminescence characterization of Dy^{3+} and Eu^{2+} ion in $M_5(PO_4)_3F$ (M = Ba, Sr, Ca) phosphors. *Journal of Alloys and Compounds*, *481*(1-2), 632–638. doi:10.1016/j.jallcom.2009.03.069

Park, J. K., Lim, M. A., Kim, C. H., Park, H. D., Park, J. T., & Choi, S. Y. (2003). White light-emitting diodes of GaN-based Sr_2SiO_4:$EuSr_2SiO_4$:Eu and the luminescent properties. *Applied Physics Letters*, *82*(5), 683–685. doi:10.1063/1.1544055

Park, S. (2014). Ce^{3+}–Mn^{2+} cooperative $Ba_9Y_2Si_6O_{24}$ orthosilicate phosphors. *Materials Letters*, *135*, 59–62. doi:10.1016/j.matlet.2014.07.134

Park, S., & Vogt, T. (2009). Luminescent phosphors, based on rare earth substituted oxyfluorides in the $A(1)_{3-x}A(2)_xMO_4F$ family with A(1)/A(2) = Sr, Ca, Ba and M=Al, Ga. *Journal of Luminescence*, *129*(9), 952–957. doi:10.1016/j.jlumin.2009.04.005

Phillips, J. M., Coltrin, M. E., Crawford, M. H., Fischer, A. J., Krames, M. R., Mueller-Mach, R., ... Tsao, J. Y. (2007). Research challenges to ultra-efficient inorganic solid-state lighting. *Laser & Photonics Reviews*, *1*(4), 307–333. doi:10.1002/lpor.200710019

Poort, S. H. M., Van Krevel, J. W. H., Stomphorst, R., Vink, A. P., & Blasse, G. (1996). Luminescence of Eu^{2+}in Host Lattices with Three Alkaline Earth Ions in a Row. *Journal of Solid State Chemistry*, *122*(2), 432–435. doi:10.1006/jssc.1996.0137

Prodjosantoso, A. K., Kennedy, B. J., Vogt, T., & Woodward, P. M. (2003). Cation and anion ordering in the layered oxyfluorides $Sr_{3-x}A_xAlO_4F$ (A=Ba, Ca). *Journal of Solid State Chemistry*, *172*(1), 89–94. doi:10.1016/S0022-4596(02)00128-7

Puchalska, M., & Zych, E. (2012). Effect of Na^+ co-dopant and activator concentration on luminescent properties of $CaGa_4O_7$:Eu^{3+}. *Journal of Luminescence*, *132*(11), 2879–2884. doi:10.1016/j.jlumin.2012.06.007

Qi, S. Y., Huang, Y. L., Li, Y. D., Cai, P. Q., Kim, S. I. I., & Seo, H. J. (2014). Probe spectrum measurements of Eu^{3+} ions as a relevant tool for monitoring *in-vitro* hydroxyapatite formation in a new borate biomaterial. *Journal of Materials Chemistry. B, Materials for Biology and Medicine*, *2*(37), 6387–6396. doi:10.1039/C4TB00871E

Qi, S. Y., Huang, Y. L., Tsuboi, T., Huang, W., & Seo, H. J. (2014). Versatile luminescence of $Eu^{2+,3+}$-activated fluorosilicate apatites $M_2Y_3[SiO_4]_3F$ (M= Sr, Ba) suitable for white light emitting diodes. *Optical Materials Express*, *4*(2), 396–402. doi:10.1364/OME.4.000396

Que, M. D., Ci, Z. P., Wang, Y. H., Zhu, G., Xin, S. Y., Shi, Y. R., & Wang, Q. (2013). Crystal structure and luminescence properties of a cyan emitting $Ca_{10}(SiO_4)_3(SO_4)_3F_2$:$Eu^{2+}$ phosphor. *CrystEngComm*, *15*(32), 6389–6394. doi:10.1039/c3ce40482j

Rack, P. D., & Holloway, P. H. (1998). The structure, device physics, and material properties of thin film electroluminescent displays. *Materials Science and Engineering R Reports*, *21*(4), 171–219. doi:10.1016/S0927-796X(97)00010-7

Redhammer, G. R., & Roth, G. (2003). Lithium and sodium yttrium orthosilicate oxyapatite, $LiY_9(SiO_4)_6O_2$ and $NaY_9(SiO_4)_6O_2$, at both 100 K and near room temperature. *Acta Crystallographica. Section C, Crystal Structure Communications*, *59*(12), i120–i124. doi:10.1107/S0108270103023321 PMID:14671334

Reid, A. F., & Ringwood, A. E. (1969). Six-coordinate silicon: High pressure strontium and barium aluminosilicates with the hollandite structure. *Journal of Solid State Chemistry*, *1*(1), 6–9. doi:10.1016/0022-4596(69)90002-4

Robertson, J. M., Van Tol, M. W., Smits, W. H., & Heynen, J. P. H. (1981). Color shift of the Ce^{3+} emission in monocrystalline epitaxially grown garnet layers. *Philips Journal of Research*, *36*, 15–30.

Ryan, F. M., Warren, R. W., Hopkins, R. H., & Murphy, J. (1978). Selective Site Laser Excitation and ESR Studies of Nd^{3+} Ions in $Ca_5(PO_4)_3F$. *Journal of the Electrochemical Society*, *125*(9), 1493–1498. doi:10.1149/1.2131702

Saradhi, M. P., & Varadaraju, U. V. (2006). Photoluminescence studies on Eu^{2+}-activated Li_2SrSiO_4 a potential orange-yellow phosphor for solid-state lighting. *Chemistry of Materials*, *18*(22), 5267–5272. doi:10.1021/cm061362u

Sato, M., Tanaka, T., & Ohta, M. (1994). Photostimulated Luminescence and Structural Characterization of $Ba_5(PO_4)_3Cl:Eu^{2+}$ Phosphors. *Journal of the Electrochemical Society*, *141*(7), 1851–1855. doi:10.1149/1.2055016

Sato, Y., Kuwahara, H., Kato, H., Kobayashi, M., Masaki, T., & Kakihana, M. (2015). Large redshifts in emission and excitation from Eu^{2+}-activated Sr_2SiO_4 and Ba_2SiO_4 phosphors induced by controlling Eu^{2+} occupancy on the basis on crystal-site engineering. *Optics and Photonics Journal*, *5*(11), 326–333. doi:10.4236/opj.2015.511031

Schubert, E. F., & Kim, J. K. (2005). Solid-state light sources getting smart. *Science*, *308*(5726), 1274–1278. doi:10.1126cience.1108712 PMID:15919985

Setlur, A. A., Heward, W. J., Gao, Y., Srivastava, A. M., Gopi Chandran, R., & Shankar, M. (2006). Crystal Chemistry and Luminescence of Ce^{3+}-Doped $Lu_2CaMg_2(Si,Ge)_3O_{12}$ and Its Use in LED Based Lighting. *Chemistry of Materials*, *18*(14), 3314–3322. doi:10.1021/cm060898c

Shi, L. A., Huang, Y. L., & Seo, H. J. (2010). Emission Red Shift and Unusual Band Narrowing of Mn^{2+} in $NaCaPO_4$ Phosphor. *The Journal of Physical Chemistry A*, *114*(26), 6927–6934. doi:10.1021/jp101772z PMID:20540584

Sivakumar, V., & Kasturi, S. (2017). Eu3+ based orange - red emitting inorganic colour convertors: An overview. In Synthesis of phosphors and their applications. Singapore: Pan Stanford Publishing Pte Ltd.

Sivakumar, V., & Varadaraju, U. V. (2007). Ce^{3+} \ Eu^{2+} Energy Transfer Studies on $BaMgSiO_4$ A Green Phosphor for Three Band White LEDs. *Journal of the Electrochemical Society*, *154*(5), J167–J171. doi:10.1149/1.2714303

Sivakumar, V., & Varadaraju, U. V. (2009). Eu^{2+}, Ce^{3+} Luminescence and $Ce^{3+} \rightarrow Eu^{2+}$ Energy-Transfer Studies on Sr_2LiSiO_4F: A White Light-Emitting Phosphor. *Journal of the Electrochemical Society*, *156*(7), J179–J184. doi:10.1149/1.3122663

Stenberg, L., Sellar, J. R., & Hyde, B. G. (1986). Incommensurately modulated α'-Sr_2SiO_4. *Nature, 320*(6061), 428–429. doi:10.1038/320428a0

Sun, W., Jia, Y., Pang, R., Li, H., Ma, T., Li, D., ... Li, C. (2015). $Sr_9Mg_{1.5}(PO_4)_7$:Eu^{2+}: A Novel Broadband Orange-Yellow-Emitting Phosphor for Blue Light-Excited Warm White LEDs. *ACS Applied Materials & Interfaces, 7*(45), 25219–25226. doi:10.1021/acsami.5b06961 PMID:26503633

Tam, T. T. H., Du, N. V., Kien, N. D. T., Thang, C. X., Cuong, N. D., Huy, P. T., ... Nguyen, D. H. (2014). Co-precipitation synthesis and optical properties of green-emitting $Ba_2MgSi_2O_7$: Eu^{2+} phosphor. *Journal of Luminescence, 147*, 358–362. doi:10.1016/j.jlumin.2013.11.066

Tang, Y. S., Hu, S. F., Lin, C. C., Bagkar, N. C., & Liu, R. S. (2007). Thermally stable luminescence of $KSrPO_4$:Eu^{2+} phosphor for white light UV light-emitting diodes. *Applied Physics Letters, 90*(15), 151108. doi:10.1063/1.2721846

Tang, Z. B., Wang, D. Y., Khan, W. U., Du, S. S., Wang, X. C., & Wang, Y. H. (2016). Novel zirconium silicate phosphor $K_2ZrSi_2O_7$:Eu^{2+} for white light-emitting diodes and field emission displays. *Journal of Materials Chemistry. C, Materials for Optical and Electronic Devices, 4*(23), 5307–5313. doi:10.1039/C6TC01449F

Tao, Z. X., Huang, Y. L., & Seo, H. J. (2013). Blue luminescence and structural properties of Ce^{3+}-activated phosphosilicate apatite $Sr_5(PO_4)_2(SiO_4)$. *Dalton Transactions (Cambridge, England), 42*(6), 2121–2129. doi:10.1039/C2DT32117C PMID:23187727

Vogt, T., Woodward, P. M., Hunter, B. A., Prodjosantoso, A. K., & Kennedy, B. J. (1999). Sr_3MO_4F (M=Al, Ga) - A New Family of Ordered Oxyfluorides. *Journal of Solid State Chemistry, 144*(1), 228–231. doi:10.1006/jssc.1999.8172

Wang, J., Lin, H., Huang, Q., Xiao, G., Xu, J., Wang, B., ... Wang, Y. (2017). Structure and luminescence behavior of a single-ion activated single-phased $Ba_2Y_3(SiO_4)_3F$:Eu white-light phosphor. *Journal of Materials Chemistry. C, Materials for Optical and Electronic Devices, 5*(7), 1789–1797. doi:10.1039/C6TC04350J

Wang, Q., Zhu, G., Xin, S., Ding, X., Xu, J., Wang, Y., & Wang, Y. (2015). A blue-emitting Sc silicate phosphor for ultraviolet excited light-emitting diodes. *Physical Chemistry Chemical Physics, 17*(41), 27292–27299. doi:10.1039/C5CP02899J PMID:26242881

Wang, S. S., Chen, W. T., Li, Y., Wang, J., Sheu, H. S., & Liu, R. S. (2013). Neighboring-Cation Substitution Tuning of Photoluminescence by Remote-Controlled Activator in Phosphor Lattice. *Journal of the American Chemical Society, 135*(34), 12504–12507. doi:10.1021/ja404510v PMID:23941075

Wang, X., Zhao, Z., Wu, Q., Wang, C., Wang, Q., Yanyan, L., & Wang, Y. (2016). Structure, photoluminescence and abnormal thermal quenching behavior of Eu^{2+}-doped $Na_3Sc_2(PO_4)_3$: A novel blue-emitting phosphor for n-UV LEDs. *Journal of Materials Chemistry. C, Materials for Optical and Electronic Devices, 4*(37), 8795–8801. doi:10.1039/C6TC01049K

Wang, Z. Y., Xia, Z. G., Molokeev, M. S., Atuchin, V. V., & Liu, Q. L. (2014). Blue-shift of Eu^{2+} emission in $(Ba,Sr)_3Lu(PO_4)_3$:Eu^{2+} eulytite solid-solution phosphors resulting from release of neighbouring-cation-induced stress. *Dalton Transactions (Cambridge, England)*, *43*(44), 16800–16804. doi:10.1039/C4DT02319F PMID:25287659

Wen, D., Dong, Z., Shi, J., Gong, M., & Wu, M. (2013). Standard White-Emitting $Ca_8MgY(PO_4)_7$:Eu^{2+},Mn^{2+} Phosphor for White-Light-Emitting LEDs. *ECS Journal of Solid State Science and Technology*, *2*(9), R178–R185. doi:10.1149/2.019309jss

Wen, D., Kuwahara, H., Kato, H., Kobayashi, M., Sato, Y., Masaki, T., & Kakihana, M. (2016). Anomalous Orange Light-Emitting $(Sr,Ba)_2SiO_4$: Eu^{2+} Phosphors for Warm White LEDs. *ACS Applied Materials & Interfaces*, *8*(18), 11615–11620. doi:10.1021/acsami.6b02237 PMID:27090069

Wickleder, M. S. (2002). Inorganic Lanthanide Compounds with Complex Anions. *Chemical Reviews*, *102*(6), 2011–2088. doi:10.1021/cr010308o PMID:12059261

Wierzbicka-Wieczorek, M., Kolitsch, U., & Tillmanns, E. (2010). Synthesis and structural study of five new trisilicates, $BaREE_2Si_3O_{10}$ (REE = Gd, Er, Yb, Sc) and $SrY_2Si_3O_{10}$, including a review on the geometry of the Si_3O_{10} unit. *European Journal of Mineralogy*, *22*, 245–258. doi:10.1127/0935-1221/2010/0022-1969

Wu, W., & Xia, Z. G. (2013). Synthesis and color-tunable luminescence properties of Eu^{2+} and Mn^{2+}-activated $Ca_3Mg_3(PO_4)_4$ phosphor for solid state lighting. *RSC Advances*, *3*(17), 6051–6057. doi:10.1039/c3ra40313k

Xia, Y., Chen, J., Liu, Y., Molokeev, M. S., Guan, M., Huang, Z., & Fang, M. (2016). Crystal structure evolution and luminescence properties of color tunable solid solution phosphors $Ca^{2+}xLa_8-x(SiO_4)_6-x(PO_4)xO_2$:$Eu^{2+}$. *Dalton Transactions (Cambridge, England)*, *45*(3), 1007–1015. doi:10.1039/C5DT03786G PMID:26648132

Xia, Z., Molokeev, M. S., Im, W. B., Unithrattil, S., & Liu, Q. (2015). Crystal Structure and Photoluminescence Evolution of $La_5(Si_2+xB1-x)(O_{13}-xNx)$:$Ce^{3+}$ Solid Solution Phosphors. *The Journal of Physical Chemistry C*, *119*(17), 9488–9495. doi:10.1021/acs.jpcc.5b01211

Xia, Z., Zhang, Y., Molokeev, M. S., & Atuchin, V. V. (2013). Structural and Luminescence Properties of Yellow-Emitting $NaScSi_2O_6$:Eu^{2+} Phosphors: Eu^{2+} Site Preference Analysis and Generation of Red Emission by Codoping Mn^{2+} for White-Light-Emitting Diode Applications. *The Journal of Physical Chemistry C*, *117*(40), 20847–20854. doi:10.1021/jp4062225

Xia, Z. G., Li, Q., & Sun, J. Y. (2007). Luminescence properties of $Ba_5SiO_4(F,Cl)_6$:Eu^{2+} phosphor. *Materials Letters*, *61*(8-9), 1885–1888. doi:10.1016/j.matlet.2006.07.149

Xia, Z. G., & Liu, Q. L. (2016). Progress in discovery and structural design of color conversion phosphors for LEDs. *Progress in Materials Science*, *84*, 59–117. doi:10.1016/j.pmatsci.2016.09.007

Xia, Z. G., & Liu, R. S. (2012). Tunable Blue-Green Color Emission and Energy Transfer of $Ca_2Al_3O_6F$:Ce^{3+}, Tb^{3+} Phosphors for Near-UV White LEDs. *The Journal of Physical Chemistry C*, *116*(29), 15604–15609. doi:10.1021/jp304722z

Xia, Z. G., Molokeev, M. S., Im, W. B., Unithrattil, S., & Liu, Q. L. (2015). Crystal Structure and Photoluminescence Evolution of $La_5(Si_{2+x}B_{1-x})(O_{13-x}N_x):Ce^{3+}$ Solid Solution Phosphors. *The Journal of Physical Chemistry C, 119*(17), 9488–9495. doi:10.1021/acs.jpcc.5b01211

Xia, Z. G., Sun, J. Y., Du, H. Y., & Zhou, W. (2006). Luminescence of Eu^{2+} in alkali earth chlorosilicate phosphor and their color-tunable properties. *Optical Materials, 28*(5), 524–529. doi:10.1016/j.optmat.2005.01.029

Xiao, W. G., Zhang, X., Hao, Z. D., Pan, G. H., Luo, Y. S., Zhang, L. G., & Zhang, J. H. (2015). Blue-emitting $K_2Al_2B_2O_7:Eu^{2+}$ phosphor with high thermal stability and high color purity for near-UV-pumped white light-emitting diodes. *Inorganic Chemistry, 54*(7), 3189–3195. doi:10.1021/ic502773t PMID:25751730

Xiao, Y., Hao, Z., Zhang, L., Xiao, W., Wu, D., Zhang, X., ... Zhang, J. (2017). Highly efficient green-emitting phosphors $Ba_2Y_5B_5O_{17}$ with low thermal quenching due to fast energy transfer from Ce^{3+} to Tb^{3+}. *Inorganic Chemistry, 56*(8), 4539–4545. doi:10.1021/acs.inorgchem.7b00085 PMID:28358516

Xiao, Y., Hao, Z., Zhang, L., Xiao, W., Wu, D., Zhang, X., ... Zhang, J. (2017). Highly Efficient Green-Emitting Phosphors $Ba_2Y_5B_5O_{17}$ with Low Thermal Quenching Due to Fast Energy Transfer from Ce^{3+} to Tb^{3+}. *Inorganic Chemistry, 56*(8), 4538–4544. doi:10.1021/acs.inorgchem.7b00085 PMID:28358516

Xie, M., Li, D., Pan, R., Zhou, X., & Zhu, G. (2015). Crystallographic sites of Eu2+ and Eu^{2+}–Tb^{3+} energy transfer in Sr_2LiSiO_4F. *RSC Advances, 5*(29), 22856–22862. doi:10.1039/C5RA00585J

Xie, R. J., Hirosaki, N., Mitomo, M., Yamamoto, Y., Suehiro, T., & Sakuma, K. (2004). Optical Properties of Eu^{2+} in α-SiAlON. *The Journal of Physical Chemistry B, 108*(32), 12027–12031. doi:10.1021/jp048295g

Xin, S. Y., Wang, Y. H., Zhu, G., Zhang, F., Gong, Y., Wen, Y., & Liu, B. T. (2013). Tunable white light emitting from mono Ce^{3+} doped $Sr_5(PO_4)_2SiO_4$ phosphors for light emitting diodes. *Materials Research Bulletin, 48*(4), 1627–1631. doi:10.1016/j.materresbull.2013.01.008

Xu, H., Liu, Q., Huang, L., He, Z., Wang, S., Jiang, C., & Peng, Q. (2017). Luminescence and energy transfer of color-tunable $Na_6Ca_3Si_6O_{18}:Ce^{3+},Tb^{3+}$ phosphors for WLED. *Optics & Laser Technology, 89*, 151–155. doi:10.1016/j.optlastec.2016.09.035

Yang, P., Yu, X., Yu, H., Jiang, T., Zhou, D., & Qiu, J. (2012). Effects of crystal field on photoluminescence properties of $Ca_2Al_2SiO_7:Eu^{2+}$ phosphors. *Journal of Rare Earths, 12*(12), 1208–1212. doi:10.1016/S1002-0721(12)60207-5

Yang, Z., Lin, P. C., Guo, C. F., & Liu, W. R. (2015). Color-tunable luminescence and energy transfer properties of Eu^{2+}- and Mn^{2+}-activated $BaCa_2MgSi_2O_8$ phosphor for ultraviolet light-emitting diodes. *RSC Advances, 5*(17), 13184–13191. doi:10.1039/C4RA14848G

Yeh, K. Y., Lin, C. H., Maggay, I. V. B., & Liu, W. R. (2016). Novel green-emitting $K_2Ba_5Si_{12}O_{30}:Eu^{2+}$ phosphors with excellent thermal quenching for white light-emitting diodes. *Optical Materials, 59*, 8–14. doi:10.1016/j.optmat.2016.04.026

Yi, H., Wu, L., Wu, L., Zhao, L., Xia, Z., Zhang, Y., ... Xu, J. (2016). Crystal. Structure of High-Temperature Phase β-$NaSrBO_3$ and Photoluminescence of β-$NaSrBO_3$:Ce^{3+} I*norgan.ic. Chemistry*, *55*, 6487–6495. PMID:27299933

Yu, R., Zhong, S., Xue, N., Lia, H., & Ma, H. (2014). Synthesis, structure, and peculiar green emission of $NaBaBO_3$:Ce^{3+} phosphors. *Dalton Transactions (Cambridge, England)*, *43*(28), 10969–10976. doi:10.1039/C4DT00859F PMID:24905051

Yu, R. J., Li, H. J., Ma, H. L., Wang, C. F., & Wang, H. (2014). A New Blue-Emitting Phosphor of Ce^{3+}-Activated Fluorosilicate Apatite $Ba_2Y_3[SiO_4]_3F$. *Journal of the American Ceramic Society*, *97*(4), 1151–1156. doi:10.1111/jace.12765

Yuan, B., Huang, Y., Yu, Y. M., & Seo, H. J. (2012). Luminescence and structure of Eu^{2+}-doped $Ba_2CaMg_2Si_6O_{17.}$ *Ceramics International*, *38*(3), 2219–2223. doi:10.1016/j.ceramint.2011.10.070

Zhai, D. Y., Ning, L. X., Huang, Y. C., & Liu, G. K. (2014). Ce–O Covalence in Silicate Oxyapatites and Its Influence on Luminescence Dynamics. *The Journal of Physical Chemistry C*, *118*(29), 16051–16059. doi:10.1021/jp5049293

Zhang, J., Wang, Y., Zhang, F., & Huang, Y. (2011). Single-Phase White-Emitting $Ca_8MgGd(PO_4)_7$:Ln^{3+}, Mn^{2+} ($Ln^{3+} = Ce^{3+}$, Tb^{3+}, and Dy^{3+}) for Mercury-Free Lamps. *Journal of the Electrochemical Society*, *158*(4), J110–J114. doi:10.1149/1.3551508

Zhang, S., Hu, Y., Chen, L., Fu, Y., & Ju, G. (2016). $La_3GaGe_5O_{16}$:Cr^{3+} phosphor: the near-infrared persistent luminescence. *Optical Materials Express*, *4*(4), 1247–1255. doi:10.1364/OME.6.001247

Zhang, S. Y., Huang, Y. L., Nakai, Y., Tsuboi, T., & Seo, H. J. (2011). The Luminescence Characterization and Thermal Stability of Eu^{2+} Ions-Doped $NaBaPO_4$ Phosphor. *Journal of the American Ceramic Society*, *94*(9), 2987–2992. doi:10.1111/j.1551-2916.2011.04481.x

Zhang, Y., Li, G. G., Geng, D. L., Shang, M. M., Peng, C., & Lin, J. (2012). Color-tunable emission and energy transfer in $Ca_3Gd_7(PO_4)(SiO_4)_5O_2$:$Ce^{3+}$/$Tb^{3+}$/$Mn^{2+}$ phosphors. *Inorganic Chemistry*, *51*(21), 11655–11664. doi:10.1021/ic3015578 PMID:23075230

Zheng, Z., & Tang, W. (2016). Efficient sensitization of Eu^{2+}/Mn^{2+} emissions by Ce^{3+} doping in NaMgPO4 host under UV excitation. *Applied Physics. A, Materials Science & Processing*, *122*(3), 229. doi:10.100700339-016-9784-5

Zhong, J., Chen, D., Yuan, Y., Chen, L., Yu, H., & Ji, Z. (2017). Synthesis and spectroscopic investigation of $Ba_3La_6(SiO_4)_6$:Eu^{2+} green phosphors for white light-emitting diodes. *Chemical Engineering Journal*, *309*, 795–801. doi:10.1016/j.cej.2016.10.109

Zhong, J., Zhuang, W., Xing, X., Liu, R., Li, Y., Liu, Y., & Hu, Y. (2015). Synthesis, Crystal Structures, and Photoluminescence Properties of Ce^{3+}-Doped $Ca_2LaZr_2Ga_3O_{12}$: New Garnet Green-Emitting Phosphors for White LEDs. *The Journal of Physical Chemistry C*, *119*(10), 5562–5569. doi:10.1021/jp508409r

Zhou, J., & Xia, Z. (2014). Synthesis, luminescence properties and energy transfer behavior of $Na_2CaMg(PO_4)_2$:Eu^{2+}, Mn^{2+} phosphors. *Journal of Luminescence*, *146*, 22–26. doi:10.1016/j.jlumin.2013.09.031

Zhou, J., Xia, Z., Yang, M., & Shen, K. (2012). High efficiency blue-emitting phosphor: Ce^{3+}-doped $Ca_{5.45}Li_{3.55}(SiO_4)_3O_{0.45}F_{1.55}$ for near UV-pumped light-emitting diodes. *Journal of Materials Chemistry*, *22*(41), 21935–21941. doi:10.1039/c2jm34146h

Zhou, W., Pan, F., Zhou, L., Hou, D., Huang, Y., Tao, Y., & Liang, H. (2016). Site Occupancies, Luminescence, and Thermometric Properties of $LiY_9(SiO_4)_6O_2$:Ce^{3+} Phosphors. *Inorganic Chemistry*, *55*(20), 10415–10424. doi:10.1021/acs.inorgchem.6b01656 PMID:27700068

Zhu, G., Shi, Y. R., Mikami, M., Shimomura, Y., & Wang, Y. H. (2014). Electronic structure and photo/cathodoluminescence properties investigation of green emission phosphor $NaBaScSi_2O_7$:Eu^{2+} with high thermal stability. *CrystEngComm*, *16*(27), 6089–6097. doi:10.1039/C4CE00523F

Chapter 7
Synthesis of Magnesium Based Nanophosphors and Nanocomposites by Different Techniques:
Silicates and Ferrites

Ramachandra Naik
New Horizon College of Engineering, India

Revathi V.
New Horizon College of Engineering, India

S. C. Prashantha
East West Institute of Technology, India

H. Nagabhushana
Tumkur University, India

ABSTRACT

This chapter consists of systematic synthesis of magnesium silicate nanoparticles by solution combustion method using ODH fuel. Compared to other fuels, ODH has the better advantages of obtaining crystalline nanomaterials without post calcinations. By doping different rare earth ions to magnesium silicate, the product can be used for luminescent applications. Synthesis of magnesium ferrite nanoparticle is done using molten salt method followed by the synthesis of polymer composite in the form of fibers by the process of electrospinning.

INTRODUCTION

Nanomaterials find wide range of applications due to their unique chemical, physical, electrical, magnetic, optical and mechanical properties. Because of these properties, they were useful as catalyst, sensors, coating materials, tunable lasers, and memory devices. The unique spectroscopic properties of rare-

DOI: 10.4018/978-1-5225-5170-6.ch007

earth ions in different host lattices was prompted the development of rare- earth luminescent materials for lamps, cathode ray tubes, radiation monitoring systems, lasers, scintillators, bio sensors and white light-emitting diodes (LEDs) Cho. et.al (2010), Wang.et.al (2009) and Prashantha et.al (2011). White light-emitting diodes (LEDs) were the potential materials for significantly improving lighting efficiency, resulting in reduction of the excitation energy and also reduction in pollution from fossil fuel power plants Li et.al (2011). In order to enhance the quality of white-light, one of the current academic interest is the research for phosphors of single-component and cost effective preparation methods. The demand for developing efficient luminescent materials such as rare earth activated powders were attracted researchers due to their possible photonic applications, good luminescent characteristics, stability in high vacuum, biosensors and absence of corrosive gas emission under electron bombardment when compared to currently used sulfide based phosphors. Among all the rare earth ions, Eu^{3+} was usually adapted as a red emitting center because of its unique $4f^6$ structure that can be activated by ultraviolet rays effectively and emit high purity color of the red light Zhang et.al (2010), Brito et.al(2012) and Maldiney et.al(2012). Near-ultraviolet (NUV) conversion was the most suitable method which can achieve WLED. In this method, red/green/blue tricolor phosphors were pumped by NUV to generate white light. So phosphors play a crucial role in these solid-state lighting devices. However, the luminescent efficiency was low in this system owing to the strong re-absorption of the blue light by the red and green phosphors Chen et.al(2010), Li et.al(2012) and Cho et.al(2010).

The silicates are the largest, the most interesting and the most complicated class of minerals. The richness in crystal structures endows silicates affluent chemistry, physics and materials content including Geolites, Mesoporous materials, inorganic-organic composites etc. Exploration upon new type nano structures of silicates with unexpected properties seems especially technologically and economically important, simply due to their low cost and richest resources Hao et.al(1998), Zhao et.al(1999). From ancient times minerals have been a convenient object for investigating the nature of luminescence. In the last decades, research have been carried out on almost exclusively on artificial analogs of the minerals CaF_2, ZnS, $Ca_3\ (PO_4)_2$ .F, Zn_2SiO_4, Al_2O_3, and Mg_2SiO_4. The search of materials suitable for solid-state radiation dosimeters has been the aim of many researchers after the discovery of TL Nagabhushana(2002).

Forsterite is a member of the Olivine family of crystals. It has orthorhombic crystalline structure. Forsterite is a crystalline magnesium silicate which has extremely low electrical conductivity that makes it an ideal substrate material in electronics. On the other hand it shows good refractoriness, low thermal expansion, good chemical stability and excellent insulation properties even at high temperatures Saberi et.al (2006). Forsterite is known to be an insulator at high frequencies. In particular Cr^{3+} doped Mg_2SiO_4 found to be an excellent tunable laser material in the near IR range of 1.1-1.3 μm Tsai (2002). Forsterite is known to be a good Bioceramics material. Therefore it is significant to develop new type of bioceramics Ni et.al (2007).

In recent years, materials with porous architecture and high surface area being developed for numerous potential applications in nanotechnology. The need for such materials is essentially because of the technological progress Gopichandran and Patil (1992) and Saravanapavan and Hench (2003). These materials have been applied in the field of catalysis, Lasers, sensors and solar cells. Recently, Mesoporous silicate materials have been proposed for biomedical applications such as drug delivery systems and bone tissue regeneration materials Xia and Chang (2008). Nanotechnology has emerged as an attractive field for the development of novel materials having unusual properties and has provided different path ways to solve many unresolved issues in various fields including ceramic, silicate materials Fu et.al (2007) and huang et.al (2002).

It is a material of interest due to good refractoriness, low dielectric permittivity, low thermal expansion, high chemical stability and excellent insulation property even at elevated temperature Chesters (1973) and Diesperova (1977). Furthermore, Forsterite has been reported as a new bio material, and might be used in dental and orthopedic prostheses. Mg_2SiO_4 has been synthesized by different methods such as solid state Kosanovic et.al (2005), sol gel Retricevic (1998) etc. The production of Mg_2SiO_4 via solid state method usually requires high temperature, long reaction time and difficulty in achieving sample homogeneity Maliavski (1997). With these features Forsterite (Mg_2SiO_4) has been chosen as a system for the present investigation.

COMBUSTION SYNTHESIS

Low temperature solution combustion (LCS) technique is a versatile process leading to synthesis of single phase, composites, solid solutions as well as complex compound oxide phases in homogeneous form. This process has the advantage of choice of a wide variety of fuels, rapid cooling leading to nucleation of crystallites without any growth and also has the potential to scale up. Because of the gas evolution, large particles or agglomerates can be disintegrated during the process and the products formed are of high purity. The resulting product is very fine particulates of friable agglomerates that can easily be ground to obtain a much finer particle size Patil (1997), Patil et.al (2002), Bera et.al(1999), Bera (2002) and Aruna (1998).

Characteristics of LCS Process

1. Fast heating rates
2. Short reaction time (few seconds to ~5 minutes)
3. The particles are in nano size
4. The process is fast, instantaneous and results in high crystalline products.
5. Due to evolution of large amounts of gases produced during combustion process, nanosize, porous and foamy product is obtained
6. Silicate samples, which are prepared by LCS method have higher crystallinity even though it requires less energy than the solid–state reaction method
7. The temperature instantaneously shoots upto > 1000 °C with flame and combustion of the fuel during combustion process.

Some of the Disadvantages of Other Methods to Synthesize Silicate Phosphors

Solid State Method

The doped and undoped silicates are commonly prepared by high-temperature ceramic method, involving the grinding and firing of constituent precursors. In this method there is difficulty in achieving sample homogeneity. Also, it requires high temperature furnace (~1500ºC) and at the same time high temperature crucible are required.

Co-Precipitation Method

This method can produce more homogeneous powders than conventional mixed-oxide methods. However, the required stoichiometry can be lost during the filtering steps when there is a large difference of solubility among the precipitates. Also, as the number of components increases, it becomes more difficult in finding a suitable precipitating agent.

Sol-Gel Method

Sol gel process can provide molecular level of mixing and high degree of homogeneity which leads to reduction in crystallization temperature and prevent from phase segregation during heating. However, multi component silicate systems the hydrolysis and condensation rates are different with in silica and other alkoxides which may cause non uniform precipitation and chemical in homogeneity of the gels, and also results in higher crystallization temperature and undesired phases. On the other hand the starting chemicals like tetraethyl ortho silicate used in this method are high cost.

Molten Salt Synthesis

It is one of the traditional methods for preparing ceramic powders. In this a molten salt is used as the medium for preparing complex oxides from their constituent materials (oxides and carbonates). In the process of synthesis, a large amount of salt is used as the solvent to control the size, shape and other characteristics of the product. The molten salt synthesis is different from the flux method, where the salt is added to enhance the reaction rate.

Principle of Propellant Chemistry

The stoichiometry of the redox mixture for combustion is calculated based on the thermochemical concepts of propellant chemistry. In the combustion synthesis, the energy released from the exothermic reaction between the nitrates and the fuel which is usually ignited at a temperature much lower than the actual phase transformation temperature can rapidly heat the system to a high temperature and sustain it long enough even in the absence of an external heat source for the synthesis to occur. The basis of the combustion synthesis technique comes from the thermo chemical concepts used in the field of propellants and explosives. One approach that can be used is the examination of gun powder reaction. Gun powder is generally a fine milled mixture of potassium nitrate, sulfur and charcoal with the weight ratio of 75:10:15. Once ignited, it burns quickly, a strongly exothermic reaction accompanied by a release of smoke. When complete combustion is established, the overall reaction can be represented as follows.

$$2KNO_3 + S + 3C \rightarrow K_2S(s) + 3CO_2(g) + N_2(g) + \text{Energy} \quad (1)$$

i.e., gun powder $\rightarrow$ solid product + gaseous product + heat

It should be noted that the three ingredients of gunpowder are solid, among them potassium nitrate acts as an oxidant whereas both sulfur and charcoal act as fuels. This triggered an idea to develop a novel combustion method for synthesizing oxide ceramic powders. The concept is that, if mixtures composed of metal nitrates with a suitable fuel in an appropriate ratio are ignited and burnt to form ceramic oxides. Gunpowder generates a lot of gases when burned. These gases are conducive to heat loss through convection, resulting in rapid reduction of product temperature. On a practical level, it is suggested that fine powders could be produced due to the prevention from both the growth of grains and the agglomeration of particles. As stated above, a theory for improving the combustion synthesis method has been established accordingly. Use of suitable raw material in combustion synthesis ensures stability of the chemical composition and high quality of products. More importantly, it must be readily available and convenient to use. In addition, it should not react violently and should only produce non-toxic gases.

The Factors Influencing the Combustion Process

The mechanism of the combustion reaction is quite complex. Processing parameters such as type of fuel, amount of fuel, the amount of gaseous products released during combustion, water content, flame temperature (T_f) that is the ignition temperature, addition of extra oxidizer, fuel-to-oxidizer ratio, furnace temperature etc play an important role in deciding material properties and phase formation.

The Type of Fuel and Its Role in LCS Process

In LCS process for the synthesis of various Silicate materials different fuels are used, based on the number of review articles on combustion synthesis, the fuels so far employed in LCS are Urea, Oxalyldihydrazine (ODH), Carbohydrazine (CH), Tetraformaltrisazine (TFTA), Diformylhydrazine (DFH), Hexamethylenetetramine (HMT), Glucose, Citric acid etc. By employing these fuels a large number of Silicates have been synthesized Chandran et.al(1994) and Chandran and Patil (1990).

The Importance of Gaseous Products

The combustion-derived products are generally porous and nanocrystalline because of large number of gaseous products like H_2O, CO_2 and N_2 generate and released during the reaction. The number of moles of gaseous products released during reaction affects the product very much. In fact they increase the surface area of the powder product by creating micro and nano porous regions.

Water Content

In LCS process precursor water was required for complete atomic mixing and to prevent selective crystallization. When the excess of water is left at the time of ignition, the water may reduce the flame temperature, causing incomplete crystallization of the product. When the ignition temperature is too low, then the reaction may not initiate. It was determined that specific surface area decreases with an increase in precursor water content. This was attributed to less heat being available for particle growth due to the heat of vaporization for water Nagbhushana (2007).

Fuel-to-Oxidizer Ratio

The combustion synthesis reaction could be controlled by varying the fuel-to-oxidizer ratio (F/O). It is found that as the amount of fuel increases, the reaction temperature also increases and facilitates crystal growth. High crystalline combustion products could be synthesized by either in equilibrium state (Fuel/Oxidizer = 1) or fuel rich combustion method. It is noted that the reaction temperature and the reaction time are two important factors that control the phase development during the process. The optimal temperature and reaction time vary from material to material and are determined primarily by the intrinsic phase transformation temperature that is characteristic of each system. It is important to note that, the fuel lean mixture undergoes decomposition (not combustion) producing amorphous products. The incandescent combustion with the appearance of flame was observed only with stoichiometric and near stoichiometric system Jain et.al (1981), Kingsley Pederson (1993), Kingsley Pederson (1993), Chandrappa et.al(1995), Venkatachari (1995) and Zhang (1994).

Mg_2SiO_4 nanophosphors were prepared by the low temperature (350 °C) solution combustion route in a very short time (less than 5 min). In this route, suitable amount of magnesium nitrate (Mg $(NO_3)_2.6H_2O$ (Sigma Aldrich) and fumed silica (SiO_2 (Sigma Aldrich) were mixed in stoichiometric amounts with laboratory prepared oxalyl dihydrazide (ODH; $C_2H_6N_4O_2$) fuel and dissolved in a minimum quantity of doubled distilled water in a cylindrical pyrex dish and mixed thoroughly using magnetic stirrer for about 5 min. The details of the stoichiometric calculations of the redox mixtures (oxidizer and fuel) were given elsewhere Sunitha et.al (2012). The pyrex dish containing this solution was placed in a pre-heated muffle furnace maintained at 350±10 °C. The solution boiled resulting in a transparent gel. The gel then formed white foam, which expanded to fill the vessel. Thereafter, the reaction was initiated somewhere in the interior a flame appeared on the surface of the foam and proceeded rapidly throughout the entire volume, leaving a white powder with an extremely porous structure. The energy released from the reaction was produced temperature greater than 1200 °C. The reaction was self-propagating and able to sustain high temperature to form Mg_2SiO_4 phosphor. The entire combustion process was over less than ~5 min. The final product was grinded into a fine powder by agate and mortar. Theoretical equation assuming complete combustion of the redox mixture used for the synthesis of Mg_2SiO_4 can be written as Naik. et.al (2014)

$$2Mg\,[NO_3]_2 + SiO_2 + 2C_2H_6N_4O_2\,[ODH] \rightarrow Mg_2SiO_4 + 6N_2 + 6H_2O + 4CO_2 \text{ ------} \quad (2)$$

Molten Salt Method

One of the modest and well known mechanisms adopted for the preparation of ferrite ceramic powders is the molten salt method. Particles of size ranging from a few nanometers to a few micrometers can be synthesised by prudent handling of the conditions during synthesis. This method involves the usage of molten salt as a medium for deriving complex oxides from the corresponding constituent carbonates and oxides. There are enormous numbers of techniques available for the synthesis of these ferrite nanoparticles Rahaman (2003) in bulk amount, of which powder metallurgical method is a generally adopted method. Molten salt method is an altered version of powder metallurgical method. In this method, a suitable salt preferably possessing a low melting point is mixed with the component reactants. When the temperature of this mixture reaches the salt's melting point, the salt which is in the molten state, dons the roll of a solvent.

The main variation of molten salt method from that of any solid state reactions is in the amount of salt added to enhance the rate of reaction. In contrast to solid state reactions where very feeble percentage of molten salt is added, in molten salt synthesis huge quantity of salt is added so as to control the size and shape of the resultant particle. Commonly used salts like chlorides and sulphates may be replaced by eutectic mixtures of salts so as to reduce the liquid formation temperature.

For example, eutectic composition of 0.5NaCl–0.5KCl has a melting point of 650 °C, which is much more lower than the melting points of NaCl and KCl which are 801 °C and 770 °C respectively, thereby preferring the former than the individual salts. Similarly among salts of sulphates popularly used eutectic mixture is $0.635Li_2SO_4$–$0.365Na_2SO_4$ whose melting point is 594 °C. When ordinary solvents are used, constituents may not dissolve completely whereas in this method all the reactants get dissolved in the molten salt and the product particle is obtained from the homogeneous solution after precipitation.

In molten salt synthesis, to get particles of appropriate size a large number of nuclei are necessary as the product particles nucleate heterogeneously on the surfaces of the reactant particles. Due to this, it is not mandatory that all reactants must be highly soluble. Another parameter which has an influence on the particle size is the cooling rate, because a vast number of particles are already present before cooling and the materials dissolved in the molten salt precipitate on the surfaces of the already existing solid particles. In this synthesis, the role of the molten salt is to

- Enhance the rate of reaction
- Reduce the reaction temperature
- Magnify the degree of homogeneity
- Govern the product particle size
- Handle particle shape and
- Handle the state of agglomeration

The salient features of the salt chosen are preferably, it should be,

- Stable
- Readily available
- Cost effective
- Easily washed away with water
- Low melting point
- Low vapour pressure
- Non-reactive with reactants and products

It is always favourable to select the salt which acts only as solvent and does not react with the component particles. Hence at most care must be taken to choose the salt. An example for an unsuitable salt is LiCl which is used for the preparation of $LiFe_5O_8$ Wickham (1971).

The desired precursor materials otherwise known as reactants are mixed with the chosen salt with the required characteristics. The salt is chosen based on the required nature of the resultant product. If we desire to synthesise the product in nano form then it is necessary that suitable surfactants shall be added Mao. et.al (2003). The required reactants are taken in stoichiometric ratio.

Care must be paid when carbonates are used as reactants as they have high solubility compared to oxides thereby causing deviations in the stoichiometry of the product. A typical amount of salt added to the reactant mixture is around 80-120 wt% of the reactant mixture, which is decided by the requirement so as to enable the salt to considerably spread through the space between the reactant particles and to overlay the surface of the reactants.

It is also a fact that the quantity of salt has an influence on the size of the product particle Yoon et.al (1998). Too small the amount of salt may result in less liquid phase. An extremely large amount may lead to estrangement of the reactant particles by sedimentation Kimura et.al (1980). This may happen due to the difference in the densities and sizes of the reactant particles which results in different sedimentation rates, thus causing split up of the reactant particles thereby reducing the rate of reaction. Yet another problem caused by the presence of large amount of salt is the laboriousness during treatment.

When the amount of salt is excessive, the gaps between the reactants find difficulty in holding the molten salt, leading to the loss of molten salt from a gob of reactants. Oozed molten salt never acts as a solvent. Furthermore, the molten salt adheres to the wall of the crucible and changes to a hard lump after solidification. The dissolution of hard salt in the lumps is quite laborious. Also when the quantity of salt and the temperature is low, the system tends to move on to double phase region (Solid-liquid) because of which large number of particles nucleates disparately on the reactant surface.

Contrary to this when both the temperature and the quantity of salt is more the system retains a single phase. The particles homogeneously nucleate in the liquid phase. Because the cooling rate is relatively high compared to that employed in the single-crystal growth, many nuclei form but the number of nuclei is less than that nucleated heterogeneously on the surfaces of the reactants leading to the formation of particles of larger size.

The required amount of salt together with the constituent precursors are taken ground well and put in a closed crucible or container. The container is kept in a furnace and heated to the desired temperature. The temperature and the tenure for which the mixture has to be heated are decided based on the properties of the product. As the heating temperature is enhanced, the rate of salt evaporation also gets increased. The duration for which the mixture needs to be heated is decided based on the size and shape of the product particles. Normal duration is around 1hr and temperature around 800°C to 1100°C with durations between 30 and 60 min. Also it is found that the rate of heating has a strong effect on the particle size Yoon et.al(1998). The product is taken out after heating and washed with deionised water to remove excess of salt, as the salts used in this method are mostly water soluble. The procedure of synthesis is represented by a flow chart given in Figure 1.

As salts of chlorides and sulphates dissolve easily in water, washing the mixture with water for few times shall remove the salt content in the mixture. The mixture is washed with deionised water multiple times and after removing the afloat water the filtered powder is dried with hand drier. To prevent agglomeration the powder is again washed with a solvent of low surface tension and dried again. There are two steps by which the product is obtained. They are,

- Reaction Stage
- Particle – Growth Stage

In the former stage, the reaction occurs in the existence of solid constituent particles as they are not completely dissolved by the molten salt. The resultant particle in the powder form is precipitated as all constituent members of the precursor dissolves in the molten salt. At the end of this stage, the product specks are the only solid material in the molten salt.

Figure 1. Flow Chart of Molten Salt Method

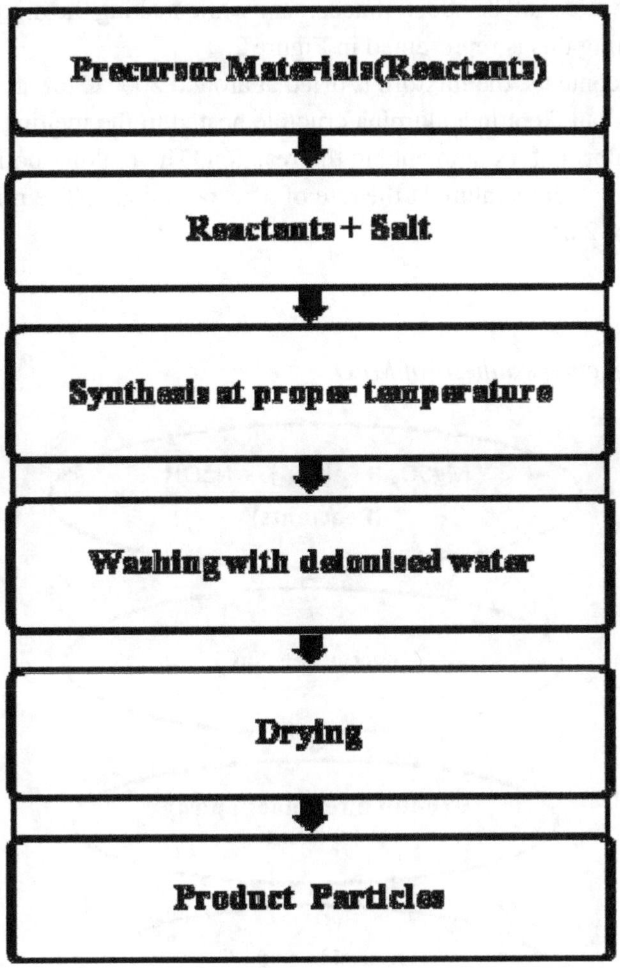

The later stage starts once the heating commences, where in a condition of low degree of super saturation, the product particles change their size and shape. The shape of the product can also be tweaked. The size and shape of particles changes in the course of the stages of particle formation. It is determined by the relative dissolution rate of the reactant particles into the molten salt, and the dissolution rate in turn is controlled by the solubility and particle size of the reactant. In the particle-growth stage, the particle size is increased and the particle shape approaches the equilibrium form. The process of molten salt synthesis is used widely since 2005, for preparing from ferroelectric and ferromagnetic materials to materials for Li-ion batteries Santhanam and Rambabu (2010), semiconductors Huang et.al (2010), phosphors Yan and Lei (2010) and photo catalysts Arney et.al (2008), especially to obtain materials of nano dimension Mao (2007).

The nanosized particles of magnesium ferrite are prepared following molten salt route. The precursors are taken in desired ratio. To obtain nanoparticles of magnesium ferrite – $MgFe_2O_4$, we have taken magnesium sulphate – $MgSO_4$, ferrous nitrate – $Fe(NO_3)_3$ and sodium hydroxide - NaOH along with the chosen salt sodium chloride – NaCl in the ratio 1:2:8:10. The ingredients are taken in an agate motor. They are then grounded well in to smooth paste for around 90 minutes to enable proper mixing of constituents. Deionised water shall be added if necessary while making the mixture into a smooth paste.

The flow chart depicting this is represented in Figure 2.

To remove the water contents, the mixture is dried at around 200 °C for an hour. The dried mixture is powdered again, taken out, kept in a alumina crucible heated to the melting point of the salt, which in this case is sodium chloride for commencing the reaction. The melting point is around 800 °C and the mixture is heated to that temperature at the rate of 5 °C per minute. The mixture is maintained at a temperature of 800 °C for 2 hours.

Figure 2. Flow Chart depicting synthesis of MFO

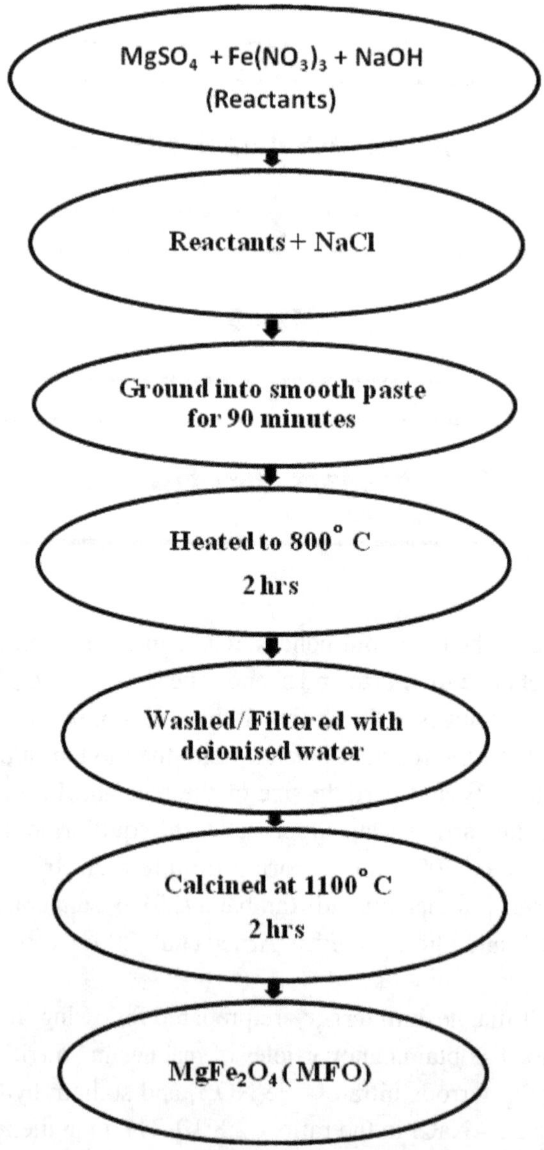

The mixture is taken out after cooling. Powdered well and the powder is kept in a filter paper. The mixture is then washed slowly and throughly with deionised water multiple number of times. This is done to ensure that any excess content of salt present is washed away. To prevent any agglomeartion, the mixture is also washed with acetone. The filtered water is decanted and the residual powder is again kept in a alumina crucible and calcined to 1100°C for 2 hours at the same rate to get the resultant particle magnesium ferrite. The calcined powder is taken out after cooling, powdered again using agate motor kept in a clean container, for all related studies.

NANOCOMPOSITES

"Nature is a master chemist with incredible talent", states an article titled 'Nano sandwiches' Maeda and Armes (1995), in 1998. Nature has mastered the use of nanocomposites and we researchers are the students who learn from nature. Nature makes strong composites such as bones, shells with the use of natural reagents and polymers like carbohydrates, lipids and proteins. A nanocomposite is a material obtained by mixing two or more phases such as particles, layers or fibers, where the dimension of one of the phases is in the nano range that is, in the order of 1-100nm. Depending on the nature of the components and the method of preparation, significant differences in the composite properties may be obtained.

The combination of nanomaterials with polymer is very attractive not only to reinforce polymer but also to introduce new electronic properties based on the morphological modification or electronic interaction between the two components. There have been great efforts to fabricate nanocomposite to obtain unique physical properties since these properties are increasingly size dependent at low dimension. It may be noted that the formation and the properties of these composites depend not only on the properties of the components but also on the size of the filler materials which are added to the polymers Chiang and Popielarz (2002). It also depends on the degree of mixing and method of preparation. When the sizes of these fillers are large enough, only little interaction is expected between them and the polymer material. When the size of the filler material is reduced, then the interaction is found to be effective due to which the performance of the composite is increased to a larger extent Akamatsu and Deki (1997), Roco (1999), Chiu (2002), Park et.al (2001) and Lau et.al (2006).

Organic-inorganic composites are prepared by various methods. The primitive and the simplest of all involve the direct blending of nanoparticle and the polymer matrix. In this method it is easy to control the size of the nanoparticle. Here the nanoparticles which have to be incorporated into the polymer matrix are prepared well in advance. The advantageousness of this technique is its simplicity. However, due to the tendency of the nanoparticles to agglomerate, owing to their large specific surface energy, there is an uneven distribution of the nanoparticles in to the polymer matrix, resulting in a loss of their function.

The 1D composite structure which attracts more importance by the researchers can have a homogeneous uniformly miscible inorganic and organic phase or can have a phase separation. Based on the degree of mixing of the organic and inorganic materials, the 1D structure can be a core-shell type, scattered type, multi layered type and so on.

The 1D composite structure in the form of fibres can be prepared by various techniques. There are many techniques available to synthesis composite structures in the form of fibers. They are:

- Self-Assembly, Zhang et.al (2005) and Kimizuka (2008)
- Phase Separation Zhao et.al (2001)

- Dry Spinning Anitha (2011)
- Mechanical Drawing Nain et.al (2005)
- Template Synthesis Wang (2010) and
- Electrospinning Teo et.al (2006).

ELECTROSPINNING

Electrospinning is a technique used to form one dimensional, continuous fiber of materials, whose schematic representation is given in Figure 3.

As the name implies, fibers of desired materials are produced, whose thickness range from 50 nm to 1000 nm by the process of spinning under the effect of electric field Reneker and Chun (1996). This method is employed to obtain fibers of polymer composites, used for various applications.

Of the above mentioned processes, electrospinning is the most widely studied and attractive technique for producing one dimensional fibers. Fibers produced by the process of electrospinning are thin compared to those obtained by mechanical drawing. Also the chances of contamination are less in fibers produced by electrospinning as it is a contactless procedure. It is less complex than self-assembly and can be used for a wide range of materials unlike phase separation.

Figure 3. Schematic representation of the process of Electrospinning

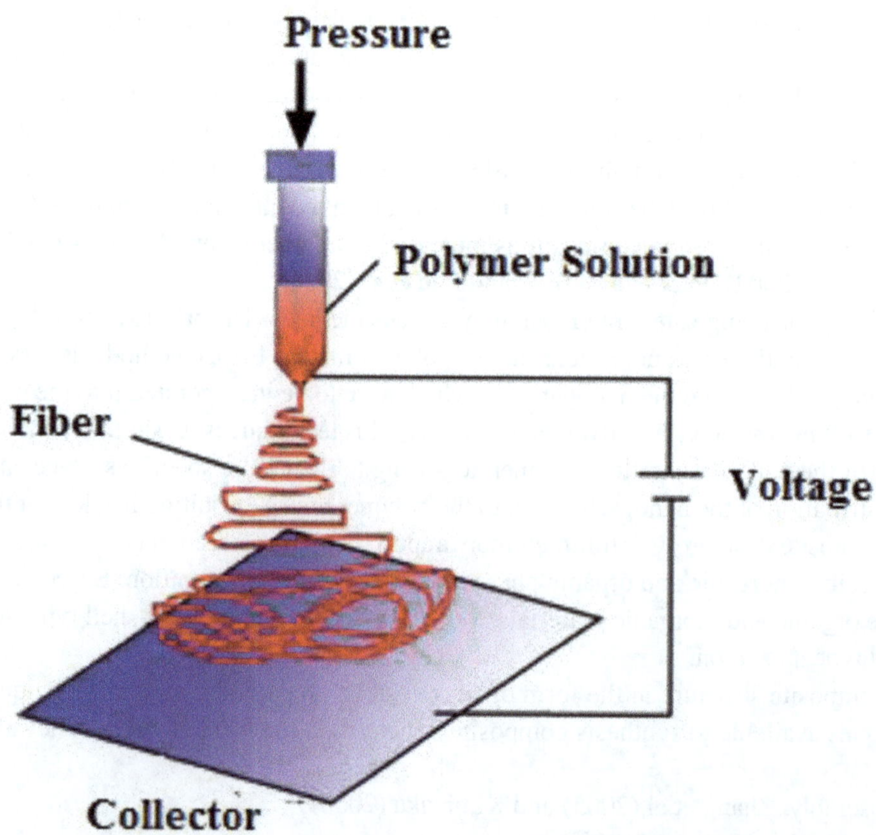

The one dimensional fibers obtained by the process of electrospinning have the following characteristics:

- High surface to volume ratio
- Porosity
- Flexibility
- Weightless Fibers

In addition to the above the fibers obtained can be made suitable for specific application. The fibers obtained are continuous, uniform and can be produced on a larger scale. Fibers with cores, hollow fibers and porous fibers are successfully prepared. The process of electrospinning is cost effective when compared to other methods of fiber fabrication. Due to these electrospinning has attracted increased attention in the past few years in a wide range of biomedical and industrial applications due to the ease of forming fibers with a broad range of properties.

Even though, electrospinning is a relatively robust and simple technique to produce fibers from a wide variety of polymers; it has its own limitations like broad range of fiber diameter, randomly oriented fibers with low mechanical properties. The fibers obtained possess dimensions in microscopic scale and when woven as mats possess dimension in macroscopic scale. This makes these fiber mats more application oriented compared to those prepared conventionally.

Electrospun fibers are broadly applied in biomedical applications, as tissue engineering scaffolds, in wound healing, drug delivery, filtration, as affinity membrane, in immobilization of enzymes, small diameter vascular graft implants, healthcare, biotechnology, environmental engineering, defence and security, in textiles as smart and protective clothing like fire retardant, chemical retardant fabrics, as actuators, in electromagnetic interference (EMI) shielding, used as sensors like gas sensors, chemical sensors, piezoelectric sensors, fluorescence sensors and as ion exchange membranes in filtration. In addition to these they are used in energy storage and generation and in various researches that are underway. Pictorial representations of these applications are shown in Figure 4.

Owing to the application of these fibers in diverse fields, a lot of research and developments are happening in the fields of electrospinning, which is represented graphically in Figure 16.

Electrospinning is a highly adaptable method to form one dimensional, continuous fiber of materials whose dimension ranges in the order few micrometers to nanometers Chronakis (2005), Ramakrishna (2006) and Reneker (2008). The fibers obtained using these processes have a wide range of application in all fields of science and technology Jingwei (2008), Moghe et.al (2009), Huang(2006), Moghe (2008) and Ramakrishna et.al (2005).

The fibers are formed by subjecting the required polymer/composite viscous solution to an electric field. In this technique when the polymer solution is subjected to an electric field, it gets charged and forms a cone shaped droplet at the tip of the needle or spinneret Baumgarten (1971), Taylor(1969) and hayati et.al (1986) . The drop form to a cone shaped structure and extends in the form of jet towards the collector which is grounded. This happens if and only if the electrostatic force of attraction overcomes the force due to surface tension of the polymer solution. The solvent gets evaporated on the way and the polymer composite in the form of fiber gets collected in the collector.

The basic electrospinning setup is represented in Figure 5, where both vertical and horizontal setup is shown.

The composite solution, which is to be formed into fibers, is kept in a 2 ml syringe with a needle fitted in its end. The tip of the needle is cut and made uniform. The needle also acts as an electrode. When

Figure 4. Applications of Electrospun Fibers

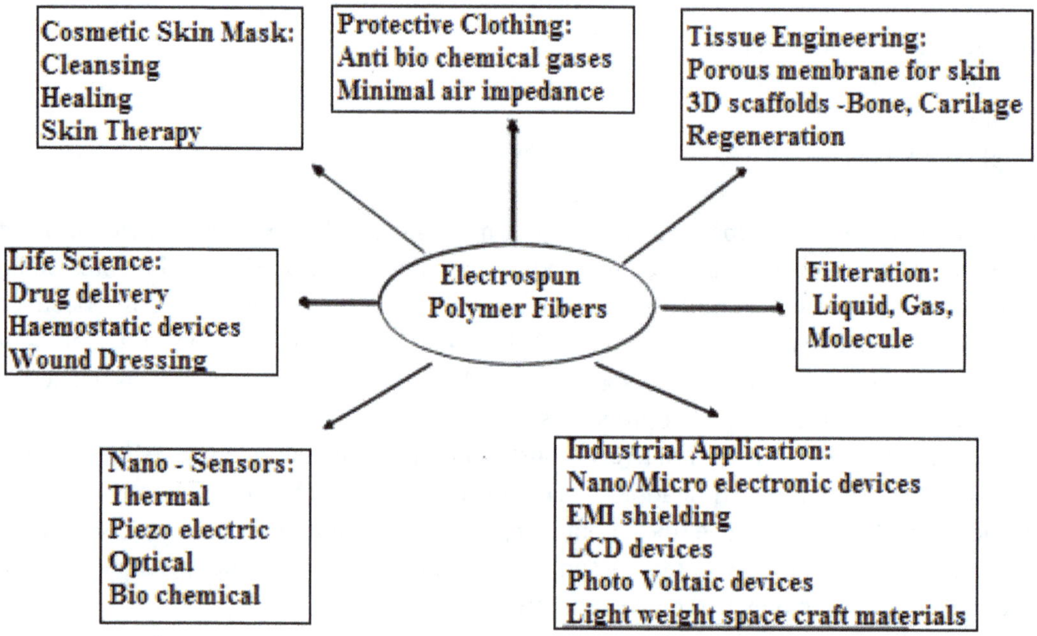

Figure 5. Typical Electrospinning Setup

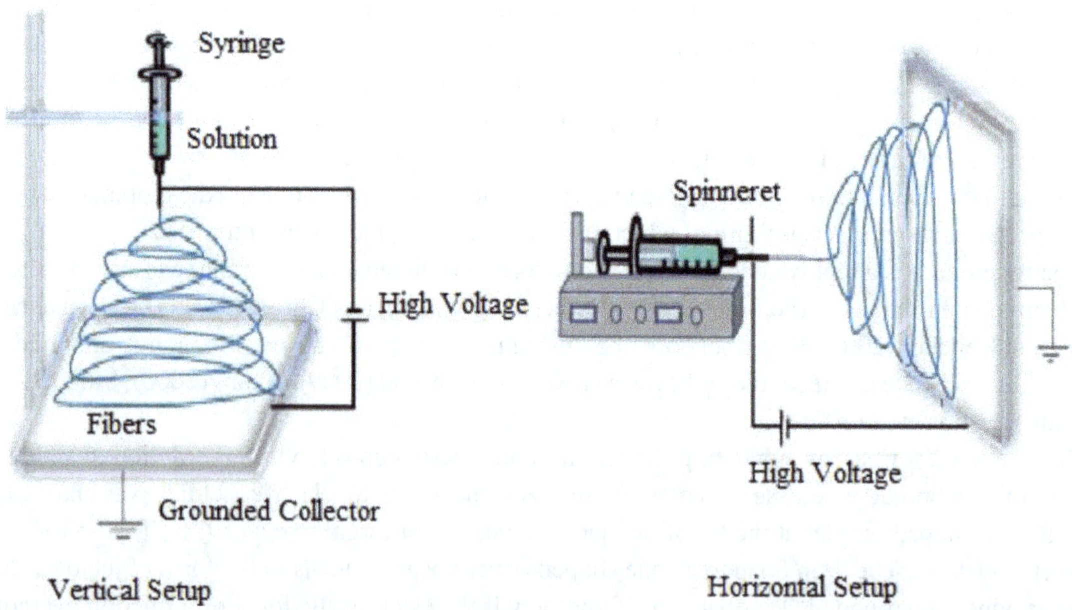

an electric field of required amount is applied across the needle and the grounded collector, the charges accumulate on the surface of the solution drop which is formed at the tip of the needle. When the force of repulsion between the similar charges on the droplet exceeds the force due to surface tension the drop extends to a cone called Taylor cone, which extends further to form a jet of solution.

Under the effect of columbic forces the jet of solution stretches towards the collector plate and gets collected while the solvent in the solution evaporates on the way. To collect dry fibers the distance between the collector and the pin tip is adjusted. The polymer chain entanglements within the solution prevents the electrospinning jet from breaking up and depending on the stability of the electrospinning process fibres with lengths of up to many hundreds of metres can be obtained. We shall now try understanding the history of electrospinning and the various factors which influence the formation and the characteristics of the fibers involved in the process.

The polymer composite of PVDF with the ferrite nanoparticles - prepared by the method of molten salt method was obtained in one dimensional fiber form by the method of electrospinning. The solution of the composite which has to be electrospun was obtained by preparing two separate solutions one with polymer and the other with ferrite.

The base polymer PVDF is mixed with acetone and DMF which are used as solvents; the ferrite powder of the desired percentage by weight with respect to PVDF is dissolved in DMF. The two solu-

Figure 6. Flow Chart for formation of composite fiber

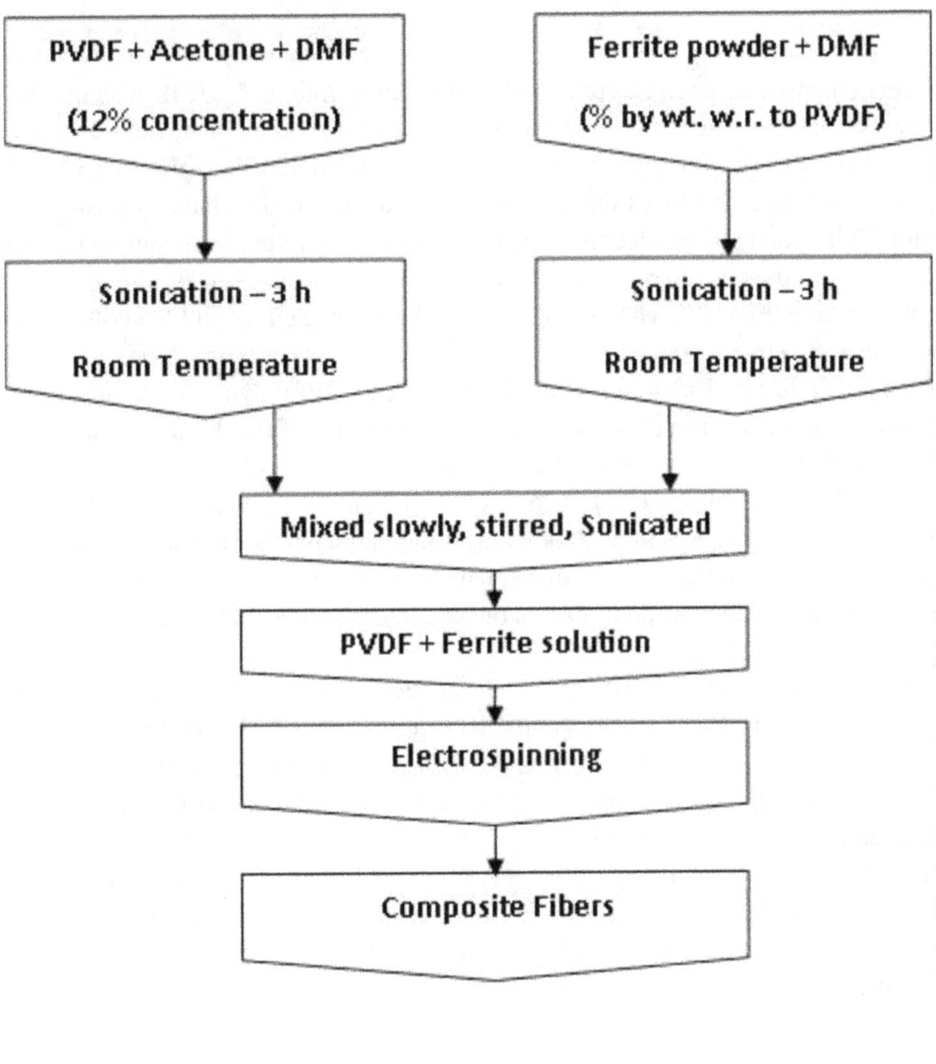

tions are independently sonicated for 3 h at room temperature. The two solutions are then mixed slowly, stirred continuously and sonicated to get the required polymer composite solution. Care should be taken to maintain the concentration of the composite solution at 12%. This solution is made in to one dimensional fiber by the process of electrospinning.

COMPOSITE FIBERS

The polymer composite solution obtained by the process mentioned in the flow chart Figure 3.6 is spun into one dimensional fiber by the process of electrospinning. Fibers of dimension ranging from micro to nanometers are obtained by the application of suitable electric field from the corresponding polymer composite solution.

The instrument used for the process comprises of 3 main parts such as,

- power supply
- pump to drive the syringe and
- Collector

The polymer composite solution was prepared with a concentration of 12% by weight. The solution is taken in a 2ml disposable syringe fitted with a 0.5 mm diameter needle (Dispovan – India) whose tip is severed in such a way that the tapering end is removed. This facilitates the polymer composite solution to form spherical droplets at the tip of the needle. The needle tip made relatively positive with respect to the collector. This is done by connecting the positive polarity of the power supply to the needle and the negative polarity to the collector. The photograph of the apparatus used for the process of electrospinning is represented in Figure 7. The photographs of the syringe pump and the collector are given in Figure 8 and 9 respectively.

A suitable high voltage of 1 kV/cm produced a sustained jet of polymer composite solution which is held at the tip of the needle. The charges induced on the surface of the droplet formed at the tip repel each other and travel towards the grounded collector. It is found that as the strength of the electric field is increased from 0 kV, the droplet extends to Taylor cone and forms a jet which travels towards the grounded collector when the right value of the voltage is applied. On its way the solvent gets evaporated and the fibers get collected on the aluminium foil which is rolled and fixed on the cylindrical collector. This aluminium foil serves dual purpose, one as the counter electrode and the other as the substrate on which the fibers are collected.

The distance between the tip of the needle and the collector was varied and finally standardised as 10 cm so that smooth fibers are formed. The applied voltage is kept as 10 kV and hence the electric field applied is 1 kV/cm. The complete process was performed at room temperature. The flow rate of the polymer composite solution was optimised as 0.7 mL/hr. The parameters used for the fiber formation throughout this study are given in Table 1

To begin with a desired amount of PVDF was taken and dissolved in proper proportion of acetone and dimethylformamide (DMF) which are used as solvents. The mixture is sonicated for 3 h to get a homogeneous solution. The solution is then loaded into the syringe and kept in to the spinning apparatus. The PDVF solution prepared is shown in Figure 10.

Figure 7. Electrospinning Apparatus

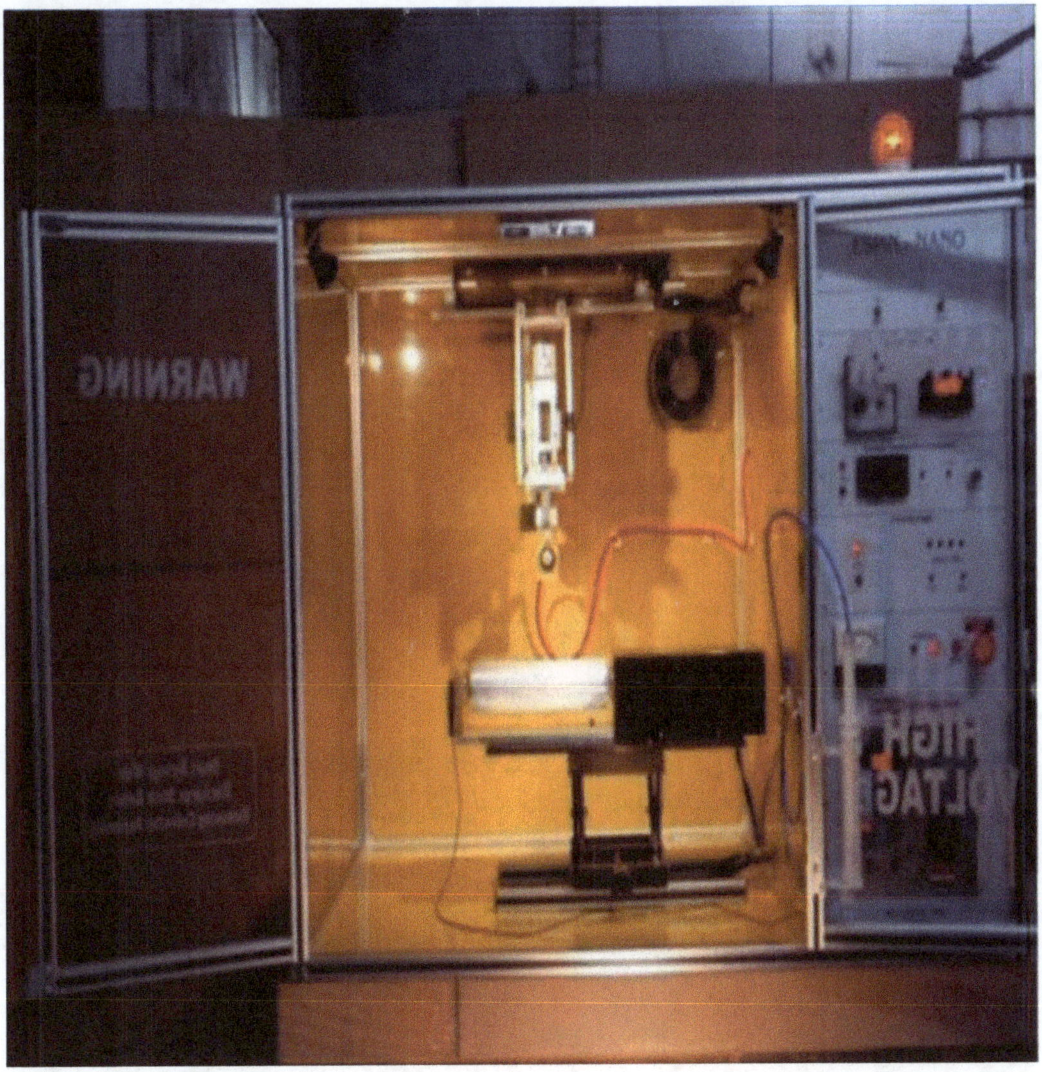

By keeping the process parameters and system parameters to the optimum value, fibers of pure PVDF were spun. The SEM of the pure PVDF is as shown in Figure 11. The fiber diameter is found to be 90 nm.

The ferrite loaded PVDF fiber is obtained by taking desired percentage weight of MFO which is dissolved in DMF. This solution is sonicated for around 2 to 3 h. The required amount of PVDF is taken and the measured amount of DMF and acetone are added to it. This solution is also sonicated for almost the same time. The two solutions are mixed slowly and thoroughly and then sonicated for 3 hrs to get the composite solution which is loaded in the 2 ml syringe. The solution is spun in to fibers by keeping the optimum spinning conditions. The fibers are spun for 1%, 2%, 3%, 5% and 10% percentage weight of MFO in PVDF. Similarly BMFO – PVDF composite fibers are also prepared. Table 2 shows the compositions for which the fibers are prepared Revathi (2014).

Figure 8. Syringe Pump

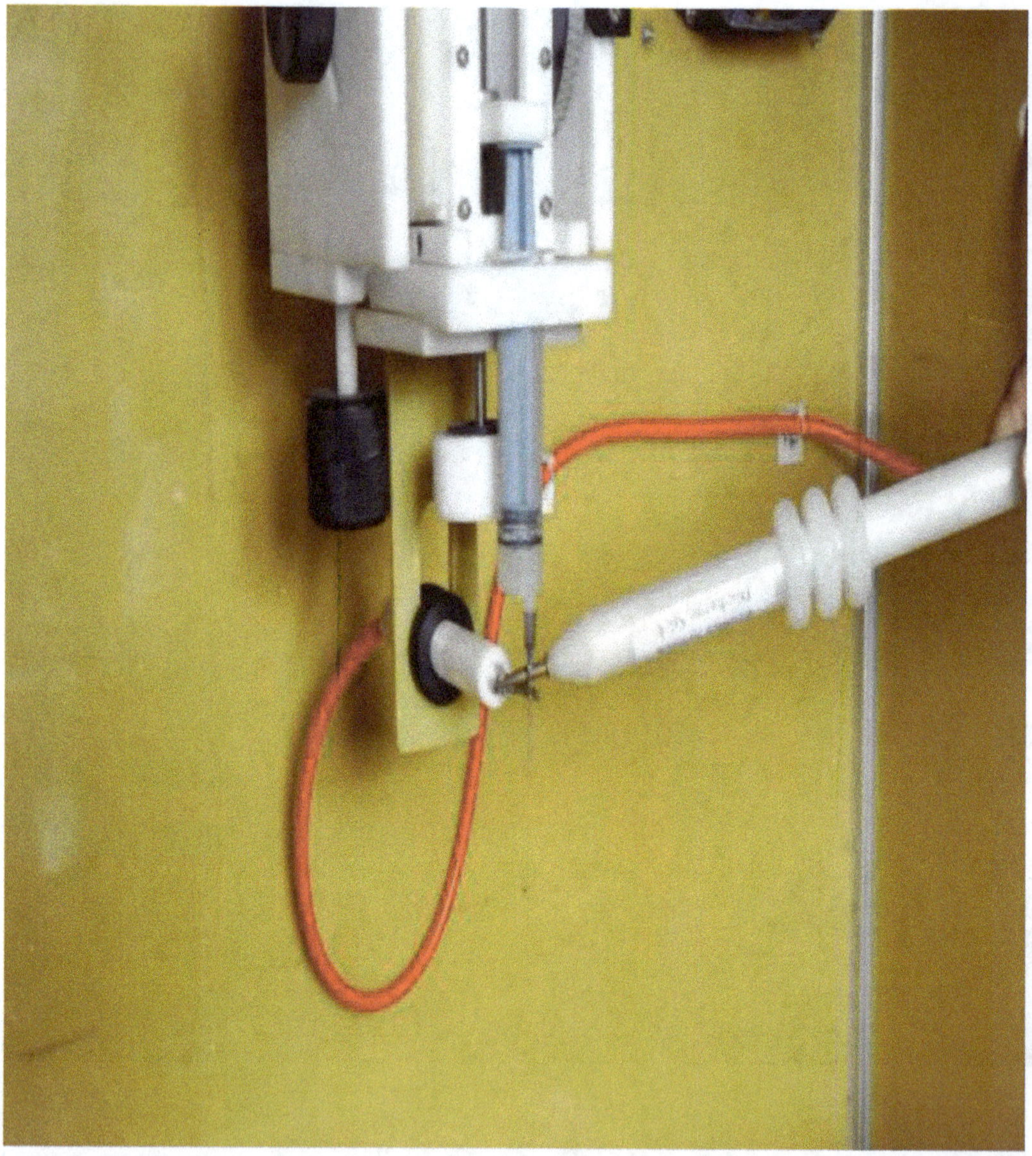

The scanning electron microgram (SEM) of the electrospun PVDF fibers, PVDF - MFO and PVDF – BMFO fibers are taken. Using SEM the dimension, morphology and quality of the electrospun fibers can be obtained. The SEM picture of the pure PVDF electrospun fiber is shown in Figure 12, where in the comparison with the MFO loaded PVDF fiber is also shown.

The thickness of the pure PVDF fiber is around 90 nm and that of PVDF- MFO composite fiber is around 150 nm. The dimension of the fiber remains almost same for 1% to 10% loading of MFO by weight with PVDF. The increase in the thickness of the MFO loaded fiber can be due to the encapsulation of the MFO particles by PVDF. This indicates a typical core shell formation, where the core is the ferrite material and the shell is the polymer.

Figure 9. Collector Drum

Table 1. Parameters for electrospinning process

Polymer	Solvent	Concentration	Collector	Voltage	Electrode Separation	Flow Rate
PVDF	DMF & Acetone	12% by wt	Aluminium foil	10 kV	10 cm	0.7 mL/hr

The size of the MFO nanoparticle is found to be around 30 nm using the Scherrer formula. This supports the fact the resultant fiber diameter of the ferrite loaded fiber is increased due to the encapsulation of the ferrite by the polymer. The diameter of the fibers is obtained and a histogram to represent the number of fibers with different diameter ranges is made. The histogram showing the number of fibers with fiber diameter is shown in Figure 13. From the histogram it is clear that majority of the fibers have diameter around 150 nm.

Figure 10. PVDF solution

Figure 11. SEM of Electrospun PVDF Fibers

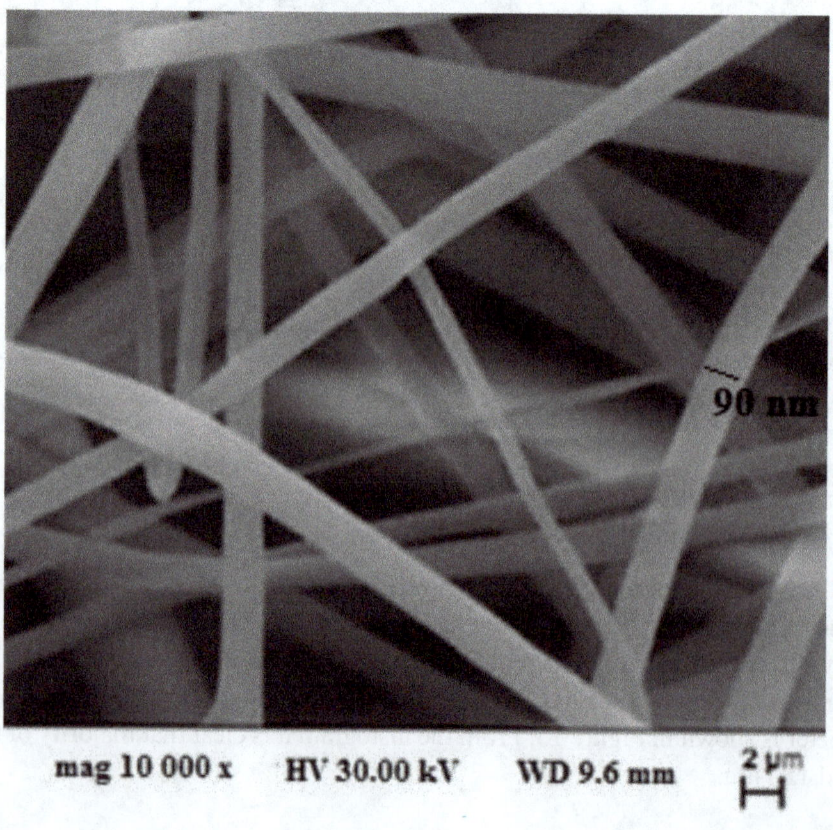

Table 2. Composition of ferrite with PVDF

$Ba_xMg_{(1-x)}Fe_2O_4$	Ferrite % by Weight + PVDF				
	1%	2%	3%	5%	10%
x = 0	√	√	√	√	√
x = 0.05	√	√	√	√	√
x = 0.1	√	√	√	√	√
x = 0.2	√	√	√	√	√

Figure 12. PVDF Fiber & MFO loaded PVDF Fibers

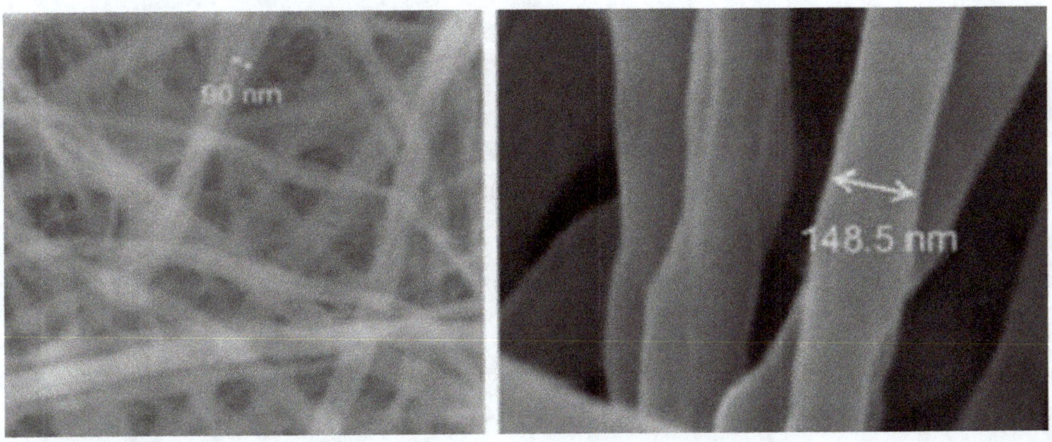

Figure 13. Histogram of PVDF–MFO electrospun fiber diameter in nm

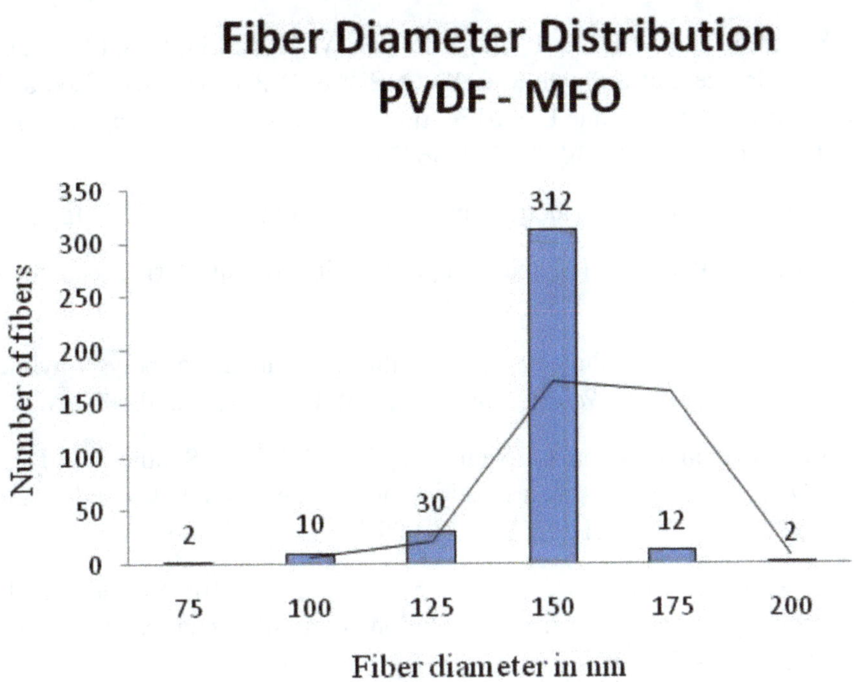

The magnesium ferrite nanoparticles were synthesized by molten salt method. Also Barium substituted magnesium ferrite nanoparticles were also synthesized. The particle size was found to be around 45nm. A composite fiber of ferrite with the base polymer PVDF was prepared by the process of electrospinning. Smooth fibers of the composite were obtained on the aluminium foil by optimizing the various parameters in the electrospinning process. The MFO – PVDF fibers were prepared for 1%, 2%, 3%, 5% and 10% by wt of MFO. The SEM micrograph of the pure PVDF fibers and the ferrite loaded PVDF fibers show that the diameter of the ferrite loaded fiber is larger than that of the pure PVDF fiber, which conclude the fact that core shell composite fibers with ferrite material as core and polymer as shell are formed.

REFERENCES

Akamatsu, K., & Deki, S. (1997). Nanoscale Metal Particles Dispersed in Polymer Matrix. *Nanostructured Materials*, *8*(8), 1121–1129. doi:10.1016/S0965-9773(98)00044-0

Anitha, S. (2011). *Nanofibrous composite membrane of ZnO/polymer for UV ssensor, superhydrophoobic and antibacterial applications* (Ph.D Thesis). SRM University, Chennai, India.

Arney, D., Porter, B., Greve, B., & Maggard, P. A. (2008). New Molten-Salt Synthesis and Photocatalytic Properties of La2Ti2O7 Particles. *Journal of Photochemistry and Photobiology A Chemistry*, *199*(2-3), 230–235. doi:10.1016/j.jphotochem.2008.06.005

Baumgarten, P. K. (1971). Electrostatic Spinning of Acrylic Microfibers. *Journal of Colloid and Interface Science*, *36*(1), 71–79. doi:10.1016/0021-9797(71)90241-4

Brito, H. F., Hölsä, J., Laamanen, T., Lastusaari, M., Malkamäki, M., & Rodrigues, L. C. V. (2012). Persistent luminescence mechanisms: Human imagination at work. *Optical Materials Express*, *2*(4), 371–381. doi:10.1364/OME.2.000371

Chen, L., Chen, K. J., Lin, C. C., Chu, C. I., Hu, S. F., Lee, M. H., & Liu, R. S. (2010). Combinatorial Approach to the Development of a Single MassYVO_4:Bi^{3+},Eu^{3+} Phosphor with Red and Green Dual Colors for High Color Rendering White Light-Emitting Diodes. *Journal of Combinatorial Chemistry*, *12*(4), 587–594. doi:10.1021/cc100063x PMID:20560596

Chesters, J. H. (1973). Refractories production and properties. The Iron and Steel Institute.

Chiang, C. K., & Popielarz, R. (2002). Polymer composites with high dielectric constant. *Ferroelectrics*, *275*(1), 1–9. doi:10.1080/00150190214285

Chiu, H. J. (2002). Spherulitic Morphology and crystallization kinetics of poly (vinylidene fluoride)/ poly(vinyl acetate) blends. *Journal of Polymer Research*, *9*(3), 169–174. doi:10.1023/A:1021387524243

Cho, S., Lee, R., Lee, H., Kim, J., Moon, C., Nam, S., & Park, J. (2010). Synthesis and characterization of Eu^{3+} doped Lu_2O_3 nanophosphor using a solution-combustion method. *Journal of Sol-Gel Science and Technology*, *53*(2), 171–175. doi:10.100710971-009-2073-4

Cho, S., Lee, R., Lee, H., Kim, J., Moon, C., Nam, S., & Park, J. (2010). Synthesis and characterization of Eu^{3+}doped Lu_2O_3 nanophosphor using a solution-combustion method, J. Sol-Gel. *Sciences et Techniques (Paris)*, *53*, 171–175.

Chronakis, I. S. (2005). Novel nanocomposites and nanoceramics based on polymer nanofibers using electrospinning process--A review. *Journal of Materials Processing Technology*, *167*(2-3), 283–293. doi:10.1016/j.jmatprotec.2005.06.053

Davis, M. E. (2002). Ordered porous materials for emerging applications. *Nature*, *417*(6891), 813–821. doi:10.1038/nature00785 PMID:12075343

Diesperova, M. I., Bron, V. A., & Boriskova, T. I. (1977). Alekseeva and E.I.Kelareva. *Refractories and Industrial Ceramics*, *18*(5-6), 278.

Fu, Z., Zhou, S., Pan, T., & Zhang, S. (2007). Preparation and luminescent properties of cubic Eu3+:Y2O3 nanocrystals and comparison to bulk Eu3+:Y2O3. *Journal of Luminescence*, *124*(2), 213–216. doi:10.1016/j.jlumin.2006.02.020

Gopichandran, R., & Patil, K. C. (1992). Articles. *Materials Letters*, *12*, 437. doi:10.1016/0167-577X(92)90209-3

Hao, R. R., Fang, X. Y., & Niu, S. C. (1998). *Series of Inorganic Chemistry* (Vol. 3). Beijing: Science Press.

Hayati, I., Bailey, A. I., & Tadros, T. F. (1986). Mechanism of stable jet formation in electro hydrodynamic atomization. *Nature*, *319*(6048), 41–43. doi:10.1038/319041a0

Huang, H., Xu, G. Q., Chin, W. S., Gan, L. M., & Chew, C. H. (2002). Synthesis and characterization of Eu:Y2O3 nanoparticles. *Nanotechnology*, *13*(3), 318–323. doi:10.1088/0957-4484/13/3/316

Huang, Z., Li, B., & Liu, J. (2010). Molten-Salt Synthesis of Oxyapatite $La_{9.33}Si_6O_{26}$ Powders as Electrolytes for Intermediate Temperature Solid Oxide Fuel Cells. *Physica Status Solidi. A, Applications and Materials Science*, *207*(10), 2247–2251. doi:10.1002/pssa.200925605

Huang, Z. M., He, C.-L., Yang, A., Zhang, Y., Han, X.-J., Yin, J., & Wu, Q. (2006). Encapsulating drugs in biodegradable ultrafine fibers through co-axial electrospinning. *Journal of Biomedical Materials Research. Part A*, *77A*(1), 169–179. doi:10.1002/jbm.a.30564 PMID:16392131

Jain, S. R., Adiga, K. C., & Vernerker, V. R. P. (1981). A new approach to thermochemical calculations of condensed fuel-oxidizer mixtures. *Combustion and Flame*, *40*, 71–79. doi:10.1016/0010-2180(81)90111-5

Jingwei, X., Xiaoran, L., & Younan, X. (2008). Putting Electrospun Nanofibers to Work for Biomedical Research. *Macromolecular Rapid Communications*, *9*(22), 1775–1792. PMID:20011452

Kimizuka, N. (2008). Self-Assembled Nanomaterials. *Advances in Polymer Science*, *219*, 1–26.

Kimura, T., Takahashi, T., & Yamaguchi, T. (1980). Preparation and Characteristics of Ni- Ferrite Powders obtained in the presence of Fused Salts. *Journal of Materials Science*, *15*(6), 1491–1497. doi:10.1007/BF00752130

Kingsley, J. J., & Pederson, L. R. (1993). Combustion synthesis of perovskite LnCrO3 powders using ammonium dichromate. *Materials Letters*, *18*(1-2), 89–96. doi:10.1016/0167-577X(93)90063-4

Kingsley, J. J., & Pederson, L. R. (1993). Energetic Materials in Ceramics Synthesis. *Proceedings of the Materials Research Society*, *296*, 361. doi:10.1557/PROC-296-361

Lau, K. T., Gu, C., & Hui, D. (2006). A critical review on nanotube and nanotube/nanoclay related polymer composite materials. *Composites. Part B, Engineering*, *37*(6), 425–436. doi:10.1016/j.compositesb.2006.02.020

Li, G. G., Geng, D. L., Shang, M. M., Zhang, Y., Peng, C., Cheng, Z. Y., & Lin, J. (2011). Color tuning luminescence of $Ce^{3+}/Mn^{2+}/Tb^{3+}$-triactivated $Mg_2Y_8(SiO_4)_6O_2$ via energy transfer: Potential single-phase white-light-emitting phosphors. *The Journal of Physical Chemistry C*, *115*(44), 21882–21892. doi:10.1021/jp204824d

Li, G. G., Zhang, Y., Geng, D. L., Shang, M. M., Peng, C., Cheng, Z. Y., & Lin, J. (2012). Single-Composition Trichromatic White-Emitting $Ca_4Y_6(SiO_4)_6O$: $Ce^{3+}/Mn^{2+}/Tb^{3+}$ Phosphor: Luminescence and Energy Transfer. *ACS Applied Materials & Interfaces*, *4*(1), 296–305. doi:10.1021/am201335d PMID:22136252

Maeda, S., & Armes, S. (1995). Preparation and characterisation of polypyrrole – Tin (IV) oxide nanocomposite colloids. *Chemistry of Materials*, *7*(1), 171–178. doi:10.1021/cm00049a026

Maldiney, T., Viana, B., Gourier, D., Richard, C., Scherman, D., Bessodes, M., ... Smet, P. F. (2012). *In vivo* optical imaging with rare earth doped $Ca_2Si_5N_8$ persistent luminescence nanoparticles. *Optical Materials Express*, *2*(3), 261–268. doi:10.1364/OME.2.000261

Mao, Y., Banerjee, S., & Wong, S. S. (2003). Large Scale Synthesis of Single – Crystalline Perovskite Nanostructures. *Journal of the American Chemical Society*, *125*(51), 15718–15719. doi:10.1021/ja038192w PMID:14677942

Mao, Y., Park, T. J., Zhang, F., Zhou, H., & Wong, S. (2007). S., "Environmentally Friendly Methodologies of Nanostructure Synthesis. *Small*, *3*(7), 1122–1139. doi:10.1002mll.200700048 PMID:17554768

Moghe, A. K., & Gupta, B. S. (2008). Co-axial Electrospinning for Nanofiber structures: Preparation and Applications. *Polymer Reviews (Philadelphia, Pa.)*, *48*(2), 353–377. doi:10.1080/15583720802022257

Moghe, A. K., & Gupta, B. S. (2009). Hybrid Nanofiber Structures for Tissue Engineering. *AATCC Review*, *9*(10), 43–47.

Nain, A., Amon, C., & Sitti, M. (2005). Proximal Probes based Nanorobotic Drawing of Polymer Micro/Nanofibers. *IEEE Transactions on Nanotechnology*, *5*(5), 499–510. doi:10.1109/TNANO.2006.880453

Ni, S., Chou, L., & Chang, J. (2007). Preparation and characterization of forsterite (Mg2SiO4) bioceramics. *Ceram. Interna.*, *33*(1), 83–88. doi:10.1016/j.ceramint.2005.07.021

Park, C., Park, O., Lim, J., & Kim, H. (2001). The fabrication of syndiotactic polystyrene/organophilic clay nanocomposites and their properties. *Polymer*, *42*(17), 7465–7475. doi:10.1016/S0032-3861(01)00213-0

Prashantha, S. C., Lakshminarasappa, B. N., & Nagabhushana, B. M. (2011). Photoluminescence and thermoluminescence studies of Mg_2SiO_4:Eu^{3+} nano phosphor. *Journal of Alloys and Compounds*, *509*(42), 10185–10189. doi:10.1016/j.jallcom.2011.03.148

Rahaman. (2003). *Ceramic Processing and sintering* (2nd ed.). Marcel Dekker.

Ramachandra Naik, S. C. (2014). Article. *Biological Chemistry*, *195*, 140–149.

Ramakrishna, S., Fujihara, K., Teo, W. E., Lim, C. T., & Ma, Z. (2005). *An introduction to Electrospinning and Nanofibers, Basics related to Electrospinning*. Singapore: World Scientific Publishing Co. Pvt. Ltd. doi:10.1142/5894

Ramakrishna, S., Fujihara, K., Teo, W.-E., Yong, T., Ma, Z., & Ramaseshan, R. (2006). Electrospun nanofibers: Solving global issues. *Materials Today*, *9*(3), 40–50. doi:10.1016/S1369-7021(06)71389-X

Reneker, D. H., & Chun, I. (1996). Nanometre diameter fibres of polymer produced by electrospinning. *Nanotechnology*, *7*(3), 216–223. doi:10.1088/0957-4484/7/3/009

Reneker, D. H., & Yarin, A. L. (2008). Electrospinning jets and polymer nanofibers. *Polymer*, *49*(10), 2387–2425. doi:10.1016/j.polymer.2008.02.002

Retricevic, V., & Gayen, S. K. (1998). Article. *Applied Physics Letters*, *52*, 1040.

Revathi, V. (2014). Article. *Chin Shu Hsueh Pao*, *27*, 557–562.

Roco, M. C. (1999). Nanoparticles and nanotechnology research. *Journal of Nanoparticle Research*, *1*(1), 1–6. doi:10.1023/A:1010093308079

Santhanam, R., & Rambabu, B. (2010). Research Progress in High Voltage Spinel $LiNi_{0.5}Mn_{1.5}O_4$ Material. *Journal of Power Sources*, *195*(17), 5442–5451. doi:10.1016/j.jpowsour.2010.03.067

Saravanapavan & Hench. (2003). J. Non-Crys. *Soli.*, *318*, 14.

Sunitha, D. V., Nagabhushana, H., & Fouran Singh, B. M. (2012). Thermo, Iono and photoluminescence properties of 100 MeV Si^{7+} ions bombarded $CaSiO_3$:Eu^{3+} nanophosphor. *Journal of Luminescence*, *132*(8), 2065–2071. doi:10.1016/j.jlumin.2012.03.019

Taylor, G. (1969). Electrically driven jets. *Proceedings of the Royal Society of London. Series A, Mathematical and Physical Sciences*, *313*(1515), 453–475. doi:10.1098/rspa.1969.0205

Teo, W. E., & Ramakrishna, S. (2006). A review on electrospinning design and nanofibre assemblies. *Nanotechnology*, *17*(14), R89–R106. doi:10.1088/0957-4484/17/14/R01 PMID:19661572

Tsai, M. T. (2002). Synthesis of nanocrystalline forsterite fiber via a chemical route. *Materials Research Bulletin*, *37*(13), 2213–2226. doi:10.1016/S0025-5408(02)00926-1

Venkatachari, K. R., Huang, D., Ortrander, S. P., Schulze, W. A., & Stangle, G. C. (1995). A combustion synthesis process for synthesizing nanocrystalline zirconia powders. *Journal ofMaterials Research*, *10*(03), 748–755. doi:10.1557/JMR.1995.0748

Wang, J., & Xu, Y. (2009). Mirrabos Hojamberdiev, Jianhong Peng, Gangqiang Zhu, A facile route to synthesize luminescent YVO_4:Eu^{3+}. *J. Non-Cryst. Solid.*, *355*, 903–907.

Wang, Y., Zheng, M., Lu, H., Feng, S., Ji, G., & Cao, J. (2010). Template Synthesis of Carbon Nanofibers Containing Linear Mesocage Arrays. *Nanoscale Research Letters*, *5*(6), 913–916. doi:10.100711671-010-9562-9 PMID:20671793

Wickham. (1971). The Preparation of Ferrites with the Aid of Fused Salts. *Proceedings of the International Conference*, 105-107.

Xia, W., & Chang, J. (2008). Preparation and the phase transformation behavior of amorphous mesoporous calcium silicate. *Microporous and Mesoporous Materials*, *108*(1-3), 345–351. doi:10.1016/j.micromeso.2007.04.013

Yan, B., & Lei, F. (2010). Molten Salt Synthesis, Characterization and Luminescence of $ZnWO_4:Eu^{3+}$ Nanoparticles. *Journal of Alloys and Compounds*, *507*(2), 460–464. doi:10.1016/j.jallcom.2010.07.203

Yoon, K. H., Cho, Y. S., & Kang, D. H. (1998). Molten Salt Synthesis of Lead-based Relaxors. *Journal of Materials Science*, *33*(12), 2977–2984. doi:10.1023/A:1004310931643

Zhang, C. M., Huang, S. S., Yang, D. M., Kang, X. J., Shang, M. M., Peng, C., & Lin, J. (2010). Tunable luminescence in Ce^{3+}, Mn^{2+}-codoped calcium fluorapatite through combining emissions and modulation of excitation: A novel strategy to white light emission. *Journal of Materials Chemistry*, *20*(32), 6674–6680. doi:10.1039/c0jm01036g

Zhang, S., Gelainb, F., & Zhao, X. (2005). Designer self-assembling peptide nanofiber scaffolds for 3D tissue cell cultures. *Seminars in Cancer Biology*, *15*(5), 413–420. doi:10.1016/j.semcancer.2005.05.007 PMID:16061392

Zhang, Y. (1994). Article. *Journal of Materials Research*, *9*, 1997. doi:10.1557/JMR.1994.1997

Zhao, D., Feng, F., Quo, Q., Melosh, N., Fredrickson, G. H., Chmelka, B. F., & Stucky, G. D. (1999). Triblock Copolymer Syntheses of Mesoporous Silica with Periodic 50 to 300 Angstrom Pores. *Science*, *279*(5350), 548–552. doi:10.1126cience.279.5350.548 PMID:9438845

Zhao, J., Han, W., Chen, H., Tu, M., Zeng, R., Shi, Y., ... Zhou, C. (2001). Preparation, structure and crystallinity of chitosan nano-fibers by a solid–liquid phase separation technique. *Carbohydrate Polymers*, *83*(4), 1541–1546. doi:10.1016/j.carbpol.2010.10.009

Chapter 8

Perspectives of Hydrothermal Synthesis of Fluorides for Luminescence Applications: Fluorides Phosphors for Luminescence

Nilesh Sureshrao Ugemuge
Priyadarshini College of Engineering, India

Snajay J. Dhoble
R. T. M. Nagpur University, India

ABSTRACT

Hydrothermal synthesis is an easy, portable, less-hazardous, and low-cost synthesis method. Various researchers across the globe are worked on the synthesis of different materials via this route. Practically, fluorides are difficult to synthesize due to their hygroscopic nature by conventional methods. But, the hydrothermal synthesis is used to prepare several compositionally optimized fluoride-based materials using closed-system physical and chemical processes in an aqueous solution at low temperatures and pressures. The silent features of the hydrothermal method over conventional methods of materials processing are a crystallization of materials, crystal growth, in the processing of a wide range of materials not only the bulk crystals but fine particles with a controlled size and morphology. Therefore, in order to place its numerous recent developments, past and current research come together in this chapter. This chapter is a recent clocking update for synthesis, materials, and their applications.

INTRODUCTION

The term Hydrothermal was coined from two words, "hydro" means water and "thermal" means heat and describe the action of water at elevated pressure and temperature (Byrappa, 2001). It is a method used to produce different materials by closed-system physical and chemical processes in an aqueous solution at temperatures above 100 °C and pressures above 1 atm. However, there is no definite lower limit for the temperature and pressure conditions. Hydrothermal synthesis is an advanced material pro-

DOI: 10.4018/978-1-5225-5170-6.ch008

cessing technique, which has numerous advantages than other methods. With the help of aforementioned method, one can prepare high-quality crystals of materials which are even unstable near the melting point. Homogeneous materials of pure phase are obtained, offers high diffusivity, low viscosity, and mass transport. By tailoring the chemical environment, it gives control over the shape and size of the crystals, consumes less energy, less reaction time and is a safe method with no environmental hazards as it's a closed system processing technique. There are various other terms preferred by chemists like solvothermal which means a chemical reaction in the presence of a non-aqueous solvent or solvent in supercritical or near supercritical conditions. Similarly, several other terms like glycothermal, alcothermal, ammonothermal, carbonothermal and lyothermal are also used (Yoshimura, 2008). Among the various synthesis techniques, a hydrothermal synthesis route is a simple approach for synthesizing various inorganic materials with controllable morphologies and architecture (Jia, 2009; Xu, 2010). The temperature in hydrothermal synthesis is low compared to most traditional melt techniques of synthesis. A great variety of materials like metal oxides, silicates, hydroxides, sulfides, carbonates, phosphates, tellurides, nitrides, selenides, $KTiOPO_4$ (Bierlein, 1976), tungstates (Reis,1990), Ti- superconductors (Chen, 1994), layered compounds (Sugita, 1990), artificial gems (Zhao, 1997; Hosaka, 1991), intercalation compounds (Whittingham, 1995) and zeolites (Barrer, 1982) have been synthesized using the hydrothermal technique. A wide variety of combinations of aqueous and solvent mixture systems can be used in hydrothermal processing. Moreover, compared to solid state processes, the possibility for acceleration of diffusion, higher dispersion, controlled morphology, better nucleation, high reactivity, adsorption, reaction rate and crystallization in the liquid is more (Yoshimura, 2000). The hydrothermal process is a method of forming ceramic powders by heating and pressurizing solutions or suspensions of metal salts, oxides, hydroxides or metal powders in water or in other solvents. The process is conducted in a pressure vessel called an autoclave. Normally, as the mixture is heated in this closed system, the pressure rises and these conditions result in the formation of submicron particles with controlled size and shapes. This route has been used to prepare a variety of monodispersed powders, materials with specific particles with controlled size and shapes, crystalline phases and frequently at temperatures lower than those required in normal processing procedures. The formation of monodispersed spheres by providing the appropriate conditions needed for homogeneous nucleation which only assisted by the hydrothermal process. For example, using heat and pressure in an autoclave, monodispersed metal oxide powders are formed by homogeneous nucleation of metal hydroxide particles, which are produced by forced hydrolysis of metal alkoxides. By controlling the release of precipitating ion, forced hydrolysis is normally achieved. The advantage of the hydrothermal process is that low-quality reactants can be used because metallic impurities remain in solution. Also, a variety of reaction conditions such as pH, temperature, and pressure controls the particle size. The relatively low reaction temperature, around 300 ^{o}C and ability to use impure reactants are attractive features for commercial exploitation of this technique. In recent times, hydrothermal synthesis has been used for preparing fluoride phosphors (You, 2002).

Hydrothermal synthesis is typically the chemical reactions in water/solvents conducted above the ambient temperature and pressure in a sealed or closed system. It is a special type of chemical transport reaction that relies on the liquid-phase transport of reactants to form the desired product. Under autogenous conditions, water functions both as a pressure-transmitting medium and as a solvent. In a sealed vessel, as the temperature increases, the vapor pressure of water increases below its critical temperature but above its boiling temperature (Rabenau, 1969). The hydrothermal technique has its place in several branches of science and technology today where its roots attached to the appearance of several other related techniques. In the last few years many luminescent materials have been discovered and applied

Figure 1. Number of papers/patents on hydrothermal research in materials (Internet sources)

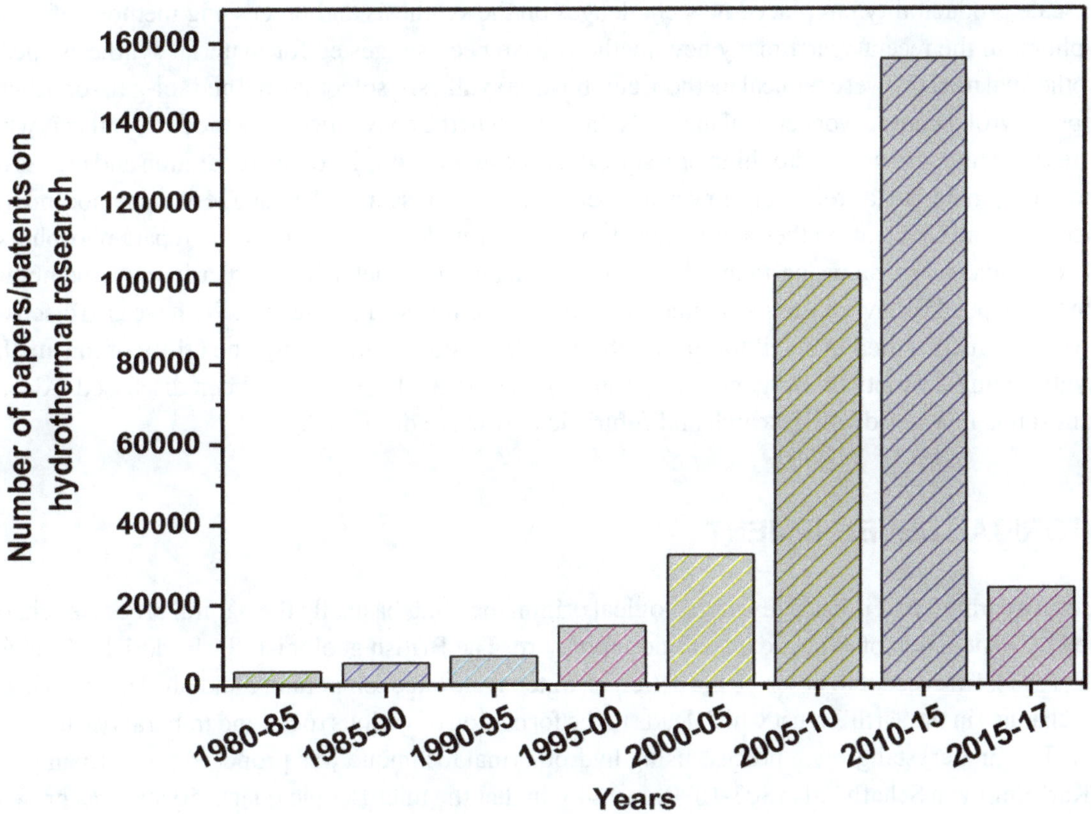

Figure 2. Contribution of states in hydrothermal research in material science (You, 2002)

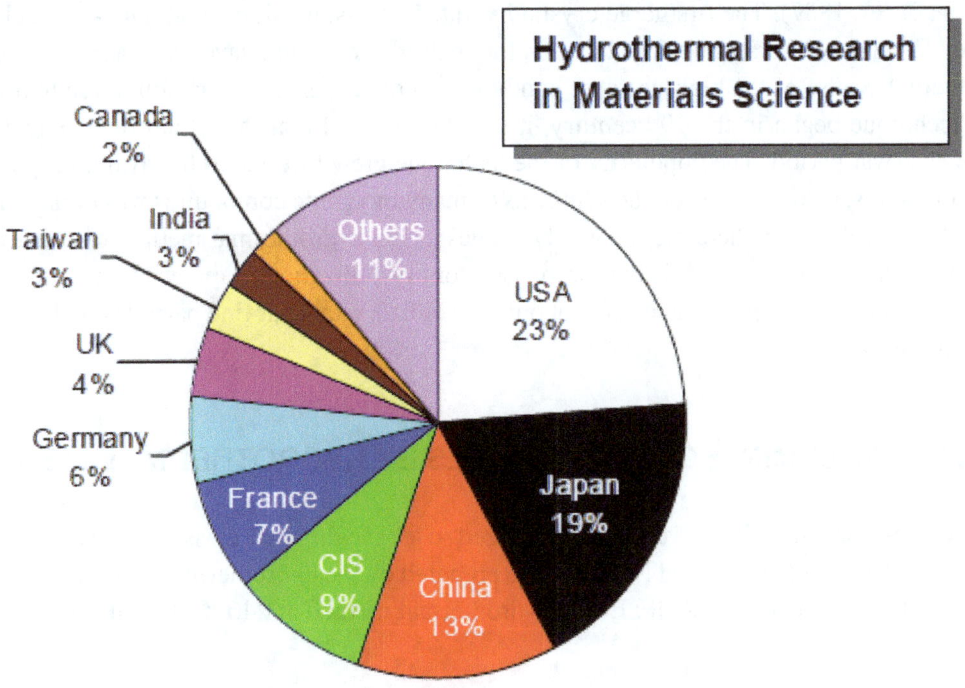

in various applications but the demand for high efficiency, high quality, longer lifetime, low cost and large scale producibility has placed new challenges on the synthesis and processing methods of known phosphors. In the recent years, many new methods have been suggested for material synthesis such as hydrothermal method, wet chemical method, combustion synthesis, solvothermal and sol-gel, sonochemical, spray pyrolysis and cryochemical methods. In contrast to the conventional method of synthesis which requires high temperatures and/or high pressure, these novel methods involve preparation and processing of inorganic materials at relatively lower temperatures and pressures. The suggested method provide milder route and efficient synthesis of materials but also provide an opportunity to prepare phosphors of stable colloidal suspensions, particles in the submicron and nanometer scale, microporous membranes etc. In this paper. We have discussed synthesis methods of luminescent materials. We have also discussed the hydrothermal method in detail for the synthesis of fluorides. Additionally, the utility of this method for synthesizing a variety of inorganic compounds including the fluorides has been discussed. Also, its current status in a worldwide research and futuristic outfits are discussed.

HISTORICAL DEVELOPMENT

The term hydrothermal is basically of geological origin, meaning basically the natural processes characterized by joint action of heat and water under pressure. The British geologist, Sir Roderick Murchison (1792-1871), first used it to describe the action of water at elevated temperature and pressure in bringing about changes in the earth's crust which leads to the formation of various rocks and minerals (Murchison, 1840). The first crystal growth process using hydrothermal technique was proposed by German geologist Karl Emil von Schafhäutl (1803-1890) in 1845, in that the microscopic quartz crystal was grown in a pressure cooker (Schafhäutl, 1845). By 1900, more than 150 mineral species were synthesized using hydrothermal technique including diamond (Morey, 1913). Commercial application of the hydrothermal technique began in 1908 when K.J. Bayer used bauxite mineral under hydrothermal conditions to obtain Aluminium (Bayer, 1887). The first large crystal hydrated potassium silicate, about 2–3 mm long, was obtained by Friedel and Sarasin (1881). Further, the hydrothermal autoclave was named hydrothermal bomb by them, because of the high-pressure conditions in the experiment. Although the hydrothermal synthesis technique began in the 19th century, it should be noted that the actual development occurs during the post-war period. The popularity of the technique grew fast due to the availability of the improved equipment's, as knowledge on the synthesis of many inorganic compounds was obtained through other established routes. As there are many advantages of the hydrothermal method over conventional methods of materials processing, it is popularly used for crystallization of materials, crystal growth, in the processing of a wide range of materials not only the bulk crystals but fine particles with a controlled size and morphology.

SYNTHESIS OF LUMINESCENT MATERIALS BY HYDROTHERMAL METHOD

Several fluoride-based materials have been synthesized so far and these are classified as well as discussed in this section. NaGdF and Eu-doped NaGdF were prepared by the hydrothermal method (Kang, 2001). Youa et al.,(2002) synthesized NaGdF: Eu by hydrothermal method at 240 °C. $LiYbF_4$: Er^{3+} octahedral microcrystals have been successfully prepared through an EDTA (ethylenediamine tetraacetic acid) assisted facile hydrothermal method (Lu, 2011). The hydrothermal method was employed to prepare Pr^{3+} ions doped $LiYF_4$ (Wang, 2007). By using the hydrothermal technique, $BaYF_5:Yb^{3+}$, Er^{3+} (BYF) upconversion (UC) luminescence nanoparticles have been prepared (Du, 2011). Synthesis of BaF_2 nano-crystal and Ce doped BaF_2 nanoparticles has been reported by hydrothermal reaction (Gao, 2006; Zhang, 2008). The complex fluorides KMF_3 (M = Mg, Ca, Zn, Mn, Ni and Co) have been prepared by a hydrothermal technique (Li, 2000). Uniform and well-dispersed SrF_2 hierarchical microspheres have been successfully synthesized by a facile and friendly hydrothermal method (Zhang, 2010). $BaMgF_4:RE^{3+}$ (RE=Eu, Tb, Sm, Dy, Yb-Er/Tm) micro phosphors had firstly prepared by the hydrothermal process (Yan, 2013). In the presence of different surfactants, micro and nanocrystals cubic-phase $SrF_2:Yb^{3+}/Er^{3+}$ upconversion luminescence phosphors were synthesized via a facile hydrothermal route (Jiayue, 2011). Fujihara et al. reported the aqueous synthesis of $BaMgF_4$ particles. This route was also employed to prepare $SrAlF_5$ (Hua, 2002) Since the 1970s, $LiBaF_3$ prepared with solid state reaction by LiF-BaF_2 as a raw material (Yan, 2008). The complex fluorides KMF_3 (M = Mg, Ca, Zn, Mn, Ni, and Co) prepared by the conventional solid-state reaction (Somiya, 1981; Dzik, 2000). A new non- linear crystal Ba(Al, Ga)BO_3F_2 has been synthesized by doping with Ga in $BaAlBO_3F_2$ (Wang, 2011) The high-performance fluoride ion conductor $BaSnF_4$ has been prepared by solid-state reaction (Patro, 2011). A persistent phosphor of $KY_3F_{10}:Tb^{3+}$ synthesized by solid-state reaction for x-ray or cathode-ray tubes reported by Jinsu Zhang et al. (2009). A well-known laser host $LiYF_4$ was also prepared by Boyer et al. (2004). Chemical co-precipitation has been used to fabricate oxide-based and fluoride-based phosphors (Zhang, 2010; Pan, 2003; Zhou, 2002). $BaYF_5:Yb^{3+}$, Er^{3+} (BYF) upconversion (UC) luminescence nanoparticles have been prepared using co-precipitation techniques (Zhang, 2008). BaF_2:Ce nano phosphors were synthesized using a chemical precipitation method with oleic acid as the capping ligands (Kang, 2010). Nanocrystalline LiF:Mg,Cu, Phosphor material have been prepared by the chemical co-precipitation method (Sahare, 2010). For obtaining oxide films, acetates of different cations are widely used as substrates, although also sulfides can be obtained. Using the spray pyrolysis method, many TL materials, both in film and powder, have been prepared successfully (Azorin, 2002; Hipolito, 2001; Azorin, 1998; Lee, 2001). High sensitive $Y_2O_3:Eu^{3+}$, $Gd_2O_3:Eu^{3+}$, $(Y_xGd_{1-x})O_3:Eu^{3+}$, $YAG(Y_3A_{15}O_{12})$: Eu^{3+} and $(Y_xGd_{1-x})BO_3:Eu^{3+}$ luminescence display phosphors have been made by pyrolysis route followed by sintering at a relatively low temperature of 1000 °C (Sivakumar, 2012). Well-dispersed commonly used LED yellow phosphors YAG:Ce can be synthesized by spray pyrolysis method (Purwanto, 2007; Mancic, 2007). Synthesis of BaF_2 nanocrystal and Ce doped BaF_2 nanoparticles has been reported by microemulsion technique (Hua, 2003; Lian, 2004). The complex fluorides KMF_3 (M = Mg, Ca, Zn, Mn, Ni and Co) have been prepared by microemulsion method (Hua, 2007; Cao, 2004). Efficient laser in Nd^{3+} doped $LiGdF_4$ has been reported (Zhang, 1993), visible quantum cutting was observed in $LiKGdF_5:Eu^{3+}$, (Kodama, 2005) $LiGdF_4$:Eu (Wegh, 1999) was synthesized by ammonium complex method.

In a typical preparation of $NaGdF_4$ phosphor by hydrothermal synthesis, the starting materials NaF, Gd_2O_3, Eu_2O_3, NH_4HF_2, H_2O were taken in appropriate ratio and reaction mixture at pH = 4 was sealed in a Teflon lined stainless steel autoclave and heated at 240 °C for 2 days. After the autoclave was cooled and depressurized the products are washed, dried and then heated at 950 °C for 1hr. The resultant powder indicates the formation of hexagonal $NaYF_4$.

METHODS OF SYNTHESIS OF LUMINESCENT MATERIALS

There are various methods for the synthesis of luminescent materials, looking at the chemical formula of most of the phosphors, one may feel that the synthesis of the luminescent materials should be straightforward as the host materials are well known. However, synthesis of phosphors is not straightforward as it appears from their chemical formula. There are many constraints in the synthesis of phosphor while using it for any particular application. The constraints mostly are due to the physical and chemical properties of initial ingredients used, limitations of a particular method of synthesis, incorporation of the activator at the desired sites, elimination of the unwanted impurities, specific grain size and morphology suitable for the application, cost of production etc.

SOLID STATE REACTION (SSR)

The solid-state reaction (SSR), which occurs between powders in the solid state, has been the most widely and intensively used synthesis approach for the synthesis of phosphors including oxides, (oxy) nitrides, and fluorides because it is suitable for mass production of polycrystalline bulk materials and comparatively simple than other routes of synthesis. Since solids do not react with constituents at room temperature, it is necessary to heat them at elevated temperatures as high as up to 1500^{0}C for the proper reaction to take place at an appreciable rate. In SSR method, the solid reactants react chemically without the presence of any solvent at high temperatures and form the stable product. Also, synthesis of oxide-based phosphors like YAG by conventional solid-solid reaction requires temperature higher than 1000 °C, because of the refractory nature of rare-earth oxides and alumina (Ye, 2010). Atom diffusion between different raw materials controls the solid-state reaction. Hence, repeated grinding and repeated heating are required. A controlled atmosphere is also required to master the stoichiometry of the host lattice and valence of the activator. For the synthesis of complex fluorides, the solid-state reaction is commonly used to avoid possible contamination from oxygen, this method requires complicated set-up and fluoric atmosphere (Wegh, 1997). The desired properties of synthesized products depend upon the final sintering temperature. By the conventional solid-state method, single phase compound is normally difficult to prepare. Hence, low concentration doping of the order of 1–3% of activator has always been delicate in SSR.

Wet Chemical Method

The wet-chemical method is the solution-based chemical synthesis, which is defined as a technique which does not comprise of the normal mixing of product, calculations, and grinding (Rao, 1994; Zhang, 2007). Simple oxide phosphor synthesis such as Eu^{3+} activated Y_2O_3 red phosphor is possible via wet-chemical technique. Wet-chemical methods such as co-precipitation, sol-gel, hydrothermal, evaporative decomposition, reverse micellar, surface modifier etc. have been employed to synthesize lamp phosphors activated with rare earth and received considerable attention since they offer the possibilities for controlling homogeneity, purity of phase, size distribution, surface area, and microstructural uniformity of the phosphors. Sol-gel and hydrothermal methods have been successfully employed to synthesize various phosphors. These wet-chemical methods dope rare earth activators uniformly. But, calcination is required to get crystalline (required) phosphors. To prevent the hydrolysis, use of chlorides as a starting material or using 'freshly' prepared fluorides in the initial procedure has been suggested. It will be highly desirable to have the complex fluoride powders as starting materials for crystal growth. X-ray diffraction studies of some of the complex fluorides like $LiCaAlF_6$, $CaAlF_5$, $BaAlF_5$ etc. shows that these compounds formed immediately after precipitation during the wet chemical method. It is 'surprising' that such 'complex' fluoride systems were formed at room temperature itself with a simple method of preparation. This may be due to the formation of 'molecules' during the precipitation process itself. The wet chemical processes can produce high-quality final products in the form of powders; the production costs are much higher than in calculations as wet chemical methods require stringent drying conditions, high pH sensitivity, use of complex equipment and expensive precursors/ disposal of the by-product.

SOL-GEL SYNTHESIS

Sol-gel synthesis is a solution based chemical process of synthesizing wide range of materials especially mixed oxides, which is used due to its advantages of low temperature synthesis, flexible nature, etc. The control on stoichiometry of the resultant product is benefit of this method. A sol is a stable dispersion of colloidal particles or polymers in a solvent. The particles may be crystalline or amorphous in nature. A gel consists of a three dimensional network, which can enclose the liquid phase, in a colloidal form of gel, the network is consists of agglomeration of colloidal particles. The sol-gel process is an efficient method, which uses the same type of reactions among the precursors as used in other methods, followed by condensation to yield a polymeric network. The sol or gel is applied to the substrate and then thermal treatment is given to remove undesired organic matter from the product (Sahare, 2010). In such a procedure the decomposition of complexes is one of the convenient ways to release the necessary precursors to particle precipitation. Sol-gel process offers many advantages for producing fluoride materials: (i) the level of purity can be controlled rather than being limited to a fluoride precursor, (ii) viscous sintering below the crystallization temperature should eliminate formation of crystalline defects, (iii) thin films, fibers and monoliths can be made, giving it broader range of applications.

CHEMICAL CO-PRECIPITATION

In this method, all the raw materials are dissolved in solutions and all the cations are co-precipitated by adding certain reagents. Final precipitates are separated and heat treated to decompose the precursor and finally annealed to crystallize the powder. Co-precipitation synthesis is a simple method for the preparation of complex fluorides because it does not require expensive and high-purity starting materials as RE fluorides and results in an inhomogeneous distribution of dopant ion and smaller grains. However, repeated washing is required to eliminate the anions coming from the precursor salts used, this makes the process time-consuming and complicated. Calcination is usually required to get fine crystalline phosphors. Since the precipitants are formed slowly and uniformly throughout the solution during the precipitation process, particles of uneven size can be prepared. In co-precipitation, control over crystallite size, particle size distribution, morphology is difficult to achieve. The difficulty in maintaining chemical homogeneity is serious and often produces a detrimental effect on the properties of the phosphors. Another problem is that the final product is generally agglomerated. Therefore, grinding, dry milling or wet milling with water or non-aqueous liquid is used for particle size reduction. These processes can introduce impurities from grinding media.

SPRAY PYROLYSIS

Spray pyrolysis is a technique to synthesize pure and homogeneous ceramic materials, wherein the principle of this method is that when a droplet of the spray solution reaches the hot substrate, fine powders would be deposited due to the pyrolytic decomposition of the solution. By this method, homogeneous powders with a particle size less than 100 nm of oxide may be produced, which allows tailoring materials on a nano-scale. As a result of solvent vaporization and thermal decomposition, powders with the complex composition can be synthesized. In this case, a spray formed by a carrier gas (inert to the system) and the solution containing the substance to decompose on the surface of the substrate to yield the desired film, is impinged on a substrate of a material which is heated up to the appropriate and desired temperature. It must be used with substances which decompose at relatively low temperatures compared to others and do not let other solid waste than those must form the film (Gurav, 1993). Although, spray pyrolysis method is a fast and economical technique, but has the certain disadvantage of requiring precise control of concentration, flow, spray height, substrate temperature, etc. parameters. In powder and film form, many thermo luminescence (TL) materials, can be prepared successfully using this method.

MICROEMULSION METHOD

Microemulsions are isotropic, macroscopically homogeneous and thermodynamically stable solutions containing at least three components, namely a nonpolar phase (usually oil), polar phase (usually water), and a surfactant (Hu, 2009). On a microscopic level, the surfactant molecules form an interfacial film separating the non-polar and the polar domains from each other. This interfacial layer forms different microstructures ranging from droplets of oil dispersed in a continuous water phase (O/W-microemulsion) over a bicontinuous ''sponge'' phase to water droplets dispersed in a continuous oil phase (W/O-microemulsion). Unique properties of microemulsion method are a large interfacial area, thermodynamic stability, ultralow interfacial tension and the ability to solubilize otherwise immiscible liquids. The microemulsion technique controls the particle properties such as mechanisms of particle size control, morphology, geometry, surface area, and homogeneity (Malik, 2012).

REACTIVE ATMOSPHERIC PROCESSING (RAP)

The stoichiometry of fluorides can be preserved by preparing the crystals in 'reactive' atmosphere instead of the inert atmosphere for example fluoride crystals prepared in fluorine atmosphere will prohibit the substitution of F^- by other impurity ions. In the simplified, modified RAP, the phosphor ingredients are heated with an appropriate RAP agent like Teflon, NH_4F, HF, NH_4Cl, CCl_4 etc. in a closed glass tube at about 450 °C for 1 hour. After heating, it is then transferred into a graphite crucible for air quenching, and post-heated to a suitable temperature. The reacted ingredients or molten phosphor are then quickly quenched to room temperature. In this process, a vacuum is not required but still, it produces phosphors, which are as good as those produced by other sophisticated techniques. The procedure was found suitable for producing Cu^+ doped phosphors, which could not be prepared by preparation in the air or inert atmosphere.

AMMONIUM COMPLEX METHOD

Ammonium complex method includes fluorination of oxides of rare earths using ammonium hydrogen fluoride at room temperature, which forms some ammonium complex and decomposition of this complex at high temperature results into complex fluorides like alkaline rare earth fluorides, trifluorides and alkali rare earth fluorides. The above described method needs a careful controlling of temperature during synthesis as the slight variation in temperature can form different phase of ammonium complex.

COMBUSTION SYNTHESIS

The principle concept of this technique is that, once initiated, a highly exothermic reaction can become self- sustaining and will propagate in the form of a combustion wave through the reactant mixture. For this to happen, a chemical reaction must have relatively high activation energy and must also generate a sufficient amount of heat. The synthesis of refractory high-temperature materials (e.g. nitrides, borides, carbides, and silicides) from elemental constituents exhibits both of the characteristics mentioned above.

SELF-PROPAGATED HIGH TEMPERATURE SYNTHESIS (SHS)

Self-propagated high-temperature synthesis (SHS) synthesis reaction occurs generally through ignition, propagation of combustion wavefront and cooling. Various heat sources used to ignite samples are, for instance, tungsten coil (Yeh, 2004), electric arc, electrically heated Mo filament, a magnesium ribbon, laser pulse (Bertolino, 2003), highly exothermic reaction, rapid heating in a furnace or microwave energy (Ganesh, 2005). Combustion reaction can be induced either by a self-propagating mode or by a thermal explosion mode. The SHS process with the advantages of energy and time savings has been recognized as an alternative method to the conventional methods of producing advanced materials, such as carbides, borides, nitrides, hydrides, and intermetallics etc. (Cao, 2008).

MICROWAVE INITIATED SSM

The result of microwave initiated ignition usually occurs at the center of the body, with the radially outward propagation of combustion wavefront. This unconventional result leads to a number of differences compared with conventionally ignited SSM reactions by other means. These include crude control of the wavefront propagation, ignition in both weakly exothermic systems and denser green bodies and the generation of different microstructures owing to dissimilar time–temperature and spatial temperature profiles. This technology extends to the range of materials and compositions that can be produced in a self-propagating manner by other methods. Commercially important developments are likely to be those that utilize these features of microwave assisted SSM to produce tailored microstructures for niche applications (Peng, 2002).

HYDROTHERMAL SYNTHESIS

Any heterogeneous reaction in the presence of aqueous solvents or mineralizers under high pressure and temperature conditions to dissolve and recrystallize (recover) materials that are relatively insoluble under ordinary conditions, is known as hydrothermal processing (Byrappa, 2007). Chemical reactions, precipitation, crystal growth, and modification of solid structures by ion exchange in the presence of water as an active solvent at high-pressure conditions lead to the formation of materials with unique properties which cannot be easily obtained by other methods (Hlavacek, 1996). The technique requires short reaction time. The advanced material, whether it is bulk single crystals, or fine particles, or nanoparticles, can be processed by the hydrothermal technique. Hydrothermal treatment is advantageous in the preparation of various inorganic micro and nanomaterials with less pollution, high homogeneity, and high purity (Chen, 2009). Feng et al., (1996) proposed a hydrothermal approach for the synthesis of complex fluorides. Activated carbon (AC) composite with HKUST-1 metal-organic framework (AC-HKUST-1 MOF) was prepared by the ultrasonically assisted hydrothermal method (Azad, 2016).

HYDROTHERMAL SOLVENTS

At the critical point and at above 100 °C temperature and above 1 atm. pressure, the hydrothermal solvents exhibit different properties. The properties of the solvent under hydrothermal temperature and pressure conditions must be understood to comprehend the hydrothermal reactions. Water is an important solvent which has remarkable properties as a reaction medium under hydrothermal conditions, where it acts very differently at standard conditions. Water is very volatile, therefore, removal from the product is quite easy. It is cheap compared to other solvents and is non-toxic as well as environment-friendly. Water is thermodynamically stable and is nonflammable which makes it advantageous. In nature, also, water plays an important role in the formation of various rocks and minerals and in the creation of life (origin of life). Water is a polar solvent and the advantage of using water over other solvents is that temperature and pressure can modify and control its polarity. A supercritical fluid is defined as the liquid, whose density decreases with increase in temperature and a critical point, its density is same as gas. Above critical point, it is neither liquid nor gas any longer and becomes a supercritical fluid. For hydrothermal synthesis, supercritical fluids and water are excellent media as density and dielectric constants can be varied with pressure and temperature which facilitates controlled particle size and increased reaction rates. Figure 3 shows the kinetics of hydrothermal synthesis around the critical point due to the variation of properties of water due to change in temperature. Figure 4 shows a variation of dielectric constant of water with temperature and pressure (Uematsu, 1976). The parameters like solubility and dielectric constant vary with temperature. At the critical point, the decrease of solubility is due to the lower density and dielectric constant (Schafhäutl, 1845).

The density of water in the supercritical state above the critical temperature (374 °C) and pressure (22.1MPa) varies greatly with minute variations in temperature and pressure. The change in density results in a change in all the fluid properties around the critical point, including dielectric constant which controls the rate of reaction, equilibrium, and solubility. At room temperature, the dielectric constant of water is nearly 78, but it goes on decreasing as the temperature rises around the critical point, to around 2 to 10, which is similar to that of solvents like ethanol, acetone, etc., which are organic in nature (Adschiri, 2010). As a green solvent, water plays an important role in atom- economical reactions

Figure 3. Kinetic of supercritical hydrothermal synthesis [Adschiri, internet source]

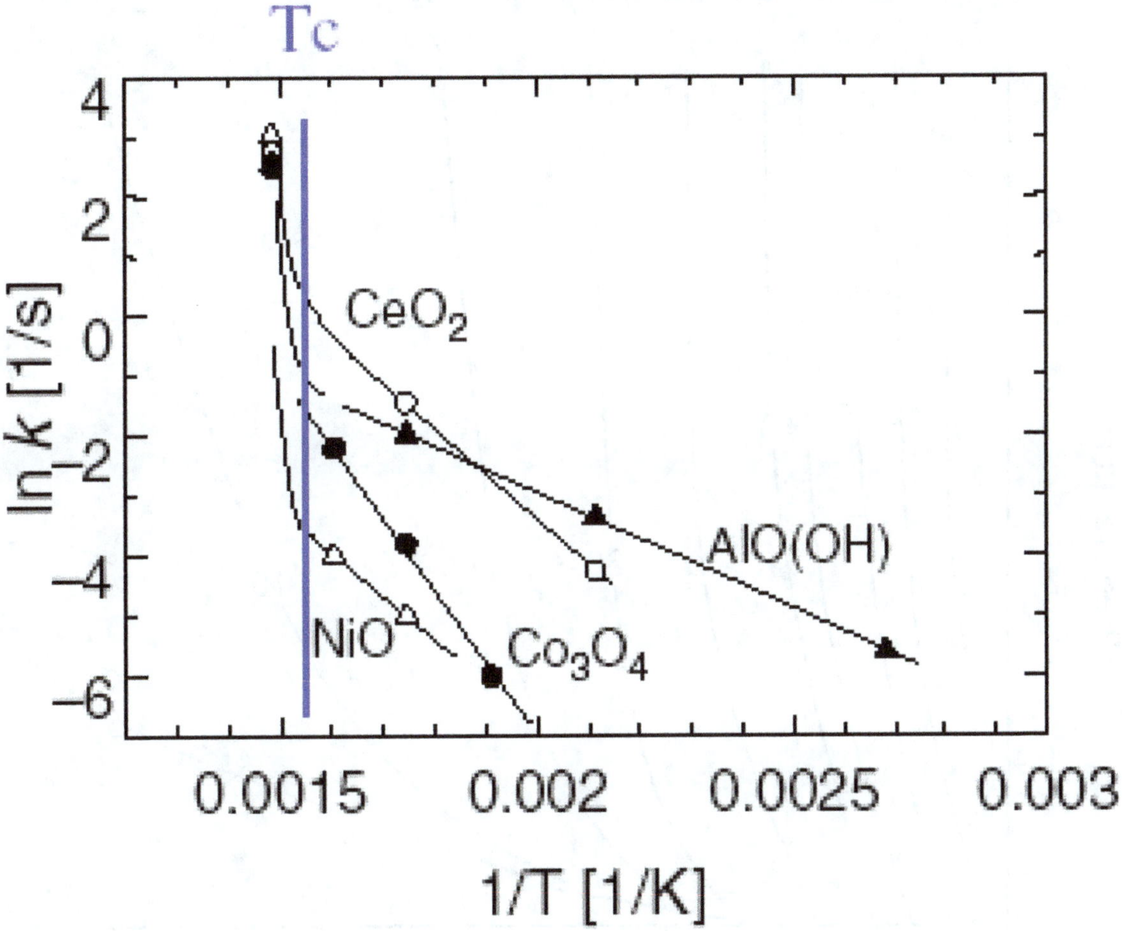

and typically chemical transport and soluble reactants of these complexes at elevated temperatures lead to crystallite growth. In aqueous solution, the behavior of solid metal oxides and their corresponding cations is complex, but nonetheless, it is important to understand metal speciation under a wide range of aqueous conditions, including hydrothermal conditions. Further, the comparison of solvents used and their actions are summarized in Table 1 and Table 2, respectively.

AUTOCLAVE FOR HYDROTHERMAL SYNTHESIS

An autoclave is a pressure vessel capable of sustaining and sterilizing solvents at high temperature and pressure for longer time duration (Black, 1993). These autoclaves are designed for use in hydrothermal synthesis at very high pressures and temperatures. Their cone/cone sealing makes them safe for withstanding temperatures of up to 250 °C and are lined with Teflon insert. These autoclaves should withstand highly corrosive solvents for reaction. The pressure and temperature conditions and corrosive nature of the solvent in the experiment should be considered before selecting proper autoclave for synthesis. Due

Figure 4. Variation of the dielectric constant of water with temperature and pressure, where marked numbers denote the dielectric constant [Adschiri, internet source]

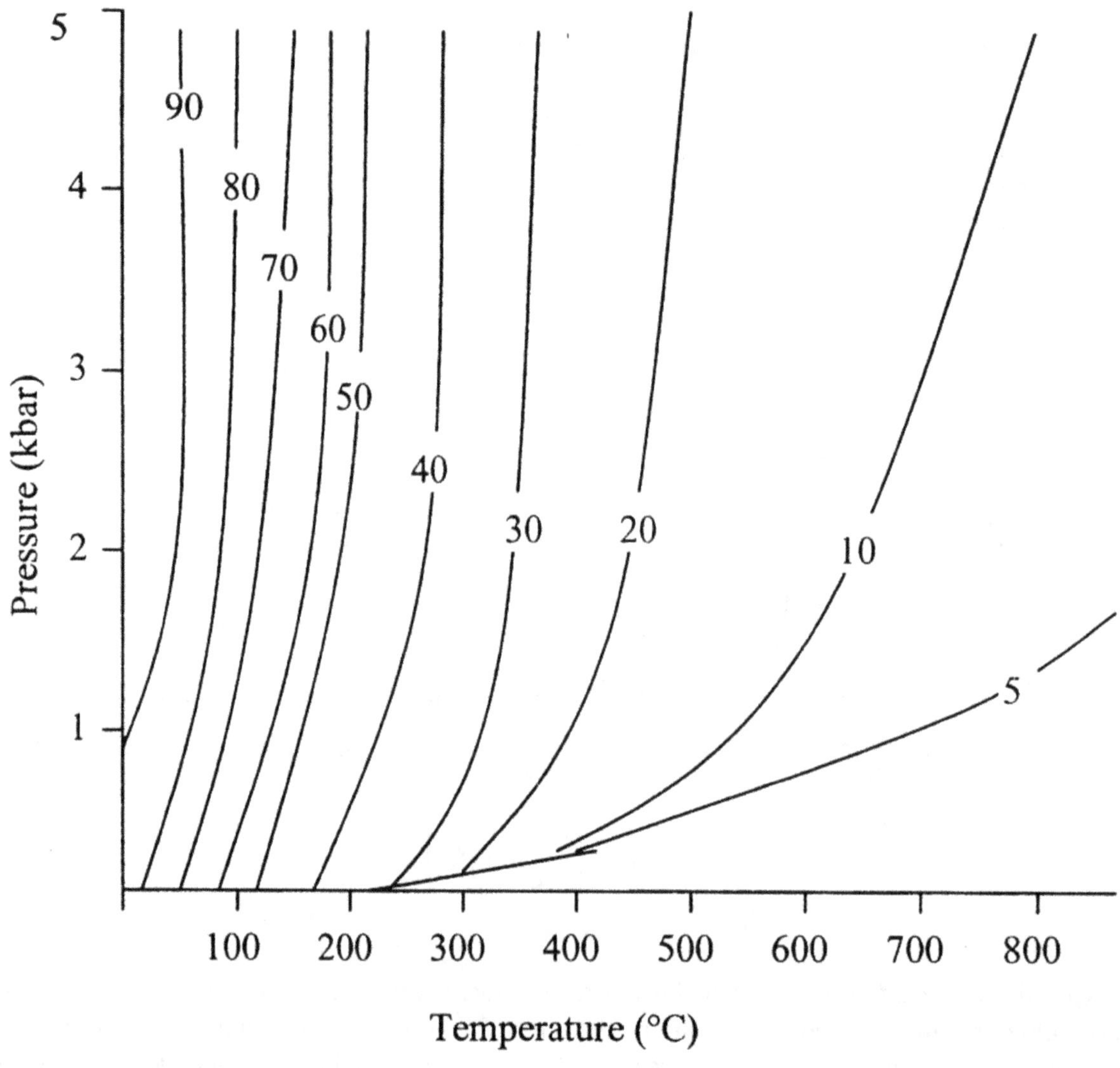

to corrosive nature of the solvents used in hydrothermal reactions, the autoclave requires a lining for the inner wall or separate liners placed in. For temperatures below 300 °C, Teflon is the popularly used liner which should sit exactly inside the autoclave without any gap. With the increase in temperature, Teflon expands and hermitic sealing can be obtained. Above 300 °C, Teflon cannot be used as a liner due to its dissociation and the coat getting torn. Figure 5 shows the schematic representation of an autoclave used for hydrothermal processing. Hydrothermal Autoclave should be corrosion resistant, should withstand high temperature and pressure for a long time duration, should not react with acids and bases, should be easy to assemble and operate. Most of the autoclaves are fitted with fixed pressure gauges for measurement or pressure can also be calculated using Kennedy's PVT relation (Kennedy, 1950). Usually, after completion of the reaction, the autoclave is allowed to cool in the furnace.

Table 1. Various solvents used in hydrothermal synthesis with temperature and duration

Temperature	Solvent	Duration
150 °C	Methanol	24 h
80 -200 °C	Aqueous	24h - 36 h
110 °C	ODE	1 h
193 °C	Chloride Solution	36 h
160 °C	Oleic acid	30 min
150 °C	Ethanol and water Mixture	10 min
200 °C	1,4-butanediol	30 min
180-250 °C	Ethylenediamine–Ethanol	5 h to 20 h
190 °C	Ethanol	24 h
150 °C	Ethylene Glycol	24 h
200 °C	Mixture of oleic acid–glycol–ethanol	24 h
220 °C	Pentanol and oleic acid	24 h
340 °C	Oleyl amine	2 h
240 °C	H_2O	3-5 days
230⁰C	Trifluoroacetic acid solution	24 h
220 °C	Nitric Acid	24 h
120 °C	HNO_3 aqueous solution	12 h
180 °C	Dilute HNO_3	12 h
160-200 °C	NH_4HF_2,and distilled water	4-5 days
220 °C	Aqueous solution of $SrCl_2$_$6H_2O$ (A.R.), $NaAlO_2$ (C.P.) and $Eu(NO_3)_3$	72 h
220 °C	H_2O	5 days
220 °C	EDTA dissolved in deionised water	20 h
120 °C	NaOH acqueous solution	12h
210 °C	Isopropyl alcohol	24 h

Table 2. Action of hydrothermal fluid (high-temperature–high-pressure aqueous solution/ vapor) on solid state materials

Sr. No.	Classified	Action	Application
1.	Transfer medium	Transfer of kinetic energy, heat and pressure forming, etc.	Erosion, machining abrasion, HIP
2.	Adsorbate	Adsorption/desorption at the surface	Dispersion, surface diffusion, catalyst, crystallization, sintering, ion exchange, etc.
3.	Solvent	Dissolution/precipitation	Synthesis, growth, purification, extraction, modification, degradation, Etching, corrosion, etc.
4.	Reactant	Reaction	Formation/decomposition (hydrates, hydroxides, oxides, corrosion, etc.)

Figure 5. Schematic representation of an autoclave

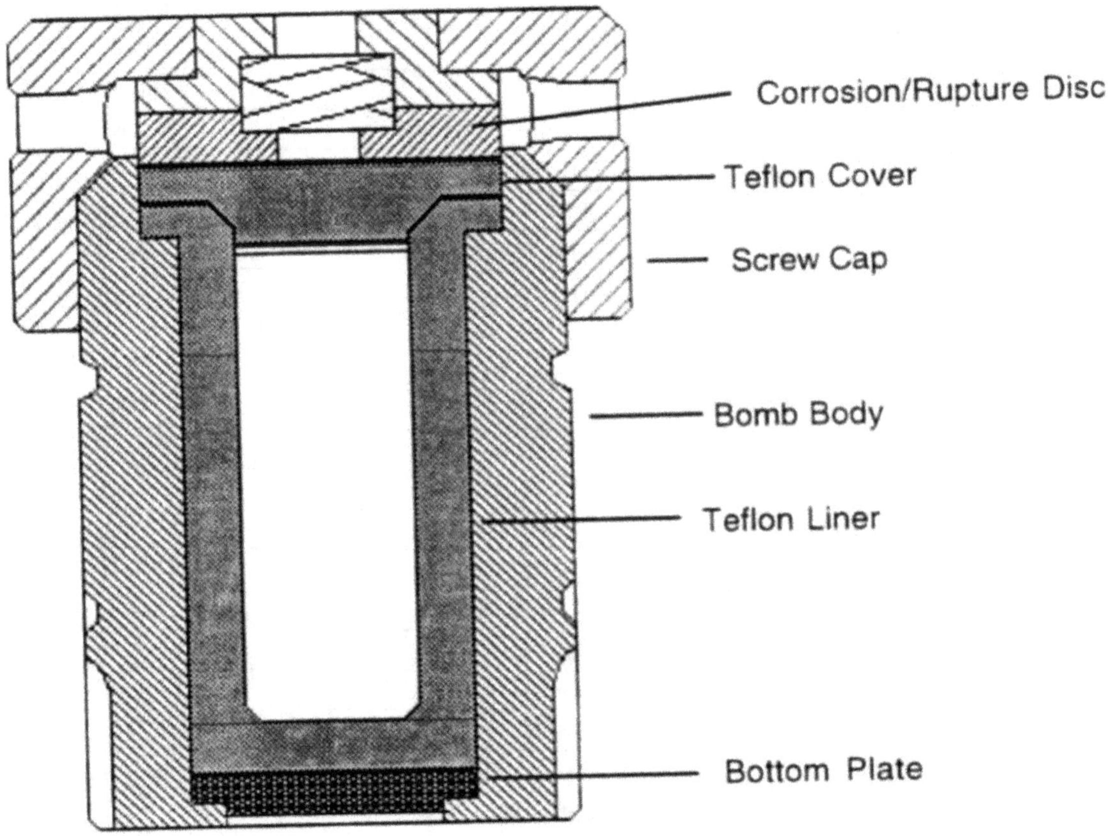

The ideal characteristics of the hydrothermal autoclave are; i). Inert to acids, bases and oxidizing agents, ii) it should be easily assembled and dissembled, iii) it should have sufficient length to obtain a desired temperature gradient, iv) it should be leak-proof at desired temperature and pressure, and v) it should bear high pressure and temperature for the long duration of time.

RECENT TRENDS IN HYDROTHERMAL SYNTHESIS

Recently, researchers are dealing with multi-energy processing techniques of materials like microwave heating into hydrothermal synthesis. Temperature, pressure and chemical potential are the three parameters, which are important in materials processing in the hydrothermal technique. The advanced material processing has complicated thermodynamic relations due to additional energy variables. The hybrid hydrothermal techniques include hydrothermal–microwave synthesis, mechanochemical synthesis, electrochemistry, SONAR, hot pressing and photochemical synthesis. The materials synthesized by these hybrid hydrothermal techniques are the basis of the nanotechnology because of controlled morphologies and particle size quantization (Ekimov, 1985).

HYDROTHERMAL CRYSTAL GROWTH

The hydrothermal method was used initially to grow crystals of quartz following which several other minerals in bulk form like carbonates, phosphates, sulfates, zeolite, ruby, corundum etc. were grown. The hydrothermal technique is also extensively used to prepare single crystals and polycrystals of piezoelectric, magnetic, optic, ceramic and host of other materials. The single crystals of very high purity, controlled size with fewer defects and dislocation can be prepared by hydrothermal method. The hydrothermal processing is in a closed system, which facilitates phase equilibrium studies due to the influence of temperature, pressure and compositional variations individually. The material synthesized by the hydrothermal technique is agglomeration free, fine-grained with controlled morphology, of high purity and has single crystals of narrow size distribution. There are many other advantages of hydrothermal technique like minimum expense, high rate of reaction, an environment free, no need for very highly sophisticated instruments and good dispersion. The production of bulk single crystals using hydrothermal technique is still restricted to quartz, emerald, corundum, ruby etc. until today. From 1970 onwards, the use of hydrothermal technology for the growth of other large single crystals declined significantly. In the recent years, there is again a growing interest in the use of hydrothermal technology for the single crystal growth. The supercritical hydrothermal synthesis method can be used for the nanosize crystals of metal oxides. Single crystals of specific characteristics can be synthesized. The products obtained by low- temperature wet method usually need heat treatment to show results, but phosphor nanoparticles show luminescence without any heat treatment. The organic and inorganic hybrid nanoparticles can be synthesized by the supercritical hydrothermal method.

NANOTECHNOLOGY AND HYDROTHERMAL METHOD

There are many hundreds of materials at the nanoscale prepared using hydrothermal technology. There are nearly 8000 publications dealing with advanced nanomaterial processing in the last 8 years. The properties of metallic nanoparticles are structural and strongly dependent upon their morphology. The nanostructures of these metal alloys form under non-equilibrium conditions (Tian, 2003). For the preparation of these nanoparticles, hydrothermal techniques are being used. The nanomaterials have the unique surface area, mechanical, chemical, physical, thermal, electrical, optical and magnetic properties. These unique properties define them as nanostructures, nanoelectronics, nanophotonics, nanobiotechnology, etc. The control over particle size and morphology with dispersion and chemical homogeneity of the particles produced becomes the biggest advantages of using hydrothermal technology for nanomaterials processing. Using hydrothermal processing, advanced nanomaterials like nanocomposites, metal oxides, semiconductors, silicates, hydroxides, tungstates, titanates, carbon, zeolites, sulfides, ceramics, etc. have been prepared. To modify the physical and chemical properties of nanocrystals, capping agents and surfactants help. The hydrothermal technique is an attractive method to produce highly homogeneous nanoparticles, nanocomposites, and hybrids.

FLUORIDES AS PHOSPHORS

Fluorides are very attractive materials for many potential applications such as display monitors, photonics, imaging, light amplification, as well as precursors for transparent ceramics processing. Compared to other oxide materials, fluoride has additional advantages such as highly transparent in a wide wavelength from the vacuum ultraviolet (VUV) to the infrared (IR) region and a low phonon energy that decreases the nonradiative transition probability of the active ions. However, there is a dearth of synthesis and characterization of rare-earth-doped nano fluorides (Pandurangappa, 2012).

Fluoride is a perfect fluorescent host matrix owing to their low vibrational energy, and subsequently minimizes the quenching of the excited state of the rare-earth ions. Furthermore, it owns adequate environmental and thermal stability as well as large solubility without deleterious effects for all lanthanide ions. These are essential to most optical applications (Stouwdam, 2002; Yan, 2005; Weber, 1967; Stouwdam, 2003).

Fluoride materials are being used for different applications such as phosphor materials (after doping with suitable impurities), optical window materials (in its pure form), etc. Single CaF_2 crystals are widely applicable in various optical devices operating in the UV and visible spectral ranges (Sun, 2009). $KMgF_3$ is useful as an optical window material in the VUV region (Babin, 2004). It has also been studied as a UV scintillator and a radiation dosimetry phosphor, etc. (Gektin, 1996; Masson, 2001; Martini, 1988).

Due to their various technological applications, especially optical, fluoride materials have attracted the attention of several researchers. When rare earth ions are doped with fluoride materials, they show favorable properties for optical applications such as high quantum efficiencies. Fluorides are of significant interest because of their wide range of applications as a laser material, X-ray imaging and storage phosphor, thermoluminescence detector, scintillator, a quantum cutter for VUV applications etc. Due to peculiar luminescent properties of fluorides like the high optical quality of crystals and their wide transmission range, they are well suited for the study of spectroscopic properties. The significant amount of laser emissions have been obtained from fluoride crystals with a wavelength ranging from 285nm in the UV to 4.34μm in the IR (Nakajima, 2000).

Under excitation by X-rays or γ-rays radiations or particles, the number of emitted photons decrease with increasing bandgap, fluorides have lower efficiencies than other halides or oxides. However, the high optical quality, chemical stability, and radiation hardness make them well suited for application in fast detectors for high energy particles as scintillator crystals. Fluorides can be also used in the process of ionization and hole burning. Photoionization is responsible for the quenching of the luminescence of cation with reducing character in some lattices and two-step ionization affects the efficiency of lasers. To obtain high capacity optical memories (hole burning), the ionization is used.

Some fluorides are used as energy storage phosphors. In this field, most of the work has been devoted to photostimulated phosphors for digital x-ray imaging system. Most of them contain fluorine which brings the chemical stability required for storage of energy. LiF:Mg,Cu,P and LiF:Mg,Cu,Na,Si are commonly used for thermoluminescence dosimetry materials. $LiCaAlF_6$:Ce found application as scintillator material (Gektin, 2002). $LiCaAlF_6$: Ce and $LiSrAlF_6$: Ce is reported as tunable solid-state lasers in the UV region (Dubinskii, 1993; Marshall, 1994). Cr doped $LiCaAlF_6$, $LiSrAlF_6$ are also good tunable laser materials and can be pumped by a diode laser or flash lamp (Wang, 1996). BaFBr:Eu^{2+} is a well-known photostimulable phosphor that has been widely used for X-ray imaging and as a storage phosphor (Crawford, 1991). $LiCaAlF_6$:Eu^{2+} is useful for detection of the X-ray, gamma -radiation as well as thermal neutrons.[102] Use of $KMgF_3$:Eu^{2+} phosphor as a UV dosimetry material is also suggested

(Gektin, 1998). $LiBaF_3:Eu^{2+}$ material is found suitable for X-ray storage phosphor.[104] Pr^{3+} doped YF_3, LuF_3, $BaMgF_4$, $LiCaAlF_6$, $LiSrAlF_6$ and $KMgF_3$ are materials, which exhibit quantum cutting under VUV (Kuck, 2003).

Fluoride compounds are the most appropriate materials for solid-state lasers with specific wavelengths because of optical quality and chemical stability of crystals, low probability of nonradiative processes and wide transmission range. Thus the 4f-4f line emissions of lanthanide ions have been used in order to obtain infrared laser radiation up to 4.34 μm and blue or green radiation by upconversion pumping. Tunable laser operation in the ultraviolet has been demonstrated using the broad 5d-4f emission of Ce^{3+}. Fluoride crystals are also suitable for fast detectors for high energy radiations. Scintillators based on the Ce^{3+} emission have been particularly investigated. Fluorides or Fluorine-rich compounds are used in data storage. Research on Samarium doped fluorides has lead in recent years to materials with the broad inhomogeneous band and permanent hole burning at room temperature.

Doping of rare earth ions in fluorides produces efficient emission in UV region leading these materials for important applications. However, preparation of fluorides is difficult due to their high reactivity and susceptibility to hydrolysis, which is detrimental to the luminescence properties. In order to prevent hydrolysis, these materials were prepared by methods which need hazardous and tedious arrangements such as passing hot HF over the initial constituent fluorides, using the reactive atmosphere of argon or nitrogen at high temperature (>600 °C) etc. These expensive methods to synthesize fluoride phosphor made the devices based on these phosphors costlier. Hence some new simple low-cost methods for the synthesis of fluorides are needed.

Preparation of complex fluorides is not a simple task. Fluorides are highly susceptible to hydrolysis. At high temperatures, fluorides are known to incorporate hydroxyl ion and oxygen causing distortion in the lattice. The growth of high quality and optically clear fluorides crystals for laser applications require high purity oxygen free starting materials and rigorous attention to keeping treatment and growth furnaces free of oxygen and water contamination (Doulan, 2003). Otherwise, oxides, oxyfluorides, and other complicated complexes may easily form giving rise to detrimental scattering and absorption centers. Presence of OH^- is always suspected in samples prepared by solid state reaction, co-precipitation or hydrothermal methods (Karbowiak, 2005; Hua, 2006). Karbowiak et al., (2005) observed that when the samples were prepared with constituent fluorides as the starting materials, this was a major issue. Hydrolysis of fluoride has a great impact on their luminescence properties and is responsible for the deterioration of luminescence. The conventional solid state methods of preparing these materials are quite difficult and rather involved. The above-reported synthesis methods, in general, involved tedious, time-consuming procedures required proper care to control the reaction condition, longer reaction times, etc. A simplified procedure that overcomes the disadvantages of conventional methods and further improves the potential usage of the prepared materials is desirable. The conventional ceramic method of synthesizing these complex compound is both tedious and time-consuming owing to the low diffusion rates involved in the solid-state reactions (Sharma, 1992; Goodenough, 1992; Vaidhyanathan, 1998).

RARE EARTH FLUORIDE NANO/MICROCRYSTALS

In recent years, a lot of research is carried out on rare-earth materials because of their novel properties like electronic, optical, and chemical due to the 4f electrons of rare earth elements, which refer to the lanthanide (Ln) series, yttrium, and scandium (He, 2012; Cho, 2010; Ye, 2010; Zhang, 2008; Park, 2009;

Liu, 2007). Intense research attention is given to rare-earth nanocrystals with controllable shapes and sizes due to their wide range of applications in optics, biological labeling, catalysis fields, etc. In rare-earth nanocrystals, the movement of electrons and photons would be confined in two or all directions which change the properties (Murray, 1993; Peng, 2000; Murray, 1995; Chan, 1998; Bruchez, 2000). Rare earth fluorides are being studied extensively owing to their unusual spectroscopic characteristics, but these crystals are grown as small nanosize crystals rather than the bulk single crystals. The use of hydrothermal solutions provides the study of the behavior of the whole series of rare earth elements in the presence of F^- ions under the same conditions. For example, fluorides possess very low vibrational energies as compared to oxygen-based systems, the quenching of the rare earth ion's excited states will be minimum (Bender, 2000). Fluorides can be the host for luminescent lanthanide ions due to their thermal and environmental stability. It is easy to synthesize fluorides of compositions $ALnF_4$, A_3LnF_6, A_2LnF_5 (A= Li, Na, K, Rb, Cs), amongst others using the hydrothermal conditions. The surface properties need to be engineered to get nanocrystals of high quality. The growth dynamics modulation in certain cases of nanocrystals is important. Various synthetic routes which are solution based have been used to process many important compounds into monodisperse nanocrystals. Fluorides form a large group of inorganic compounds like fluorocarbonates, oxyfluorides, fluoroarsenates, fluorophosphates, etc. The hydrothermal method helps in developing new fluoride crystal structures, which are difficult to obtain by other conventional methods particularly the active elements like Nd^{3+}, and Eu^{3+} which have good luminescent properties. The other transitional metals like fluorides bearing Fe, Mn exhibit good magnetic properties. Transition metal fluorides and fluoro carbonates/ fluorophosphates show ionic conductivity, optical and magnetic properties. The microporous structure of oxyfluorides facilitates uses as gas absorbers and catalyst. Rare-earth compounds are distinguished for their luminescence properties, and different doping modes may lead to quite a different emission behavior, which is appealing to applications such as biological labeling and optics (Heer, 2004). Especially nanometer-sized up-conversion phosphors with uniform size and high luminescence efficiency are desirable as sensitive biological labeling material for the detection of biomolecules such as DNA, RNA, or proteins (Yi, 2004; Yan, 2005). In the recent past, several successful examples have been demonstrated for the synthesis of rare-earth compound nanocrystals with up-conversion properties (Somiya, 1981; Dzik, 2000; Wang, 2011). The hydrothermal process is extensively employed in the synthesis of rare earth ions activated inorganic compounds, such as yttrium vanadate, lanthanum fluoride, lanthanum phosphate and yttrium oxide.

ADVANTAGES OF HYDROTHERMAL METHOD

The hydrothermal method has the ability to synthesize crystals of substances which are unstable near the melting point, and the ability to synthesize large crystals of high quality. Hydrothermal/ solvothermal synthesis demand a sealed autoclave environment to form high temperature and pressure to obtain good crystal morphology. Hydrothermal synthesis is environment-friendly because it consumes less energy, less solid waste/liquid waste /gases, no hazardous process materials, high selectivity, closed system processing. The main advantages of the hydrothermal method are: (a) kinetics of the reaction are increased many-fold with a small increase in temperature, (b) new metastable products can be formed, (c) single crystals are obtained generally, (d) from impure feed-stocks, high purity products can be obtained, (e) the process is cost effective, as no precipitants are needed in many cases, (f) pollution is minimized because of the closed system conditions and reagents can be recycled, (g) hydroxylated clays and zeolite molecular sieves cannot be synthesized by any other method (Komarneni, 2003). It accelerates interactions between solid and fluid species resulting in phase pure and homogeneous materials with good compositions and good reaction kinetics. It is also categories as environment-friendly having higher diffusivity, lower viscosity, controlled mass transport and solvents have higher dissolving power. It provides highly crystalline particles with a better morphology and controlled particle size. Hydrothermal synthesis has a strong influence on the particle size and formation of graphene quantum dots which can be used in electronics, solar cell, Li-ion battery, and pollution treatment, cell imaging, drug delivery, DNA intercalator, and so on. There are several advantages of the hydrothermal method over conventional solid-state reaction. For example, compounds that have elements with an unusual oxidation state can be synthesized such as the formation of ferromagnetic chromium (IV) oxides (Rabenau, 1985).

Synthesis of low-temperature phases and metastable compounds by using simple quartz ampoules are possible by hydrothermal method. Although hydrothermal synthesis is considered as a high-temperature technique, in reality, these temperatures are low compared to most conventional melt techniques. There are several important advantages of low-temperature crystal growth. For example, it allows the growth of low-temperature polymorphs like α-quartz that are often difficult or impossible to prepare by other synthetic methods. α-quartz is the most intensively studied material grown hydrothermally in the industry for its piezoelectric properties and application in electronic devices. Piezoelectric α -quartz is only stable below 580 °C and must be grown below this temperature. This is a problem for conventional flux crystal growth or melting method. In the hydrothermal method, less time is required for synthesis compared to conventional methods. For instance, although a solid-state reaction can be performed in a few weeks, the hydrothermal reaction can be done in a few days.

The rapid growth rates because of the rapid diffusion processes are also the advantage of hydrothermal synthesis. In this technique, crystal growth is carried out from a liquid solution, but the viscosity of the liquid is lower. For conventional methods, the solubility of a liquid should be high or there will be a rate problem. Under hydrothermal conditions, because of low viscosity of materials in comparison to the viscosity at ambient temperature, diffusion is not a problem. The inadequacies of this method are the requirement of high-pressure and also difficulty in gathering data. Because of fast diffusion under hydrothermal conditions, during the process of crystal growth, supersaturation occurs, which can result in the dendritic growth of crystal that increases the chance of impurities and can reduce crystal quality. Even for materials that have low solubility at ambient temperature, the diffusion is fast. For example, quartz cannot be grown at low temperature, however, under hydrothermal conditions; quartz crystal can be grown at relatively high rates (Eanes, 2000).

CONCLUSION

Hydrothermal technology has made its place as a most powerful tool in materials processing because of its various advantages over other conventional methods of synthesis. It is the green technology, being environmentally benign. It is highly suitable for advanced materials processing from bulk single crystals to superfine crystals and nanocrystals or nanoparticles. The physical chemistry and solution chemistry of the hydrothermal media leads to the modeling of the hydrothermal reactions having great assistance in the intelligent engineering of the materials under much reduced pressure-temperature conditions. For the growth of large bulk crystals of quartz, berlinite, gallium phosphate, zeolites, and other clay materials, the hydrothermal technique is the only best suitable technique. The use of multi-energy systems like microwave-hydrothermal, mechanochemical hydrothermal, sonar-hydrothermal, electrochemical-hydrothermal, etc. drives this technology to a new and totally unexplored avenue in the 21st century. Hydrolysis and/or oxygen-containing fluoride has a great impact on their luminescence properties and is responsible for the deterioration of luminescence, but hydrothermal process overcomes these limitations and retains of good luminescence in oxygen-free fluorides.

REFERENCES

Azorın, J., Esparza, A., Falcony, C., Rivera, T., Garcıa, M., & Martınez, E. (2002).. . *Radiation Protection Dosimetry*, *100*, 277–279. doi:10.1093/oxfordjournals.rpd.a005867 PMID:12382877

Azorin, J., Rivera, T., Martínez, E., & Garcia, M. (1998). Thermoluminescence of Eu-doped ZrO2 thin films exposed to ultraviolet and visible light. *Radiation Measurements*, *29*(3-4), 315–317. doi:10.1016/S1350-4487(98)00013-4

Babin, V., Oskam Koert, D., Vergeer, P., & Meijerink, A. (2004). Article. *Radiation Measurements*, *38*, 463–466. doi:10.1016/j.radmeas.2004.01.036

Barrer, R. M. (1982). *Hydrothermal Chemistry of Zeolites*. London: Academic Press.

Bender, C. M., Burlitch, J. M., Barber, D., & Pollock, C. (2000). Synthesis and Fluorescence of Neodymium-Doped Barium Fluoride Nanoparticles. *Chemistry of Materials*, *12*(7), 1969–1976. doi:10.1021/cm9904741

Bierlein, J., & Geir, T. (n.d.). *U.S. Patent 3,1976,949,323*. US Patent Office.

Boyer, D., & Mahiou, R. (2004). Powders and Coatings of $LiYF_4$:Eu^{3+} Obtained via an Original Way Based on the Sol–Gel Process. *Chemistry of Materials*, *16*(13), 2518–2521. doi:10.1021/cm040143a

Brincker, C. J., & Scherer, G. W. (1990). *Sol–Gel Science*. New York: Academic Press.

Bruchez, M. Jr, Moronne, M., Gin, P., Weiss, S., & Alivisatos, A. P. (1998). Semiconductor Nanocrystals as Fluorescent Biological Labels. *Science*, *281*(5385), 2013–2016. doi:10.1126cience.281.5385.2013 PMID:9748157

Byrappa, K., & Adschiri, T. (2007). Hydrothermal technology for nanotechnology. *Progress in Crystal Growth and Characterization of Materials*, *53*(2), 117–166. doi:10.1016/j.pcrysgrow.2007.04.001

Byrappa, K., & Yoshimura, M. (2001). *Handbook of Hydrothermal Technology*. Noyes Publications.

Cao, M., Wang, Y., Qi, Y., Guo, C., & Hu, C. (2004). Synthesis and characterization of MgF2 and KMgF3 nanorods. *Journal of Solid State Chemistry*, *177*(6), 2205–2209. doi:10.1016/j.jssc.2004.01.005

Chan, W. C. W., & Nie, S. M. (1998). Quantum Dot Bioconjugates for Ultrasensitive Nonisotopic Detection. *Science*, *281*(5385), 2016–2018. doi:10.1126cience.281.5385.2016 PMID:9748158

Chen, Q., Qian, Y., Chen, Z., Tang, K., Zhou, G., & Zhang, Y. (1994). Preparation of a Tl-based superconductor by a hydrothermal method. *Physica. C, Superconductivity*, *224*(3-4), 228–230. doi:10.1016/0921-4534(94)90258-5

Chen, X. P., Zhang, Q. Y., Yang, C. H., Chen, D. D., & Zhao, C. (2009). Comparative investigation on structure and luminescence properties of fluoride phosphors codoped with Er3+/Yb3+. *Spectrochimica Acta. Part A: Molecular Spectroscopy*, *74*(2), 441–445. doi:10.1016/j.saa.2009.06.040

Cho, H. S., Dong, Z., Pauletti, G. M., Zhang, J., Xu, H., Gu, H., ... Shi, D. (2010). Fluorescent, Superparamagnetic Nanospheres for Drug Storage, Targeting, and Imaging: A Multifunctional Nanocarrier System for Cancer Diagnosis and Treatment. *ACS Nano*, *4*(9), 5398–5404. doi:10.1021/nn101000e PMID:20707381

Crawford, M. K., & Brixner, L. H. (1991). Photostimulable phosphors for X-ray imaging: Applications and mechanism. *Journal of Luminescence*, *48*, 37–42. doi:10.1016/0022-2313(91)90073-5

Doulan, J. L., & Moncorge, R. (2003). Article. *Ann. Chim. Sci. Mat.*, *28*, 5.

Du, H., Zhang, W., & Sun, J. (2011). Structure and upconversion luminescence properties of BaYF5:Yb3+, Er3+ nanoparticles prepared by different methods. *Journal of Alloys and Compounds*, *509*(7), 3413–3418. doi:10.1016/j.jallcom.2010.12.101

Dubinskii, M. A., Semashko, V. V., Naumov, A. K., Abdulsabirov, R. Y., & Korableva, S. L. (1993). Article. *Laser Physics*, *3*, 216.

Dzik, G. D., Sokolska, I., Golab, S., & Baluka, M. (2000). Article. *Journal of Alloys and Compounds*, *300*, 254–260. doi:10.1016/S0925-8388(99)00779-3

Eanes, M. (2000). *Synthesis and characterization of alkali silver chalcogenides and alkali rare earth germanates by supercritical fluids, Doctor of Philosophy*. Clemson University.

Ekimov, A., Efros, A., & Onushchenko, A. (1985). Quantum size effect in semiconductor microcrystals. *Solid State Communications*, *56*(11), 921–924. doi:10.1016/S0038-1098(85)80025-9

Ganesh, I., Johnson, R., Rao, G. V. N., Mahajan, Y. R., Madavendra, S. S., & Reddy, B. M. (2005). Microwave-assisted combustion synthesis of nanocrystalline MgAl2O4 spinel powder. *Ceramics International*, *31*(1), 67–74. doi:10.1016/j.ceramint.2004.03.036

Gao, P., Xie, Y., & Li, Z. (2006). Controlling the Size of BaF2 Nanocubes from 1000 to 10 nm. *European Journal of Inorganic Chemistry*, *16*(16), 3261–3265. doi:10.1002/ejic.200600237

Gektin, A., Komar, V., Shlyahturov, V., & Shiran, N. (1996). Pure and Ce-doped KMgF/sub 3/-scintillators. *IEEE Transactions on Nuclear Science*, *43*(3), 1295–1297. doi:10.1109/23.506681

Gektin, A., Shiran, N., Neichev, S., Gavrilyuk, V., Bensalah, A., Fukuda, T., & Shimamura, K. (2002). LiCaAlF6:Ce crystal: a new scintillator. *Nucl. Instru. Methods A*, *486*(1-2), 274–277. doi:10.1016/S0168-9002(02)00717-9

Gektin, A. V., Krasovitskaya, I. M., & Shiran, N. V. (1998). UV-dosimetric material based on KMgF3 perovskite. *Radiation Measurements*, *29*(3-4), 337–340. doi:10.1016/S1350-4487(98)00054-7

Goodenough, J. B., Manthiram, A., Paranthaman, M., & Zhen, Y. S. (1992). Oxide ion electrolytes. *Materials Science and Engineering B*, *12*(4), 357–364. doi:10.1016/0921-5107(92)90006-U

Gurav, A., Kodas, T., Pluym, T., & Xiong, Y. (1993). Aerosol Processing of Materials. *Aer. Sci. Tech.*, *19*(4), 411–452. doi:10.1080/02786829308959650

He, F., Niu, N., Zhang, Z. G., Zhang, X., Wang, D., Bai, L., ... Yang, P. P. (2012). Morphology-controllable synthesis and enhanced luminescence properties of β-NaLuF4:Ln (Ln = Eu, Tb and Ce/Tb) microcrystals by solvothermal process. *RSC Advances*, *2*(19), 7569–7577. doi:10.1039/c2ra20675g

Heer, S., Kopme, K., Gudel, H. U., & Haase, M. (2004). Highly Efficient Multicolour Upconversion Emission in Transparent Colloids of Lanthanide-Doped $NaYF_4$ Nanocrystals. *Advanced Materials*, *16*(23-24), 2102–2105. doi:10.1002/adma.200400772

Hipolito, G., Falcony, M., Aguilar-Frutis, C., & Azorín, M. A. (2001). Article. *J. Appl. Phys. Lett.*, *79*, 4369–4371. doi:10.1063/1.1428110

Hlavacek, V., & Puszynski, J. A. (1996). Chemical Engineering Aspects of Advanced Ceramic Materials. *Industrial & Engineering Chemistry Research*, *35*(2), 349–377. doi:10.1021/ie9501034

Hu, A., Yao, Z., & Yu, X. (2009). Phase behavior of a sodium dodecanol allyl sulfosuccinic diester/ *n* -pentanol/methyl acrylate/butyl acrylate/water microemulsion system and preparation of acrylate latexes by microemulsion polymerization. *Journal of Applied Polymer Science*, *113*(4), 2202–2208. doi:10.1002/app.30284

Hua, R., Jia, Z., & Shi, C. (2007). Preparation of KMgF3 and Eu-doped KMgF3 nanocrystals in water-in-oil microemulsions. *Materials Research Bulletin*, *42*(2), 249–255. doi:10.1016/j.materresbull.2006.06.004

Hua, R., Jia, Z., Xie, D., & Shi, C. (2002). Solvothermal synthesis of the complex fluorides KMgF3 and KZnF3 with the Perovskite structures. *Materials Research Bulletin*, *37*(6), 1189–1195. doi:10.1016/S0025-5408(02)00732-8

Hua, R., & Jiang, H. (2006). Article. *Materials Letters*, *60*, 3822–3825. doi:10.1016/j.matlet.2006.03.121

Hua, R., Zang, C., Shao, C., Xie, D., & Shi, C. (2003). Synthesis of barium fluoride nanoparticles from microemulsion. *Nanotech.*, *14*(6), 588–591. doi:10.1088/0957-4484/14/6/304

Jia, G., Zheng, Y. H., Liu, K., Song, Y. H., You, H. P., & Zhang, H. J. (2009). Article. *Journal of Physical Chemistry*, *C113*, 153–158.

Jiayue, S., Jianbo, X., & Zhang, X. (2011). Du Haiyan. *Journal of Rare Earths*, *29*(1), 32–38.

Kang, Z., Barta, M., Nadler, J., Wagner, B., Rosson, R., & Kahn, B. (2010). Synthesis of BaF2:Ce nanophosphor and epoxy encapsulated transparent nanocomposite. *Journal of Luminescence*, *131*(10), 2140–2143. doi:10.1016/j.jlumin.2011.05.030

Kang, Z. J., Wang, Y. X., You, F. T., & Lin, J. H. (2001). Hydrothermal Syntheses and Crystal Structure of NH4Ln3F10 (Ln=Dy, Ho, Y, Er, Tm). *Journal of Solid State Chemistry*, *158*(2), 358–362. doi:10.1006/jssc.2001.9131

Karbowiak, M., Mech, A., Bednarkiewicz, A., Strek, W., & Kepinski, L. (2005). Comparison of different NaGdF4:Eu3+ synthesis routes and their influence on its structural and luminescent properties. *Journal of Physics and Chemistry of Solids*, *66*(6), 1008–1019. doi:10.1016/j.jpcs.2005.01.002

Kennedy, G. C. (1950). Pressure-volume-temperature relations in water at elevated temperatures and pressures. *American Journal of Science*, *248*(8), 540–564. doi:10.2475/ajs.248.8.540

Kodama, N., & Oishi, S. (2005). Visible quantum cutting through downconversion in KLiGdF5:Eu3+ crystals. *Journal of Applied Physics*, *98*(10), 103515–103520. doi:10.1063/1.2135893

Komarneni, S. (2003). Article. *Current Science*, *85*(12), 1730–1734.

Kuck, S., Sokolska, I., & Henke, M. (2003). Article. *Journal of Luminescence*, *102*, 176–181.

Le Masson, N. J. M., Bos, A. J. J., Winkelman, A. J. M., & van Eijk, C. W. E. (2001). Optically stimulated luminescence in KMgF/sub 3/:Ce/sup 3+/ comparison of dosimetric characteristics with Al/sub 2/O/sub 3/:C. *IEEE Transactions on Nuclear Science*, *48*(4), 1143–1147. doi:10.1109/23.958739

Lee, J. H., Kang, Y. C., Park, S. B., & Park, H. D. (2001). Luminescence Characteristics of Eu-doped Strontium Halophosphate Phosphor Particles Prepared by Spray Pyrolysis. *Japanese Journal of Applied Physics*, *40*(Part 1), 3222–3226. doi:10.1143/JJAP.40.3222

Li, H., Jia, Z., & Shi, C. (2000). The Mild Hydrothermal Synthesis of Complex Fluorides of $AZnF_3$ (A = Na, K). *Chemistry Letters*, *9*(9), 1106–1110. doi:10.1246/cl.2000.1106

Lian, H., Liu, J., Ye, Z., & Shi, C. (2004). Synthesis and photoluminescence properties of erbium-doped BaF2 nanoparticles. *Chemical Physics Letters*, *386*(4-6), 291–294. doi:10.1016/j.cplett.2004.01.036

Liu, R. S., Liu, Y. H., Bagkar, N. C., & Hu, S. F. (2007). Enhanced luminescence of SrSi2O2N2:Eu2+ phosphors by codoping with Ce3+, Mn2+, and Dy3+ ions. *Applied Physics Letters*, *91*(6), 61119. doi:10.1063/1.2768916

Lu, C., Huang, W., Ni, Y., & Xu, Z. (2011). Hydrothermal synthesis and luminescence properties of octahedral LiYbF4: Er3+ microcrystals. *Materials Research Bulletin*, *46*(2), 216–221. doi:10.1016/j.materresbull.2010.11.008

Malik, M. A., Wani, M. Y., & Hashim, M. A. (2012). Microemulsion method: A novel route to synthesize organic and inorganic nanomaterials. *Arabian Journal of Chemistry*, *5*(4), 397–417. doi:10.1016/j.arabjc.2010.09.027

Mancic, L., Del Rosario, G., Stanojevic, Z., & Milosevic, O. (2007). Phase evolution in Ce-doped yttrium–aluminum-based particles derived from aerosol. *Journal of the European Ceramic Society*, *27*(13-15), 4329–4332. doi:10.1016/j.jeurceramsoc.2007.02.153

Marshall, C. D., Payne, S. A., Spaeth, J. A., Krupke, W. F., Quarles, G. J., Castillo, V., & Chai, B. H. T. (1994). Ultraviolet laser emission properties of Ce^3+-doped LiSrAlF_6 and LiCaAlF_6. *Journal of the Optical Society of America. B, Optical Physics*, *11*(10), 2054–2065. doi:10.1364/JOSAB.11.002054

Martini, M., Meinardi, F., & Scacco, A. (1988). Impurity-induced thermally stimulated luminescence of KMgF3:Ce3+ crystals. *Chemical Physics Letters*, *293*(1-2), 43–46. doi:10.1016/S0009-2614(98)00764-7

Morey, G. W., & Niggli, P. (1913). The Hydrothermal Formation of Silicates, A Review. *Journal of the American Chemical Society*, *35*(9), 1086–1130. doi:10.1021/ja02198a600

Murray, C. B., Kagan, C. R., & Bawendi, M. G. (1995). Self-Organization of CdSe Nanocrystallites into Three-Dimensional Quantum Dot Superlattices. *Science*, *270*(5240), 1335–1338. doi:10.1126cien ce.270.5240.1335

Murray, C. B., Norris, D. J., & Bawendi, M. G. (1993). Synthesis and characterization of nearly monodisperse CdE (E = sulfur, selenium, tellurium) semiconductor nanocrystallites. *Journal of the American Chemical Society*, *115*(19), 8706–8715. doi:10.1021/ja00072a025

Nakajima, T., Zemva, B., & Tressaud, T. (2000). *Adv. Inorganic Fluorides: Synthesis*. Characterization & Applications, Elsevier Series S.A.

Pan, Y. X., Wu, M. M., & Su, Q. (2003). Synthesis of Eu3+-doped calcium and strontium carbonate phosphors at room temperature. *Materials Research Bulletin*, *38*(9-10), 1537–1544. doi:10.1016/S0025-5408(03)00174-0

Pandurangappa, C., & Lakshminarasappa, B. N. (2012). Optical studies on lanthanum-doped calcium fluoride. *Journal of Materials Science*, *47*(2), 892–897. doi:10.100710853-011-5869-x

Park, Y. I., Kim, J. H., Lee, K. T., Jeon, K. S., Na, H. B., Yu, J. H., ... Lee, N. (2009). Article. *Advanced Materials*, *21*, 4467–4471. doi:10.1002/adma.200901356

Patro, L. N., & Hariharan, K. (2011). Influence of synthesis methodology on the ionic transport properties of BaSnF4. *Materials Research Bulletin*, *46*(5), 732–737. doi:10.1016/j.materresbull.2011.01.010

Peng, J. H., Binner, J., & Bradshaw, S. (2002). Microwave initiated self-propagating high temperature synthesis of materials. *Materials Science and Technology*, *18*(12), 1419–1427. doi:10.1179/026708302225006142

Peng, X. G., Manna, L., Yang, W. D., Wickham, J., Scher, E., Kadavanich, A., & Alivisatos, A. P. (2000). Shape control of CdSe nanocrystals. *Nature*, *404*(6773), 59–61. doi:10.1038/35003535 PMID:10716439

Purwanto, A., Wang, W. N., Lenggoro, I. W., & Okuyama, K. (2007). Formation and Luminescence Enhancement of Agglomerate-Free YAG:Ce[sup 3+] Submicrometer Particles by Flame-Assisted Spray Pyrolysis. *Journal of the Electrochemical Society*, *154*(3), 91–96. doi:10.1149/1.2426874

Rabenau, A., & Rau, H. (1969). Article. *Philips Tech. Rev.*, *30*, 89–96.

Rao, C. N. R. (1994). *Chemical Approaches to the Synthesis of Inorganic Materials*. New York: John Wiley.

Reis, K., Ramanan, A., & Whittingham, M. (1990). Hydrothermal synthesis of sodium tungstates. *Chemistry of Materials*, *2*(3), 219–221. doi:10.1021/cm00009a003

Sahare, P. D., Bakare, J. S., Dhole, S. D., Ingale, N. B., & Rupasov, A. A. (2010). Synthesis and luminescence properties of nanocrystalline LiF:Mg,Cu,P phosphor. *Journal of Luminescence*, *130*(2), 258–265. doi:10.1016/j.jlumin.2009.08.018

Sahare, P. D., Bakare, J. S., Dhole, S. D., Ingale, N. B., & Rupasov, A. A. (2010). Synthesis and luminescence properties of nanocrystalline LiF:Mg,Cu,P phosphor. *Journal of Luminescence*, *130*(2), 258–265. doi:10.1016/j.jlumin.2009.08.018

Schafhäutl, G. A. (1845). Article. *Royal Bavarian Academy of Sciences*, *20*, 557.

Sharma, V., Shukla, A., & Gopalakrishnan, K. (1992). Effect of aliovalent-cation substitution on the oxygen-ion conductivity of Bi4V2O11. *J. Solid State Ionics*, *58*(3-4), 359–362. doi:10.1016/0167-2738(92)90141-B

Sivakumar, V., Lakshmanan, A., Satheesh Kumar, R., Kalpana, S., Sangeetha Rani, R., & Jose, M. T. (2012).. . *Indian Journal of Pure and Applied Physics*, *50*, 123–128.

Somiya, S., Hirano, S., Yoshimura, M., & Yanagisawa, K. (1981). Hydrothermal crystal growth of perovskite-type fluorides. *Journal of Materials Science*, *16*(3), 813–816. doi:10.1007/BF02402800

Stouwdam Jan, W., Hebbink Gerald, A., Jurriaan, H., & van Veggel Frank, C. J. M. (2003). Article. *Chemistry of Materials*, *15*, 4604–4616. doi:10.1021/cm034495d

Stouwdam Jan, W., & van Veggel Frank, C. J. M. (2002). Article. *Nano Letters*, *2*, 733–737. doi:10.1021/nl025562q

Sugita, M., Tsuji, M., & Abe, M. (1990). Synthetic Inorganic Ion-Exchange Materials. LVIII. Hydrothermal Synthesis of a New Layered Lithium Titanate and Its Alkali Ion Exchange. *Bulletin of the Chemical Society of Japan*, *63*(7), 1978–1984. doi:10.1246/bcsj.63.1978

Tian, Z., Liu, J., Voigt, J. A., Xu, H., & Mcddermott, M. J. (2003). Dendritic Growth of Cubically Ordered Nanoporous Materials through Self-Assembly. *Nano Letters*, *3*(1), 89–92. doi:10.1021/nl025828t

Vaidhyanathan, B., Balaji, K., & Rao, K. J. (1998). Microwave-Assisted Solid-State Synthesis of Oxide Ion Conducting Stabilized Bismuth Vanadate Phases. *Chemistry of Materials*, *10*(11), 3400–3401. doi:10.1021/cm980092f

Wang, D. W., Huang, S. H., You, F. T., Qi, S. Q., Fu, Y. B., Zhang, G. B., ... Huang, Y. (2007). Vacuum ultraviolet spectroscopic properties of Pr3+ in MYF4 (M=Li, Na, and K) and LiLuF4. *Journal of Luminescence*, *122*, 450–456. doi:10.1016/j.jlumin.2006.01.203

Wang, J., Yue, Y., Yao, J., & Hu, Z. (2011). Growth and characterization of Ba(Al,Ga)BO3F2 crystal. *Journal of Crystal Growth*, *318*(1), 962–965. doi:10.1016/j.jcrysgro.2010.10.223

Wang, Y., Huang, C., Chen, L., & Fang, Z. (1996). Crystal growth of Cr3+:LiCaAlF6 by Bridgman technique. *Journal of Crystal Growth*, *167*(1-2), 176–179. doi:10.1016/0022-0248(96)00250-3

Weber, M. J. (1967). Probabilities for Radiative and Nonradiative Decay of Er 3 + in La F 3. *Physical Review*, *157*(2), 262–271. doi:10.1103/PhysRev.157.262

Wegh, R. T., Donker, H., Meijerink, A., Lamminmaki, R. J., & Holsa, J. (1997). Vacuum-ultraviolet spectroscopy and quantum cutting for Gd 3 + in LiYF 4. *Physical Review B: Condensed Matter and Materials Physics*, *56*(21), 13841–13848. doi:10.1103/PhysRevB.56.13841

Wegh, R. T., Donker, H., Oskam, K. D., & Meijerink, A. (1999). Visible Quantum Cutting in $LiGdF_4:Eu^{3+}$ Through Downconversion. *Science*, *283*(5402), 663–666. doi:10.1126cience.283.5402.663 PMID:9924020

Whittingham, M., Guo, J. D., Chen, R., Chirayil, T., Janauer, G., & Aavalij, P. (1995). The hydrothermal synthesis of new oxide materials. *Solid State Ionics*, *75*, 257–268. doi:10.1016/0167-2738(94)00220-M

Xu, Z. H., Li, C. X., Li, G. G., Chai, R. T., Peng, C., Yang, D. M., & Lin, J. (2010). Article. *Journal of Physical Chemistry, C*, 114, 2573–2582.

Yan, R. X., & Li, Y. D. (2005). Down/Up Conversion in Ln3+-Doped YF3 Nanocrystals. *Advanced Functional Materials*, *15*(5), 763–770. doi:10.1002/adfm.200305044

Yan, R. X., & Li, Y. D. (2005). Down/Up Conversion in Ln3+-Doped YF3 Nanocrystals. *Advanced Functional Materials*, *15*(5), 763–770. doi:10.1002/adfm.200305044

Yan, Z. Y., Yan, B., & Jia, L. P. (2013). Article. *Materials Research Bulletin*, *1*, 39.

Ye, S., Xiao, F., Pan, Y. X., Ma, Y. Y., & Zhang, Q. Y. (2010). Phosphors in phosphor-converted white light-emitting diodes: Recent advances in materials, techniques and properties. *Materials Science and Engineering R Reports*, *71*(1), 1–34. doi:10.1016/j.mser.2010.07.001

Ye, X., Collins, J. E., Kang, Y., Chen, J., Chen, D. T. N., Yodh, A. G., & Murray, C. B. (2010). Morphologically controlled synthesis of colloidal upconversion nanophosphors and their shape-directed self-assembly. *Proceedings of the National Academy of Sciences of the United States of America*, *107*(52), 22430–22435. doi:10.1073/pnas.1008958107 PMID:21148771

Yeh, C. L., & Sung, W. Y. (2004). Synthesis of NiTi intermetallics by self-propagating combustion. *Journal of Alloys and Compounds*, *376*(1-2), 79–88. doi:10.1016/j.jallcom.2003.12.016

Yi, G. S., Lu, H. C., Zhao, S. Y., Yue, G., & Yang, W. J. (2004). Article. *Nano Letters*, *4*, 2191–2196. doi:10.1021/nl048680h

Yoshimura, M., & Byrappa, K. (2008). Hydrothermal processing of materials: Past, present and future. *Journal of Materials Science*, *43*(7), 2085–2103. doi:10.100710853-007-1853-x

Yoshimura, M., Suchanek, W. L., & Byrappa, K. (2000). Soft Solution Processing: A Strategy for One-Step Processing of Advanced Inorganic Materials. *Materials Research Bulletin*, *25*(09), 17–25. doi:10.1557/mrs2000.175

You, F., Wang, Y., Lin, J., & Tao, Y. (2002). Hydrothermal synthesis and luminescence properties of NaGdF4:Eu. *Journal of Alloys and Compounds*, *343*(1-2), 151–155. doi:10.1016/S0925-8388(02)00203-7

Youa, F., Wanga, Y., Lina, J., & Tao, Y. (2002). Article. *Journal of Alloys and Compounds*, *343*, 151–155. doi:10.1016/S0925-8388(02)00203-7

Zhang, C., Hou, Z., Chai, R., Cheng, Z., Xu, Z., Li, C., ... Lin, J. (2010). Article. *Journal of Physical Chemistry, C*, 114, 6928–6936.

Zhang, F., Wan, Y., Shi, Y., Tu, B., & Zhao, D. (2008). Ordered Mesostructured Rare-Earth Fluoride Nanowire Arrays with Upconversion Fluorescence. *Chemistry of Materials*, *20*(11), 3778–3784. doi:10.1021/cm800489e

Zhang, J., Hao, Z., Zhang, X., Luo, Y., Ren, X., Wang, X., & Zhang, J. (2009). Color tunable phosphorescence in KY3F10:Tb3+ for x-ray or cathode-ray tubes. *Journal of Applied Physics*, *106*(3), 034915–034916. doi:10.1063/1.3190511

Zhang, Q. Y., & Huang, X. Y. (2010). Recent progress in quantum cutting phosphors. *Progress in Materials Science*, *55*(5), 353–427. doi:10.1016/j.pmatsci.2009.10.001

Zhang, Q. Y., Yang, G. F., & Jiang, Z. H. (2007). Cooperative downconversion in GdAl3(BO3)4:RE3+, Yb3+ (RE=Pr, Tb, and Tm). *Applied Physics Letters*, *91*(5), 051903. doi:10.1063/1.2757595

Zhang, X., Quan, Z., Yang, J., Yang, P., Lian, H., & Lin, J. (2008). Solvothermal synthesis of well-dispersed MF_2 (M = Ca,Sr,Ba) nanocrystals and their optical properties. *Nanotech.*, *19*(7), 075603–075608. doi:10.1088/0957-4484/19/7/075603

Zhang, X. X., Bass, M., Villaverde, A. B., Lefaucheur, J., Pham, A., & Chai, B. H. T. (1993). Efficient laser performance of Nd:$GdLiF_4$: A new laser crystal. *Applied Physics Letters*, *62*(11), 1197–1199. doi:10.1063/1.108732

Zhao, C., Feng, S., Chao, Z., Shi, C., Xu, R., & Ni, J. (1996). Hydrothermal synthesis of the complex fluorides LiBaF3 and KMgF3 with perovskite structures under mild conditions. *Chemical Communications (Cambridge)*, (14): 1641–1642. doi:10.1039/cc9960001641

Zhao, X., Roy, R., Cherian, K., & Badzian, A. (1997). Article. *Science*, *385*, 513–515.

Zhou, Y. H., Lin, J., Yu, M., Wang, S. B., & Zhang, H. J. (2002). Synthesis-dependent luminescence properties of Y3Al5O12:Re3+ (Re=Ce, Sm, Tb) phosphors. *Materials Letters*, *56*(5), 628–636. doi:10.1016/S0167-577X(02)00567-0

Section 4

Emerging Synthesis Techniques of Luminescent Materials for OLED's Mechano and Electroluminescence Devices

Thermo-luminescence, photo-luminescence, electro-luminescence, and mechano-luminescence studies are discussed. Organic light-emitting diodes are overviewed and discussed.

Chapter 9
Phenanthroimidazole-Based Organic Fluorophores for Organic Light-Emitting Diodes:
An Overview

Sivakumar Vaidyanathan
National Institute of Technology Rourkela, India

Jairam Tagare
National Institute of Technology Rourkela, India

ABSTRACT

Considering the imminent global energy crisis and inefficient energy utilization, energy-efficient organic light-emitting diodes (OLEDs) are considered one of the most competitive candidates for displays and particularly for future energy-saving lighting sources. Full color displays require all primary colors: red, green, and blue (RGB). In recent decades, numerous phenanthroimidazole (PI) RGB-emitting materials have been developed for efficient OLEDs. In organic electronics, considerable interest is shown on PI, due to ease in fluorophore modification. This chapter focuses on the design and synthesis of PI-based materials and their applications in OLEDs. At first, some nondoped blue, green, and yellow fluorescent materials are comprehensively studied. Then attention has been paid for typical blue, green, yellow, orange, and red PhOLEDs of PI-based fluorophores as a host materials are briefly presented. The molecular design concept, general synthetic routes for PI materials, and the applications of fluorophores in fluorescent OLEDs and host materials in PhOLEDs are reviewed.

INTRODUCTION

Organic light-emitting diodes (OLEDs) are competitive candidates for the next generation flat-panel displays and solid state lighting sources since the pioneering work by Tang *et al.*, in 1987 (Tang, & VanSlyke, 1987). Extensive research has been carried out to promote OLEDs into commercial applications as solid-state lighting resources and smart displays (Kamtekar, Monkman, & Bryce, 2010). A

DOI: 10.4018/978-1-5225-5170-6.ch009

full-color display requires red, green, and blue emission of relatively equal stability, color purity and efficiency are imperative for achieving efficient white OLEDs. In full-color OLEDs, the blue emitter can not only efficiently decrease power consumption of the devices but also be utilized to produce light of other colours by energy cascade to lower energy fluorescent or phosphorescent dopants. However, the progress in highly efficient blue, green and red emitters has remained a formidable challenge to date (Ulla, et al., 2014, Lee, et al., 2015 Wang, et al., 2014, Tang, et a., 2015). Especially, the performance of a blue emitting device is often inferior to that of green and red devices for the intrinsic wide band gap of the blue emitting material. Spin-statistics suggest that the phosphorescent OLEDs (PhOLEDs) can achieve internal quantum efficiency (η_{int}) of 100% compared to that of fluorescent OLEDs (which can only reach up to η_{int} of 25%). By harvesting singlet and triplet excitons through intersystem crossing (heavy metal ion effect), one can achive100% internal quantum efficiency (IQE). A great deal of effort has been devoted to develop efficient phosphors that can emit color covering the whole visible region. Red, yellow and green, phosphorescent OLEDs (PhOLEDs) have achieved at very high efficiency with good device lifetime during the past decades. However, the design of efficient blue phosphors still remains an alternative word challenge as discussed in recent review (Chi, 2010). Since the innovative work of Forrest and Thompson on phosphorescent OLEDs, (Baldo, et al., 1998) there have been several studies to achieve 100% internal quantum efficiency in OLEDs and most studies on high efficiency OLEDs have been focussed on developing organic materials and device structures for phosphorescent OLEDs. Currently, a number of red, green, and blue phosphorescent OLEDs are reported by several research groups (Chopra, et al., 2008; Zhu, et al., 2011; Xiao, Su, Agata, Lan, &Kido, 2009,S. Su, et al., 2008; Bin, Cho, & Hong, 2012,S. Kim, et al., 2013; D. Kim, et al., 2013; Lee, & Lee, 2013).However, PhOLEDs have distinct disadvantages, which includes, the relatively long lifetime of phosphorescent heavy metal complexes leading to dominant triplet–triplet (T1–T1) annihilation at high currents, and may also cause a long range of exciton diffusion (>100 nm) that could get quenched in the adjacent layers of materials in OLEDs (T. Lee, et al., 2009) and they have lower electroluminescence (EL) efficiency under high current densities (Baldo, Adachi, & Forrest, 2000) rather low reliability in the blue region of the visible spectrum for practical applications and more expensive. However, to achieve an optimized device performance, careful control of the dopant concentration is required (Wu, Yeh, Chan, & Chen, 2002). On the other hand, performance degradation due to phase separation upon heating is also a problem encountered in these devices (Chang, He, Chen, Guo, & Yang, 2001, G. Zhong, et al., 2002; Gong, Wan, Lei, &Bai, 2005). Fluorescent OLEDs have continued to attract interest because of their long operational lifetimes, high colour purity of EL and potential to be manufactured at low cost in next-generation full-colour display and lighting applications. Fluorescent OLED devices have exhibited a good efficiency (Shih, Chuang, Chien, Diau,&Shu,2007,Y. Yuan, et al.,2011; K. Wu,et al., 2008; M. Wu, et al.,2007; H. Huang,et al., 2011; Xiao, Su, Agata, Lan, Kido, 2009; Chou, & Cheng, 2010).However, in fluorescent OLEDs, the IQE is limited to 25% due to the deactivation of triplet excitons (Baldo, Thompson, & Forrest, 2000; H. Nakanotani, e al., 2014).To overcome these disadvantages of both phosphorescent OLEDs and Fluorescent OLEDs, intensive research is going onto develop the efficient OLEDs (Kamtekar, Monkman, & Bryce, 2010; Wang,& Ma,2010; Zhu, Peng, Cao, & Roncali, 2011; Tao, Yang, &Qin, 2011; Jeon, & Lee, 2012; Yook, & Lee, 2012; Zhu, &Yang, 2013; Yang, Huang, Li, & Li, 2016).

Recently, a new concept was introduced by Leo *et al.*, (G. Schwartz, et al., 2009) to fabricate white OLEDs by combining blue fluorescent emitters with green, red and/or orange phosphorescent emitters. This methodology permits the harvesting of both the fluorescence from the host of blue emitters and the complementary phosphorescence from the triplet dopants simultaneously to generate white emission. In this context, highly efficient blue fluorescent materials have been considered as a promising alternative to promote the commercialization of OLEDs. However, the performance of blue emitting devices is often underdeveloped by the intrinsic wide band gap of blue emitting materials, which makes it hard to inject charges into emitters. Solving this problem is a materials related issue. Significant efforts have been made to design versatile blue fluorescent materials aiming at further improving device efficiency, chromaticity and lifetime. The output is very prosperous. For instance, Adachi et al. fabricated pure blue OLEDs by efficient thermally activated delayed fluorescence materials, and achieved amazing external quantum efficiencies of 9.9% and Commission Internationale del'Eclairage (CIE) coordinates of (0.15, 0.07), which is nearly double the theoretical upper limit (5%) for fluorescent devices (Q. Zhang, et al., 2012).

Recently some blue, green, yellow, orange and red fluorophores with high luminescent efficiency based on pyrene, anthracene, phosphine oxide, fluoranthene, pyrazolo-quinoline, oxadiazoles, naphthalimide, phenanthroimidazole have been developed (Z. Wang, et al., 2010; Z. Wang,et al., 2011; Y. Sun, et al., 2011; C. Wu, et al., 2012; Y. Lee, et al., 2013; Ulla, Garudachari, Satyanarayan, Umesh,&Isloor,2014; Jeon, Jang, Son, &Lee, 2011; H. Inomata, et al., 2004; W. Yuan,et al., 2012; Y. Zhang, et al., 2012). However, phenanthroimidazole continue to show considerable interest in organic electronics, due to ease in fluorophore modification at N1 and C2 position. On the other hand, these materials have the rigid skeleton, show high carrier mobilities, high thermal stability, as well as good film forming ability, leading to promising applications as green, yellow and blue emitters in high-performance fluorescent OLEDs (Subeesh, Shanmugasundaram, Sunesh, Nguyen, & Choe, 2015; X. Tang, et al., 2015; W. Qin, et al., 2015; Y. Yuan, et al., 2013).On the other hand these materials have high triplet energy, hence, these materials are useful as host materials for blue, green, yellow, orange, and red high performance PhOLEDs (D. Liu, et al., 2015; M. Chen, et al., 2015; B. Wang, et al., 2016; C. Li, et al., 2015).Su*et al.,* developed two phenanthroimidazole based blue fluorophores with excellent hybridized local and charge-transfer (HLCT) and "hot exciton" properties for fluorescent blue OLEDs and the host for orange-red phosphor to achieve highly efficient hybrid white organic light-emitting diodes (WOLEDs) in a single-emissive-layer single-dopantarchitecture even at a high concentration of phosphorescent dopant (X. Ouyang, et al., 2015).

This book chapter focuses on the design and synthesis of phenanthroimidazole based blue, green, yellow and red fluorescent and phosphorescent materials and their applications in OLEDs. At first, some nondoped blue, green, and yellow fluorescent materials endowed with unipolar and bipolar-transporting abilities are comprehensively studied. Then some typical blue, green, yellow, orange and red PhOLEDs of phenanthroimidazole based fluorophores as hosts are briefly presented. The molecular design concept, general synthetic routes for PI materials, thermal stabilities, and the applications of fluorophores in fluorescent OLEDs and host materials in PhOLEDs are reviewed. We expect this book chapter provides valuable insights into the design criteria for materials towards efficient OLEDs.

Luminescence

Luminescence is the production of light from a compound (either organic or inorganic). Organic compounds that are suitable for OLED applications are expected to display photoluminescence and electroluminescence. In order for a material to emit energy in the form of light energy, it needs to be in a higher energy (excited) state which can return back to the ground state by releasing energy in the form of a photon or through a nonradiative pathway. The excited state can be populated by photoexcitation (photoluminescence) or electronic excitation (electroluminescence).

Photoluminescence

In photoluminescence, an incident photon excites an electron from the highest occupied molecular orbital (HOMO) to the lowest unoccupied molecular orbital (LUMO). A molecular excited state is formed from which the electron can then decay back to the HOMO. This process can be radiative or non-radiative, i.e.,a photon can be released or a non-radiative process can aid the relaxation of the electron to the HOMO. The energy of the photon will be equal to the energy difference between the HOMO and LUMO. The photoluminescence (fluorescence and phosphorescence) phenomenon can be explained by Jablonski diagram (Figure 1).

Electroluminescence

Electroluminescence can be defined as the production of light from a material when an electric field is applied across it. OLED's main working principle is electroluminescence. Organic light-emitting diodes are double charge injection devices, which require the simultaneous supply of both holes and electrons to the emissive layer (Kulkarni, Tonzola, Babel, & Jenekhe,2004). OLED device is sandwich between

Figure 1. Photoluminescence phenomena (schematic diagram)

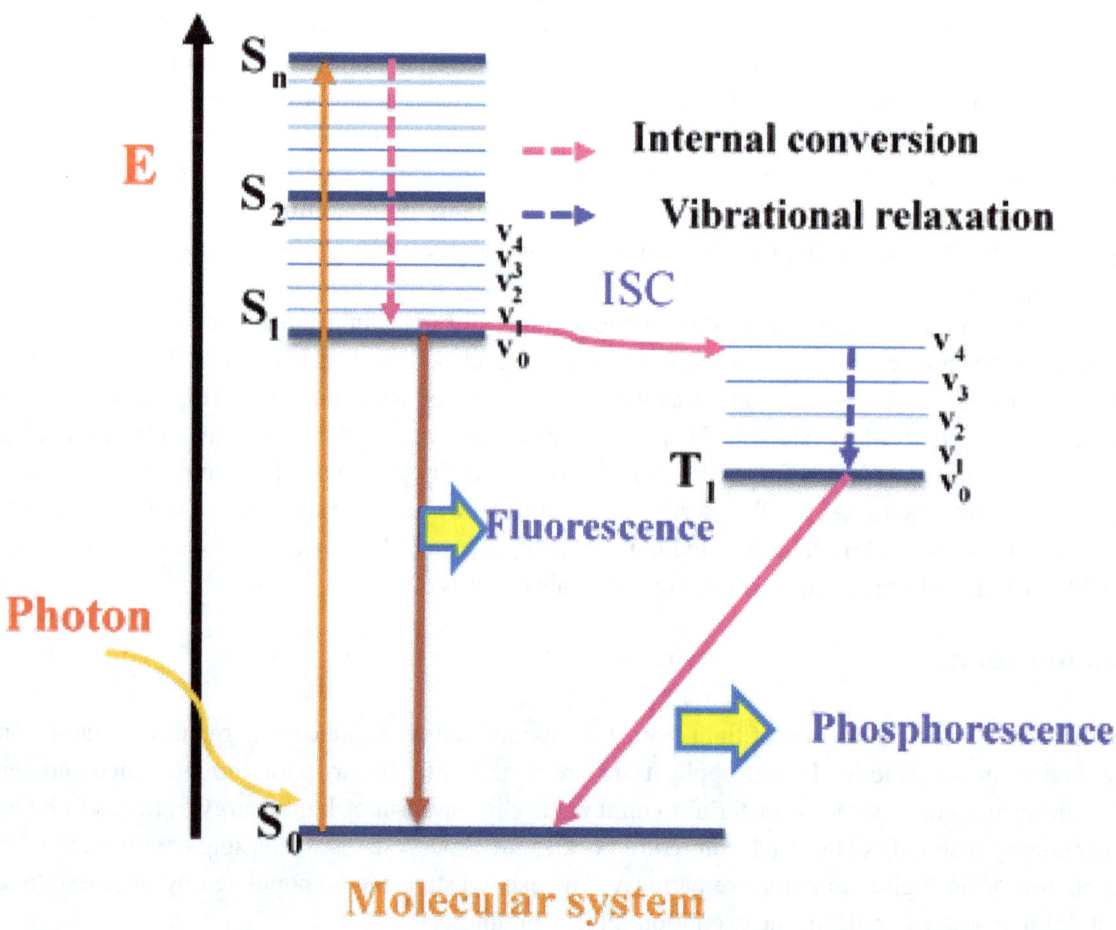

two electrodes which are cathode and anode. When avoltage is applied to it, electrons are injected from the cathode into the LUMO and holes are injected from the anode into the HOMO. This forms radical anions with an electron (negatively charged) on the LUMO and radical cations with a hole (positively charged) on the HOMO (Figure 2). If the charges combine within the emissive material they can form a molecular excited state. These excited states are called 'excitons', which can release photons to decay to the ground state. A schematic of Electroluminescence can be seen in below Figure 2.

A successful OLED generally needs facile and balanced charge transport as well as a high conversion efficiency of excitons to light (Lundin, Blackman, Gordon, & Officer,2006). To fulfill these requirements, efficient OLEDs have a tendency to have multilayer device configurations, an electron transport layer (ETL), a hole transport layer (HTL) and an emissive layer (EML), some of the devices also have an electron injection layer and a hole-injection layer to improve the device performance. Generally, the charge injection/transport layers are used to inject and transport holes and electrons to the EML, where the charges recombine and form the excitons. Schematic energy-level alignment of host–dopant systems is shown in Figure 3. Generally, in host/guest (dopant) systems, two routes can lead to dopant emission (Baldo, & Forrest,2000). (a) The singlet excitons formed in the host under electrical excitation can be transferred to dopant via Forster energy transfer (Figure 4.) and then singlet excitons in the dopant are converted to triplet excitons by efficient intersystem crossing. (b) The triplet excitons formed in the host

Figure 2. Electroluminescence phenomena

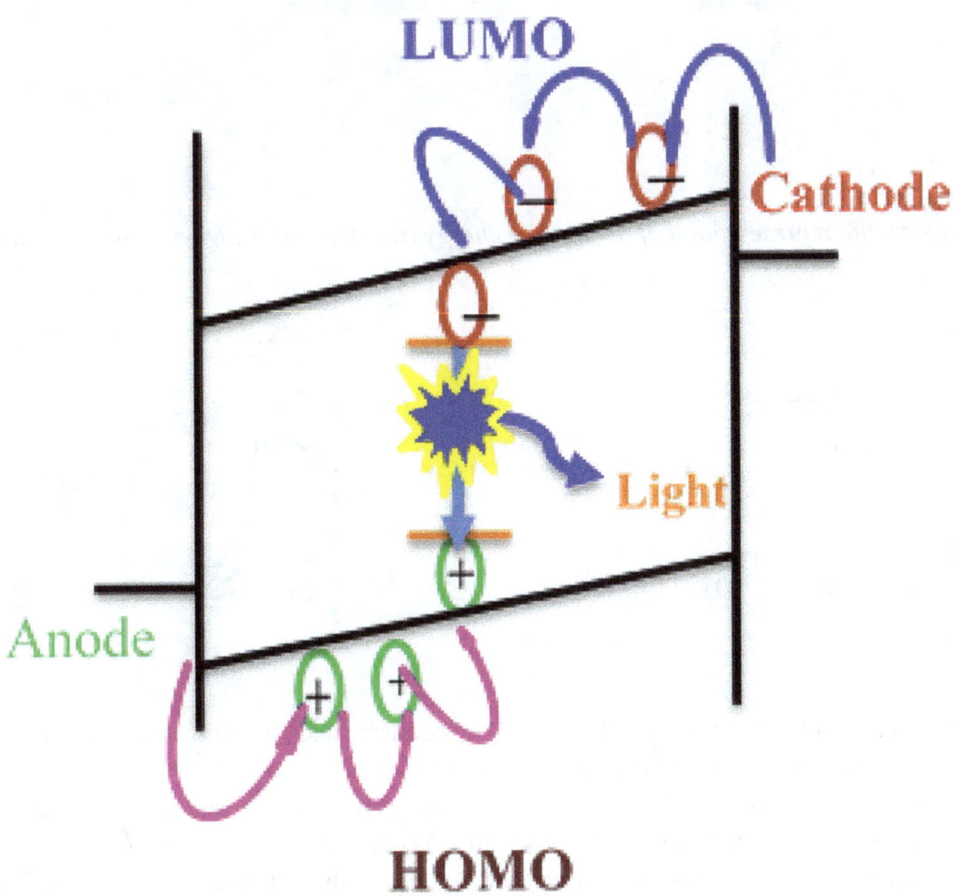

Figure 3. Schematic energy-level alignment of host–dopant systems

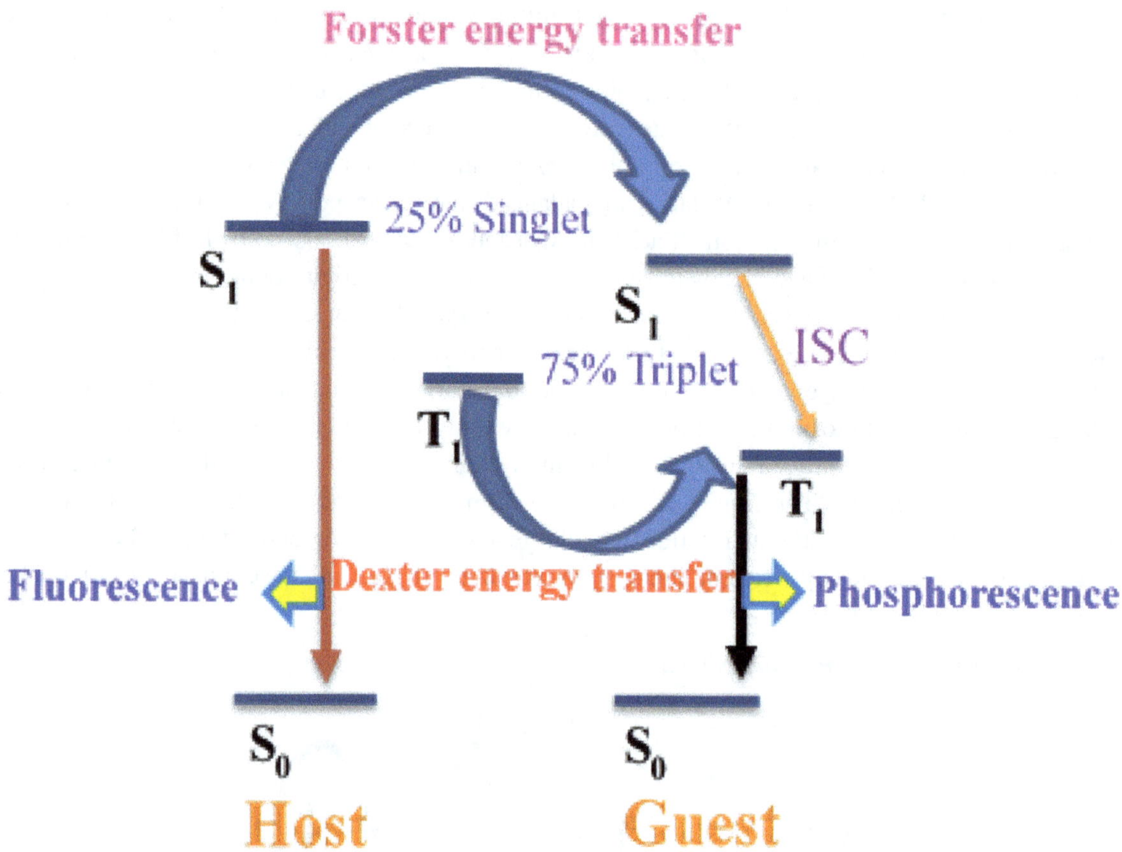

Figure 4. Schematic representation of (a) Forster energy transfer and (b) Dexter energy transfer

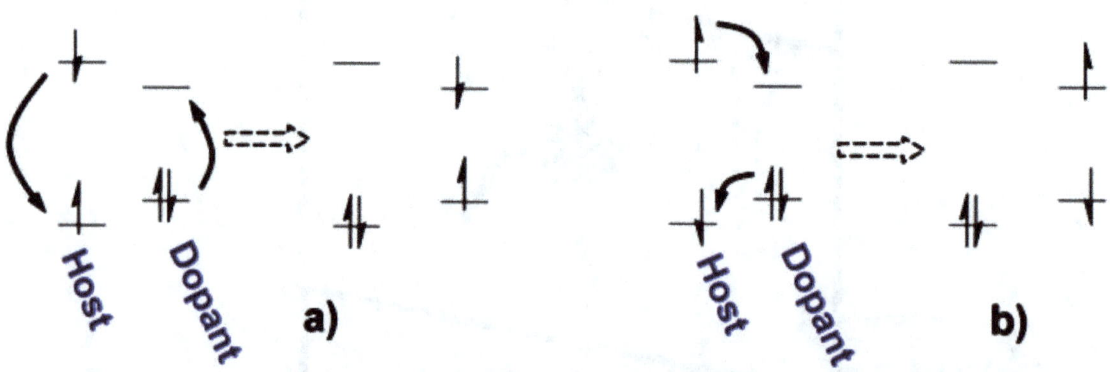

can be transferred to phosphor through Dexter energy transfer and then the radiative decay of triplet excitons (phosphorescence) in the phosphor gives light emission.

Forster energy transfer is a Coulombic interaction between the host exciton and the dopant, which is a fast and long-range process (up to 10 nm) while Dexter energy transfer is an electron exchange interaction between the host exciton and the dopant, which is a short distance process (1.5–2.0 nm) (Y. Huang, et al., 2010).

OLED Device Structures

A schematic representation of OLED and its mode of operation are depicted in Figure5 It consists of an anode, typically indium tin oxide (ITO), the hole transporting layer (HTL), the emissive layer (EML), in fluorescence OLED the EML containing only emissive materials whereas in phosphorescent OLED the EML containing a host material and the emissive dopant, the hole blocking layer (HBL), the electron-transporting layer (ETL), and the cathode. Generally, anode should have high work function than the cathode and is transparent in order to allow the output of the electro generated light. The cathode is made of a low work function metal that facilitates electron injection. The thickness of each active layer is generally comprised between a few and a few tens of nanometres in order to allow the development of a high electric field upon application of a moderate voltage across the device. The energy difference between the highest occupied molecular orbital (HOMO) and lowest unoccupied molecular orbital (LUMO) of the materials, namely its band gap, determines its most relevant properties and in particular the energy of the emitted photons and hence the color of the generated light. When a forward bias is applied, electrons are injected from the cathode into the LUMO of the emitter to form negative charge carriers (radical anions/electrons (-)) and holes are injected from the anode into the HOMO to form the positive charge carriers (radical cations/ holes (+)) (1). The charge transfer is recognized via a hopping-mechanism between the localized p-electron states. These charge carriers migrate in the emitting layer (3) and their recombination on the same emitting molecule leads to the formation of excitons, which can then undergo radiative decay to give rise to light emission(4) (fluorescence OLEDs) (Kraft, Grimsdale, & Holmes,1998), whereas, in Phosphorescent OLEDs the hole and electron accumulate at the LUMO and HOMO of the emissive dopant (3). Then the electron–hole pair formation occurs leading to an exciton generation (4) which can then undergo radiative decay to give rise to light emission.

Basic Properties of Phenanthroimidazole (PI)

Phenanthroimidazole consist of electron deficient Phenontroline moiety as well as five member heterocyclic compounds (imidazole), phenontroline moiety is electron deficient in nature, it has been proved that phenontroline derivatives such as BCP and 4,7-diphenyl-1,10-phenanthroline (BPhen) are well-known as an electron transporting and hole blocking materials (Kijima, Asai, & Tamura, 1999; O'Brien, Baldo, Thompson, & Forrest, 1999; Naka, Okada, & Onnagawa,2000; Huang, Pfeiffer, Werner, Blochwitz, &Leo,2002).Imidazole and Phenontroline are electron deficient molecules, so phenanthroimidazole-groupin the fluorophores (unipolar, bipolar) acting as an electron deficient group will efficiently transfer the electron in the OLEDs. The PIgroup consists of two highly electronegative nitrogen atoms, which makes the PI group highly polar and electron with drawing making it advantageous for its application as host in PHOLEDs (Naka, Okada, & Onnagawa, 2000). Phenanthroimidazole continues to show considerable interest in organic electronics, due to ease in fluorophore modification at N1 and C2 position. On the other hand, these materials have the rigid skeleton, show high carrier mobilities, high thermal stability, as well as good film forming ability, leading to promising applications in high-performance OLEDs (M. Subeesh, et al., 2015; C. Tang, et al., 2015).On the other hand these materials, having high triplet energy, are used as host materials for high-performance PhOLEDs. The electron withdrawing properties of the PI moiety can be improved with different substitution (electron withdrawing groups) at N1 and C1 position of the PI group. The PIgroup can be effectively combined with the hole transport units (carbazole, diphenylamine and triphenylamine) to obtain bipolar charge transport properties in the fluorophores for organic light emitting diodes.

Figure 5. (A) Schematic representation of a multi-layered Organic light emitting diode (OLED). (B) Schematic diagram of the OLED structure.

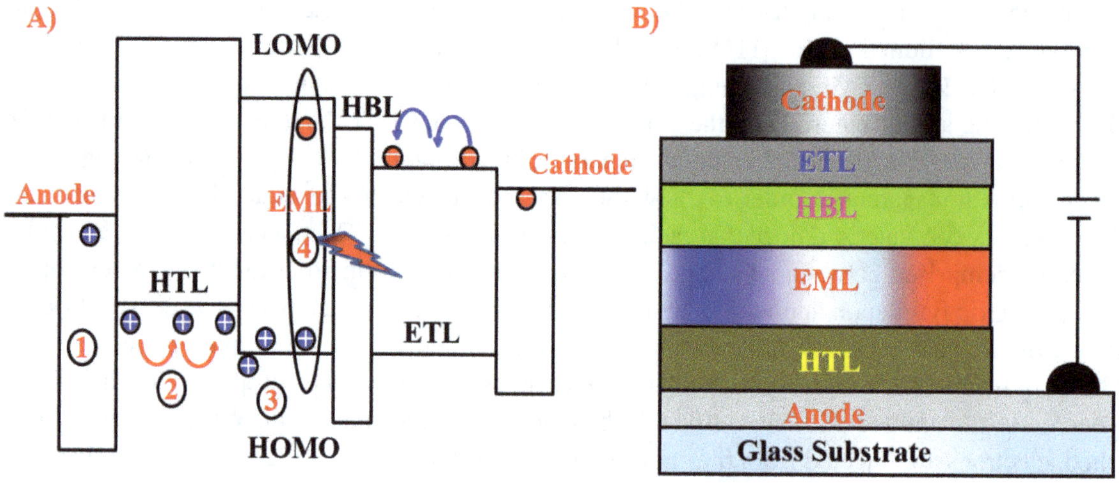

Molecular Design of Phenanthroimidazole Based Fluorophores

PI compounds/derivatives can be designed/chemical modification by attaching R and R[1] group (electron rich or electron deficient or both). In this molecular design strategy, the Phenontroline moiety is acting as a core group (D. Liu, et al., 2015; M. Chen, et al., 2015). R groups should be amine containing alkyl chain, phenyl, phenyl substituted, fluorene, naphthalene, naphthalene substituted, anthracene, anthracene substituted, pyrene, carbazole, and triphenylamine groups (Q. Zhang, et al., 2012). R[1] groups should be aldehyde or boronic acid containing, phenyl, phenyl substituted, double bond spacer phenyl, naphthalene, naphthalene substituted, anthracene, anthracene substituted, pyrene, carbazole, and triphenylamine groups (B. Wang, et al., 2016).These R and R[1] can act as electron rich (carbazole/triphenylamine), electron deficient (pyridine, isoquinoline, electron withdrawing group substituted phenyl…etc.,) and spacer group in the molecular structures. When these group combining through the chemical reaction (condensation and Suzuki coupling reaction), five member hetero cyclic imidazole moiety with different substituting at N1 and N2 position was obtained. This formed imidazole ring can act as electron deficient group and will easily transfer the electron to the device. The electron deficiency of the imidazole moiety can be increased by adding different electron withdrawing group in the N1 and C1 position of the imidazole moiety. By using this method, one can produce a number of various unipolar and bipolar phenanthroimidazole fluorophores for organic light emitting diodes. The molecular design of selected PI derivatives is outlined in Figure 6. Here, we have clearly shown the design strategy for the phenanthroimidazole derivatives.

Synthetic Routes for the PI Derivatives

The PI derivatives were synthesized by following the reaction schemes (Figure 7). The targeted fluorophore 'I' was synthesized via one step reaction by condensation. The condensation occurs among the diketone (phenantrolene-9,10-dione core), amine (amine with R group) and aldehyde (aldehyde with theR[1]group) in the presence of ammonium acetate and acetic acid. The synthetic strategy to obtain the

Figure 6. Schematic representation of the molecular design of PI derivatives

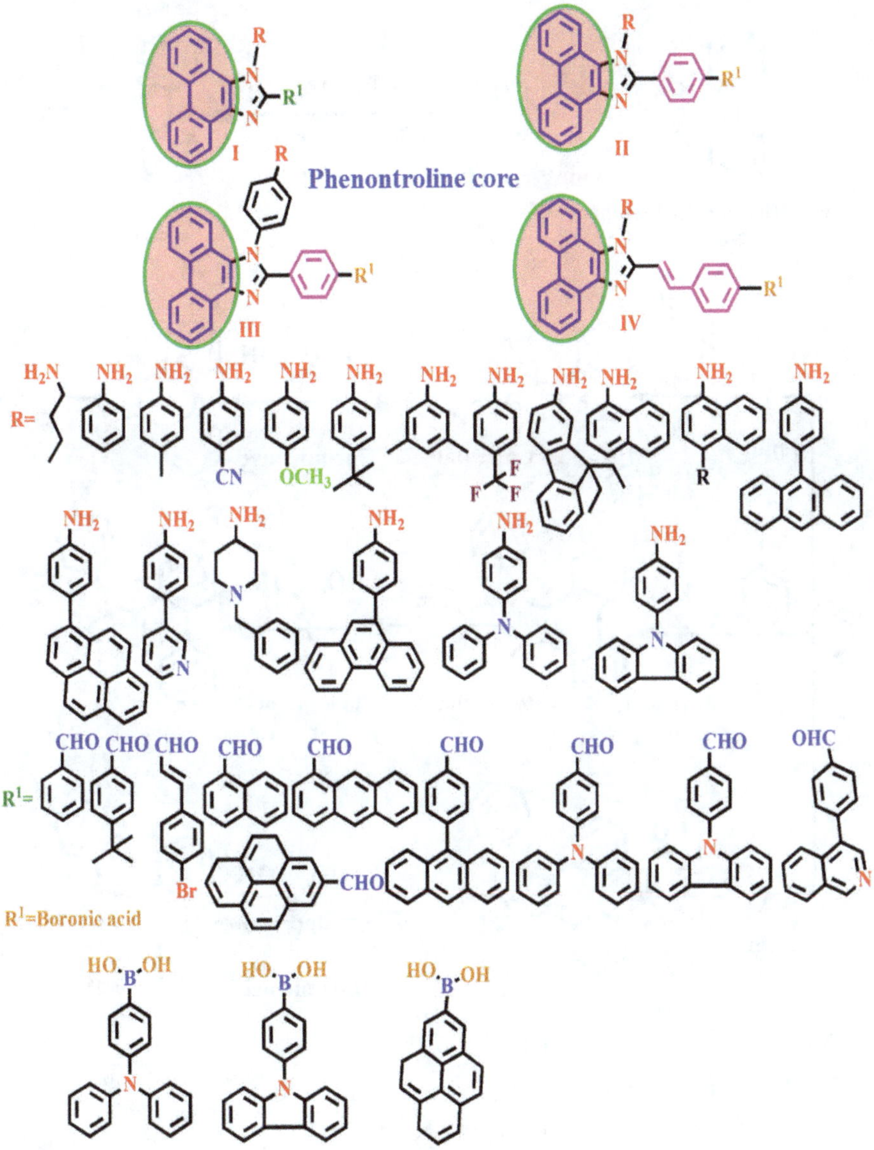

target PI fluorophores Iis illustrated in Scheme-1 in Figure 7. The targeted fluorophores II, III and IV was synthesized via two step reaction by condensation and Suzuki coupling reaction. The condensation occurs among diketone (phenantrolene-9, 10-dione core), amine (amine with R group) and bromo-benzaldehyde/ bromo-acrylaldehyde in the presence of ammonium acetate and acetic acid is to be carried out. Further, the obtained product was reacted with boronic acid (R^1=boronic acid) by Suzuki coupling reaction to achieve the desired PI fluorophores. The synthetic strategy to obtain the target PI fluorophores II, III, and IV are illustrated in Scheme-2 in Figure 7.

Figure 7. Synthetic routes for the Phenanthroimidazole derivatives

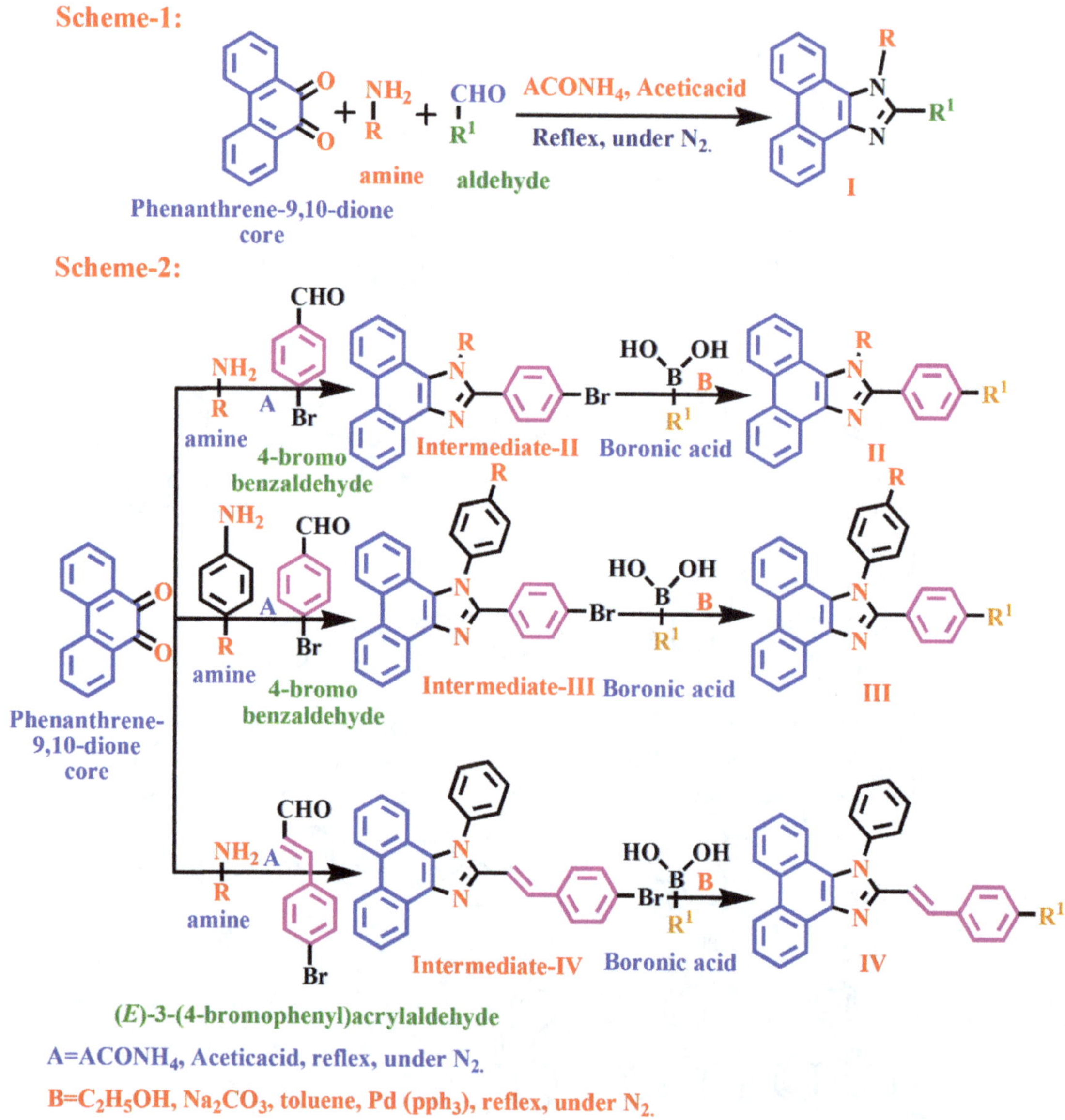

Phenanthroimidazole Based Materials for Fluorescent OLEDs

In recent decades phenanthroimidazole based materials are used as emissive materials in the fluorescent OLEDs. These materials have the potential to produce blue, green and yellow colors fluorescent OLEDs. These materials are generally used to produce deep blue color OLEDs, due to ease in fluorophore modification at N1 and C2 position (X. Ouyang, et al., 2015; A. Kulkarni, et al., 2004; N. Lundin, et al., 2006; Baldo, & Forrest, S. 2000; Y. Huang, et al., 2010; Kraft, Grimsdale, &Holmes,1998; Kijima, Asai,&Tamura,1999; O'Brien, et al., 1999; Naka, Okada,& Onnagawa, 2000; J. Huang, et al.,2002; Y. Zhang, et al., 2016; M. Chen, et al.,2015; Jayabharathi, Prabhakaran, Thanikachalam, &Sundharesan,

2016; T. Shan, et al., 2017). On the other hand, these materials have rigid skeleton, show high carrier mobilities, high thermal stability, as well as good film forming ability, leading to promising applications in high-performance fluorescent OLEDs (S. Zhang, et al., 2012; M. Subeesh, et al., 2015).In this section, we will comprehensively discuss about unipolar and bipolar fluorophores based on PI for fluorescent OLEDs.

Unipolar PI Fluorophores for Fluorescent OLEDs

Unipolar materials consist of electron deficient or electron rich groups, thermally stable and photophysical active groups (naphthalene, anthracene, and pyrene.... etc.). A desirable unipolar fluorescent material should have good charge carrier transport ability, good film forming ability and good thermal stability. Thus, the molecular structure of unipolar fluorescent materials usually incorporates hole or electron transport units and thermally stable groups. Various unipolar phenanthroimidazole based emissive materials are reported till now (T. Shan, et al., 2017; Zhuang, 2012; Y. Zhang, et al., 2013; S. Zhuang, et al., 2014; Z. Wang, et al., 2014; J. Wang, et al., 2015). Here we want to discuss about phenanthroimidazole (only), anthracene, naphthalene, pyrene, and pyridine containing unipolar fluorescent materials for fluorescent OLEDs.

N-Type Phenanthroimidazole (Only) Based Fluorophores for Fluorescent OLEDs

Liu *et al.*, designed a series of n-type phenanthroimidazole fluorophores (1PPIP, 2TPIP and 3APIP) as emissive materials for deep blue nondoped OLEDs (C. Kuo, et al., 2009).The design tactic used to develop these fluorophores is by connecting two phenanthroimidazole group shaving different N1 substitutions with the biphenyl group through a chemical reaction. Nondoped devices based on 1PPIP as the emissive layer turned on at very low voltages of <3.0 V and exhibited pure-blue light with CIE coordinates of x=0.15, y=0.14. The 1 PPIP-based device-A-C (Figure 8.) can reach not only a high external quantum efficiency of 6.31% but also a high power efficiency of 7.30 lm/W. Furthermore, the power efficiency remains as high as 4.94 lm/W at a practical brightness of 200 cd/m^2. When using 2TPIP or 3APIP as the emissive layer, the devices also exhibited high external quantum efficiencies of more than 5%and highly saturated blue light can be obtained from the 2TPIP-based device with $CIE_y \leq 0.10$.The superiority of phenanthroimidazole attracted considerable attention.

Ma *et al.,* designed and synthesized a series of linear and branched blue-emitting PI derivatives (4PPI, 5BPPI and 6TPPI) (Z. Wang, et al., 2011). The synthetic routes of PI derivatives shown in scheme-3. The double-layered devices exhibit blue emission peaked at around 451 nm for 4 PPI,468 nm for 5 BPPI and 454 nm for 6 TPPI. The 5 BPPI based device (identical to that of 1 PPIP reported by Liu *et al.,*) achieved a higher luminance and a lower turn-on voltage than the other device. The results suggest that the phenanthroimidazole unit is an excellent building block for tuning the carrier injection properties as well as blue emission.

Scheme 3. The synthetic routes to 4 PPI, 5 BPPI and 6 TPPI. (a. bis(pinacolato)diboron, $Pd(PPh_3)_4$, sodium carbonate, THF and distilled water, refluxed for 48 h; b. 4-bromobenzaldehyde, bis(pinacolato) diboron, $Pd(dppf)_2Cl_2$, potassium acetate in DMSO, 80^{0}C, 24 h; c. $Pd(PPh_3)_4$, sodium carbonate in THF and distilled water, refluxed for 48 h under nitrogen; d. in CH_3COOH and ammonium acetate, refluxed for 2 h.)(Wang, 2011).

Figure 8. A) Luminance Vs Voltage characteristics of device A-C (1 PPIP). Inset: the EL spectra at 8 V and B) the EL spectra of device B (1 PPIP) at various applied voltages.

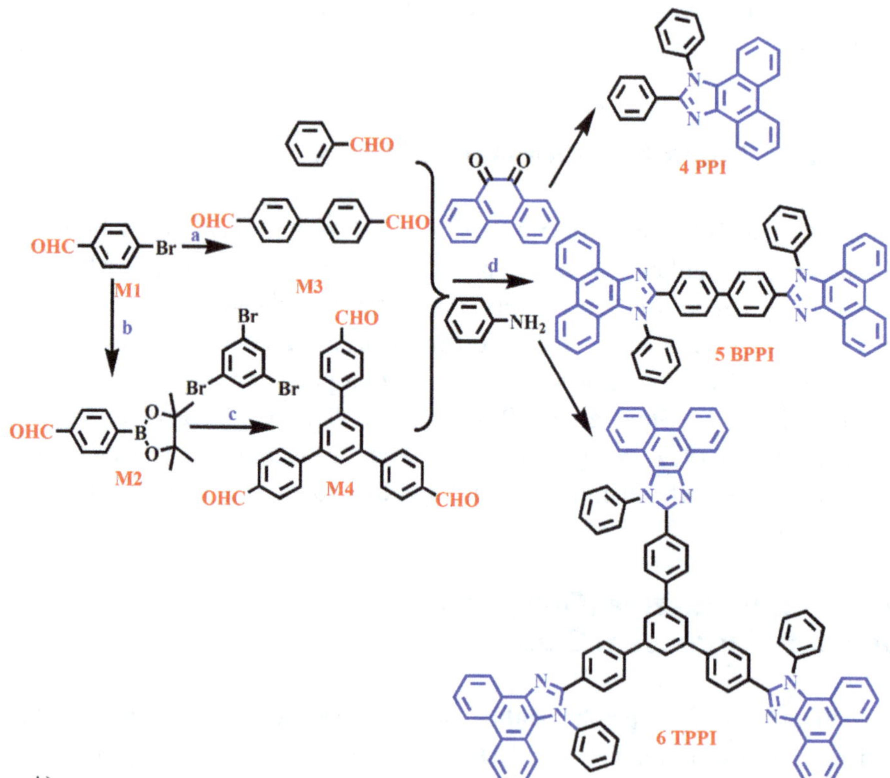

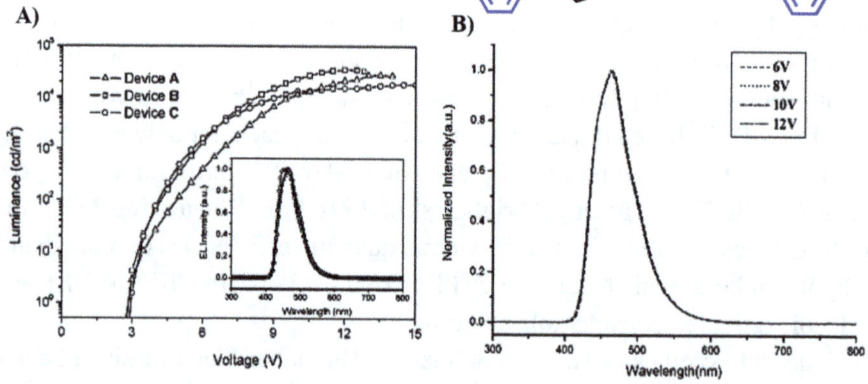

Y. Wang *et al.*, synthesized four PI derivatives (7 a, 8 b, 9 c and 10 d) and were carefully studied their single crystal structures, photophysical, electrochemical and mobility properties (Y. Yuan, et al., 2011). The highly efficient Alq_3based electroluminescence devices have been achieved by employing compounds7 a - 10 d as an organic buffer layer inserted between NPB and Alq_3 layers. A maximum luminescence efficiency (LE) of 8.1 cd/A was found with a maximum brightness of 65130 cd/m^2 at 15 V for 10 d based device, which exhibited much higher efficiencies compared to the device of [ITO/NPB/Alq_3/LiF/Al] without 7 a - 10 d. This improvement is due to the comparatively weak hole transport (HT) ability of the phenanthroimidazole derivatives. This can reduce excessive hole to transport into the

EML, thus improve the balance of the hole and electron and realize the higher efficiency. The results of this study not only demonstrated an alternative idea to design novel HT materials but also provided a convenient way to improve the performance of the NPB/Alq_3based devices by inserting a suitable organic buffer layer in the device.

Wang *et al.,* prepared a new PI based material 11 CN-BPPI (Z. Wang, et al., 2012). The 11 CN-BPPI showed similar absorption and emission spectra of 5 BPPI. Compared with the model compound 5 BPPI, the cyano-group inserted did not influence or effect the blue emission. However, the material showed comprehensive enhancement in basic characteristics involving purification costs, fluorescent efficiency, thermal stability, and charge injection properties. Similar to 5 BPPI, the device of 11 CN-BPPI exhibited blue emission peaked around 468 nm. The device based on 11 CN-BPPI showed better performance with a maximum luminance of 1400 cd/m^2, a maximum current efficiency of 0.78 cd/A and a maximum luminance efficiency of 0.26 lm at lower current density than 5 BPPI, which should be due to the more balanced proportion of holes to electrons (excitons). It was proved that 11 CN-BPPI could depress excessive holes to transport in the emission-layer and could realize the tuning of carrier mobilities. Wang *et al.,* synthesized two isomers of 5 BPPI, 12 L-BPPI and 13 Z-BPPI by tuning the substituted position (chemical modification of structure) on the biphenyl from para- to meta-coupling (Q. Wang, et al., 2014). Since it can help to change the conjugated degree at the C2-position in the PI block. In order to weaken the change in the degree of conjugation at the C2-position, the para-coupling of biphenyl at the C2-position was replaced by the alternative meta-coupling, which was employed because of its weak conjugation ability. By tuning the substituted position on the biphenyl center, the fluorescence of 12 L-BPPI (433 nm) and 13 Z-BPPI (402 nm) were obviously blue-shifted in comparison with5 BPPI (468 nm) in the films, and the non-doped devices based on 12 L-BPPI and 13 Z-BPPI as emitters exhibited stable deep blue emissions with CIE coordinates of (0.16, 0.10) and (0.16, 0.11), respectively (Figure 9). The new arrival (12 L-BPPI and 13 Z-BPPI) maintained higher thermal stability, balanced charge injection and proper carries barriers properties like BPPI, and had a certain suppressed aggregation ability. As a successful example, the structure properties relationship of 12 L-BPPI or 13 Z-BPPI reveals not only more information about the C2 substituted group, but also new ideas for the design of other PI derivatives.

Lee *et al.,* designed and synthesized two PI derivatives (14 BBTPI and 15 mBBTPI) by linking two phenanthroimidazole units via the para and meta position of a freely rotatable phenyl bridge (M. Chen, et al., 2015). In this design, the freely rotatable benzene ring in the molecular center plays a vital role in confining p-conjugation due to its tetrahedral electronic conformation (J. Ye, et al., 2013). The design strategy provides a suitable level of conjugation between the two phenanthroimidazole units such that fluorescence is strengthened over the single unit while a violet-blue emission can be maintained by limiting the amount of red shift behavior. By using a suitable device structure, OLED was fabricated based on 15 mBBTPI as a non-doped EML, which exhibits decent performance with current efficiency (CE) up to 1.99 cd/A, power efficiency (PE) up to 1.81 lm/W and a color index of (0.16, 0.06). The obtained results are comparable to the reported efficient non-doped violet blue devices.

By using molecular twisting and bulky side group substitution concept, Lee *et al.,* was developed two deep-blue emitters, 17 BiPI-1 and 18 BiPI-2 (Z. Zhu, 2016). These two twisted linear molecules maintain the excellent photophysical and electrical properties as compared to that of planaranalog and remarkably suppress/limit the redshifts in solid state emission. In particular, 17 BiPI-1 showed deep-blue emission in both solution and solid phase, and high carrier mobilities. Standard blue electroluminescence (color purity: (0.15, 0.08)) and high device efficiency (CE: 4.62 cd/A, EQE: 6.18%) were observed in a

Figure 9. The EL spectra of the devices at 9 V

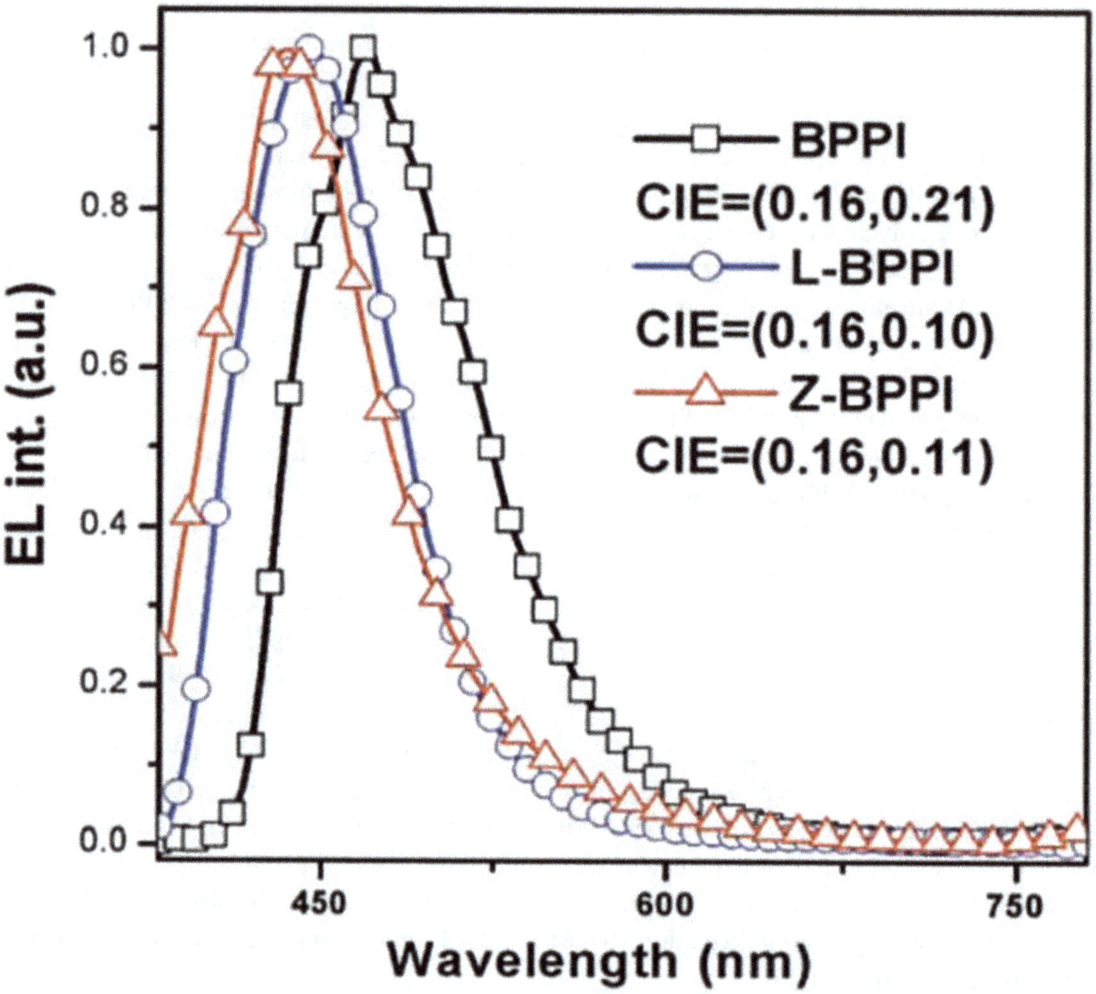

non-doped OLED of 17 BiPI-1. This research shows a successful example of deep-blue molecule design by twisting linear molecular systems to tune color purity without scarifying photophysical properties and overall efficiency. Lee *et al.*, designed a phenanthroimidazole (PI) as donor, a donor–acceptor type deep-blue fluorophore (20 BITPI) (W. Chen, 2017) 20 BITPI displays a mild charge transfer (CT) feature as evidenced by theoretical simulation and photophysical analysis. Results from UV-aging test on neat films of 20 BITPI, demonstrated that PI is indeed a more stable donor comparing to common arylamineanalogs. An OLED using 20 BITPI as an emitter was fabricated and exhibits good device performances EQE over 7% (Figure 10.) with stable deep-blue emission (color index: (0.15, 0.13)) and longer operation lifetime than the similarly structured device using arylamine-based emitters (M. Lai, 2008; M. Liu, 2015). Single-organic layer device based on 20 BITPI also shows superior performances, which is comparable to the best results from the arylaminebased donor–acceptor emitters (Feng, 2014; Huang, 2006), suggesting that PI is a stable donor with good hole transport/injection capability.

Figure 10. Performances of the BITPI-based OLED

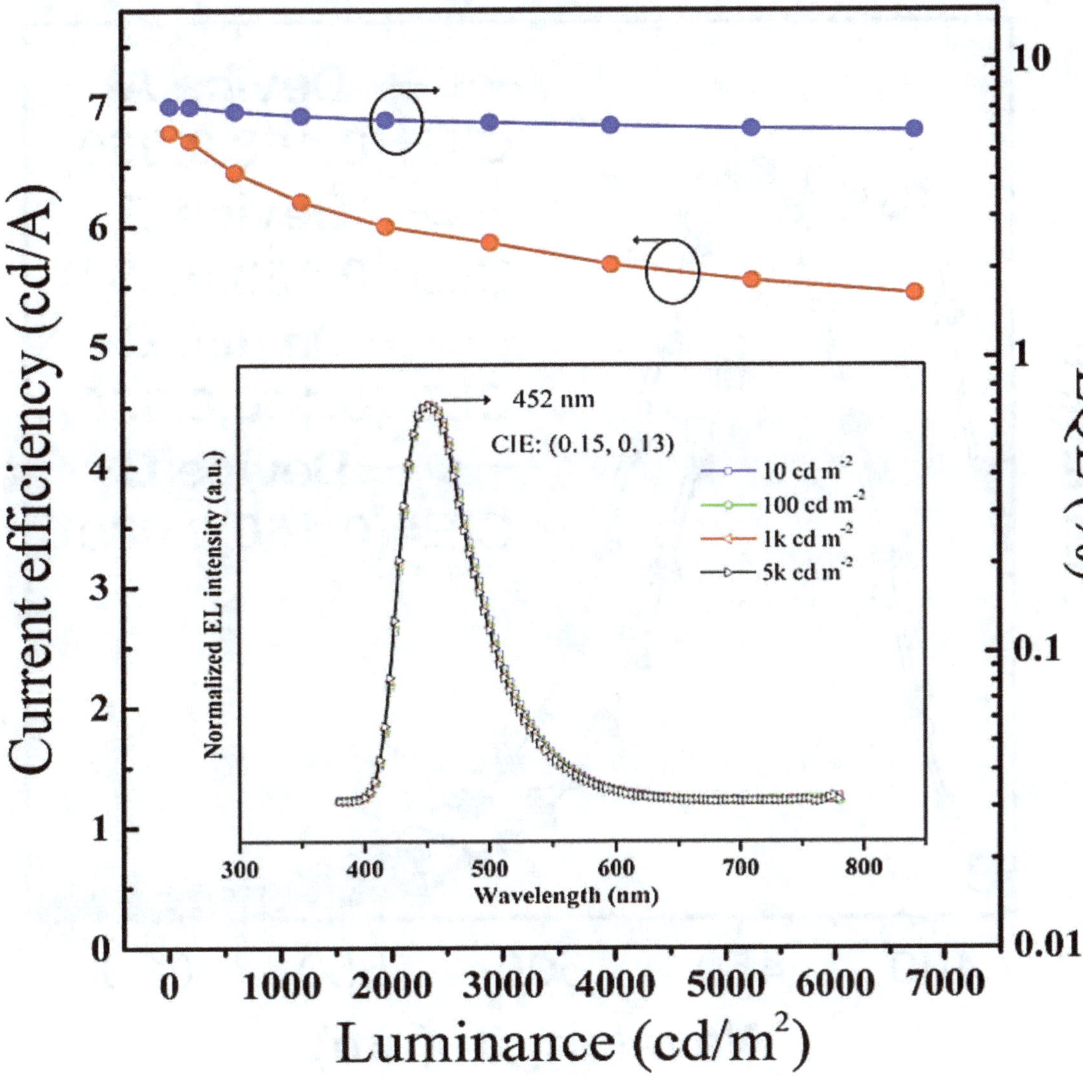

Wang et al., designed and synthesized two deep-blue fluorophores, 21 mm-BPPI and 22 mm-CN-BPPI, by integrating two 1,2-diphenyl-1H-phenanthro[9,10-d]imidazole (PPI) groups in a meta–meta pattern to the benzene moiety as the bridge (Z. Wang, et al., 2016).Owing to this meta–meta linking, the dimer conjugation degree of 21 mm-BPPI was effectively limited. Thus 21 mm-BPPI exhibited very similar photophysical and electrochemical properties of PPI. Due to this meta–meta connection in the mm-BPPI, thermal stability is improved as compared to that of PPI blue fluorophores overcoming the limitation of using the PPI based fluorophores for practical application. Non-doped optimized devices were fabricated with 21 mm-BPPI and 22 mm-CN-BPPI, and exhibited stable deep-blue emission with CIE coordinates of 0.156, 0.050 (21 mm-BPPI) and 0.150, 0.080 (22 mm-CN-BPPI), respectively (Figure 11). After inserting cyano groups in the basic PPI structure, higher glass-transition temperature and a quasi-reversible redox process were observed in 22 mm-CN-BPPI. Additionally, the lower efficiency

Figure 11. The EL spectra of the devices and the CIE at 6V

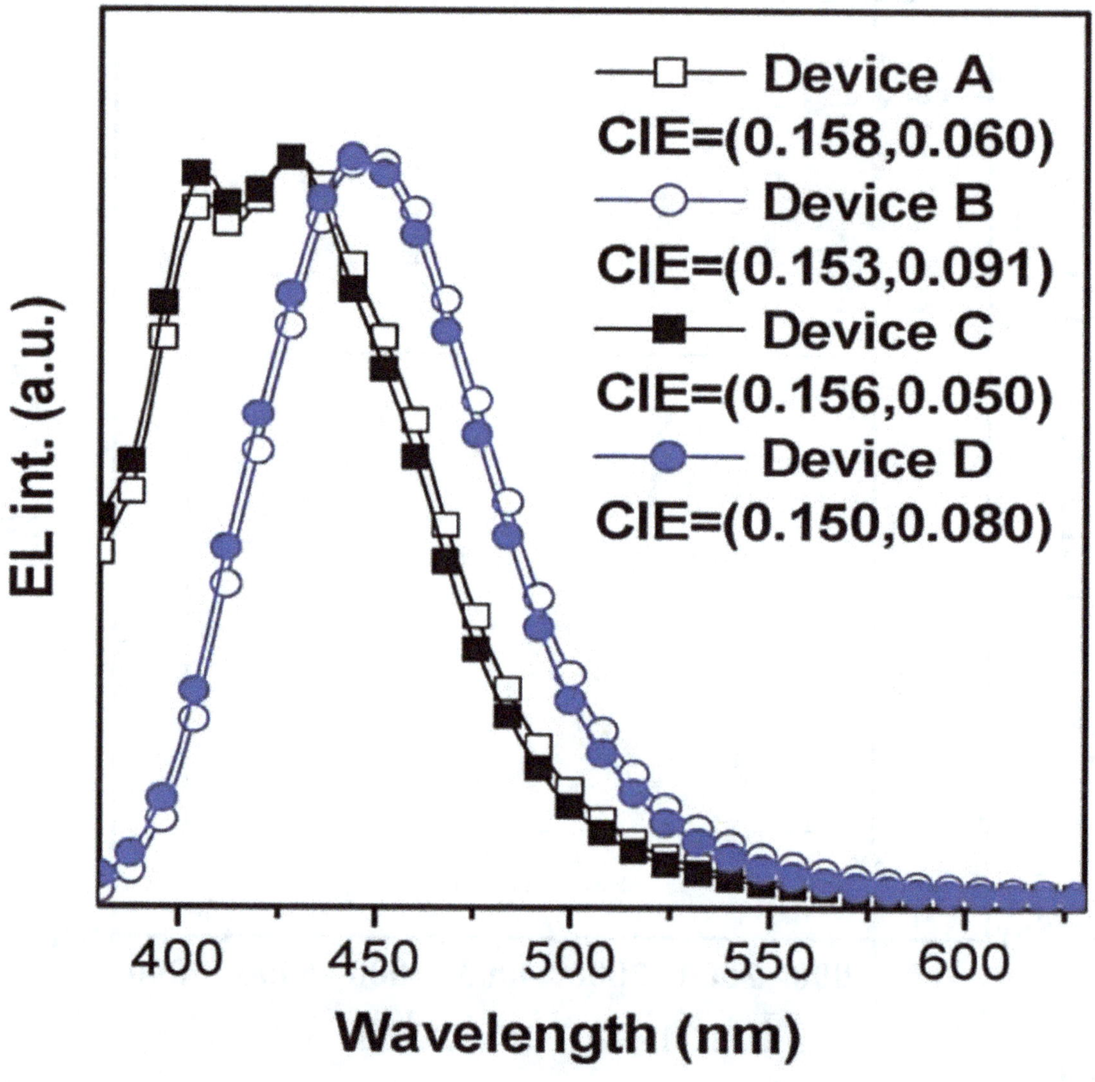

roll-off was realized in its non-doped devices. This study demonstrates an effective strategy towards stable deep blue emitters with higher stability, better efficiency and the lower roll-off ratio by chemical modification (meta–meta coupling or cyano group insertion) in non-doped deep-blue-emitting devices based on aryl-substituted phenanthroimidazole blocks.

Among all the N-type phenanthroimidazole fluorophores (1-22), 20 BITPI shown good device performances EQE over 7% with stable blue emission (color index: (0.15, 0.13)), due to more twisting nature of the compound (Table 1). Thus, twisting nature of the fluorophores effectively Coase blue shift in the PL and EL emissions. Additionally, PI fluorophores with twisting nature and high mass are showing good thermal stability (Z. Wang, et al., 2014; J. Li, et al., 2013). The device performances of selected fluorophores are listed in Table 1.

N-Type Anthracene, Naphthalene and Pyrene Integrated PI Fluorophores for Fluorescent OLEDs

Anthracene is intrinsically a blue fluorescent fluorophore in the solid state (Zhang, *et al.*, 2016). Tremendous efforts have been made to investigate various kinds of PI based fluorophores with anthracenemoieties to improve their electroluminescence (EL) properties. Intrigued by the good PL properties of anthracene in solution, much effort has been devoted to developing anthracene PI derivatives as deep-blue emitters with reduced luminescent quenching in films (Zhang, *et al.*, 2016; Zhuang, *et al.*, 2014; F. Zhang, *et al.*, 2016). Naphthalene group has been widely used in blue PI fluorophores (H. Gao, 2010). Attaching the naphthalene group to the PI fluorophores, substantially improve the electron mobility in the device, since, the dominant contribution of naphthalene groups to the lowest unoccupied molecular orbital (LUMO), which is beneficial to the LUMO orbital overlap among molecules, and thus boosting the electron mobility in the device (Jayabharathi, et al., 2016; T. Shan, et al., 2017; 73, W. Chen, et al., 2015).

Three asymmetric phenanthroimidazole-substituted anthracene derivatives (23 ACPI, 24 1-NaCPI, and 25 2-NaCPI) was designed and synthesized by Zhu *et al.,* (Zhu *et al.,*2012). The introduction of PI moieties with anthracene, which effectively increases the electron injection and transport ability, and further reduces the hole injection barrier at the interface of the hole-transporting layer (HTL) and the emissive layer and balanced recombination (Y. Zhang, et al., 2011). To explore the electroluminescence properties of these fluorophores, single layer, two layers and three layers, devices were fabricated. With respect to the three layer device 4 using 23 ACPI as the emitting layer, its maximum current efficiency reaches 4.36 cd/A with CIE coordinates of (0.156, 0.155). In the single layer device 10 based on 23 ACPI, maximum current efficiency reaches 1.59 cd/A with CIE coordinates of (0.169, 0.177). Interestingly, both the devices (4 and 10) have low turn on voltage and negligible efficiency roll off at high current densities. These materials showed perfect thermal stability and excellent EL performance with low onset voltages, which indicate that these new materials are a promising class of blue emitters for practical applications in panel display. The same group (Zhu *et al.*,) were synthesized anthracene and pyrene based three blue-emissive fluorophores (26 ANPI, 27 2-NaNPI and 28PNPI) and studied their OLED characteristics (S. Zhuang, et al., 2014). The non-doped three layers device using 28PNPI was fabricated and exhibits a maximum current efficiency of 4.99 cd/A with CIE color coordinates of ($x = 0.15$, $y = 0.17$). When doped with sky blue fluorescent material BUBD-1, all the blue host 26 ANPI, 27 2-NaNPI, 23 ACPI, 24-1-NaCPI and 25-2-NaCPI containing anthracene moiety shown impressive luminous current efficiency around 12 cd/A. The primary results suggest that the efficient energy transfer between fluorescence host and dopant are affected by the Förster energy transfer mechanism. The structural engineering of the phenanthroimidazole by changing the conjugated position and moiety can affect the EL device performance of the materials obtained.

Ge *et al.*, synthesized two novel twisted anthracene-PI derivatives 29p-PABPI and 30 m-PABPI (J. Wang, et al., 2015). The non-doped fluorescent OLEDs were fabricated by using anthracene PI derivatives as the emissive fluorophores. The maximum current efficiencies were achieved to be 3.98 and 1.32 cd/A and the maximum power efficiencies are 2.80 and 1.14 lm/W, respectively. The EQE is 3.61% and 1.33% for 29p-PABPI and 30m-PABPI, respectively (Figure 13.). It is particularly interesting to compare the electroluminescent properties of 29 p-PABPI and 30 m-PABPI, both have the same constructed structural unit. When the molecular configuration of these fluorophores is changed, their EL devices exhibited similar blue emission but electroluminescent efficiency is varied almost three

Figure 12. N-type phenanthroimidazole (only) based fluorophores for fluorescent OLEDs 1-22

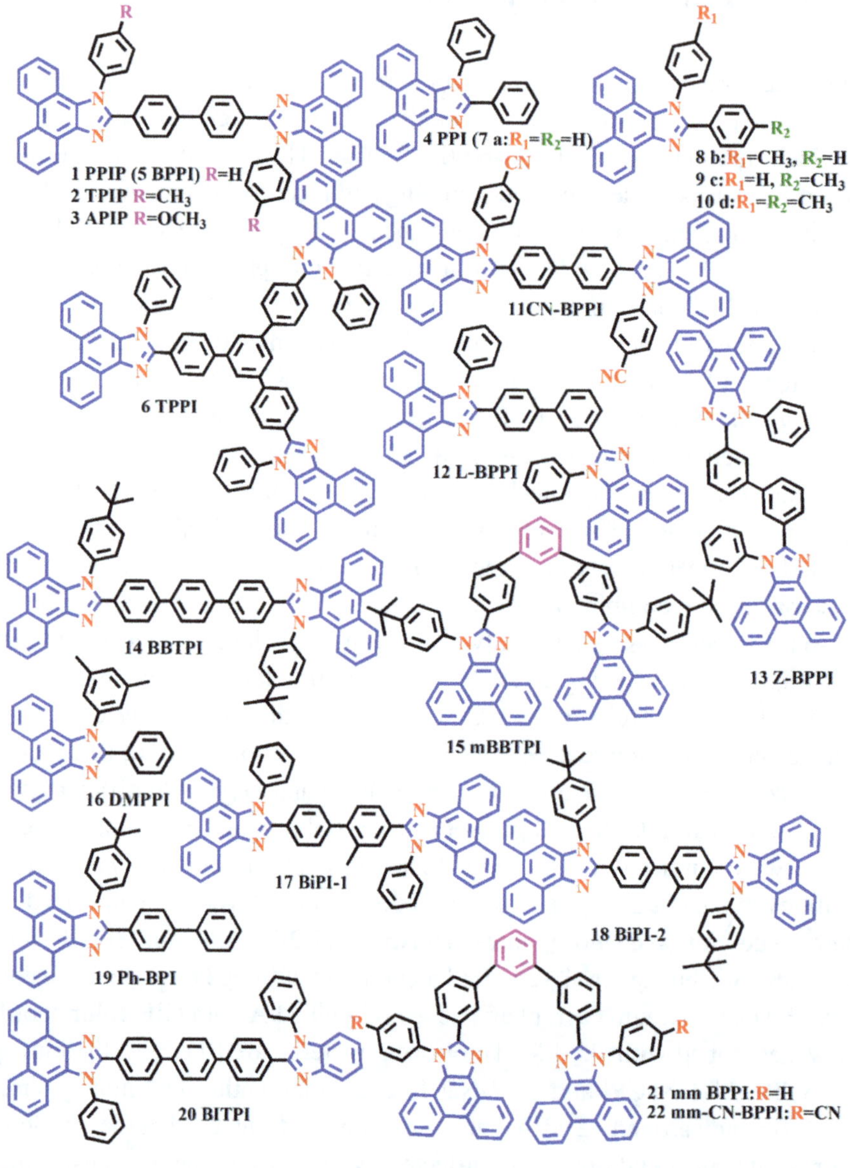

times. The results that are discovered a new rule of molecular design based on anthracene derivatives for obtaining high-performance blue emission materials.

Lee *et al.*, were designed and synthesized an efficient binaphthyl-bridged PI blue fluorophores, 31NBTPI and322NBTPI (W. Chen, et al., 2015). Comparing with its mononaphthyl counterpart 31NBTPI, 322NBTPI exhibits a higher thermal stability, more twisted geometric structure as well as more spatially separated HOMO and LUMO. Thus it shows deeper blue emission, more balanced charge transport properties, which could give rise to a lower efficiency roll-off. Comparing to its counterparts, the 32 2NBTPI-based device exhibits much less efficiency roll-off (Figure 14). For instance, at a current density of 100 cd/m^2, the efficiency roll-off of the 322NBTPI-based device is evidently lower than

Table 1. Device performances of N-type phenanthroimidazole (only) based fluorophores for fluorescent OLEDs 1-22

Fluorophores	Voltage [a] [V]	EQE [b] (%)	C. E. [c]/cd A^{-1}	P. E. [d]/lm W^{-1}	Color Index	Ref.
1 PPIP	3.0	6.31	7.47	7.30	0.15, 0.14	80
2 TPIP	3.9	5.43	4.69	2.71	0.15, 0.09	
3 APIP	4.1	5.26	6.40	3.84	0.15, 0.15	
4 PPI	4.7	-	-	-	-	81
5 BPPI	2.8	4.0	6.87	6.2	0.16, 0.21	
6 TPPI	4.2	-	-	-	-	
7 a	5.0	-	6.10	-	0.33,0.55	82
8 b	4.9	-	4.07	-	0.32,0.55	
9 c	4.1	-	4.16	-	0.33,0.55	
10 d	3.8	-	8.03	-	0.35,0.56	
11 CN-BPPI	5.6	-	0.78	0.26	0.17, 0.22	83
12 L-BPPI	4.5	-	0.68	-	0.16, 0.10	78
13 Z-BPPI	5.0	-	0.31	-	0.16, 0.11	
14 BBTPI	2.7	5.77	5.48,	4.77	0.15, 0.10	84
15 mBBTPI	3.2	3.63	1.99	1.81	0.16, 0.06	
16 DMPPI	18.1	13.2	19.2	9.6	-	97
17 BiPI-1	2.8	6.18	4.62	4.55	0.15, 0.08	86
18 BiPI-2	2.9	4.52	3.38	3.44	0.15, 0.08	
19 Ph-BPI	3.0	0.68	0.62	0.55	0.15, 0.10	71
20 BITPI	*2.6*	*7.11*	*6.78*	*7.10*	*0.15, 0.13*	*87*
21 mm BPPI	3.5	2.36	0.95	0.93	0.156, 0.05	95
22 mm-CN-BPPI	4	1.17	0.89	0.66	0.15, 0.08	

Voltage [a]: turn on voltage, EQE [b]: quantum efficiency, C. E. [c]: current efficiency, P. E. [d]: power efficiency.

those of the 14 BBTPI-(W. Chen, et al., 2015).and the 31 NBTPI-based devices. The performance of the 322NBTPI-based device is comparable to the state-of-the-art non-doped deep-blue OLEDs with CIE y coordinates lower than 0.10. All the above evidence demonstrate that the molecular design strategy towards high efficiency and deep-blue OLEDs used here is effective.

Jayabharathi *et al.*, first employed PI derivatives (16 DMPPI, 33TMPPI and 34 NMPPI) as nondoped emitters in hybrid organic–inorganic LEDs (HyLEDs) (Jayabharathi, et al., 2016). Efficient HyLEDs with an electron injection layer of 3.8 nm sized zinc oxide nanomaterials and their different emissive layers of PI derivatives have been used to fabricate the OLED. Among the nondoped emissive layers such as 16 DMPPI, 33TMPPI, and 34NMPPI, a combination of 26 nm thickness of ZnO as electron injection layer (EIL) and 33TMPPI as an emissive layer exhibits maximum luminance (L) of 17345 cd/m^2 with external quantum efficiency (η_{ex}) of 13.8% and current (η_c) and power (η_p) efficiencies of 19.5 cd/A and 9.9 lm/W, respectively. This high efficiency could be due to the presence of an electron withdrawing effect of the trifluoromethyl group at the p-position of the aldehydic ring in the emissive layer.This

Figure 13. A) Electroluminescence spectra of devices, B) The external quantum efficiency curves

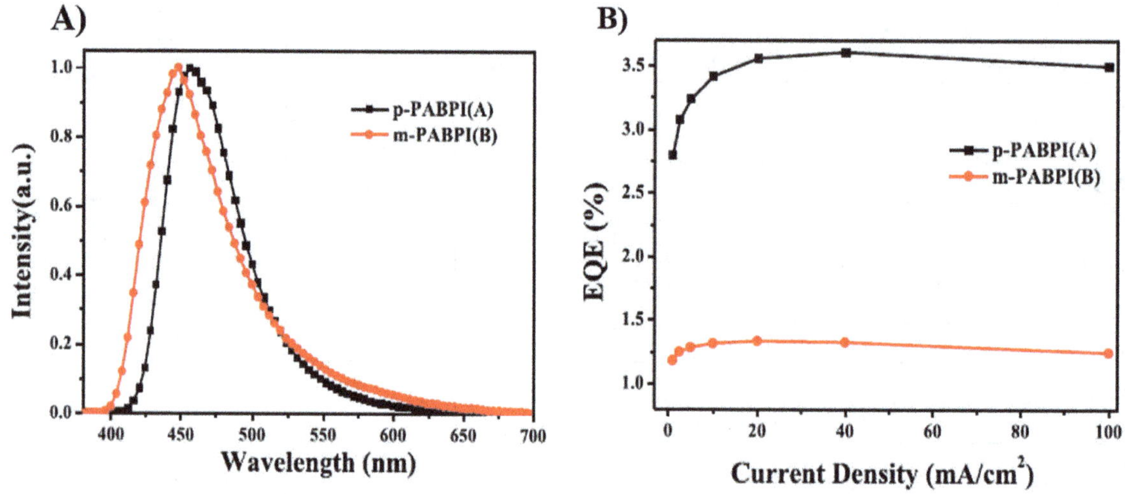

lowers the HOMO energy (−5.4 eV) and thereby lowers the energy barrier so that the leakage of hole

Figure 14. Current efficiency and EQE vs. current density of the 14 BBTPI-, the 31NBTPI- and the 322NBTPI-based devices

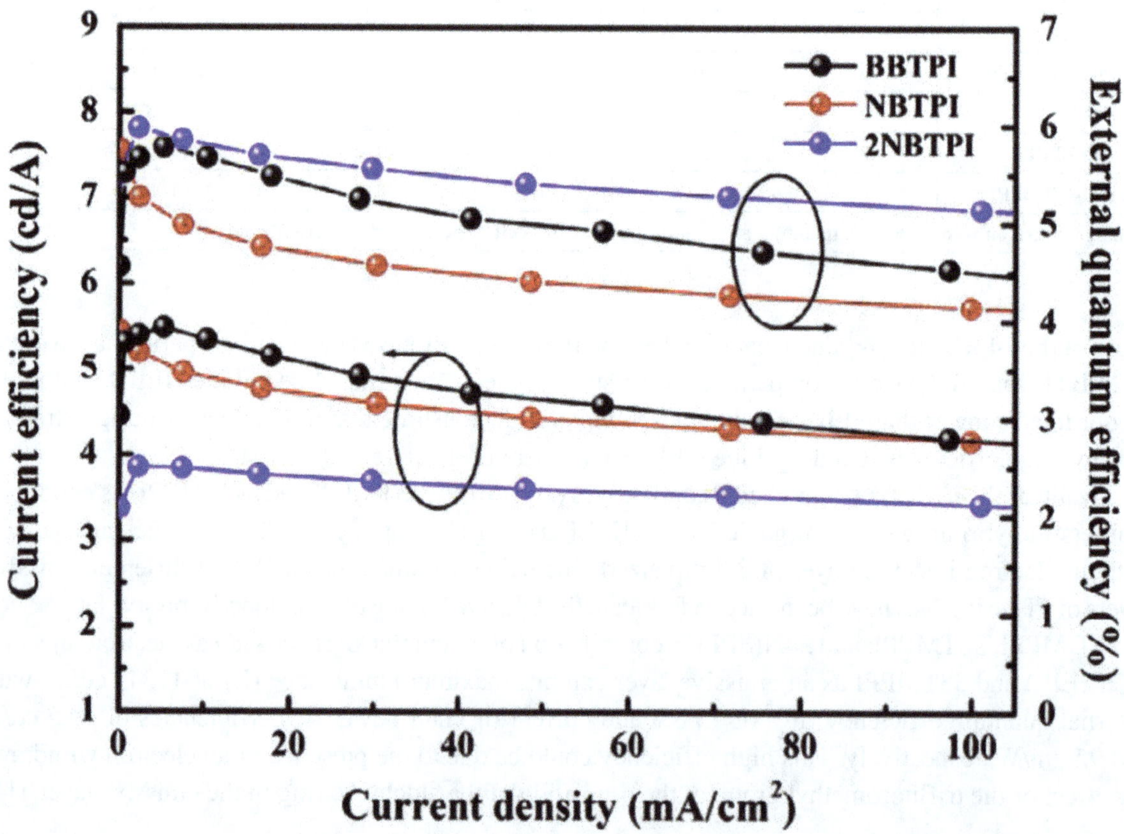

is avoided, as larger numbers of electrons are injected into the emissive layer from the 26 nm thickness of ZnO, e− - h+ balance becomes improved which results in higher efficiencies.

Lu *et al.*, synthesized two imidazole-based fluorophores, 35PIMNA and 36PyINA by coupling naphthalene with Phenanthreneimidazole (PIM) and pyreneimidazole (PyI), respectively (T. Shan, et al., 2016). The work has proved that the phenanthroimidazolePIM and PyI have great potentials to produce deep blue nondoped OLEDs. 35PIMNA exhibited deep blue emission with the smallest CIEy of 0.034 ever reported innondoped OLEDs (Figure 15). 36PyINA also displays deep-blue emission (0.157, 0.084) and great EL performance (EQE = 3.68%) with negligible efficiency roll-off. To take advantage of high photoluminescence QE of 35PIMNA and a large fraction of singlet exciton formation of 36PyINA, doped devices were fabricated by dispersing 36PyINA into 35PIMNA. A significantly improved maximum external quantum efficiency (EQE) of 5.05% is obtained through very effective energy transfer between host to guest, with CIE coordinates of (0.156, 0.060) (Figure 15). The EQE remains 4.67% at 1000 cd/m^2, which is among the best of deep blue OLEDs reported matching stringent EBU standard well. This study provides novel ideas for designing high-efficiency deep blue emitters and fabricating OLEDs with excellent performance.

Zhai *et al.,* designed and synthesized two blue emitting dual core materials consisting of phenanthroimidazole and anthracene or pyrene chromophores (37-1 and 38-2) (F. Zhang, et al., 2016). The electron accepting character of the phenanthroimidazole group is expected to increase the electron-transporting mobility of the emitters and to further improve devices performance. Devices based on 37-1 and 38-2 show low turn-on voltages 2.6 and 2.5 V and stable blue emissions with CIE of (0.16, 0.17) and (0.15, 0.12), respectively. The maximum current efficiencies of 1.15 and 0.90 cd/A and maximum external quantum efficiencies of 0.83% and 0.85% with little efficiency roll-off at high current densities were achieved, respectively. As depicted in Figure 16, the EL efficiencies of devices decrease gradually with a further increase of luminance, which is the so-called EL efficiency roll-off (B. Liang, et al., 2007; Y. Zhu, et al., 2011).

Tong *et al.*, synthesized a series of PI derivatives (39 2N-PI, 40 anthracene-PI, 19 Ph BPI, 41 1N-BPI, 42 2N-BPI, 43 2N-API and 44Py-BPI) with N1 and C2 position modification have been characterized

Figure 15. A) EL spectra of MA (35 PIMNA) and MB (36PyINA) devices at 100 cd m−2. B) CIE chromaticity coordinates of NTSC, EBU standard blue, and the three devices

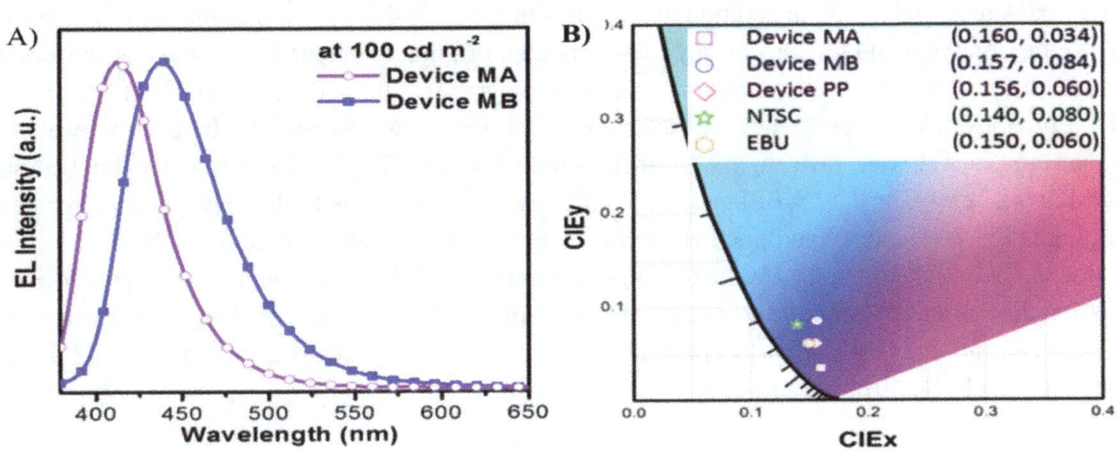

Figure 16. Current efficiency–current density–external quantum efficiency characteristics of the devices based on compounds 37-1 and 38-2

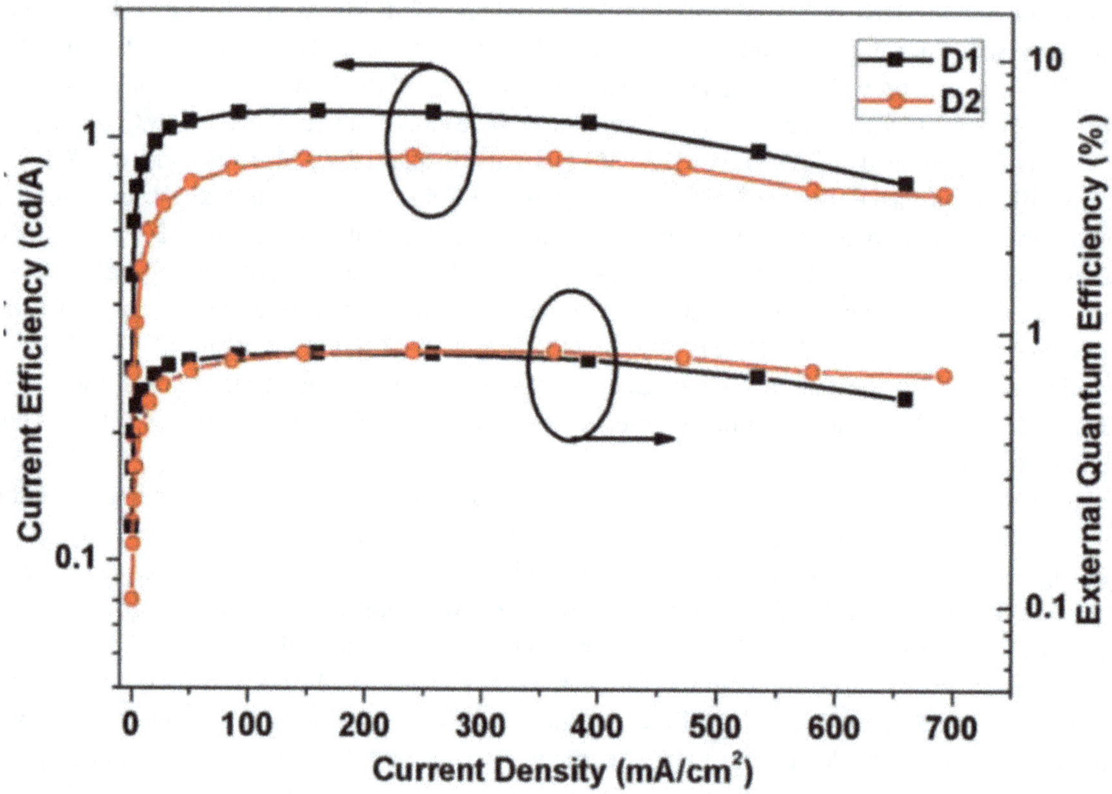

and applied as non-doped emitters in OLEDs (Y. Zhang, et al., 2016). The emitters were found to give deep-blue emissions with peaking wavelength ranging from 395 nm to 439 nm in thin solid film state, while attaching bulky aromatic moieties such as anthracene and pyrene leads redshift of the emission wavelength (466 nm and 451 nm). The emitters show high fluorescent quantum yield as high as 87.51% in the solid state. Non-doped deep-blue OLEDs with the simple three-layer structure using these emitters were observed to show deep-blue emission with CIEy coordinate lesser than 0.12 and best efficiency of 1.51 cd/A and 1.64 lm/W. In addition to low turn-on (i.e. 2.5–4.0 V), the present devices also show stable emission color and little efficiency roll-off at high brightness output. It is noted that the isomers with the different moiety (naphthalene, anthracene and pyrene) at the different position of the PI group exhibit obvious different properties. For example 432N-API shows deeper HOMO energy level (5.59 eV) and emits at a shorter wavelength (399 nm) compared with 42 2N-BPI(439 nm) in the thin film state (HOMO: 5.27 eV). The 42 2N-BPI and 432N-API based device showed relatively better performance with high device efficiency and good color purity. From single crystal structure analysis, it is found that the modification position and dihedral angle are important factors for obtaining efficient deep blue emission and efficient luminescent. Deep blue emission can be obtained attaching aromatic moieties through benzene ring (e.g.: 422N-BPI) other than attaching bulky moieties directly to the C-2 position of PI (e.g.: 392N-PI, 40anthracene-PI).

Jayabharathi *et al.*, synthesized PI based blue-light-emitting materials (45 NPIP, 46 MeNPIP and 47 MeONPIP) for non-doped OLEDs (Jayabharathi, et al., 2016). The synthetic routes of PI derivatives shown in scheme-4. The non-doped devices with PI derivatives as the emissive layer show the maximum external quantum efficiency and current efficiency are 5.20, 6.19 and 6.92% and 6.81, 6.92 and 7.80 cd/A, respectively (45 NPIP, 46MeNPIP and 47 MeONPIP). These results could be attributed to the more balanced charge-transporting properties within the emissive layer achieved by better charge injection and confinement provided by the hole-transport layer. The electroluminescent performances reveal that 47 MeONPIP is a potential non-doped blue-emitting material.

The PI fluorophores with the naphthalene, anthracene and pyrene moieties exhibit different photophysical properties and these fluorophores showing good thermal stability (Jayabharathi, et al., 2016; J. Wan, et al., 2015). These moieties efficiently increases the electron injection and transport ability and further reduces the hole injection barrier at the interface of the hole-transporting layer (HTL) and the emissive layer and balanced recombination (Y. Zhang, et al., 2011).Among all the N-type anthracene, naphthalene and pyrene based fluorophores (23-47), 47 MeONPIP display good device performances EQE over 6.92%, due to more balanced charge-transporting properties within the emissive layer (Table 2).The device performances of selected fluorophores are listed in Table 2.

Figure 17. Scheme-4. Synthetic route of phenanthroimidazole derivatives.

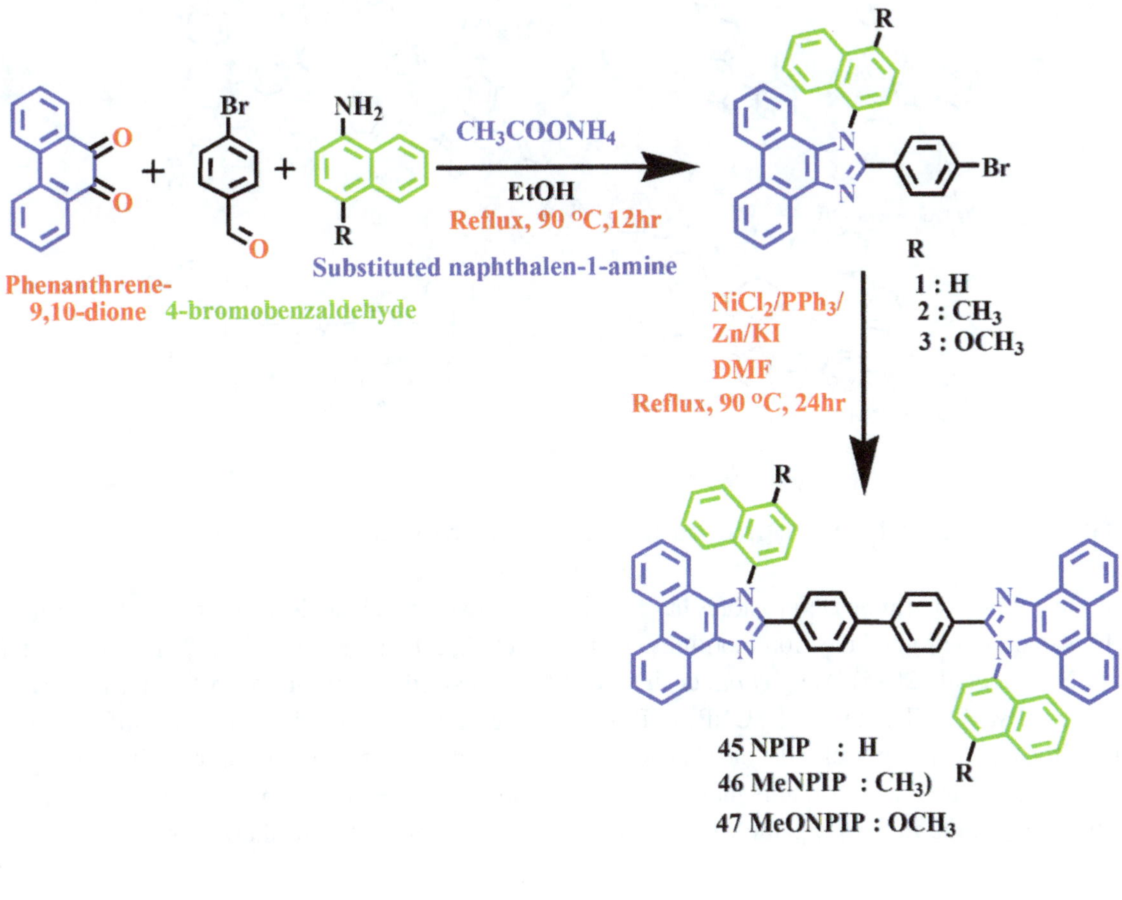

Figure 18. N-type anthracene, naphthalene and pyrene based PI fluorophores for fluorescent OLEDs 23-47

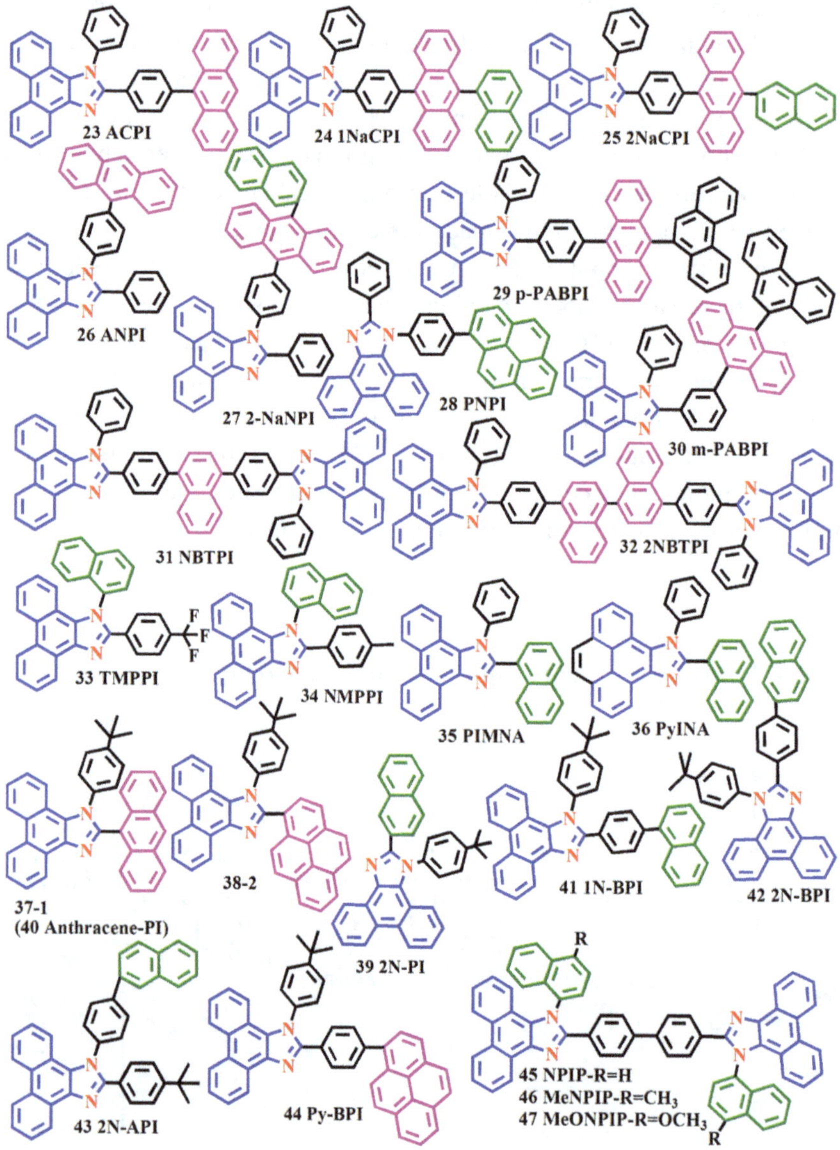

N-Type Pyridine PI Fluorophores for Fluorescent OLEDs

Electron deficient pyridine fluorophores have been proved to be efficient electron transport materials in OLEDs due to their good electron motilities (Ichikawa, et al., 2010; Togashi, et al., 2012; D. Ye, et al., 2014; N. Lin, et al., 2014).Wang *et al.*, designed and synthesized electron-transport materials (ETMs) (48 CPI-p3Py, 49 NPI-p3Py and 50 CNPI-p3Py), by incorporating the pyridine ring to the different positions of the rigid phenanthroimidazole skeleton (B. Wang, et al. 2015).The electronmobilities in organic semiconductors are lower than hole mobilities, the development of ETMs with high electronmobility is a bottle neck in the field of organic electronics (Hughes, &Bryce, 2005, Kulkarni, Tonzola, Babel, &

Table 2. Device performances of N-type anthracene, naphthalene and pyrene based PI fluorophores for fluorescent OLEDs 23-47

Fluorophores	Voltage[a] [V]	EQE[b] (%)	C. E.[c]/cd A^{-1}	P. E.[d]/lmW^{-1}	Color Index	Ref.
23 ACPI	2.83	2.72	4.30	2.29	0.156, 0.216	75
24 1NaCPI	2.73	2.82	4.32	2.46	0.165, 0.199	
25 2NaCPI	2.71	3.59	4.63	2.48	0.150, 0.160	
26 ANPI	3.74	1.56	2.41	1.01	0.166, 0.213	77
27 2-NaNPI	3.32	2.71	3.24	1.51	0.154, 0.147	
28 PNPI	3.59	2.82	3.64	1.76	0.153, 0.163	
29 p-PABPI	-	3.61	3.98	2.80	0.15,0.13	79
30 m-PABPI	-	1.33	1.32	1.14	0.16,0.12	
31 NBTPI	2.8	5.73	5.43,	-	0.15, 0.11	91
32 2NBTPI	2.9	5.95	3.93	-	0.15, 0.09	
33 TMPPI	16	13.8	19.5	9.9	-	97
34 NMPPI	17.9	6.5	15.3	7.2	-	
35 PIMNA	3.8	2.43	0.513	-	0.160, 0.034	98
36 PyINA	2.9	3.68	2.62	-	0.157, 0.084	
37-1	2.6	0.83	1.15	0.71	0.16, 0.17	93
38-2	2.5	0.85	0.90	0.49	0.15, 0.12	
39 2N-PI	2.8	0.95	0.89	0.78	0.15, 0.11	71
40 Anthracene-PI	3.2	0.80	1.33	0.97	0.16, 0.24	
41 1N-BPI	3.2	0.60	0.53	0.55	0.14, 0.10	
42 2N-BPI	3.0	1.61	1.37	1.44	0.14, 0.11	
43 2N-API	3.4	1.47	1.2	1.07	0.15, 0.11	
44 Py-BPI	2.5	2.07	3.27	3.17	0.15, 0.18	
45 NPIP	2.7	5.20	6.81	5.19	-	73
46 MeNPIP	2.9	6.10	6.92	5.86	-	
47 MeONPIP	*3.1*	*6.92*	*7.80*	*6.10*	-	

Voltage [a]: turn on voltage, EQE [b]: quantum efficiency, C. E. [c]: current efficiency, P. E. [d]: power efficiency.

Jenekhe, 2004, Sasabe, & Kido,2011). Generally, high-performance ETMs should have the low-lying highest occupied molecular orbital (HOMO) and the lowest unoccupied molecular orbital (LUMO) energy level to block hole leakage from the emissive layer (EML)(avoid unnecessary emission) and facilitate electron injection from the cathode to the electron-transport layer (N. Li, 2010, S. Su, Sasabe, Y. Pu, Nakayama,& Kido,2010).Wang *et al.,* used a new strategy to develop efficient ETMs. The pyridine unit was linked through the C2-phenyl (48 CPI-p3Py),and expected to have high electron mobility for the efficient conjugation in the para-position of C2-phenyl. However, when the pyridine group was conjugated to the N1-phenyl (49 NPI-p3Py), shows desired low-lying HOMO/LUMO energy levels and a high triplet energy level. Based on the new strategy for electron mobility/energy-level trade-off optimization, 50 CNPI-p3Py with pyridine rings at para-positions of the C2-phenyl and N1-phenyl of the phenanthroimidazole exhibits not only a higher electron mobility than 48 CPI-p3Py, also lower-lying HOMO/LUMO energy levels compared with that of 49 NPI-p3Py. Simultaneously maintain a relatively high triplet energy level which will help to enhance the overall efficiency. The 50 CNPI-p3Py fluorophore exhibits not only lower-lying HOMO/LUMO energy levels, but also a relatively high electron mobility (around 10^{-3} cm^2 V^{-1} s^{-1}), which is 2-fold greater than that of the widely used BPhen material (Naka, Okada, & Onnagawa,2000). More interestingly, the blue fluorescent OLEDs with 48 CPI-p3Py or 50 CNPI-p3Py as an electron-transporting layer (ETL) exhibited superior performance compared to that using BPhen, unusually lowering the driving voltage and improving efficiencies.

In particular, the device with 50 CNPI-p3Py as an ETL showed a maximum power efficiency of 10.65 lm/W, a maximum current efficiency of 15.17 cd/A, a maximum EQE of 7.75% and low efficiency roll-off even at an illumination-relevant luminance of 10 000 cd/m^2. By the efficient utilization of the up-conversion of a triplet excited state, which are among the most efficient ETMs and proposes a new strategy for the molecular design of high-performance ETMs.The device performances of selected fluorophores are listed in Table 3.

Bipolar PI Based Materials for Fluorescent OLEDs

The electroluminescence (EL) performances of fluorescent unipolar materials often suffer from the charge injection and transportation in the emission layer. High efficiencies of OLEDs are usually obtained through optimizing the charge balance in multilayer devices. In terms of the materials, an emission layer featured with improved charge injection and transport characteristics could effectively elevate the device performances. Consequently, except for good thermal stabilities, desirable fluorescent emitters required better to have good charge carrier transport ability to provide more balance in hole and electron fluxes in the simple device structure. In this context, hole and electron transport units/moieties have been incorporated into the fluorescent backbone of the luminophores. The carbazole (CBZ), diphenylamine (DPA) and triarylamine (TPA) groups are well-known for its good hole-transporting ability via the radical cation species with high stability (Holmes, et al., 2003, Tsai, et al., 2006, J. Li, et al., 2005, Shih, Chien, Wu, & Shu,2007). They have been incorporated into fluorophores to improve hole-injection, hole-transporting properties and increase the thermal stability. Most of the developed fluorescent materials exhibit low electron affinities (EAs), leading to inefficient electron injection into the fluorescent emitters. Poor electron injection can be compensated by using additional electron-transporting groups such as triazines, oxadiazoles, pyridine, phosphine oxide and benzimidazole, etc., (Kang, et al., 2007, Xia, et al., 2007, S. Su, Takahashi, Chiba, Takeda, & Kido, 2009, Chien, et al., 2009, Kondakova, et al., 2008). In this part, bipolar transport fluorescent materials will be introduced according to different electron releasing groups, with the order of carbazole and triphenylamine

Carbazole (CBZ) Based Bipolar Materials for Fluorescent OLEDs

The carbazole is a typical hole transporting unit due to the electron donating character and it is commonly modified with electron transport units in various ways to construct bipolar fluorophores. By using carbazole group various research group has been produced several bipolar materials.

Lu *et al.*, synthesized four PI derivatives 51 PSC, 52 PDC, 53 PCC and 54 PPP bearing terminal carbazole or pyrene group linked by different bridges (such as, phenylene for 51 PSC, phenylvinyl for 52 PDC, phenylvinylcarbazolylvinyl for 53 PCC and phenylvinyl for 54 PPP)in order to obtain blue-

Figure 19. N-type pyridine PI fluorophores for fluorescent OLEDs 48-50

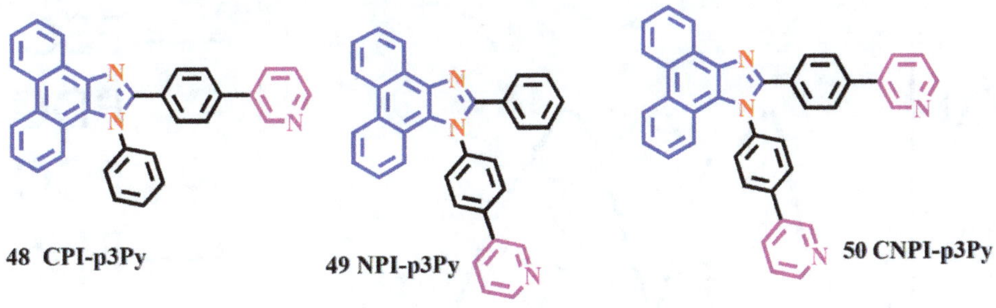

Table 3. Device performances of N-type pyridine PI fluorophores for fluorescent OLEDs 48-50

Fluorophores	Voltage [a] [V]	EQE [b] (%)	C. E. [c]/cd A^{-1}	P. E. [d]/lm W^{-1}	Color Index	Ref.
48 CPI-P3Py	2.9	7.44	14.94	10.07	0.15, 0.32	105
49 NPI-P3Py	3.8	5.25	10.26	3.08	0.14, 0.31	
50 CNPI-P3Py	*2.8*	*7.75*	*15.17*	*10.65*	*0.14, 0.31*	
BPhen	3.6	5.74	12.49	4.89	0.15, 0.34	
Voltage [a]: turn on voltage, EQE [b]: quantum efficiency, C. E. [c]: current efficiency, P. E. [d]: power efficiency.						

emitting fluorophores (J. Peng, et al., 2014).The synthetic routes of PI derivatives shown in scheme-5. FluorescentOLEDs was fabricated by using 51 PSC, 52 PDC, 53 PCC and 54 PPP fluorophores as emitting materials. The blue emission (Figure 19) with CIEx,y of (0.16, 0.17) was achieved from the device based on 51 PSC as emissive layer. In addition, the devices based on other phenanthroimidazole derivatives could emit bluish green or green light. The turn-on voltage, current efficiency and maximum luminance of the devices based on 51 PSC, 52 PDC, 53 PCC and 54 PPP were 3.4 V, 3.4 V,3.8 V, 3.4 V and 0.92 cd/A, 1.83 cd/A, 1.28 cd/A, 0.32 cd/A, as well as1190 cd/m^2, 1930 cd/m^2, 1270 cd/m^2, 1149 cd/m^2, respectively. It is suggesting that the phenanthroimidazole derivatives might become candidates as emitting materials employed in OLEDs and other related fields.

Lu *et al.,* designed and synthesized a series of highly efficient deep blue fluorophores containing carbazole as a hole transporting and phenanthroimidazole (PI) as an electrontransporting moieties (55 M1, 56 M2, 57 M3, 58 M4 and 59 M5) (Z. Gao, et al., 2014). The phenanthroimidazole unit is committed to the 3, 6 or 9 positions of carbazole, respectively, to achieve fluorophores with limited conjugation length. The ratio of carbazole and phenanthroimidazole is designed to be 2:1 or 1:1 to gain insight into the relationship between the bipolar molecular structures. OLEDs using these fluorophores were fabricated to deliver a deep blue CIE of y < 0.08. Among them, 56M2 displays the highest external quantum efficiency of 3.02%, CIE y of 0.056 and negligible efficiency roll-off. By taking the advantage of these deep blue fluorophores, they are further served as good hosts for phosphorescent dyes. High performance green phosphorescent device based on 56 M2 is attained with a maximum current efficiency of 33.35 cd/A, a power efficiency of 22.99 lm/W and a maximum external quantum efficiency of 9.47%. When the 55 M1 fluorophore, doped with an orange fluorescent (CNDPASDB) material, upon careful tuning the doping proportion, the two emitting-component (55 M1/CNDPASDB) white OLED is successfully fabricated (device-11) with a maximum current efficiency of 5.53 cd/A and CIE coordinates of (0.313,

Figure 20. Electroluminescence spectra of the devices

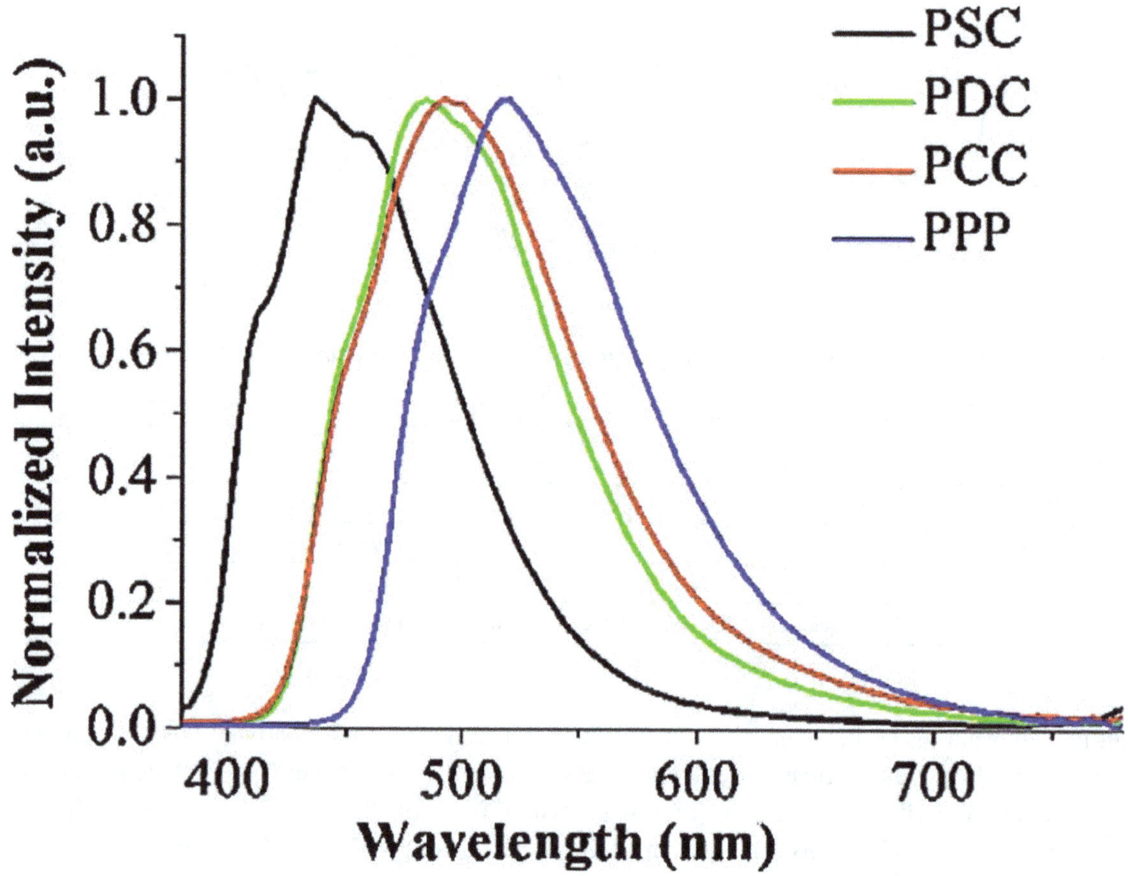

0.305). (Figure 22). All these results give us a new foreground for the design of blue fluorophores for OLED.

The strategy for creating blue-emitting fluorophores could be the incorporation of high-energy emissive fragments through twisted arrangements. The twisted structure can decrease conjugation between these fragments and consequently benefit for blue molecular emission. Furthermore, the twisted structure is unfavorable for intermolecular π...π stacking that usually leads to red-shifted and weakened the emission in the aggregation process of fluorescent molecules (C. Li, et al., 2016). As a consequence, the twisted structure facilitates the realization of efficient solidstate blue emission which is the key requirement for the emitting materials in non-doped blue OLEDs. By using above strategy, Wang *et al.*, were synthesized two new sky-blue fluorophores, 60 PCz-An-PPI and 87 TPA-An-PPI, with the integration of phenyl phenanthroimidazole (PPI) as an electron transporting and triphenylamine (TPA)/phenylcarbazole (PCz) as a hole transporting groups, being attached to the 9,10-positions of anthracene (C. Li, et al., 2017).The twisted arrangement of the anthracene and PPI groups is inferred from the steric bulk of the PPI unit, while that of the anthracene and TPA/PCz groups is deduced from the known H–H repulsion between anthracene and the benzene ring attached to it (B. Kim, et al., 2013, S. Lee, et al., 2017). Structurally simple non-doped OLEDs with 87 TPA-An-PPI as the emitting layer showed high performance, with the maximum luminance (L_{max}) of 51 390 cd/m^2, the maximum power efficiency (PE_{max}) of 10.42 lm/W,

Figure 21. Scheme-5. The synthetic routes for the phenanthroimidazole derivatives.

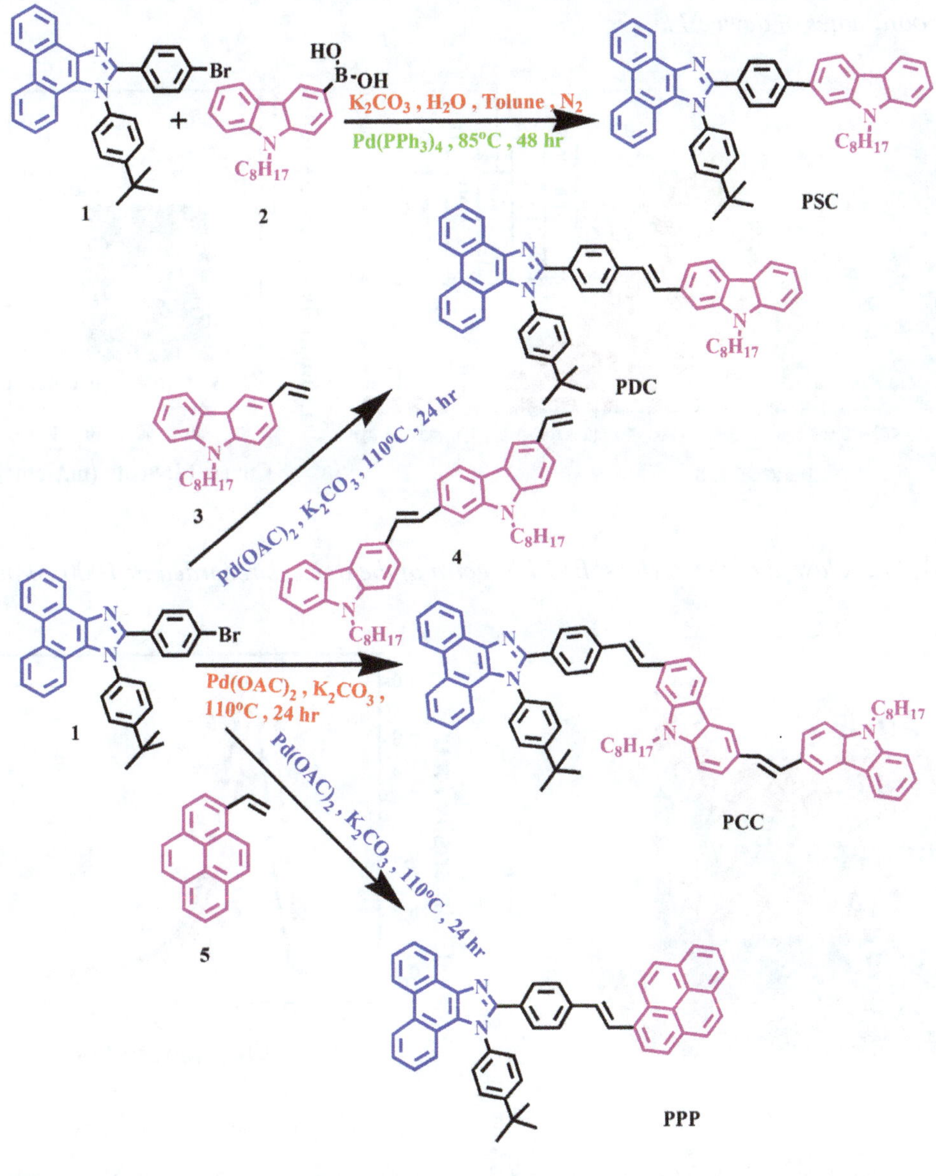

and the maximum external quantum efficiency (EQE_{max}) of 4.61%. More importantly, at a practical high luminance of 1000 cd/m², the PE and EQE retained high values of 8.48 lm/W and 4.60%, respectively, and the corresponding driving voltage was as low as 3.6 V. Both devices D1 (87 TPA-An-PPI) and D2 (60 PCz-An-PPI) emitted sky-blue light with the emission maxima located at 492 and 472 nm, respectively (Figure21.). The CIE coordinates of D1 and D2 recorded at 1000 cd/cm² were (0.16, 0.38) and (0.15, 0.23), respectively. Which suggests the potential of 60 PCz-An-PPI and 87 TPA-An-PPI and for practical application.

Venugopal *et al.*, synthesized donor-acceptor PI derivatives, 61 SPI-Cz, 62 SPICN-Cz, 88 SPI-TPA and 89 SPICN-TPA using strong donor triphenylamine (TPA) and weak donorcarbazole (Cz), respectively, and benzonitrile substituted phenanthroimidazole (SPICN) as acceptor (Figure 22.) (Venugopal,

Figure 22. (a) EL spectra for device 11 at 10 V, inset: energy level diagram and chemical structure of the materials used and (b) power efficiency and current efficiency vs. current density for device 11, inset: CIE coordinates of device 11

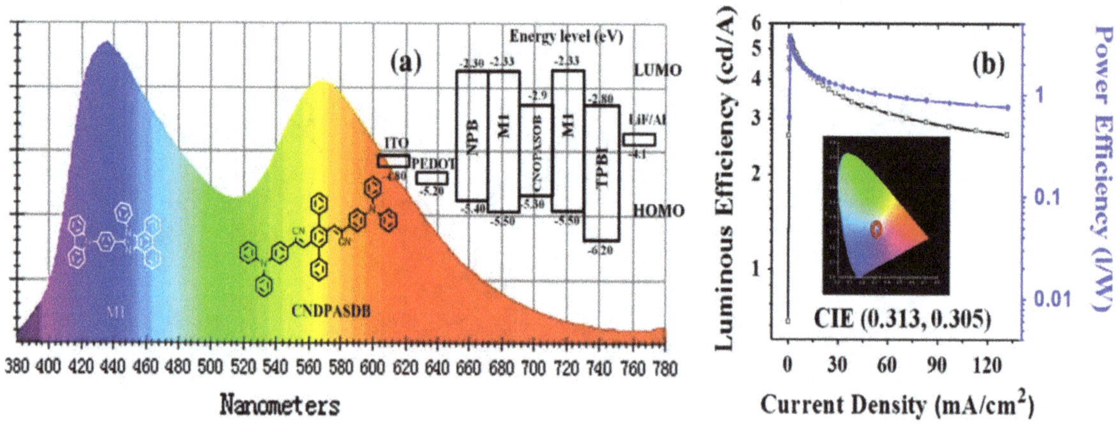

Figure 23. A) Stack layout of the devices, B) EL spectra of the devices recorded at 1000 cd/m²

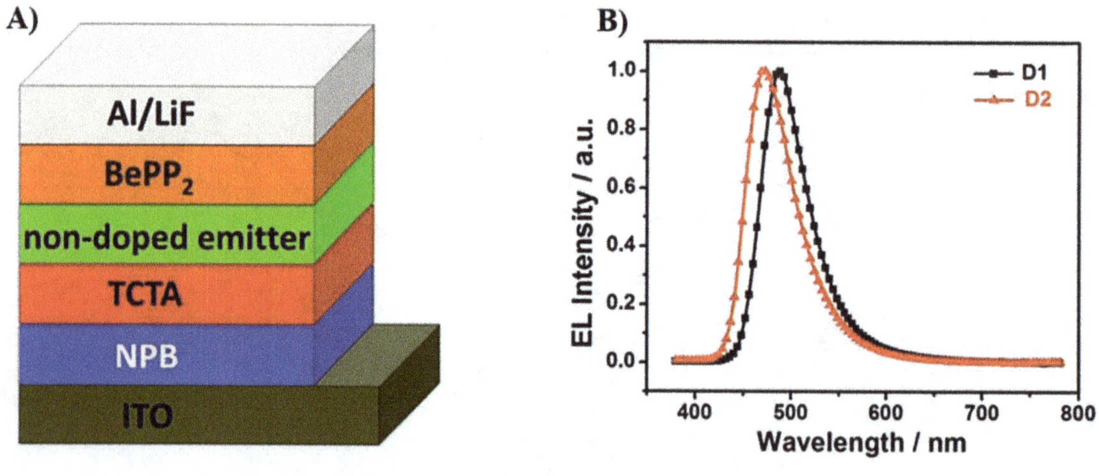

Palanivel, & Jayaraman,2017). The nondoped OLED device based on 62 SPICN-Cz exhibit better electroluminescent performances than those of 89 SPICN-TPA–based device: high external quantum efficiency of 2.58%, current efficiency of 2.90 cd/A, power efficiency of 2.26 lm/W and high external quantum efficiency of 2.58% with CIE coordinates of (0.15, 0.12). The donor-acceptor system could be useful to design low-cost, high-efficiency fluorescent OLED materials.

Among all the carbazole based bipolar fluorophores (51-63), 60 PCz-An-PPI shown blue emission with CIE coordinates of (0.15, 0.23) and high EQE over 3.99%, due to bipolar nature of the fluorophore. 56 M2based device delivers a deep blue emission with CIE y of 0.056 compared with the 60 PCz-An-PPI device. The 56 M2 also displays the high external quantum efficiency of 3.02% and negligible efficiency roll-off. The design strategy of limited conjugation length of the fluorophores is very useful for producing deep blue emitting fluorophores for OLED applications (Z. Gao, et al., 2014). The device performances of selected carbazole based bipolar fluorophores are listed in Table 4.

Diphenylamine (DPA) and Triphenylamine (TPA) Based Bipolar Materials for Fluorescent OLEDs

The diphenylamine and triarylamine group is well-known for its good hole transporting and injection ability via the radical cation species with high stability. Triphenylamine is having strong donor nature than the carbazole (Venugopal, Palanivel, & Jayaraman, 2017). Both of them can be incorporated into molecules or luminophores to improve hole-injection and -transporting properties, and to increase bulky volumes and thermal stability.

Utilizing the donor–π–acceptor approach, Tong *et al.,* designed three new blue fluorophores (64 BPA-BPI, 65 TPA-BPI and 66 PATPA) by using phenanthroimidazole as the acceptor for its good electron-

Figure 24. Carbazole based bipolar materials for fluorescent OLEDs 51-63

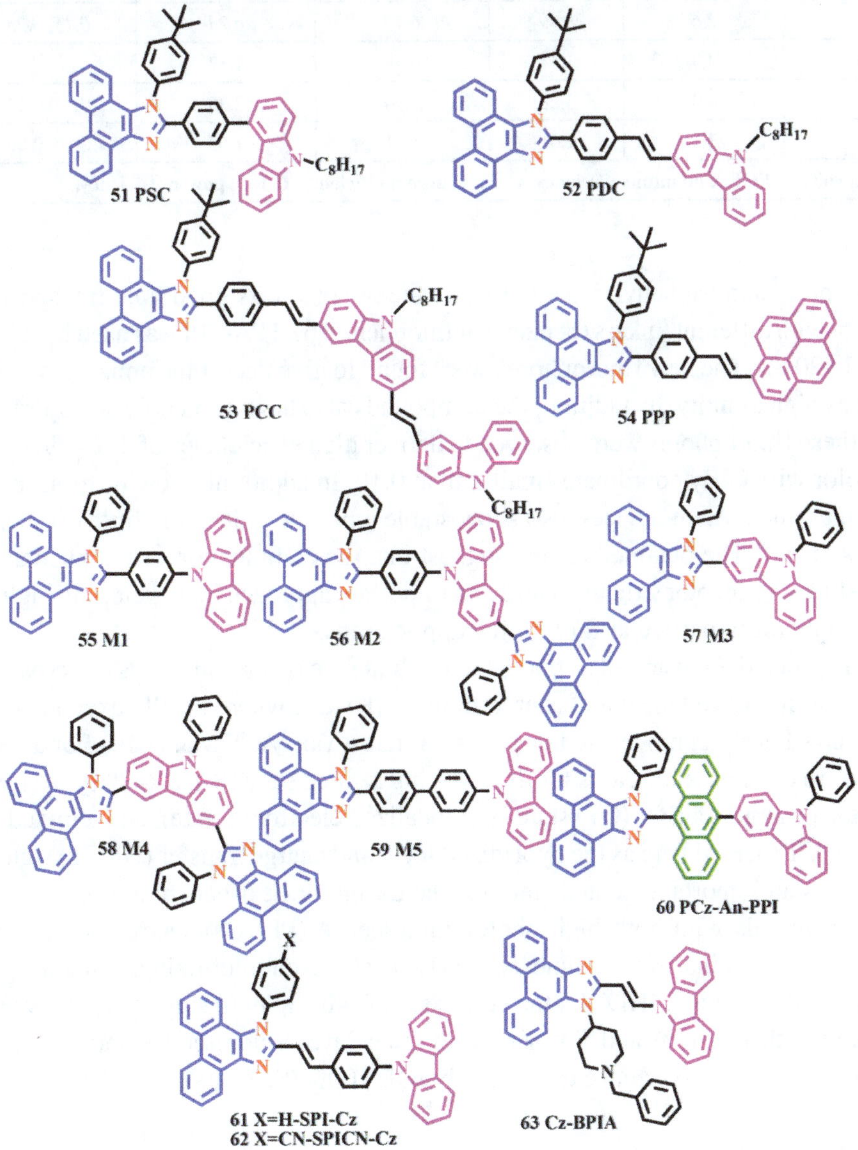

Table 4. Device performances of Carbazole based bipolar fluorophores for fluorescent OLEDs 51-63

Fluorophores	Voltage [a] [V]	EQE [b] (%)	C. E. [c]/cd A^{-1}	P. E. [d]/lm W^{-1}	Color index	Ref.
51 PSC	3.4	-	0.92	0.85	0.16, 0.17	121
52 PDC	3.4	-	1.83	1.38	0.21, 0.36	
53 PCC	3.8	-	1.28	1.05	0.23, 0.38	
54 PPP	3.4	-	0.32	0.24	0.33, 0.53	
55 M1	4.5	1.94	0.65	0.48	0.165,0.050	122
55 M1 + CNDPASDB	3.6	2.79	5.53	3.56	0.313,0.305	
56 M2	3.4	3.02	1.53	0.86	0.167,0.056	
57 M3	4.0	1.51	0.57	0.32	0.165,0.053	
58 M4	3.2	2.11	1.25	0.76	0.169,0.078	
59 M5	6.2	1.46	0.88	0.30	0.170,0.077	
60 PCz-An-PPI	*2.6*	*3.99*	*6.43*	*6.23*	*0.15, 0.23*	*124*
61 SPI-Cz	4.3	1.41	0.58	1.12	-	127
62 SPICN-Cz	4.0	3.28	2.90	2.26	0.15, 0.12	
63 Cz-BPIA	3.3	2.61	1.91	1.63	0.16, 0.09	143

Voltage [a]: turn on voltage, EQE [b]: quantum efficiency, C. E. [c]: current efficiency, P. E. [d]: power efficiency.

transporting mobility and triphenylamine (TPA) as the donor for its good hole transport mobility and connecting them with different linkers (benzene or thiophene), 67 TPA-TPI was used here for comparison (Y. Zhang, et al., 2012). The new fluorophores were found to give deep-blue emissions with fluorescent quantum yield as high as unity. In addition, the compounds also have good hole-transporting properties. OLEDs using these fluorophores were observed to deliver high efficiencies of 2.63 cd/A and 2.53 lm/W at deep blue color with CIEy coordinate smaller than 0.10. In addition to low turn-on (e.g., 2.8 V) and operation voltages, the present devices also show stable emission colors and little efficiency roll-off at high brightness output. The obtained results suggest that the combination of the phenanthroimidazole and the donor–linker–acceptor structure can be an important approach for developing high-performance deep-blue fluorophores in particular for lighting applications.

Recently, Ma group demonstrated a concept to facilitate the reverse inter-system crossing (RISC, T1 – S1) by means of hybridized local and charge-transfer (HLCT, where locally excited (LE) and charge transfer (CT) states fused in one) excited or emissive states(Yang, &Y.Ma, 2013).For a donor–acceptor molecule, there are three possible cases for its low-lying excited state (S_1): a π–π^* state (LE) (a localized electronic state), a charge-transfer (CT) state (a delocalized electronic state), and a mixed or hybridized state of π–π^* and CT (named here as the hybridized local and charge transfer (HLCT) state) (Figure 25.) The HLCT state is an important excited state for the design of next-generation organic light-emitting diode (OLED) materials with both high photoluminescence (PL) efficiency and a large fraction of singlet exciton generation in electroluminescence (EL). The exciton utilizing efficiency (EUE) can be enhanced by using this channel (HLCT), the conversion from triplet to singlet is relatively fast (unique feature compared to that of TTA and TADF), avoid the delayed fluorescence and triplet concentration quenching and it is possible to achieve theoretically IQE 100% (S. Zhang, et al., 2013).

Figurer 25. Illustration of the three possible cases of the LE and CT state energy levels in D–A molecules: a) and b) independent energy levels for their large energy gaps, and c) possible state mixing for the small energy gap between the LE and CT states

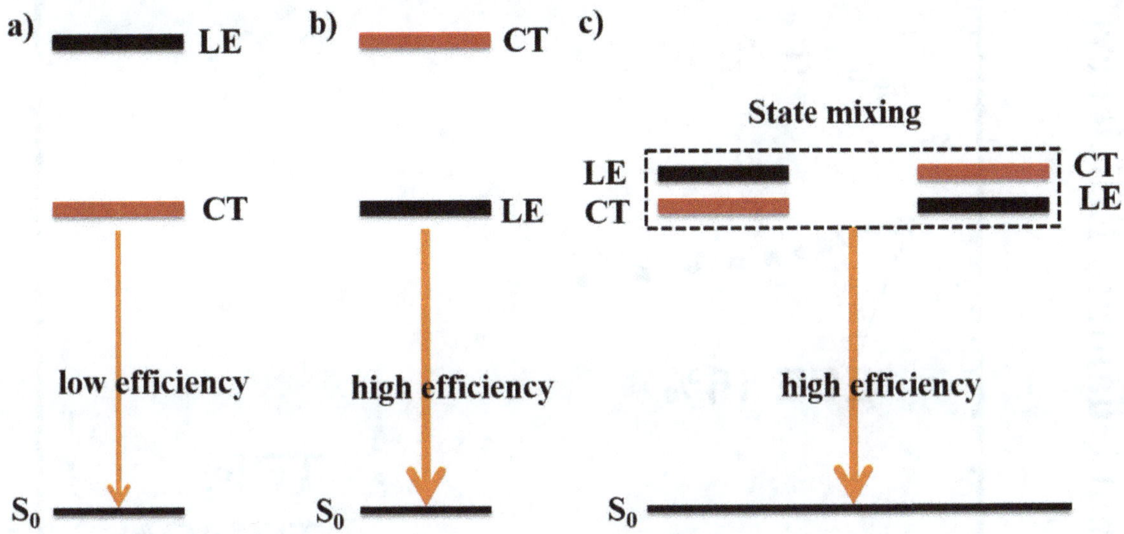

To improve the exciton utilizing efficiency (EUE), Ma *et al.*, was synthesized two PI fluorophores (68 TPM and 69 TPMCN).[131] 68 TPM is constructed with directly linked triphenylamine (TPA) and phenanthroimidazole (PM) moieties, while 69 TPMCN is substituted with a cyano group on the phenyl side group along an orthogonal direction to the TPM backbone. To evaluate the EUE of 68 TPM and 69 TPMCN, non-doped electroluminescent devices with these two compounds as a light emitting layer are fabricated and characterized. The non-doped OLEDs device of 69 TPMCN showed a blue peak at 463 nm, CIE coordinates (0.15, 0.18), with a measured PLE of 13%, a greatly enhanced EUE of 85% is estimated in the EL device (Figure 26).In contrast, though 68 TPM possesses a higher PLE in films (35%), the 68 TPM EL device with a same structure as 69 TPMCN exhibits a very low EUE (16%, far less than the 25% limit of spin statistics), as a result of the consumption of singlet excitons along the ISC process from singlet to triplet excitons, regardless of the occurrence of any photochemical process. It is inspiring that the maximum EUE of 69 TPMCN is over three times higher than the 25% limit of spin statistics. The enhanced EUE effect in 69 TPMCN can be attributed to the HLCT character in the emissive state. The orthogonal cyano-substitution acts as a statecontroller to promote the CT proportion in the HLCT emissive state so that a very high EUE could be achieved in EL devices without doping.

Ma *et al.,* synthesized fully twisted donor–acceptor fluorophore (71 TPA–PIM) based on the five member imidazole ring (W. Li, et al., 2014), which is a derivative of the previously reported excellent deep-blue fluorophore N, N-diphenyl-40-(1-phenyl-1H-phenanthro[9,10-d]imidazol-2-yl)biphenyl-4-amine (70 TPA–PPI) [W. Li, et al 2012), obtained by breaking a chemical bond in the phenanthrene part of phenanthro-[9,10-d]imidazole (Figure 26). Surprisingly, an extraordinarily narrow EL emission with a FWHM of only 35 nm and a CIE coordinate of (0.161, 0.046) was observed in the OLED using 71 TPA–PIM as the emissive fluorophore (Figure 26.), the y value of which is the smallest ever reported. Simultaneously, the 71 TPA–PIM device also showed a high quantum efficiency of 3%The fully twist-

Figure 26. External quantum efficiencies at different current densities in the EL device

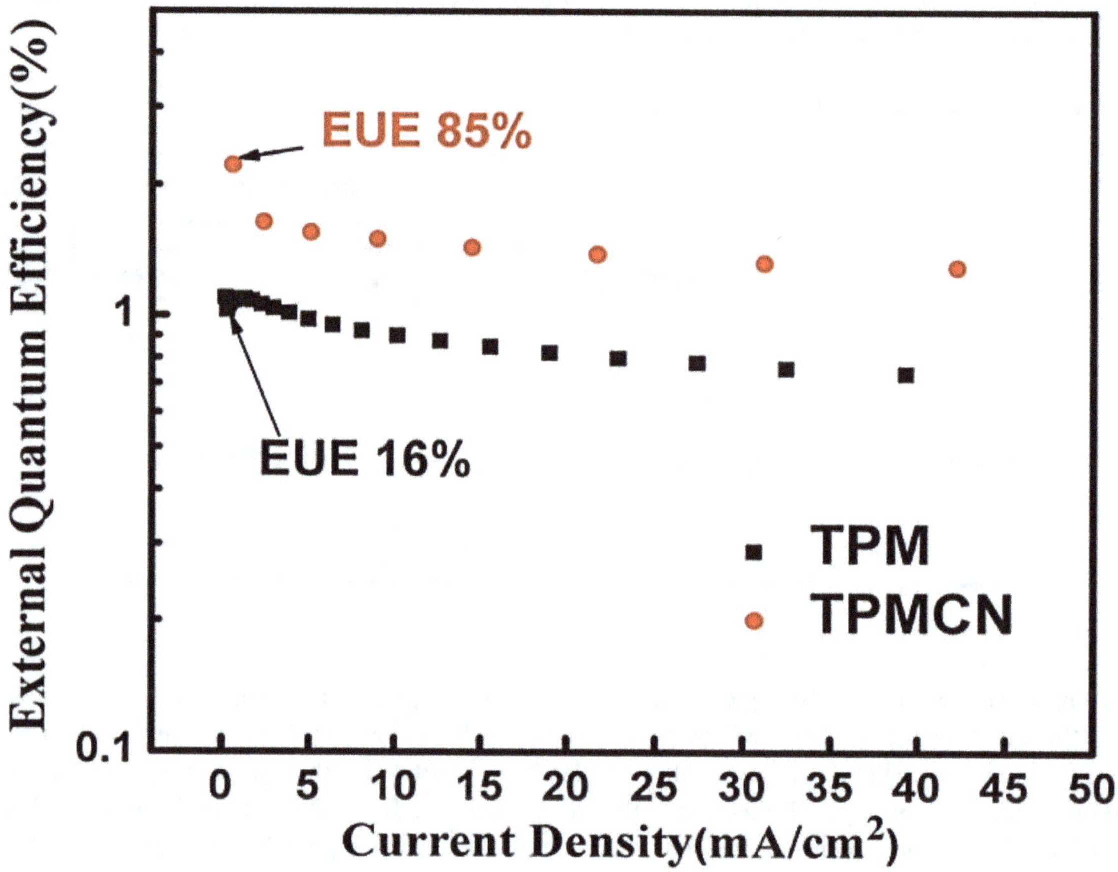

ing structural 71 TPA–PIM molecule maybe useful as an excellent color saturation deep-blue emitter inOLEDs.

Su *et al.*, designed and synthesized two Phenanthrene derivatives (72 p-PTPAPI and 73 m-PTPAPI), by introducing strong D-π-A (donor-π-acceptor) structure strategy to manage effectively their intramolecular charge transfer processes for the fluorophores (Ouyang, et al., 2015). Which contain the strong electron-donating triphenylamine (TPA) unit and electron-accepting imidazole unit. Both fluorophores have same molecular structure only different in the connecting mode of p-conjugating linkers between the D and A units from para-to meta-position. The devices based on Phenanthrene fluorophores exhibit deep-blue light emissions with CIE coordinates of (0.15, 0.15) for p-PATPA and violet-blue (0.16, 0.05) for m-PATPA at 100 cd/m^2, respectively. The m-PATPA based device shows good performance with EQE, current efficiency (CE), and power efficiency (PE) of 2.39%, 0.78 cd/A, and 0.72 lm/W, respectively. Furthermore, at the high luminescence, the device still indicated good performance with relatively low-efficiency roll-off. The obtained results here successfully provide an efficient design strategy with strong donor and acceptor for violet-blue emitters.

Wang *et al.* successfully designed and synthesized two Phenanthroimidazole fluorophores (74 DPAA-PPI and 75 tBuDPAA-PPI), by integrating the respective advantages of the phenanthroimidazole (PI), anthracene (AN) and diphenylamino (DPA) groups (C.Li, et al., 2016).PI is a wide band gap semicon-

Figure 27. Electroluminescence spectra of 70 TPA–PPI and 71 TPA–PIM, together with their molecular structures, FWHM and 1931 CIE coordinate values

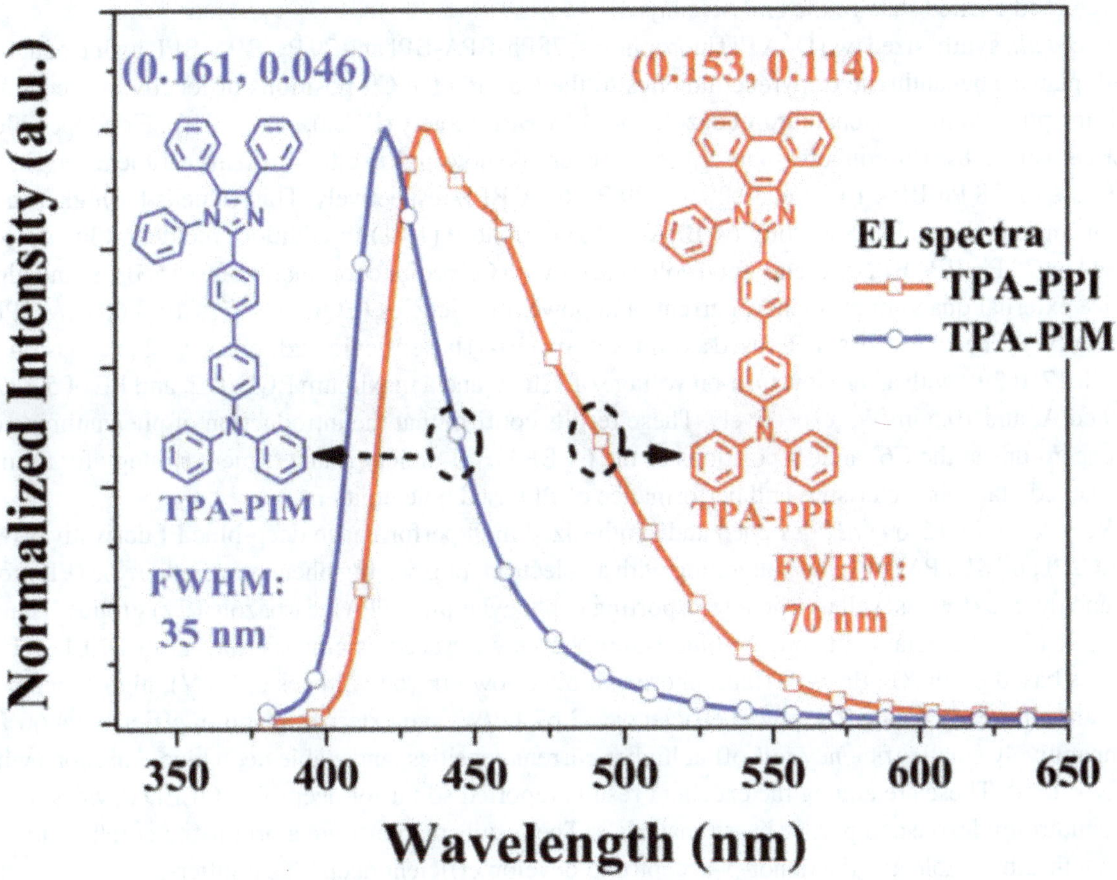

ductor with thegood electron-transporting ability and high thermal stability (Misra, Jadhav, Dhokale, & Mobin, 2014, Kwon, Park,& Park,2013). A DPA substituted anthracene moiety possesses high fluorescence quantum yield and the good hole transporting properties. Based on the simple fabrication process and device structure without a doped thin film, hole-injection layer and light out-coupling enhancement technology, the optimized OLEDs using 74 DPAA-PPI and 75 tBuDPAA-PPI as non-doped emitting layers showed very low turn-on voltages of 2.4 V and 2.5 V, respectively. The device based on 74 DPAA-PPI as an emissive component, which exhibits the highest power efficiency (PE) of 13.9 lm/W, EQE of 5.2% and brightness over 100 000 cd/m^2, respectively. The rational combination between PI, AN and DPA should be an efficient strategy to achieve high performance organic electroluminescent materials.

Venugopal *et al.* have used the donor–linker–acceptor methodology to design and synthesized non-doped blue fluorophores (76 TPA–MPI and 77 TPA–MPS), having an electron transporting acceptor PI moiety and a hole transporting donor TPA fragment with different linkers of phenyl (76 TPA–MPI) and styryl (77 TPA–MPS) (T. Venugopal, et al., 2016).138] Influence of the linker in the synthesized fluorophores on thermal properties, optical properties and device performances were investigated. The DFT optimization geometry has shown that less tilting of the styryl spacer in 77 TPA–MPS increases the intramolecular charge transfer (CT) due to reduced conjugation. Thus shows higher efficiency with low turn-on voltage. The nondoped device based on 77 TAP–MPS shows higher efficiencies, 1.73 cd/A (ηc), 1.46 lm/W (ηp), and2.11% (η_{ex}). The higher efficiency is attributed to the commission from the intercrossed excited state of the emissive layer.

Lee *et al.,* synthesized two D–A PI fluorophores, 78Ph-BPA-BPI and 79 Py-BPA-BPI, by introducing rigid, planar phenanthrene or pyrene moieties to the C6- and the C9-positions of locally excited-state emitting phenylamine–phenanthroimidazole (64 BPA-BPI) moiety (B. Liu, et al., 2016). Nondoped blue OLEDs using these fluorophores as an emissive materials show high exciton-utilizing efficiencies (EUE) of 36.2% for 78 Ph-BPA-BPI and 39.2% for 79 Py-BPA-BPI, respectively. This value is far higher than that using the original LE-emitting 64 BPA-BPI fluorophore (19%).In addition, the deep-blue device (based on 78 Ph-BPABPI) presents good color purity with CIE color coordinates of (0.15, 0.08) and high device external quantum efficiency, current, and power efficiencies (EQE: 4.56%, CE: 3.60 cd/A, PE: 3.66 lm/W). The 79 Py-BPA-BPI (used as emissive material) based device exhibits sky-blue emission of CIE (0.17, 0.29), with a very low turn-on voltage of 2.15 V and a maximum EQE, CE, and PE of 5.64%, 10.9 cd/A, and 10.5 lm/W, respectively. These results confirm that the introduction of phenanthrene or pyrene groups at the C6- andC9-positions of the 64 BPA-BPI moiety is an efficient strategy for tuning the excited-state characteristics and performance of PI based blue emitters.

Very recently, Liao *et al.,* designed and synthesized high performance deep-blue PI derivatives (80 TPACzPI and 81 TPAPyPI), by integrating with an electron transporting phenanthroimidazole (PI) moiety and pyrene (Py), as well as a hole transporting triphenylamine (TPA)/ carbazole (Cz) groups (Islam, et al., 2017). The obtained two deep-blue fluorophores were used as emissive layers in OLEDs. The devices based on the81TPAPyPI fluorophore exhibited low turn-on voltages (3.36 V), high efficiency (current efficiency 4.24 cd/A, power efficiencies 3.67 lm/W, and external quantum efficiency4.66%), comparatively small efficiency roll-off at higher current densities, and stable deep blue emissions with CIEy < 0.10. These are among the excellent results reported so far for deep-blue OLED devices using phenanthroimidazole and pyrene based materials. The results demonstrate a promising combination of phenanthroimidazole and the donor-p–acceptor to develop efficient deep blue emitters.

The p-conjugation of the PI derivatives also influences the blue emission. In order to adjust the p-conjugation length, the introduction of a steric hindrance group to create a twisted conformation seems to be a rational strategy (Z. Zhu, et al., 2016, W. Chen, et al., 2014). Recently, Tong *et al.,* were reported nondoped high-efficiency devices based on 65 TPA-BPI,[128] but the color purity of CIE (0.15, 0.09) did not meet the demands of National Television Standards Committee (NTSC) and High Definition Television (HDTV). The same group was designed three D–p–A fluorophores, 82 mtp, 83 Tmtp and 84 Cmtp, as deep-blue fluorophores for OLED applications.[142] Compared to 65 TPA-BPI, a methyl group was added to the biphenyl bridge, leading to an improved twist angle between the two benzene planes and a shortened conjugation length. The non-doped 82 mtp-based device emitted deep-blue light of 436 nm and CIE of (0.15, 0.05), and it exhibited a maximum EQE of 3.89%. The 84 Cmtp-based device emitted blue light of 445 nm and CIE of (0.15, 0.07). Especially, the 83 Tmtp-based device showed a saturated and much deeper CIE color coordinate of (0.15, 0.04), and maximal EQE of 3.00% was achieved. (Figure 29)

Venugopal *et al.,* were synthesized three blue emissive donor–spacer–acceptor (D-spacer -A) PI fluorophores 85 TPA-BDIS, 86 TPA-BPIS and 63 Cz-BPIS using styryl as an aromatic spacer.[143] The blue fluorophore 85 TPA-BDIS was obtained by breaking a chemical bond in the phenanthrene part of phenanthroimidazole of 86 TPA-BPIS. The fully twisted 85 TPA-BDIS based OLED exhibits a narrow electroluminescence emission at 441 nm (peaking wavelength) with a full-width at half maximum of

Figure 28. Diphenylamine (DPA) and Triphenylamine (TPA) based bipolar materials for fluorescent OLEDs 64-77

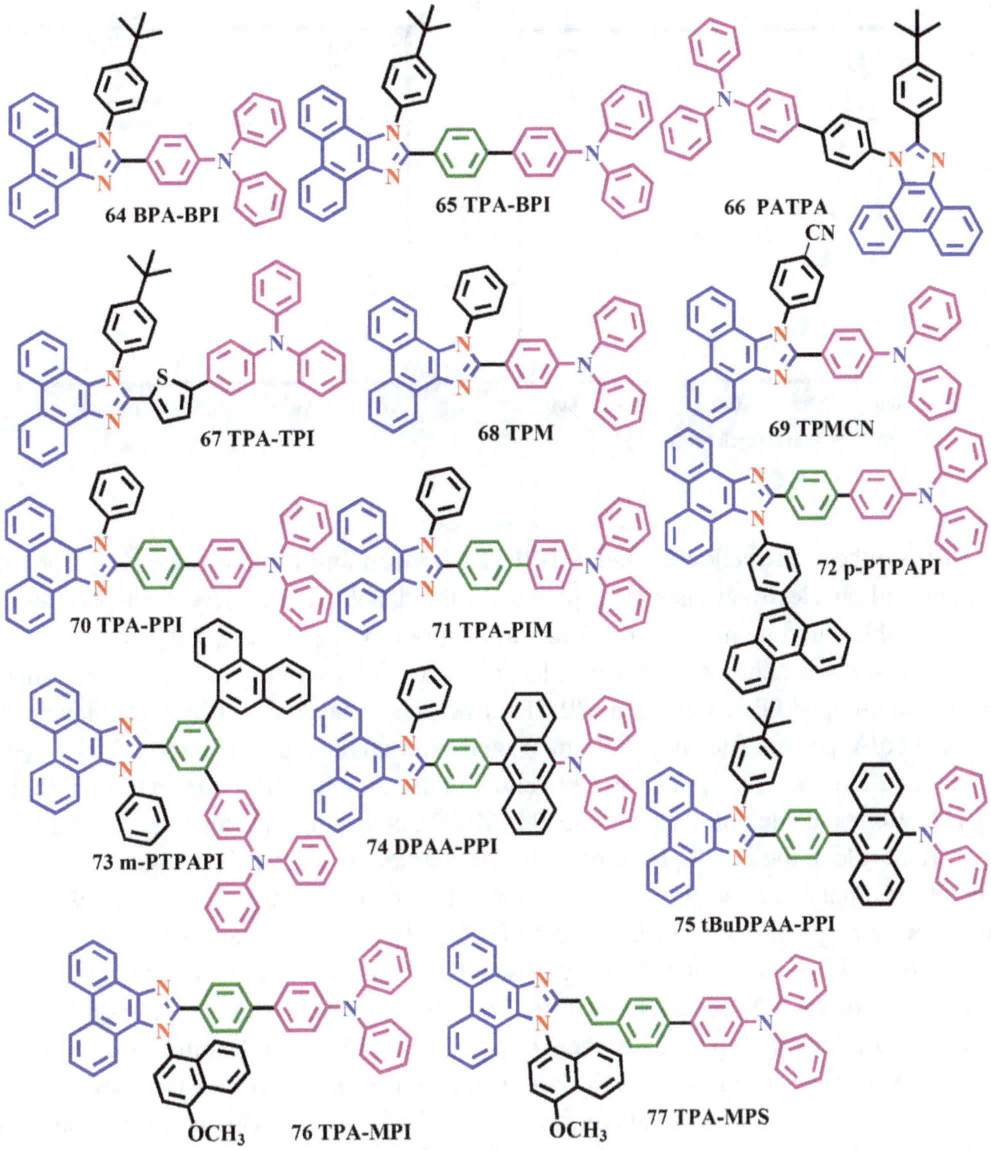

40 nm, as well as it shows a brightness of 1942 cd/m², current efficiency of 1.48 cd/A, power efficiency of 1.31 lm/W and external quantum efficiency of 1.98%.The electroluminescence performance of the non-doped device based on 86 TPA-BPIS shows a high brightness of 2852 cd/m², a current efficiency of 1.88 cd/A, a power efficiency of 1.51 lm/W and an external quantum efficiency of 2.53%. Compared with 85 TPA-BDIS, the weak donor carbazole substituted 63 Cz-BPIS exhibits an excellent performance: a blue emission with CIE coordinates of (0.16, 0.09), a current density of 1.91 cd/A, a power efficiency of 1.63 lm/W and an external quantum efficiency of 2.61%. These results are helpful in the design of next generation efficient low cost fluorescent OLED materials.

Figure 29. (a) EL spectra of the 82 mtp, 83 Tmtp and 84 Cmtp devices. (b) The EQE and luminance characteristics of the devices

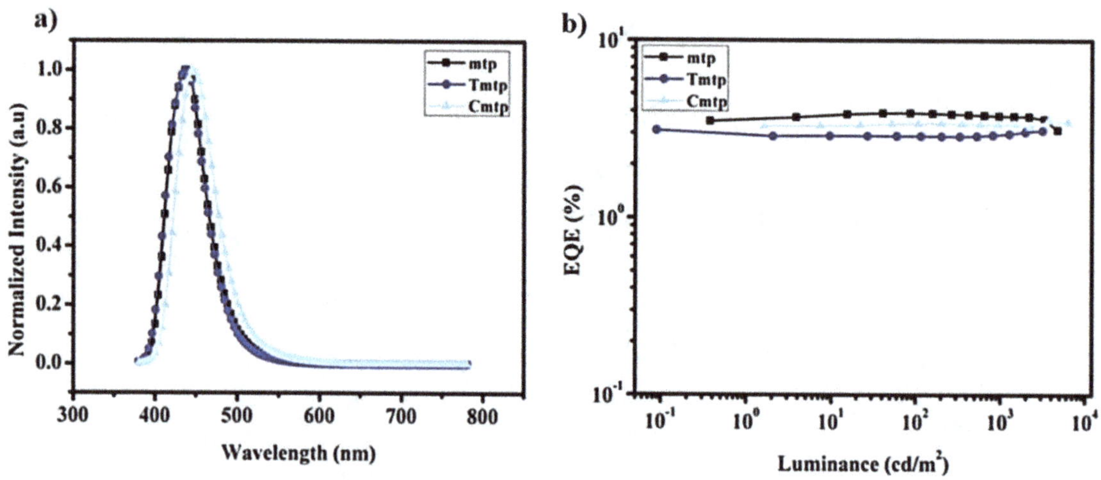

Jairam*et al.*, synthesized a yellow emitter (90 PIFTPA), containing a hole-transporting triphenylamine (TPA) moiety and an electrontransporting phenanthroimidazole (PI) moiety with photo-chemically and thermally stable fluorene moiety in N1-position (Jairam, Ulla, Kajjam, Satyanarayan, & Vaidya-nathan, 2017). The emitter shows good thermal stability as well as high glass transition temperatures. The fabricated non-doped OLED using 90 PIFTPA as emissive layer shows high efficiencies (current efficiency-6.45 cd/A, power efficiency-4.39 lm/W, and external quantum efficiency-3.01%) with a low turn-on voltage of 2.87 V. The device showed yellow emission with CIE coordinates of (0.38, 0.524) (Figure 30). These results demonstrate that the 90 PIFTPA emitter is a promising candidate and could play an important role in the development of white OLEDs and solid state lighting.

Among all the diphenylamine and triphenylamine based bipolar materials (64-92), 79 Py-BPA-BPI shown good device performances, EQE, CE, and PE of 5.64%, 10.9 cd/A, and 10.5 lm/W, respectively. Compared to other TPA fluorophores 83 Tmtp-based device showed a saturated and much deeper CIE color coordinate of (0.15, 0.04), and maximal EQE of 3.00% was achieved. These results confirm that the introduction of methyl group to the biphenyl bridge of 65 TPA-BPI, leading to an improved twist angle between the two benzene planes and a shortened conjugation length. So, more twisting nature of the fluorophores are promising materials for deep blue OLEDs. The device performances of selected DPA and TPA based bipolar materials are listed in Table 5.

Other Bipolar PI Based Materials for Fluorescent OLEDs

Lee *et al.* designed and synthesized a series of bipolar phenanthroimidazole derivatives, 91TTP–TPI, 92DPT–TPI and 93DPF–TPI, by incorporating polyaromatic hydrocarbon groups to phenanthroimidazole through a phenyl bridge (Y. Yuan, et al., 2013).The synthetic routes of PI derivatives shown in scheme-6. The bulky polyaromatic hydrocarbon units endow the molecules with high glass-transition temperatures, nonplanar twisty structures which reduce molecular aggregations in the solid state. Theoretical calculations and single-carrier devices confirmed their bipolar transporting nature. Nondoped OLEDs using these PI derivatives as emitters show blue-violet to sky-blue emissions with CIE coordinates of (0.16, 0.05)

Figure 30. (a) Energy-level diagram of the materials used in this study, (b) Chromaticity diagram (inset shows the image of the operational OLED)

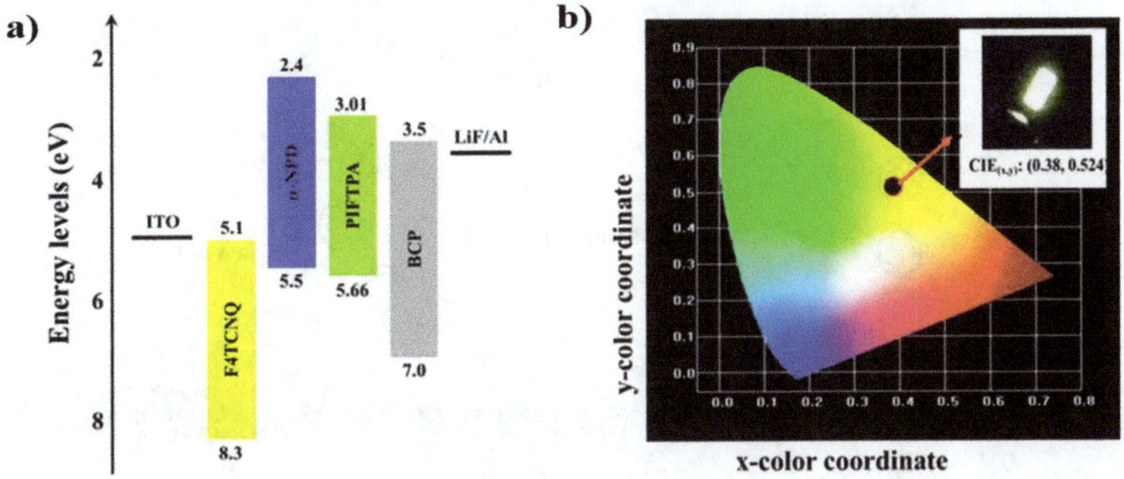

for 91TTP-TPI, (0.16, 0.07) for 92DPT-TPI, and (0.17, 0.24) for 93DPF-TPI. These nondoped devices show impressive external quantum efficiencies (EQE) of 5.02, 5.25, 4.85%, respectively (corresponding current efficiencies: 2.10, 3.13, and 8.41 cd/A). These devices also show small efficiency roll-off at high brightness (1000 cd/m^2), 93 DPF-TPI based device can achieve a higher brightness of 10000 cd/m^2 with EQE maintaining at 4.49%. These values are the best or among the best ones compared to those of reported nondoped OLEDs (G. Li, et al., 2017).

Fang *et al.*, design and synthesized two blue fluorescent fluorophores (94PhImFD and 95PhImTD), by attaching a hole-transporting dibenzofuran or dibenzothiophene and an electrontransporting phenanthroimidazole moiety (Chen, Wu, Zhao, & Fang,2015).Non-doped OLEDs are fabricated by employing the fluorophores94 PhImFD and 95PhImTD as emitters and the device exhibited promising performances. The devices show a deep blue emission with CIE coordinates of (0.15, 0.11) for 94PhImFD and (0.15, 0.10) for 95PhImTD, respectively. The 95 PhImTD emitter based device (B), with well-fitting energy levels and increased electron transport ability, possesses favorable efficiencies of 1.34 cd/m^2 for CE, 0.82 lm/W for PE and 1.63% for EQE. Meanwhile, WOLED devices using 94PhImFD or 95PhImTD as the host and yellow PO-01 as the dopant were realized with a maximum CE of 8.12 cd/A and CIE color coordinates of (0.339,0.330). The results demonstrated not only that the phenanthroimidazole unit is an excellent building block to construct deep blue emission materials, but also that chemical structure modification by the introduction of a suitable electron-donor substituent could influence the performance of devices. Very recently, the same group (Fang *et al.*,) synthesized a novel deep-blue emitter 96 PhImPOTD based on phenathroimidazole, which is integrated by an electron-donating dibenzothiophene unit and electron-withdrawing PI and diphenylphosphine oxide group (Chen, Wu, Hu. Zhao, & Fang, 2017). Non-doped OLEDs (Device-A) based on 96PhImPOTD emitter display a low turn-on voltage of 3.6 V, a favorable efficiency of 1.13 cd/A and a deep blue emission (Figure 30 a and b) with CIE color coordinates of (0.15, 0.08). The CIE is very close to the NTSC (National Television Standards Committee) blue standard (CIE: 0.14, 0.08). The 96PhImPOTD is also utilized as host for a yellow emitter (PO-01) to fabricate white organic light-emitting diodes (Device-B). This device gives maximum CE

Figure 31. Diphenylamine (DPA) and Triphenylamine (TPA) based bipolar materials for fluorescent OLEDs 78-92

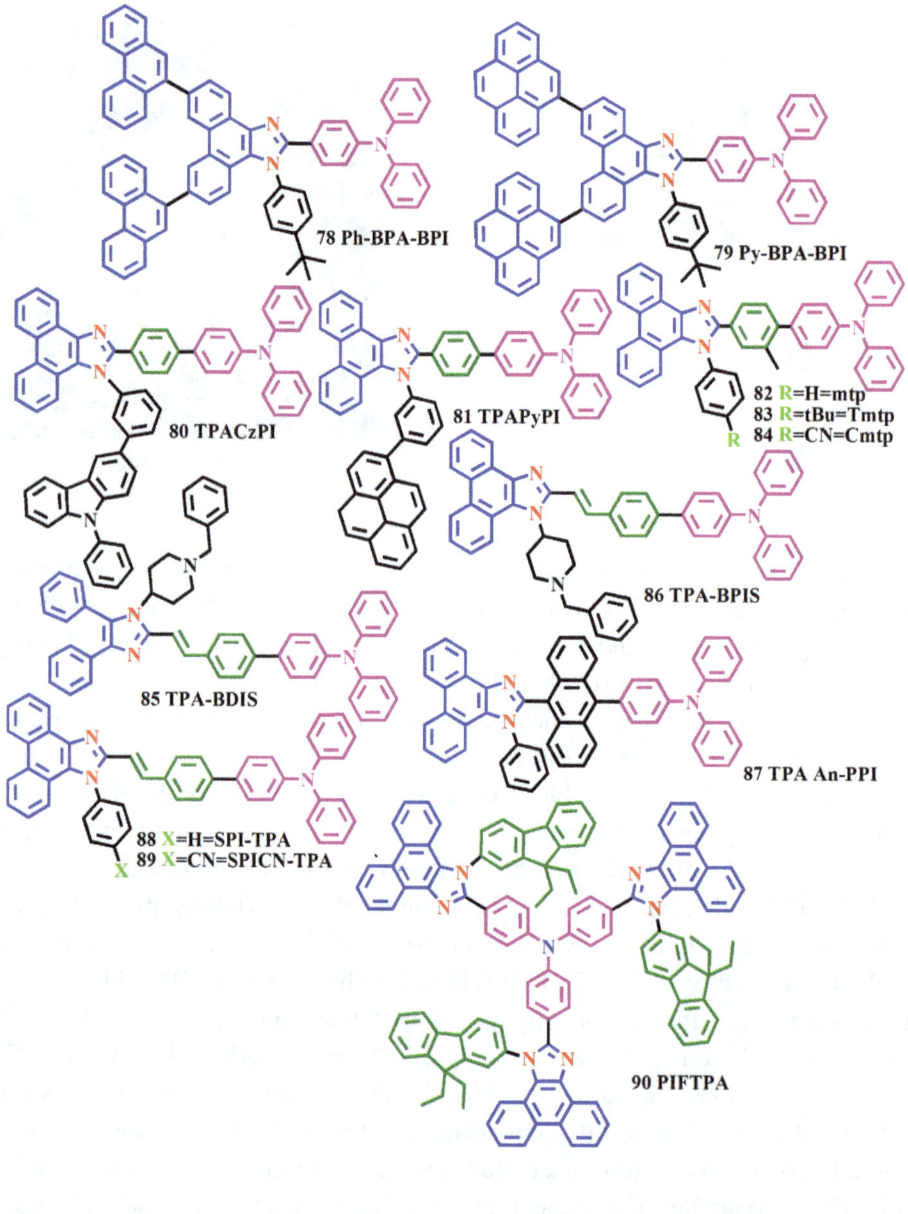

of 4.83 cd/A and CIE color coordinates of (0.32, 0.32) at the luminance of 1000 cd/m^2 (Figure 30. c and d). This work demonstrates not only that the phenanthroimidazole unit is an excellent building block to construct deep blue emission materials, but also the introduction of a diphenylphosphine oxide deprotonation substituent is an effective tactic for harvesting deep-blue emitting devices.

The phenanthroimidazole based UV fluorescent material (97 SiBPI), by attacking butyltriphenylsilane to the PPI group was synthesized by Lu *et al.,* (Gao, Liu, Li, Cheng, & Lu, 2016). Attached butyltriphenylsilane group could effectively increase the flexibility of the compound to help conserve better morphological stability with convenient solution-processibility. In addition, the tetrahedralgeometry of

Table 5. Device performances of Diphenylamine and Triphenylamine based bipolar PI materials for fluorescent OLEDs 64-92

Fluorophores	Voltage [a] [V]	EQE [a] (%)	C. E. [b]/cd A^{-1}	P. E. [c]/lm W^{-1}	Color Index	Ref.
64 BPA-BPI	2	0.55	0.65	0.68	0.15, 0.05	128
65 TPA-BPI	2.8	3.08	2.63	2.53	0.15, 0.09	
66 PATPA	3.8	0.72	0.34	0.24	0.15, 0.06	
67 TPA-TPI	2.6	2.8	4.6	4.68	0.19, 0.39	
68 TPM	4	1.05	0.4	0.86	0.18,0.14	131
69 TPMCN	3.5	2.2	2.69	1.96	0.15, 0.18	
70 TPA–PPI	-	-	-	-	0.153, 0.114	133
71 TPA–PIM	-	3.28	1.14	0.79	0.161, 0.046	132
72 p-PTPAPI	3.2	3.71	3.73	3.47	0.15, 0.15	134
73 m-PTPAPI	3.4	2.39	0.78	0.72	0.15, 0.05	
74 DPAA-PPI	2.4	5.2	14.3	13.9	0.20,0.55	135
75 tBuDPAA-PPI	2.5	4.1	11.3	11.4	0.19,0.55	
76 TPA-MPI	5.1	1.30	1.53	1.26	-	138
77 TPA-MPS	4.6	2.11	1.73	1.46	-	
78 Ph-BPA-BPI	2.52	4.56	3.60	3.66	0.15, 0.08	139
79 Py-BPA-BPI	*2.15*	*5.64*	*10.9*	*10.5*	*0.17, 0.29*	
80 TPACzPI	3.67	3.11	2.84	3.02	0.15, 0.08	140
81 TPAPyPI	3.20	3.99	2.64	2.61	0.15, 0.09	
82 mtp	3.0	3.89	2.11	2.09	0.15, 0.05	142
83 Tmtp	3.3	3.0	1.24	1.14	0.15, 0.04	
84 Cmtp	3.3	3.41	2.23	2.04	0.15, 0.07	
85 TPA-BDIS	4.1	1.98	1.48	1.31	0.16, 0.09	143
86 TPA-BPIS	3.5	2.53	1.88	1.51	0.16, 0.12	
87TPA An-PPI	2.4	4.61	9.83	10.42	0.16, 0.38	124
88 SPI-TPA	4.5	1.21	0.43	0.93	-	127
89 SPICN-TPA	4.2	2.35	2.81	2.05	0.15, 0.14	
90 PIFTPA	2.87	3.01	6.45	4.39	0.38, 0.524	144

Voltage [a]: turn on voltage, EQE [b]: quantum efficiency, C. E. [c]: current efficiency, P. E. [d]: power efficiency.

silane segment could increase the sterichindrance, suppress intermolecular interactions and enhance the efficiency in th esolid state. Compared with 4 PPI (0.40) (Y. Yuan, et al., 2011), very high PL quantum yield of 0.77 for 97 SiBPI is observed in thin films. The solution-processed device based on 97 SiBPI exhibits good performance with the maximum external quantum efficiency of 1.76%, current efficiency of 0.57 cd/A, and CIE coordinates of (0.158, 0.042), which is comparable to the performance of its vacuum-deposited device. It was among the best results ever reported for solution-processed UV light-emitting diodes, which inspired its application in the future.

Figure 32. Scheme-6. Synthetic Routes for TTP-TPI, DPT-TPI, and DPF-TPI

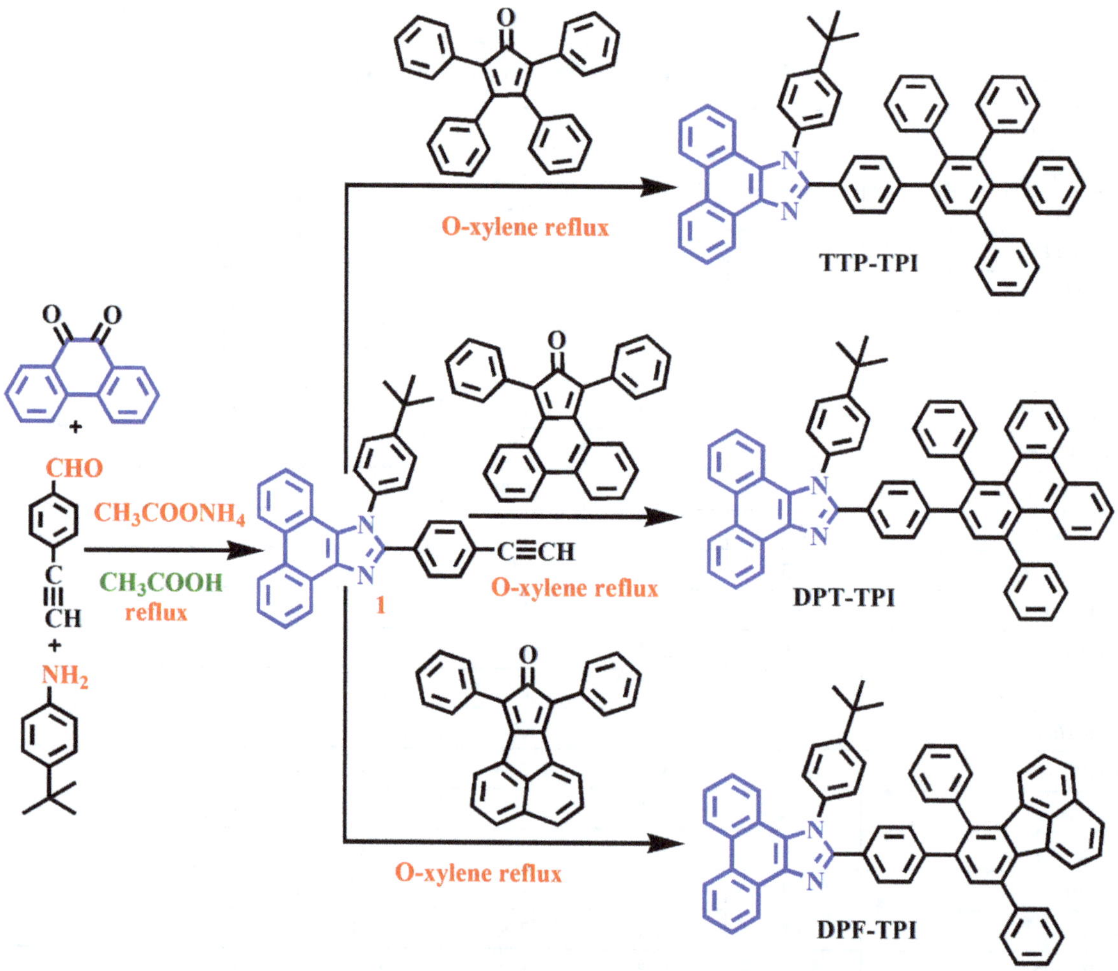

Wang *et al.,* synthesized three novel bipolar deep-blue fluorophores (98 DPACTPI, 99DPACPhTPI and 100 DPACFPPI) employing the donor 9,9-diphenyl-9,10-dihydroacridine (DPAC) and the acceptor, phenanthroimidazole group (Z. Huang, et al., 2017).The highly twisted conformations between DAPC and phenanthroimidazole in the molecules efficiently interrupt molecular π-conjugation and inhibit π-π intermolecular interactions, resulting in good thermal stability and efficient deep-blue emission. Non-doped OLEDs based on the three novel fluorophores show deep-blue light emissions. The nondoped OLED device based on 99DPACPhTPI showed excellent EL performance with CIE coordinates of (0.156, 0.047), low turn-on voltage of 3.0 V, high external quantum efficiency (EQE) of 3.5%, current efficiency (CE) of 1.38 cd/A, and power efficiency (PE) of 1.40 lm/W. This result is comparable to the reported high-efficiency OLEDs with similar color.

Very recently, Lu *et al.,* synthesized two deep blue PI fluorophores, 101PITO and 102PISF (X. Tang, et al., 2017), which are having same chemical structures but exhibit different photophysical and device properties. The synthetic routes of PI derivatives shown in scheme-7.These two fluorophores consist of phenanthroimidazole and dibenzothiophene analogs. The dissimilarity of their chemical structures lies

Figure 33. a) EL spectra for device A at different voltages, b) Current density–voltage–luminance characteristics for devices A, c) EL spectra for device B at different voltages, and d) Current density–voltage–luminance characteristics for device B

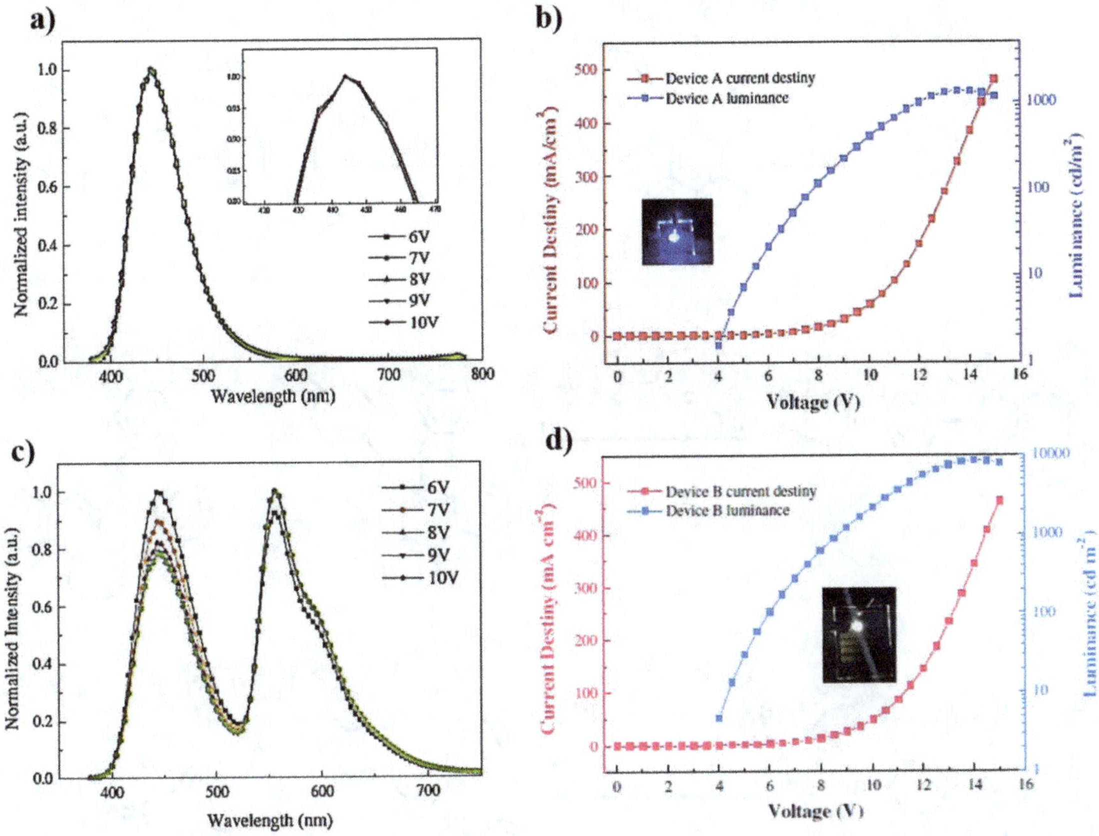

in the different oxidation states of the S atom. For 101 PITO, the S atom is oxidized and the resulting structure dibenzothiophene S,S-dioxide becomes electron deficient. Hence, the 101PITO shows significant solvatochromism, implying a charge transfer (CT) excited state formed between the donor (D) phenanthroimidazole and acceptor (A) dibenzothiophene S,S-dioxide. In addition, the 101PITO exhibits more balanced carrier transport. However, the 104 PISF is capable of emitting in the shorter wavelength region which is beneficial to obtain better color purity. The nondoped device of 101 PITO shows maximum EQE is 3.88% and CIE color coordinates of (0.17, 0.31.). On the other hand, the 102PISF displays blue EL but a little lower efficiency. The CIE coordinates of 102 PISF nondoped device is (0.16, 0.09) and the maximum EQE is 2.60%. The device performances of other bipolar PI materials are listed in Table 6.

Among all the other bipolar PI fluorophores (91-102). 92 DPT–TPI based device shown blue emission (0.16, 0.07) with high EQE (5.25%). The bulky polyaromatic hydrocarbon units endow the molecules with high glass-transition temperatures, nonplanar twisty structures which reduce molecular aggregations in the solid state. The nondoped OLED device based on 99 DPACPhTPI showed deep blue emission (0.156, 0.047) than the other bipolar materials, due to the highly twisted conformations between DAPC and phenanthroimidazole group. Resulting in good thermal stability and efficient deep-blue emission.

Figure 34. Scheme-7. Synthetic procedures of 101 PITO and 102 PISF. (a) bis(pinacolato)diboron, Pd(dppf)Cl_2, KOAc, 1,4-dioxane, 90 °C, N2 protection, 72 h; (b) 4-bromobenzaldehyde, CH_3COONH_4, aniline, acetic acid, 120 °C, N2 protection, overnight. (c) Pd(PPh_3)$_4$, K_2CO_3 (aq, 2 M), toluene, 90 °C, N2 protection, 72 h.

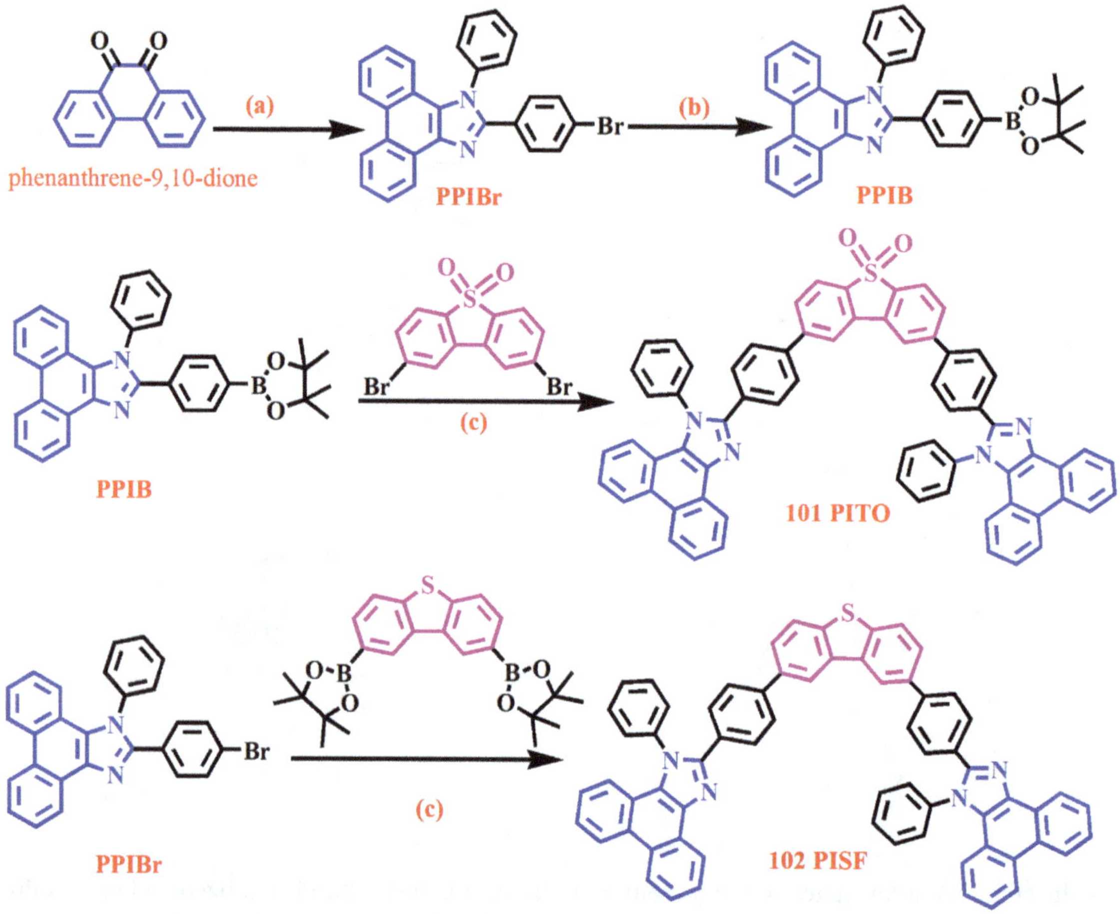

Host Materials for Phosphorescent OLEDs

Host materials are composed of two or more groups such as electron releasing and electron withdrawing groups. Host materials can be classified as hole-transport-type (HT-type), electron transport-type (ET-type) and bipolar transport hosts according to charge carrier transporting feature. Carbazole (CBZ) or triphenylamine (TPA) derivatives have been widely used as unipolar hole transporting host materials in PhOLEDs due to their sufficiently large T_1 (3.02 eV and 3.04eV) and good hole-transporting ability. CBP and mCP (Figure 35.) are most widely used as a host for the phosphorescent OLEDs. CPB and mCPwas used as a hole transporting host in the initial work of electrophosphorescence (Holmes, D'Andrade, & Forrest, 2003). OLED with these molecules show EQE of 2.9% for CBP and 4.6% for mCP, respectively. The obtained low EQE for CBP can be attributed to the lower T_1 energy of CBP (T_1: 2.56 eV), which is lower than that of the blue triplet emitters (>2.65 eV), resulting in an inefficient energy transfer from host to guest leads to poor device efficiency. However mCP having very low thermal

Figure 35. Other bipolar PI materials for fluorescent OLEDs 91-102

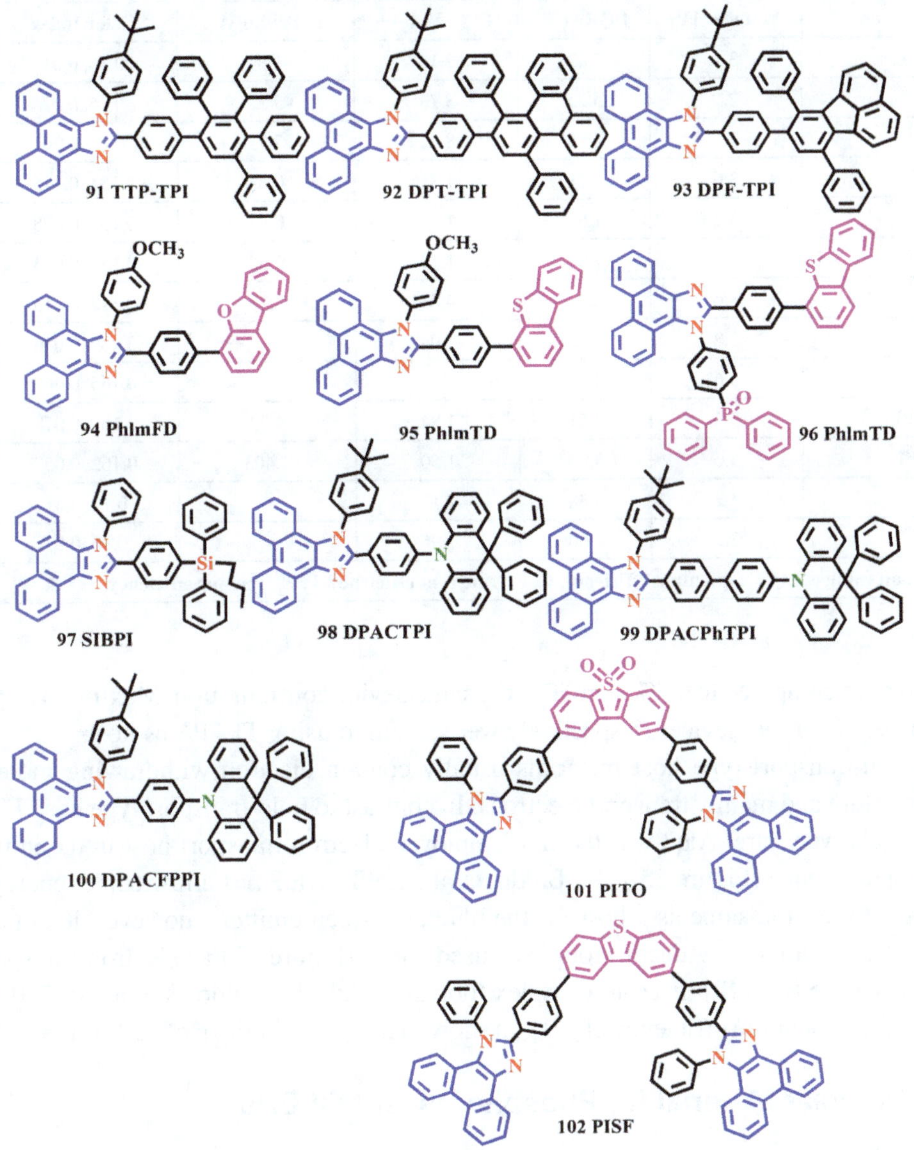

and morphological stability, mCP exhibits a low Tg at 60 °C, which is an obvious drawback. Therefore, to improve the overall efficiency of PhOLEDs, there is a need for the development new or novel host materials with high triplet energy and good thermal and morphological stabilities. The facile synthesis of a fluorene/triarylamine hybrid, tris[4-(9-phenylfluoren-9-yl)phenyl]amine (TFTPA) was reported through a simple Friedel–Crafts reaction (Figure 35) (Shih, Chien, Wu, & Shu,2007). The triplet energy of TFTPA remains as high as 2.89 eV, it also shows a relatively high glass transition temperature (Tg) of 186 °C. By using TFTPA as host with FIrpic, $Ir(ppy)_3$ and $(piq)_2Ir(acac)$ as emitters, Blue-, green- and red-electrophosphorescent devices have been made, a maximum EQE of 13.1% was attained in TFTPA

Table 6. Device performances of other bipolar materials for fluorescent OLEDs 91-102

Fluorophores	Voltage [a] [V]	EQE [a] (%)	C. E. [b]/cd A^{-1}	P. E. [c]/lm W^{-1}	Color Index	Ref.
91 TTP–TPI	3.1	5.02	2.10	1.88	0.16, 0.05	145
92 DPT–TPI	*2.9*	*5.25*	*3.13*	*3.22*	*0.16, 0.07*	
93 DPF–TPI	2.7	4.85	8.41	7.23	0.17, 0.24	
94 PhImFD	3.6	1.21	1.31	0.88	0.151, 0.115	146
95 PhImTD	3.6	1.63	1.34	0.82	0.151, 0.098	
96 PhImPOTD	3.6	1.00	1.13	0.50	0.151, 0.084	147
96PhImPOTD +PO01	3.6	2.2	4.83	2.15	0.317, 0.315	
97 SiBPI	3.8	2.05	1.04	-	0.166, 0.046	148
98 DPACTPI	3.6	2.31	1.84	1.75	0.165, 0.068	149
99 DPACPhTPI	3.3	3.51	1.84	4.02	0.160, 0.078	
100 DPACFPPI	3.6	3.03	1.50	1.403	0.162, 0.057	
101 PITO	3.2	3.88	7.17	-	0.17, 0.31	150
102 PISF	3.2	2.60	1.96	-	0.16, 0.09	

Voltage [a]: turn on voltage, EQE [b]: quantum efficiency, C. E. [c]: current efficiency, P. E. [d]: power efficiency.

hosted blue device compared to 6.7% of mCP at the same device configuration. Maximum EQE of 12.0% for green and 9.6% for red devices, respectively were acquired using TFTPA as host.

The electron-transport-type host materials usually contain electron-withdrawing moiety in their molecular structure and mainly transport electrons. In contrast to hole-transport type host, ET-type host materials are relatively rare. Alq3 was the first employed electron transport host in the initial work of electro phosphorescence (Figure 35.) (M. Baldo, et al., 1998). Alq3 has a low triplet energy (2.0 eV), which restricts to use the same as a host for the blue and green emitters, however, it can be used for red to near-infrared emitters. Adachi group was used TPBI (Figure 35) as electron transport host for red phosphorescence from $Btp_2Ir(acac)$ complex (dopant), (Adachi, Baldo, & Forrest. 2001), with the maximum external quantum efficiency of 6.9%, it shows a relatively high Tg of 122 °C then CPB (62°C).

PI Based Unipolar Material for Phosphorescent OLEDs

The n-type imidazole moiety has been extensively employed as an electron-transporting material (Kulkarni, Tonzola, Babel, & Jenekhe,2004, Tamao, Uchida, Izumizawa, Furukawa, & Yamaguchi, 1996, Ge, et al., 2008, Takizawa, Montes, & Anzenbacher,2009, Hung, et al., 2010, Gong, et al., 2010, Gong, et al., 2011, Chen, et al., 2011). Moreover, Ma and coworkers additionally demonstrated that the phenanthro-imidazolyl unit is an excellent electron-injection structure (Z. Wang, et al., 2011).

Cheng et al., design and synthesized two new imidazole derivatives 103 PPIE and 104 TPIE (H. Chou, et al., 2012). Both 103PPIE and 104TPIE adopt a stilbene group as the core structure end-capped with two 1-phenyl-1 H -phenanthro[9,10-d]imidazol-2-yl and 1,4,5-triphenyl-1 H -imidazol-2-yl groups, respectively. To know the suitability of these n-type diimidazolylstilbenes as the electroluminescent(EL) materials, the device A–D was fabricated by utilizing 103PPIE and 104TPIE as the dopants and DMPPP as the host. The devices A and B exhibited high EQEs of 8.1 and 6.9%, current efficiencies of 10.0 and

Figure 36. Typical hole transport and electron transport materials

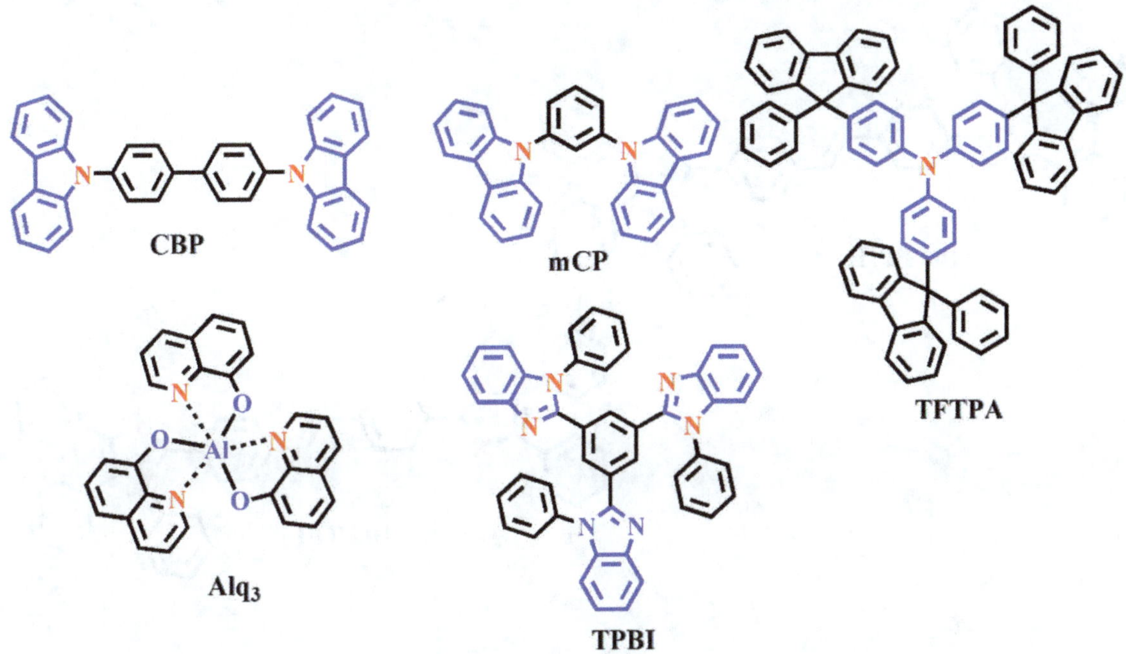

6.8 cd/A, and CIE color coordinates of (0.14, 0.14) and (0.15, 0.10), respectively. The performance of 103PPIE and 106 TPIE-based devices can be improved by inserting a thin layer of 4,4′,4′′-tri(9-carbazoyl) triphenylamine (TCTA) (10 nm) between NPB and the emitting layer. Devices C and D were fabricated, which emit deep-blue light with CIE color coordinates of (0.14, 0.15) and (0.15, 0.11) very efficiently. Extremely high EQEs of 7.8 and 7.7% and current efficiencies of 10.4 and 7.9 cd/A, respectively. To compare the results of these materials, with a common p-type deep-blue dopant, a device E with common p-type deep-blue dopant BCzVBi is fabricated. The performance of device E showed an EQE of 5.7%, a current efficiency of 7.1 cd/A, and CIE coordinates of (0.14, 0.14). The result demonstrates that the n-type dopants have higher device efficiency than the p-type one. In addition, compared with the performance of previously reported blue-fluorescent devices with true blue CIE coordinates, the extremely high efficiencies displays for the103PPIE- and 104TPIEbased devices. These n-type materials provide an alternative for dopant emitters in OLEDs other than the well-known p-type ones. The device performances of PI based unipolar material for phosphorescent OLEDs are listed in Table 7.

Bipolar Host Material for Phosphorescent OLEDs

Bipolar transport hosts materials are constructed with both electron transporting and as well as hole transporting groups. A suitable bipolar transport host in PhOLEDs should meet certain intrinsic requirements (Chaskar, Chen, & Wong,2011).

1) Bipolarity to facilitate both hole and electron injection and transport. The electron-donating moieties are normally comprised of aryl amines, such as carbazole, diphenylamine, and triphenylamine etc., while electron-withdrawing moieties often consist of phenanthroimidazole, oxadiazoles, triazine, benzimidazole, pyridine, triazine, phosphine oxide and so on. 2) Good thermal and morphological stabilities to

Figure 37. PI based unipolar material for phosphorescent OLEDs 103-104

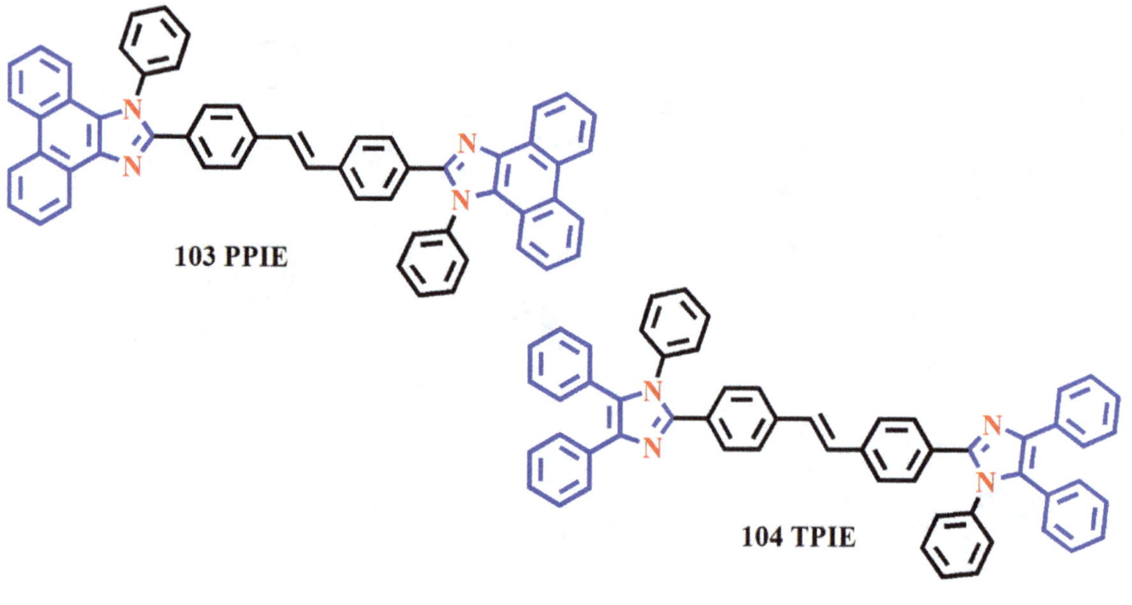

ensure the device stability. 3) Matching HOMO and LUMO energy levels with those of adjacent layers to reduce charge injection barriers and device driving voltages. 4) Higher triplet (T_1) energy than that of the dopant to effectively prevent exothermic reverse energy transfer from the guest emitter to the host.

Most of the previously reported unipolar host materials are capable of preferentially transporting holes or electrons in the device. Generally, the recombination zone tends to occur close to the emissive layer/ electron transporting layer (EML/ETL) interface when using a hole-transport host (Adachi, Baldo, & Forrest. 2001). Similarly, the recombination zone is likely to be located at the EML/HTL interface in the device, when the electron-transport host was used (Takizawa, Montes, & Anzenbacher, 2009). The poor carrier mobility and unbalanced charges in the emitting layer have been proven to be detrimental to the efficiency of OLEDs. Furthermore, PhOLEDs with narrow charge recombination zone could expedite triplet–triplet (T–T) annihilation, resulting in efficiency roll off due to a local accumulation of high density of triplet excitons, especially under high current densities (Zeng, Lee, Merkel, & Chen,2009). To this effect, the strategy by using both hole- and electron-transporting hosts (bipolar) has been demonstrated

Table 7. Devices Performances of PI based unipolar material for phosphorescent OLEDs 105-106

Host Material	Dopant	Devices	V_{on}	EQE [a] (%)	C. E. [b]/cd A^{-1}	P. E. [c]/lm W^{-1}	Color Index	Ref.
DMPPP	***103 PPIE***	A	*3.0*	*8.1*	*10.0*	*4.8*	*(0.14, 0.14)*	161
DMPPP	**104 TPIE**	B	3.1	6.9	6.8	2.8	(0.15, 0.10)	
DMPPP	**103 PPIE**	C	2.8	7.8	10.4	6.4	(0.14, 0.15)	
DMPPP	**104 TPIE**	D	3.2	7.7	7.9	3.1	(0.15, 0.11)	
DMPPP	**BCzVBi**	E	3.5	5.7	7.1	3.4	(0.14, 0.14)	
EQE [a]: quantum efficiency, C. E. [b]: current efficiency, P. E. [c]: power efficiency.								

to broaden the recombination zone and improve the device efficiency (Kondakova, et al., 2008, Kim, Jang, & Lee,2007, Leung, et al., 2007). Very recently, bipolar transport host materials have aroused great attention in the area of phosphorescent OLEDs because they can provide more balance in electron and hole fluxes and simplify device structure (M. Guan, et al., 2006, a. Y. Tao, et al., 2010, b. Tao, Y., et al., 2010, C. Chen, et al., 2009, Chi, Hung, Chiu, & Wong,2009).

Carbazole (CBZ) Based Bipolar Host Forphosphorescent OLEDs

Carbazole moiety was the most widely used group forhigh triplet energy host materials due to high triplet energy and versatile chemical modification inspite of good hole transportproperties (R. Holmes, et al., 2003). Carbazolehas a high triplet energy of 3.02 eV which is high enough for the blue host material.

Yang *et al.*, synthesized a series of bipolar phosphorescent hosts (105pPhBICP, 106mPhBICP, 107pPhBINCP and 108mPhBINCP), by conjugating carbazole (CBZ) moiety to the different positions of the rigid skeleton 1,2-diphenyl-1H-phenanthro[9,10-d]imidazole (PI) (H. Huang, et al., 2012). The photophysical and electrochemical properties of these hosts can be tuned through the different linkage between the electronic donor (CBZ) and acceptor (PI) components. These novel bipolar phosphorescent hosts were found to possess sufficiently large triplet energy levels for the green phosphorescent emitter $Ir(ppy)_3$ and the yellow phosphorescent emitter $(fbi)_2Ir(acac)$. Peak wavelengths of the EL spectra (Figure 28. a) of green devices A–D exhibit an emission similar to the phosphorescent emitter $Ir(ppy)_3$. This suggests that the emission peaks are indeed derived from the dopants and the triplet energy transfer from the hosts to the dopant are complete. Green device D using 108mPhBINCP as host exhibits high efficiencies of 77.6 cd/A and 80.3 lm/W ($\eta_{EQE,\ max}$ = 21% (Figure 38. b)) at the normal $Ir(ppy)_3$ emission. Even the yellow device E doping the $(fbi)_2Ir(acac)$ into host 105pPhBICP exhibits $\eta_{c,max}$ of 57.2 cd/A, $\eta_{EQE,max}$ of 19.3%, and $\eta_{p,max}$ of 59.8 lm/W, respectively. The high efficiency could be attributed to the following aspects: 1) The bipolar properties of the hosts could balance the charge transfer and improve the carrier recombination ratio in the emissive layer;2)The high triplet energies of these hosts could efficiently suppress adverse energy back-transfer from the guest to the host, consequently resulting in the good performances of these devices. Finally, the good thermal stability restrains the strong bimolecular interactions of the phosphorescent Ir(ppy)3 emitter to reduce the triplet–triplet annihilations. The performances of PhOLEDs based on these phosphorescent hosts are attractive compared to Ir(ppy)3- and (fbi)2Ir(acac)-based PhOLEDs with similar configurations reported recently (Y. Tao, et al., 2008).

In order to improve the thermal stability of above mentioned bipolar host materials (107pPhBINCP and 108mPhBINCP are only around 110 ^{0}C), the same group (Yang *et al.*,) designed and synthesized a series of bipolar host materials (109 PhBIDpCP, 110 PhBIDPmCP, 111 PhBIDmpCP and 112 PhBID-mCP) (H. Huang, 2013), by the conjugation of two carbazole units to one phenanthroimidazole moiety through different linkage modes. As expected, these new compounds exhibit high thermal stabilities, with the decomposition temperatures (Td, 5 wt% weight loss) and the glass transition temperatures (Tg) in the range of 468–514 ^{0}C and 143–282 ^{0}C, respectively. The device C (111 PhBIDmpCP) shows a low onset voltage of 2.8 V, high brightness of 42 160 cd/m^2 at 10.0 V, corresponding to a maximum current efficiency of 74.3 cd/A and a maximum power efficiency of 74.4 lm/W, as well as a maximum external quantum efficiency (EQE_{max}) of 20.2%.Devices A (109 PhBIDpCP), B (110PhBIDpmCP) and D (112 PhBIDmCP) exhibit maximum current efficiencies of 59.6 cd/A, 45.7 cd/A, 60.3 cd/A and power efficiencies of 59.4 lm/W, 45.6 lm/W, 59.2 lm/W, respectively.

Figure 38. a) EL spectra of the devices A–D based on the phenanthroimidazole derivatives, b) External quantum efficiency vs. current density for devices A–D

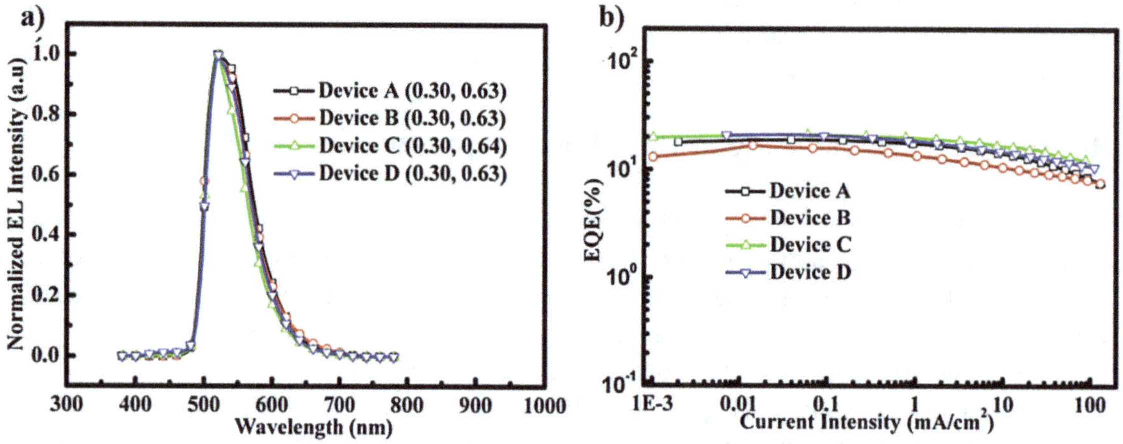

Zhang *et al.*, designed and synthesized two novel host materials, 113 DCzBPI and 118DTPABPI,by integrating electron-donating arylamine (carbazole/triphenylamine) and electron-accepting phenanthroimidazole (PI) moieties (X. Zhang, et al., 2013).The synthetic routes of PI derivatives shown in scheme-8. In the fluorophores (113 and 118), the arylamine and phenanthroimidazole moieties are extended to different planes in order to decrease the interaction of donor and acceptor units. Both the fluorophores show high triplet energy levels. Their EL properties were investigated by using 113DCzBPI or 118 DTPABPI as host materials. The devices with 118DTPABPI host, achieve maximum external quantum efficiency of 17.1% and a maximum current efficiency of 43.2 cd/A. While the devices with113DCzBPI host, the maximum external quantum efficiency is up to 21.2% with the CE_{max} of 53.8 cd/A. The maximal brightness of these devices 63,610 cd/m^{2}for 113DCzBPI and 40,040 cd/m^{2}for 118DTPABPI is achieved. These results provide a new approach for designing highly efficient phosphorescence host materials for electroluminescent applications, in particular, for greenPhOLEDs.

Wang *et al.,* designed and synthesized two deep blue emitting/host fluorophores119PPI-PPITPA and 114PPI-PPIPCz, by integrating with two phenyl-phenanthroimidazole (PPI-acceptor) and one triphenylamine (TPA-donor) or phenyl-carbazole (PCz-donor) groups. (C. Li, et al., 2015).The deep blue emission and good ambipolar carrier transport characteristics of 119PPI-PPITPA (device-B1) and 114 PPI-PPIPCz (device-B2) ensure that the OLEDs with the two fluorophores as non-dopant EMLs show an extremely high EQE of 7.7% and 8.1% with deep blue CIE (0.15,0.08) and (0.15,0.07).Moreover, the small singlet–triplet splitting and good carrier transport properties allow 119PPI-PPITPA (devices-R1, Y1 and G2) and 114PPI-PPIPCz (devices-R2, Y2 and G2) to be used as a host to fabricate red, yellow and green (RYG) along with phosphorescent dopant for PhOLEDs. Device G2 exhibits a low turn-on voltage of 2.8 V. The maximum EQE of 17.8% and PE of 65.7 lm/W are among the best values for green phosphorescent OLEDs with blue emitters as hosts (Chou, & Cheng,2010, Su, Cai, & Kido, 2011, X. Liu, et al., 2014). However, device G1 shows a much lower EL efficiency than device G2. The reason for low EL efficiency can be attributed to the lower triplet energy level (T1 = 2.40 eV) of 119PPI-PPITPA, which may cause back transfer of energy from the green phosphorescent guest triplet states to the host, resulting in a loss of efficiency. Yellow PhOLEDs with excellent performance was also achieved by using these two fluorophores as host materials. The device Y1 exhibits the maximum EQE of 19.1%, CE of 61.8 cd/A, and PE of 65.5 lm/W, and device Y2 shows a similar behavior with the maximum EQE of18.1%, CE of 56.5 cd/A, and PE of 58.8 lm/W. Finally, deep red PhOLEDs were fabricated using 119PPI-PPITPA and 114PPI-PPIPCzas host materials. Device R2 exhibits the maximum luminance of24 810 cd/m^2 at 13.5 V and excellent EL efficiencies (18.0%,21.3 cd/A, and 20.9 lm/W), whereas device R1 exhibits the maximum luminance of 18 020 cd/m^2 at 11.0 V and ELefficiencies (16.8%, 19.8 cd/A, and 21.5 lm/W) (Figure 39). The PE values of devices R1 and R2 are comparable to recently reported highly efficient deep red PhOLEDs (Y. Tao, et al., 2009, H. Sasabe, et al., 2008, F. Hsu, et al., 2009, C. Chen, et al., 2014, Fan, Sun, Su, & Cheng, 2011). These results demonstrate an efficient strategy to achieve full-color OLEDs by developing the deep blue organic luminescent materials with an ideal D–A fluorophores.

Figure 39. Scheme-8. The synthetic routes for 113 DCzBPI and 118 DTPABPI fluorophores

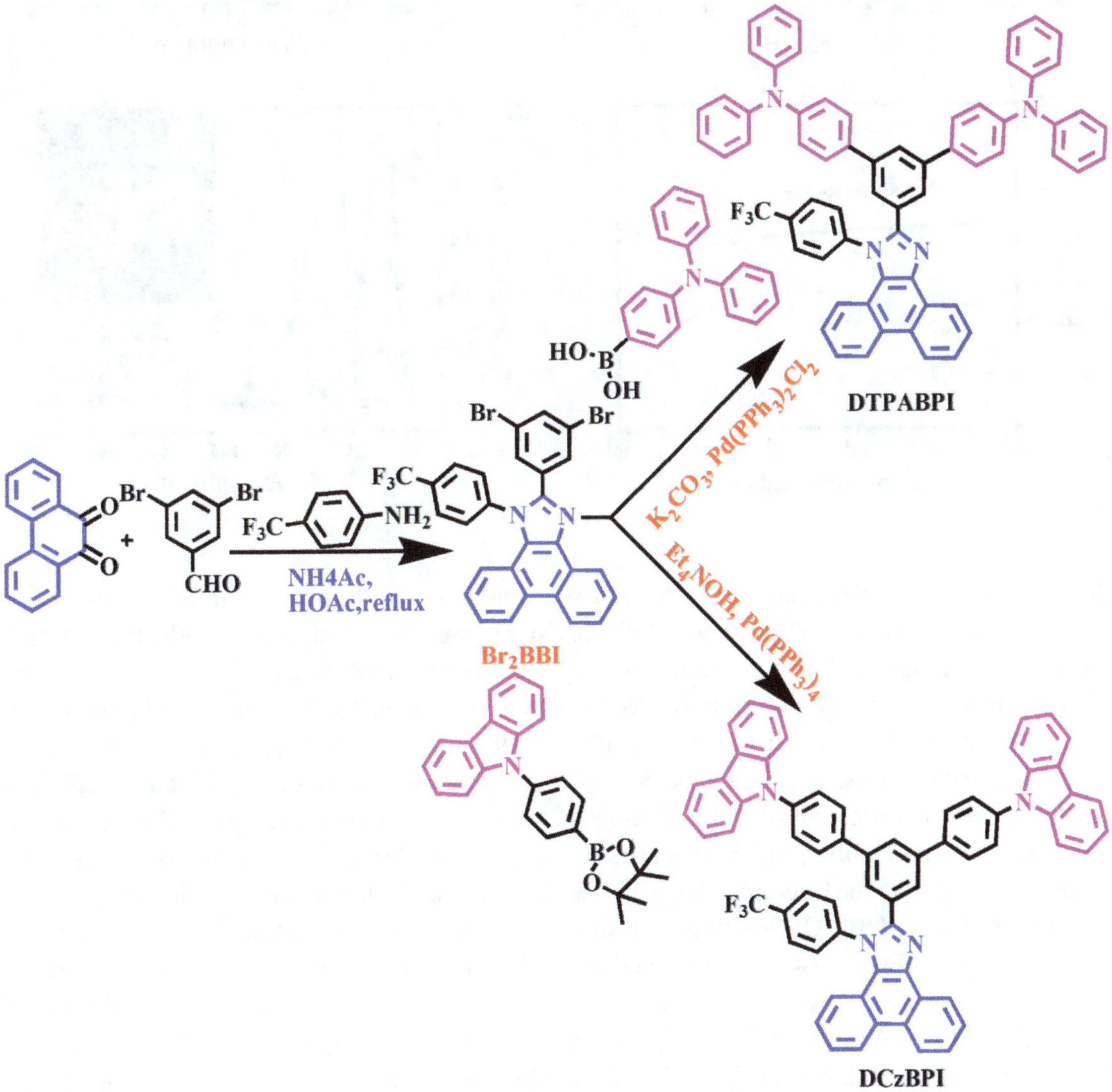

Figure 40. a) External quantum efficiency vs. luminance characteristics; b) EL spectra and CIE1931 coordinates of devices B1, G1, Y1 and R1 based on 119PPI-PPITPA; c) External quantum efficiency versus luminance characteristics; and d) EL spectra and CIE1931 coordinates of devices B2, G2, Y2 and R2 based on 114 PPI-PPIPCz

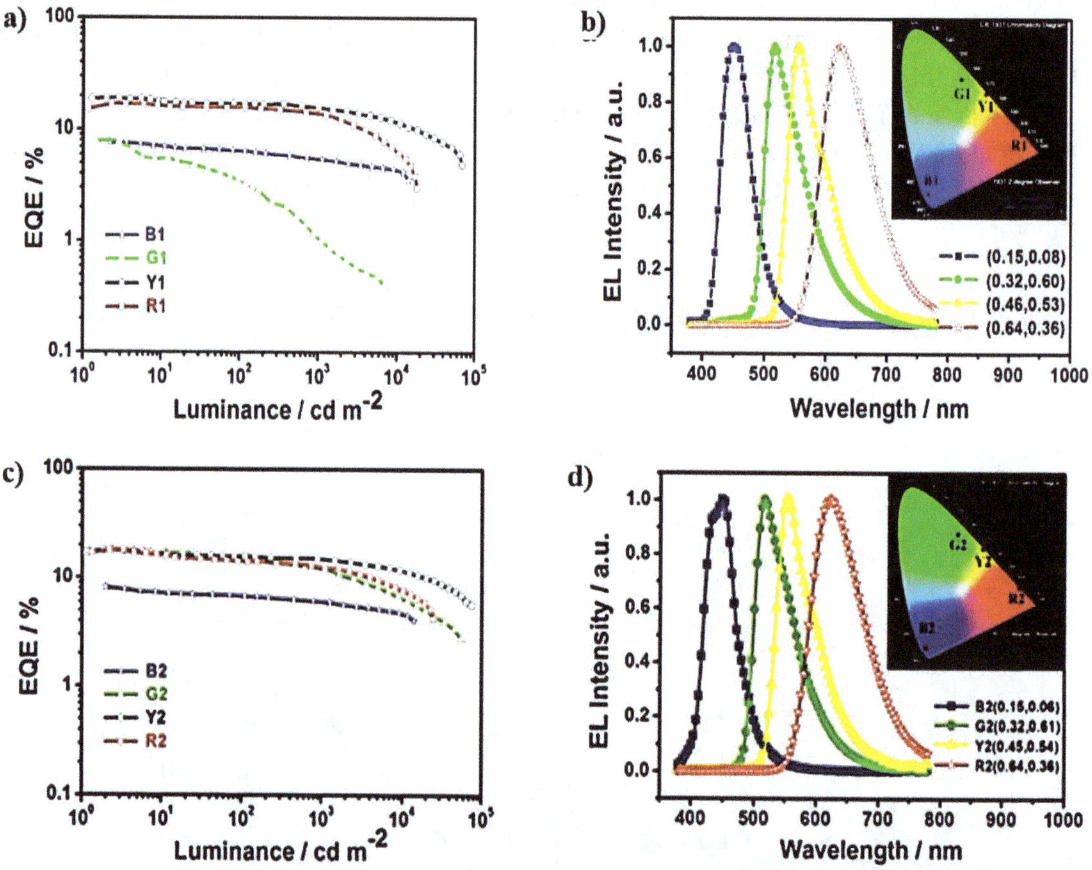

Recently, Wang *et al.*, designed and synthesized two para-linked bipolar phenanthroimidazole–diazacarbazolefluorophores (115PPI-pCNCz and 116PPI-pNNCz), by incorporating the novel electron-accepting 1,5-diazacarbazole into different positions on a rigid phenanthroimidazole (PI) skeleton (B. Wang, et al., 2016). The synthetic routes of PI derivatives shown in scheme-9. The C-connected fluorophore 115 PPI-pCNCz has a high glass-transition temperature of 191 ^{0}C and a relatively narrow bandgap of 3.04 eV, which contribute to decreasing the device operating voltage and improving the power efficiency. Meanwhile, the N-connected analog 116 PPI-pNNCz retains a high triplet energy of 2.63 eV and balanced carrier-transport ability, which favors the electron–hole recombination within the emission layer. More interestingly, both the fluorophores can be used to construct high-performance three-primary color (red, green and blue (RGB)) OLEDs through a simple material system and a similar device architecture. In particular, for 116PPI-pNNCz, the undoped deep-blue fluorescent device (B2) exhibits an external quantum efficiency of 2.97% with CIE coordinates of (0.15, 0.06). Highly efficient green (device-G2-116 PPI-pNNCz) (79.8 cd/A, 88.3 lm/W and 22.6%) and red (device-R2-116 PPI-pNNCz) (24.7 cd/A, 22.5 lm/W and 19.3%)were achieved. (Figure 40.) PhOLEDs are also achieved by using it as the universal

Figure 41. a) External quantum efficiencies versus current density curves, b) EL spectra for devices G1, R1 based on 115PPI-pCNCz and devices G2, R2 based on 116PPI-pNNCz, respectively

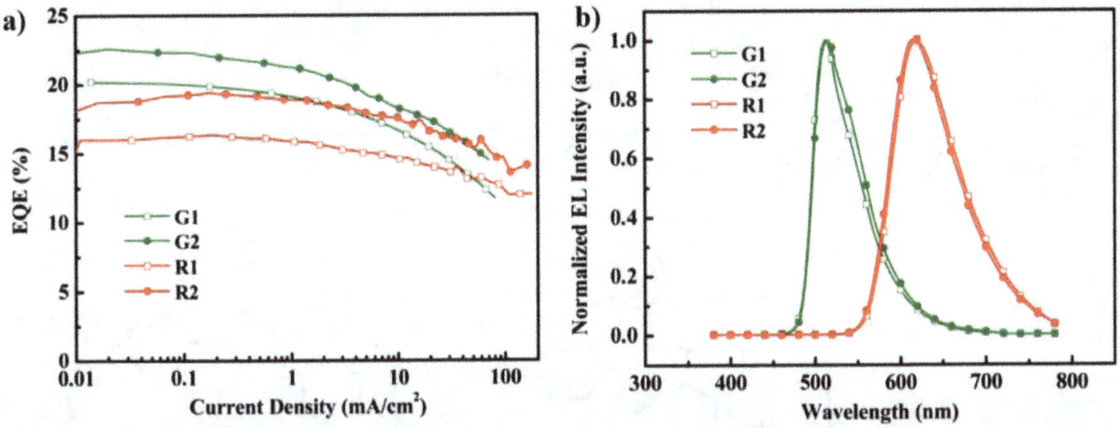

host material. More interestingly, all the devices exhibit a low-efficiency roll-off at the normal operation brightness of 1000 cd/m^2, which are among the best results for reported full-color OLEDs to date. The device performances of carbazole based bipolar host for PhOLEDs are listed in Table 8.

Triphenylamine (TPA) Based Bipolar Host Forphosphorescent OLED

Like carbazole, triphenylamine (TPA) also having a high triplet energy of 3.04 eV.[115] Several triphenylamine derivatives, such as TPD, NPB, TAPC and TCTA (structures are given in Figure 43) are well-known good hole-transport materials. Additionally, most triphenylamine derivatives have higher HOMO level (ca. 5.3 eV) than carbazole materials, which is near to the work function of ITO or the widely used hole transporting material NPB, thus allowing efficient hole injection (L. Tao, et al., 2009). These characteristics put the TPA derivatives close to the desired criteria for electrophosphorescent host materials.

Tong *et al.*, designed and synthesized a new bipolar fluorophore, 120 PPI-F-TPA, consisting of an electron-withdrawing phenanthroimidazole (PI) moiety and an electron-donating triphenylamine group, with indirect linkage (B. Liu, et al., 2016). The sp^3-hybridized C9 atom of the fluorene linkage efficiently interrupts molecular conjugation and inhibits π–π intermolecular interactions, resulting in efficient violet-blue emission, excellent thermal stability and high triplet energy. The 120PPI-F-TPA fluorophore shows impressive performance as the emitting layer in non-doped OLEDs, which reached an external quantum efficiency of 3.11% with a CIE coordinate of (0.16, 0.05). Moreover, the high triplet energy allows 120PPI-F-TPA to be used as a host for PhOLEDs. High performance green (device 2) and orange-red (device 3) PhOLEDs with the maximum EQEs, current efficiencies (CE) and power efficiencies (PE) of 15.6% and 12.5%, 57 cd/A and 27 cd/A, 60 lm/W and 28.3 lm/W, respectively (Figure 44.), have been successfully obtained. More importantly, all the devices exhibit low-efficiency roll-off; in particular, that of the orange-red PhOLEDs is extremely small. The orange-red PhOLED has a decay rate of EQE less than 1% at 1000 cd/m^2, 13.6% at 10000 cd/m^2 and 29.5% even at 50000 cd/m^2, which is very rare among orange or orange-red PhOLEDs at such high brightness.

Very recently, Grigalevicius *et al.*, described a series of new phenanthroimidazole based fluorophores for the bipolar emitting host (117 H1, 121H2 and 122 H3) (Tavgeniene, et al., 2017). The synthetic routes of

Figure 42. Scheme-9. Synthetic routes and chemical structures of the materials. Reagents and conditions: (a) Pd(PPh3)2Cl2, K2CO3, toluene, EtOH, H2O; (b) PPh3, 1,2-dichlorobenzene; (c) CuI, K2CO3, 18-crown-6, DMPU

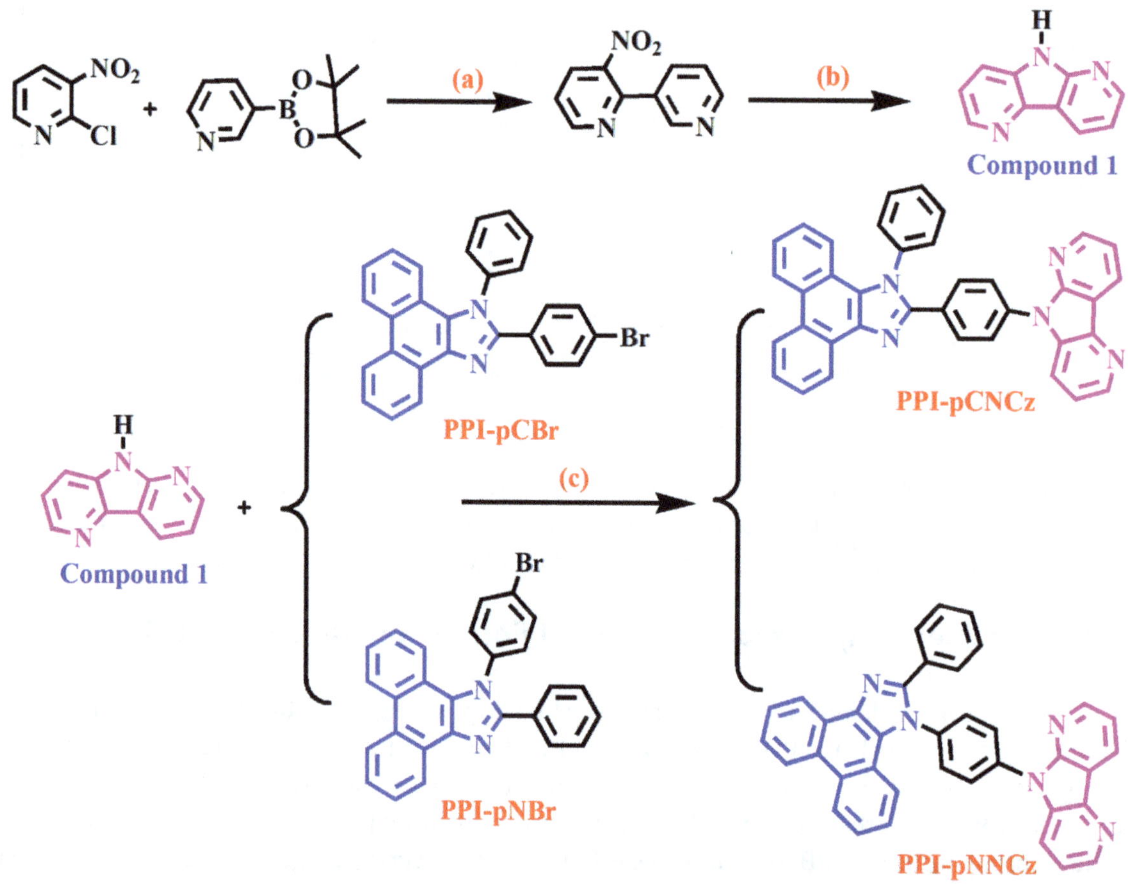

PI derivatives shown in scheme-10. All the synthesized fluorophores were used as hosts in red PhOLEDs and their properties were compared with commercially available host materials (TCTA, PYD-2Cz and CBP).Device E with the 121H2 host gave a higher current density than device D (117H1), indicating that replacing the 9-hexylcarbazolyl fragment with the triphenylamine moiety in 121H2 improves the carrier transport. Moreover, based on the molecular structure of fluorophore121H2, fluorophore122H3 has an additional imidazole moiety, giving it an even higher carrier transport capability. Since the energy bandgaps of 121H2 and 122H3 are much lower than those of the other tested hosts, the respective turn on voltages of 121H2 and 122H3 are as low as 2.8 V and 2.6 V, respectively. Furthermore, the maximum luminance of device F reached 15000 cd/m^2, which exceeded that of device C with a bipolar CBP host (~13500 cd/m^2).Devices E and F exhibited EL efficiencies peaks of up to 15.9% (21.5 cd/A and 29.9 1 m/W) and 12.4% (18.3 cd/A and 23.9 1 m/W), respectively (Figure 45).These results indicate that the gradually improved carrier transport capabilities for 121H2 and 122H3 could result in devices easily achieving carrier balance in a simplified trilayer architecture. The device performances of triphenylamine based bipolar host for phOLEDs are listed in Table 9.

Figure 43. Carbazole based bipolar host for phosphorescent OLEDs 105-117

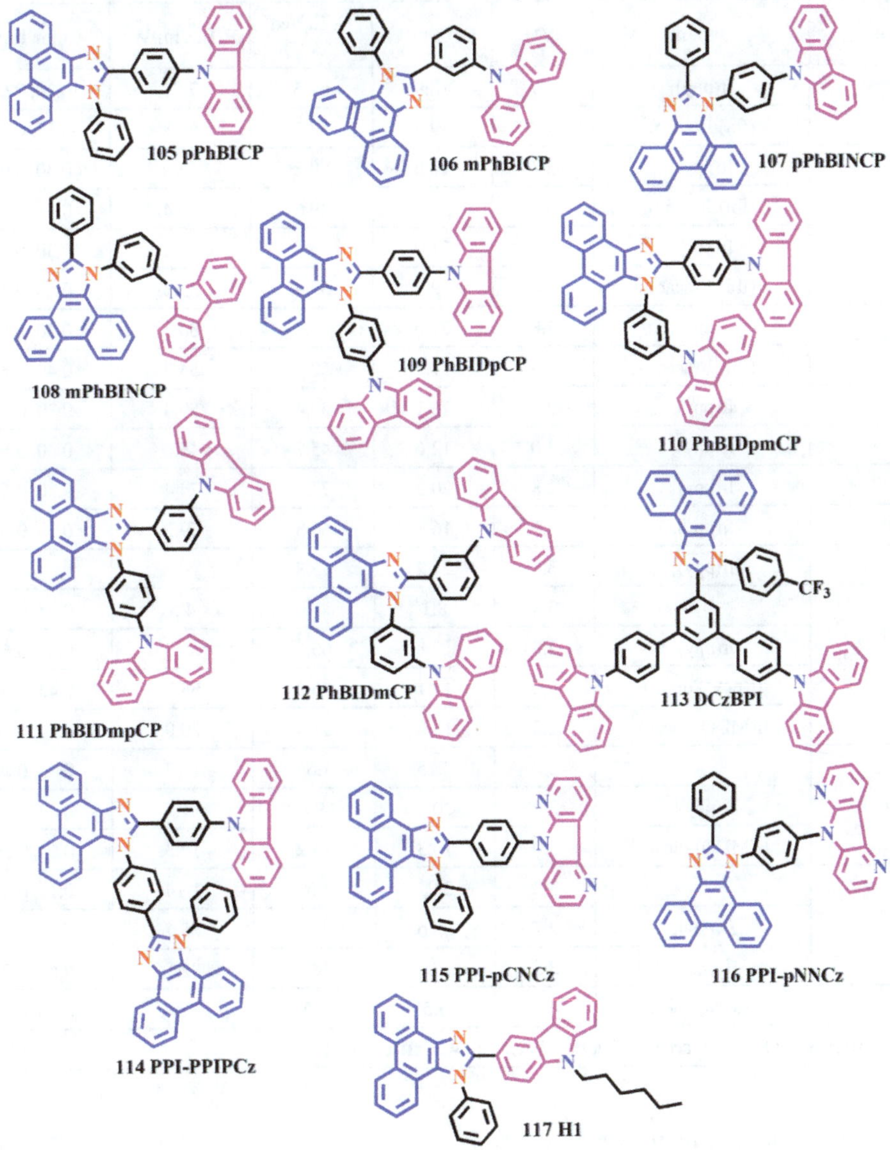

Other Bipolar Host for Phosphorescent OLEDs

Other than phenanthroimidazole group, dimesitylborane, phosphine oxide derivatives are also used to improve the thermal stability, electrochemical stability and carrier injection/transport ability of fluorophores (Wakamiya, Taniguchi, & Yamaguchi, 2006, M. Lin, et al., 2012, H. Shi, et al., 2014, Padmaperuma, Sapochak, & Burrows,2006, D. Yu, et al., 2012, C. Han, et al., 2011, Wada, et al., 2013, J. Zhao, et al., 2011, W. Jiang, et al., 2011, D. Yu, et al., 2012, Jeon, Yook, Jogo, & Lee,2010).The synthetic routes of PI derivatives shown in scheme-6. Tetraarylsilane compounds have attracted an intensive attention as organic EL materials, attributed to their wide bandgap and the high thermal stability resulting from

Table 8. Device performances of carbazole host materials for phosphorescent OLEDs105-117

Host Materials	Dopant	V_{on}	EQE [a] (%)	C. E. [b]/cd A^{-1}	P. E. [c]/lm W^{-1}	Color Index	Ref.
105 pPhBICP	$Ir(ppy)_3$	2.7	18.9	69.5	74.7	0.30, 0.63	161
	$(fbi)_2Ir(acac)$	2.8	19.3	57.2	59.8	0.51, 0.49	
106 mPhBICP	$Ir(ppy)_3$	3.1	16.6	60.2	54.0	0.30, 0.63	
	$(fbi)_2Ir(acac)$	3.1	16.9	49.6	46.4	0.51, 0.48	
107 pPhBINCP	$Ir(ppy)_3$	2.7	21.0	76.6	79.5	0.30, 0.64	
	$(fbi)_2Ir(acac)$	3.0	19.3	57.3	55.3	0.50, 0.48	
108 mPhBINCP	$Ir(ppy)_3$	2.8	21.0	77.6	80.3	0.30, 0.63	
	$(fbi)_2Ir(acac)$	2.8	17.5	52.9	54.7	0.49, 0.49	
109 PhBIDpCP	$Ir(ppy)_3$	2.8	16.1	59.4	59.4	0.30, 0.63	172
110 PhBIDPmCP	$Ir(ppy)_3$	3.0	12.6	45.7	45.6	0.30, 0.63	
111 PhBIDmpCP	$Ir(ppy)_3$	2.8	20.2	74.3	74.4	0.30, 0.64	
112 PhBIDmCP	$Ir(ppy)_3$	3.0	16.3	60.3	59.2	0.30, 0.63	
113 DCzBPI	$Ir(ppy)_3$	3.4	21.2	53.8	-	-	175
114 PPI-PPIPCz	-	3.4	8.1	4.8	4.5	0.15, 0.07	176
	$Ir(ppy)_3$	2.8	17.8	60.6	65.7	0.32, 0.61	
	$Ir(ppy)_2(dipig)$	3.0	18.1	56.5	58.8	0.45, 0.54	
	$Ir(MDQ)_2(acac)$	2.3	18.0	21.3	20.9	0.64, 0.36	
115 PPI-pCNCz	-	3.0	2.15	1.63	1.17	0.15, 0.08	185
	$Ir(ppy)_3$	2.6	20.2	69.8,	81.2	0.27, 0.63	
	$Ir(MDQ)_2(acac)$	3.0	16.3	19.2	19.3	0.64, 0.36	
116 PPI-pNNCz	-	3.3	2.97	1.69	1.14	0.15, 0.06	
	$Ir(ppy)_3$	*2.7*	*22.6*	*79.8*	*88.3*	*0.28, 0.63*	
	$Ir(MDQ)_2(acac)$	*3.3*	*19.3*	*24.7*	*22.5*	*0.63, 0.36*	
117 H1	$Ir(MDQ)_2acac$	4.8	5.5	8.5	6.5	0.60, 0.39	188

EQE [a]: quantum efficiency, C. E. [b]: current efficiency, P. E. [c]: power efficiency.

Figure 44. The typical hole transporting materials

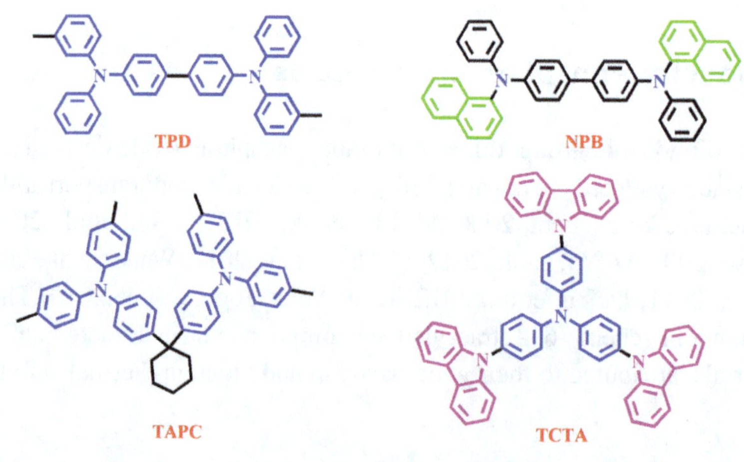

Figure 45. a) The EL spectra of device 2 and device 3. d) The EQE and luminance characteristics of the devices.

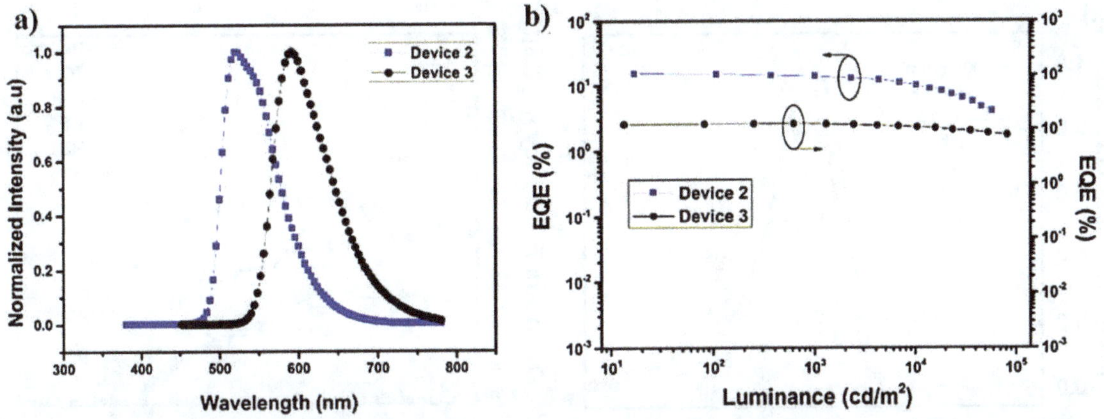

their non-conjugated tetrahedral molecular geometry (X. Ren, et al., 2004, S. Gong, et al., 2011, H. Liu, et al., 2014, D. Sun, et al., 2014).

Wang *et al.*, designed and synthesized two novel bipolar host materials (123 p-BPPI and 124 m-BPPI) containing phenanthroimidazole (PI)/dimesitylborane (Mes_2B) with para- and meta-linkage (K. Wang, et al., 2014). The synthetic routes of PI derivatives shown in scheme-11. The attaching Mes_2B moiety to the PI group improves the electron injection/transport ability, thermal and electrochemical stability of both synthesized fluorophores. The fabricated series of PhOLEDs displays high efficiencies of 66.3 cd/A, 63.1 lm/W, 18.2% for green emission (device-G), 55.2 cd/A, 66.6 lm/W, 14.5% for yellow emission (device-Y2) and 20.1 cd/A, 20.4 lm/W, 13.5% for red emission (device-R1) by using 123 p-BPPI and 124 m-BPPI as the universal host materials, respectively (Figure 48). More importantly, all these high-efficiency values are obtained at practical brightness 100–2000 cd/m^2. The results indicate that bipolar host 123p-BPPI and 124m-BPPI have high potential in fabricating various color OLEDs for displays and lighting applications.

The Same group (Wang *et al.,)* designed and synthesized two novel hybrids of an electron-accepting phosphine oxide moiety attached to a phenanthroimidazole (PI) group (125 DPO-PPI and 126 DPO-2PPI) (Wang, et al., 2014). The P=O group is used as a point of saturation between the PPI moiety and the outer phenyl groups, so the high triplet energy of PPI is preserved to act as a host for red and green phosphorescent dopants. The strong intermolecular interactions and steric effect of the diphenylphosphine oxide (DPO) moiety endows the films with high quantum yields in the deep-blue emission region. The fabricated non-doped devices B1 (125 DPO-PPI) and B2 (126 DPO-2PPI) exhibit deep blue emissions with CIE coordinates of (0.16, 0.09) and (0.16, 0.08), respectively. Furthermore, the efficient green (G1 and G2 devices) and red (R1 and R2 devices) PhOLEDs are fabricated by using 125 DPO-PPI and 126 DPO-2PPI fluorophores. Device G1 using 125 DPO-PPI as the host achieves a maximum current efficiency ($\eta_{c,\ max}$) of 62.6 cd/A, a maximum power efficiency ($\eta_{p,max}$) of 70.2 lm/W, and a maximum external quantum efficiency ($\eta_{ext,max)}$ of 17.2%. In comparison, device G2 hosted by 126 DPO-2PPI exhibits slightly higher performance with $\eta_{c,max}$ of 65.4 cd/A, $\eta_{p,max}$ of 73.3 lm/W, and $\eta_{ext,max}$ of 18.0%. Device R2 exhibited a higher maximum brightness of 29 245 cd/m^2, at 9 V, and maximum EL efficiencies (19.0 cd/A, 21.3 lm/W, 13.5%) with CIE coordinates of (0.63, 0.36). Device R1 achieved a maximum bright-

Figure 46. a) Normalized EL spectra at a luminance of 103 cd/m², b) external quantum efficiency vs. luminance, for Devices A (TCTA), B (PYD-2Cz), C (CBP), D (117 H1), E (121 H2) and F (122 H3)

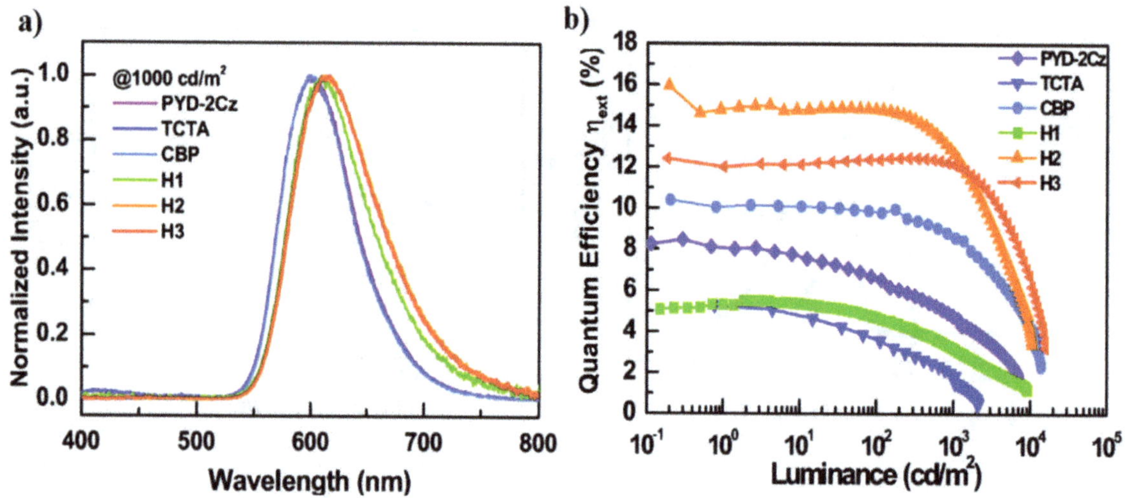

Figure 47. Scheme-10. The synthetic pathway of PI derivatives H1-H3.

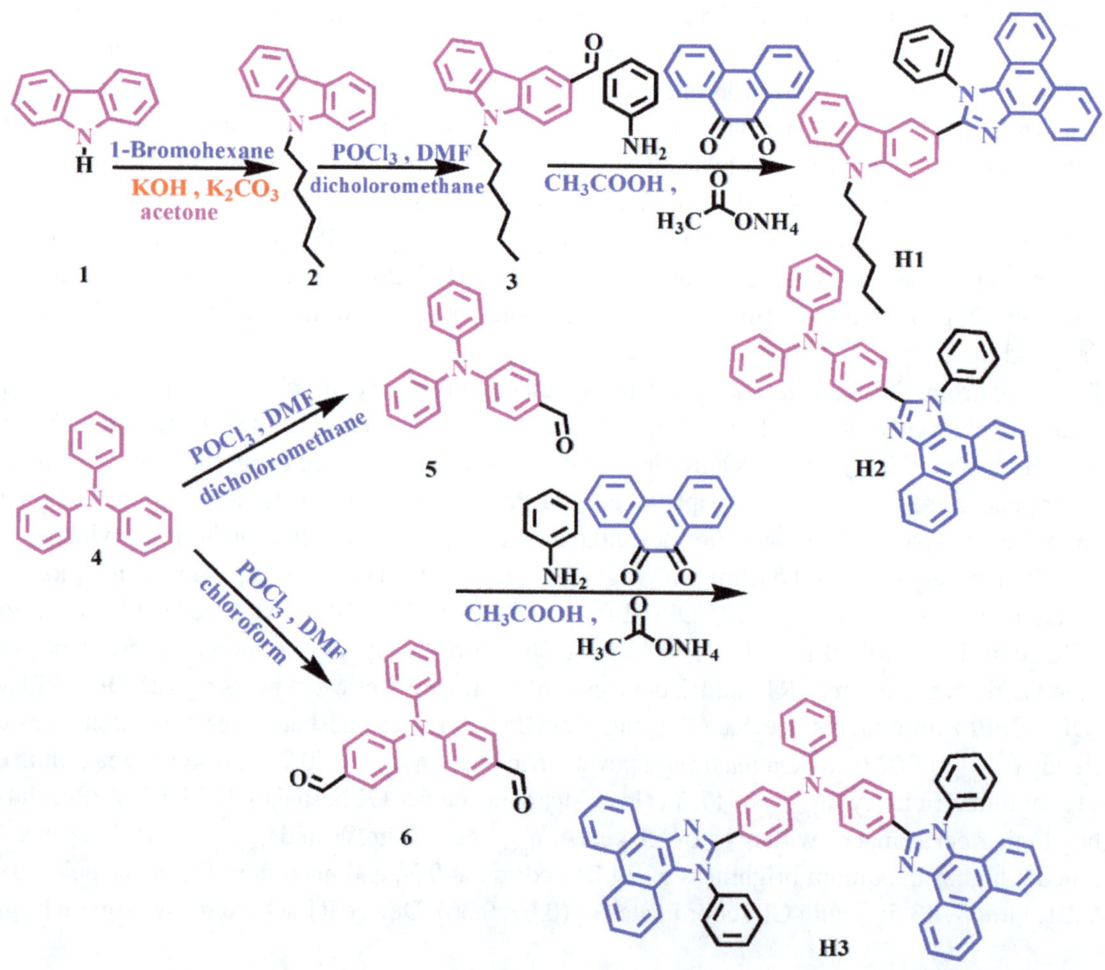

Figure 48. Triphenylamine (TPA) based bipolar host for phosphorescent OLEDs 118-122

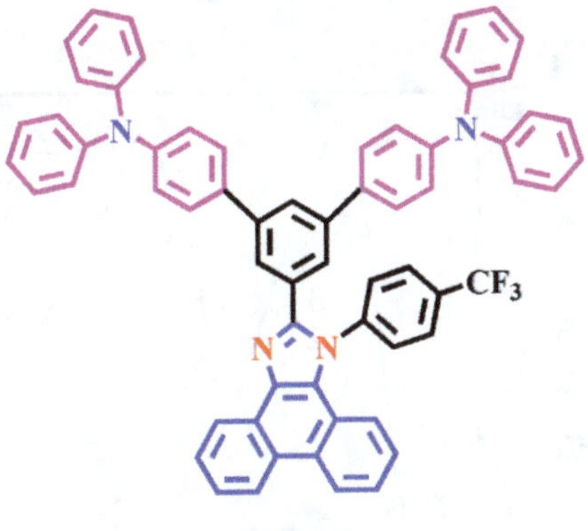

118 DTPABPI

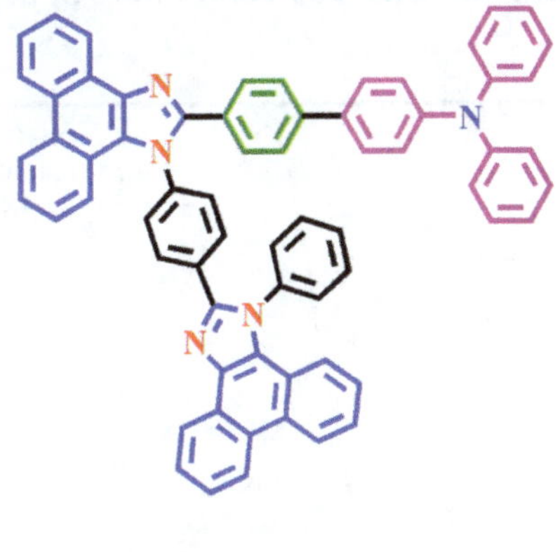

119 PPI-PPITPA

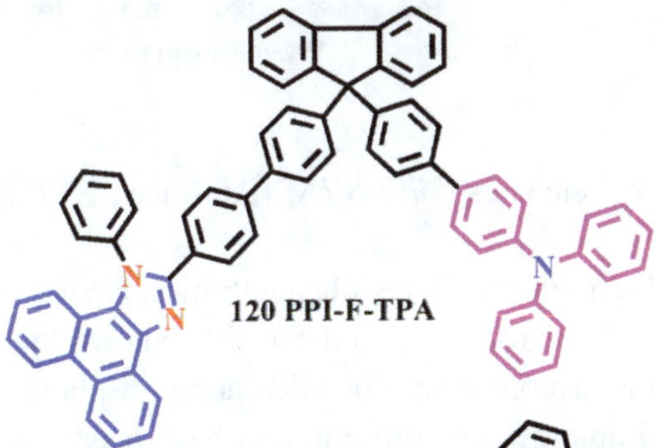

120 PPI-F-TPA

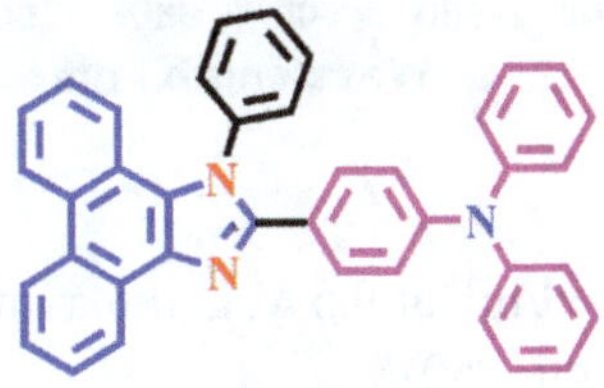

121 H2

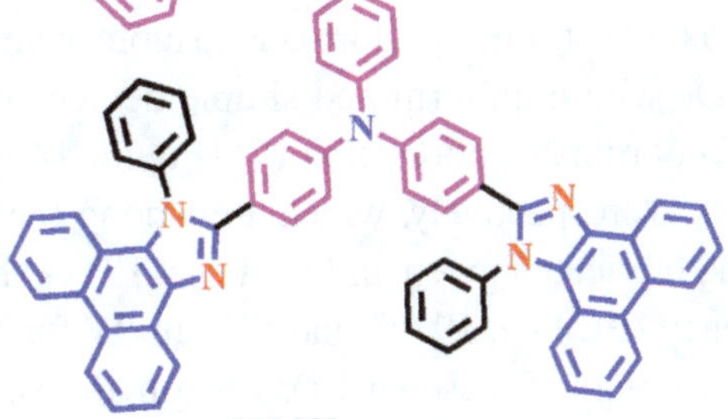

122 H3

Table 9. Device performances of triphenylamine based bipolar host for phosphorescent OLEDs 118-122

Host Materials	Dopant	V_{on}	EQE [a] (%)	C. E. [b]/cd A^{-1}	P. E. [c]/lm W^{-1}	Color Index	Ref.
118 DTPABPI	$Ir(ppy)_3$	2.8	17.1	43.2	-	-	175
119 PPI-PPITPA	-	3.2	7.7	5.6	5.5	0.15, 0.08	176
	$Ir(ppy)_3$	2.8	7.9	26.6	29.9	0.32, 0.60	
	$Ir(ppy)_2(dipig)$	2.9	19.1	61.8	65.5	0.46, 0.53	
	$Ir(MDQ)_2(acac)$	2.8	16.8	19.8	21.5	0.64, 0.36	
120 PPI-F-TPA	-	4.2	3.11	1.35	1.0	0.16, 0.05	187
	$Ir(ppy)_3$	2.7	15.6	57.0	60.0	0.34, 0.62	
	$Ir(2phq)_3$	2.8	12.5	27.0	28.3	0.58, 0.42	
121 H2	$Ir(MDQ)_2acac$	2.8	15.9	21.5	29.9	0.62, 0.38	188
122 H3	$Ir(MDQ)_2acac$	2.6	12.4	18.3	23.9	0.62, 0.38	
EQE [a]: quantum efficiency, C. E. [b]: current efficiency, P. E. [c]: power efficiency.							

Figure 49. a) EL spectra and CIE1931 coordinates of devices Y1 and R1 at the brightness of 100 cd/m², b) EL spectra and CIE1931 coordinates of devices G, Y2 and R2 at the brightness of 100 cd/m²

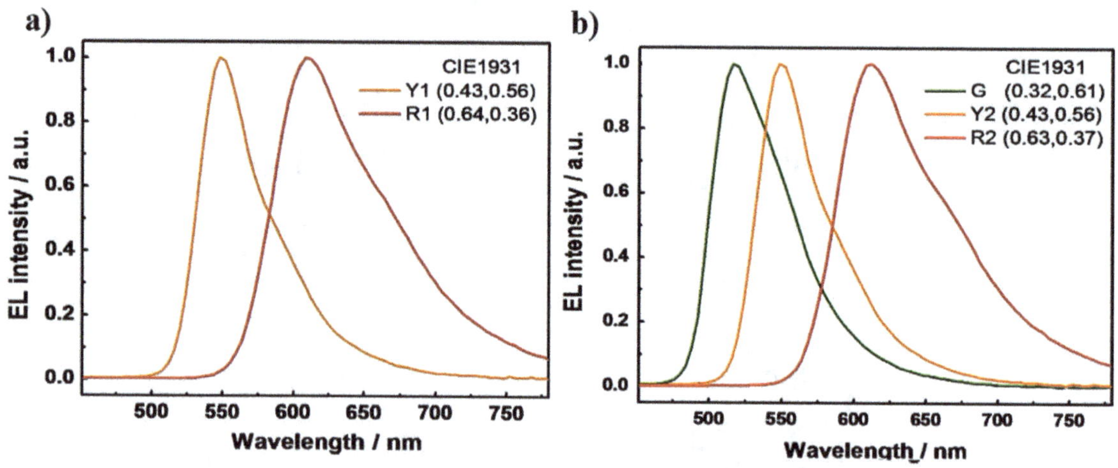

ness of 26 239 cd/m², at 9.5 V, and maximum EL efficiencies (17.9 cd/A, 19.9 lm/W, 12.7%) with CIE (0.63, 0.36) (Figure 50).

Wang et al., designed and synthesized a novel tetraphenylsilane–phenanthroimidazole hybrid fluorophore127 Si(PPI)$_2$ (D. Liu, et al., 2015). The synthetic routes of PI derivatives shown in scheme-12. By using (127 Si(PPI)$_2$) as a host, high-performance fluorescent blue (FB) and phosphorescent green (PG) and red (PR) OLEDs with a uniform and simple device configuration have been achieved. The blue device with the CIE coordinates of (0.18, 0.17) (Figure 44.) exhibited the maximum EQE and PE values of 6.1% and 8.0 lm/W, respectively, which are among the highest EL efficiencies ever reported for pure-blue devices. The fabricated green and red PhOLEDs exhibit high external quantum efficiency (EQE) and power efficiency (PE), i.e. 19.2% and 51.1 lm/W for PG and 12.0% and 15.6 lm/W for PR. Furthermore, the emission colors of these OLEDs remain almost unchanged within the whole range of driving voltages.

Among all the bipolar host materials (carbazole, triphenylamine and other host materials) 105-127, 116 PPI-pNNCzfluorophore shown good device performances, EQE, CE, and PE of 22.6%, 79.8 cd/A, 88.3 lm/W (green device) and 19.3%, 24.7 cd/A, 22.5 lm/W (red device), respectively. These results are attributed due to incorporating the novel electron-accepting 1,5-diazacarbazole into different positions

Figure 50. Scheme-11. Synthetic procedure of 123 p-BPPI and 124 m-BPPI

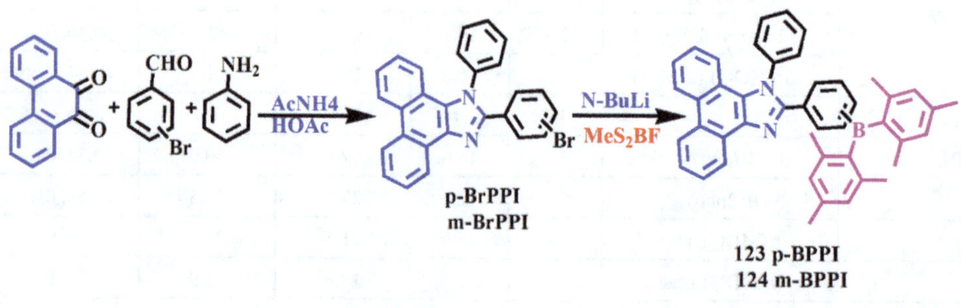

Figure 51. a) Electroluminescence spectra and CIE1931 coordinates for devices B1, G1 and R1 based on 125 DPO-PPI, b) Electroluminescence spectra and CIE1931 coordinates for devices

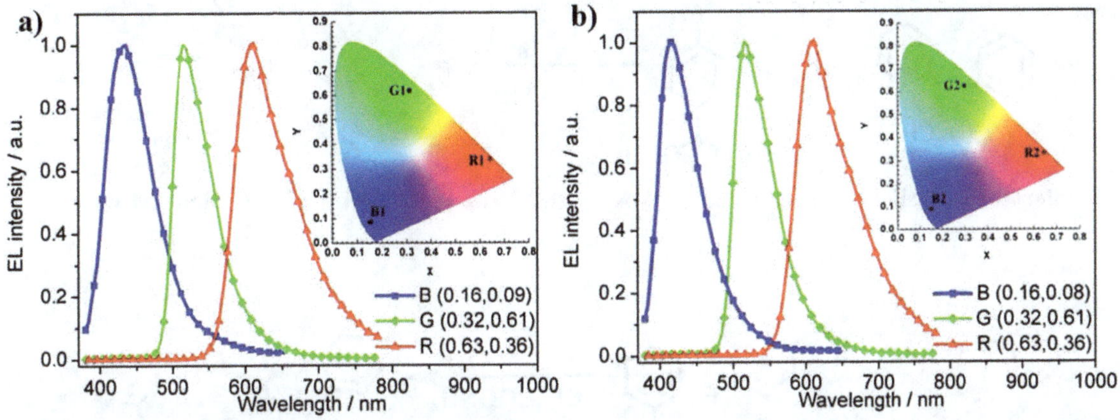

Figure 52. EL spectra of devices FB, PG and PR at the luminance of 1000 cd/m²

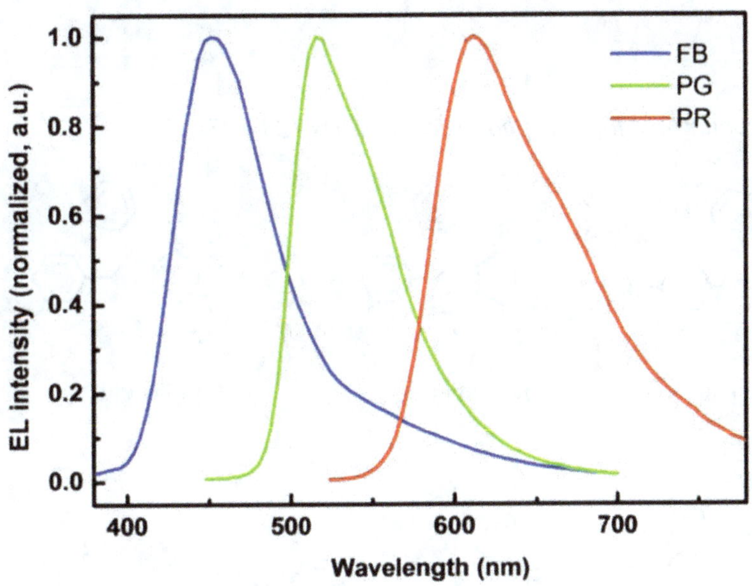

on a rigid phenanthroimidazole (PI) skeleton (115 PPI-pCNCz and 116 PPI-pNNCz), which favors the balanced carrier-transport ability and the electron–hole recombination within the emission layer. However, bipolar host materials can be used as emissive materials in fluorescent OLEDs. Wang *et al.*, have been used 119 PPI-PPITPA and 114 PPI-PPIPCz fluorophores for deep blue fluorescent OLEDs. These fluorophores sowed good thermal and ambipolar carrier transport characteristics. The non-dopant OLEDs of these fluorophores show good extremely high EQE of 7.7% and 8.1% (119 PPI-PPITPA and 114 PPI-PPIPCz) with deep blue CIE (0.15, 0.08) and (0.15, 0.07), respectively.

Figure 53. Scheme-12. Synthesis procedure and molecular structure of 127 PPI-F-TPA

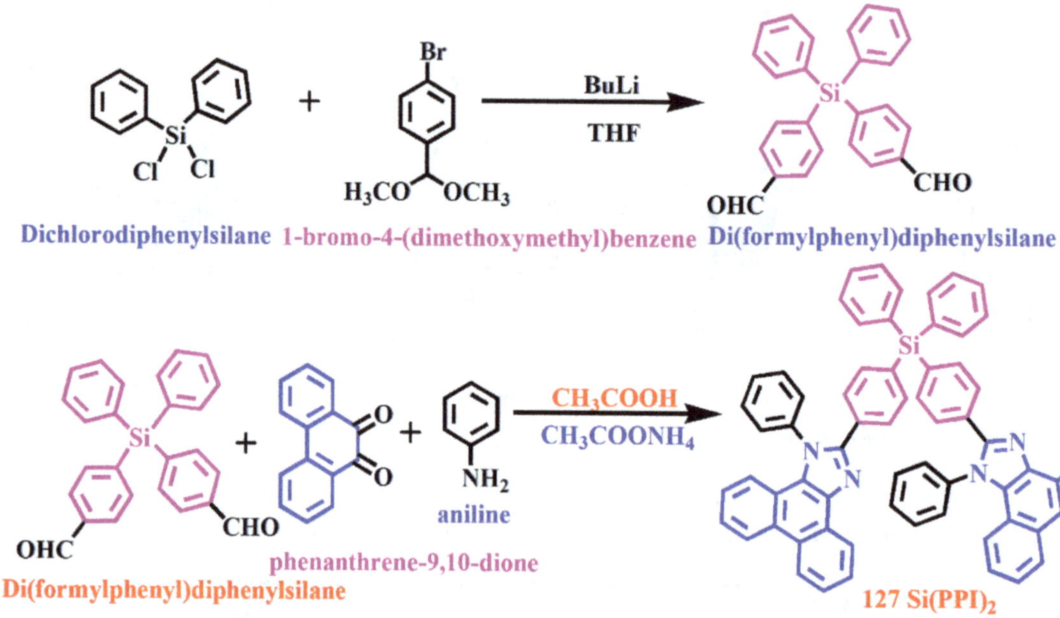

Figure 54. Other bipolar host for phosphorescent OLEDs 123-127

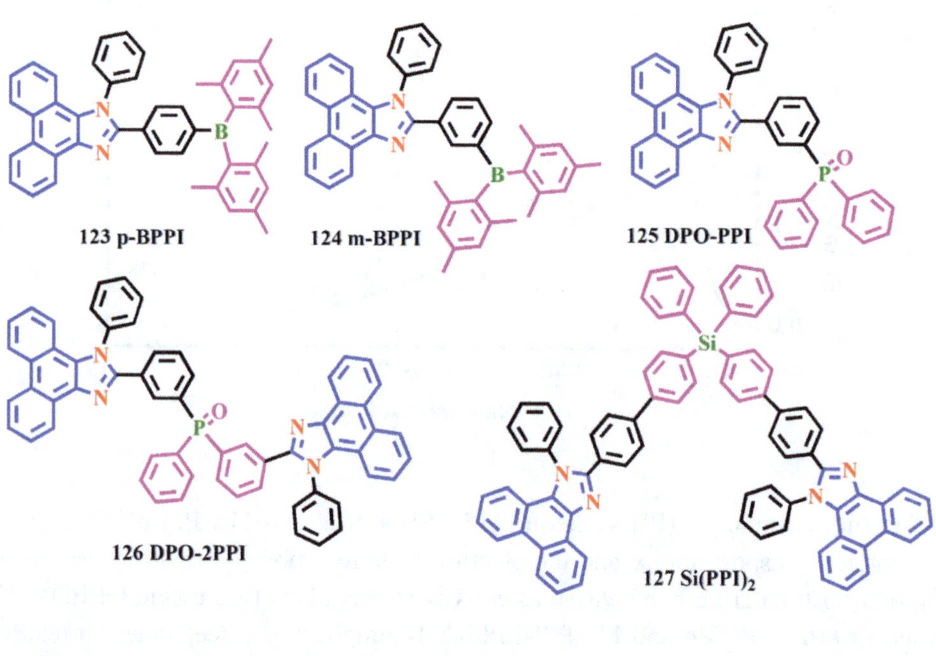

CONCLUSION

In this book chapter, we reviewed the recent developments of phenanthroimidazole based emissive fluorophores and host materials for fluorescent and phosphorescent OLEDs with detailed consideration paid to the design, synthesis, physical properties and performances of fluorescent and phosphorescent

Table 10. Device performances of other bipolar host for phosphorescent OLEDs 123-127

Host Materials	Dopant	V_{on}	EQE [a] (%)	C. E. [b]/ cd/A	P. E. [c]/ lm/W	Color Index	Ref.
123 p-BPPI	$(ppy)_2Ir(tipg)$	2.4	13.1	52.7	68.9	0.43, 0.56	204
	$(bt)_2Ir(dipba)$	2.4	13.5	20.1	20.04	0.64, 0.36	
124 m-BPPI	$Ir(ppy)_3$	2.6	18.2,	66.3	63.1	0.32, 0.61	
	$(ppy)_2Ir(tipg)$	2.6	14.5	55.2	66.6	0.43, 0.56	
	$(bt)_2Ir(dipba)$	2.5	12.6	19.8	19.5	0.63, 0.37	
125 DPO-PPI	-	3.4	1.88	1.51	1.29	0.16, 0.09	205
	$Ir(ppy)_3$	2.6	17.2	62.6	70.2	0.32, 0.61	
	$(bt)_2Ir(dipda)$	2.6	12.7	17.9	19.9	0.63, 0.36	
126 DPO-2PPI	-	3.4	2.24	1.75	1.53	0.16, 0.08	
	$Ir(ppy)_3$	2.6	18.0	65.4	73.3	0.30, 0.63	
	$(bt)_2Ir(dipda)$	2.6	13.5	19.0	21.3	0.63, 0.36	
127 Si(PPI)2	-	3.1	6.1	-	8.0	0.18, 0.17	206
	$Ir(ppy)_3$	3.4	19.2	-	51.1	0.32, 0.61	
	$(bt)_2Ir(dipba)$	3.3	12.0	-	15.6	0.64, 0.36	

EQE [a]: quantum efficiency, C. E. [b]: current efficiency, P. E. [c]: power efficiency.

OLEDs. Several strategies have been discussed for the design of phenanthroimidazole fluorophores (unipolar, bipolar materials with a spacer and without a spacer). PI derivatives are found to be very effective as emissive fluorophores and host materials, help in enhancing the overall quantum efficiency of the device and lowering the driving voltage of OLEDs. Among all the reported phenanthroimidazolebased fluorophores, the bipolar fluorophores are of significance in improving device performances with deep blue color coordinates. The bipolar fluorophores could not only broaden the exciton-formation zone, consequently reduce efficiency roll-off. In particular, phenanthroimidazole/TPA based fluorophores showed high quantum efficiency, little efficiency roll-off and low driving voltage due to high T_1 (triple energy) and bipolar charge transport properties. Therefore, the phenanthroimidazole fluorophores with the TPA functional group are promising materials for deep blue fluorescent OLEDs. PI fluorophores consist of strong electron withdrawing (phenanthroline/imidazole) group were also effectively used for hole blocking (Bphen) and as ETL (TPBI) in fluorescent OLEDs and PhOLEDs. Compared with other emitters, the PI fluorophores are having some advantages, which are ease in fluorophore modification at N1 and C2 position and these materials have the rigid skeleton, show high carrier mobilities, high thermal stability, as well as good film forming ability, leading to promising applications in high-performance fluorescent OLEDs and PhOLEDs. We believe that systematic engineering is required to design the efficient fluorophores to achieve highly efficient and stable fluorescent OLEDs.

To achieve high-efficiency PhOLEDs, the appropriate host materials are of equal importance to the phosphorescent emitters. This book chapter also provides an overview of the recent progress of phenanthroimidazole based bipolar host materials for PhOLEDs. In general, to design a host for a blue phosphor, a high enough triplet energy is main concern; to design a host for green or red phosphor, relatively narrow energy gap is desirable. The good charge transporting ability and high glass transition temperatures of host materials benefit the device performance. However, phenanthroimidazole based materials have high triplet energy, hence, these materials are useful as host materials for blue, green, yellow, orange, and red high performance PhOLEDs. Among all the bipolar host materials (carbazole, triphenylamine and other host materials), 116 PPI-pNNCzfluorophore shown good device performances, EQE, CE, and PE of 22.6%, 79.8 cd/A, 88.3 lm/W (green device) and 19.3%, 24.7 cd/A, 22.5 lm/W (red device), respectively. These results are attributed due to incorporating the novel electron-accepting 1,5-diazacarbazole into different positions on a rigid phenanthroimidazole (PI) skeleton (115 PPI-pCNCz and 116 PPI-pNNCz), which favors the balanced carrier-transport ability and the electron–hole recombination within the emission layer. Therefore, the organic materials with PI functional groups are promising as the emissive materials and host in various fluorescent OLEDs and PhOLEDs, and further development of new PI derivatives is actively ongoing to improve the device performances of fluorescent OLEDs and PHOLEDs.

ACKNOWLEDGMENT

The authors gratefully acknowledge the Board of Research in Nuclear Sciences (BRNS), Department of Atomic Energy (DAE), and Inspire faculty award, INSPIRE (Innovation in Science Pursuit for Inspired Research) and SERB (Science and Engineering Research Board), Department of Science and Technology (DST), India for financial support.

REFERENCES

Adachi, C., Baldo, M. A., Forrest, S. R., Lamansky, S., Thompson, M. E., & Kwong, R. C. (2001). High-efficiency red electrophosphorescence devices. *Applied Physics Letters*, *78*(11), 11. doi:10.1063/1.1355007

Baldo, M. A., Adachi, C., & Forrest, S. R. (2000). Transient analysis of organic electrophosphorescence. II. Transient analysis of triplet-triplet annihilation. *Physical Review B: Condensed Matter and Materials Physics*, *62*(16), 10967–10977. doi:10.1103/PhysRevB.62.10967

Baldo, M. A., & Forrest, S. R. (2000). Transient analysis of organic electrophosphorescence: I. Transient analysis of triplet energy transfer. *Physical Review B: Condensed Matter and Materials Physics*, *62*(16), 10958–10966. doi:10.1103/PhysRevB.62.10958

Baldo, M. A., O'Brien, D. F., You, Y., Shoustikov, A., Sibley, S., Thompson, M. E., & Forrest, S. R. (1998). Highly efficient phosphorescent emission from organic electroluminescent devices. *Nature*, *395*(6698), 151–154. doi:10.1038/25954

Baldo, M. A., Thompson, M. E., & Forrest, S. R. (2000). High-efficiency fluorescent organic light-emitting devices using a phosphorescent sensitizer. *Nature*, *403*(6771), 750–753. doi:10.1038/35001541 PMID:10693799

Baldo, M. A. O., Brien, D. F., You, Y., Shoustikov, A., Sibley, S., Thompson, M. E., & Forrest, S. R. (1998). Highly efficient phosphorescent emission from organic electroluminescent devices. *Nature*, *395*(6698), 151–154. doi:10.1038/25954

Bin, J. K., Cho, N. S., & Hong, J. I. (2012). New Host Material for High-Performance Blue Phosphorescent Organic Electroluminescent Devices. *Advanced Materials, 24*(21), 2911–2915. doi:10.1002/adma.201200972 PMID:22549891

Chang, S. C., He, G., Chen, F. C., Guo, T. F., & Yang, Y. (2001). Degradation mechanism of phosphorescent-dye-doped polymer light-emitting diodes. *Applied Physics Letters, 79*(13), 2088–2090. doi:10.1063/1.1404995

Chaskar, A., Chen, H. F., & Wong, K. T. (2011). Bipolar Host Materials: A Chemical Approach for Highly Efficient Electrophosphorescent Devices. *Advanced Materials, 23*(34), 3876–3895. doi:10.1002/adma.201101848 PMID:21805507

Chen, C. H., Hsu, L. C., Rajamalli, P., Chang, Y. W., Wu, F. L., Liao, C. Y., ... Cheng, C. H. (2014). Highly efficient orange and deep-red organic light emitting diodes with long operational lifetimes using carbazole–quinoline based bipolar host materials. *Journal of Materials Chemistry. C, Materials for Optical and Electronic Devices, 2*(30), 6183–6191. doi:10.1039/C4TC00523F

Chen, C. H., Huang, W. S., Lai, M. Y., Tsao, W. C., Lin, J. T., Wu, Y. H., ... Wu, C. C. (2009). Versatile, Benzimidazole/Amine-Based Ambipolar Compounds for Electroluminescent Applications: Single-Layer, Blue, Fluorescent OLEDs, Hosts for Single-Layer, Phosphorescent OLEDs. *Advanced Functional Materials, 19*(16), 2661–2670. doi:10.1002/adfm.200900561

Chen, M., Yuan, Y., Zheng, J., Chen, W. C., Shi, L. J., Zhu, Z. L., ... Lee, C. S. (2015). Novel Bipolar Phenanthroimidazole Derivative Design for a Nondoped Deep-Blue Emitter with High Singlet Exciton Yields. *Adv. Opt. Mater, 3*(9), 1215–1219. doi:10.1002/adom.201500258

Chen, M., Yuan, Y., Zheng, J., Chen, W. C., Shi, L. J., Zhu, Z. L., ... Lee, C. S. (2015). Novel Bipolar Phenanthroimidazole Derivative Design for a Nondoped Deep-Blue Emitter with High Singlet Exciton Yields. *Adv. Optical Mater, 3*(9), 1215–1219. doi:10.1002/adom.201500258

Chen, S., Wu, Y., Hu, S., Zhao, Y., & Fang, D. (2017). Non-Doped Deep Blue and Doped White Electroluminescence Devices Based on Phenanthroimidazole Derivative. *Journal of Fluorescence, 27*(2), 451–461. doi:10.100710895-016-1970-5 PMID:27900511

Chen, S., Wu, Y., Zhao, Y., & Fang, D. (2015). Deep blue organic light-emitting devices enabled by bipolar phenanthro[9,10-d]imidazole derivatives. *RSC Advances, 5*(88), 72009–72018. doi:10.1039/C5RA13814K

Chen, W. C., Wu, G. F., Yuan, Y., Wei, H. X., Wong, F. L., Tong, Q. X., & Lee, C. S. (2015). A meta-molecular tailoring strategy towards an efficient violet-blue organic electroluminescent material. *RSC Advances, 5*(23), 18067–18074. doi:10.1039/C4RA16954A

Chen, W. C., Yuan, Y., Ni, S. F., Zhu, Z. L., Zhang, J. Z., Jiang, Q., ... Lee, C. S. (2017). Highly Efficient Deep-Blue Electroluminescence from a Charge-Transfer Emitter with Stable Donor Skeleton. *ACS Applied Materials & Interfaces, 9*(8), 7331–7338. doi:10.1021/acsami.6b14638 PMID:28185447

Chen, W. C., Yuan, Y., Wu, G. F., Wei, H., Ye, J., Chen, M., ... Lee, C. S. (2015). Molecular modification on bisphenanthroimidazole derivative for deep-blue organic electroluminescent material with ambipolar property and high performance. *Organic Electronics, 17*, 159–166. doi:10.1016/j.orgel.2014.11.024

Chen, W. C., Yuan, Y., Wu, G. F., Wei, X. H., Tang, L., Tong, Q. X., ... Lee, C. S. (2014). Staggered Face-to-Face Molecular Stacking as a Strategy for Designing Deep-Blue Electroluminescent Materials with High Carrier Mobility. *Adv. Opt. Mater*, *2*(7), 626–631. doi:10.1002/adom.201400078

Chen, Y. M., Hung, W. Y., You, H. W., Chaskar, A., Ting, H. C., Chen, H. F., ... Liu, Y. H. (2011). Carbazole–benzimidazole hybrid bipolar host materials for highly efficient green and blue phosphorescent OLEDs. *Journal of Materials Chemistry*, *21*(38), 14971. doi:10.1039/c1jm12598b

Chi, L. C., Hung, W. Y., Chiu, H. C., & Wong, K. T. (2009). A high-efficiency and low-operating-voltage green electrophosphorescent device employing a pure-hydrocarbon host material. *Chemical Communications (Cambridge)*, (26): 3892. doi:10.1039/b904795f PMID:19662243

Chi, Y., & Chou, P. T. (2010). Transition-metal phosphors with cyclometalating ligands: Fundamentals and applications. *Chemical Society Reviews*, *39*(2), 638–655. doi:10.1039/B916237B PMID:20111785

Chien, C. H., Chen, C. K., Hsu, F. M., Shu, C. F., Chou, P. T., & Lai, C. H. (2009). Multifunctional Deep-Blue Emitter Comprising an Anthracene Core and Terminal Triphenylphosphine Oxide Groups. *Advanced Functional Materials*, *19*(4), 560–566. doi:10.1002/adfm.200801240

Chopra, N., Lee, J., Zheng, Y., Eom, S. H., Xue, J., & So, F. (2008). High efficiency blue phosphorescent organic light-emitting device. *Applied Physics Letters*, *93*(14), 143307. doi:10.1063/1.3000382

Chou, H. H., Chen, Y. H., Hsu, H. P., Chang, W. H., Chen, Y. H., & Cheng, C. H. (2012). Synthesis of Diimidazolylstilbenes as n-Type Blue Fluorophores: Alternative Dopant Materials for Highly Efficient Electroluminescent Devices. *Advanced Materials*, *24*(43), 5867–5871. doi:10.1002/adma.201202222 PMID:22915130

Chou, H. H., & Cheng, C. H. (2010). A Highly Efficient Universal Bipolar Host for Blue, Green, and Red Phosphorescent OLEDs. *Advanced Materials*, *22*(22), 2468–2471. doi:10.1002/adma.201000061 PMID:20446307

Chou, H. H., & Cheng, C. H. (2010). A Highly Efficient Universal Bipolar Host for Blue, Green, and Red Phosphorescent OLEDs. *Advanced Materials*, *22*(22), 2468–2471. doi:10.1002/adma.201000061 PMID:20446307

Fan, C. H., Sun, P. P., Su, T. H., & Cheng, C. H. (2011). Host and Dopant Materials for Idealized Deep-Red Organic Electrophosphorescence Devices. *Advanced Materials*, *23*(26), 2981–2985. doi:10.1002/adma.201100610 PMID:21567483

Farinola, G. M., & Ragni, R. (2011). Electroluminescent materials for white organic light emitting diode. *Chemical Society Reviews*, *40*(7), 3467–3482. doi:10.1039/c0cs00204f PMID:21437308

Feng, X. J., Chen, S. F., Ni, Y., Wong, M. S., Lam, M. M. K., Cheah, K. W., & Lai, G. Q. (2014). Fluorene Derivatives for Highly Efficient Non-Doped Single-Layer Blue Organic Light-Emitting Diodes. *Organic Electronics*, *15*(1), 57–64. doi:10.1016/j.orgel.2013.10.019

Gao, H. Z. (2010). Theoretical investigation into charge mobility in 4, 4'-bis(1-napthylphenylamino) biphenyl. *Theoretical Chemistry Accounts*, *127*(5-6), 759–763. doi:10.100700214-010-0804-9

Gao, Z., Liu, F., Li, J., Cheng, G., & Lu, P. (2016). Solution-processed UV light emitting diode based on butyltriphenylsilane modified phenanthro[9,10-d]imidazole with high efficiency. *RSC Advances*, *6*(85), 81744–81749. doi:10.1039/C6RA18755B

Gao, Z., Wang, Z., Shan, T., Liu, Y., Shen, F., Pan, Y., ... Ma, Y. (2014). High-efficiency deep blue fluorescent emitters based on phenanthro[9,10-d]imidazole substituted carbazole and their applications in organic light emitting diodes. *Organic Electronics*, *15*(11), 2667–2676. doi:10.1016/j.orgel.2014.07.019

Ge, Z., Hayakawa, T., Ando, S., Ueda, M., Akiike, T., Miyamoto, H., ... Kakimoto, M. (2008). Solution-Processible Bipolar Triphenylamine-Benzimidazole Derivatives for Highly Efficient Single-Layer Organic Light-Emitting Diodes. *Chemistry of Materials*, *20*(7), 2532–2537. doi:10.1021/cm7035458

Gong, J. R., Wan, L. J., Lei, S. B., Bai, C. L., Zhang, X.-H., & Lee, S.-T. (2005). Direct evidence of molecular aggregation and degradation mechanism of organic light-emitting diodes under joule heating: An STM and photoluminescence study. *The Journal of Physical Chemistry B*, *109*(5), 1675–1682. doi:10.1021/jp046509o PMID:16851142

Gong, S., Chen, Y., Luo, J., Yang, C., Zhong, C., Qin, J., & Ma, D. (2011). Bipolar Tetraarylsilanes as Universal Hosts for Blue, Green, Orange, and White Electrophosphorescence with High Efficiency and Low Efficiency Roll-Off. *Advanced Functional Materials*, *21*(6), 1168–1178. doi:10.1002/adfm.201002066

Gong, S., Chen, Y., Yang, C., Zhong, C., Qin, J., & Ma, D. (2010). De Novo Design of Silicon-Bridged Molecule Towards a Bipolar Host: All-Phosphor White Organic Light-Emitting Devices Exhibiting High Efficiency and Low Efficiency Roll-Off. *Advanced Materials*, *22*(47), 5370–5373. doi:10.1002/adma.201002732 PMID:20972980

Gong, S. L., Fu, Q., Wang, Q., Yang, C. L., Zhong, C., Qin, J. G., & Ma, D. G. (2011). Highly Efficient Deep-Blue Electrophosphorescence Enabled by Solution-Processed Bipolar Tetraarylsilane Host with Both a High Triplet Energy and a High-Lying HOMO Level. *Advanced Materials*, *23*(42), 4956–4959. doi:10.1002/adma.201102758 PMID:21959864

Guan, M., Chen, Z., Bian, Z., Liu, Z., Gong, Z., Baik, W., ... Huang, C. (2006). The host materials containing carbazole and oxadiazole fragment for red triplet emitter in organic light-emitting diodes. *Organic Electronics*, *7*(5), 330–336. doi:10.1016/j.orgel.2006.04.006

Han, C., Xie, G., Xu, H., Zhang, Z., Yu, D., Zhao, Y., ... Liu, S. (2011). Towards Highly Efficient Blue-Phosphorescent Organic Light-Emitting Diodes with Low Operating Voltage and Excellent Efficiency Stability. *Chemistry (Weinheim an der Bergstrasse, Germany)*, *17*(2), 445–449. doi:10.1002/chem.201001981 PMID:21207558

Holmes, R. J., D'Andrade, B. W., Forrest, S. R., Ren, X., Li, J., & Thompson, M. E. (2003). Efficient, deep-blue organic electrophosphorescence by guest charge trapping. *Applied Physics Letters*, *83*(18), 3818–3820. doi:10.1063/1.1624639

Holmes, R. J., Forrest, S. R., Tung, Y. J., Kwong, R. C., Brown, J. J., Garon, S. M., & Thompson, E. (2003). Blue organic electrophosphorescence using exothermic host–guest energy transfer. *Applied Physics Letters*, *82*(15), 2422–2424. doi:10.1063/1.1568146

Hsu, F. M., Chien, C. H., Hsieh, Y. J., Wu, C. H., Shu, C. F., Liu, S. W., & Chen, C. T. (2009). Highly efficient red electrophosphorescent device incorporating a bipolar triphenylamine/bisphenylsulfonyl-substituted fluorene hybrid as the host. *Journal of Materials Chemistry*, *19*(42), 8002. doi:10.1039/b910292b

Huang, H., Fu, Q., Zhuang, S., Liu, Y., Wang, L., Chen, J., & Yang, C. (2011). Novel Deep Blue OLED Emitters with 1,3,5-Tri(anthracen-10-yl)benzene-Centered Starburst Oligofluorenes. *The Journal of Physical Chemistry C*, *115*(11), 4872–4878. doi:10.1021/jp110652y

Huang, H., Wang, Y., Wang, B., Zhuang, S., Pan, B., Yang, X., ... Yang, C. (2013). Controllably tunable phenanthroimidazole–carbazole hybrid bipolar host materials for efficient green electrophosphorescent devices. *Journal of Materials Chemistry. C, Materials for Optical and Electronic Devices*, *1*(37), 5899. doi:10.1039/c3tc30832d

Huang, H., Wang, Y., Zhuang, S., Yang, X., Wang, L., & Yang, C. (2012). Simple Phenanthroimidazole/Carbazole Hybrid Bipolar Host Materials for Highly Efficient Green and Yellow Phosphorescent Organic Light-Emitting Diodes. *The Journal of Physical Chemistry C*, *116*(36), 19458–19466. doi:10.1021/jp305764b

Huang, J., Pfeiffer, M., Werner, A., Blochwitz, J., Leo, K., & Liu, S. (2002). Low-voltage organic electroluminescent devices using pin structures. *Applied Physics Letters*, *80*(1), 139–141. doi:10.1063/1.1432110

Huang, T. H., Lin, J. T., Chen, L. Y., Lin, Y. T., & Wu, C. C. (2006). Dipolar Dibenzothiophene S,S-Dioxide Derivatives Containing Diarylamine: Materials for Single-Layer Organic Light-Emitting Devices. *Advanced Materials*, *18*(5), 602–606. doi:10.1002/adma.200502078

Huang, Y. K., Jen, T. H., Chang, Y. T., Yang, N. J., Lu, H. H., & Chen, S. A. (2010). Enhancing Shielding of Triplet Energy Transfer to Poly(p-phenylene)s from Phosphor Dopant by Addition of Branched Alcohol for Highly Efficient Electrophosphorescence. *ACS Applied Materials & Interfaces*, *2*(4), 1094–1099. doi:10.1021/am900878f PMID:20380421

Huang, Z., Wang, B., Zhang, Q., Xiang, S., Lv, X., Ma, L., ... Wang, L. (2017). Highly twisted bipolar emitter for efficient nondoped deep-blue electroluminescence. *Dyes and Pigments*, *140*, 328–336. doi:10.1016/j.dyepig.2017.01.028

Hughes, G., & Bryce, M. R. (2005). Electron-transporting materials for organic electroluminescent and electrophosphorescent devices. *Journal of Materials Chemistry*, *15*(1), 94. doi:10.1039/b413249c

Hung, W. Y., Chi, L. C., Chen, W. J., Chen, Y. M., Chou, S. H., & Wong, K. T. (2010). A new benzimidazole/carbazole hybrid bipolar material for highly efficient deep-blue electrofluorescence, yellow–green electrophosphorescence, and two-color-based white OLEDs. *Journal of Materials Chemistry*, *20*(45), 10113. doi:10.1039/c0jm02143a

Ichikawa, M., Wakabayashi, K., Hayashi, S., Yokoyama, N., Koyama, T., & Taniguchi, Y. (2010). Bi- or ter-pyridine tris-substituted benzenes as electron-transporting materials for organic light-emitting devices. *Organic Electronics*, *11*(12), 1966–1973. doi:10.1016/j.orgel.2010.09.004

Inomata, H., Goushi, K., Masuko, T., Konno, T., Imai, T., Sasabe, H., ... Adachi, C. (2004). High-Efficiency Organic Electrophosphorescent Diodes Using 1,3,5-Triazine Electron Transport Materials. *Chemistry of Materials*, *16*(7), 1285–1291. doi:10.1021/cm034689t

Islam, A., Wang, Q., Zhang, L., Lei, T., Hong, L., Yang, R., ... Ge, Z. (2017). Efficient non-doped deep blue organic light emitting diodes with high external quantum efficiency and a low efficiency roll-off based on donor-acceptor molecules. *Dyes and Pigments*, *142*, 499–506. doi:10.1016/j.dyepig.2017.03.061

Jairam, T., Ulla, H., Kajjam, A. B., Satyanarayan, M. N., & Vaidyanathan, S. (2017). Star-Shaped Phenanthroimidazole-Triphenylamine-Based Yellow Organic Emitter for Organic Light Emitting Diodes. *ChemistrySelect*, *2*(8), 2611–2620. doi:10.1002lct.201700056

Jayabharathi, J., Prabhakaran, A., Thanikachalam, V., Jeeva, P., & Sundharesan, M. (2016). Air Stable HyLEDs Using Efficient Electron Injection and Emitting Materials. *Industrial & Engineering Chemistry Research*, *55*(29), 8087–8095. doi:10.1021/acs.iecr.6b01151

Jayabharathi, J., Prabhakaran, A., Thanikachalam, V., & Sundharesan, M. (2016). Highly efficient non-doped blue electroluminescent materials for organic lightemitting devices. *RSC Advances*, *6*(67), 62208–62217. doi:10.1039/C6RA11439C

Jeon, S. O., Jang, S. E., Son, H. S., & Lee, J. Y. (2011). External Quantum Efficiency Above 20% in Deep Blue Phosphorescent Organic Light-Emitting Diodes. *Advanced Materials*, *23*(12), 1436–1441. doi:10.1002/adma.201004372 PMID:21433109

Jeon, S. O., & Lee, J. Y. (2012). Phosphine oxide derivatives for organic light emitting diodes. *Journal of Materials Chemistry*, *22*(10), 4233–4243. doi:10.1039/C1JM14832J

Jeon, S. O., Yook, K. S., Jogo, C. W., & Lee, J. Y. (2010). High-Efficiency Deep-Blue-Phosphorescent Organic Light-Emitting Diodes Using a Phosphine Oxide and a Phosphine Sulfide High-Triplet-Energy Host Material with Bipolar Charge-Transport Properties. *Advanced Materials*, *22*(16), 1872–1876. doi:10.1002/adma.200903321 PMID:20512965

Jiang, W., Duan, L., Qiao, J., Dong, G., Wang, L., & Qiu, Y. (2011). Tuning of Charge Balance in Bipolar Host Materials for Highly Efficient Solution-Processed Phosphorescent Devices. *Organic Letters*, *13*(12), 3146–3149. doi:10.1021/ol201039n PMID:21591625

Kamtekar, K. T., Monkman, A. P., & Bryce, M. R. (2010). Recent Advances in White Organic Light-Emitting Materials and Devices (WOLEDs). *Advanced Materials*, *22*(5), 572–58. doi:10.1002/adma.200902148 PMID:20217752

Kamtekar, K. T., Monkman, A. P., & Bryce, M. R. (2010). Recent Advances in White Organic Light-Emitting Materials and Devices (WOLEDs). *Advanced Materials*, *22*(5), 572–582. doi:10.1002/adma.200902148 PMID:20217752

Kang, J. W., Lee, D. S., Park, H. D., Park, Y. S., Kim, J. W., Jeong, W. I., ... Kim, J. J. (2007). Silane- and triazine-containing hole and exciton blocking material for high-efficiency phosphorescent organic light emitting diodes. *Journal of Materials Chemistry*, *17*(35), 3714. doi:10.1039/b705741e

Kijima, Y., Asai, N., & Tamura, S. (1999). A Blue Organic Light Emitting Diode. *Japanese Journal of Applied Physics*, *38*(Part 1, No. 9A), 5274–5277. doi:10.1143/JJAP.38.5274

Kim, B., Park, Y., Lee, J., Yokoyama, D., Lee, J. H., Kido, J., & Park, J. (2013). Synthesis and electroluminescence properties of highly efficient blue fluorescence emitters using dual core chromophores. *Journal of Materials Chemistry. C, Materials for Optical and Electronic Devices*, *1*(3), 432–440. doi:10.1039/C2TC00185C

Kim, D. H., Cho, N. S., Oh, H. Y., Yang, J. H., Jeon, W. S., Park, J. S., ... Kwon, J. H. (2013). Highly Efficient Red Phosphorescent Dopants in Organic Light-Emitting Devices. *Advanced Materials*, *23*(24), 2721–2726. doi:10.1002/adma.201100405 PMID:21495090

Kim, S. H., Jang, J., & Lee, J. Y. (2007). Triplet host engineering for triplet exciton management in phosphorescent organic light-emitting diodes. *Applied Physics Letters*, *91*, 083511. doi:10.1063/1.2773941

Kim, S. Y., Jeong, W. I., Mayr, C., Park, Y. S., Kim, K. H., Lee, J. H., ... Kim, J. J. (2013). Organic Light-Emitting Diodes with 30% External Quantum Efficiency Based on a Horizontally Oriented Emitter. *Advanced Functional Materials*, *23*(31), 3896–3900. doi:10.1002/adfm.201300104

Kondakova, M. E., Pawlik, T. D., Young, R. H., Giesen, D. J., Kondakov, D. Y., Brown, C. T., ... Klubek, K. P. (2008). High-efficiency, low-voltage phosphorescent organic light-emitting diode devices with mixed host. *Journal of Applied Physics*, *104*(9), 094501. doi:10.1063/1.3000046

Kraft, A., Grimsdale, A. C., & Holmes, A. B. (1998). Electroluminescent Conjugated Polymers-Seeing Polymers in a New Light. *Angewandte Chemie International Edition*, *37*(4), 402–428. doi:

Kulkarni, A. P., Tonzola, C. J., Babel, A., & Jenekhe, S. A. (2004). Electron Transport Materials for Organic Light-Emitting Diodes. *Chemistry of Materials*, *16*(23), 4556–4573. doi:10.1021/cm049473l

Kulkarni, A. P., Tonzola, C. J., Babel, A., & Jenekhe, S. A. (2004). Electron Transport Materials for Organic Light-Emitting Diodes. *Chemistry of Materials*, *16*(23), 4556–4573. doi:10.1021/cm049473l

Kuo, C. J., Li, T. Y., Lien, C. C., Liu, C. H., Wub, F. I., & Huang, M. J. (2009). Bis(phenanthroimidazolyl) biphenyl derivatives as saturated blue emitters for electroluminescent devices. *Journal of Materials Chemistry*, *19*(13), 1865–1871. doi:10.1039/b816327h

Kwon, J. E., Park, S., & Park, S. Y. (2013). Realizing Molecular Pixel System for Full-Color Fluorescence Reproduction: RGB-Emitting Molecular Mixture Free from Energy Transfer Crosstalk. *Journal of the American Chemical Society*, *135*(30), 11239–11246. doi:10.1021/ja404256s PMID:23876082

Lai, M. Y., Chen, C. H., Huang, W. S., Lin, J. T., Ke, T. H., Chen, L. Y., ... Wu, C. C. (2008). Benzimidazole/Amine-Based Compounds Capable of Ambipolar Transport for Application in Single-Layer Blue-Emitting OLEDs and as Hosts for Phosphorescent Emitters. *Angewandte Chemie International Edition*, *47*(3), 581–585. doi:10.1002/anie.200704113 PMID:18058878

Lee, C. W., & Lee, J. Y. (2013). Above 30% External Quantum Efficiency in Blue Phosphorescent Organic Light-Emitting Diodes Using Pyrido[2,3-*b*]indole Derivatives as Host Materials. *Advanced Materials*, *25*(38), 5450–5454. doi:10.1002/adma.201301091 PMID:23788128

Lee, J. H., Cheng, S. H., Yoo, S. J., Shin, H., Chang, J. H., Wu, C. I., ... Kim, J. J. (2015). An Exciplex Forming Host for Highly Efficient Blue Organic Light Emitting Diodes with Low Driving Voltage. *Advanced Functional Materials*, *25*(3), 361–366. doi:10.1002/adfm.201402707

Lee, S., Kim, B., Jung, H., Shin, H., Lee, H., Lee, J., & Park, J. (2017). Synthesis and electroluminescence properties of new blue dual-core OLED emitters using bulky side chromophores. *Dyes Pigm*, *136*, 255–261. doi:10.1016/j.dyepig.2016.08.010

Lee, T. W., Noh, T., Shin, H. W., Kwon, O., Park, J. J., Choi, B. K., ... Kim, Y. R. (2009). Characteristics of Solution-Processed Small-Molecule Organic Films and Light-Emitting Diodes Compared with their Vacuum-Deposited Counterparts. *Advanced Functional Materials*, *19*(10), 1625–1630. doi:10.1002/adfm.200801045

Lee, Y. H., Wu, T. C., Liaw, C. W., Wen, T. C., Feng, S. W., Lee, J. J., ... Guo, T. F. (2013). Non-doped active layer, benzo[k]fluoranthene-based linear acenes, for deep blue- to green-emissive organic light-emitting diodes. *Organic Electronics*, *14*(4), 1064–1072. doi:10.1016/j.orgel.2013.01.021

Leung, M., Yang, C. C., Lee, J. H., Tsai, H. H., Lin, C. F., Huang, C. Y., ... Chiu, C. F. (2007). The Unusual Electrochemical and Photophysical Behavior of 2,2'-Bis(1,3,4-oxadiazol-2-yl)biphenyls, Effective Electron Transport Hosts for Phosphorescent Organic Light Emitting Diodes. *Organic Letters*, *9*(2), 235–238. doi:10.1021/ol062668+ PMID:17217273

Li, C., Li, Z., Yan, X., Zhang, Y., Zhang, Z., & Wang, Y. (2017). Structurally simple non-doped sky-blue OLEDs with high luminance and efficiencies at low driving voltages. *Journal of Materials Chemistry. C, Materials for Optical and Electronic Devices*, *5*(8), 1973–1980. doi:10.1039/C6TC05639C

Li, C., Wang, S., Chen, W., Wei, J., Yang, G., Ye, K., ... Wang, Y. (2015). High performance full color OLEDs based on a class of molecules with dual carrier transport channels and small singlet–triplet splitting. *Chemical Communications (Cambridge)*, *51*(53), 10632–10635. doi:10.1039/C5CC03492B PMID:26040328

Li, C., Wang, S., Chen, W., Wei, J., Yang, G., Ye, K., ... Wang, Y. (2015). High performance full color OLEDs based on a class of molecules with dual carrier transport channels and small singlet–triplet splitting. *Chemical Communications (Cambridge)*, *51*(53), 10632–10635. doi:10.1039/C5CC03492B PMID:26040328

Li, C., Wei, J., Song, X., Ye, K., Zhang, H., Zhang, J., & Wang, Y. (2016). Non-doped luminescent material based organic light-emitting devices displaying high brightness under very low driving voltage. *Journal of Materials Chemistry. C, Materials for Optical and Electronic Devices*, *4*(29), 7013–7019. doi:10.1039/C6TC01975G

Li, C. L., Wei, J., Han, J., Li, Z., Song, X., Zhang, Z., ... Wang, Y. (2016). Efficient deep-blue OLEDs based on phenanthro[9,10-d]imidazole-containing emitters with AIE and bipolar transporting properties. *Journal of Materials Chemistry. C, Materials for Optical and Electronic Devices*, *4*(42), 10120–10129. doi:10.1039/C6TC03923E

Li, G., Zhao, J., Zhang, D., Zhu, J., Shi, Z., Tao, S., ... Tong, Q. (2017). Non-doped deep blue emitters based on twisted phenanthroimidazole derivatives for organic light-emitting devices (CIE y E 0.04). *New Journal of Chemistry*, *41*(12), 5191–5197. doi:10.1039/C7NJ00155J

Li, J., Liu, D., Li, Y., Lee, C. S., Kwong, H. L., & Lee, S. (2005). A High Tg Carbazole-Based Hole-Transporting Material for Organic Light-Emitting Devices. *Chemistry of Materials*, *17*(5), 1208–1212. doi:10.1021/cm034731k

Li, N., Wang, P., Lai, S. L., Liu, W., Lee, C. S., Lee, S. T., & Liu, Z. (2010). Synthesis of multiaryl-substituted pyridine derivatives and applications in non-doped deep-blue OLEDs as electron-transporting layer with high hole-blocking ability. *Advanced Materials*, *22*(4), 527–530. doi:10.1002/adma.200902430 PMID:20217747

Li, W., Liu, D., Shen, F., Ma, D., Wang, Z., Feng, T., ... Ma, Y. (2012). A Twisting Donor-Acceptor Molecule with an Intercrossed Excited State for Highly Efficient, Deep-Blue Electroluminescence. *Advanced Functional Materials*, *22*(13), 2797–2803. doi:10.1002/adfm.201200116

Li, W., Pan, Y., Yao, L., Liu, H., Zhang, S., Wang, C., ... Ma, Y. (2014). A Hybridized Local and Charge-Transfer Excited State for Highly Efficient Fluorescent OLEDs: Molecular Design, Spectral Character, and Full Exciton Utilization. *Adv. Opt. Mater*, *2*(9), 892–901. doi:10.1002/adom.201400154

Li, W., Yao, L., Liu, H., Wang, Z., Zhang, S., Xiao, R., ... Ma, Y. (2014). Highly efficient deep-blue OLED with an extraordinarily narrow FHWM of 35 nm and a y coordinate <0.05 based on a fully twisting donor–acceptor molecule. *Journal of Materials Chemistry. C, Materials for Optical and Electronic Devices*, *2*(24), 4733–4736. doi:10.1039/C4TC00487F

Liang, B., Wang, L., Xu, Y. H., Shi, H. H., & Cao, Y. (2007). High-Efficiency Red Phosphorescent Iridium Dendrimers with Charge-Transporting Dendrons: Synthesis and Electroluminescent Properties. *Advanced Functional Materials*, *17*(17), 3580–3589. doi:10.1002/adfm.200700222

Lin, M., Chi, L., Chang, H., Huang, Y., Tien, K., Chen, C., ... Chi, Y. (2012). A diarylboranesubstituted carbazole as a universal bipolar host material for highly efficient electrophosphorescence devices. *Journal of Materials Chemistry*, *22*(3), 870–876. doi:10.1039/C1JM13323C

Lin, N., Qiao, J., Duan, L., Xue, J., & Wang, L. (2014). Rational Design of Chelated Aluminum Complexes toward Highly Efficient and Thermally Stable Electron-Transporting Materials. *Chemistry of Materials*, *26*(12), 3693–3700. doi:10.1021/cm5011604

Liu, B., Yuan, Y., He, D., Huang, D. Y., Luo, C. Y., Zhu, Z. L., ... Lee, C. S. (2016). High-Performance Blue OLEDs Based on Phenanthroimidazole Emitters via Substitutions at the C6- and C9-Positions for Improving Exciton Utilization. *Chemistry (Weinheim an der Bergstrasse, Germany)*, *22*, 1–9. PMID:27412733

Liu, B., Zhao, J., Luo, C., Lu, F., Tao, S., & Tong, Q. (2016). A novel bipolar phenanthroimidazole derivative host material for highly efficient green and orange-red phosphorescent OLEDs with low efficiency roll-off at high brightness. *Journal of Materials Chemistry. C, Materials for Optical and Electronic Devices*, *4*(10), 2003–2010. doi:10.1039/C5TC04393J

Liu, D., Du, M., Chen, D., Ye, K., Zhang, Z., Liu, Y., & Wang, Y. (2015). A novel tetraphenylsilane–phenanthroimidazole hybrid host material for highly efficient blue fluorescent, green and red phosphorescent OLEDs. *Journal of Materials Chemistry. C, Materials for Optical and Electronic Devices*, *3*(17), 4394–4401. doi:10.1039/C5TC00333D

Liu, D., Du, M., Chen, D., Ye, K., Zhang, Z., Liu, Y., & Wang, Y. (2015). A novel tetraphenylsilane–phenanthroimidazole hybrid host material for highly efficient blue fluorescent, green and red phosphorescent OLEDs. *Journal of Materials Chemistry. C, Materials for Optical and Electronic Devices*, *3*(17), 4394–4401. doi:10.1039/C5TC00333D

Liu, H., Chen, P., Hu, D. H., Tang, X. Y., Pan, Y. Y., Zhang, H. H., ... Ma, Y. G. (2014). Separation of Electrical and Optical Energy Gaps: Selectively Adjusting the Electrical and Optical Properties for a Highly Efficient Blue Emitter. *Chemistry (Weinheim an der Bergstrasse, Germany)*, *20*(8), 2149–2153. doi:10.1002/chem.201304544 PMID:24478181

Liu, M., Li, X. L., Chen, D. C., Xie, Z., Cai, X., Xie, G., ... Cao, Y. (2015). Study of Configuration Differentia and Highly Efficient, Deep-Blue, Organic Light-Emitting Diodes Based on Novel Naphtho[1,2-D] imidazole Derivatives. *Advanced Functional Materials*, *25*(32), 5190–5198. doi:10.1002/adfm.201502163

Liu, X. K., Zheng, C. J., Lo, M. F., Xiao, J., Lee, C. S., Fungc, M. K., & Zhang, X. H. (2014). A multifunctional phosphine oxide–diphenylamine hybrid compound as a high performance deep-blue fluorescent emitter and green phosphorescent host. *Chemical Communications (Cambridge)*, *50*(16), 2027. doi:10.1039/c3cc48531e PMID:24413567

Lundin, N. J., Blackman, A. G., Gordon, K. C., & Officer, D. L. (2006). Cover Picture: Synthesis and Characterization of a Multicomponent Rhenium(I) Complex for Application as an OLED Dopant. *Angewandte Chemie International Edition*, *45*(16), 2582–2584. doi:10.1002/anie.200504532 PMID:16555347

Misra, R., Jadhav, T., Dhokale, B., & Mobin, S. M. (2014). Reversible mechanochromism and enhanced AIE in tetraphenylethene substituted phenanthroimidazoles. *Chemical Communications (Cambridge)*, *50*(65), 9076. doi:10.1039/C4CC02824D PMID:24983656

Naka, S., Okada, H., Onnagawa, H., & Tsutsui, T. (2000). High electron mobility in bathophenanthroline. *Applied Physics Letters*, *76*(2), 197–199. doi:10.1063/1.125701

Nakanotani, H., Higuchi, T., & Furukawa, T. Masui1, K.,Morimoto, K.,Numata, M.,Tanaka, H.,Sagara, Y., Yasuda,T., &Adachi, C.(2014).High-efficiency organic light-emitting diodes with fluorescent emitters. *Nature Communications*. doi:10.1038/ncomms5016

O'Brien, D. F., Baldo, M. A., Thompson, M. E., & Forrest, S. R. (1999). Improved Energy Transfer in Electrophosphorescent Devices. *Applied Physics Letters*, *74*(3), 442–444. doi:10.1063/1.123055

Ouyang, X., Li, X. L., Ai, L., Mi, D., Ge, Z., & Su, S. J. (2015). Novel Hot Exciton Blue Fluorophores for High Performance Fluorescent/Phosphorescent Hybrid White Organic Light-Emitting Diodes with Superhigh Phosphorescent Dopant Concentration and Improved Efficiency Roll-Off. *ACS Applied Materials & Interfaces*, *7*(15), 7869–7877. doi:10.1021/am5081106 PMID:25826288

Ouyang, X., Li, X. L., Zhang, X., Islam, A., Ge, Z., & Su, S. J. (2015). Effective management of intramolecular charge transfer to obtain from blue to violet-blue OLEDs based on a couple of Phenanthrene isomers. *Dyes and Pigments*, *122*, 264–271. doi:10.1016/j.dyepig.2015.06.036

Padmaperuma, A. B., Sapochak, L. S., & Burrows, P. E. (2006). New Charge Transporting Host Material for Short Wavelength Organic Electrophosphorescence: 2,7-Bis(diphenylphosphineoxide)-9,9-dimethylfluorene. *Chemistry of Materials*, *18*(9), 2389–2396. doi:10.1021/cm0600677

Peng, J., Ye, K., Zhang, G., Zhan, Y., Jia, J., Xue, P., & Lu, R. (2014). Synthesis, photophysical and electroluminescent properties of phenanthroimidazole derivatives with terminal carbazole or pyrene. *Synthetic Metals*, *193*, 94–101. doi:10.1016/j.synthmet.2014.04.004

Qin, W., Yang, Z., Jiang, Y., Lam, J. W. Y., Liang, G., Kwok, H. S., & Tang, B. Z. (2015). Construction of Efficient Deep Blue Aggregation-Induced Emission Luminogen from Triphenylethene for Nondoped Organic Light-Emitting Diodes. *Chemistry of Materials*, *27*(11), 3892–3901. doi:10.1021/acs.chemmater.5b00568

Ren, X. F., Li, J., Holmes, R. J., Djurovich, P. I., Forrest, S. R., & Thompson, M. E. (2004). Ultrahigh Energy Gap Hosts in Deep Blue Organic Electrophosphorescent Devices. *Chemistry of Materials*, *16*(23), 4743–4747. doi:10.1021/cm049402m

Sasabe, H., Chiba, T., Su, S. J., Pu, Y. J., Nakayama, K. I., & Kido, J. (2008). 2-Phenylpyrimidine skeleton-based electron-transport materials for extremely efficient green organic light-emitting devices. *Chemical Communications (Cambridge)*, (44): 5821. doi:10.1039/b812270a PMID:19009093

Sasabe, H., & Kido, J. (2011). Multifunctional Materials in High-Performance OLEDs: Challenges for Solid-State Lighting. *Chemistry of Materials*, *23*(3), 621–630. doi:10.1021/cm1024052

Schwartz, G., Reineke, S., Rosenow, T. C., Walzer, K., & Leo, K. (2009). Triplet Harvesting in Hybrid White Organic Light-Emitting Diodes. *Advanced Functional Materials*, *19*(9), 1319–1333. doi:10.1002/adfm.200801503

Shan, T., Gao, Z., Tang, X., He, X., Gao, Y., Li, J., ... Ma, Y. (2017). Highly efficient and stable pure blue nondoped organic light-emitting diodes at high luminance based on phenanthroimidazole-pyrene derivative enabled by triplei-triplet annihilation. *Dyes and Pigments*, *142*, 189–197. doi:10.1016/j.dyepig.2017.03.032

Shan, T., Liu, Y., Tang, X., Bai, Q., Gao, Y., Gao, Z., ... Ma, Y. (2016). Highly Efficient Deep Blue Organic Light-Emitting Diodes Based on Imidazole: Significantly Enhanced Performance by Effective Energy Transfer with Negligible Efficiency Roll-off. *ACS Applied Materials & Interfaces*, *8*(42), 28771–28779. doi:10.1021/acsami.6b10004 PMID:27723979

Shi, H., Xin, D., Dong, X., Dai, J., Wu, X., Miao, Y., ... Choi, M. (2014). A star-shaped bipolar host material based on carbazole and dimesitylboron moieties for fabrication of highly efficient red, green and blue electrophosphorescent devices. *Journal of Materials Chemistry. C, Materials for Optical and Electronic Devices*, *2*(12), 2160–2168. doi:10.1039/C3TC32236J

Shih, P. I., Chien, C. H., Wu, F. I., & Shu, C. F. (2007). A Novel Fluorene-Triphenylamine Hybrid That is a Highly Efficient Host Material for Blue-, Green-, and Red-Light-Emitting Electrophosphorescent Devices. *Advanced Functional Materials, 17*(17), 3514–3520. doi:10.1002/adfm.200700191

Shih, P. I., Chuang, C. Y., Chien, C. H., Diau, E. W. G., & Shu, C. F. (2007). Highly Efficient Non-Doped Blue-Light-Emitting Diodes Based on an Anthrancene Derivative End-Capped with Tetraphenylethylene Groups. *Advanced Functional Materials, 17*(16), 3141–3146. doi:10.1002/adfm.200700355

Su, S. J., Cai, C., & Kido, J. (2011). RGB Phosphorescent Organic Light-Emitting Diodes by Using Host Materials with Heterocyclic Cores: Effect of Nitrogen Atom Orientations. *Chemistry of Materials, 23*(2), 274–284. doi:10.1021/cm102975d

Su, S. J., Sasabe, H., Pu, Y. J., Nakayama, K., & Kido, J. (2010). Pyridine-Containing Bipolar Host Materials for Highly Efficient Blue Phosphorescent OLEDs. *Advanced Materials, 22*, 3311. doi:10.1002/adma.200904249 PMID:20552601

Su, S. J., Takahashi, Y., Chiba, T., Takeda, T., & Kido, J. (2009). Relationship Structure–Property of Pyridine-Containing Triphenyl Benzene Electron-Transport Materials for Highly Efficient Blue Phosphorescent OLEDs. *Advanced Functional Materials, 19*, 1260. doi:10.1002/adfm.200800809

Su, S. J., Tanaka, D., Li, Y. J., Sasabe, H., Takeda, T., & Kido, J. (2008). Novel Four-Pyridylbenzene-Armed Biphenyls as Electron-Transport Materials for Phosphorescent OLEDs. *Organic Letters, 10*(5), 941–944. doi:10.1021/ol7030872 PMID:18237184

Subeesh, M. S., Shanmugasundaram, K., Sunesh, C. D., Nguyen, T. P., & Choe, Y. (2015). Phenanthroimidazole Derivative as an Easily Accessible Emitter for Non-Doped Light-Emitting Electrochemical Cells. *The Journal of Physical Chemistry C, 119*(41), 23676–23684. doi:10.1021/acs.jpcc.5b07871

Sun, D. M., Zhou, X. K., Li, H. H., Sun, X. L., Zheng, Y. H., Ren, Z. J., ... Yan, S. K. (2014). A versatile hybrid polyphenylsilane host for highly efficient solution-processed blue and deep blue electrophosphorescence. *Journal of Materials Chemistry. C, Materials for Optical and Electronic Devices, 2*(39), 8277–8284. doi:10.1039/C4TC01467G

Sun, Y., Duan, L., Zhang, D., Qiao, J., Dong, G., Wang, L., & Qiu, Y. (2011). A Pyridine-Containing Anthracene Derivative with High Electron and Hole Mobilities for Highly Efficient and Stable Fluorescent Organic Light-Emitting Diodes. *Advanced Functional Materials, 21*(10), 1881–1886. doi:10.1002/adfm.201002691

Takizawa, S., Montes, V. A., & Anzenbacher, P. (2009). Phenylbenzimidazole-Based New Bipolar Host Materials for Efficient Phosphorescent Organic Light-Emitting Diodes. *Jr. Chemistry of Materials, 21*(12), 2452–2458. doi:10.1021/cm9004954

Takizawa, S. Y., Montes, V. A., & Anzenbacher, P. (2009). Phenylbenzimidazole-Based New Bipolar Host Materials for Efficient Phosphorescent Organic Light-Emitting Diodes. *Jr. Chem. Mater, 21*(12), 2452–2458. doi:10.1021/cm9004954

Tamao, K., Uchida, M., Izumizawa, T., Furukawa, K., & Yamaguchi, S. (1996). Silole Derivatives as Efficient Electron Transporting Materials. *Journal of the American Chemical Society, 118*(47), 11974–11975. doi:10.1021/ja962829c

Tang, C., Bi, R., Tao, Y., Wang, F., Cao, X., Wang, S., ... Huang, W. (2015). A versatile efficient one-step approach for carbazole–pyridine hybrid molecules: Highly efficient host materials for blue phosphorescent OLEDs. *Chemical Communications (Cambridge)*, *51*(9), 1650–1653. doi:10.1039/C4CC08335K PMID:25500635

Tang, C. W., & VanSlyke, S. A. (1987). Organic electroluminescent diodes. *Applied Physics Letters*, *51*(12), 913–915. doi:10.1063/1.98799

Tang, X., Bai, Q., Peng, Q., Gao, Y., Li, J., Liu, Y., ... Ma, Y. (2015). Efficient Deep Blue Electroluminescence with an External Quantum Efficiency of 6.8% and CIEy < 0.08 Based on a Phenanthroimidazole–Sulfone Hybrid Donor–Acceptor Molecule. *Chemistry of Materials*, *27*(20), 7050–7057. doi:10.1021/acs.chemmater.5b02685

Tang, X., Shan, T., Bai, Q., Ma, H., He, X., & Lu, P. (2017). Efficient Deep Blue Electroluminescence Based on Phenanthroimidazole-Dibenzothiophene Derivatives with Different Oxidation States of Sulfur Atom. *Chemistry, an Asian Journal*, *12*(5), 552–560. doi:10.1002/asia.201601626 PMID:28070976

Tao, L., Li, J., Yu, Y., Jiang, Y., Zhou, C. S., Lee, S. T., ... Kwon, O. (2009). Bipolar Molecule as an Excellent Hole-Transporter for Organic-Light Emitting Devices. *Chemistry of Materials*, *21*(7), 1284–1287. doi:10.1021/cm803087c

Tao, Y., Wang, Q., Ao, L., Zhong, C., Qin, J., Yang, C., & Ma, D. (2010). Molecular design of host materials based on triphenylamine/oxadiazole hybrids for excellent deep-red phosphorescent organic light-emitting diodes. *Journal of Materials Chemistry*, *20*(9), 1759. doi:10.1039/b920227g

Tao, Y., Wang, Q., Ao, L., Zhong, C., Yang, C., Qin, J., & Ma, D. (2010). Highly Efficient Phosphorescent Organic Light-Emitting Diodes Hosted by 1,2,4-Triazole-Cored Triphenylamine Derivatives: Relationship between Structure and Optoelectronic Properties. *The Journal of Physical Chemistry C*, *114*(1), 601–609. doi:10.1021/jp908886d

Tao, Y., Wang, Q., Shang, Y., Yang, C. L., Ao, L., Qin, J., ... Shuai, Z. (2009). Multifunctional bipolar triphenylamine/oxadiazole derivatives: Highly efficient blue fluorescence, red phosphorescence host and two-color based white OLEDs. *Chemical Communications (Cambridge)*, (1): 77–79. doi:10.1039/B816264F PMID:19082004

Tao, Y., Wang, Q., Yang, C., Wang, Q., Zhang, Z., Zou, T., ... Ma, D. (2008). A Simple Carbazole/Oxadiazole Hybrid Molecule: An Excellent Bipolar Host for Green and Red Phosphorescent OLEDs. *Angewandte Chemie International Edition*, *47*(42), 8104–8107. doi:10.1002/anie.200803396 PMID:18798180

Tao, Y., Yang, C., & Qin, J. (2011). Organic host materials for phosphorescent organic light-emitting diodes. *Chemical Society Reviews*, *40*(5), 2943–2970. doi:10.1039/c0cs00160k PMID:21369622

Tavgeniene, D., Krucaite, G., Baranauskyte, U., Wu, J. Z., Su, H. Y., Huang, C. W., ... Grigalevicius, S. (2017). Phenanthro[9,10-d]imidazole based new host materials for efficient red phosphorescent OLEDs. *Dyes and Pigments*, *137*, 615–621. doi:10.1016/j.dyepig.2016.11.003

Togashi, K., Nomura, S., Yokoyama, N., Yasuda, T., & Adachi, C. (2012). Low driving voltage characteristics of triphenylene derivatives as electron transport materials in organic light-emitting diodes. *Journal of Materials Chemistry*, *22*(38), 20689. doi:10.1039/c2jm33669c

Tsai, H., Lin, H. W., Su, H. C., Ke, T. H., Wu, C. C., Fang, F. C., ... Wu, C. I. (2006). Incorporation of a CN group into mCP: A new bipolar host material for highly efficient blue and white electrophosphorescent devices. *Advanced Materials*, *18*, 1216–1220. doi:10.1002/adma.200502283

Ulla, H., Garudachari, B., Satyanarayan, M. N., Umesh, G., & Isloor, A. M. (2014). Blue organic light emitting materials: Synthesis and characterization of novel 1,8-naphthalimide derivatives. *Optical Materials*, *36*(3), 704–711. doi:10.1016/j.optmat.2013.11.017

Ulla, H., Raveendra Kiran, M., Garudachari, B., Satyanarayan, M. N., Umesh, G., & Isloor, A. M. (2014). Blue emitting halogen–phenoxy substituted 1,8-naphthalimides for potential organic light emitting diode applications. *Optical Materials*, *37*, 311–321. doi:10.1016/j.optmat.2014.06.016

Venugopal, T., Jeeva, P., Jayabharathi, J., Ramanathan, P., Prabhakaran, A., & Purani, E. S. (2016). Nondoped Blue Organic Light Emitting Devices with Donor–π–Acceptor Derivatives as the Emissive Material. *Industrial & Engineering Chemistry Research*, *55*(36), 9639–9647. doi:10.1021/acs.iecr.6b02549

Venugopal, T., Palanivel, J., & Jayaraman, J. (2017). Nondoped blue fluorescent OLED based on cyano-phenanthrimidazole-styryl-triphenylamine/carbazole materials. *Journal of Physical Organic Chemistry*, *30*(12), 3695. doi:10.1002/poc.3695

Venugopal, T., Sarojpurani, E., Jayabharathi, J., & Jeeva, P. (2017). Efficient phenanthroimidazole-styryltriphenylamine derivatives for blue OLEDs: A combined experimental and theoretical study. *New Journal of Chemistry*, *41*(6), 2443–2457. doi:10.1039/C6NJ03801H

Wada, A., Yasuda, T., Zhang, Q., Yang, Y. S., Takasu, I., Enomoto, S., & Adachi, C. (2013). A host material consisting of a phosphinic amide directly linked donor–acceptor structure for efficient blue phosphorescent organic light-emitting diodes. *Journal of Materials Chemistry. C, Materials for Optical and Electronic Devices*, *1*(13), 2404. doi:10.1039/c3tc00939d

Wakamiya, A., Taniguchi, T., & Yamaguchi, S. (2006). Intramolecular B–N coordination as a scaffold for electron-transporting materials: Synthesis and properties of boryl-substituted thienylthiazoles. *Angewandte Chemie International Edition*, *45*(19), 3170–3173. doi:10.1002/anie.200504391 PMID:16570334

Wang, B., Lv, X., Tan, J., Zhang, Q., Huang, Z., Yi, W., & Wang, L. (2016). Bipolar phenanthroimidazole–diazacarbazole hybrids with appropriate bandgaps for highly efficient and low roll-off red, green and blue electroluminescent devices. *Journal of Materials Chemistry. C, Materials for Optical and Electronic Devices*, *4*(36), 8473–8482. doi:10.1039/C6TC02683D

Wang, B., Lv, X., Tan, J., Zhang, Q., Huang, Z., Yi, W., & Wang, L. (2016). Bipolar phenanthroimidazole–diazacarbazole hybrids with appropriate bandgaps for highly efficient and low roll-off red, green and blue electroluminescent devices. *Journal of Materials Chemistry. C, Materials for Optical and Electronic Devices*, *4*(36), 8473–8482. doi:10.1039/C6TC02683D

Wang, B., Mu, G., Tan, J., Lei, Z., Jin, J., & Wang, L. (2015). Pyridine-containing phenanthroimidazole electron-transport materials with electron mobility/energy-level trade-off optimization for highly efficient and low roll-off sky blue fluorescent OLEDs. *Journal of Materials Chemistry. C, Materials for Optical and Electronic Devices*, *3*(29), 7709–7719. doi:10.1039/C5TC01436K

Wang, J., Lou, X., Liu, Y., Zhao, G., Islam, A., Wang, S., & Ge, Z. (2015). Controllable molecular configuration for significant improvement of blue OLEDs based on novel twisted anthracene derivatives. *Dyes and Pigments*, *118*, 137–144. doi:10.1016/j.dyepig.2015.03.005

Wang, K., Wang, S., Wei, J., Chen, S., Liu, D., Liu, Y., & Wang, Y. (2014). New multifunctional phenanthroimidazole–phosphine oxide hybrids for high-performance red, green and blue electroluminescent devices. *Journal of Materials Chemistry. C, Materials for Optical and Electronic Devices*, *2*(33), 6817–6826. doi:10.1039/C4TC00749B

Wang, K., Wang, S., Wei, J., Miao, Y., Liu, Y., & Wang, Y. (2014). Novel diarylborane–phenanthroimidazole hybrid bipolar host materials for high-performance red, yellow and green electrophosphorescent devices. *Organic Electronics*, *15*(11), 3211–3220. doi:10.1016/j.orgel.2014.08.062

Wang, Q., & Ma, D. (2010). Management of charges and excitons for high-performance white organic light-emitting diodes. *Chemical Society Reviews*, *39*(7), 2387–2398. doi:10.1039/b909057f PMID:20358089

Wang, Q., Oswald, I. W. H., Yang, X., Zhou, G., Jia, H., Qiao, Q., ... Gnade, B. E. (2014). A Non-Doped Phosphorescent Organic Light-Emitting Device with Above 31% External Quantum Efficiency. *Advanced Materials*, *26*(48), 8107–8113. doi:10.1002/adma.201402947 PMID:25219957

Wang, Y., Zhang, X., Han, B., Peng, J., Shiyou, H. H., Huang, Y., ... Lu, Z. (2010). The synthesis and photoluminescence characteristics of novel blue light-emitting naphthalimide derivatives. *Dyes and Pigments*, *86*(2), 190–196. doi:10.1016/j.dyepig.2010.01.003

Wang, Z., Feng, Y., Li, H., Gao, Z., Zhang, X., Lu, P., ... Liu, S. (2014). Dimeric phenanthroimidazole for blue electroluminescent materials: The effect of substituted position attached to biphenyl center. *Physical Chemistry Chemical Physics*, *16*(22), 10837–10843. doi:10.1039/C4CP00209A PMID:24759956

Wang, Z., Li, X., Xue, K., Li, H., Zhang, X., Liu, Y., ... Chen, P. (2016). Towards stable deep-blue emission and low efficiency roll-off in OLEDs based on phenanthroimidazole dimers. *Journal of Materials Chemistry. C, Materials for Optical and Electronic Devices*, *4*(9), 1886–1894. doi:10.1039/C5TC04048E

Wang, Z., Lu, P., Chen, S., Gao, Z., Shen, F., Zhang, W., ... Ma, Y. (2011). Phenanthro[9,10-d]imidazole as a new building block for blue light emitting materials. *Journal of Materials Chemistry*, *21*(14), 5451. doi:10.1039/c1jm10321k

Wang, Z., Xu, C., Wang, W., Dong, X., Zhao, B., & Ji, B. (2011). Novel pyrene derivatives: Synthesis, properties and highly efficient non-doped deep-blue electroluminescent device. *Dyes and Pigments*, *92*(1), 732–736. doi:10.1016/j.dyepig.2011.06.025

Wang, Z. M., Song, X. H., Gao, Z., Yu, D. W., Zhang, X. J., Lu, P., ... Ma, Y. G. (2012). Tuning of the electronic and optical properties of 4,4′-bis(1-phenylphenanthro[9,10-d]imidazol-2-yl)biphenyl via cyano substitution in un-conjugated phenyl. *RSC Advances*, *2*(25), 9635–9642. doi:10.1039/c2ra21054a

Wu, C. A., Chou, H. H., Shih, C. H., Wu, F. I., Cheng, C. H., Huang, H. L., ... Tseng, M. R. (2012). Synthesis and physical properties of meta-terphenyloxadiazole derivatives and their application as electron transporting materials for blue phosphorescent and fluorescent devices. *Journal of Materials Chemistry*, *22*(34), 17792–17799. doi:10.1039/c2jm33376g

Wu, K. C., Ku, P. J., Lin, C. S., Shih, H. T., Wu, F. I., Huang, M. J., ... Cheng, H. (2008). The Photophysical Properties of Dipyrenylbenzenes and Their Application as Exceedingly Efficient Blue Emitters for Electroluminescent Devices. *Advanced Functional Materials, 18*(1), 67–75. doi:10.1002/adfm.200700803

Wu, M. F., Yeh, S. J., Chen, C. T., Murayama, H., Tsuboi, T., Li, W. S., & Wang, J. K. (2007). The Quest for High-Performance Host Materials for Electrophosphorescent Blue Dopants. *Advanced Functional Materials, 17*(12), 1887–1895. doi:10.1002/adfm.200600800

Wu, W. C., Yeh, H. C., Chan, L. H., & Chen, C. T. (2002). Red Organic Light-Emitting Diodes with a Non-doping Amorphous Red Emitter. *Advanced Materials, 14*(15), 1072–1075. doi:

Xia, H., Li, M., Lu, D., Zhang, C., Xie, W., Liu, X., ... Ma, Y. (2007). Host selection and configuration design of electrophosphorescent devices. *Advanced Functional Materials, 17*(11), 1757–1764. doi:10.1002/adfm.200600882

Xiao, L., Su, S. J., Agata, Y., Lan, H., & Kido, J. (2009). Nearly 100% Internal Quantum Efficiency in an Organic Blue-Light Electrophosphorescent Device Using a Weak Electron Transporting Material with a Wide Energy Gap. *Advanced Materials, 21*(12), 1271–1274. doi:10.1002/adma.200802034

Xiao, L., Su, S. J., Agata, Y., Lan, H., & Kido, J. (2009). Nearly 100% Internal Quantum Efficiency in an Organic Blue-Light Electrophosphorescent Device Using a Weak Electron Transporting Material with a Wide Energy Gap. *Advanced Materials, 21*(12), 1271–1274. doi:10.1002/adma.200802034

Yang, B., & Ma, Y. G. (2013). Progress in next-generation organic electroluminescent materials: Material design beyond exciton statistics. *Science in China. Series B, Chemistry, 43*, 1457.

Yang, J., Huang, J., Li, Q., & Li, Z. (2016). Blue AIEgens: Approaches to control the intramolecular conjugation and the optimized performance of OLED devices. *Journal of Materials Chemistry. C, Materials for Optical and Electronic Devices, 4*(14), 2663–2684. doi:10.1039/C5TC03262H

Ye, D., Chen, M., Liu, S., Su, J., Wang, Y. F., Lo, C. C., ... Kido, J. (2014). Pyridine-Containing Electron-Transport Materials for Highly Efficient Blue Phosphorescent OLEDs with Ultralow Operating Voltage and Reduced Efficiency Roll-Off. *Advanced Functional Materials, 24*(21), 3268–3275. doi:10.1002/adfm.201303785

Ye, J., Chen, Z., Fung, M. K., Zheng, C., Ou, X., Zhang, X., ... Lee, C. S. (2013). Carbazole/Sulfone Hybrid D-π-A-Structured Bipolar Fluorophores for High-Efficiency Blue-Violet Electroluminescence. *Chemistry of Materials, 25*(13), 2630–2637. doi:10.1021/cm400945h

Yook, K. S., & Lee, J. Y. (2012). Organic Materials for Deep Blue Phosphorescent Organic Light-Emitting Diodes. *Advanced Materials, 24*(24), 3169–3190. doi:10.1002/adma.201200627 PMID:22641461

Yu, D., Zhao, F., Han, C., Xu, H., Li, J., Zhang, Z., ... Yan, P. (2012). Ternary Ambipolar Phosphine Oxide Hosts Based on Indirect Linkage for Highly Efficient Blue Electrophosphorescence: Towards High Triplet Energy, Low Driving Voltage and Stable Efficiencies. *Advanced Materials, 24*(4), 509–514. doi:10.1002/adma.201104214 PMID:22213078

Yu, D., Zhao, F., Zhang, Z., Han, C., Xu, H., Li, J., ... Yan, P. (2012). Insulated donor–p–acceptor systems based on fluorene-phosphine oxide hybrids for non-doped deep-blue electroluminescent devices. *Chemical Communications (Cambridge)*, *48*(49), 6157. doi:10.1039/c2cc31066j PMID:22590706

Yuan, W. Z., Gong, Y., Chen, S., Shen, X. Y., Lam, J. W. Y., Lu, P., ... Tang, B. Z. (2012). Efficient Solid Emitters with Aggregation-Induced Emission and Intramolecular Charge Transfer Characteristics: Molecular Design, Synthesis, Photophysical Behaviors, and OLED Application. *Chemistry of Materials*, *24*(8), 1518–1528. doi:10.1021/cm300416y

Yuan, Y., Chen, J. X., Lu, F., Tong, Q. X., Yang, Q. D., Mo, H. W., ... Lee, C.-S. (2013). Bipolar Phenanthroimidazole Derivatives Containing Bulky Polyaromatic Hydrocarbons for Nondoped Blue Electroluminescence Devices with High Efficiency and Low Efficiency Roll-Off. *Chemistry of Materials*, *25*(24), 4957–4965. doi:10.1021/cm4030414

Yuan, Y., Chen, J. X., Lu, F., Tong, Q. X., Yang, Q. D., Mo, H. W., ... Lee, C. S. (2013). Bipolar Phenanthroimidazole Derivatives Containing Bulky Polyaromatic Hydrocarbons for Nondoped Blue Electroluminescence Devices with High Efficiency and Low Efficiency Roll-Off. *Chemistry of Materials*, *25*(24), 4957–4965. doi:10.1021/cm4030414

Yuan, Y., Li, D., Zhang, X., Zhao, X., Liu, Y., Zhang, J., & Wang, Y. (2011). Phenanthroimidazole-derivative semiconductors as functional layer in high performance OLEDs. *New Journal of Chemistry*, *35*(7), 1534–1540. doi:10.1039/c1nj20072k

Yuan, Y., Li, D., Zhang, X., Zhao, X., Liu, Y., Zhang, J., & Wang, Y. (2011). Phenanthroimidazole-derivative semiconductors as functional layer in high performance OLEDs. *New Journal of Chemistry*, *35*(7), 1534–1540. doi:10.1039/c1nj20072k

Zeng, L., Lee, T. Y. H., Merkel, P. B., & Chen, S. H. (2009). A new class of non-conjugated bipolar hybrid hosts for phosphorescent organic light-emitting diodes. *Journal of Materials Chemistry*, *19*(46), 8772. doi:10.1039/b909787b

Zhang, F., Li, W., Wei, D., Wei, X., Li, Z., Zhang, S., ... Zhai, B. (2016). Synthesis, photophysical and electroluminescent properties of blue-emitting dual core imidazole–anthracene/pyrene derivatives. *RSC Advances*, *6*(65), 60264–60270. doi:10.1039/C6RA04958C

Zhang, Q., Li, J., Shizu, K., Huang, S., Hirata, S., Miyazaki, H., & Adachi, C. (2012). Design of Efficient Thermally Activated Delayed Fluorescence Materials for Pure Blue Organic Light Emitting Diodes. *Journal of the American Chemical Society*, *134*(36), 14706–14709. doi:10.1021/ja306538w PMID:22931361

Zhang, S., Li, W., Yao, L., Pan, Y., Shen, F., Xiao, R., ... Ma, Y. (2013). Enhanced proportion of radiative excitons in non-doped electro-fluorescence generated from an imidazole derivative with an orthogonal donor–acceptor structure. *Chemical Communications (Cambridge)*, *49*(96), 11302. doi:10.1039/c3cc47130f PMID:24158601

Zhang, X., Lin, J., Ouyang, X., Liu, Y., Liu, X., & Ge, Z. (2013). Novel host materials based on phenanthroimidazole derivatives for highly efficient green phosphorescent OLEDs. *Journal of Photochemistry and Photobiology A Chemistry*, *268*, 37–43. doi:10.1016/j.jphotochem.2013.06.012

Zhang, Y., Lai, S. L., Tong, Q. X., Chan, M. Y., Ng, T. W., Wen, Z. C., ... Lee, C. S. (2011). Synthesis and characterization of phenanthroimidazole derivatives for applications in organic electroluminescent devices. *Journal of Materials Chemistry*, *21*(22), 8206. doi:10.1039/c1jm10326a

Zhang, Y., Lai, S. L., Tong, Q. X., Lo, M. F., Ng, T. W., Chan, M. Y., ... Lee, C. S. (2012). High Efficiency Nondoped Deep-Blue Organic Light Emitting Devices Based on Imidazole-π-triphenylamine Derivatives. *Chemistry of Materials*, *24*(1), 61–70. doi:10.1021/cm201789u

Zhang, Y., Lai, S. L., Tong, Q. X., Lo, M. F., Ng, T. W., Chan, M. Y., ... Lee, S. (2012). High Efficiency Nondoped Deep-Blue Organic Light Emitting Devices Based on Imidazole-π-triphenylamine Derivatives. *Chemistry of Materials*, *24*(1), 61–70. doi:10.1021/cm201789u

Zhang, Y., Ng, T. W., Lua, F., Tong, Q. X., Lai, S. L., Chan, M. Y., ... Lee, C. S. (2013). A pyrene-phenanthroimidazole derivative for non-doped blue organic light-emitting devices. *Dyes and Pigments*, *98*(2), 190–194. doi:10.1016/j.dyepig.2013.01.022

Zhang, Y., Wang, J. H., Han, G., Lu, F., & Tong, Q. X. (2016). Phenanthroimidazole derivatives as emitters for non-doped deep-blue organic light emitting devices. *RSC Advances*, *6*(75), 70800–70809. doi:10.1039/C6RA13605B

Zhao, J., Xie, G., Yin, C., Xie, L., Han, C., Chen, R., ... Huang, W. (2011). Harmonizing Triplet Level and Ambipolar Characteristics of Wide-Gap Phosphine Oxide Hosts toward Highly Efficient and Low Driving Voltage Blue and Green PHOLEDs: An Effective Strategy Based on Spiro-Systems. *Chemistry of Materials*, *23*(24), 5331–5339. doi:10.1021/cm201654c

Zhong, G. Y., Xu, Z., He, J., Zhang, S. T., Zhan, Y. Q., Wang, X. J., ... Hou, X. Y. (2002). Aggregation and permeation of 4-(dicyanomethylene)-2-methyl-6-(p-dimethylaminostyryl)-4H-pyran molecules in $Alq_{3.}$ *Applied Physics Letters*, *81*(6), 1122–1124. doi:10.1063/1.1497438

Zhu, M., & Yang, C. (2013). Blue fluorescent emitters: Design tactics and applications in organic light-emitting diodes. *Chemical Society Reviews*, *42*(12), 4963. doi:10.1039/c3cs35440g PMID:23436004

Zhu, X. H., Peng, J., Cao, Y., & Roncali, J. (2011). Solution-processable single-material molecular emitters for organic light-emitting devices. *Chemical Society Reviews*, *40*(7), 3509–3524. doi:10.1039/c1cs15016b PMID:21483933

Zhu, Y. C., Zhou, L., Li, H. Y., Xu, Q. L., Teng, M. Y., Zheng, Y. X., ... You, X. Z. (2011). Highly Efficient Green and Blue-Green Phosphorescent OLEDs Based on Iridium Complexes with the Tetraphenylimidodiphosphinate Ligand. *Advanced Materials*, *23*(35), 4041–4046. doi:10.1002/adma.201101792 PMID:21800382

Zhu, Y. C., Zhou, L., Li, H. Y., Xu, Q. L., Teng, M. Y., Zheng, Y. X., ... You, X. Z. (2011). Highly Efficient Green and Blue-Green Phosphorescent OLEDs Based on Iridium Complexes with the Tetraphenylimidodiphosphinate Ligand. *Advanced Materials*, *23*(35), 4041–4046. doi:10.1002/adma.201101792 PMID:21800382

Zhu, Z. L., Chen, M., Chen, W. C., Ni, S. F., Peng, Y. Y., Zhang, C., ... Lee, C. S. (2016). Removing shortcomings of linear molecules to develop high efficiencies deep-blue organic electroluminescent materials. *Organic Electronics*, *38*, 323–329. doi:10.1016/j.orgel.2016.09.004

Zhu, Z. L., Chen, W. C., Zhang, L. D., Liu, X. L., Tong, Q. X., Wong, F. L., ... Lee, C. S. (2016). A pyridine based meta-linking deep-blue emitter with high conjugation extent and electroluminescence efficiencies. *Journal of Materials Chemistry. C, Materials for Optical and Electronic Devices*, *4*(26), 6249–6255. doi:10.1039/C6TC01689H

Zhuang, S.,Shangguan, R., Huang, H.,Tu, G., Wang, L., &Zhu, X. (2014).Synthesis, characterization, physical properties, and blue electroluminescent device applications of phenanthroimidazole derivatives containing anthracene or pyrene moiety. *Dyes and Pigments,101,* 93-102.

Zhuang, S., Shangguan, R., Jin, J., Tu, G., Wanga, L., Chen, J., ... Zhu, X. (2012). Efficient nondoped blue organic light-emitting diodes based on phenanthroimidazole-substituted anthracene derivatives. *Organic Electronics*, *13*(12), 3050–3059. doi:10.1016/j.orgel.2012.08.032

APPENDIX

Table 11. Abbreviations

Alq_3,	tris-(8-hydroxyquinoline)aluminium;
BCP,	2,9-dimethyl-4,7-diphenyl-1,10-phenanthroline;
Bphen,	4,7-diphenyl-1, 10-phenanthroline;
CBP	4,4′-bis(carbazol-9-yl)-biphenyl
CBZ	Carbazole
CE,	current efficiency;
CIE,	Commission internationale de L'Eclairage;
CT,	charge transfer;
CV,	cyclic voltammetry;
DFT	Density function theory
DPA,	diphenylamine;
DPO	Diphenyl phosphine oxide
D–π–P,	donor–π–acceptor;
EAs	electron affinities
EBU	European Broadcasting Union
EIL	Electron injection layer
EL,	electroluminescent;
EML,	Emissive layer;
EQE,	external quantum efficiency;
ET,	energy transfer;
ETL,	electron-transport layer;
EUE	exciton utilization efficiency
FMOs,	frontier molecular orbitals;
FWHM,	full width at half maximum;
HBL,	Hole blocking layer;
HDAC	High Definition Television
HIL,	hole-injecting layer;
HLCT,	hybridized local and charge transfer
HOMO,	highest occupied molecular orbital;
HTL,	hole-transporting layer;
HyLEDs,	hybrid organic–inorganic LEDs
IC,	internal conversion;
IQE	internal quantum efficiency;
ISC,	intersystem crossing;
ITO,	indium tin oxide;
LE,	Luminescence efficiency;

continued on following page

Table 11. Continued

LED,	light-emitting diode;
LUMO,	lowest unoccupied molecular orbital;
NPB,	N,N'-bis-(1-naphthyl)-N,N'-diphenyl-1,1 -diphe nyl-4;
NPD	N,N′-bis(1-naphthyl)-N,N′-diphenyl-1,1′-biphenyl-4,4′-diamine
NTSC	National Television Standards Committee
OLED,	organic light-emitting diode;
PE,	power efficiency;
PI,	phenanthroimidazole;
PL,	photoluminescence;
PLQE,	photoluminescent quantum efficiency;
PO,	phosphine oxide;
TAPC	1,1-bis((di-4- tolylamino)phenyl)cyclohexane
TBPI	2,2′,2′′-(1, 3, 5-benzenetriyl) tris-[1-phenyl-1H- benzimidazole
TCTA	4,4′,4′′-tri(9-carbazoyl)triphenylamine
Td,	decomposition temperature;
Tg,	glass transition temperature;
THF,	tetrahydrofuran;
Tm,	melting point temperature;
TPA,	triphenylamine;
TPBI,	1,3,5-tri(1-phenyl-1H-benzo[d] imidazole-2-yl)phen;
TPD,	tri-methylphenyl diamine;
TPPO,	triphenylphosphine oxide;
TPTZ,	2,4,6- tripyridyltriazine;
TTA,	triplet–triplet annihilation
UV	Ultra Violet
$V_{on,}$	turnon-voltage;
WOLEDs,	white organic light emitting diodes
ZnO	Zink oxide
TADF	thermally activated delayed fluorescence
PhOLEDs	Phosphorescent OLEDs
Mes2B	dimesitylborane

Chapter 10
Thermo, Photo and Mechanoluminescence Studies of Eu^{3+} Doped $Y_4Al_2O_9$ Phosphors:
TL, PL, and ML Analysis of YAM: Eu^{3+}

Shubha Tripathi
Dr. C. V. Raman University, India

Manish Kumar Mishra
Bhilai Institute of Technology Raipur, India

Vikas Kumar Jain
Government Engineering College, India

Ratnesh Tiwari
Bhilai Institute of Technology Raipur, India

Neha Dubey
Government V. Y. T. Post Graduate Autonomous College, India

ABSTRACT

The chapter reports synthesis and characterization of europium-doped Y4Al2O9 phosphor for display and dosimetric applications. The europium-activated Y4Al2O9 (YAM) phosphor is synthesized via solid state reaction method. Synthesized phosphors were characterized by powder x-ray diffraction (PXRD) techniques, scanning electron microscopy (SEM) technique, and transmission electron microscopy (TEM) technique. Particle size calculated from TEM analysis and crystallite size was calculated by Scherer's formula. All synthesized phosphor for different concentration of europium ion were studied by photo, thermo, and mechanoluminescence study. It is found that for photoluminescence analysis of Eu3+ doped phosphor has prominent spectra in red region and electric dipole transition (5D0 → 7F2) dominant over the magnetic dipole transition (5D0 → 7F1) due to non-centro symmetry between rare earth ions. Broad excitation spectra found for photoluminescence study.

DOI: 10.4018/978-1-5225-5170-6.ch010

INTRODUCTION

During the past decades, nanostructured materials have attracted considerable attention for their novel and enhanced properties, e.g., the Mn doped ZnS phosphor can yield both high luminescent efficiencies and short lifetime Bhargava et al. (1994) and Zhang et al. (1998). Nanostructured materials may be developed to form a novel type of luminescent materials for display applications.

The Y_2O_3–Al_2O_3 system is a promising material for refractory coatings and for ceramic and semiconductor processing technology Lo et al. (1998) and Aguilar et al. (2000). Doped yttrium aluminium garnet (YAG) is widely used as a laser host material Lupei et al. (2002) and Vodopyanov et al. (1998), and yttrium aluminumperovskite (YAP) used as scintillation host material Glodo et al. (2000). In addition, rare earth doped YAG is also employed as a phosphor Lu et al. (2002) and Choe et al. (2001). However, there are few reports on rare earth doped $Y_4Al_2O_9$ (YAM). It has been reported that the space group for the crystal structure of $Y_4Al_2O_9$ is $P2_1/c$ of monoclinic Yamane et al. (1998 a & b). The Al atoms are coordinated to four oxygen atoms, the Y atoms are coordinated to either six or seven oxygen atoms Yamane et al. (1998 a & b), and its site symmetry is C1 Rabinovitch et al. (2004). There are four formula units in the unit cell of the room temperature phase of $Y_4Al_2O_9$ and four different rare earth sites in the asymmetric unit. Xia et al. (2005) have synthesized the $Y_4Al_2O_9$:Eu^{3+} phosphor through a sol–gel combustion method, where the $Y_4Al_2O_9$ phase can form through sintering at 800 ^{0}C. However, higher doping concentration could be realized in $Y_4Al_2O_9$:Eu^{3+}nano-crystal host lattice. Shengli Liu et al. have also prepared the $Y_4Al_2O_9$:(Tb^{3+}, Eu^{3+}) by sol–gel process Shengli and Su (1997) Wang et al. (2006), Kaur et al. (2012) and Dubey et al. (2013) have studied the VUV excitation and photoluminescence characteristics of this $Y_{3.8}Al_2O_9$:$Re_{0.2}$ (Re = Tb^{3+}, Eu^{3+}) phosphor synthesized via a citric-gel method. This phosphor shows strong absorption in VUV region. The results indicate that this phosphor could be one of the potential candidates for PDPs applications.

The rare-earth ions show abundant emission colors based on their 4f–4f or 5d–4f transitions. Eu^{3+} show red/orange emission, respectively. Y_2O_3:Eu^{3+}phosphor emit red emission and have excellent chemical stability. This phosphor is the only existing red/green phosphor used in three-band fluorescent lamps. For Y_2O_3:Eu^{3+} nanoparticles, the intensity of $^5D_1$ emission of Eu^{3+} ions from the C3i site (S6) is significantly less compared to the emission from Eu^{3+} ions in the C_2 site for these samples Jeong et al. (2003). The Eu–O charge transfer band resulting from an electron transfer from the ligand O^{2-}(2p) orbit to the empty states of the $4f^6$ configuration (Eu^{3+}–O^{2-} transition) and the energy transfer efficiency between the Y_2O_3 host and Eu^{3+} ions were investigated by Kelley et al. (2006).

Plasma display panels (PDPs) dominate the segment of the next generation flat panel displays which uses VUV excitation of Red, Green and Blue (RGB) phosphors for image display. For full coloured large area flat display, PDP is a very promising technique due to the progress made in technology by way of improvements in cost, resolution, lifetime, power consumption, high performance and luminescence efficiency which results in the reductions of thickness and weight Tian et al. (2006) and Rao et al. (2006). Motivated by the advances in PDP technique, the demand of highly efficient vacuum ultraviolet (VUV) phosphors has increased tremendously in the past decade Yang et al. (2006). The demand became crucial as phosphors play a very important role in PDPs in terms of augmented performance such as higher efficiency for lower power consumption and higher reliability for longer lifetime. Luminescence characteristics of phosphors and their behaviour under panel making process, energetic discharge ions, electrons and solarization from VUV generated by the Xe-Ne gas plasma are important factors for PDP. In PDP's plasma resonance, vacuum ultraviolet (VUV) radiation lines of Xe atoms at 147 nm and a molecular excimer Xe^2 at 172 nm are used for excitation of phosphors to emit visible luminescence in red, green and blue. All the three red, blue and green phosphors have shown degradation in luminescence intensity due to the thermal treatment of PDP manufacturing process and this remains a major problem for PDP phosphors Zhang et al. (2003), Zeng et al. (2006), Bizarri and Moine et al. (2005) and Okazaki et al. (2000). Another challenge in PDP technology today is to improve the purity of red phosphor so as to obtain monochromatic red emission with good color coordinates Ronda (1997), Sohn et al. (2000). Commercially used $(Y,Gd)BO_3:Eu^{3+}$ ($YGB:Eu^{3+}$) red phosphor has poor color purity with a dominant orange component and the CIE chromaticity x value (0.65) Shionoya and Yen (1999) falls short of declared standard value of 0.67 by National TelevisionStandard Committee (NTSC). For viewing high speed motor sports, video gaming and other high definition programs, decay time of the phosphor should be as low as possible to avoid image overlap. Decay time of $YGB:Eu^{3+}$ phosphor is long (12 msec) which makes this phosphor unsuitable for next generation PDP sets. It is, therefore, necessary to develop highly efficient, baking resistant, high color purity red-emitting phosphor with shorter decay time. As an alternative, the search for new efficient red-emitting phosphor other than $YGB:Eu^{3+}$ is being conducted and reviewed by many researchers Han et al. (2009) and Kang et al. (1999). Europium-doped Yttrium Aluminium Garnet (YAG) has attracted substantial attention in recent years because of its high resistance to electron irradiation which makes Yttria – Alumina system promising candidates in Cathode Ray Tubes, Field Emission Display, Vacuum Fluorescent Displays Ching et al. (1999) etc. The Yttria – Alumina system has several phases, including $YAlO_3$ (YAP: Yttrium Aluminium Polymorphs), $Y_4Al_2O_9$ (YAM: Yttrium Aluminium Monoclinic), and $Y_3Al_5O_{12}$ (YAG: Yttrium Aluminium Garnet). Commonly, YAP and YAM form as an intermediate product of YAG phase in solid state reactions. Even if YAG is synthesized with a stochiometric mixture of Y_2O_3 and Al_2O_3, two detrimental phases, YAP and YAM, often coexist as by-products Lupei et al. (2002). We have deliberately synthesised single phase $YAM:Eu^{3+}$ by conventional solid state reaction technique and investigated the luminescence properties in detail to see the effect of changed crystal field in Eu3+ emission characteristics. The luminescence properties of $YAM:Eu^{3+}$ under UV excitation was reported earlier by many researchers for different applications Lu et al. (2002) and Choe et al. (2001). However, very few reports Xia et al. (2005) and Wang et al. (2006) are available on the VUV excitation properties of $YAM:Eu^{3+}$ for PDP application so far. To the best of our knowledge, we report for the first time Eu3+ doped YAM that is highly efficient, degradation controlled, pure red-emitting phosphor with additional feature of short decay time suitable for advanced PDPs.

In this context it is first time reported the thermo, photo and mechanoluminescence study of YAM phosphor doped with europium ion and detail study of spectroscopic parameters. The chapter describes the synthesis and characterization of YAM phosphor doped with europium ions with variable concentration. Kinetic parameters were calculated using CGCD technique for TL studies. Photoluminescence excitation and emission spectra explained in details. Mechanoluminescence study for UV induced sample has been reported in the chapter.

Experimental

To prepare $Y_4Al_2O_9$ with various concentrations of europium (0.2mol % to 2mol %), consists of heating stoichiometric amounts of reactant mixture is taken in alumina crucible and is fired in air at 1100°C for 1 hour in a muffle furnace. Every heating is followed by intermediate grinding using agate mortar and pestle. The Eu^{3+} activated $Y_4Al_2O_9$phosphor was prepared via high temperature modified solid state diffusion. The starting materials were as follows: Y_2O_3, Al_2O_3 Eu_2O_3 and H_3BO_3 (as a flux) in molar ratio (0.2% to 2 mol%.of Eu^{3+}) were used to prepare the phosphor .The mixture of reagents was ground together to obtain a homogeneous powder. After being ground thoroughly in stoichiometric ratios by using an agate mortar by dry grinding for nearly 45 minutes, to ensure the best homogeneity and reactivity, powder was transferred to alumina crucible, and then heated in a muffle furnace at 1350 °C for 4 hour Dubey et al. (2014). The phosphor materials were cooled to room temperature naturally.

The sample was characterized using XRD, FTIR, EDX (Energy dispersive X-ray analysis) FEGSEM and HRTEM. The XRD measurements were carried out using Bruker D8 Advance X-ray diffractometer. The X-rays were produced using a sealed tube and the wavelength of X-ray was 0.154 nm (Cu K-alpha). The X-rays were detected using a fast counting detector based on Silicon strip technology (BrukerLynxEye detector).Observation of particle morphology was investigated by FEGSEM (field emission gun scanning electron microscope) (JEOL JSM-6360). The photoluminescence (PL) emission and excitation spectra were recorded at room temperature by use of a Shimadzu RF-5301 PC spectrofluorophotometer. The excitation source was a xenon lamp. Thermally stimulated luminescence glow curves were recorded at room temperature by using TLD reader I1009 supplied by Nucleonix Sys. Pvt. Ltd. Hyderabad Dubey ety al. (2013), Jeong et al. (2003), Kelley et al. (2006). The obtained phosphor under the TL examination is given UV radiation using 254nm UV source. Heating rate used for TL measurement is 5^0Cs^{-1}. It is well known that ML is produced during deformation of crystal, a correlation between the ML intensity and the crystal deformation is expected. Various techniques have been used for deformation of the crystal for ML studies Saluja et al. (2016). ML Devices, which we have used, was designed by Chandra (1983), Chandra and Deshmukh (1985) to improve the performance of earlier designed devices Tiwari et al. (2015). The ML intensity were measured at different impact velocities by dropping a load of particular mass and shape from different heights. The photomultiplier tube and its housing are put in the box through the lid. The box is perfectly dark for accurate ML measurements. A hollow cylindrical aluminium pipe of diameter 2.1 cm was mounted on the box by the clamps. At the top of hollow pipe a friction less pulley is fixed with the help of wooden cavity. For the measurement of different heights a scale parallel to guided hollow cylinder is fixed. The ML is excited impulsively by dropping a load of particular mass from different heights. The phosphor was placed on a transparent Lucite plate and was covered with aluminium foil with the help of adhesive taps. A RCA 931A photomultiplier tube was connected to an EHT supply of -700 V and output of the tube was supplied to a digital storage oscilloscope Tiwari et al. (2014), Kaur et al. (2015), Kaur et al. (2010a) and kaur et al. (2010b)

RESULTS AND DISCUSSION

Structural and Morphology of YAM:Eu^{3+} Phosphor

Figure 1 shows the XRD pattern of1.5mol% Eu^{3+} doped YAM phosphor. The phosphors doped with other concentration shows same behaviour and it indicates that the small amount of doping does not affect the crystal structure.The crystal system of the prepared samples was identified as a monoclinic structure.

The size of the crystals has been computed from the full width half maximum (FWHM) of the intense peak using Scherer formula Lu et al. (2002). Formula used for calculation is

Figure 1. XRD pattern of $Y_4Al_2O_9$: Eu^{3+} doped phosphor

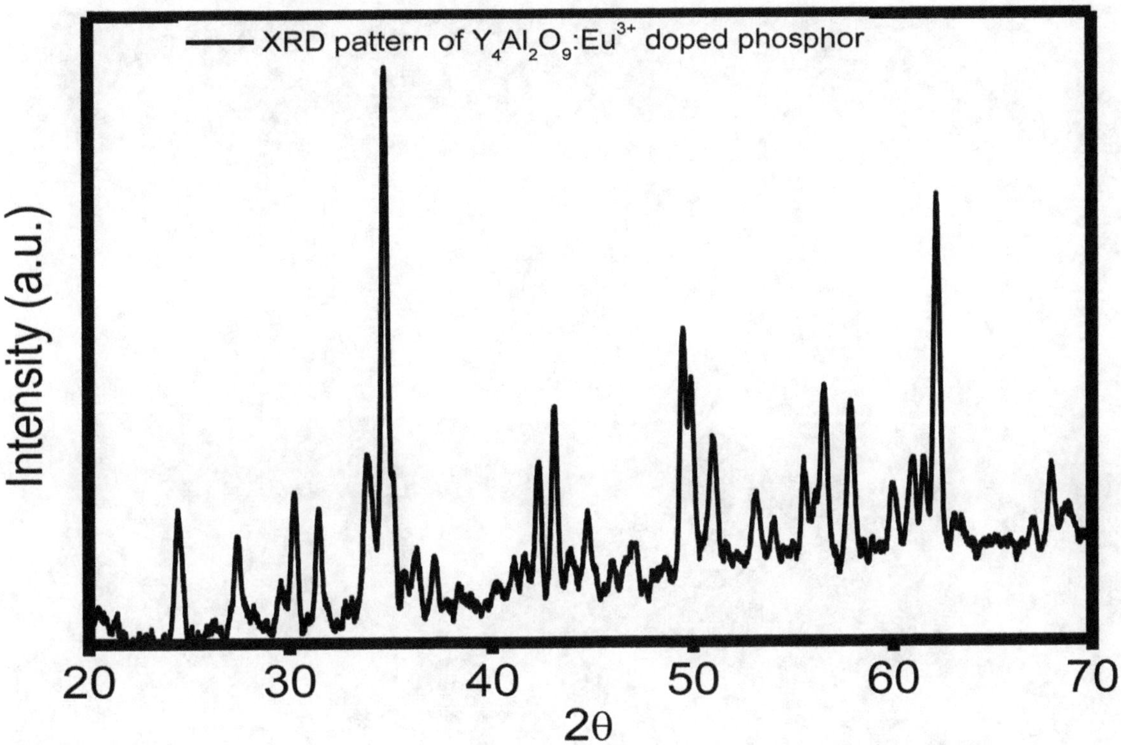

$$D = \frac{0.9\lambda}{\beta \cos\theta} \tag{1}$$

Here D is crystallite size; β is FWHM (full width half maximum), λ is the wavelength of X-ray source and θ is angle of diffraction. The average crystallite size is ~ 56 nm.

Morphology

Figure 2 shows the morphology of YAM:Eu^{3+} (1.5 mol%) phosphor by FEGSEM technique. The synthesized phosphor shows good morphology with particles of 2-5 ~μm size. The particles are distinct in nature which is the main criteria for the lighting application.

TEM Analysis of Prepared Phosphor

Particle size were analysed using TEM images of prepared phosphor (Figure 3 a & b), had a spherical sized structure and exact particle size varies from 66 nm -111 nm.

Figure 2. FEGSEM image of prepared phosphor

Figure 3. (a) TEM image of prepared phosphor low resolution (b) high resolution

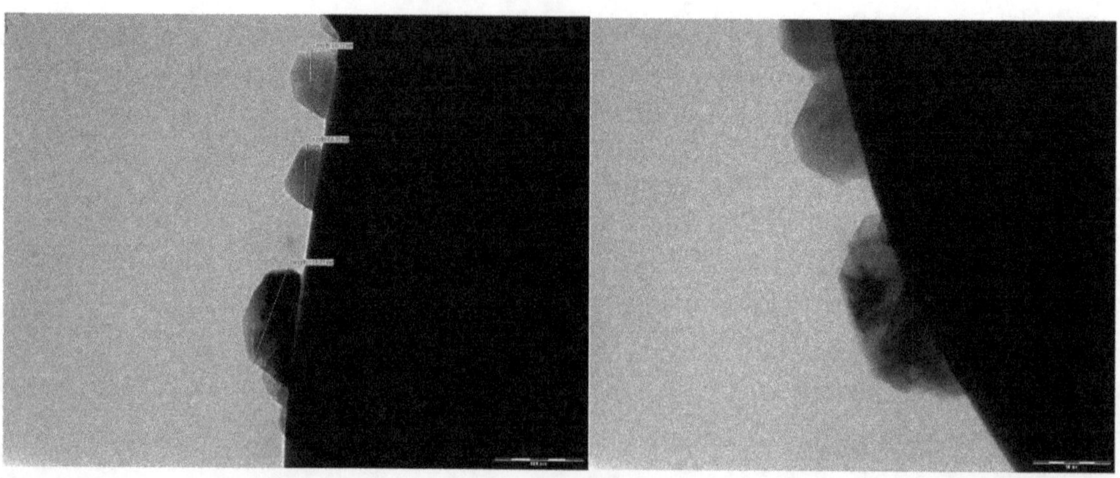

FTIR Analysis

FTIR Spectrum of $Y_4Al_2O_9$:Eu^{3+} powder is shown in the figure 4. This spectrum expresses a strong peak at 588 cm^{-1} and 727 cm^{-1} which are the characteristics of Y-O vibrations. The peak found at 852 cm^{-1} is probably because of Al-O Vibrations. Presence of B-O gives rise to an IR peak within a range of 1016 – 1194 cm^{-1}; we also have a peak at 1299 cm^{-1} so we predict this peak as a result of B-O vibration because we have used H_3BO_3 as flux Rao et al. (2000) and Yang et al. (2006). All these discussed peak found together confirms the formation of $Y_4Al_2O_9$ phosphor.

Photoluminescence Study of YAM: Eu^{3+} Phosphor

Figure 5 shows the excitation spectra of YAM:Eu^{3+} phosphor. The excitation spectra of $Y_4Al_2O_9$:Eu^{3+} phosphor mainly consists of the charge transfer and (CTB) of Eu^{3+} located in 220–280 nm. The CTB of crystalline phosphor shows blue shift. The energy position of Eu^{3+}CTB is closely related to the covalency of Eu–O bond and the coordination number of Eu^{3+}. The covalency of Eu^{3+}–O^{2-} bond is strongly influenced by the next nearest cation M^{3+} ($M=Y^{3+}$, Al^{3+}). Stronger chemical bond between Eu–O bond and M^{3+}ions forms for the crystalline phosphor. Therefore, the CTB of crystalline $Y_4Al_2O_9$:$Eu^{3+}$is located

Figure 4. FTIR spectra of YAM: Eu^{3+} doped phosphor

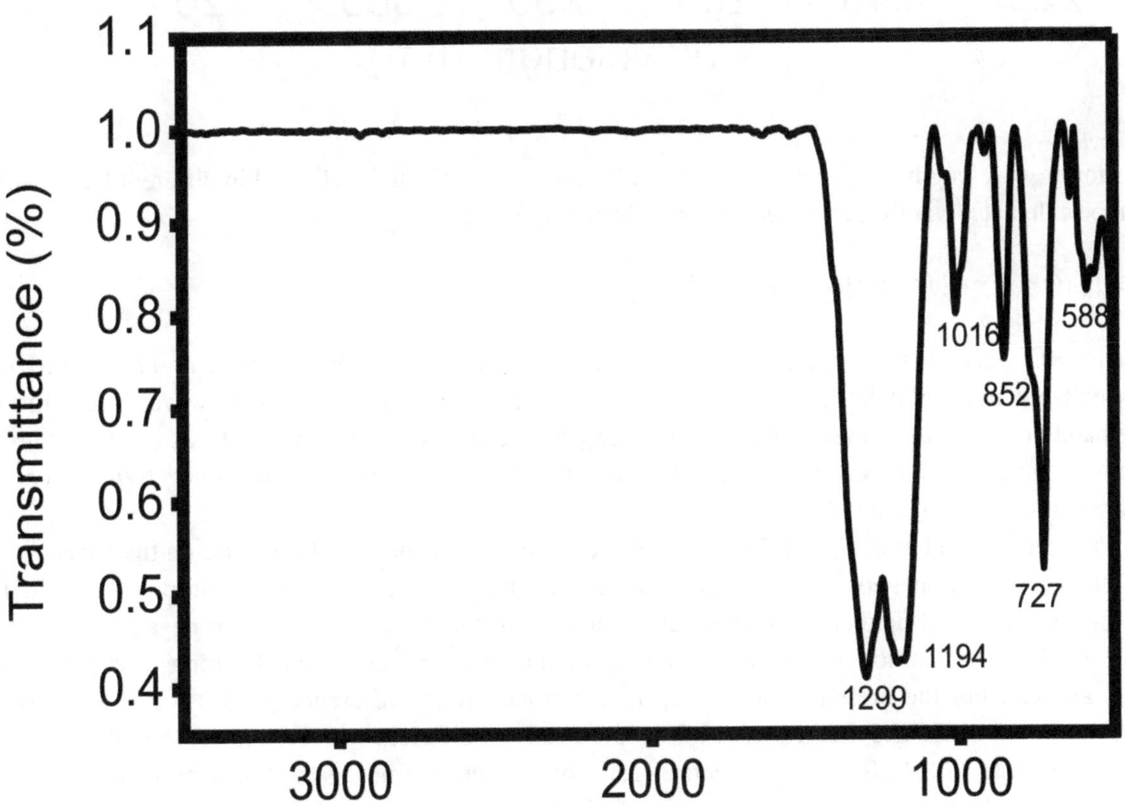

Figure 5. PL excitation spectra of $Y_4Al_2O_9$: Eu^{3+} (0.5 mol%) doped phosphor excited with 613nm

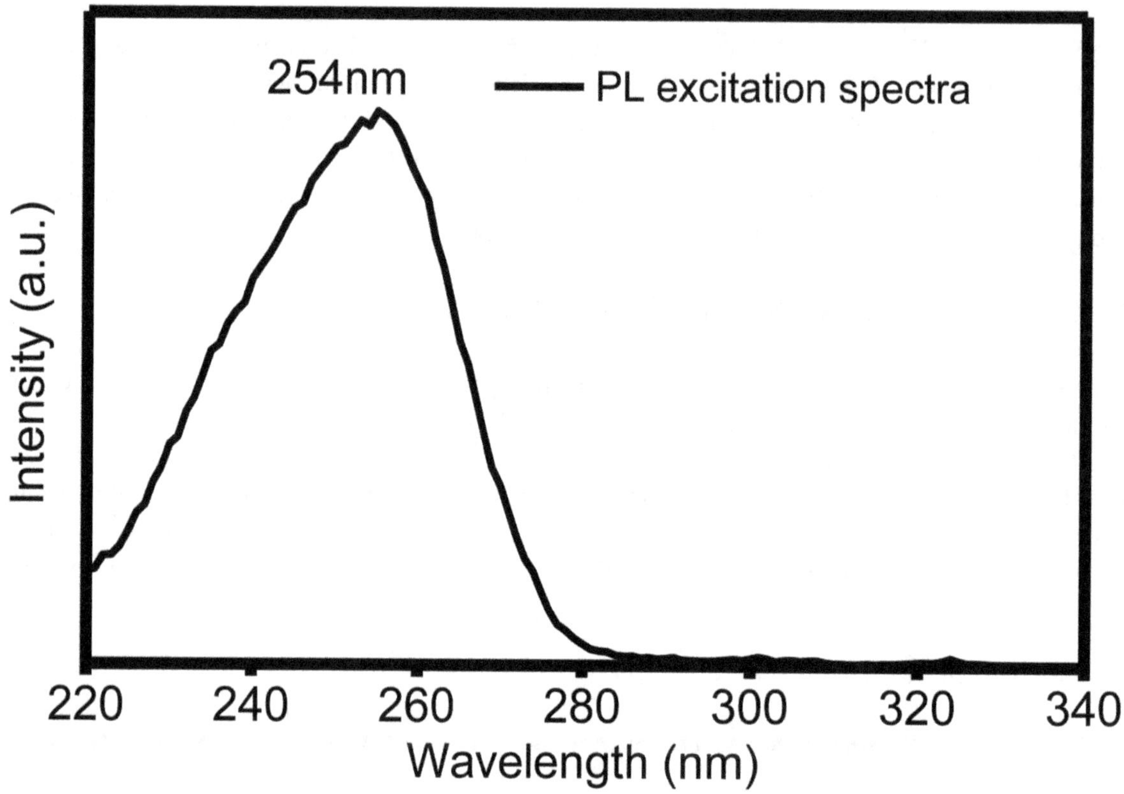

at shorter wavelength. The accurate position of the energy σ of Eu-O CT band in different host matrix can be calculated via the Jorgensen formula Huang et al. (2012):

$$\sigma = [\chi\,(O^{2-}) - \chi\,(Eu^{3+}\ \text{ion})] \times 30000\ \text{cm}^{-1} \tag{2}$$

where, $\chi(O^{2-})$ and $\chi(Eu^{3+}$ ion) are the electronegativity of the ligand anion O^{2-} and central metal ion Eu^{3+} respectively. In general the Eu^{3+}- O^{2-} band located at ~254nm Som et al. (2014).However, a red shift in the band position is observed in the present case which can be explained on the basis of two different factors as mentioned by X. Liu *et al.* (2014). These two factors comprise the electronegetivity and bond characteristics of Eu-O bond.

Previously, Parchur et al. (2012) explained the large shift of the Eu–O CT band on the basis of the difference in electronegetivity by using the Jørgensen equation. According to their proposed model, the change in the crystal field environments can modify the difference in optical electronegativity between Eu^{3+} and O^{2-}. The position of energy is dependent on the difference in optical electronegativity between Eu^{3+} and O^{2-}. For the present host matrix, the electronegativity difference $\chi\,(O^{2-}) - \chi\,(Eu^{3+}$ ion), is observed to be 1.25. The present value of electronegativity difference is much lower than the reported value Som et al. (2014). Such lower electronegativity difference indicates the lesser value of CT energy σ. Therefore, the red shift of CT band can be attributed to the lowering of CT energy.

Emission Transitions of Eu^{3+} in YAM Host Matrix

Figure 6 shows the emission spectra of $Y_4Al_2O_9$:Eu^{3+}phosphor with different concentration of Eu^{3+}. The emission spectra are characteristic $^4f_6$ energy level transition emission. They are mainly due to two dipole transitions. One is $^5D_0$ to $^7F_1$ magnetic dipole transition, and another is $^5D_0$ to$^7F_2$ forced electric dipole transition. The intensity ratio of $^5D_0$ to $^7F_2$ to $^5D_0$ to $^7F_1$ can be viewed as a clue concerning the nature of the chemical surroundings of the luminescent center and its symmetry Zhang et al. (2003).As seen from Figure 6, the sharp and distinct emission lines, lying between 580–640 nm and observed due to $^5D_0 \rightarrow ^7F_J$ (J = 0–2) levels of Eu^{3+} ions, depend upon the location of Eu^{3+} ion. If the Eu^{3+} ion is located at the crystallographic site without inversion symmetry, owing to which the electric-dipole transition $^5D_0 \rightarrow ^7F_2$ red emission is observed to be dominated. If Eu^{3+} ion is situated at the site with inversion centre, the orange emission ($^5D_0 \rightarrow ^7F_1$) should be dominated Lee et al. (2002). In the present situation, the dominance of ~613 nm peak ($^5D_0 \rightarrow ^7F_2$) over the orange emission~594 nm ($^5D_0 – ^7F_1$)transition indicates that Eu^{3+} ions occupy a low symmetry site in YAM host.

Figure 6. PL emission spectra of YAM: Eu^{3+} doped phosphor recorded under various concentration of Eu^{3+} ion

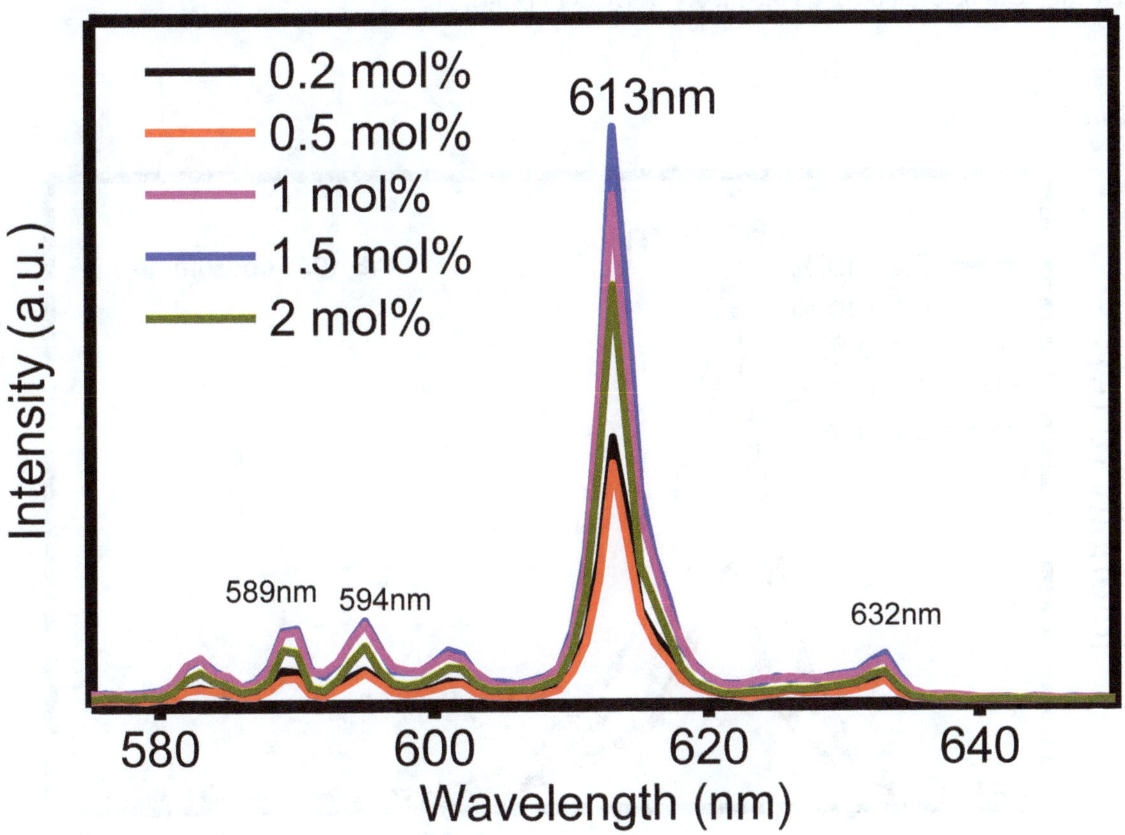

Concentration Quenching

With the increase in Eu^{3+} concentrations, the emission intensity increased systematically without any change in the peak positions, as shown in Figure 7. The inset (a) (CIE coordinate for red emission) depicted the variation of emission intensity with Eu^{3+} concentration, indicating that the concentration quenching occurs at 1.5 mol% of Eu^{3+} ions concentration (613 nm emission).

The energy transfer among two Eu^{3+} ions may be take place via exchange interaction or multipole–multipole interaction. The type of interaction can be identified via the calculation of the critical distance (R_c) between two Eu^{3+} in YAM host as per the formula of Blasse (1986):

$$R_c \approx 2\left[\frac{3V}{4\pi CN}\right]^{1/3} \tag{3}$$

where C is the critical concentration of the Eu^{3+}, N is the number of cations in the unit cell and V is the volume of the unit cell. In the present case, R_c is estimated to be 11.83 Å. The value of R_c is greater than 5 Å in the present host indicating that the exchange interaction cannot take place. Therefore, multipolar interaction should be responsible for concentration quenching. The interaction type may be dipole–dipole (d–d), dipole–quadrupole (d–q), and quadrupole–quadrupole (q–q), etc. The exact type of multipolar interaction can be evaluated via the theory of Dexter (1954):

Figure 7.

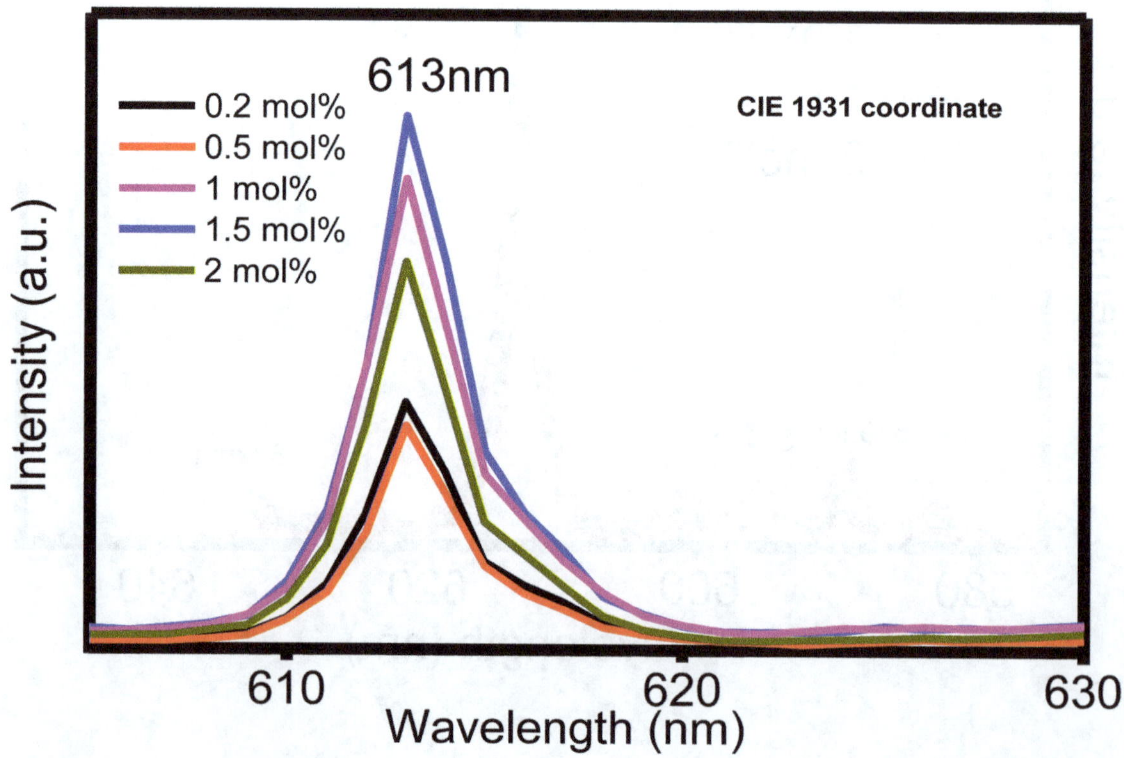

$$\frac{I}{C} = \frac{k_1}{\beta C^{s/3}} \tag{4}$$

where *I* is the emission intensity, *C is* the activator concentration involved in self-concentration quenching, k_1 and β are constants for each interaction in the same excitation conditions for a given host lattice.

The character of multipolar interaction can be estimated as d–d, d–q and q–q when the values of s are 6, 8 and 10, respectively. The value of s can be obtained by the slop (s/3) of the linear plot of ln(I/C) vslnC, as shown in Figure 8. In the present case, the value of s is obtained to be 5.97 (~6 for the electric d–d interaction) indicating that the d–d interaction is the main mechanism for the concentration quenching transition of Eu^{3+} in YAM phosphors Liu et al. (2012) and Ofelt (1962).

Strength of Eu-O Bond Covalency and Spectral Characterization

Various spectroscopic parameters of Eu^{3+} in YAMhost are estimated via Judd- Ofelt theory in order to examine the exact site symmetry and luminescence behaviour. The behaviour of the radiative potential of Eu^{3+} ions in YAM host are obtained by estimating different intensity parameters Ω_t (t = 2, 4, 6) Judd (1962) and Som et al. (2015). Herein, Ω_2 parameter indicates the asymmetric nature of Eu^{3+} in YAM crystal ligand. Other two parameters $\Omega_{4,6}$ depend on long range effects Kodaira et al. (2003). If $I_{0\text{-}J}$ (J=

Figure 8. Logarithmic plot for the emission intensity per Eu^{3+} ion as a function of its concentration

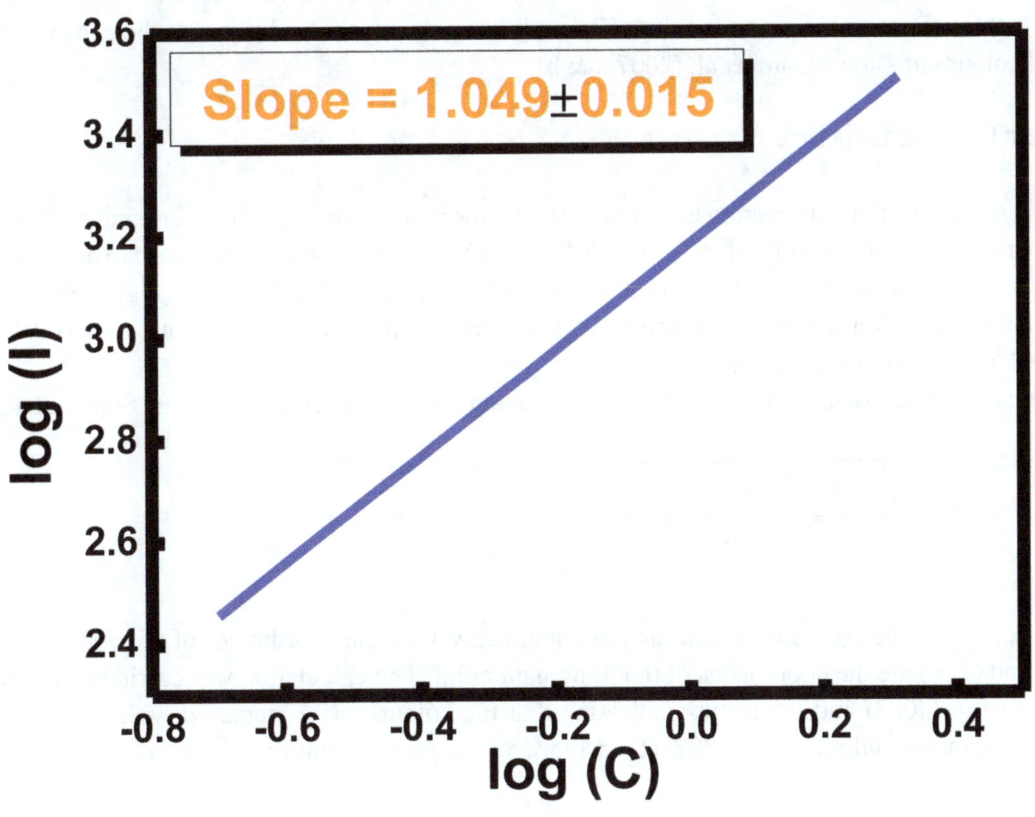

0, 1, 2, 4) denotes the integrated emission intensities for different transitions and $h\upsilon_{0\text{-}J}$ (J= 0, 1, 2, 4) stands for the energy of the corresponding transitions, then the radiative emission ($A_{0\text{-}J}$ (J= 0, 1, 2, 4)) rates can be evaluated as Feng et al. (2013):

$$\frac{A_{0-2,4}}{A_{0-1}} = \frac{I_{0-2,4}}{I_{0-1}} \frac{h\nu_{0-1}}{h\nu_{0-2,4}} \tag{5}$$

Since the magnetic dipole transition cannot be affected by the crystal ligand, the magnetic dipole radiative emission rate A_{0-1} has a constant value and it is ~50 s^{-1}. Moreover, these radiative emission rates ($A_{0-2,4}$) can be evaluated via the JO intensity parameters as follows Feng et al. (2013):

$$A_{0-J} = \frac{64\pi^4 (\nu_{0-2,4})^3 e^2}{3hc^3} \frac{1}{4\pi\varepsilon_0} \chi \sum_{J=2,4,6} \Omega_J \langle ^5D_0 | U^{(J)} | ^7F_{2,4} \rangle^2 \tag{6}$$

where, χ is the Lorentz local field correction factor. Reportedly, it was observed that the matrix elements are $\langle ^5D_0 | U^{(2)} | ^7F_2 \rangle^2$ =0.0032 and $\langle ^5D_0 | U^{(4)} | ^7F_4 \rangle^2$ =0.0023 Feng et al. (2013). From Eq. 5 and 6, the Judd–Ofelt parameters have been examined. In the present case the emission spectra of YAM phosphor doped with 2 mol% of Eu^{3+} has been used to evaluate the intensity parameters. The parameter Ω_2 is estimated to be 2.4 × 10^{-24} cm^2 indicating the dependence on the covalency between Eu^{3+} ions and ligands anions. The asymmetric nature of Eu^{3+} has been revealed by the Ω_2 value. The value Ω_2 and Ω_4 parameters indicate that the $^5D_0 \rightarrow {}^7F_2$ transition is more efficient in the present case and supports the enhancement of the red colour emission Mishra et al. (2007 a & b).

Colour Characteristics

Figure 9 represents the CIE coordinate of the present phosphor for optimized concentration of Eu^{3+}. The CIE coordinates for the optimized (1.5mol% Eu^{3+} doped YAM) phosphor are estimated to be (0.62, 0.37) which is very nearer to the commercial red phosphor Y_2O_2S: Eu^{3+} (0.667, 0.326) (Figure 8)According to National Television Standard Committee (NTSC) system the ideal red chromaticity is (0.67, 0.33) Lee et al. (2002) and Blasse (1986)

The colour purity of the emitted colour was calculated using Wang et al. (2006) and Som et al. (2015):

$$ColourPurity = \frac{\sqrt{(x_s - x_i)^2 + (y_s - y_i)^2}}{\sqrt{(x_d - x_i)^2 + (y_d - y_i)^2}} \times 100\% \tag{7}$$

where (x_s, y_s) are the coordinates of a sample point, (x_d, y_d) are the coordinates of the dominant wavelength and (x_i,y_i) are the coordinates of the illuminant point. The calculation was carried out as per ref. Kodaira et al. (2003) and the results indicating that the colour purity increased with the increase in Eu^{3+}concentration and reached to 89% at 1.5 mol% of Eu^{3+} concentration. Therefore, YAM:Eu^{3+} (1.5

Figure 9. CIE coordinate for YAM: Eu^{3+} phosphor (1.5 mol% of Eu^{3+})

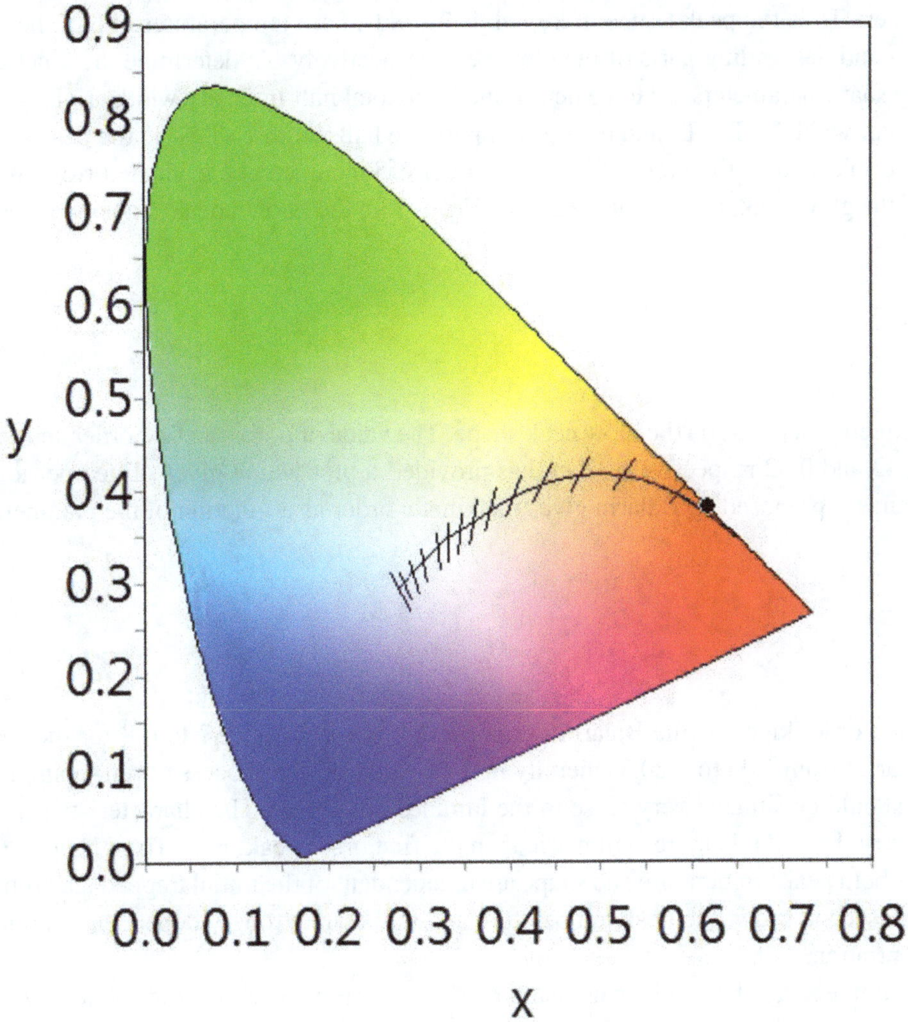

mol%) phosphor, which exhibits intense red emission with good red chromaticity and high colour purity, have promising applications in solid-state lighting.

Thermoluminescence Glow Curve Analysis

Thermoluminescence (TL) phosphors generally exhibit glow curves with one or more peaks when the charge carriers are released. The glow curve is characteristic of the different trap levels that lie in the band gap of the material. The traps are characterized by certain physical parameters that include trap depth (E) and frequency factor(s). For many TL applications, a clear knowledge of these physical parameters is essential. In the study of relatively deep trapping effect-states in various solid state materials as well as TL dating, a detailed analysis of TL glow curves is indispensable.

In this work, Chen's method Dubey et al. (2016) in his article was used to determine the kinetic energy parameters of the glow peak of the TL materials. This method is mainly based on the temperature T_m, T_1 and T_2, where T_m is the peak temperature, while T_1 and T_2 are temperatures at half the intensity on the ascending and descending parts of the glow peak respectively. To determine the kinetic parameter the following shape parameters are to be determined: the total half intensity width $\omega = T_2 - T_1$. The high temperature half width $\delta = T_2 - T_m$ and the low temperature half width $\tau = T_m - T_1$ the peak shape method is mainly used to calculate the order of kinetics. Order of kinetic can be evaluated from the symmetry factor (μ_g) of the glow peak. μ_g is calculated using Eq. 1 from the known peak shape parameters δ andω.

$$\mu_g = - = \frac{T_{2-T_m}}{T_{2-T_1}} \tag{8}$$

Order of kinetics depends on the glow peak shape. The value of μ_g for the first order and second order kinetics is 0.42 and 0.52 respectively. Chen has provided a plot which gives of geometric factor (μ_g). Another parameter proposed by Balarin gives the kinetic order as a function of the parameter.

$$\gamma = - = \frac{T_{2-T_m}}{T_{m-T_1}} \tag{9}$$

For the first order kinetics, the Balarin parameter (γ) ranges from 0.7 to 0.8 for the second order kinetics (γ) varies from 1.05 to 1.20. Generally in the first order, the process of retrapping is negligible and the trap should be situated very close to the luminescent center. The characteristics of the second order peak are wider and it is more symmetric than the first order peak. For a fixed heating rate, in first order kinetics both peak temperature and shape are independent of the initial trapped electron concentration but in the second order, the peak temperature and shape are strongly dependent on initial trapped charge concentration.

The activation energy (E) can be calculated by the general expressions formulated by Chen and is given by

$$E = C_\alpha \frac{kT_m^2}{\alpha} - b_\alpha 2kT_m \tag{10}$$

where k is Boltzmann constant. T_m is peak temperature. The constant $C_\alpha$and b_α was also calculated by the Chen's equation.

Herein the TL glow curve for γ - irradiated (0.5 – 2 kGy) sample shows well resolved peak (only concentrated for high temperature peak) centred at 283^0 C had high temperature peak (Figure 10 and Figure 11) which is responsible for TL dosimetric application. Figure 10 shows the TL glow curves of the YAM:Eu^{3+} (2 mol%) γ irradiated for doses of 0.5 kGy up to 2 kGy. These TL glow curves showed two prominent well resolved TL glow peaks at 283^0C. The results indicated the creation of trapping centres that increased with an increasing γ – ray exposure. The TL intensity of the 283^0C glow peak as a function of irradiated dose is shown in Figure 11. It also showed a linear relation with the γ exposure

Figure 10. TL glow curve of YAM:Eu^{3+} (1.5 mol%) doped phosphor with the variation of γ exposure

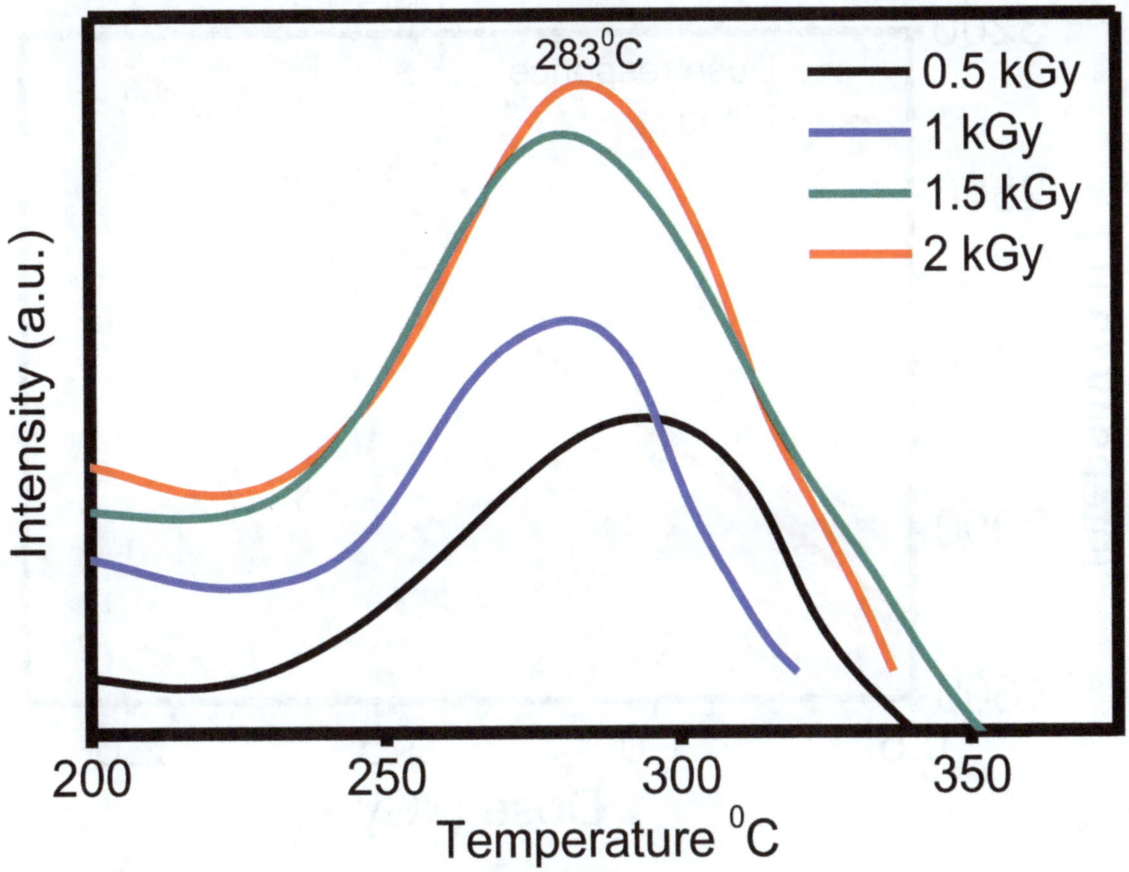

(Figure 11). Hence this behaviour of the sample might be very useful for γ-dosimetry applications Sohn et al. (2002).

Computerized Glow Curve Deconvolution of YAM: Eu^{3+} Phosphor for γ-Ray Induced Phosphor

The recorded glow curves of the prepared phosphors exposed to γ exposure with different doses seem to be composite in nature as they exhibit a broad peak. So, CGCD method was used to deconvolute and calculate the kinetic parameters. This computer programme (Glowfit) is based on Halperin and Barner equations which describe the flow of the charges between the various energy levels during a trap emptying by thermal heating. The kinetic parameters of trap levels were determined for each deconvoluted peak by this programme Dubey et al. (2016), Singh et al. (2008), Baldinozzi et al. (1998), Shrivastava et al. (2014) and Dubey et al. (2013). The fitted TL glow curves are shown in Figure 12 for YAM:Eu^{3+} and the values of trap depths and frequency factors of trapped charges calculated by CGCD method are summarized in Table 1. The CGCDof the YAM:Eu^{3+} phosphor for the fixed optimized concentration of Eu^{3+} (1.5mol%) with γ exposure of 2 kGy is shown in Figure 12. The corresponding kinetic parameters such as activation energy, order of kinetic and frequency factors were calculated using the peak shape

Figure 11. TL intensity vs. γ dose plot for YAM: Eu (1.5mol%)

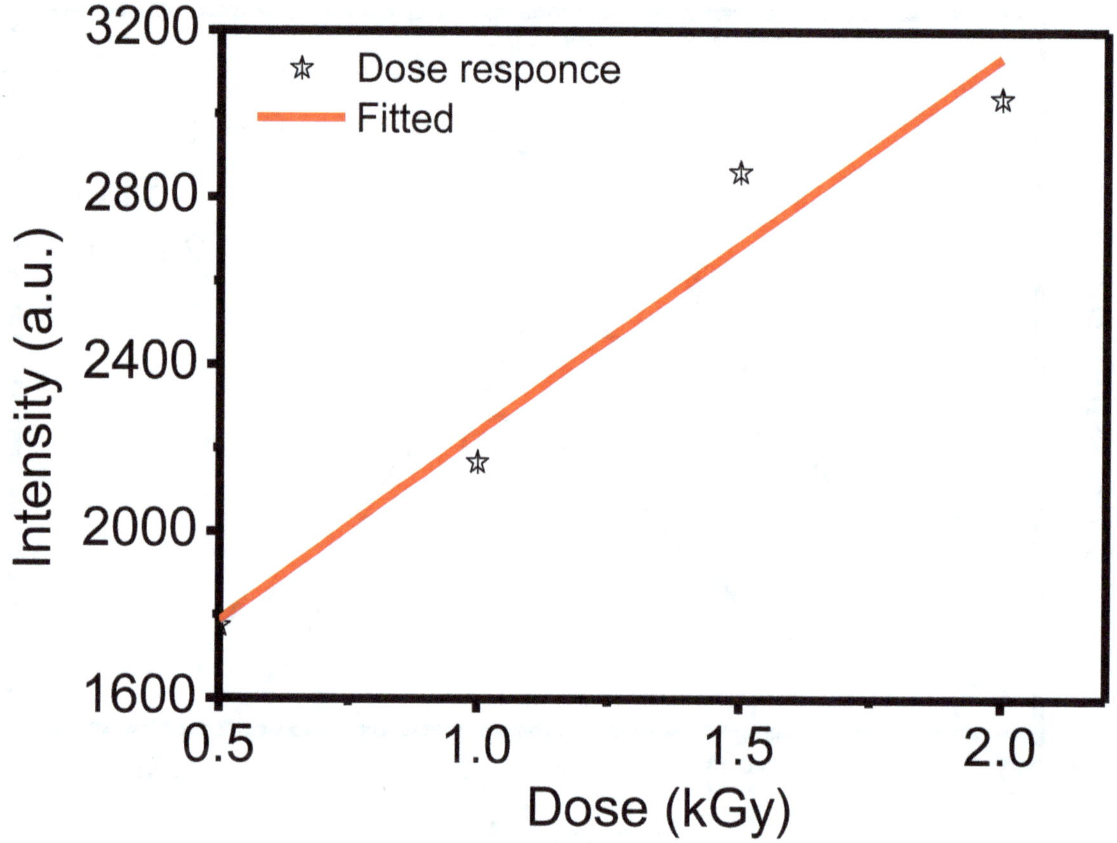

method for the individual peaks (Table 1). The five peaks that were fitted for the YAM:Eu^{3+} phosphor (1.5mol%) showed nearly a second order of kinetic and the activation energy was high for the high temperature peak. It pointed towards the formation of deeper traps in the prepared sample during γ ray irradiation. When sample were irradiated by the γ rays, which is a high energy source, deep levels were created and the electrons got trapped in the deep levels and the required energy for escaping one electron is high which means the activation energy was high for the γ irradiated samples. Figure of merit (FoM) of all fitted curve (~1.83) found less than or equal 2% which is best fit by CGCD technique.

Mechanoluminescence Study of YAM: Eu^{3+} Doped Phosphor

Mechanoluminescence characteristics of YAM: Eu^{3+} doped phosphor with variable UV exposure time was measured (Figure 13). In ML intensity versus time curve single peak is observed. The presence of a broad single peak indicates some charge transfer process involved in ML emission. The luminescence intensity depends upon UV exposure time. The experiment was carried out for different UV exposure time (5-15 min). ML emission is mostly due to the deformation of sample. In present study when phosphor is mechanically deformed, the electric field introduced during the deformation either ionizes the electrons from the defect centre or from the valence band, electron thus obtained in conduction band may either

Figure 12. CGCD curve for the YAM: Eu^{3+} (1.5mol%) doped phosphor with γ exposure of 2 kGy

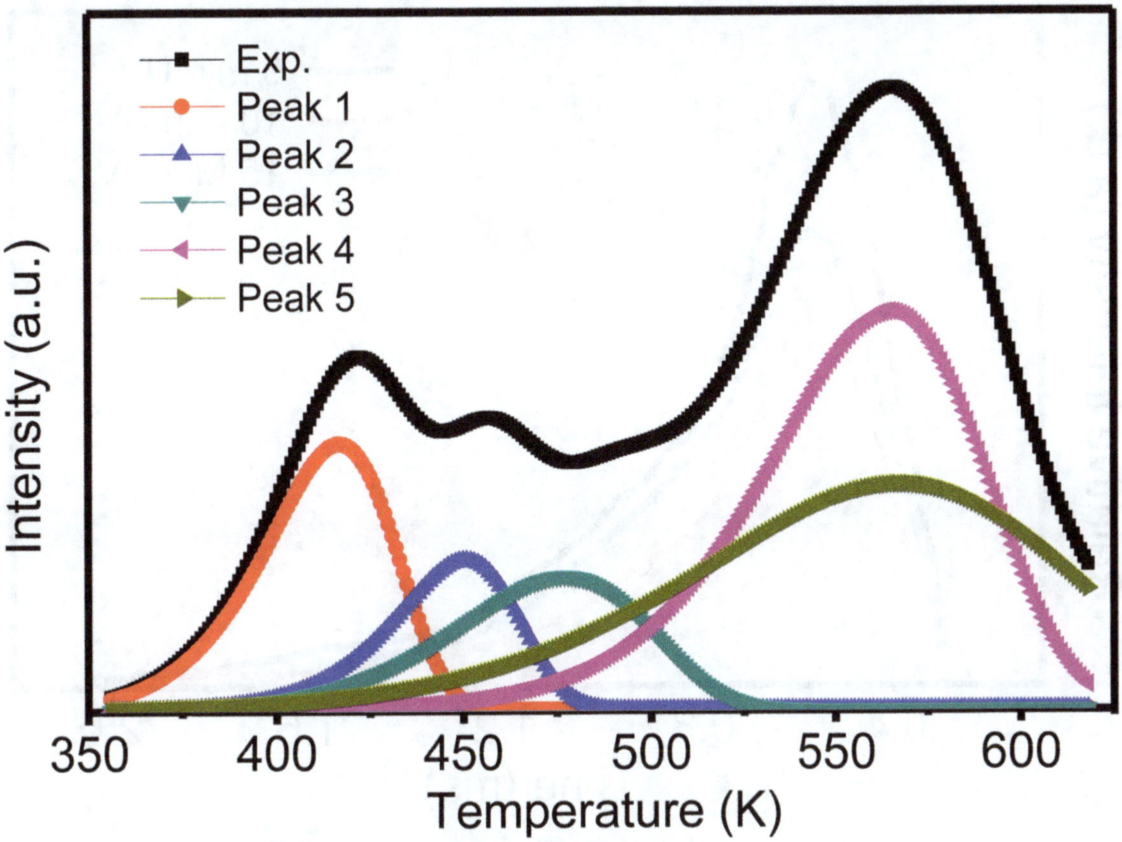

Table 1. Corresponding kinetic parameters calculated from CGCD curve for the YAM: Eu^{3+} (1.5mol%) doped phosphor after 2 kGyγ radiation

Peaks	T_1	T_m	T_2	τ	δ	ω	μ = δ / ω	Activation Energy	Frequency Factor
1	389	417	434	28	17	45	0.37	2.17	1×10^{17}
2	429	450	466	21	16	37	0.43	3.21	5×10^{23}
3	447	478	505	31	27	58	0.46	2.35	8×10^{16}
4	527	566	593	39	27	66	0.40	2.31	8×10^{14}
5	517	566	617	49	51	100	0.51	1.86	1×10^{12}

recombine directly with the holes present in the luminescence centre or recombination centre, or they may be trapped for some time in the trapping levels present near the conduction band.

Figure 15 shows that ML intensity increases linearly with $t\text{-}t_m$. In Figure 16 it is shown that peak time t_m decreases with increasing impact velocity because deformation occurs earlier at higher impact velocity. Linear response is shown in ln1000/Vo versus tm plot (Figure 17).

Figure 13. ML pattern of YAM: Eu^{3+} with different UV exposure time (5-15 min)

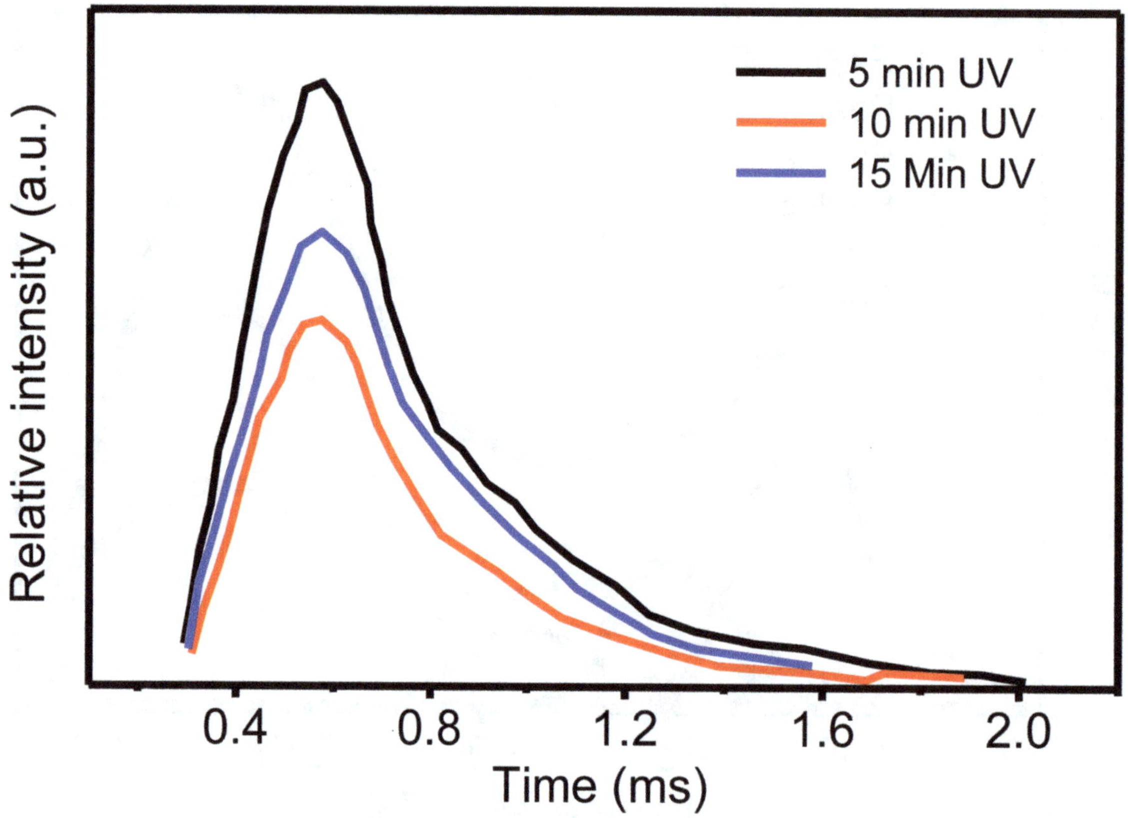

Figure 14. Dose vs. intensity plot

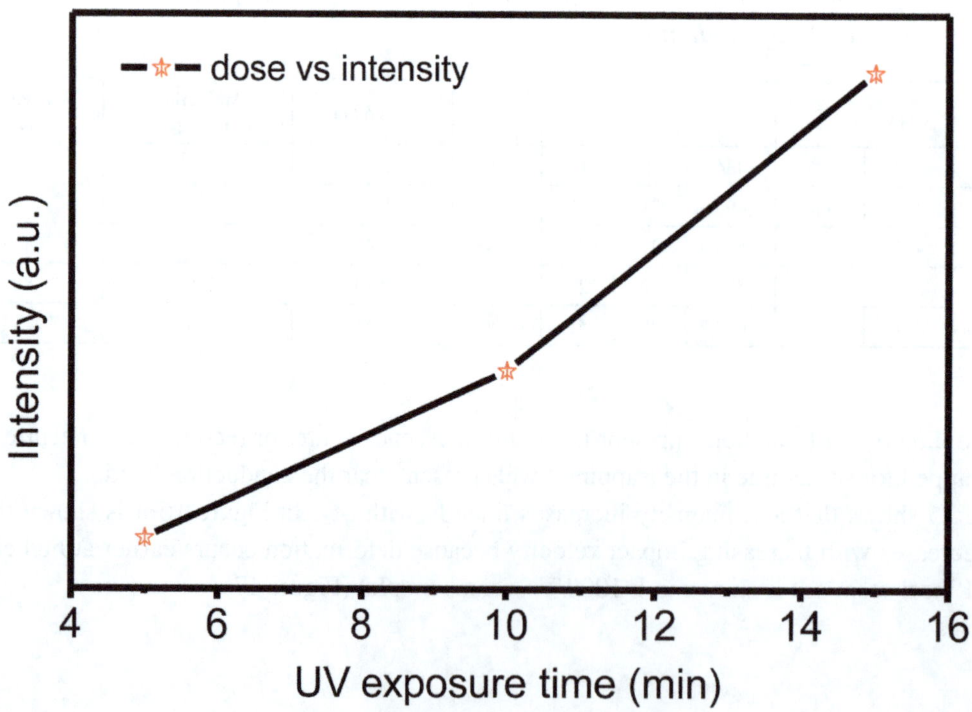

Figure 15. (t-t_m) vs. ML intensity

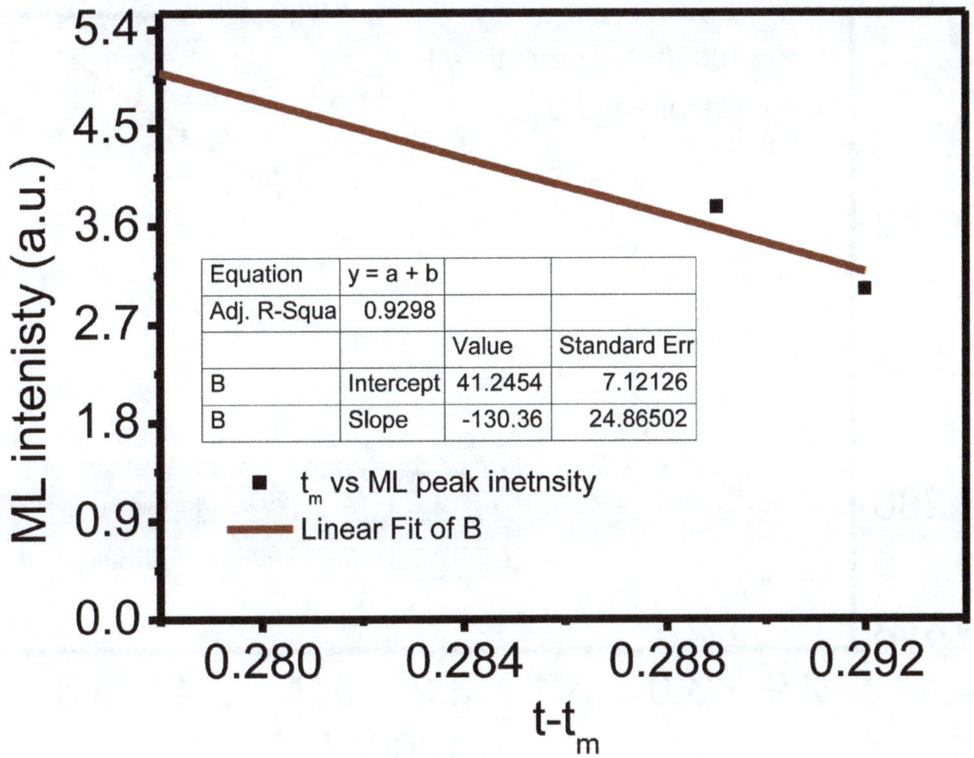

Figure 16. ML peak time(t_m) vs. impact velocity

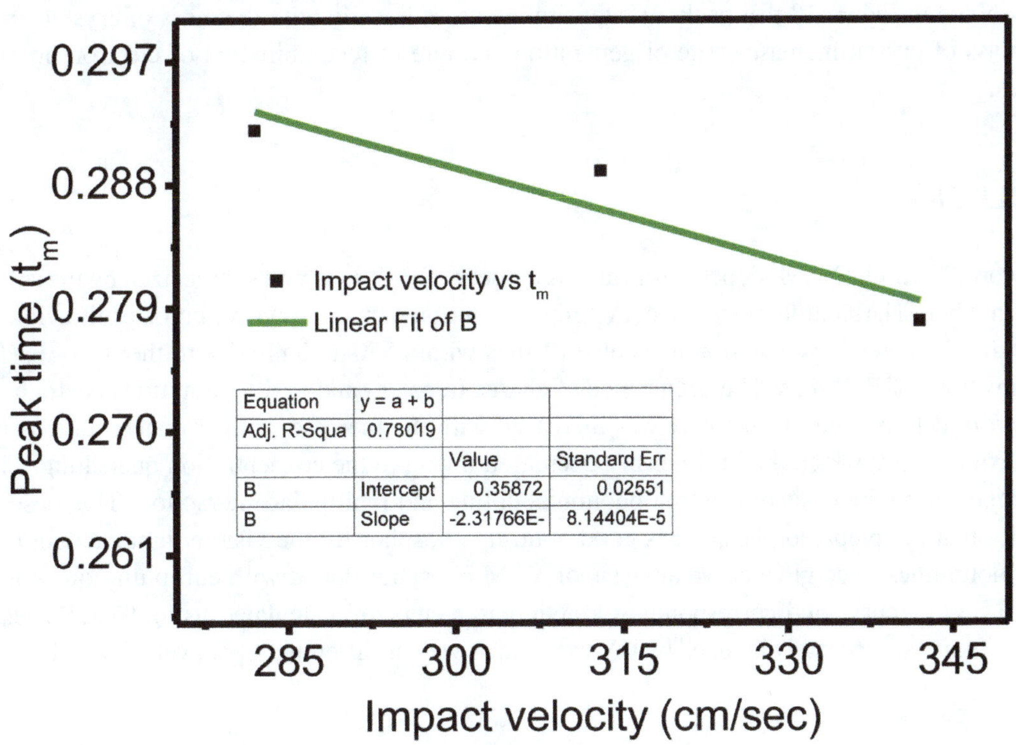

Figure 17. ln 1000/V_0 vs. peak time t_m

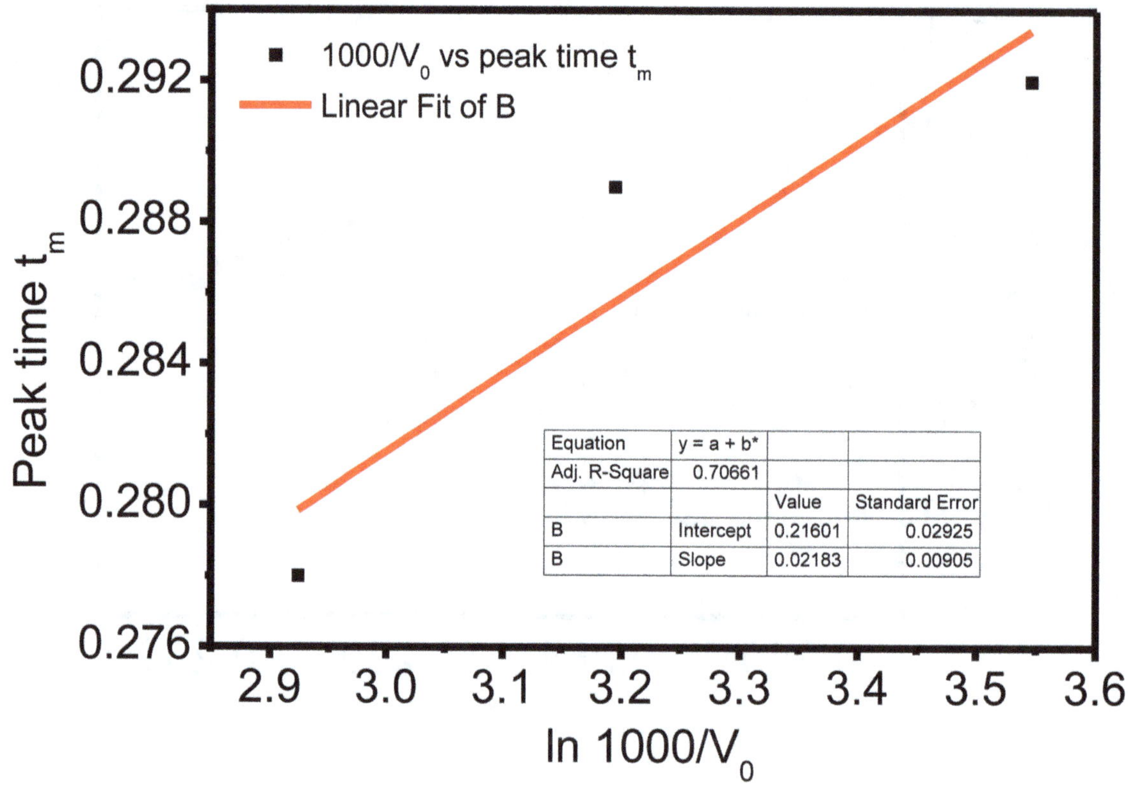

It is shown in Figure 18 that peak ml intensity increases linearly with the mass of crystals, because as the mass of crystal increases, rate of generation and rate of recombination of electron-hole pair is increases.

CONCLUSION

In this work, $Y_4Al_2O_9$ (YAM) doped with various amount of Eu^{3+} ions was synthesized by the solid state reaction method. The photoluminescence (PL) emission measurements upon excitation in ultraviolet range showed the characteristics emission lines of Eu^{3+} ions within 550–650 nm due to the $^5D_0 \rightarrow {}^7F_J$ (J=0, 1, 2, 3) transitions of Eu^{3+} ions. The occurrence of charge transfer band in the excitation spectra has been explained in details. The PL intensity was increased with the increase of Eu^{3+} concentrations up to a certain extent above which the PL intensity decreased owing to the concentration quenching. This has been explained on the basis of electron phonon coupling and multipolar interaction. This observation concludes that the prepared phosphor is good emitting phosphor for the visible light excitations. Also for thermoluminescence glow curve analysis of YAM phosphor doped with europium ion shows well resolve TL glow curve and corresponding kinetic parameters are calculated using CGCD technique. Mechanoluminescence study induced by UV exposed as well as different impact velocity and it is useful

Figure 18. Dependence of peak ml intensity on the mass of crystals

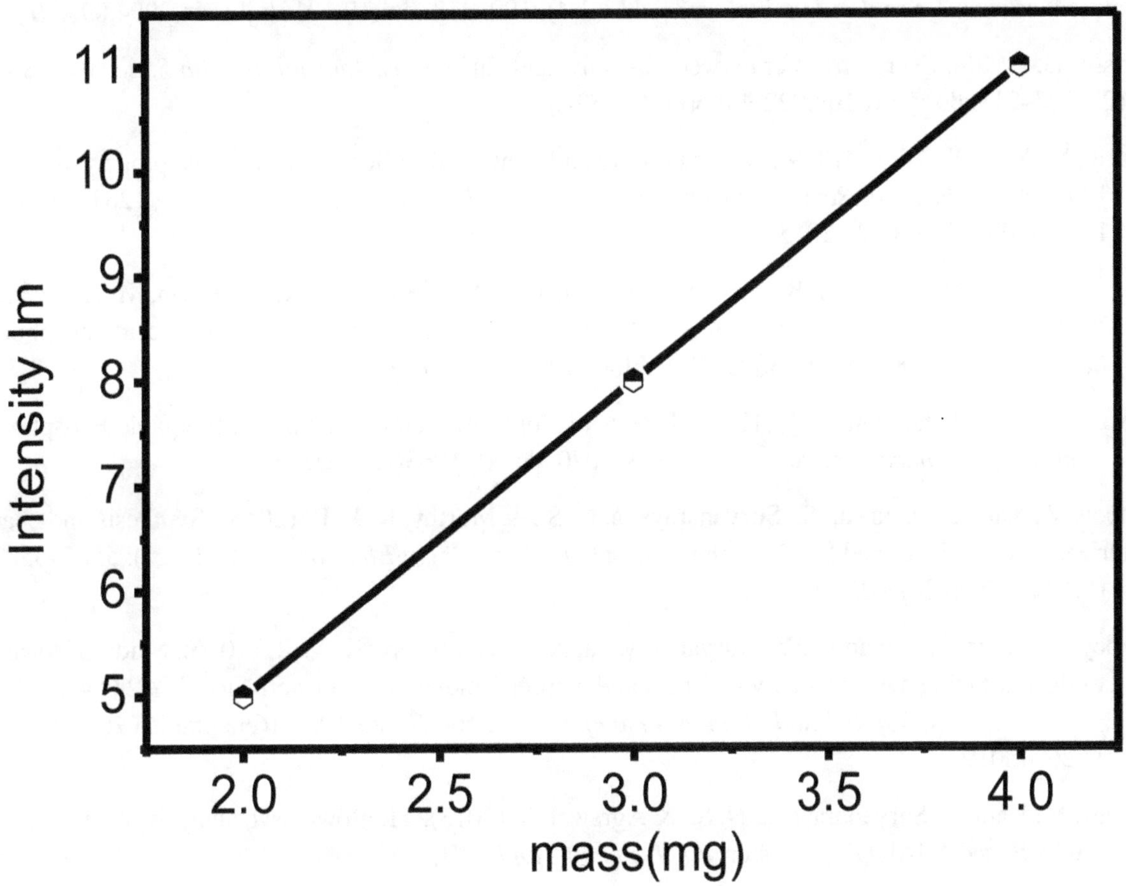

for so many applications such as mechanoluminescence (ML) dosimeter, fracture sensor, damage sensor, biological sensors etc. also our aim to calculate some spectroscopic parameters of prepared phosphor for display device applications.

REFERENCES

Aguilar, E. A., & Drew, R. A. L. (2000). Melt extraction processing of structural Y2O3–Al2O3 fibers. *Journal of the European Ceramic Society*, *20*(8), 1091–1098. doi:10.1016/S0955-2219(99)00279-4

Baldinozzi,, G., & Berar,, J-F., & Calvarin, G. (1998). Calvarin G. Rietveld refinement of two-phase Zr-doped Y_2O_3. *Materials Science Forum*, *278*, 680. doi:10.4028/www.scientific.net/MSF.278-281.680

Bhargava, R. N., Gallagher, D., Hong, X., & Nurmikko, A. (1994). Optical properties of manganese-doped nanocrystals of ZnS. *Physical Review Letters*, *72*(3), 416–419. doi:10.1103/PhysRevLett.72.416 PMID:10056425

Bizarri, G., & Moine, B. (2005). On $BaMgAl_{10}O_{17}$: Eu^{2+} phosphor degradation mechanism: thermal treatment effects. *Journal of Luminescence*, *113*(3-4), 199–213. doi:10.1016/j.jlumin.2004.09.119

Blasse, G. (1986). Energy transfer between inequivalent Eu2+ ions. *Journal of Solid State Chemistry*, *62*(2), 207–211. doi:10.1016/0022-4596(86)90233-1

Ching, W. Y., & Xu, Y. N. (1999). Nonscalability and nontransferbility in the eletronic properties of the Y-Al-O systems. *Physical Review B: Condensed Matter and Materials Physics*, *59*(20), 12815–12821. doi:10.1103/PhysRevB.59.12815

Choe, J. Y., Ravichandran, D., Blomquist, S. M., Morton, D. C., Kirchner, K. W., Ervin, M. H., & Lee, U. (2001). Alkoxy sol-gel derived $Y_{3-x}Al_5O_{12}$:Tbx thin films as efficient cathodoluminescent phosphors. *Applied Physics Letters*, *78*(24), 3800–3802. doi:10.1063/1.1378313

Dexter, D. L., & Schulman, J. H. (1954). Theory of Concentration Quenching in Inorganic Phosphors. *The Journal of Chemical Physics*, *22*(6), 1063–1070. doi:10.1063/1.1740265

Dubey, V., Kaur, J., Agrawal, S., Suryanarayana, N. S., & Murthy, K. V. R. (2013). Synthesis and characterization of Eu3+ doped SrY2O4 phosphor, *Optik– Int. J.Light Electron Opt.*, *124*(22), 5585–5587. doi:10.1016/j.ijleo.2013.03.153

Dubey, V., Kaur, J., Parganiha, Y., Suryanarayana, N. S., & Murthy, K. V. R. (2016). Study of formation of deep trapping mechanism by UV, beta and gamma irradiated Eu 3+ activated SrY 2 O 4 and Y 4 Al 2 O 9 phosphors. *Applied Radiation and Isotopes*, *110*, 16–27. doi:10.1016/j.apradiso.2015.12.047 PMID:26748019

Dubey, V., Kaur, J., Suryanarayana, N. S., & Agrawal, S. (2013). TL glow curve analysis of beta irradiated Eu3+ doped Y4Al2O9 phosphor. *Int. J. Lumin. Appl*, *3*(II), 98.

Dubey, V., Tiwari, R., Tamrakar, R. K., Kaur, J., Dutta, S., Das, S., ... Som, S. (2016). Estimation of spectroscopic parameters and colour purity of the red-light-emitting YBa 3 B 9 O 18 phosphor: Judd–Ofelt approach. *Journal of Luminescence*, *180*, 169–176. doi:10.1016/j.jlumin.2016.08.029

Dutta, S., Som, S., & Sharma, S. K. (2015). Excitation spectra and luminescence decay analysis of K^+ compensated Dy^{3+} doped $CaMoO_4$ phosphors. *RSC Advances*, *5*(10), 7380–7387. doi:10.1039/C4RA12447B

Feng, J., & Zhang, H. (2013). Hybrid materials based on lanthanide organic complexes: A review. *Chemical Society Reviews*, *42*(1), 387–410. doi:10.1039/C2CS35069F PMID:22977887

Glodo, J., & Wojtowicz, A.J. (2000). Thermoluminescence and scintillation properties of LuAP and YAP. *J. Alloys. Compd.*, *300-301*, 289.

Han, B., Liang, H., Ni, H., Su, Q., Yang, G., Shi, J., & Zhang, G. (2009). Intense red light emission of Eu^{3+}-doped $LiGd(PO_3)_4$ for mercury-free lamps and plasma display panels application. *Optics Express*, *17*(9), 7138–7144. doi:10.1364/OE.17.007138 PMID:19399090

Huang, K.-W., Chen, W.-T., Chu, C.-I., Hu, S.-F., Sheu, H.-S., Cheng, B.-M., ... Liu, R.-S. (2012). Controlling The Activator Site To Tune Europium Valence in Oxyfluoride Phosphors. *Chemistry of Materials*, *24*(11), 2220–2227. doi:10.1021/cm3011327

Jeong, J. H., Moon, B. K., Seo, H. J., Bae, J. S., Yi, S. S., Kim, I. W., & Park, H. L. (2003). Stark-shift modulation absorption spectroscopy of single quantum dots. *Applied Physics Letters*, *83*, 1346. doi:10.1063/1.1602169

Judd, B. R. (1962). Optical Absorption Intensities of Rare-Earth Ions. *Physical Review*, *127*(3), 750–761. doi:10.1103/PhysRev.127.750

Kang, Y. C., Chung, Y. S., & Park, S. B. (1999). Preparation of YAG:Europium Red Phosphors by Spray Pyrolysis Using a Filter-Expansion Aerosol Generator. *Journal of the American Ceramic Society*, *82*(8), 2056–2060. doi:10.1111/j.1151-2916.1999.tb02040.x

Kaur, J., Dubey, V., & Suryanarayana, N. S. (2015). Comparative study of ML and PL spectra of different impurity-doped (Zn, Cd) S mixed phosphors. *Research on Chemical Intermediates*, *39*(9), 4337–4349. doi:10.100711164-012-0950-x

Kaur, J., Suryanarayana, N. S., & Dubey, V. (2010a). Effect of temperature on the ML of Au doped (Zn, Cd) S mixed phosphors. *Chinese Chemical Letters*, *22*(6), 709–712. doi:10.1016/j.cclet.2010.12.006

Kaur, J., Suryanarayana, N. S., & Dubey, V. (2010b). ML spectra of gold doped (Zn, Cd) S mixed phosphors. *Chalcogenide Letters*, *7*(5), 329–334.

Kaur, J., Suryanarayana, N. S., Dubey, V., Agrawal, S., Sriwastawa, R., Jaykumar, B., ... Murthy, K. V. R. (2012). TL glow curve analysis of beta irradiated Eu3+ doped Y4Al2O9 phosphor. *Int. Jour. Lumi. Appl.*, *2*(I), 60–62.

Kelley, T. W., Diaz, A. L., & Dang, T. A. (2006). Quantitative Determination of Nonradiative Host-to-Activator Energy Transfer Efficiencies in YBO_3:Eu^{3+} and Y_2O_3:Eu^{3+} under Vacuum Ultraviolet Excitation. *Chemistry of Materials*, *18*(13), 3130–3136. doi:10.1021/cm0603831

Kodaira, C. A., Brito, H. F., Malta, O. L., & Serra, O. A. (2003). Luminescence and energy transfer of the europium (III) tungstate obtained via the Pechini method. *Journal of Luminescence*, *101*(1-2), 11–21. doi:10.1016/S0022-2313(02)00384-8

Lee, K.-G., Yu, B.-Y., Pyun, C.-H., & Mho, S.-I. (2002). Vacuum ultraviolet excitation and photoluminescence characteristics of (Y,Gd)Al3(BO3)4/Eu3+. *Solid State Communications*, *122*(9), 485–488. doi:10.1016/S0038-1098(02)00195-3

Liu, L., Zhang, B., Zhang, F., Pan, S., Zhang, F., Zhang, X., ... Yang, Z. (2015). $Pb_6Ba_2(BO_3)_5X$ (X = Cl, Br): New borate halides with strong predicted optical anisotropies derived from Pb^{2+} and $(BO_3)^{3-}$. *Dalton Transactions (Cambridge, England)*, *44*(15), 7041–7047. doi:10.1039/C5DT00641D PMID:25785911

Liu, Q., Liu, Y., Yang, Z., Han, Y., Li, X., & Fu, G. (2012). Multiwavelength excited white-emitting phosphor Dy3+-activated Ba3Bi(PO4)3. *Alloys Compd.*, *515*, 16–19. doi:10.1016/j.jallcom.2011.11.114

Liu, S., & Su, Q. (1997). Sol-gel synthesis and luminescence of $Y_4Al_2O_9$:RE^{3+} (RE = Eu, Tb). *Journal of Alloys and Compounds*, *255*(1-2), 102–105. doi:10.1016/S0925-8388(96)02866-6

Liu, X., Li, L., Noh, H. M., Moon, B. K., Choi, B. C., & Jeong, J. H. (2014). Chemical bond properties and charge transfer bands of O(2-)-Eu(3+), O(2-)-Mo(6+) and O(2-)-W(6+) in Eu(3+)-doped garnet hosts Ln3M5O12 and ABO4 molybdate and tungstate phosphors. *Dalton Transactions (Cambridge, England)*, *43*(23), 8814–8825. doi:10.1039/c4dt00674g PMID:24781975

Lo, J. R., & Tseng, T. Y. (1998). Phase development and activation energy of the Y_2O_3-Al_2O_3 system by a modified sol-gel process. *Materials Chemistry and Physics*, *56*(1), 56–62. doi:10.1016/S0254-0584(98)00139-4

Lu, C. H., & Jagannathan, R. (2002). Cerium-ion-doped yttrium aluminium garnet nanophosphors prepared through sol-gel pyrolysis for luminescent lighting. *Applied Physics Letters*, *80*(19), 3608–3610. doi:10.1063/1.1475772

Lupei, V., Pavel, N., & Taira, T. (2002). Highly efficient continuous-wave 946-nm Nd:YAG laser emission under direct 885-nm pumping. *Applied Physics Letters*, *81*(15), 2677–2679. doi:10.1063/1.1511811

Mishra, Y. K., Avasthi, D. K., Kulriya, P. K., Singh, F., Kabiraj, D., & Tripathi, A. (2007), Controlled growth of gold nanoparticles induced by ion irradiation: An in situ x-ray diffraction study. *Appl. Phys. Lett.*, *90*, 073110 (1-3).

Mishra, Y. K., Kabiraj, D., Avasthi, D. K., & Pivin, J. C. (2007). Swift heavy ion-induced dissolution of gold nanoparticles in silica matrix. *Radiation Effects and Defects in Solids*, *162*(3-4), 207–213. doi:10.1080/10420150601132883

Ofelt, S. (1962). Intensities of Crystal Spectra of Rare-Earth Ions. *The Journal of Chemical Physics*, *37*(3), 511–520. doi:10.1063/1.1701366

Okazaki, C., Shiiki, M., Suzuki, T., & Suzuki, K. (2000). Luminescence saturation properties of PDP phosphors. *Journal of Luminescence*, *87-89*, 1280–1282. doi:10.1016/S0022-2313(99)00575-X

Parchur, A. K., & Ningthoujam, R. S. (2012). Behaviour of electric and magnetic dipole transitions of Eu^{3+}, $^5D_0 \rightarrow {}^7F_0$ and Eu–O charge transfer band in Li^+ co-doped YPO_4:Eu^{3+}. *RSC Advances*, *2*(29), 10859–10868. doi:10.1039/c2ra22144f

Rabinovitch, Y., Moune, O. K., Tetard, D., & Faucher, M. D. (2004). Synthesis, Optical Absorption, and Site-Selective Excitation of the $^3P_0$ Levels in $Y_4Al_2O_9$:Pr^{3+}. *The Journal of Physical Chemistry A*, *108*(40), 8244–8255. doi:10.1021/jp048123j

Rao, R. P., & Devine, D. J. (2000). RE-activated lanthanide phosphate phosphors for PDP applications. *J. Lumin. 87-89*, 1260–1263.

Reddy, A. A., Das, S., Ahmad, S., Babu, S. S., Ferreira, J. M. F., & Prakash, G. V. (2012). Influence of the annealing temperatures on the photoluminescence of $KCaBO_3$:Eu^{3+} phosphor. *RSC Advances*, *2*(23), 8768–8776. doi:10.1039/c2ra20866k

Ronda, C. R. (1997). Recent achievements in research on phosphors for lamps and displays. *Journal of Luminescence*, *49*, 72.

Saluja, J. K., Parganiha, Y., Tiwari, N., Dubey, V., Tiwari, R., & Prabhath, A. (2016). Mechano and photoluminescence spectra of cadmium sulphide and cadmium selenide doped phosphors. *Optik-International Journal for Light and Electron Optics*, *127*(19), 7958–7966. doi:10.1016/j.ijleo.2016.05.011

Shrivastava, R., Kaur, J., Dubey, V., & Jaykumar, B. (2014). Photoluminescence, trap states and thermoluminescence decay process study of Ca2MgSi2O7: Eu2+, Dy3+ phosphor. *Bulletin of Materials Science*, *37*(4), 925–929. doi:10.100712034-014-0027-0

Singh, I. R., Ningthoujam, R. S., Sudarsan, V., Shrivastav, I., Singh, S. D., Dey, G. K., & Kulshreshtha, S. K. (2008). Luminescence study on Eu^{3+} doped Y_2O_3 nanoparticles: Particle size, concentration and core–shell formation effects. *Nano Tech.*, *19*, 1.

Sohn, K. S., Kim, S., & Park, H. D. (2002). Luminescence quenching in thermally-treated barium magnesium aluminate phosphor. *Applied Physics Letters*, *81*(10), 1759–1761. doi:10.1063/1.1504866

Som, S., Choubey, A., & Sharma, S. K. (2015). Spectral and trapping parameters of Eu in Gd_2O_2S nanophosphor. *Journal of Experimental Nanoscience*, *10*(5), 350–370. doi:10.1080/17458080.2013.837972

Som, S., Chowdhury, M., & Sharma, S. K. (2015). Kinetic parameters of γ-irradiated Y2O3 phosphors: Effect of doping/codoping and heating rate. *Radiation Physics and Chemistry*, *110*, 51–58. doi:10.1016/j.radphyschem.2015.01.015

Som, S., Kunti, A. K., Kumar, V., Kumar, V., Dutta, S., Chowdhury, M., ... Swart, H. C. (2014). Defect correlated fluorescent quenching and electron phonon coupling in the spectral transition of Eu3+ in CaTiO3 for red emission in display application. *Journal of Applied Physics*, *115*(19), 193101. doi:10.1063/1.4876316

Tian, Z., Liang, H., Lin, H., Su, Q., Guo, B., Zhang, G., & Fu, Y. (2006). Luminescence of $NaGdFPO_4:Ln^{3+}$ after VUV excitation: A comparison with $GdPO_4:Ln^{3+}$ (Ln=Ce, Tb). *Journal of Solid State Chemistry*, *179*(5), 1356–1362. doi:10.1016/j.jssc.2006.01.044

Tiwari, N., Dubey, V., & Kuraria, R. K. (2015). Mechanoluminescence Study of Europium Doped CaZrO3 Phosphor. *Journal of Fluorescence*, *26*(4), 1309–1315. doi:10.100710895-016-1817-0 PMID:27184974

Tiwari, R., Dubey, V., & Chandra, B. P. (2014). Exact model for the elastico mechanoluminescence of II-VI phosphors. *Materials Physics and Mechanics*, *19*(1), 25–38.

Vodopyanov, K. L., Shori, R., & Stafsudd, O. M. (1998). Generation of **Q**-switched Er:YAG laser pulses using evanescent wave absorption in ethanol. *Applied Physics Letters*, *72*(18), 2211–2213. doi:10.1063/1.121325

Wang, D. Y., & Wang, Y. H. (2006). Photoluminescence of $Y_4Al_2O_9$:Re (Re= Tb^{3+}, Eu^{3+}) under VUV excitation. *Journal of Alloys and Compounds*, *425*(1-2), L5–L7. doi:10.1016/j.jallcom.2006.01.033

Wang, D.-Y., & Wang, Y.-H. (2006). Photoluminescence of Y4Al2O9:Re (Re=Tb3+, Eu3+) under VUV excitation. *Journal of Alloys and Compounds*, *425*(1-2), L5–L7. doi:10.1016/j.jallcom.2006.01.033

Xia, G., Zhou, S., Zhang, J., Wang, S., Hong, W., & Jun, X. (2005). Sol–gel combustion synthesis and luminescence of Y4Al2O9:Eu3+ nanocrystal. *Journal of Non-Crystalline Solids*, *351*(37-39), 2979–2982. doi:10.1016/j.jnoncrysol.2005.07.007

Xia, G., Zhou, S., Zhang, J., Wang, S., Wang, H., & Xu, J. (2005). Sol–gel combustion synthesis and luminescence of $Y_4Al_2O_9$:Eu^{3+} nanocrystal. *Journal of Non-Crystalline Solids*, *351*(37-39), 2979–2982. doi:10.1016/j.jnoncrysol.2005.07.007

Yadav, R., Khan Atif, F., Yadav, A., Chander, H., Haranath, D., Gupta, B. K., ... Chawla, S. (2009). Intense red-emitting $Y_4Al_2O_9$:Eu^{3+}phosphor with short decay time and high color purity for advanced plasma display panel. *Optics Express*, *17*(24), 22023. doi:10.1364/OE.17.022023 PMID:19997447

Yamane, H., Nagasawa, T., Murakami, Y., Kamata, T., Shindo, D., Shimada, M., & Endo, T. (1998b). Crystal Structure and Thermal Expansion of Ca2Y2Si2O9. *Materials Research Bulletin*, *33*(6), 845–853. doi:10.1016/S0025-5408(98)00059-2

Yamane, H., Shimada, M., & Hunter, B. A. (1998a). High-Temperature Neutron Diffraction Study of Y4Al2O9. *Journal of Solid State Chemistry*, *141*(2), 466–474. doi:10.1006/jssc.1998.7998

Yang, H. M., Shi, J. X., Liang, H. B., & Gong, M. L. (2006). Novel red phosphor Mg_2GeO_4 doped with Eu^{3+} for PDP applications. *J. Mater. Sci. Eng. B*, *127*(2-3), 276–279. doi:10.1016/j.mseb.2005.10.014

Zeng, Q., Tanno, H., Egoshi, K., Tanamachi, N., & Zhang, S. (2006). Ba_5SiO_4:Eu^{2+}: An intense blue emission phosphor under vacumm ultraviolet and near-ultraviolet excitation. *Applied Physics Letters*, *88*(5), 051906. doi:10.1063/1.2168492

Zhang, S., Hou, Y., Fujii, H., Onishi, T., Kokubu, M., Obata, M., Tanno, H., Kono, T., & Uchiike, H. (2003). Effect of Nonstoichiometry on the Deterioration of Eu^{2+} -Doped Hexagonal Aluminate Phosphor for Plasma Display Applications. *Jpn. J. Appl. Phys.*, *42*(Part 1, No. 2A), 477–480.

Zhang, W. P., Xie, P. B., Duan, C. K., Yan, K., Yin, M., Lou, L. R., ... Krupa, J. C. (1998). Preparation and size effect on concentration quenching of nanocrystalline Y2SiO5:Eu. *Chemical Physics Letters*, *292*(1-2), 133–136. doi:10.1016/S0009-2614(98)00656-3

Chapter 11
Electroluminescence Principle and Novel Electroluminescent Materials

Vikas Lahariya
Amity University, India

Meera Ramrakhiani
Rani Durgavati Vishwavidyalaya, India

ABSTRACT

Over the past few decades, many efforts have been put to optimize the properties of electroluminescent devices such as electroluminescence (EL) panel, organic light emitting diode (OLED), flat panel display, EL lamps, etc. In this chapter, the authors provide a comprehensive review of the state-of-the-art research activities related to EL phenomena, the principle of electroluminescence, different types of EL, fabrication of devices, studies on novel electroluminescent materials, their characteristics and potential applications. The authors begin with a historical background of electroluminescence, description of its structure, working principle and parameters with mechanism. They discuss the experiments determining the electrical, optical, and physical properties of the powder as well as thin film EL cells. The recent progress on the improvement of their characteristics and finding novel structures, such as the latest achievements in using various semiconductor nanostructures in polymer matrix as emitting layer in EL devices, are summarized.

INTRODUCTION

Although electroluminescence (EL) has been known for more than a century, its utilization has become practical only within the last few decades. The major attributes of EL devices are compactness, ruggedness, and long operational life. But, progress remains to be achieved mostly with respect to efficiency and cost, and, for large-area displays, compatibility with solid-state integrated circuits.

Electroluminescence (EL) is an optical and electrical phenomenon where a material emits light in response to an electric current passed through it, or to an electric field. The radiation is usually in the

DOI: 10.4018/978-1-5225-5170-6.ch011

visible region of the electromagnetic spectrum. However, the same processes may yield infrared or ultraviolet radiation. It was first discovered by a French scientist Geroge Distrau in 1936, (Destriau, 1936) who observed a feeble light from ZnS:Cu powder embedded in transparent insulating binding medium with alternating electric field (a.c.) field. This is distinct from light emission resulting from heat (incandescence) or from the action of chemicals (chemoluminescence), sound (sonoluminescence), or other mechanical action (mechanoluminescence).

Electroluminescence is the result of radiative recombination of electron and holes in a material (usually a semiconductor). The excited electrons release their energy as photons i.e., light. Electrons and holes are separated either as a result of doping of the material to form a p-n junction (in semiconductor electroluminescent devices such as LEDs), or through excitation by impact of high-energy electrons accelerated by a strong electric field (as with the phosphors in electroluminescent displays ELDs). In this chapter, we will provide a comprehensive review of the state-of--art research activities related to EL phenomena, the principle of electroluminescence, different types of EL, fabrication of devices, studies on novel electroluminescent materials, their characteristics and potential applications. We begin with a historical background of Electroluminescence, description of its structure, working principle and parameters with mechanism. The recent progress on the improvement of their characteristics and finding novel structures, such as the latest achievements in using various semiconductor nanostructures in polymer matrix as emitting layer in EL devices, will be discussed.

ELECTROLUMINESCENCE DEVICES

The electroluminescence (EL) devices include light emitting diodes, which are discrete devices that produce light when a current is passed through a p-n junction of a semiconductor, as well as EL displays (ELDs) which are matrix-addressed devices that can be used to display text, graphics, and other computer images. EL is also used in lamps and backlights. EL lamps are low power and emit a soft light without any irritating glare (Figure 1). They are invariably thin, almost like ribbons or sheets of paper, and they can be produced in a variety of colors. EL provides a cool, continuous, even light that can be shaped into stripes, strips, and geometric figures (Figure 2). Strips can run for several meters, with only one connection to a power source. Shapes can be illuminated with different colors and with flashing, animated effects. The EL panels are lamps, which radiate gently, without glare, and can be used for lighting purpose.

EL can be used with furniture, wall hangings, room trim and fixtures to uniquely enhance the appearance of any space. Buildings can be outlined. Entire walls can glow with light. Ceilings can become skies. Staircases can become shining sets of steps (Figure 3). Windowless rooms can suddenly have windows --- with views. Adding the EL panels that flash, blink, stack up or stack down and the creative possibilities are endless. EL becomes an imaginative, artistic medium offering color, motion and light as tools to inspire, educate and entertain. Relevant advertising applications include electroluminescent billboards and signs (Figure 4). EL manufacturers are able to control precisely which areas of an electroluminescent sheet illuminate, and when. This has given advertisers the ability to create more dynamic advertising which is still compatible with traditional advertising spaces. EL may be used to highlight a product or advertising message more effectively than any other form of light. Company logos or advertising displays may leap towards customers, compelling their attention to the image and the message. Information illuminated by EL will command attention in any environment. Warnings are

Figure 1. An electroluminescent nightlight in operation (uses 0.08 W at 230 V, and dates from 1960; lit diameter 59 mm)

Figure 2. Animated effects include flash or blinking where one or more lamps light up intermittently. When done in sequence, the human eye will detect movement in an image.

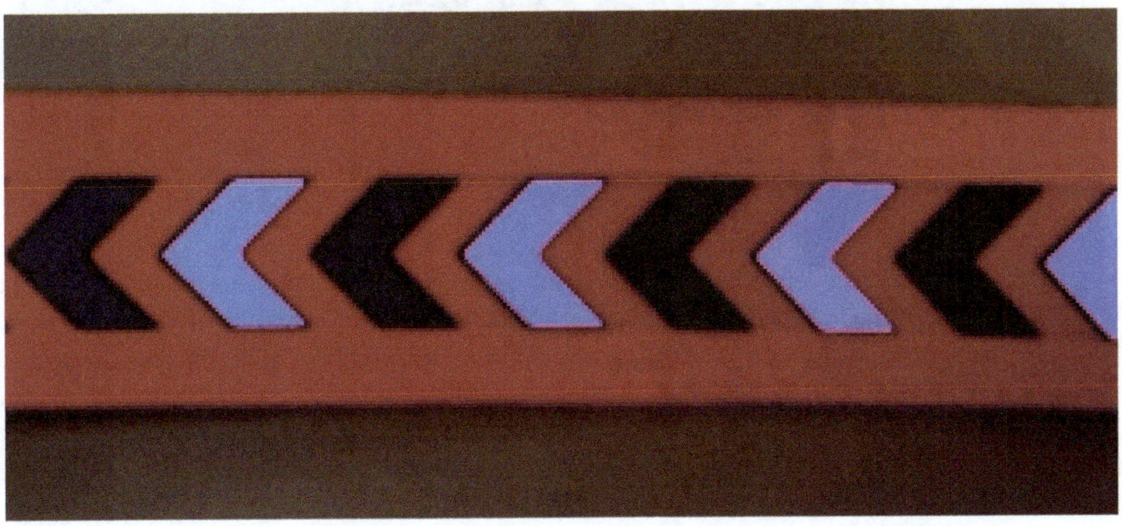

clear in an emergency. Instructions stand out to anyone in a crowd. Where power supplies are limited, EL illumination can operate for months with a small battery. EL designs may be used in products such as toys, novelties, and jewelry and fashion accessories. EL patches or panels can be sewn onto shirts,

Figure 3. Structure of EL device

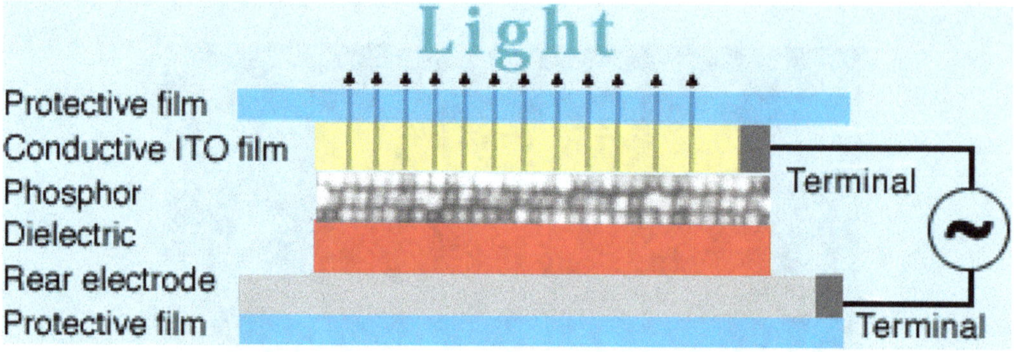

Figure 4. Construction of electroluminescent device

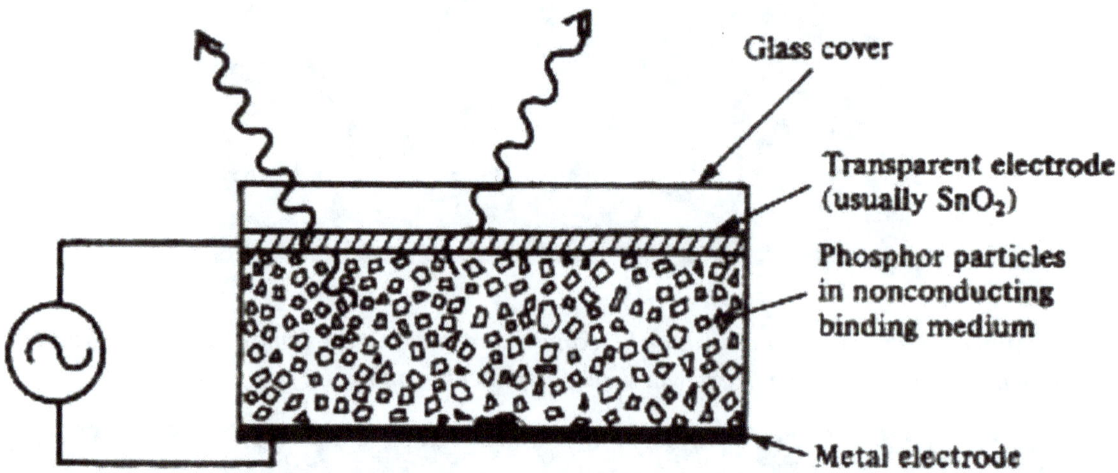

skirts, pants and hats. They can be attached to backpacks, tote bags and purses. A small EL panel can be used to illuminate any small, dark area such as a locker, a purse or a make-up kit, a tool box, a tackle box or every corner of a storage compartment

The variety of luminescent materials and phenomena is so great that sometimes the user is puzzled by the available choice. If the light is to activate an electronic system, the choice of the light source should be guided by the optimal match between the characteristics of the light source and those of the detector. However, if the emitted light is to interact with the human eye (for example in a display panel) one must be aware of the psychophysical characteristics of the eye as a detector. Recently, blue, red, and green emitting thin film electroluminescent materials have been developed that offer the potential for long life and full color electroluminescent displays. In principle, EL lamps can be made in any color. However, the commonly used greenish color closely matches the peak sensitivity of human vision, producing the greatest apparent light output for the least electrical power input. Electroluminescent technologies have low power consumption compared to competing lighting technologies, such as neon or fluorescent lamps. This, together with the thinness of the material, has made EL technology valuable to the industry and day to day applications.

DEVICES STRUCTURE

An EL film is a so-called Lambertian radiator: unlike with neon lamps, filament lamps, or LEDs, the brightness of the surface appears the same from all angles of view. Electroluminescent light is not directional and therefore hard to compare with (thermal) light sources measured in lumens or lux. The light emitted from the surface is perfectly homogeneous and is well-perceived by the eye. EL film produces single-frequency (monochromatic) light that has a very narrow bandwidth, is absolutely uniform and visible from a great distance.

The most common electroluminescent (EL) devices are composed of either powder (primarily used in lighting applications) or thin films (for information displays.)

Powder phosphor-based electroluminescent panels are frequently used as backlights to liquid crystal displays. They readily provide a gentle, even illumination to the entire display while consuming relatively little electric power. This makes them convenient for battery-operated devices such as pagers, wristwatches, and computer-controlled thermostats, and their gentle green-cyan glow is a common sight in the technological world. They do, however, require relatively high voltage (between 60 and 600 volts). For battery-operated devices, this voltage must be generated by a converter circuit within the device; this converter often makes an audible whine or siren sound while the backlight is activated. For line-voltage-operated devices, it may be supplied directly from the power line. Electroluminescent nightlights operate in this fashion. Brightness per unit area increases with increased voltage and frequency.

Thin film phosphor electroluminescence was first commercialized during the 1980s by Sharp Corporation in Japan, Finlux (Oy Lohja Ab) in Finland, and Planar Systems in the USA. Here, bright, long-life light emission is achieved in thin film yellow-emitting manganese-doped zinc sulfide material. Displays using this technology were manufactured for medical and vehicle applications where ruggedness and wide viewing angles were crucial, and liquid crystal displays were not well developed.

In EL devices the EL material is a phosphors mixture. The EL material must be enclosed between two electrodes and at least one electrode must be transparent to allow the escape of the produced light. Glass coated with indium tin oxide is commonly used as the front (transparent) electrode. It has high surface conductivity and high optical transmission while the back electrode is coated with reflective metal. Additionally, other transparent conducting materials, such as carbon nanotube (CNT) coatings or PEDOT can be used as the front electrode (Figure 5). This layer has a strong ability of transferring holes, which have been injected from the ITO anode.

An alternating current (a.c.) voltage is necessary for continuous light emission from the microcrystals suspended in an insulating medium or from luminescent materials containing electron injecting particles. Also, EL emission has been observed from thin-film luminescent materials sandwiched between two neutral insulating layers separating the film from metallic contacts, which were excited by a.c. voltage waves or pulses Light emission occurring from reverse-biased p–n junctions is an example of periodic (unidirectional) EL. Direct current dc-type EL as well as their combination can be observed by applying injecting electrodes to single crystals and thin films; carrier injection into a luminescent material under lies emission of light. Either the injection of holes and/or electrons from metallic contacts into a luminescent material or the injection of minority charge carriers at a semiconductor p–n junction produce EL under ac- and dc-excitation conditions. If the emitting excited states are produced in a recombination process of oppositely charged carriers recombination EL (REL). The injection-controlled (IC) or volume-controlled (VC) EL can be distinguished depending upon the relation between the recombination and the average transit time of the charge carriers within the sample.

Figure 5. Mechanism for EL emission involving quantum mechanical tunneling

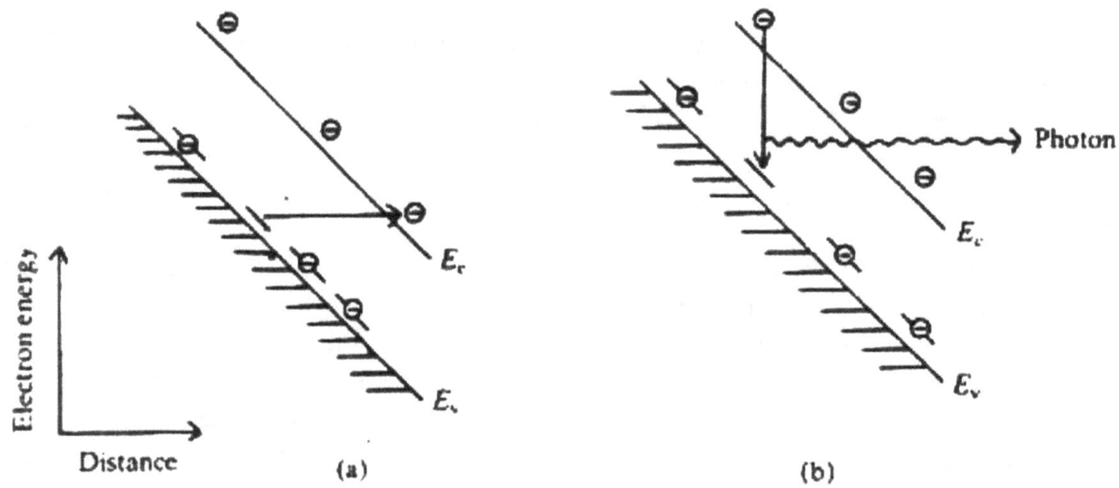

An alternating current (a.c.) / voltage is necessary for continuous light emission from the microcrystals suspended in an insulating medium or from luminescent materials containing electron injecting particles. The effect was observed for the first time with inorganic phosphors by Destriau (Destriau 1936). Also, EL emission has been observed from thin-film luminescent materials sandwiched between two insulating layers separating the film from metallic contacts, which were excited by a.c. voltage waves or pulses (Kobayshi, 1993; Mach et al 1993). Light emission occurring from reverse-biased p–n junctions is an example of periodic (unidirectional) EL. The injection EL is more common. Direct current dc-type EL as well as their combination can be observed by applying injecting electrodes to single crystals and thin films; carrier injection into a luminescent material underlies emission of light. Either the injection of holes and/or electrons from metallic contacts into a luminescent material or the injection of minority charge carriers at a semiconductor p–n junction produce EL under ac- and dc-excitation conditions. If the emitting excited states are produced by a recombination process of oppositely charged carriers, it is recombination EL (REL). The injection-controlled (IC) or volume-controlled (VC) EL can be distinguished depending upon the relation between the recombination and the average transit time of the charge carriers within the sample.

PRINCIPLE OF ELECTROLUMINESCENCE

In a simple electroluminescence device, a thin film or powder form of luminescent material is sandwiched between two electrodes. The cathode typically consists of a thin metallic film such as Ca or Ba, which can be incorporated with the emissive layer. The most commonly used anode is a semi- transparent indium tin oxide (ITO) coated glass. Both organic and inorganic materials have been used as emitting layer for EL devices. The electroluminescence depends upon a number of factors such as the concentration of activators and co-activators, the nature of applied field and the temperature during measurement. The excitation of electroluminescence involves the production of additional charge carriers.

There are four steps necessary to produce electroluminescence in ELDs:

1. Electrons tunnel from electronic states at the insulator/phosphor interface;
2. Electrons are accelerated by high electric fields in the phosphor;
3. The energetic electrons impact-ionize the luminescent center or create electron-hole pairs that lead to the activation of the luminescent center; and
4. The luminescent center relaxes toward the ground state and emits photon.

All ELDs have the same basic structure shown in Figure 6. The first layer is a base plate (usually a rigid insulator like glass), the second is a conductor, the third is a layer of phosphors, and the fourth is an insulator, and the fifth is another conductor. The insulator layers are necessary to prevent arcing between the two conductive layers. An alternating current (a.c.) is generally used to drive an ELD because the light generated by the current decays when a constant voltage is applied. Light emission is achieved by the application of large a.c. voltage to the layered structure such that electrons are injected into the high field phosphor layer from interface states at the insulator semiconductor interface due to tunneling. With sufficiently high field, electrons in the phosphor layer may gain sufficient energy to impact excite electrons in the luminescent impurities from the ground to excited state, which subsequently under go radiative decay emitting photons. As a result, the characteristic light would be emitted.

MECHANISM OF ELECTROLUMINESCENCE

The electroluminescence process categorizes on the basis of excitation and emission. Electroluminescence (EL) involves the excitation of luminescence as a result of existence of an applied electric potential difference in the phosphor. The general EL phenomena can be classified according to criteria based on three fundamental processes: (a) electrical energy supply, (b) excitation mode of emitting states and (c) light generation mechanism itself. There are several possible functions of the energy supply modes in promoting EL. The energy supply mode can determine possible mechanisms of excitation of the radiative system. For example, supplying electrical energy with insulating or non-injecting electrodes imposes

Figure 6. Mechanism for EL emission involving avalanche process

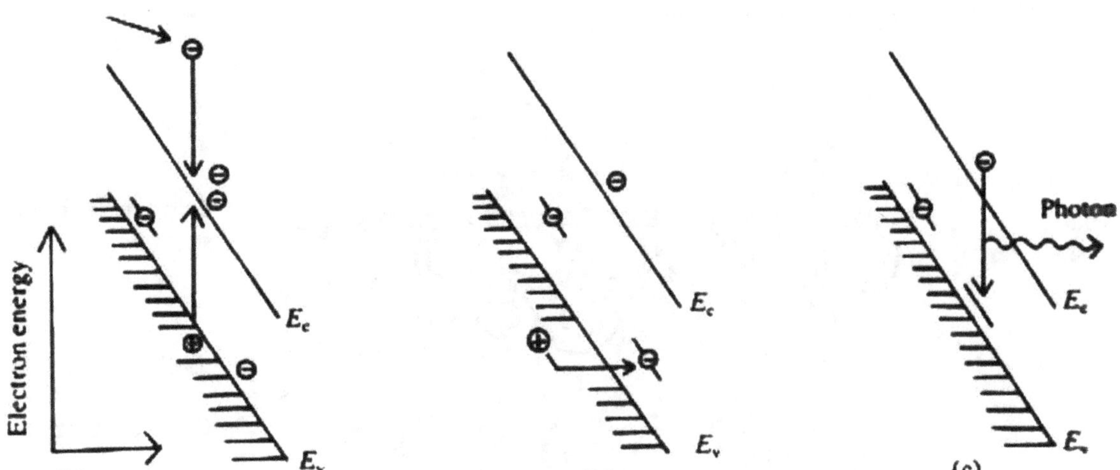

field-induced creation of excited states (high-field EL) or the generation of charge carriers inside the EL material, leading to charge-carrier-mediated EL.

Four main types of EL devices may be distinguished depending on the type of drive (a.c. or d.c.) and the character of active layer (powder or thin film). The first electroluminescent device to be extensively studied was the a.c. powder device, proposed in 1936 (Distriau,1936). In this a phosphor powder (usually ZnS:Cu) is suspended in a transparent insulating binding medium of high dielectric constant and is sandwiched between two electrodes (one of which is transparent) as shown in Figure 7. Usually there is no complete conducting path between the electrodes, so that d.c excitation is not possible. When an alternating voltage, $V_0 \cos(2\pi ft)$, is applied across the cell, however, light is emitted in the form of short bursts which last about 10^{-3} s and occur once every half cycle. It is found that the integrated light output power P can be written in the form

$$P=P_0(f)\exp[-(v_1/v_0)^{\wedge}1/2] \tag{1}$$

where V_1 is a constant and $P_0(f)$ is a function of frequency. The strongest emission from within the phosphor grain is found to take place from the side temporarily facing the cathode. Several possible emission mechanisms have been proposed; it is generally agreed, however, that there will be a high electric field within the phosphor particle. It is then possible that this field is sufficiently strong to enable electrons from occupied acceptor levels to 'tunnel' to states of the same energy in the conduction band (Figure 8a). Other electrons in the conduction band are then able to fall into these vacated levels and emit radiation (Figure 8b).

Another possibility is that an electron moving in the electric field may acquire sufficient energy to enable it to excite an electron from the valance band to the conduction band. The resulting hole quickly becomes trapped at an impurity acceptor site effectively emptying it of an electron. An electron in the conduction band can make a radiative transition by falling into the empty acceptor level (Figure 9). In phosphors containing manganese, there is evident that the Mn^{2+} ions themselves may be directly excited by the high energy electron, radiation being emitted when the ions subsequently undergoes de-excitation. The powder devices usually require several hundreds of volts (r.m.s) to drive them. They exhibit lumi-

Figure 7. Construction of d.c. electroluminescent device

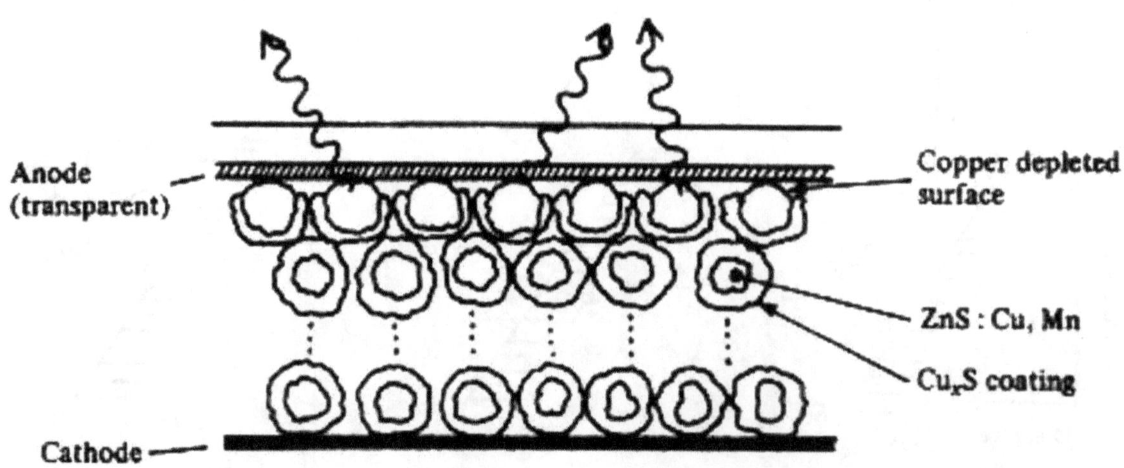

Figure 8. Types of Electroluminescence

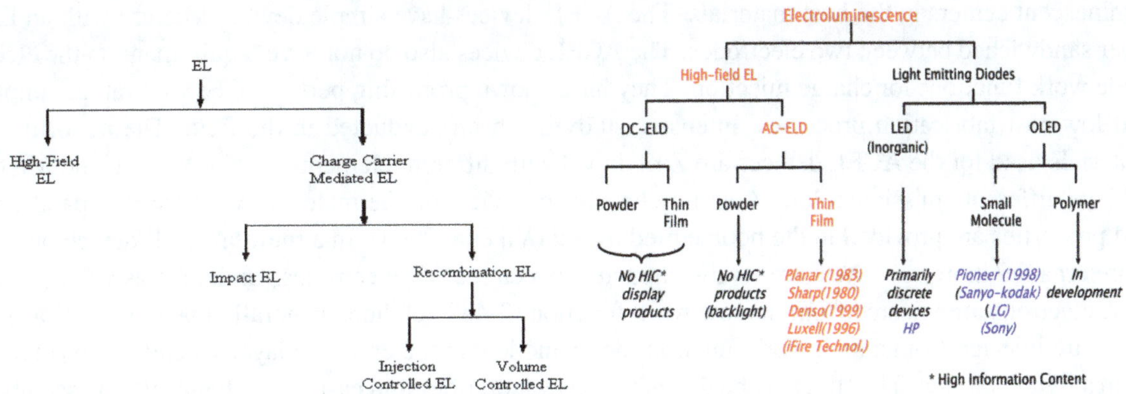

Figure 9. Electroluminescence by radiative recombination

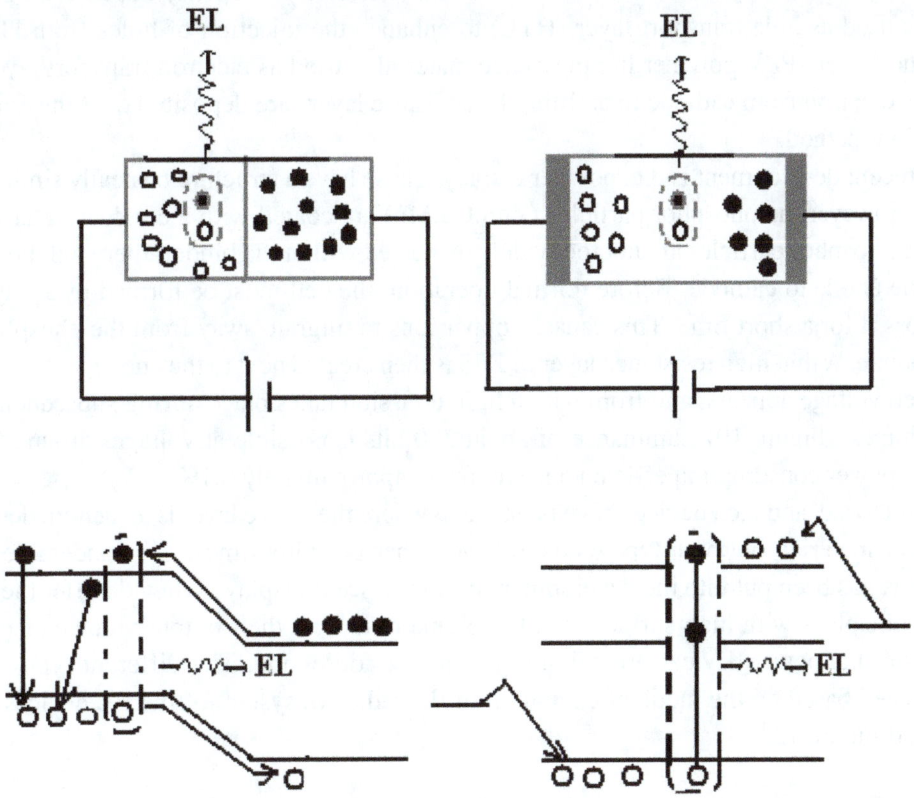

nance of about 40 nits, have power efficiencies (i.e. the ratio of optical power out to electrical power in) of about 1% and have lifetimes of about 1000h. By using different phosphor powders, red, green, yellow and blue displays are possible.

In AC Electroluminescence (ACEL) the emission behavior includes multiple steps in sequence: charge injection into the phosphor layers, acceleration of the charge carriers under high electrical field, impact excitation of the luminescent centers by high-energy electrons, and radiative relaxation of the luminescent centers in the host materials. The ACEL devices have simple device structure with an EL layer sandwiched between two electrodes. The ACEL devices also do not have requirement on the electrode work functions for charge injection. They have shown promising performance with rather simple and low-cost fabrication processes. Intensive study has been conducted in the field. The majority of materials used for the ACEL devices are ZnS doped with different elements such as Cu, Al, and Mn to achieve different emission colors. A comprehensive overview on the materials with their preparations and properties are provided in the book edited by Vij (Vij et al.2004). In a multilayer EL device one or more organic/ inorganic layers are sandwiched in between the two electrodes. Under a bias voltage 2V-30V, electrons are injected from the low work function (3-4eV) cathode generally made by Al /Ca and holes are injected from a high work function (5eV) anode into the emissive layer where they meet and form a light emission. The film may be deposited by evaporation, spin casting or chemical self assembly (Manzooret al,2005; Nayak et al 2010, M.Gao et al 1998). The basic features required for a emitting layer are that it must be good charge transporter and efficient luminescer.

For multilayer generally PEDOT:PSS (Poly 3,4 ethylenedioxy thiopene) doped with polylstyrene sulfonate) is used as hole transport layer (HTL) to enhance the injection of holes from ITO to emitting layer and MEH –PPV polymer luminescence material is used as electron transport layer (ETL) to transport the electron from cathode to emitting layer. These layers are deposited over the emitting layer by evaporation method.

A more recent development is d.c. powder display. These have a structure basically similar to that of a.c. device. However the phosphor particles (ZnS:Cu,Mn) are coated with a conducting layer of Cu_2S. Provided the phosphor particles are not too widely dispersed within the binder, there will be conducting path from the anode to cathode. Before normal operation, the cell must be formed by applying a high voltage across it for a short time. This causes copper ions to migrate away from the phosphor surfaces next to the anode. A thin high resistance layer of ZnS is then created next to the anode, across which most of the applied voltage appears, and from which light emission takes place during subsequent operation at lower voltages. (Figure 10) Luminance of about 300 nits is possible at voltages around 100 V d.c., although the power consumption efficiencies are low at approximately 0.1%.

In addition to a.c. and d.c., devices have been made where the active layer is a vacuum deposited thin film phosphor material, but both type tends to have rather poor life times. A considerable amount of research effort has been put into the development of luminescent displays. They do offer the possibility of large area displays with high surface area to volume ratio, and their output voltage characteristics, particularly of d.c. power devices are suitable for matrix addressing. The different types of EL may be distinguished based on the mode of excitation of the radiative system of the EL devices is given in Figure 11 and Figure 12.

High-Field Electroluminescence

This type of EL is defined by the assumption that an external electric field excites emitting states directly. In quantum-mechanical terms, this excitation mode is described by the probability of excitation of localized states or field ionization from the valence band (VB) (Zener effect). Free charge carriers present in a luminescent material can be accelerated by such high fields to energies of a few electron volts, leading to collision excitation of luminescence. This is often called high-field EL.

Figure 10. Light emitting device structure

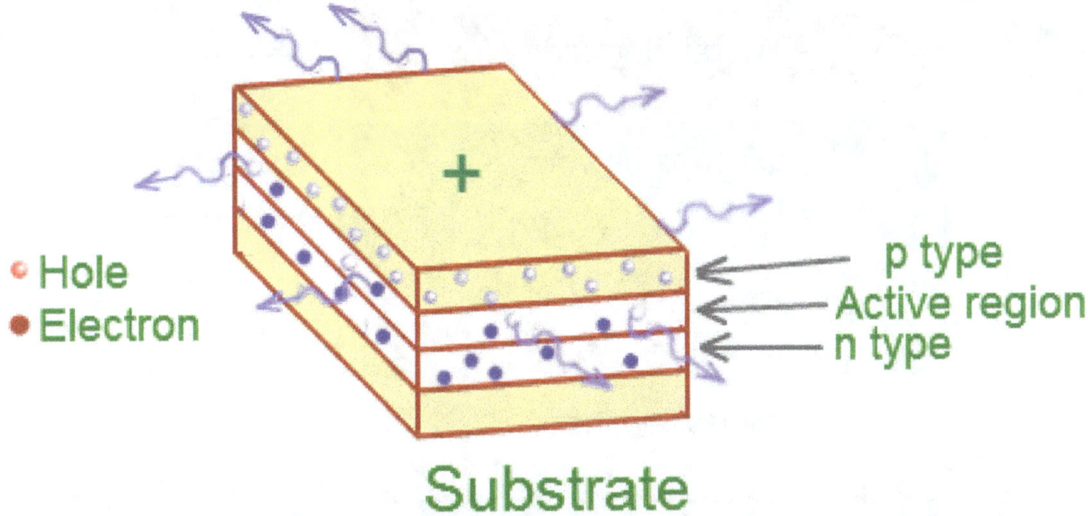

Figure 11. Standard and QD added LED bulb emission spectra
Source: QD Vision

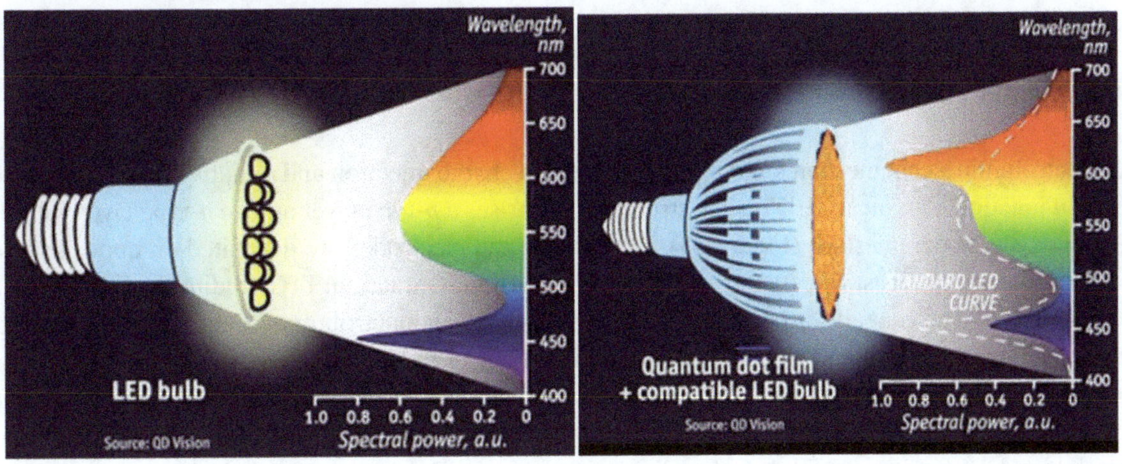

Charge-Carrier-Mediated Electroluminescence

This type of electroluminescence included the excitation by charge carriers generated in the bulk or injected at the interfaces of the luminescent material. The electric field fulfils either a charge-carrier accelerating or charge-carrier injecting functions. The charge carriers accelerated by an electric field collide with luminescent centers, chemical and physical defects or the unperturbed material molecules (atoms) exciting them to higher electronic states, including ionization. The subsequent radiative relaxation of such excited states and those produced in the opposite charge carrier recombination processes after ionization, produces impact EL (Figure 12). If electrons and holes are injected into a luminescent material at its interfaces (e.g. electrical contacts) and their recombination generates emitting excited states, this is same as recombination electroluminescence.

Figure 12. Basic Structure of Polymer LED

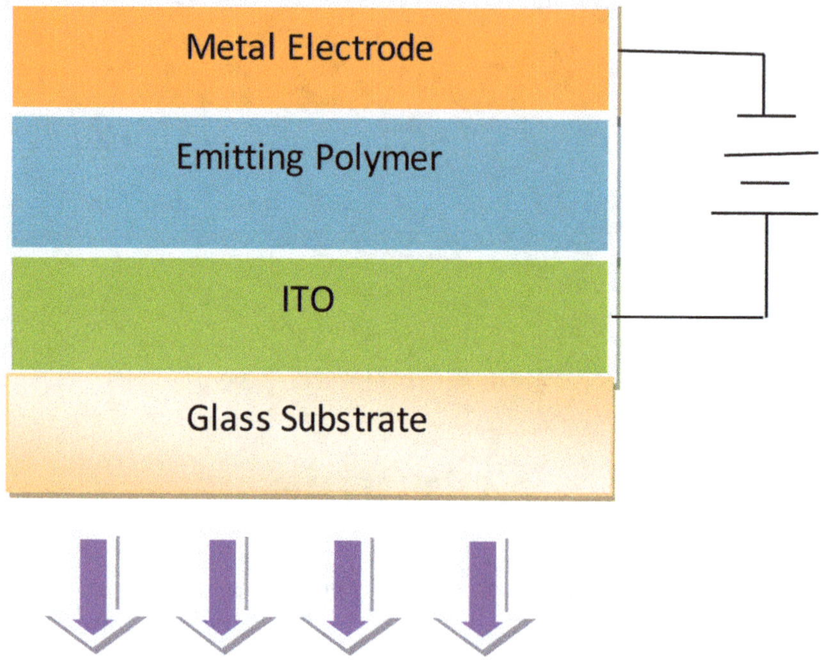

LIGHT EMITTING DIODES (LED)

The electroluminescent mechanism of LEDs is based on charge injection and radiative recombination at the $p-n$ junction. A typical LED device structure comprises a p-type layer and an n-type layer with the $p-n$ junction at the interface (Figure 13). Bottom and top electrodes are deposited on opposite sides of the device with at least one transparent electrode for light extraction. ITO (Indium-tin-oxide) glass is one of the most frequently used materials as the anode of an electroluminescent device. It has high surface conductivity and high optical transmission. Under forward bias, charges will be injected from one layer into the oppositely doped layer. The injected minority carriers will radiatively recombine with the majority carriers in the opposite side of the $p-n$ junction, leading to light emission. Both organic materials and inorganic materials have been investigated for LED devices, though the quantum efficiency and long-time stability in the OLEDs are still lower compared to the ILEDs. Another important and promising approach is to use co-polymers in combination with the emitting layer. The added units can have an enhancement in the efficiencies through Forster resonance energy transfer (FRET). In addition to p-n junction LEDs, investigations are going on following new types of LEDs for improved performance.

Types of LED

1. **Organic Light Emitting Diode (OLED):** Organic light emitting diodes (OLED) have attracted much attention due to their low driven voltage with high brightness. In the earlier stage, OLED were prepared by embedding a layer of organic fluorescent material sandwich between the two electrodes. However the efficiency of this organic electroluminescent cell was relatively low. In

Figure 13. Charge injection into Polymer LED

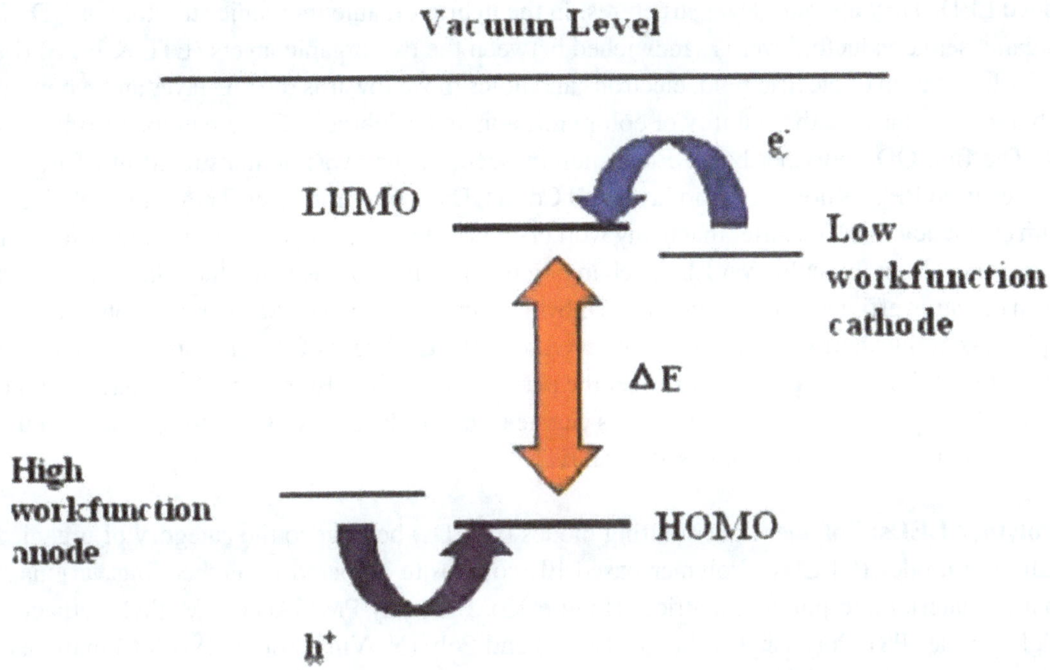

Multilayer OLEDs, two more layers (HTL & ETL) are added over the emissive layer to bring out better charge carrier transport between the polymers. OLEDs have low fabrication cost. They also have opened the way to fabricate the white light emitting diodes. The inorganic light emitting diode (ILED) requires high temperature deposition process to deposit the single crystalline inorganic layer, which is a costly process. However organic transport layers have stability problems and these layers are sensitive to air. To overcome these problems inorganic semiconductor transport layers were developed and used in QD-LEDs.

2. **Quantum Dots LED:** Quantum dots (QD) are semiconductor nanoparticle having the size less than the exciton bohr radius. Due to quantum confinement effect and large surface to volume ratio, quantum dots exhibit remarkable optical and electrical properties. They are widely used in electroluminescence application due to larger excitonic interaction between the electrons and holes, stronger carrier confinement, high quantum yield and narrow emission spectrum. Mainly the QLED have following features:
    a. Tunability of color emission can be achieved with changing the particle size of inorganic semiconductor nanoparticles.
    b. Longer lifetime and stability of color.
    c. Higher brightness and higher quantum yield.
    d. Narrow emission line width.
    e. High color purity.

By changing in the size of QDs, the emission wavelength can be changed, so the color of light emission can be changed. Therefore these advantages make colloidal QDs as promising candidates for making quantum-dot light emitting diodes (QD-LED). Figure 14 (a) represent the emission spectra of LED and QD added LED. They are multilayer structures. In the hybrid organic/inorganic structure of QD –LED, an inorganic semiconductor layer is sandwiched between the two organic layers (ETL & HTL) (Figure 14 b). On application of electric field, electrons and holes move towards the QD layer and recombine to emit photons. To enhance the stability of color emission and brightness, QD are embedded in polymer matrix. The first QD –polymer based Electroluminescent display was demonstrated in (Colvin et al, 1994).They used PPV as hole transport layer and CdSe QDs as emitting layer. They show relatively low quantum efficiency (0.01%). After that many workers fabricated QD-polymer LED using different polymer as transport layer. In multilayer EL, by changing the thickness of the constituent layer, a remarkable change in quantum efficiency has been achieved by (Coe et al, 2002) in CdSe quantum dots using TOPO as capping agent. Furthermore it has been observed that the thickness of the emissive layer and organic layer and area of the emitting layer also affect the performance of the EL device. Nowadays metal oxide NiO and ZnO are common materials to use as charge transport layers, NiO is a p-type material used as a HTL and ZnO is a n-type material used as an ETL.

3. **Polymer LEDs:** Polymer light emitting diodes (pLEDs) belongs to the category of organic light emitting diodes (OLEDs). Polymer based EL composite prepared by embedding various luminance material into polymer matrices (Figure 15). Basically PMMA(Poly Methyl methacrylate), Polysterene (PS), Poly vinyl carbozole (PVK) and Poly (N- Vinyl Alcohol) (PVA) matrices have been extensively used due to their non-conducting, long term stability and film formability. Poly (3,4 ethylenedioxy thiophene) PEDOT and PS also are used as good choice for charge transport layer in multilayer pLEDs. Generally Poly (3,4 ethylenedioxy thiophene) doped with Poly (Styrene sulfonate) PEDOT –PSS is used as hole transport layer (HTL) to enhance the injection of hole from ITO to emitting layer. In multilayer device, bathouirproinr (BCP) and tri (8- hydroxyl-quinolinate)

Figure 14. CIE diagram
Source: Meera & Vikas, 2015

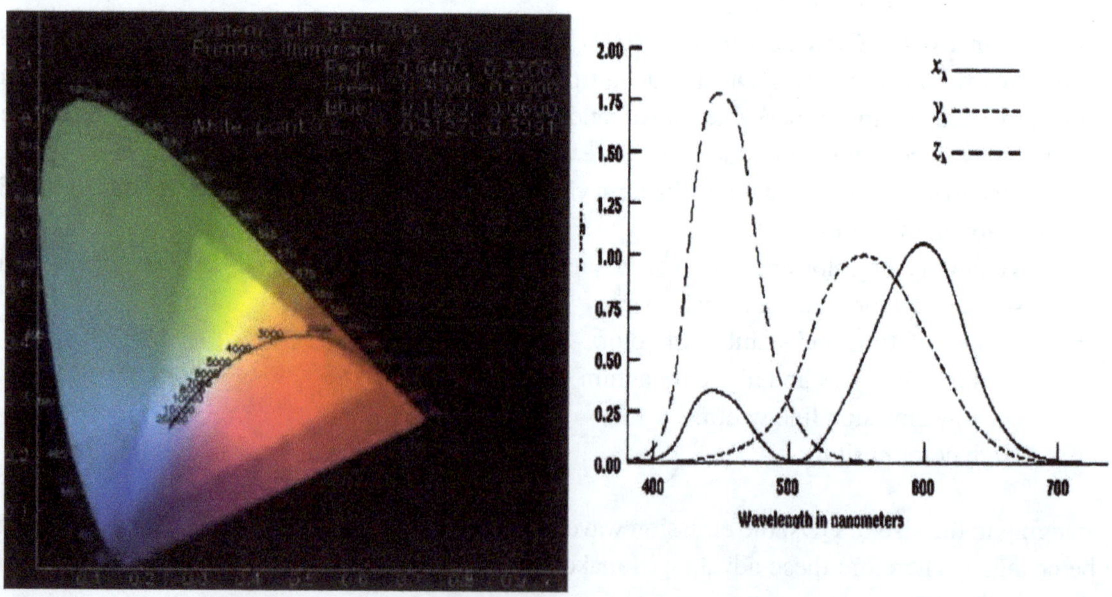

Figure 15. EL and PL Spectra of CdS nanoparticles

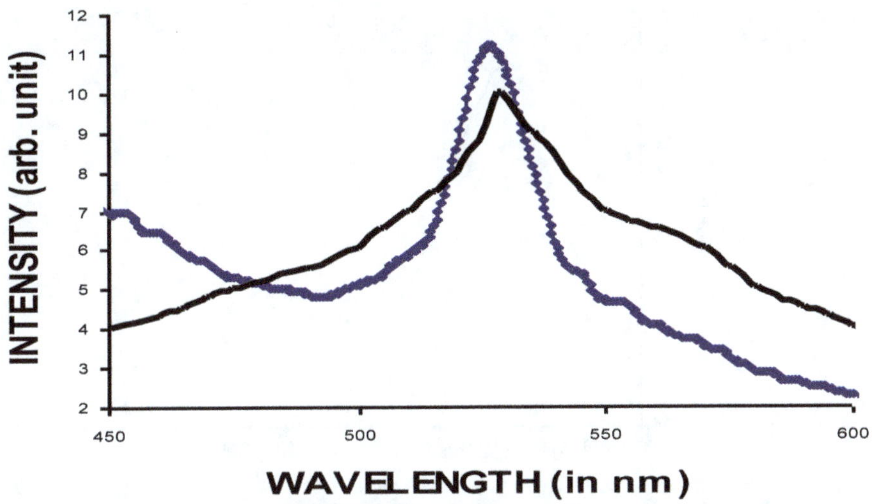

aluminum (AlQ3) used as hole blocking layer (HBL) and electron transport layer (ETL) respectively over the emissive layer (EML). The basic mechanism of pLED is based on molecular electroluminescence (EL). Electrons from the cathode and holes from the anode are injected into the emitting layer, where they recombine and release the energy in the form of light. The wavelength of light emitted depends on the band gap (Eg) of the polymer semiconductor – defined as the energy difference between the highest occupied molecular orbital (HOMO) and lowest unoccupied molecular orbital (LUMO) levels (Figure 16). pLEDs have attracted much attention due to their easy process and economically-attractive coating techniques. The first example of a polymer based OLED was published by (Burronghes et al. 1990)in which poly pheneylene vinylene (PPV) was used as the emissive layer. Since this discovery, many conjugated polymers have been developed for use in electroluminescence including poly (1,4 pheneylene) (PPP), poly(thiopene) and poly (flourene) (Vahlenkamp et al, 1994).

Characteristics of Electroluminescence

1. **Emission Brightness**: To express the Emission brightness of the ACEL devices, although many empirical and theoretical relation been proposed, the most suited relation for the voltage dependence of EL brightness may be given by

$$B = B0 \exp(-b/V^{0.5}) \quad (2)$$

where B is the emission brightness of the device, V is the applied voltage, and B0 and b are constants. The values of B0 and b are determined by the particle size of the phosphors, concentration of the EL powders in the dielectric materials, dielectric constant of the embedding medium, and the device thickness (Wilson, 1990). But voltage is not only factor that affects electroluminescent brightness. Brightness is also proportional to the frequency for a fixed applied voltage. Mostly 50 to 400 Hz frequencies are used,

Figure 16. Log B V/s 100/(V)$^{1/2}$ curve of CdS/PVA composite film (S1 5%, S2 10% loading of CdS)

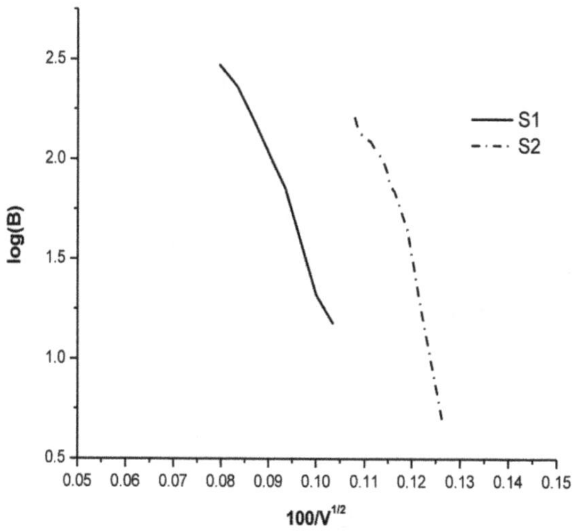

although in some cases frequencies as high as several kilohertz are used. As the frequency increases to several kilohertz, the increase in brightness tapers off.

2. **External Quantum Efficiency (EQE)**: One of the most important properties of EL device is EQE. The external quantum efficiency is defined as the ratio of the number of photons emitted in the forward direction to the number of injected charges.

$$E = \frac{N_{photons}}{Ncharges} \tag{3}$$

And the number of injected charge is given by

$$N_{charge} = I_p / e \tag{4}$$

where I_p is the injected current into the EL device and e is charge of electron.

3. **Chromaticity Coordinates**: Wyszecki and Stiles provided the important method of measuring the color property (Wyszecki et al, 1967), which is called the chromaticity co-ordinates. The chromaticity co-ordinates of a color are the ratios of each tristimulous value of the color to their sum. The most common system of chromaticity co-ordinates is the 1931 CIE- (X, Y,Z)system of color specification. It has three color matching functions, x_A, y_A and z_A. (in Figure 17).
4. **Current- Voltage Characteristics:** Thronton has shown that the current voltage characteristics of EL materials follow the normal relation for forward biased diode rectifier, which established the existence of the rectifying barrier (Thronton et al, 1961). The width of Schottky barrier d varies with the square root of the applied voltage across the emissive layer by the following relation.

$$d = (\varepsilon V/2N_0)^{1/2} \tag{5}$$

Figure 17. Voltage-Brightness curve of ZnS/PVA composite film having different loading of ZnS (A-2%, B-5%, C-10%, D-20%, E-30%, F- 40%)

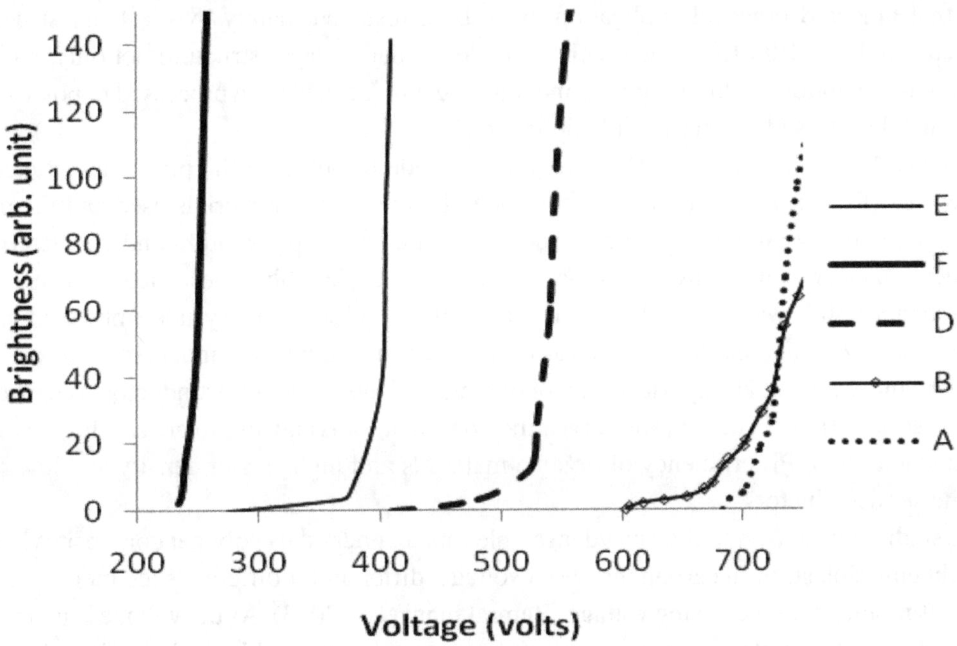

where Ɛ is the dielectric constant of the embedding medium and N_0 is the charge carrier density.

5. **Frequency Response Characteristics:** The frequency response of EL can be accounted for on the basis of relation (Ivey,1963).

$$N_e = N_\infty \left(1 - e^{-t/\tau}\right) \quad (6)$$

where N_e is the number of electron being released from trap states and reaching conduction band, N_∞ the value for infinite time, τ is effective time constant for recombination of charge carrier.

Although the light emitting properties and thickness of the layer is very important, the material choice and thickness of the electrode play important role in the device performance. It will affect the EQE and life time. Experiments show that if we compared the half life for identical devices except with different thickness of Ca and Ba cathodes, we obtain the maximum operating life for a lower thickness of approximately 30Å for both Ca and Ba; the half life decreases as the cathode thickness increases(Deng, 2011).

RECENT DEVELOPMENTS

In recent years, however, advances in electronics and material science have helped each other in accumulating new knowledge as well as in developing new process technologies and device structures. These seem to have enhanced the development of dispersion type DC-EL in recent years (Vecht & Werring, 1970) but more striking are the advent of the *Lumocen* device (Kahng,1968) and the double insulating layer structure (Russ et al,1967& Mattousi et al 1998) in which the vacuum-deposited thin phosphor film replaces the dispersion layer structure. Problems related to short life and low brightness, the major obstacles that hindered practical application, were both resolved mainly by the latter structure. The vacuum-deposited thin film EL device with a double insulating layer structure not only exhibits high brightness with exceptionally high stability and long life, but it can also be processed to possess inherent memory which lends itself to very useful applications.

Over to past few decades, more efforts have been made to optimize the properties of electroluminescent devices. The EL Properties can be controlled by the type of materials used and the method of preparation applied. By taking advantage of developments in the preparation and characterization of semiconductor nanocrystals many researchers have realized electroluminescence from nanocrystals. First EL of nanocrystals was reported (Colvin et al, 1994) on CdSe nanocrystals embedded in a matrix of poly vinyl carbazole and oxidiazole derivative. Most of the studies of nanocrystals EL are in hybrid organic/ inorganic devices. The hybridization of organic and inorganic semiconductors is expected to allow wide range selection of emitters and carrier transport materials and also to provide high performance EL devices due to high PL efficiency of organic materials and high carrier density and low resistivity of inorganic semiconductors.

The EL studies on nanocrystalline powder samples and nanocrystals polymer composites have shown that the light emission starts at certain threshold voltage, different for different specimens and then usually increases rapidly with increasing voltage (Ramrakhiani et al, 2013) .As the voltage is increased more electrons and holes are injected into the emission layer. Due to high field, bending of bands takes place and electrons are released from traps or tunnel to conduction band. The subsequent recombination of electrons and holes give light of higher intensity. The linear relation has been reported between current and voltage indicating ohmic nature. The slope of V_I lines have been found to decrease by reducing nanoparticle size and by increasing dopant concentration showing that the impedance is reduced. The EL brightness is maximum for a particular dopant concentration,(Vishwakarma et al,2011) concentration quenching occurs for higher doping. Usually EL spectrum is nearly identical to corresponding PL spectrum, (Figure 18) but EL peaks are found to be broader than PL peaks. This is probably due to local Joule heating from the large current flux and poor thermal conductivity. The correspondence between EL and PL indicates that both originate from the same states (Dabbousi et al,1995). In case of nanocomposites the emission has been found to be from deep traps also. In case of nanocrystals-polymer composites EL starts at lower threshold voltage and increases very rapidly with the voltage (Figure 19). For powder as well as nanocomposites, the relation between EL brightness (B) and applied voltage (V) has been found of the nature (Wilson & Hawkes, 1999);

$$B = Bo\ Exp(-b/\sqrt{V}) \tag{7}$$

This indicates that EL is produced by acceleration-collision mechanism. Smaller nanoparticles show enhanced electroluminescence at lesser turn-on voltage and higher brightness due to increased oscillator strength. The EL brightness is initially found to increase with increasing nanocrystalline content but beyond a certain value, saturation is observed in the EL intensity. This shows that the donor states involved in the light emission are mostly related to the surface states of the particle, which increase as size of particle is reduced since the surface area to volume is increased. The surface states need to be passivated in order to get efficient luminescence. This shows that such composites can be useful for EL devices with advantages of better flexibility and good quality films. V.Lahariya has found log B

Figure 18. Electroluminescence spectra of ZnS/ PVA composite film (voltage 312V)
Source: Ramrakhiani & Nogriya, 2012

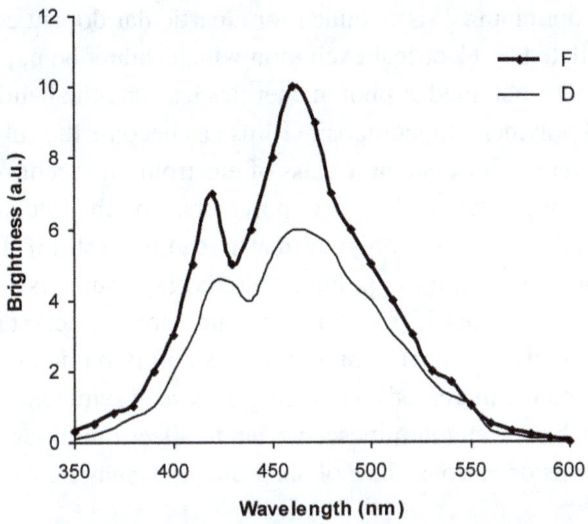

Figure 19. EL spectra of $Cd_{1-x}Zn_xSe$/PVA nanocomposite film (x= 0.0, 0.25 and 0.75)

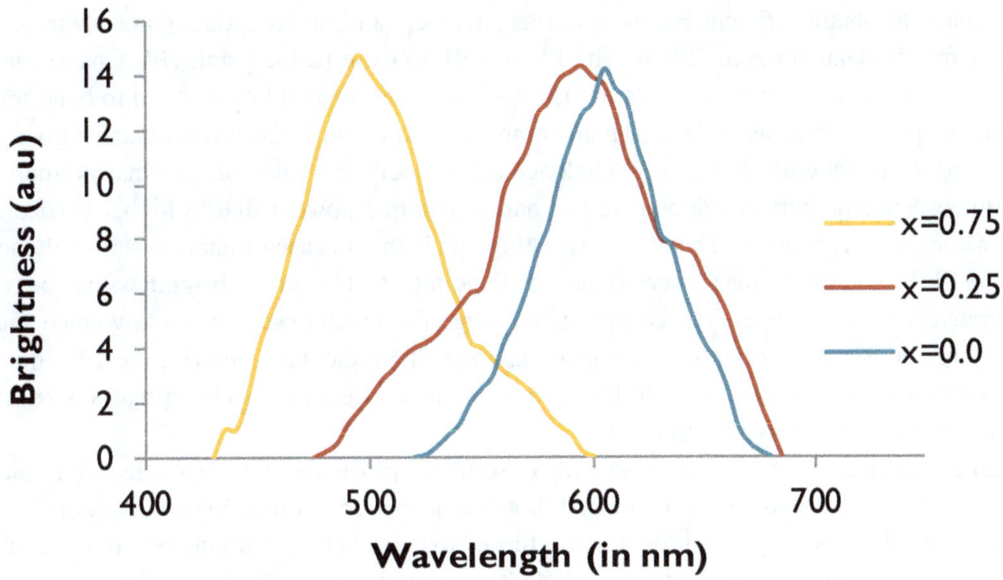

Vs 100/√V straight line with negative slope for CdS/PVA nanocomposites (Figure 18) (Lahariya et al, 2017). Experimentally it has been established that for an EL cell the relation exp $(-b/v^{1/2})$ is valid at lower voltage and relation exp(-b/v) is applicable at higher voltage (Khare et al, 2006).

Some are the inorganic semiconductor, inorganic semiconductors doped with transition metal or organic semiconductor are used as phosphor layer in electroluminescence device. Zinc Sulfide (ZnS) is most typical inorganic electroluminescent material, because the band gap is large enough to emit visible light without absorption and the efficient transport of high energy electrons. ZnS has a band gap of 3.68 eV at 300K with exciton Bhor diameter 5.2 nm. Zinc sulfide in common with most (but not all) inorganic Phosphors, must be "activated" by suitable impurities in order to produce luminescence. In case of doped samples, optimum EL is obtained for a particular dopant concentration. Usually low energy states are also populated by electrical excitation which cannot be populated by optical process. Therefore EL emission may be obtained at photon energies less that the band gap of the material.

Electroluminescence of polymer nanocomposite films has become the subject of both scientific and industrial importance in recent years. As a new class of electroluminescent materials, nanocomposites consist of semiconductor nanocrystals and organic polymers. To fabricate such a nanocomposite, the nucleation of nanoparticles directly in the polymer matrix can be useful if it is possible to control the nanoparticle dispersion and size within the polymer matrix, thus suppressing segregation. Generally conducting polymers are used as matrix for the semiconductor nanocrystals, as they provide good mechanical and optical properties conferring high kinetic stability on nanometer sized semiconductor particles. The EL efficiency can be increased by reducing the size of semiconductor crystals to nanometer range. Nanocrystallite show better photoluminescence but for electroluminescence, nanocomposites are beneficial. Different materials or various sizes of same material may be used for different color light emission in polymer matrix.

M. Ramrakhani et al reported electroluminescence in ZnS/PVA nanocomposite film with different loading of ZnS in PVA matrix in single layer El structure. Film show violet-blue light emission, which increases with higher loading of ZnS (Ramrakhiani et al, 2012) (Figure 19). Recently attempt has been made to obtain efficient EL using ternary nanocrystalline compounds and their composite in polymer matrix (Gautam et al, 2015). The EL of Cd1-xZnxSe nanocrystals /PVA nanocomposites has been studied. In case of nanocrystals the EL is obtained corresponding to band to band transition giving narrow peaks where as in case of nanocomposites emission is from defect states giving broad peak. It is observed that embedding metal chalcogenide nanocrystals into polymer matrix improves the electroluminescence properties as compared to nanocrystalline powder due to higher carries density and better transport mechanism. The most interesting fact is that in case of nanocrystals-polymer composites, the EL brightness increases very rapidly with voltage and thus high brightness can be obtained at comparatively lower voltages. The composite films require much lower threshold voltage and lower power for light emission. Varying the composition, color of EL can be tuned (Figure 19). In nanosize regime, electroluminescence is governed by size quantization effect i.e. by changing the size emission color changes and intensity can be enhanced.

Several examples have been reported recently of solution- processed multilayer electroluminescence devices incorporating semiconductor nanocrystals as the active recombination centers (doped semiconductor nanocrystals). Recently, attention has also turned to hybrid electroluminescent devices involving transition metal doped nanocrystals (Gao et al,1998 ;Que et al 1998). The EL performance of an A.C thin film EL devices is a strong function of the crystallinity of the phosphor layer, the insulator/phosphor interface properties, space charge in the phosphor, the nature of the luminescent center and their coordination in the host crystal lattice.

Another recent development is organic light-emitting diodes (OLEDs). Currently, OLEDs have reached the stage of commercialization, and there are intense efforts to use them in various applications from small- and medium-sized mobile devices to illumination equipment and large TV screens. In particular, phosphorescent materials have become core OLED materials as alternatives to the conventionally used fluorescent materials because devices made with phosphorescent materials exhibit excellent light-emitting performance. However, phosphorescent materials have several problems, such as their structure being limited to organic metal compounds containing rare metals, for example, Ir, Pt, and Os, and difficulty in realizing stable blue light emission, so the development of new materials is necessary. The next-generation OLEDs using a new light-emitting mechanism is called thermally activated delayed fluorescence (TADF). Highly efficient TADF, which was difficult to realize with conventional technologies, has been achieved by optimizing molecular structures. This has led to the realization of ultimate next-generation OLEDs that are made of common organic compounds and can convert electricity to light at an internal quantum efficiency of nearly 100%.

QD-LEDs have attracted great attention due to tunability of light by varying the composition and controlling the size and shape of the QDs. They have attracted intense research and commercialization efforts over the last decade. The combination of QDs with organic emitting materials can provide a wide range color emitting devices (Anikeeva, 2009). For solid state lighting application, white emission from polymer-QD based LEDs is very attractive. Several research studies prove that it is possible to emit white light from hybrid polymeric-inorganic diodes by blending different materials with emission in RGB (Chou et al., 2007; Huang et al., 2010; Luo et al., 2007). Luo et al. developed high efficiency white-light emitting diodes using a single copolymer. Fluorescent red, green and blue chromophores were introduced in a conjugated polymer backbone as emissive layer, resulting in a stable white light with 6.2 cd/A luminance efficiency (Luo et al., 2007). For many lighting applications, warmer white is preferred over the bluish white. QDs can be used to tune the quality of white lighting; by adding QD lumophores to the phosphors used to fabricate the LED it is possible to achieve warmer white light. Figure No. represents the emission spectra of a standard LED bulb and a QD added LED bulb.

J. Kim et at have reported a EL device with the tandem structure, particularly the device with the triple emitting layer, was six times brighter than (1460 cd/m2 at an alternating voltage application of 50 V) the structure with a single emitting layer. This tandem structure plays important roles by increasing the chance of electron–hole tunneling into the phosphor with leading to an increase in brightness by combining the light emitted from several emitting layers in a device (Kim et al, 2011).

Y. R. Park et al demonstrate that nitrogen-doped carbon nanodot (N-CD) interlayer improves electrical and luminescent properties of QD-LEDs. The N-CD were inserted as a hole transport layer (HTL) between other multilayer HTL heterojunction and the red-QD layer. The QD-LEDs with N-CD interlayer represented superior electrical rectification and electroluminescent efficiency. The N-CD interlayer acting as a FRET energy-donor layer to the QD layer, made possible by the spectral overlap and small band offset, enhanced the luminescence of the QD layer (Park et al, 2017).

Electroluminescence has found vide applications as diverse as, advertising, camera flashes and even wallpaper or ceiling, for instructions to public, warning in emergency etc. with much more impact. However its application has been limited because of low quantum yield and low response time. Although many EL devices have been investigated but still some challenges need to overcome for practical applications. It will also require wider effort to develop multicolour display, white light emission with balance in Red, Green and Blue (RGB) colour emission.

CONCLUSION

The development of intrinsic EL by means of phosphor film stagnated in the first half of 1960s, and most of the applications were taken over by LEDs. However, in recent years, the advent of the AC type thin film EL has indicated a bright prospect for practical application to various kinds of displays due to its high reliability, brightness, life and its inherent memory function. The DC-EL performance of powder type and thin film type structures has been improved several times. In a few years, it may be expected that the AC type thin film EL will become practical for character displays. It is believed that this panel is competitive with PDP in brightness, reliability and in memory function. The all solid state flat display panel has a promising viability in the field of information displays. The final application would be in the fiat panel TV display. Problems such as ortho-chromatic display, conversion efficiency and cost should be resolved before its penetration into the consumer market. The former two will be solved by further studies of the materials. As for the cost, the general trend of its reduction will be in accord with the experience curve that will be affected by the improvement in mass production technique of the panel and the number of units marketed.

REFERENCES

Anikeeva, P. O., Halpert, J. E., Bawendi, M. G., & Bulovic, V. (2009). Quantum dot lightemitting devices with electroluminescence tunable over the entire visible spectrum. *Nano Letters*, *9*(7), 2532–2536, 15306984. doi:10.1021/nl9002969 PMID:19514711

Burroughes, J. H., Bradley, D. C., Brown, A. R., Marks, R. N., Mackay, K., Friend, H. R., ... Holmes, A. B. (1990). Light emitting diodes based on conjugated polymers. *Nature*, *347*(6293), 539–541. doi:10.1038/347539a0

Chou, C. H., Yang, C. H., Hsu, C. S., & Chen, T. M. (2007). Hybrid white-light emitting-LED based on luminescent polyfluorene polymer and quantum dots. *Journal of Nanoscience and Nanotechnology*, *7*(8), 2785–2789. doi:10.1166/jnn.2007.622 PMID:17685298

Coe, S., Woo, W. K., Bawendi, M. G., & Bulovic, V. (2002). Electroluminescence from single monolayers of nanocrystals in molecular organic devices. *Nature*, *420*(6917), 800–803. doi:10.1038/nature01217 PMID:12490945

Colvin, V. L., Schlamp, M. C., & Alivisatos, A. P. (1994). Light-Emitting-Diodes made from Cadmium Selenide Nanocrystals and a Semiconductor Polymer. *Nature*, *370*(6488), 354–357. doi:10.1038/370354a0

Dabbousi, B. O., Bawendi, M. G., Onotsuka, O., & Rubner, M. F. (1995). Electroluminescence from CdSe quantum dots polymer composite. *Applied Physics Letters*, *66*(11), 1316–1318. doi:10.1063/1.113227

Deng, X. Y. (2011). Light emitting devices with conjugated polymers. *International Journal of Molecular Sciences*, *12*(12), 1575–1594. doi:10.3390/ijms12031575 PMID:21673908

Destriau, G. (1936). AC electroluminescence in ZnS. *Journal de Chimie Physique*, *33*, 587. doi:10.1051/jcp/1936330587

Gao, M., Richter, B., Kirstein, S., & Mohwald, H. (1998). Electroluminescence studies on self assembled filmof PPV and CdSe Nanoparticles. *The Journal of Physical Chemistry B*, *102*(21), 4096–4103. doi:10.1021/jp980620a

Gao, M., Richter, B., Kristen, S., & Mohwald, H. (1998). Electroluminescence studies on self assembly films of PPV and CdSe nanoparticles. *The Journal of Physical Chemistry B*, *102*(21), 4096–4103. doi:10.1021/jp980620a

Gautam, N. K., Ramrakhiani, M., Kuraria, R., & Kuraria, S. R. (2015). Electroluminescence in Organically Capped $C_{d1-x}Z_{nx}$Se Chalcogenide Nanocrystals. *Defect and Diffusion Forum*, *361*, 215–230. doi:10.4028/www.scientific.net/DDF.361.215

Hameed, S., Pradeep, P., & Baiju, M. R. (2010). Polymer Light Emitting Diodes- A review on materials and Techniques Rev. *Adv. Mater. Sc.*, *26*, 30–42.

Huang, C. Y., Huang, T. S., Cheng, C. Y., Chen, Y. C., Wan, C. T., Madhava Rao, M. V., & Su, Y. K. (2010). Three-band white light-emitting diodes based on hybridization ofpolyfluorene and colloidal CdSe-ZnS quantum dots. *IEEE Photonics Technology, Letters*, *22*(5), 305-307.

Ivey. (1963). *Electroluminescence and related effects*. New York: Academic Press.

Kahng, D. (1968). Electroluminescence of rare earth and transition metal molecules in II-VI compound via impact excitation. *Applied Physics Letters*, *13*(6), 210–212. doi:10.1063/1.1652574

Khare, A., & Bhusan, S. (2006). Electroluminescence studies of chemically deposited (Zn-Cd)S:Cu,F films. *Crystal research tech.*, *41*(7), 689-697.

Kim, J.-Y., Bae, M. J., Park, S. H., Jeong, T., Song, S., Lee, J., ... Yu, S. G. (2011). Electroluminescence enhancement of the phosphor dispersed in a polymer matrix using the tandem structure. *Organic Electronics*, *12*(3), 529–533. doi:10.1016/j.orgel.2010.12.020

Lahariya, V. (2017). Study of electroluminescence study in cadmium sulfide polymer nanocomposite films. *Journal of Nano Research*, *49*, 181-189.

Luo, J., Li, X., Hou, Q., Peng, J., Yang, W., & Cao, Y. (2007). High-efficiency white-light emission from a single copolymer: Fluorescent blue, green, and red chromophores on a conjugated polymer backbone. *Advanced Materials*, *19*(8), 1113–1117. doi:10.1002/adma.200601241

Manzoor, K., Aditya, V., Vadera, S., Kumar, N., & Kutty, T. (2005). Enhanced electroluminescence properties of doped ZnS nanorods formed by the self-assembly of colloidal nanocrystals. *Solid State Communications*, *135*(1-2), 16–20. doi:10.1016/j.ssc.2005.04.001

Mattoussi, H., Radzilowski, L. H., Dabbousi, B. O., Thomas, E. L., Bawendi, M. G., & Rubner, M. F. (1998). Electroluminescence from heterostructures of poly(phenylenevinylene) and inorganic CdSe nanocrystals. *Journal of Applied Physics*, *83*(12), 7965–7974. doi:10.1063/1.367978

Nayak, P. K., Agarwal, N., Ali, F., Patankar, M. P., Narasimhan, K. L., & Periasamy, N. (2010). Blue and white light electroluminescence in a multilayer OLED using a new aluminium complex. *Journal of Chemical Sciences*, *122*(6), 847–855. doi:10.100712039-010-0073-0

Park, Y. R., Jeong, H. Y., Soo Seo, Y., Choi, W. K., & Hong, Y. J. (2017). Quantum dots Light emitting diodes with nitrogen doped carbon nanodot hole transport and electronic energy transfer layer. *Scientific Reports*, *7*, 46422. doi:10.1038rep46422 PMID:28401912

Que, W., Zhou, Y., Lam, Y. L., Chan, Y. C., Kam, C. H., Liu, B., ... Mendis, F. V. C. (1998). Photoluminescence and Electroluminescence from copper doped Zinc sulfide nanocrystals/Polymer composites. *Applied Physics Letters, 73*(19), 2727–2729. doi:10.1063/1.122571

Ramrakhiani, M., & Nogriya, V. (2012). Chapter X Synthesis and Characterization of Zinc Sulfide nanocrystals and ZnS/PVA nanocomposite for luminescence application Polymer Processing and Characterization Apple. Academic Press Publisher.

Ramrakhiani, M., & Nogriya, V. (2013). Photo and Electroluminescence of Cadmium selenide nanocrystals and Nonocomposites. *Journal of Luminescence, 133*, 129–133. doi:10.1016/j.jlumin.2011.09.046

Russ, M. J., & Kennedy, D. I. (1967). The Effects of Double Insulating Layers on the Electroluminescence of Evaporated ZnS:Mn Films. *Journal of the Electrochemical Society, 114*(10), 1066–1071. doi:10.1149/1.2424188

Thornton, W. A. (1961). Direct Current Electroluminescence at Low Voltages. *Physical Review, 122*(1), 58–59. doi:10.1103/PhysRev.122.58

Vahlenkamp, T., & Wegner, G. (1994). Poly(2,5 dialkoxy-p-phenylene)S- synthesis and properties. *Macromolecular Chemistry and Physics, 195*(6), 1933–1952. doi:10.1002/macp.1994.021950605

Vecht, A., & Werring, N. J. (1970). Direct Current Electroluminescence in ZnS. *J. Phys. D, 3*(2), 105–120. doi:10.1088/0022-3727/3/2/202

Vij, D. (Ed.). (2004). *Handbook of Electroluminescent Materials*. Bristol, UK: Institute of Physics. doi:10.1201/9781420033410

Vishwakarma, P., Ramrakhiani, M., Singh, P., & Bisen, D. P. (2011). Synthesis and electroluminescence studies of manganese doped cadmium sulfide nanoparticles. *Open Nanoscience Journal, 5*, 34-40.

Wilson, J., & Hawkes, J. F. B. (1999). *Optoelectronics: An Introduction* (2nd ed.). Prentice-Hall of India Pvt. Ltd.

Wyszecki, G., & Stiles, W. S. (1967). *Color Science* (1st ed.). John Wiley & Sons, Inc.

Compilation of References

Abramowitz, M., & Stegun, I. A. (Eds.). (1965). *Handbook of mathematical functions*. New York, USA: Dover.

Adachi, C., Baldo, M. A., Forrest, S. R., Lamansky, S., Thompson, M. E., & Kwong, R. C. (2001). High-efficiency red electrophosphorescence devices. *Applied Physics Letters*, *78*(11), 11. doi:10.1063/1.1355007

Adikaari, A. D., Etchart, I., Guéring, P. H., Bérard, M., Silva, S. R. P., Cheetham, A. K., & Curry, R. J. (2012). Near infrared up-conversion in organic photovoltaic devices using an efficient Yb^{3+}:Ho^{3+} Co-doped Ln_2BaZnO_5 (Ln= Y, Gd) phosphor. *Journal of Applied Physics*, *111*(9), 094502. doi:10.1063/1.4704687

Agrawal, S., & Dubey, V. (2014). Article. *Jour. Rad. Res. Appl. Sci.*, *7*(4), 601–606.

Aguilar, E. A., & Drew, R. A. L. (2000). Melt extraction processing of structural Y2O3–Al2O3 fibers. *Journal of the European Ceramic Society*, *20*(8), 1091–1098. doi:10.1016/S0955-2219(99)00279-4

Akamatsu, K., & Deki, S. (1997). Nanoscale Metal Particles Dispersed in Polymer Matrix. *Nanostructured Materials*, *8*(8), 1121–1129. doi:10.1016/S0965-9773(98)00044-0

Akella, A., & Keszler, D. (1995). Sr_2LiSiO_4F: Synthesis, Structure, and Eu^{2+} Luminescence. *Chemistry of Materials*, *7*(7), 1299–1302. doi:10.1021/cm00055a005

Amara, M. B., Vlasse, M., Flem, G. L., & Hagenmuller, P. (1983). Structure of the low-temperature variety of calcium sodium orthophosphate, $NaCaPO_4$. *Acta Crystallographica. Section C, Crystal Structure Communications*, *39*(11), 1483–1485. doi:10.1107/S0108270183008963

Anikeeva, P. O., Halpert, J. E., Bawendi, M. G., & Bulovic, V. (2009). Quantum dot lightemitting devices with electroluminescence tunable over the entire visible spectrum. *Nano Letters*, *9*(7), 2532–2536, 15306984. doi:10.1021/nl9002969 PMID:19514711

Anitha, S. (2011). *Nanofibrous composite membrane of ZnO/polymer for UV ssensor, superhydrophoobic and antibacterial applications* (Ph.D Thesis). SRM University, Chennai, India.

Antic, Z., Krsmanovic, R., Dordevic, V., Dramicanin, T., & Dramicanin, M. D. (2009). Optical Properties of Y_2O_3:Eu $^{3+}$ Red Emitting Phosphor Obtained via Spray Pyrolysis. *Acta Physica Polonica A*, *4*(116), 622–624. doi:10.12693/APhysPolA.116.622

Arney, D., Porter, B., Greve, B., & Maggard, P. A. (2008). New Molten-Salt Synthesis and Photocatalytic Properties of La2Ti2O7 Particles. *Journal of Photochemistry and Photobiology A Chemistry*, *199*(2-3), 230–235. doi:10.1016/j.jphotochem.2008.06.005

Aumayr, F., Facsko, S., El-Said, A. S., Trautmann, C., & Schleberger, M. (2011).Single ion induced surface nanostructures: a comparison between slow highly charged and swift heavy ions. J. Phys.: Condens. Matter, 23, 393001 (23pp). doi:10.1088/0953-8984/23/39/393001

Auzel, F. (1966). Compteur quantique par transfert d'energie entre deux ions de terres rares dans un tungstate mixte et dans un verre. *CR Acad. Sci. Paris*, *262*, 1016–1019.

Auzel, F. (1973). Materials and devices using double-pumped-phosphors with energy transfer. *Proceedings of the IEEE*, *61*(6), 758–786. doi:10.1109/PROC.1973.9155

Auzel, F. (1990). Upconversion processes in coupled ion systems. *Journal of Luminescence*, *45*(1-6), 341–345. doi:10.1016/0022-2313(90)90189-I

Auzel, F. (1992). Upconversion process in coupled ions systems. In R. P. Rao (Ed.), *Luminescence: phenomena, materials and devices* (pp. 33–37). New York: Nova Science Publisher.

Auzel, F. (2004). Upconversion and anti-stokes processes with f and d ions in solids. *Chemical Reviews*, *104*(1), 139–174. doi:10.1021/cr020357g PMID:14719973

Avasthi, D. K. (2002). Some interesting aspects of swift heavy ions in materials science. *Current Science*, *78*, 1297–1302.

Avasthi, D. K., & Mehta, G. K. (2011). *Swift Heavy Ions for Materials Engineering and Nanostructuring* (Vol. 145). Springer Series in Materials Science. doi:10.1007/978-94-007-1229-4

Ayyangar, K., Lakshmanan, A. R., Chandra, B., & Ramadas, K. (1974). A comparison of thermal neutron and gamma ray sensitivities of common TLD materials. *Physics in Medicine and Biology*, *19*(5), 665–676. doi:10.1088/0031-9155/19/5/007 PMID:4374709

Azorin, J., Furetta, C., & Scacco, A. (1993) Preparation and properties of thermoluminescent materials. *Physica Status Solidi (A)*, *138*, 9-46.

Azorın, J., Esparza, A., Falcony, C., Rivera, T., Garcıa, M., & Martınez, E. (2002).. . *Radiation Protection Dosimetry*, *100*, 277–279. doi:10.1093/oxfordjournals.rpd.a005867 PMID:12382877

Azorin, J., Rivera, T., Martínez, E., & Garcia, M. (1998). Thermoluminescence of Eu-doped ZrO2 thin films exposed to ultraviolet and visible light. *Radiation Measurements*, *29*(3-4), 315–317. doi:10.1016/S1350-4487(98)00013-4

Babin, V., Oskam Koert, D., Vergeer, P., & Meijerink, A. (2004). Article. *Radiation Measurements*, *38*, 463–466. doi:10.1016/j.radmeas.2004.01.036

Bachmann, V., Ronda, C., & Meijerink, A. (2009). Temperature Quenching of Yellow Ce^{3+} Luminescence in YAG:Ce. *Chemistry of Materials*, *21*(10), 2077–2084. doi:10.1021/cm8030768

Baginskiy, I., & Liu, R. S. (2009). Significant Improved Luminescence Intensity of Eu^{2+}-Doped $Ca_3SiO_4Cl_2$ Green Phosphor for White LEDs Synthesized Through Two-Stage Method. *Journal of the Electrochemical Society*, *156*(5), G29–G32. doi:10.1149/1.3089366

Baldinozzi,, G., & Berar,, J-F., & Calvarin, G. (1998). Calvarin G. Rietveld refinement of two-phase Zr-doped Y_2O_3. *Materials Science Forum*, *278*, 680. doi:10.4028/www.scientific.net/MSF.278-281.680

Baldo, M. A., Adachi, C., & Forrest, S. R. (2000). Transient analysis of organic electrophosphorescence. II. Transient analysis of triplet-triplet annihilation. *Physical Review B: Condensed Matter and Materials Physics*, *62*(16), 10967–10977. doi:10.1103/PhysRevB.62.10967

Baldo, M. A., & Forrest, S. R. (2000). Transient analysis of organic electrophosphorescence: I. Transient analysis of triplet energy transfer. *Physical Review B: Condensed Matter and Materials Physics*, *62*(16), 10958–10966. doi:10.1103/PhysRevB.62.10958

Baldo, M. A., O'Brien, D. F., You, Y., Shoustikov, A., Sibley, S., Thompson, M. E., & Forrest, S. R. (1998). Highly efficient phosphorescent emission from organic electroluminescent devices. *Nature*, *395*(6698), 151–154. doi:10.1038/25954

Baldo, M. A., Thompson, M. E., & Forrest, S. R. (2000). High-efficiency fluorescent organic light-emitting devices using a phosphorescent sensitizer. *Nature*, *403*(6771), 750–753. doi:10.1038/35001541 PMID:10693799

Barrer, R. M. (1982). *Hydrothermal Chemistry of Zeolites*. London: Academic Press.

Baumgarten, P. K. (1971). Electrostatic Spinning of Acrylic Microfibers. *Journal of Colloid and Interface Science*, *36*(1), 71–79. doi:10.1016/0021-9797(71)90241-4

Baur, W. (1974). The geometry of polyhedral distortions. Predictive relationships for the phosphate group. *Acta Crystallographica. Section B, Structural Crystallography and Crystal Chemistry*, *30*(5), 1195–1215. doi:10.1107/S0567740874004560

Belik, A. A., Lazoryak, B. I., Pokholok, K. V., Terekhina, T. P., Leonidov, I. A., Mitberg, E. B., ... Kellerman, D. G. (2001). Synthesis and Characterization of New Strontium Iron(II) Phosphates, $SrFe_2(PO_4)_2$ and $Sr_9Fe_{1.5}(PO_4)_7$. *Journal of Solid State Chemistry*, *162*(1), 113–121. doi:10.1006/jssc.2001.9363

Bender, C. M., Burlitch, J. M., Barber, D., & Pollock, C. (2000). Synthesis and Fluorescence of Neodymium-Doped Barium Fluoride Nanoparticles. *Chemistry of Materials*, *12*(7), 1969–1976. doi:10.1021/cm9904741

Bhadane, M. S., Akhtar, S., Dahiwale, S. S., Hareesh, K., Asokan, K., Kanjilal, D., & Dhole, S. D. (2017). Evaluation of thermoluminescence of 200 keV carbon ion irradiated $CaSO_4$: Dy nanophosphors for medical dosimetry. *Journal of Luminescence*, *192*, 695–700. doi:10.1016/j.jlumin.2017.07.042

Bhadane, M. S., Hareesh, K., Dahiwale, S. S., Sature, K. R., Patil, B. J., Asokan, K., ... Dhole, S. D. (2016). A highly sensitive CaF_2: Dy nanophosphor as an efficient low energy ion dosimetry. *Nuclear Instruments and Methods in Physics Research Section B*, *386*, 61–69. doi:10.1016/j.nimb.2016.09.017

Bhadane, M. S., Mandlik, N. T., Patil, B. J., Dahiwale, S. S., Sature, K. R., Bhoraskar, V. N., & Dhole, S. D. (2016). $CaSO_4$:Dy microphosphor for thermal neutron dosimetry. *Journal of Luminescence*, *170*, 226–230. doi:10.1016/j.jlumin.2015.10.025

Bhargava, R. N., Gallagher, D., Hong, X., & Nurmikko, A. (1994). Optical properties of manganese-doped nanocrystals of ZnS. *Physical Review Letters*, *72*(3), 416–419. doi:10.1103/PhysRevLett.72.416 PMID:10056425

Bhattacharyya, S., Karmakar, M., Majumdar, P. S., & Singh, S. D. (2016). On the thermoluminescence glow curve analysis recorded under hyperbolic heating profile. *Int. J. Lumin. Appl.*, *6*, 194.

Bierlein, J., & Geir, T. (n.d.). *U.S. Patent 3,1976,949,323*. US Patent Office.

Biju, S., Harris, M., Vander Elst, L., Wolberg, M., Kirschhock, C., & Parac-Vogt, T. N. (2016). Multifunctional β-$NaGdF_4$: Ln^{3+} (Ln= Yb, Er, Dy) nanoparticles with NIR to visible upconversion and high transverse relaxivity: a potential bimodal contrast agent for high-field MRI and optical imaging. *RSC Advances*, *6*(66), 61443–61448. doi:10.1039/C6RA09450C

Bin, J. K., Cho, N. S., & Hong, J. I. (2012). New Host Material for High-Performance Blue Phosphorescent Organic Electroluminescent Devices. *Advanced Materials*, *24*(21), 2911–2915. doi:10.1002/adma.201200972 PMID:22549891

Bizarri, G., & Moine, B. (2005). On $BaMgAl_{10}O_{17}$: Eu^{2+} phosphor degradation mechanism: thermal treatment effects. *Journal of Luminescence, 113*(3-4), 199–213. doi:10.1016/j.jlumin.2004.09.119

Blasse, G., & Grabmaier, B. C. (1994). *Luminescent materials.* Berlin: Spring-Verlag.

Blasse, G. (1986). Energy transfer between inequivalent Eu2+ ions. *Journal of Solid State Chemistry, 62*(2), 207–211. doi:10.1016/0022-4596(86)90233-1

Bloembergen, N. (1959). Solid state infrared quantum counters. *Physical Review Letters, 2*(3), 84–85. doi:10.1103/PhysRevLett.2.84

Bos, A. J. J. (2001). High sensitivity thermoluminescence dosimetry. *Nuclear Instruments & Methods in Physics Research. Section B, Beam Interactions with Materials and Atoms, 184*(1-2), 3–28. doi:10.1016/S0168-583X(01)00717-0

Bos, A., & Dielhof, J. (1991). The analysis of Thermoluminescencent Glow Peaks in CaF_2:Tm (TLD-300). *Radiation Protection Dosimetry, 37*, 231–239.

Boyer, D., & Mahiou, R. (2004). Powders and Coatings of $LiYF_4$:Eu^{3+} Obtained via an Original Way Based on the Sol−Gel Process. *Chemistry of Materials, 16*(13), 2518–2521. doi:10.1021/cm040143a

Brincker, C. J., & Scherer, G. W. (1990). *Sol–Gel Science.* New York: Academic Press.

Brito, H. F., Hölsä, J., Laamanen, T., Lastusaari, M., Malkamäki, M., & Rodrigues, L. C. V. (2012). Persistent luminescence mechanisms: Human imagination at work. *Optical Materials Express, 2*(4), 371–381. doi:10.1364/OME.2.000371

Bruchez, M. Jr, Moronne, M., Gin, P., Weiss, S., & Alivisatos, A. P. (1998). Semiconductor Nanocrystals as Fluorescent Biological Labels. *Science, 281*(5385), 2013–2016. doi:10.1126cience.281.5385.2013 PMID:9748157

Bull, R. K. (1986). Thermoluminescence and its applications: An introduction. *Nuclear Tracks and Radiation Measurements, 11*(1-2), 105–113. doi:10.1016/1359-0189(86)90029-4

Burroughes, J. H., Bradley, D. C., Brown, A. R., Marks, R. N., Mackay, K., Friend, H. R., ... Holmes, A. B. (1990). Light emitting diodes based on conjugated polymers. *Nature, 347*(6293), 539–541. doi:10.1038/347539a0

Byrappa, K., & Adschiri, T. (2007). Hydrothermal technology for nanotechnology. *Progress in Crystal Growth and Characterization of Materials, 53*(2), 117–166. doi:10.1016/j.pcrysgrow.2007.04.001

Byrappa, K., & Yoshimura, M. (2001). *Handbook of Hydrothermal Technology.* Noyes Publications.

Campos, L. L., & Lima, M. F. (1987). Thermoluminescent $CaSO_4$:Dy Teflon Pellets for Beta Radiation Detection. *Radiation Protection Dosimetry, 18*(2), 95–97. doi:10.1093/oxfordjournals.rpd.a079889

Cao, B. S., He, Y. Y., Feng, Z. Q., Song, M., & Dong, B. (2011). Crystalline-structure-dependent green and red upconversion emissions of Er^{3+}–Yb^{3+}–Li^+ codoped TiO_2. *Optics Communications, 284*(13), 3311–3314. doi:10.1016/j.optcom.2011.03.005

Cao, M., Wang, Y., Qi, Y., Guo, C., & Hu, C. (2004). Synthesis and characterization of MgF2 and KMgF3 nanorods. *Journal of Solid State Chemistry, 177*(6), 2205–2209. doi:10.1016/j.jssc.2004.01.005

Cao, Y., Ding, J., Wang, X., & Wang, Y. (2017). Tunable white light of multi-cation-site $Na_2BaCa(PO_4)_2$:Eu,Mn phosphor: synthesis, structure and PL/CL properties. *Journal of Materials Chemistry. C, Materials for Optical and Electronic Devices, 5*(5), 1184–1194. doi:10.1039/C6TC04827G

Cao, Z. M., Chen, Y., Zhou, S., Duan, C., & Yin, M. (2015). Optical properties of Er^{3+} doped KYb_xF_{3x+1} (x=2, 3) upconverting nanoparticles. *Journal of Rare Earths, 33*(9), 911–914. doi:10.1016/S1002-0721(14)60504-4

Capobianco, J. A., Vetrone, F., Alesio, T. D., Tessari, G., Speghini, A., & Bettinelli, M. (2000). Optical spectroscopy of nanocrystalline cubic Y2O3:Er3+ obtained by combustion synthesis. *Physical Chemistry Chemical Physics*, *2*(14), 3203–3207. doi:10.1039/b003031g

Capobianco, J. A., Vetrone, F., Boyer, J. C., Speghini, A., & Bettinelli, M. (2002). Enhancement of red emission ($^4F_{9/2} \rightarrow ^4I_{15/2}$) via upconversion in bulk and nanocrystalline cubic Y_2O_3:Er^{3+}. *The Journal of Physical Chemistry B*, *106*(6), 1181–1187. doi:10.1021/jp0129582

Capobianco, J. A., Vetrone, F., Boyer, J. C., Speghini, A., & Bettinelli, M. (2002). Visible upconversion of Er^{3+} doped nanocrystalline and bulk Lu_2O_3. *Optical Materials*, *19*(2), 259–268. doi:10.1016/S0925-3467(01)00188-4

Carnall, W. T. (1978). Optical Spectroscopy of f-Element Compounds. Report Argonne National Laboratory, Chemistry Division.

Carnall, & Field, & Rajnak. (1968). Article. *J. Chem. Phys. 49*, 104253.

Catti, M., & Gazzoni, G. (1983). The β ↔ α'phase transition of Sr_2SiO_4. II. X-ray and optical study, and ferroelasticity of the β form. *Acta Crystallographica. Section B, Structural Crystallography and Crystal Chemistry*, *39*(6), 679–684. doi:10.1107/S0108768183003225

Chander, N., Sardana, S. K., Parashar, P. K., Khan, A. F., Chawla, S., & Komarala, V. K. (2015). Improving the short-wavelength spectral response of silicon solar cells by spray deposition of YVO_4: Eu^{3+} downshifting phosphor nanoparticles. *IEEE Journal of Photovoltaics*, *5*(5), 1373–1379. doi:10.1109/JPHOTOV.2015.2438633

Chang, J., Ning, Y., Wu, S., Niu, W., & Zhang, S. (2013). Effectively Utilizing NIR Light Using Direct Electron Injection from Up-Conversion Nanoparticles to the TiO_2 Photoanode in Dye-Sensitized Solar Cells. *Advanced Functional Materials*, *23*(47), 5910–5915. doi:10.1002/adfm.201301158

Chang, S. C., He, G., Chen, F. C., Guo, T. F., & Yang, Y. (2001). Degradation mechanism of phosphorescent-dye-doped polymer light-emitting diodes. *Applied Physics Letters*, *79*(13), 2088–2090. doi:10.1063/1.1404995

Chan, T. S., Liu, R. S., & Baginskiy, I. (2008). Synthesis, Crystal Structure, and Luminescence Properties of a Novel Green-Yellow Emitting Phosphor $LiZn_{1-x}PO_4$:xMn for Light Emitting Diodes. *Chemistry of Materials*, *20*(4), 1215–1217. doi:10.1021/cm7028867

Chan, W. C. W., & Nie, S. M. (1998). Quantum Dot Bioconjugates for Ultrasensitive Nonisotopic Detection. *Science*, *281*(5385), 2016–2018. doi:10.1126cience.281.5385.2016 PMID:9748158

Chaskar, A., Chen, H. F., & Wong, K. T. (2011). Bipolar Host Materials: A Chemical Approach for Highly Efficient Electrophosphorescent Devices. *Advanced Materials*, *23*(34), 3876–3895. doi:10.1002/adma.201101848 PMID:21805507

Chen, C. H., Hsu, L. C., Rajamalli, P., Chang, Y. W., Wu, F. L., Liao, C. Y., ... Cheng, C. H. (2014). Highly efficient orange and deep-red organic light emitting diodes with long operational lifetimes using carbazole–quinoline based bipolar host materials. *Journal of Materials Chemistry. C, Materials for Optical and Electronic Devices*, *2*(30), 6183–6191. doi:10.1039/C4TC00523F

Chen, C. H., Huang, W. S., Lai, M. Y., Tsao, W. C., Lin, J. T., Wu, Y. H., ... Wu, C. C. (2009). Versatile, Benzimidazole/Amine-Based Ambipolar Compounds for Electroluminescent Applications: Single-Layer, Blue, Fluorescent OLEDs, Hosts for Single-Layer, Phosphorescent OLEDs. *Advanced Functional Materials*, *19*(16), 2661–2670. doi:10.1002/adfm.200900561

Chen, D., Huang, P., Yu, Y., Huang, F., Yang, A., & Wang, Y. (2011). Dopant-induced phase transition: A new strategy of synthesizing hexagonal upconversion $NaYF_4$ at low temperature. *Chemical Communications (Cambridge)*, *47*(20), 5801–5803. doi:10.1039/c0cc05722c PMID:21491047

Chen, D., Yu, Y., Huang, F., Huang, P., Yang, A., & Wang, Y. (2010). Modifying the size and shape of monodisperse bifunctional alkaline-earth fluoride nanocrystals through lanthanide doping. *Journal of the American Chemical Society*, *132*(29), 9976–9978. doi:10.1021/ja1036429 PMID:20593808

Chen, G. Y., Liu, Y., Zhang, Y. G., Somesfalean, G., Zhang, Z. G., Sun, Q., & Wang, F. P. (2007). Bright white upconversion luminescence in rare-earth-ion-doped Y_2O_3nanocrystals. *Applied Physics Letters*, *91*(13), 133103. doi:10.1063/1.2787893

Chen, G., Somesfalean, G., Liu, Y., Zhang, Z., Sun, Q., & Wang, F. (2007). Upconversion mechanism for two-color emission in rare-earth-ion-doped ZrO_2nanocrystals. *Physical Review B: Condensed Matter and Materials Physics*, *75*(19), 195204. doi:10.1103/PhysRevB.75.195204

Cheng, L., Yang, K., Zhang, S., Shao, M., Lee, S., & Liu, Z. (2010). Highly-sensitive multiplexed in vivo imaging using PE Gylated upconversion nanoparticles. *Nano Research*, *3*(10), 722–732. doi:10.100712274-010-0036-2

Cheng-Ren, L., Shu-Feng, L., Bin, D., Jing-Chang, S., Xiao-Feng, B., & Xu-Nan, F. (2012). Intense up-conversion emissions of Yb^{3+}/Dy^{3+} co-doped Al_2O_3 nanopowders prepared by non-aqueous sol-gel method. *Chinese Physics B*, *21*(9), 097803. doi:10.1088/1674-1056/21/9/097803

Chen, J., & Zhao, J. X. (2012). Upconversion nanomaterials: Synthesis, mechanism, and applications in sensing. *Sensors (Basel)*, *12*(3), 2414–2435. doi:10.3390120302414 PMID:22736958

Chen, L., Chen, K. J., Lin, C. C., Chu, C. I., Hu, S. F., Lee, M. H., & Liu, R. S. (2010). Combinatorial Approach to the Development of a Single Mass $YVO_4:Bi^{3+},Eu^{3+}$ Phosphor with Red and Green Dual Colors for High Color Rendering White Light-Emitting Diodes. *Journal of Combinatorial Chemistry*, *12*(4), 587–594. doi:10.1021/cc100063x PMID:20560596

Chen, M., Yuan, Y., Zheng, J., Chen, W. C., Shi, L. J., Zhu, Z. L., ... Lee, C. S. (2015). Novel Bipolar Phenanthroimidazole Derivative Design for a Nondoped Deep-Blue Emitter with High Singlet Exciton Yields. *Adv. Opt. Mater*, *3*(9), 1215–1219. doi:10.1002/adom.201500258

Chen, Q., Qian, Y., Chen, Z., Tang, K., Zhou, G., & Zhang, Y. (1994). Preparation of a Tl-based superconductor by a hydrothermal method. *Physica. C, Superconductivity*, *224*(3-4), 228–230. doi:10.1016/0921-4534(94)90258-5

Chen, R. (1969). Glow Curves with General Order Kinetics. *Journal of the Electrochemical Society*, *116*(9), 1254–1257. doi:10.1149/1.2412291

Chen, R. (1969). On the Calculation of Activation Energies and Frequency Factors from Glow Curves. *Journal of Applied Physics*, *40*(2), 570–585. doi:10.1063/1.1657437

Chen, R., Kristianpoller, N., Davidson, Z., & Visocekas, R. (1981). Mixed first and second order kinetics in thermally stimulated processes. *Journal of Luminescence*, *23*(3-4), 293–303. doi:10.1016/0022-2313(81)90135-6

Chen, R., & McKeever, S. W. S. (1997). *Theory of Thermoluminescence and Related Phenomena*. Singapore: World Scientific. doi:10.1142/2781

Chen, R., & Pagonis, V. (2011). *Thermally and Optically Stimulated Luminescence: A Simulation Approach*. Chichester, UK: Wiley. doi:10.1002/9781119993766

Chen, R., & Winer, S. A. A. (1970). Effects of Various Heating Rates on Glow Curves. *Journal of Applied Physics*, *40*(13), 5227–5228. doi:10.1063/1.1658652

Chen, S., Wu, Y., Hu, S., Zhao, Y., & Fang, D. (2017). Non-Doped Deep Blue and Doped White Electroluminescence Devices Based on Phenanthroimidazole Derivative. *Journal of Fluorescence*, *27*(2), 451–461. doi:10.100710895-016-1970-5 PMID:27900511

Chen, S., Wu, Y., Zhao, Y., & Fang, D. (2015). Deep blue organic light-emitting devices enabled by bipolar phenanthro[9,10-d]imidazole derivatives. *RSC Advances*, *5*(88), 72009–72018. doi:10.1039/C5RA13814K

Chen, W. C., Wu, G. F., Yuan, Y., Wei, H. X., Wong, F. L., Tong, Q. X., & Lee, C. S. (2015). A meta-molecular tailoring strategy towards an efficient violet-blue organic electroluminescent material. *RSC Advances*, *5*(23), 18067–18074. doi:10.1039/C4RA16954A

Chen, W. C., Yuan, Y., Ni, S. F., Zhu, Z. L., Zhang, J. Z., Jiang, Q., ... Lee, C. S. (2017). Highly Efficient Deep-Blue Electroluminescence from a Charge-Transfer Emitter with Stable Donor Skeleton. *ACS Applied Materials & Interfaces*, *9*(8), 7331–7338. doi:10.1021/acsami.6b14638 PMID:28185447

Chen, W. C., Yuan, Y., Wu, G. F., Wei, H., Ye, J., Chen, M., ... Lee, C. S. (2015). Molecular modification on bisphenanthroimidazole derivative for deep-blue organic electroluminescent material with ambipolar property and high performance. *Organic Electronics*, *17*, 159–166. doi:10.1016/j.orgel.2014.11.024

Chen, W. C., Yuan, Y., Wu, G. F., Wei, X. H., Tang, L., Tong, Q. X., ... Lee, C. S. (2014). Staggered Face-to-Face Molecular Stacking as a Strategy for Designing Deep-Blue Electroluminescent Materials with High Carrier Mobility. *Adv. Opt. Mater*, *2*(7), 626–631. doi:10.1002/adom.201400078

Chen, X. P., Zhang, Q. Y., Yang, C. H., Chen, D. D., & Zhao, C. (2009). Comparative investigation on structure and luminescence properties of fluoride phosphors codoped with Er3+/Yb3+. *Spectrochimica Acta. Part A: Molecular Spectroscopy*, *74*(2), 441–445. doi:10.1016/j.saa.2009.06.040

Chen, X., Dai, P. P., Zhang, X. T., Li, C., Lu, S., Wang, X. L., ... Liu, Y. C. (2014). A Highly Efficient White Light (Sr3,Ca,Ba)$(PO_4)_3Cl:Eu^{2+}$, Tb^{3+}, Mn^{2+} Phosphor via Dual Energy Transfers for White Light-Emitting Diodes. *Inorganic Chemistry*, *53*(7), 3441–3448. doi:10.1021/ic402870p PMID:24649898

Chen, Y. M., Hung, W. Y., You, H. W., Chaskar, A., Ting, H. C., Chen, H. F., ... Liu, Y. H. (2011). Carbazole–benzimidazole hybrid bipolar host materials for highly efficient green and blue phosphorescent OLEDs. *Journal of Materials Chemistry*, *21*(38), 14971. doi:10.1039/c1jm12598b

Chesters, J. H. (1973). Refractories production and properties. The Iron and Steel Institute.

Chiang, C. K., & Popielarz, R. (2002). Polymer composites with high dielectric constant. *Ferroelectrics*, *275*(1), 1–9. doi:10.1080/00150190214285

Chien, C. H., Chen, C. K., Hsu, F. M., Shu, C. F., Chou, P. T., & Lai, C. H. (2009). Multifunctional Deep-Blue Emitter Comprising an Anthracene Core and Terminal Triphenylphosphine Oxide Groups. *Advanced Functional Materials*, *19*(4), 560–566. doi:10.1002/adfm.200801240

Chi, L. C., Hung, W. Y., Chiu, H. C., & Wong, K. T. (2009). A high-efficiency and low-operating-voltage green electrophosphorescent device employing a pure-hydrocarbon host material. *Chemical Communications (Cambridge)*, (26): 3892. doi:10.1039/b904795f PMID:19662243

Ching, W. Y., & Xu, Y. N. (1999). Nonscalability and nontransferbility in the eletronic properties of the Y-Al-O systems. *Physical Review B: Condensed Matter and Materials Physics*, *59*(20), 12815–12821. doi:10.1103/PhysRevB.59.12815

Chiu, H. J. (2002). Spherulitic Morphology and crystallization kinetics of poly (vinylidene fluoride)/poly(vinyl acetate) blends. *Journal of Polymer Research*, *9*(3), 169–174. doi:10.1023/A:1021387524243

Chi, Y., & Chou, P. T. (2010). Transition-metal phosphors with cyclometalating ligands: Fundamentals and applications. *Chemical Society Reviews*, *39*(2), 638–655. doi:10.1039/B916237B PMID:20111785

Choe, J. Y., Ravichandran, D., Blomquist, S. M., Morton, D. C., Kirchner, K. W., Ervin, M. H., & Lee, U. (2001). Alkoxy sol-gel derived $Y_{3-x}Al_5O_{12}$:Tbx thin films as efficient cathodoluminescent phosphors. *Applied Physics Letters*, *78*(24), 3800–3802. doi:10.1063/1.1378313

Cho, H. S., Dong, Z., Pauletti, G. M., Zhang, J., Xu, H., Gu, H., ... Shi, D. (2010). Fluorescent, Superparamagnetic Nanospheres for Drug Storage, Targeting, and Imaging: A Multifunctional Nanocarrier System for Cancer Diagnosis and Treatment. *ACS Nano*, *4*(9), 5398–5404. doi:10.1021/nn101000e PMID:20707381

Chopra, N., Lee, J., Zheng, Y., Eom, S. H., Xue, J., & So, F. (2008). High efficiency blue phosphorescent organic light-emitting device. *Applied Physics Letters*, *93*(14), 143307. doi:10.1063/1.3000382

Cho, S., Lee, R., Lee, H., Kim, J., Moon, C., Nam, S., & Park, J. (2010). Synthesis and characterization of Eu^{3+} doped Lu_2O_3 nanophosphor using a solution-combustion method. *Journal of Sol-Gel Science and Technology*, *53*(2), 171–175. doi:10.100710971-009-2073-4

Cho, S., Lee, R., Lee, H., Kim, J., Moon, C., Nam, S., & Park, J. (2010). Synthesis and characterization of Eu^{3+}doped Lu_2O_3 nanophosphor using a solution-combustion method, J. Sol-Gel. *Sciences et Techniques (Paris)*, *53*, 171–175.

Chou, C. H., Yang, C. H., Hsu, C. S., & Chen, T. M. (2007). Hybrid white-light emitting-LED based on luminescent polyfluorene polymer and quantum dots. *Journal of Nanoscience and Nanotechnology*, *7*(8), 2785–2789. doi:10.1166/jnn.2007.622 PMID:17685298

Chou, H. H., & Cheng, C. H. (2010). A Highly Efficient Universal Bipolar Host for Blue, Green, and Red Phosphorescent OLEDs. *Advanced Materials*, *22*(22), 2468–2471. doi:10.1002/adma.201000061 PMID:20446307

Chou, H. H., Chen, Y. H., Hsu, H. P., Chang, W. H., Chen, Y. H., & Cheng, C. H. (2012). Synthesis of Diimidazolyl-stilbenes as n-Type Blue Fluorophores: Alternative Dopant Materials for Highly Efficient Electroluminescent Devices. *Advanced Materials*, *24*(43), 5867–5871. doi:10.1002/adma.201202222 PMID:22915130

Christodoulides, C. (1985). Determination of activation energies by using the widths of peaks of thermoluminescence and thermally stimulated depolarisation currents. *Journal of Physics. D, Applied Physics*, *18*(8), 1501–1510. doi:10.1088/0022-3727/18/8/015

Chronakis, I. S. (2005). Novel nanocomposites and nanoceramics based on polymer nanofibers using electrospinning process--A review. *Journal of Materials Processing Technology*, *167*(2-3), 283–293. doi:10.1016/j.jmatprotec.2005.06.053

Chung, J. H., Lee, S. Y., Shim, K. B., & Ryu, J. H. (2012). White lighting upconversion in $Tm^{3+}/Ho^{3+}/Yb^{3+}$ co-doped $CaWO_4$. *Applied Physics Express*, *5*(5), 052602. doi:10.1143/APEX.5.052602

Chung, K. S., Choe, H. S., Lee, J. I., & Kim, J. L. (2007). A new method for the numerical analysis of thermoluminescence glow curve. *Radiation Measurements*, *42*(4-5), 731–734. doi:10.1016/j.radmeas.2007.02.028

Chung, K. S., Choe, H. S., Lee, J. I., Kim, J. L., & Chang, S. Y. (2005). A computer program for the deconvolution of thermoluminescence glow curves. *Radiation Protection Dosimetry*, *115*(1-4), 343–349. doi:10.1093/rpd/nci073 PMID:16381744

Ci, Z. P., Sun, Q. S., Sun, M. X., Jiang, X. J., Qin, S. C., & Wang, Y. H. (2014). Structure, photoluminescence and thermal properties of Ce^{3+}, Mn^{2+} co-doped phosphosilicate $Sr_7La_3[(PO_4)_{2.5}(SiO_4)_3(BO_4)_{0.5}](BO_2)$ emission-tunable phosphor. *Journal of Materials Chemistry. C, Materials for Optical and Electronic Devices*, *2*(29), 5850–5856. doi:10.1039/C4TC00217B

Coe, S., Woo, W. K., Bawendi, M. G., & Bulovic, V. (2002). Electroluminescence from single monolayers of nanocrystals in molecular organic devices. *Nature*, *420*(6917), 800–803. doi:10.1038/nature01217 PMID:12490945

Colvin, V. L., Schlamp, M. C., & Alivisatos, A. P. (1994). Light-Emitting-Diodes made from Cadmium Selenide Nanocrystals and a Semiconductor Polymer. *Nature*, *370*(6488), 354–357. doi:10.1038/370354a0

Corless, R. M., Gonnet, G. H., Hare, D. E. G., Jeffrey, D. J., & Knuth, D. E. (1996). On the Lambert W function. *Advances in Computational Mathematics*, *5*(1), 329–359. doi:10.1007/BF02124750

Corless, R. M., Jeffrey, D. J., & Knuth, D. E. (1997). A sequence of series for the Lambert W Function. In *ISSAC '97: Proceedings of the 1997 International Symposium on Symbolic and Algebraic Computation*, (pp. 197). New York: ACM Press. 10.1145/258726.258783

Crawford, M. K., & Brixner, L. H. (1991). Photostimulable phosphors for X-ray imaging: Applications and mechanism. *Journal of Luminescence*, *48*, 37–42. doi:10.1016/0022-2313(91)90073-5

Czaya, R., & Bissert, G. (1971). Die Kristallstruktur von Tricalciummonosilikatdichlorid ($Ca_2SiO_4.CaCl_2$). *Acta Crystallographica. Section B, Structural Crystallography and Crystal Chemistry*, *27*(4), 747–752. doi:10.1107/S0567740871002875

Dabbousi, B. O., Bawendi, M. G., Onotsuka, O., & Rubner, M. F. (1995). Electroluminescence from CdSe quantum dots polymer composite. *Applied Physics Letters*, *66*(11), 1316–1318. doi:10.1063/1.113227

DaCosta, M. V., Doughan, S., Han, Y., & Krull, U. J. (2014). Lanthanide upconversion nanoparticles and applications in bioassays and bioimaging: A review. *Analytica Chimica Acta*, *832*, 1–33. doi:10.1016/j.aca.2014.04.030 PMID:24890691

Dai, P., Lee, S. P., Chan, T. S., Huang, C. H., Chiang, Y. W., & Chen, T. M. (2016). $Sr_3Ce(PO_4)_3:Eu^{2+}$: a broadband yellow-emitting phosphor for near ultraviolet-pumped white light-emitting devices. *The Journal of Physical Chemistry C*, *4*, 1170–1177.

Dai, Q., Foley, M. E., Breshike, C. J., Lita, A., & Strouse, G. F. (2011). Ligand-Passivated Eu:Y_2O_3 Nanocrystals as a Phosphor for White Light Emitting Diodes. *Journal of the American Chemical Society*, *133*(39), 15475–15486. doi:10.1021/ja2039419 PMID:21863840

Davis, M. E. (2002). Ordered porous materials for emerging applications. *Nature*, *417*(6891), 813–821. doi:10.1038/nature00785 PMID:12075343

Deilhof, J. B., Bos, A. J. J., Zoetelief, J., & Broerse, J. J. (1988). Sensitivity of CaF_2 thermoluminescent materials to fast neutrons. *Radiation Protection Dosimetry*, *23*(1-4), 405–408. doi:10.1093/oxfordjournals.rpd.a080207

Deng, D., Yu, H., Li, Y., Hua, Y., Jia, G., Zhao, S., ... Xu, S. (2013). $Ca_4(PO_4)_2O:Eu^{2+}$ red-emitting phosphor for solid-state lighting: structure, luminescent properties and white light emitting diode application. *Journal of Materials Chemistry. C, Materials for Optical and Electronic Devices*, *1*(19), 3194–3199. doi:10.1039/c3tc30148f

Deng, X. Y. (2011). Light emitting devices with conjugated polymers. *International Journal of Molecular Sciences*, *12*(12), 1575–1594. doi:10.3390/ijms12031575 PMID:21673908

Destriau, G. (1936). AC electroluminescence in ZnS. *Journal de Chimie Physique*, *33*, 587. doi:10.1051/jcp/1936330587

Dexter, D. L., & Schulman, J. H. (1954). Theory of Concentration Quenching in Inorganic Phosphors. *The Journal of Chemical Physics*, *22*(6), 1063–1070. doi:10.1063/1.1740265

Dey, R., Kumari, A., Soni, A. K., & Rai, V. K. (2015). $CaMoO_4$: Ho^{3+}–Yb^{3+}–Mg^{2+} upconverting phosphor for application in lighting devices and optical temperature sensing. *Sensors and Actuators. B, Chemical*, *210*, 581–588. doi:10.1016/j.snb.2015.01.007

Dey, R., Rai, V. K., & Pandey, A. (2012). Green upconversion emission in Nd^{3+}–Yb^{3+}–Zn^{2+}: Y_2O_3 phosphor. *Spectrochimica Acta. Part A: Molecular and Biomolecular Spectroscopy*, *99*, 288–291. doi:10.1016/j.saa.2012.09.001 PMID:23085286

Di Bartolo, B., & Forte, O. (Eds.). (2006). *Advances in spectroscopy for lasers and sensing* (Vol. 231). Springer Science & Business Media. doi:10.1007/1-4020-4789-4

Dieke, G. H., & Crosswhite, H. M. (1963). The Spectra of the Doubly and Triply Ionized Rare Earths. *Applied Optics*, *2*(7), 675. doi:10.1364/AO.2.000675

Dieke, G. H., Crosswhite, H. M., & Crosswhite, H. M. (1968). *Spectra and energy levels of rare earth ions in crystals* (Vol. 5). New York: Interscience Publishers.

Diesperova, M. I., Bron, V. A., & Boriskova, T. I. (1977). Alekseeva and E.I.Kelareva. *Refractories and Industrial Ceramics*, *18*(5-6), 278.

Ding, J. Y., Wu, Q. S., Li, Y. Y., Long, Q., Wang, C., & Wang, Y. H. (2015). $Sr_{7.3}Ca_{2.7}(PO_4)_6F_2$: Eu^{2+}, Mn^{2+}: a novel single-phased white light-emitting phosphor for NUV-LEDs. *Dalton Transactions (Cambridge, England)*, *44*(20), 9630–9636. doi:10.1039/C5DT00907C PMID:25924106

Ding, X., & Wang, Y. (2017). Structure and photoluminescence properties of a novel apatite green phosphor $Ba_5(PO_4)_2SiO_4$:Eu^{2+} excited by NUV light. *Physical Chemistry Chemical Physics*, *19*(3), 2449–2458. doi:10.1039/C6CP08300E PMID:28058425

Dong, H., Sun, L. D., Wang, Y. F., Xiao, J. W., Tu, D., Chen, X., & Yan, C. H. (2016). Photon upconversion in Yb^{3+}–Tb^{3+} and Yb^{3+}–Eu^{3+} activated core/shell nanoparticles with dual-band excitation. *Journal of Materials Chemistry. C, Materials for Optical and Electronic Devices*, *4*(19), 4186–4192. doi:10.1039/C6TC00413J

Dorenbos, P. (2003). Energy of the first $4f^7 \rightarrow 4f^6 5d$ transition of Eu^{2+} in inorganic compounds. *Journal of Luminescence*, *104*(4), 239–260. doi:10.1016/S0022-2313(03)00078-4

Dosev, D., Kennedy, I. M., Godlewski, M., Gryczynski, I., Tomsia, K., & Goldys, E. M. (2006). Fluorescence upconversion in Sm-doped Gd_2O_3. *Applied Physics Letters*, *88*(1), 011906. doi:10.1063/1.2161400

Doulan, J. L., & Moncorge, R. (2003). Article. *Ann. Chim. Sci. Mat.*, *28*, 5.

Dou, Q. Q., Guo, H. C., & Ye, E. (2014). Near-infrared upconversion nanoparticles for bio-applications. *Materials Science and Engineering C*, *45*, 635–643. doi:10.1016/j.msec.2014.03.056 PMID:25491873

Downing, E., Hesselink, L., Ralston, J., & Macfarlane, R. (1996). A three-color, solid-state, three-dimensional display. *Science*, *273*(5279), 1185–1189. doi:10.1126cience.273.5279.1185

Drofenik, M., & Golic, L. (1979). Refinement of the Sr_2EuFeO_5 and Sr_3EuAlO_5 structures. *Acta Crystallographica. Section B, Structural Crystallography and Crystal Chemistry*, *B35*(5), 1059–1062. doi:10.1107/S0567740879005574

Dubey, Kaur, Parganiha, Suryanarayana, & Murthy. (2016). Article. *Appl. Rad. Isotop.s, 110*, 16-27.

Dubey, Tiwari, Tamrakar, Markande, Rathore, & Pradhan. (2015). Article. *Indian Journal of Materials Science*. 10.1155/2015/745052

Dubey, & Suryanarayana, & Kaur. (2010). Kinetics of TL Glow Peak of Limestone from Patharia of CG Basin (India). *Journal of Minerals & Materials Characterization & Engineering*, *9*(12), 1101–1111. doi:10.4236/jmmce.2010.912080

Dubey, V., & Jagjeet Kaur, N. S. (2013). Thermoluminescence and chemical characterization of natural calcite collected from Kodwa mines. *Research on Chemical Intermediates*, *39*(8), 3689–3697. doi:10.100711164-012-0872-7

Dubey, V., Kaur, J., & Agrawal, S. (2015). Effect of europium concentration on photoluminescence and thermoluminescence behavior of Y2O3:Eu3+ phosphor. *Research on Chemical Intermediates*, *41*(7), 4727–4739. doi:10.100711164-014-1563-3

Dubey, V., Kaur, J., & Agrawal, S. (2015). Effect of europium doping levels on photoluminescence and thermoluminescence of strontium yttrium oxide phosphor. *Materials Science in Semiconductor Processing*, *31*, 27–37. doi:10.1016/j.mssp.2014.10.052

Dubey, V., Kaur, J., Agrawal, S., Suryanarayana, N. S., & Murthy, K. V. R. (2013). Optik – Int. *J.Light Electron Opt.*, *124*(22), 5585–5587. doi:10.1016/j.ijleo.2013.03.153

Dubey, V., Kaur, J., Agrawal, S., Suryanarayana, N. S., & Murthy, K. V. R. (2014). Effect of Eu3+ concentration on photoluminescence and thermoluminescence behavior of YBO3:Eu3+ phosphor. *Superlattices and Microstructures*, *67*, 156–171. doi:10.1016/j.spmi.2013.12.026

Dubey, V., Kaur, J., Parganiha, Y., Suryanarayana, N. S., & Murthy, K. V. R. (2016). Study of formation of deep trapping mechanism by UV, beta and gamma irradiated Eu^{3+} activated SrY_2O_4 and $Y_4A_{12}O_9$. *Applied Radiation and Isotopes*, *110*, 16–27. doi:10.1016/j.apradiso.2015.12.047 PMID:26748019

Dubey, V., Kaur, J., Suryanarayana, N. S., & Agrawal, S. (2013). TL glow curve analysis of beta irradiated Eu3+ doped Y4Al2O9 phosphor. *Int. J. Lumin. Appl*, *3*(II), 98.

Dubey, V., Tiwari, R., Tamrakar, R. K., Kaur, J., Dutta, S., Das, S., ... Som, S. (2016). Estimation of spectroscopic parameters and colour purity of the red-light-emitting YBa3B9O18 phosphor: Judd–Ofelt approach. *Journal of Luminescence*, *180*, 169–176. doi:10.1016/j.jlumin.2016.08.029

Dubey, V., Tiwari, R., Tamrakar, R. K., Rathore, G. S., Sharma, C., & Tiwari, N. (2014). Infrared spectroscopy and upconversion luminescence behaviour of erbium doped yttrium (III) oxide phosphor. *Infrared Physics & Technology*, *67*, 537–541. doi:10.1016/j.infrared.2014.09.014

Dubinskii, M. A., Semashko, V. V., Naumov, A. K., Abdulsabirov, R. Y., & Korableva, S. L. (1993). Article. *Laser Physics*, *3*, 216.

Dufour, C., Khomenkov, V., Rizza, G., & Toulemonde, M. (2012). Ion-matter interaction: The three-dimensional version of the thermal spike model. Application to nanoparticle irradiation with swift heavy ions. *Journal of Physics. D, Applied Physics*, *45*(6), 065302. doi:10.1088/0022-3727/45/6/065302

Du, H., Zhang, W., & Sun, J. (2011). Structure and upconversion luminescence properties of BaYF5:Yb3+, Er3+ nanoparticles prepared by different methods. *Journal of Alloys and Compounds*, *509*(7), 3413–3418. doi:10.1016/j.jallcom.2010.12.101

Dutta, P., & Rai, S. (2011). Optical transitions and frequency upconversions of Ho^{3+} and Ho^{3+}/Yb^{3+} ions in $Al(NO_3)_3–SiO_2$ sol–gel glasses. *Optik (Stuttgart)*, *122*(10), 858–863. doi:10.1016/j.ijleo.2010.05.029

Dutta, S., Som, S., & Sharma, S. K. (2013). Optimization and characterization of trap level distribution in γ-irradiated doped/codoped $CaMoO_4$ phosphors. *Physica B, Condensed Matter*, *417*, 39–45. doi:10.1016/j.physb.2013.02.024

Dutta, S., Som, S., & Sharma, S. K. (2015). Excitation spectra and luminescence decay analysis of K^+ compensated Dy^{3+} doped $CaMoO_4$ phosphors. *RSC Advances*, *5*(10), 7380–7387. doi:10.1039/C4RA12447B

Dwivedi, C., & Dutta, V. (2013). Effect of seed layer on the self assembly of spray pyrolyzed Al-doped ZnO nanoparticles. *AIP Advances*, *3*(3), 032127. doi:10.1063/1.4795762

Dwivedi, Y., Bahadur, A., & Rai, S. B. (2011). Optical avalanche in Ho-Yb: Gd_2O_3 nanocrystals. *Journal of Applied Physics*, *110*(4), 043103. doi:10.1063/1.3622624

Dwivedi, Y., & Zilio, S. C. (2013). Infrared cascade and cooperative multicolor upconversion emissions in $Y_8V_2O_{17}$:Eu-Yb nanophosphors. *Optics Express*, *21*(4), 4717–4727. doi:10.1364/OE.21.004717 PMID:23482004

Dzik, G. D., Sokolska, I., Golab, S., & Baluka, M. (2000). Article. *Journal of Alloys and Compounds*, *300*, 254–260. doi:10.1016/S0925-8388(99)00779-3

Eanes, M. (2000). *Synthesis and characterization of alkali silver chalcogenides and alkali rare earth germanates by supercritical fluids, Doctor of Philosophy*. Clemson University.

Ebendor-Heidepriem, H., & Ehrt, D. (1996). Article. *Journal of Non-Crystalline Solids*, *208*, 205. doi:10.1016/S0022-3093(96)00524-8

Ekimov, A., Efros, A., & Onushchenko, A. (1985). Quantum size effect in semiconductor microcrystals. *Solid State Communications*, *56*(11), 921–924. doi:10.1016/S0038-1098(85)80025-9

Fain, J., Montret, M., & Sahraoui, L. (1980). Thermoluminescent response of CaF_2: Dy and LiF: Mg, Ti under heavy ion bombardment. *Nuclear Instruments and Methods*, *175*(1), 37–39. doi:10.1016/0029-554X(80)90245-1

Fan, C. H., Sun, P. P., Su, T. H., & Cheng, C. H. (2011). Host and Dopant Materials for Idealized Deep-Red Organic Electrophosphorescence Devices. *Advanced Materials*, *23*(26), 2981–2985. doi:10.1002/adma.201100610 PMID:21567483

Farinola, G. M., & Ragni, R. (2011). Electroluminescent materials for white organic light emitting diode. *Chemical Society Reviews*, *40*(7), 3467–3482. doi:10.1039/c0cs00204f PMID:21437308

Feng, J., & Zhang, H. (2013). Hybrid materials based on lanthanide organic complexes: A review. *Chemical Society Reviews*, *42*(1), 387–410. doi:10.1039/C2CS35069F PMID:22977887

Feng, X. J., Chen, S. F., Ni, Y., Wong, M. S., Lam, M. M. K., Cheah, K. W., & Lai, G. Q. (2014). Fluorene Derivatives for Highly Efficient Non-Doped Single-Layer Blue Organic Light-Emitting Diodes. *Organic Electronics*, *15*(1), 57–64. doi:10.1016/j.orgel.2013.10.019

Fleming, R. J. (1968). Tabulation of glow-curve characteristics. *Canadian Journal of Physics*, *46*(14), 1569–1573. doi:10.1139/p68-483

Furetta, C. (2003). *Handbook of Thermoluminescence* (2nd ed.). Singapore: World Scientific Publishing Co. Pte. Ltd. doi:10.1142/5167

Fu, X. Y., Fang, L. J., Niu, S. Y., & Zhang, H. W. (2013). Luminescence properties and energy transfer investigations of $SrMgSi_2O_6$:Ce, Tb phosphors. *Journal of Luminescence*, *142*, 163–166. doi:10.1016/j.jlumin.2013.03.055

Fu, X., Lü, W., Jiao, M., & You, H. (2016). Broadband Yellowish-Green Emitting $Ba_4Gd_3Na_3(PO_4)_6F_2$:Eu^{2+} Phosphor: Structure Refinement, Energy Transfer, and Thermal Stability. *Inorganic Chemistry*, *55*(12), 6107–6113. doi:10.1021/acs.inorgchem.6b00648 PMID:27249557

Fu, Z., Zhou, S., Pan, T., & Zhang, S. (2007). Preparation and luminescent properties of cubic Eu3+:Y2O3 nanocrystals and comparison to bulk Eu3+:Y2O3. *Journal of Luminescence*, *124*(2), 213–216. doi:10.1016/j.jlumin.2006.02.020

Ganesh, I., Johnson, R., Rao, G. V. N., Mahajan, Y. R., Madavendra, S. S., & Reddy, B. M. (2005). Microwave-assisted combustion synthesis of nanocrystalline MgAl2O4 spinel powder. *Ceramics International*, *31*(1), 67–74. doi:10.1016/j.ceramint.2004.03.036

Gao, G., Turshatov, A., Howard, I. A., Busko, D., Joseph, R., Hudry, D., & Richards, B. S. (2017). Up-Conversion Fluorescent Labels for Plastic Recycling: A Review. *Advanced Sustainable Systems*.

Gao, H. Z. (2010). Theoretical investigation into charge mobility in 4, 4[1]-bis(1-napthylphenylamino) biphenyl. *Theoretical Chemistry Accounts*, *127*(5-6), 759–763. doi:10.100700214-010-0804-9

Gao, M., Richter, B., Kirstein, S., & Mohwald, H. (1998). Electroluminescence studies on self assembled filmof PPV and CdSe Nanoparticles. *The Journal of Physical Chemistry B*, *102*(21), 4096–4103. doi:10.1021/jp980620a

Gao, P., Xie, Y., & Li, Z. (2006). Controlling the Size of BaF2 Nanocubes from 1000 to 10 nm. *European Journal of Inorganic Chemistry*, *16*(16), 3261–3265. doi:10.1002/ejic.200600237

Gao, Z., Liu, F., Li, J., Cheng, G., & Lu, P. (2016). Solution-processed UV light emitting diode based on butyltriphenylsilane modified phenanthro[9,10-d]imidazole with high efficiency. *RSC Advances*, *6*(85), 81744–81749. doi:10.1039/C6RA18755B

Gao, Z., Wang, Z., Shan, T., Liu, Y., Shen, F., Pan, Y., ... Ma, Y. (2014). High-efficiency deep blue fluorescent emitters based on phenanthro[9,10-d]imidazole substituted carbazole and their applications in organic light emitting diodes. *Organic Electronics*, *15*(11), 2667–2676. doi:10.1016/j.orgel.2014.07.019

GarcíaSolé, Bausa, & Jaque. (2005). *An Introduction to the Optical Spectroscopy of Inorganic Solids*. John Wiley and Sons Ltd.

Garlick, G. F. J., & Gibson, A. F. (1948). The Electron Trap Mechanism of Luminescence in Sulphide and Silicate Phosphors. *Proceedings of the Physical Society*, *60*(6), 574–590. doi:10.1088/0959-5309/60/6/308

Gautam, N. K., Ramrakhiani, M., Kuraria, R., & Kuraria, S. R. (2015). Electroluminescence in Organically Capped $C_{d1-x}Z_{nx}$Se Chalcogenide Nanocrystals. *Defect and Diffusion Forum*, *361*, 215–230. doi:10.4028/www.scientific.net/DDF.361.215

Gedam, S. C., Dhoble, S. J., & Park, K. (2013). Resonant energy transfer from $Ce^{3+}\rightarrow Dy^{3+}$ and Mn^{2+} in $NaZnSO_4Cl$ chlorosulphate phosphor. *Journal of Luminescence*, *139*, 47–51. doi:10.1016/j.jlumin.2013.02.010

Gektin, A. V., Krasovitskaya, I. M., & Shiran, N. V. (1998). UV-dosimetric material based on KMgF3 perovskite. *Radiation Measurements*, *29*(3-4), 337–340. doi:10.1016/S1350-4487(98)00054-7

Gektin, A., Komar, V., Shlyahturov, V., & Shiran, N. (1996). Pure and Ce-doped KMgF/sub 3/-scintillators. *IEEE Transactions on Nuclear Science*, *43*(3), 1295–1297. doi:10.1109/23.506681

Gektin, A., Shiran, N., Neichev, S., Gavrilyuk, V., Bensalah, A., Fukuda, T., & Shimamura, K. (2002). LiCaAlF6:Ce crystal: a new scintillator. *Nucl. Instru. Methods A*, *486*(1-2), 274–277. doi:10.1016/S0168-9002(02)00717-9

George, N. C., Pell, A. J., Dantelle, G., Page, K., Llobet, A., Balasubramanian, M., ... Seshadri, R. (2013). Local Environments of Dilute Activator Ions in the Solid-State Lighting Phosphor $Y_{3-x}Ce_xAl_5O_{12.}$ *Chemistry of Materials*, *25*(20), 3979–3995. doi:10.1021/cm401598n

Ge, Z., Hayakawa, T., Ando, S., Ueda, M., Akiike, T., Miyamoto, H., ... Kakimoto, M. (2008). Solution-Processible Bipolar Triphenylamine-Benzimidazole Derivatives for Highly Efficient Single-Layer Organic Light-Emitting Diodes. *Chemistry of Materials*, *20*(7), 2532–2537. doi:10.1021/cm7035458

Gibart, P., Auzel, F., Guillaume, J. C., & Zahraman, K. (1996). Below band-gap IR response of substrate-free GaAs solar cells using two-photon up-conversion. *Japanese Journal of Applied Physics*, *35*(8R), 4401–4402. doi:10.1143/JJAP.35.4401

Giri, N. K., Mishra, K., & Rai, S. B. (2011). Upconversion based tunable white-light generation in Ln: Y_2O_3 nanocrystalline phosphor (Ln= Tm/Er/Yb). *Journal of Fluorescence*, *21*(5), 1951–1958. doi:10.100710895-011-0894-3 PMID:21614491

Glasser, L. S. D., & Glasser, F. P. (1965). Silicates M_3SiO_5. I. Sr_3SiO_5. *Acta Crystallographica*, *18*(3), 453–454. doi:10.1107/S0365110X65000993

Glodo, J., & Wojtowicz, A.J. (2000). Thermoluminescence and scintillation properties of LuAP and YAP. *J. Alloys. Compd.*, *300-301*, 289.

Gnaser, H. (1999). *Introduction*. Springer Berlin Heidelberg.

Gnaser, H. (1999). *Low-Energy Ion Irradiation of Solid Surfaces*. Berlin: Springer-Verlag.

Gong, J. R., Wan, L. J., Lei, S. B., Bai, C. L., Zhang, X.-H., & Lee, S.-T. (2005). Direct evidence of molecular aggregation and degradation mechanism of organic light-emitting diodes under joule heating: An STM and photoluminescence study. *The Journal of Physical Chemistry B*, *109*(5), 1675–1682. doi:10.1021/jp046509o PMID:16851142

Gong, S. L., Fu, Q., Wang, Q., Yang, C. L., Zhong, C., Qin, J. G., & Ma, D. G. (2011). Highly Efficient Deep-Blue Electrophosphorescence Enabled by Solution-Processed Bipolar Tetraarylsilane Host with Both a High Triplet Energy and a High-Lying HOMO Level. *Advanced Materials*, *23*(42), 4956–4959. doi:10.1002/adma.201102758 PMID:21959864

Gong, S., Chen, Y., Luo, J., Yang, C., Zhong, C., Qin, J., & Ma, D. (2011). Bipolar Tetraarylsilanes as Universal Hosts for Blue, Green, Orange, and White Electrophosphorescence with High Efficiency and Low Efficiency Roll-Off. *Advanced Functional Materials*, *21*(6), 1168–1178. doi:10.1002/adfm.201002066

Gong, S., Chen, Y., Yang, C., Zhong, C., Qin, J., & Ma, D. (2010). De Novo Design of Silicon-Bridged Molecule Towards a Bipolar Host: All-Phosphor White Organic Light-Emitting Devices Exhibiting High Efficiency and Low Efficiency Roll-Off. *Advanced Materials*, *22*(47), 5370–5373. doi:10.1002/adma.201002732 PMID:20972980

Goodenough, J. B., Manthiram, A., Paranthaman, M., & Zhen, Y. S. (1992). Oxide ion electrolytes. *Materials Science and Engineering B*, *12*(4), 357–364. doi:10.1016/0921-5107(92)90006-U

Gopichandran, R., & Patil, K. C. (1992). Articles. *Materials Letters*, *12*, 437. doi:10.1016/0167-577X(92)90209-3

Green, M. A., & Bremner, S. P. (2017). Energy conversion approaches and materials for high-efficiency photovoltaics. *Nature Materials*, *16*(1), 23–34. doi:10.1038/nmat4676 PMID:27994249

Grossweiner, I. (1953). A Note on the Analysis of First-Order Glow Curves. *Journal of Applied Physics*, *24*(10), 1306–1307. doi:10.1063/1.1721152

Guan, M., Chen, Z., Bian, Z., Liu, Z., Gong, Z., Baik, W., ... Huang, C. (2006). The host materials containing carbazole and oxadiazole fragment for red triplet emitter in organic light-emitting diodes. *Organic Electronics*, *7*(5), 330–336. doi:10.1016/j.orgel.2006.04.006

Guo, C., Ding, X., Luan, L., & Xu, Y. (2010). Two-color emitting of Eu^{2+} and Mn^{2+} co-doped $Sr_2Mg_3P_4O_{15}$ for UV LEDs. *Sensors and Actuators. B, Chemical*, *143*(2), 712–715. doi:10.1016/j.snb.2009.10.023

Guo, H. (2007). Green and red upconversion luminescence in CeO_2:Er^{3+}powders produced by 785 nm laser. *Journal of Solid State Chemistry*, *180*(1), 127–131. doi:10.1016/j.jssc.2006.10.003

Guo, H., Dong, N., Yin, M., Zhang, W., Lou, L., & Xia, S. (2004). Visible upconversion in rare earth ion-doped Gd_2O_3 nanocrystals. *The Journal of Physical Chemistry B*, *108*(50), 19205–19209. doi:10.1021/jp048072q

Guo, N., Huang, Y. J., Jia, Y. C., Lv, W. Z., Zhao, Q., Lu, W., ... You, H. P. (2013). A novel orange-yellow-emitting $Ba_3Lu(PO_4)_3$:Eu^{2+}, Mn^{2+} phosphor with energy transfer for UV-excited white LEDs. *Dalton Transactions (Cambridge, England)*, *42*(4), 941–947. doi:10.1039/C2DT31657A PMID:23108262

Guo, N., Huang, Y. J., Yang, M., Song, Y. H., Zheng, Y. H., & You, H. P. (2011). A tunable single-component warm white-light $Sr_3Y(PO_4)_3$: Eu^{2+}, Mn^{2+} phosphor for white-light emitting diodes. *Physical Chemistry Chemical Physics, 13*(33), 15077–15082. doi:10.1039/c1cp20635d PMID:21773615

Guo, N., Jia, Y. C., Lü, W., Lv, W. Z., Zhao, Q., Jiao, M. M., ... You, H. P. (2013). Direct warm-white-emitting $Sr_3Sc(PO_4)_3$:Eu^{2+}, Mn^{2+} phosphor with tunable photoluminescence via efficient energy transfer. *Dalton Transactions (Cambridge, England), 28*(16), 5649–5654. doi:10.1039/c3dt32946a PMID:23429499

Guo, N., Lu, W., Jia, Y. C., Lv, W. Z., Zhao, Q., & You, H. P. (2013). Eu^{2+}& Mn^{2+}-Coactivated $Ba_3Gd(PO_4)_3$ Orange-Yellow-Emitting Phosphor with Tunable Color Tone for UV-Excited White LEDs. *ChemPhysChem, 14*(1), 192–197. doi:10.1002/cphc.201200836 PMID:23161526

Guo, N., You, H. P., Jia, C. Z., Ouyang, R. Z., & Wu, D. H. (2014). A Eu^{2+} and Mn^{2+}-co-activated fluoro-apatite-structure $Ca_6Y_2Na_2(PO_4)_6F_2$ as a standard white-emitting phosphor via energy transfer. *Dalton Transactions (Cambridge, England), 43*(32), 12373–12379. doi:10.1039/C4DT01021C PMID:24993422

Guo, N., You, H., Song, Y., Yang, M., Liu, K., Zheng, Y., ... Zhang, H. (2010). White-light emission from a single-emitting-component $Ca_9Gd(PO_4)_7$:Eu^{2+},Mn^{2+} phosphor with tunable luminescent properties for near-UV light-emitting diodes. *Journal of Materials Chemistry, 20*(41), 9061–9067. doi:10.1039/c0jm01860k

Guo, N., Zhang, X. Y., Jia, Y. C., Qiao, H., & You, H. P. (2012). Warm-white-emitting from Eu^{2+}/Mn^{2+}-codoped $Sr_3Lu(PO_4)_3$ phosphor with tunable color tone and correlated color temperature. *The Journal of Physical Chemistry C, 116*(1), 1329–1334. doi:10.1021/jp209891b

Guo, N., Zheng, Y. H., Jia, Y. C., Qiao, H., & You, H. P. (2012). A tunable warm-white-light $Sr_3Gd(PO_4)_3$:Eu^{2+}, Mn^{2+} phosphor system for LED-based solid-state lighting. *New Journal of Chemistry, 36*(1), 168–172. doi:10.1039/C1NJ20532C

Guo, W., Zheng, K., Xie, W., Sun, L., Shen, L., Liu, C., ... Zhang, Z. (2014). Efficiency enhancement of inverted polymer solar cells by doping $NaYF_4$: Yb^{3+}, Er^{3+} nanocomposites in PCDTBT: PCBM active layer. *Solar Energy Materials and Solar Cells, 124*, 126–132. doi:10.1016/j.solmat.2014.01.038

Gupta, K. K., Kadam, R. M., Dhoble, N. S., Lochab, S. P., Singh, V., & Dhoble, S. J. (2016). Photoluminescence, thermoluminescence and evaluation of some parameters of Dy^{3+} activated $Sr_5(PO_4)_3F$ phosphor synthesized by sol-gel method. *Journal of Alloys and Compounds, 688*, 982–993. doi:10.1016/j.jallcom.2016.07.114

Gurav, A., Kodas, T., Pluym, T., & Xiong, Y. (1993). Aerosol Processing of Materials. *Aer. Sci. Tech., 19*(4), 411–452. doi:10.1080/02786829308959650

Haering, R. R., & Adams, E. N. (1960). Theory and Application of Thermally Stimulated Currents in Photoconductors. *Physical Review, 117*(2), 451–454. doi:10.1103/PhysRev.117.451

Haferkorn, B., & Meyer, G. (1998). Li_2EuSiO_4, einEuropium(II)-dilithosilicat:Eu[$(Li_2Si)O_4$]. *Zeitschrift fur Anorganische und Allgemeine Chemie, 624*(7), 1079–1081. doi:

Halperin, A., & Braner, A. A. (1960). Evaluation of Thermal Activation Energies from Glow Curves. *Physical Review, 117*(2), 408–415. doi:10.1103/PhysRev.117.408

Halperin, A., Braner, A. A., Ben-Zvi, A., & Kristianpoller, N. (1960). Thermal Activation Energies in NaCl and KC1 Crystals. *Physical Review, 117*(2), 416–422. doi:10.1103/PhysRev.117.416

Hameed, S., Pradeep, P., & Baiju, M. R. (2010). Polymer Light Emitting Diodes- A review on materials and Techniques Rev. *Adv. Mater. Sc., 26*, 30–42.

Han, B., Liang, H., Ni, H., Su, Q., Yang, G., Shi, J., & Zhang, G. (2009). Intense red light emission of Eu^{3+}-doped $LiGd(PO_3)_4$ for mercury-free lamps and plasma display panels application. *Optics Express*, *17*(9), 7138–7144. doi:10.1364/OE.17.007138 PMID:19399090

Han, C., Xie, G., Xu, H., Zhang, Z., Yu, D., Zhao, Y., ... Liu, S. (2011). Towards Highly Efficient Blue-Phosphorescent Organic Light-Emitting Diodes with Low Operating Voltage and Excellent Efficiency Stability. *Chemistry (Weinheim an der Bergstrasse, Germany)*, *17*(2), 445–449. doi:10.1002/chem.201001981 PMID:21207558

Hao, R. R., Fang, X. Y., & Niu, S. C. (1998). *Series of Inorganic Chemistry* (Vol. 3). Beijing: Science Press.

Hasan, F., Kitis, G., & Charalambou, S. (1985). Regenerated thermoluminescence in $CaSO_4$-Dy. *Journal of Physics. C. Solid State Physics*, *18*(8), 1743–1751. doi:10.1088/0022-3719/18/8/020

Hayati, I., Bailey, A. I., & Tadros, T. F. (1986). Mechanism of stable jet formation in electro hydrodynamic atomization. *Nature*, *319*(6048), 41–43. doi:10.1038/319041a0

Hebert, T., Wannemacher, R., Lenth, W., & Macfarlane, R. M. (1990). Blue and green cw upconversion lasing in Er:$YLiF_4$. *Applied Physics Letters*, *57*(17), 1727–1729. doi:10.1063/1.104048

Heer, S., Kopme, K., Gudel, H. U., & Haase, M. (2004). Highly Efficient Multicolour Upconversion Emission in Transparent Colloids of Lanthanide-Doped $NaYF_4$ Nanocrystals. *Advanced Materials*, *16*(23-24), 2102–2105. doi:10.1002/adma.200400772

He, F., Niu, N., Zhang, Z. G., Zhang, X., Wang, D., Bai, L., ... Yang, P. P. (2012). Morphology-controllable synthesis and enhanced luminescence properties of β-NaLuF4:Ln (Ln = Eu, Tb and Ce/Tb) microcrystals by solvothermal process. *RSC Advances*, *2*(19), 7569–7577. doi:10.1039/c2ra20675g

Hehlen, M. P., Brik, M. G., & Krämer, K. W. (2013). 50th anniversary of the Judd–Ofelt theory: An experimentalist's view of the formalism and its application. *Journal of Luminescence*, *136*, 221–239. doi:10.1016/j.jlumin.2012.10.035

Hermus, M., Phan, P., & Brgoch, J. (2016). Ab Initio Structure Determination and Photoluminescent Properties of an Efficient, Thermally Stable Blue Phosphor, $Ba_2Y_5B_5O_{17}$:Ce^{3+}. *Chemistry of Materials*, *28*(4), 1121–1127. doi:10.1021/acs.chemmater.5b04542

Hermus, M., Phan, P., Duke, A. C., & Brgoch, J. (2017). Tunable Optical Properties and Increased Thermal Quenching in the Blue-Emitting Phosphor Series: $Ba_2(Y_{1-x}Lu_x)_5B_5O_{17}$:$Ce^{3+}$ (x = 0–1). *Chemistry of Materials*, *29*(12), 5267–5275. doi:10.1021/acs.chemmater.7b01416

He, Y., Zhang, J., Zhou, W., Han, J., Qiu, Z., Yu, L., ... Lian, S. (2014). Multicolor Emission in a Single-Phase Phosphor $Ca_3Al_2O_6$:Ce^{3+}, Li^+: Luminescence and Site Occupancy. *Journal of the American Ceramic Society*, *97*(5), 1517–1522. doi:10.1111/jace.12801

Hilderbrand, S. A., Shao, F., Salthouse, C., Mahmood, U., & Weissleder, R. (2009). Upconverting luminescent nanomaterials: Application to in vivo bioimaging. *Chemical Communications*, (28): 4188–4190. doi:10.1039/b905927j PMID:19585016

Hipolito, G., Falcony, M., Aguilar-Frutis, C., & Azorín, M. A. (2001). Article. *J. Appl. Phys. Lett.*, *79*, 4369–4371. doi:10.1063/1.1428110

Hirao, K., Tamai, K., Tanabe, S., & Soga, N. (1993). Frequency upconversion its new mechanism in Tm3+-doped fluoro aluminate glasses. *Journal of Non-Crystalline Solids*, *160*(3), 261–267. doi:10.1016/0022-3093(93)91270-D

Hlavacek, V., & Puszynski, J. A. (1996). Chemical Engineering Aspects of Advanced Ceramic Materials. *Industrial & Engineering Chemistry Research*, *35*(2), 349–377. doi:10.1021/ie9501034

Hodgson, R. T., Balgin, J. E. E., Pal, R., Neri, J. M., & Hammer, D. A. (1980). Ion beam annealing of semiconductors. *Applied Physics Letters*, *37*(2), 187–188. doi:10.1063/1.91819

Holmes, R. J., D'Andrade, B. W., Forrest, S. R., Ren, X., Li, J., & Thompson, M. E. (2003). Efficient, deep-blue organic electrophosphorescence by guest charge trapping. *Applied Physics Letters*, *83*(18), 3818–3820. doi:10.1063/1.1624639

Holmes, R. J., Forrest, S. R., Tung, Y. J., Kwong, R. C., Brown, J. J., Garon, S. M., & Thompson, E. (2003). Blue organic electrophosphorescence using exothermic host–guest energy transfer. *Applied Physics Letters*, *82*(15), 2422–2424. doi:10.1063/1.1568146

Hoppe, H. A. (2009). Recent Developments in the Field of Inorganic Phosphors. *Angewandte Chemie*, *48*(20), 3572–3582. doi:10.1002/anie.200804005 PMID:19350600

Horowitz, Y. S. (1984). *Thermoluminescence & Thermoluminescent Dosimetry*. Boca Raton, FL: CRC Press.

Horowitz, Y. S., Rosenkrantz, M., Mahajna, S., & Yosian, D. (1996). The track interaction model for alpha particle induced thermoluminescence supralinearity: Dependence of the supralinearity on the vector properties of the alpha particle radiation field. *Journal of Physics. D, Applied Physics*, *29*(1), 205–217. doi:10.1088/0022-3727/29/1/031

Hsu, C. H., & Lu, C. H. (2011). Microwave-hydrothermally synthesized $(Sr_{1-x-y}Ce_xTb_y)Si_2O_{2-\delta}N_{2+\mu}$ phosphors: Efficient energy transfer, structural refinement and photoluminescence properties. *Journal of Materials Chemistry*, *21*(9), 2932–2939. doi:10.1039/c0jm02501a

Hsu, F. M., Chien, C. H., Hsieh, Y. J., Wu, C. H., Shu, C. F., Liu, S. W., & Chen, C. T. (2009). Highly efficient red electrophosphorescent device incorporating a bipolar triphenylamine/bisphenylsulfonyl-substituted fluorene hybrid as the host. *Journal of Materials Chemistry*, *19*(42), 8002. doi:10.1039/b910292b

Hu, A., Yao, Z., & Yu, X. (2009). Phase behavior of a sodium dodecanol allyl sulfosuccinic diester/ *n* -pentanol/methyl acrylate/butyl acrylate/water microemulsion system and preparation of acrylate latexes by microemulsion polymerization. *Journal of Applied Polymer Science*, *113*(4), 2202–2208. doi:10.1002/app.30284

Huang, C. Y., Huang, T. S., Cheng, C. Y., Chen, Y. C., Wan, C. T., Madhava Rao, M. V., & Su, Y. K. (2010). Three-band white light-emitting diodes based on hybridization ofpolyfluorene and colloidal CdSe-ZnS quantum dots. *IEEE Photonics Technology, Letters*, *22*(5), 305-307.

Huang, C. H., & Chen, T. M. (2010). $Ca_9La(PO_4)_7$:Eu^{2+},Mn^{2+}: an emission-tunable phosphor through efficient energy transfer for white light-emitting diodes. *Optics Express*, *18*(5), 5089–5099. doi:10.1364/OE.18.005089 PMID:20389521

Huang, C. H., & Chen, T. M. (2011). Novel yellow-emitting $Sr_8MgLn(PO_4)_7$:Eu^{2+} (Ln= Y, La) phosphors for applications in white LEDs with excellent color rendering index. *Inorganic Chemistry*, *50*(12), 5725–5730. doi:10.1021/ic200515w PMID:21591766

Huang, C. H., Chen, T. M., Liu, W. R., Chiu, Y. C., Yeh, Y. T., & Jang, S. M. (2010). A Single-Phased Emission-Tunable Phosphor $Ca_9Y(PO_4)_7$:Eu^{2+},Mn^{2+} with Efficient Energy Transfer for White-Light-Emitting Diodes. *ACS Applied Materials & Interfaces*, *2*(1), 259–264. doi:10.1021/am900668r

Huang, C. H., Chen, Y. C., Chen, T. M., Chan, T. S., & Sheu, H. S. (2011). Near UV-pumped yellow-emitting $Sr_8MgSc(PO_4)_7$:Eu^{2+} phosphor for white-light LEDs with excellent color rendering index. *Journal of Materials Chemistry*, *21*(15), 5645–5649. doi:10.1039/c0jm04524a

Huang, C. H., Kuo, T. W., & Chen, T. M. (2010). Novel Red-Emitting Phosphor $Ca_9Y(PO_4)_7$:Ce^{3+},Mn^{2+} with Energy Transfer for Fluorescent Lamp Application. *ACS Applied Materials & Interfaces*, *2*(5), 1395–1399. doi:10.1021/am100043q PMID:20429527

Huang, C. H., Liu, W. R., & Chen, T.-M. (2010). Single-phased white-light phosphors $Ca_9Gd(PO_4)_7$: Eu^{2+}, Mn^{2+} under near-ultraviolet excitation. *The Journal of Physical Chemistry C, 114*(43), 18698–18701. doi:10.1021/jp106693z

Huang, C. H., Luo, L. Y., & Chen, T. M. (2011). An investigation on the luminescence and $Ce^{3+} \rightarrow Eu^{2+}$ energy transfer in $Ca_9Y(PO_4)_7$: Ce^{3+}, Eu^{2+} phosphor. *Journal of the Electrochemical Society, 158*(11), J341–J344. doi:10.1149/2.029111jes

Huang, H., Fu, Q., Zhuang, S., Liu, Y., Wang, L., Chen, J., & Yang, C. (2011). Novel Deep Blue OLED Emitters with 1,3,5-Tri(anthracen-10-yl)benzene-Centered Starburst Oligofluorenes. *The Journal of Physical Chemistry C, 115*(11), 4872–4878. doi:10.1021/jp110652y

Huang, H., Wang, Y., Wang, B., Zhuang, S., Pan, B., Yang, X., ... Yang, C. (2013). Controllably tunable phenanthroimidazole–carbazole hybrid bipolar host materials for efficient green electrophosphorescent devices. *Journal of Materials Chemistry. C, Materials for Optical and Electronic Devices, 1*(37), 5899. doi:10.1039/c3tc30832d

Huang, H., Wang, Y., Zhuang, S., Yang, X., Wang, L., & Yang, C. (2012). Simple Phenanthroimidazole/Carbazole Hybrid Bipolar Host Materials for Highly Efficient Green and Yellow Phosphorescent Organic Light-Emitting Diodes. *The Journal of Physical Chemistry C, 116*(36), 19458–19466. doi:10.1021/jp305764b

Huang, H., Xu, G. Q., Chin, W. S., Gan, L. M., & Chew, C. H. (2002). Synthesis and characterization of Eu:Y2O3 nanoparticles. *Nanotechnology, 13*(3), 318–323. doi:10.1088/0957-4484/13/3/316

Huang, J., Pfeiffer, M., Werner, A., Blochwitz, J., Leo, K., & Liu, S. (2002). Low-voltage organic electroluminescent devices using pin structures. *Applied Physics Letters, 80*(1), 139–141. doi:10.1063/1.1432110

Huang, K.-W., Chen, W.-T., Chu, C.-I., Hu, S.-F., Sheu, H.-S., Cheng, B.-M., ... Liu, R.-S. (2012). Controlling The Activator Site To Tune Europium Valence in Oxyfluoride Phosphors. *Chemistry of Materials, 24*(11), 2220–2227. doi:10.1021/cm3011327

Huang, T. H., Lin, J. T., Chen, L. Y., Lin, Y. T., & Wu, C. C. (2006). Dipolar Dibenzothiophene S,S-Dioxide Derivatives Containing Diarylamine: Materials for Single-Layer Organic Light-Emitting Devices. *Advanced Materials, 18*(5), 602–606. doi:10.1002/adma.200502078

Huang, Y. K., Jen, T. H., Chang, Y. T., Yang, N. J., Lu, H. H., & Chen, S. A. (2010). Enhancing Shielding of Triplet Energy Transfer to Poly(p-phenylene)s from Phosphor Dopant by Addition of Branched Alcohol for Highly Efficient Electrophosphorescence. *ACS Applied Materials & Interfaces, 2*(4), 1094–1099. doi:10.1021/am900878f PMID:20380421

Huang, Y. L., Gan, J. H., & Seo, H. J. (2011). Luminescence Investigation of Eu-Activated $Sr_5(PO_4)_2SiO_4$ Phosphor by Combustion Synthesis. *Journal of the American Ceramic Society, 94*(4), 1143–1148. doi:10.1111/j.1551-2916.2010.04190.x

Huang, Y. L., Gan, J. H., Zhu, R., Wang, X. G., & Seo, H. J. (2011). Structural Phase Formation and Tunable Luminescence of Eu^{2+}-Activated Apatite-Type $(Ca,Sr,Ba)_5(PO_4)_2(SiO_4)$. *Journal of the Electrochemical Society, 158*(11), J334–J340. doi:10.1149/2.026111jes

Huang, Y., Zhang, X., & Pan, J. (2017). Synthesis and photoluminescence of violet-blue phosphor $Ba_{10}(PO_4)_6F_2$: Eu^{2+}. *Solid State Sciences, 63*, 9–15. doi:10.1016/j.solidstatesciences.2016.11.007

Huang, Z. M., He, C.-L., Yang, A., Zhang, Y., Han, X.-J., Yin, J., & Wu, Q. (2006). Encapsulating drugs in biodegradable ultrafine fibers through co-axial electrospinning. *Journal of Biomedical Materials Research. Part A, 77A*(1), 169–179. doi:10.1002/jbm.a.30564 PMID:16392131

Huang, Z., Li, B., & Liu, J. (2010). Molten-Salt Synthesis of Oxyapatite $La_{9.33}Si_6O_{26}$ Powders as Electrolytes for Intermediate Temperature Solid Oxide Fuel Cells. *Physica Status Solidi. A, Applications and Materials Science, 207*(10), 2247–2251. doi:10.1002/pssa.200925605

Huang, Z., Wang, B., Zhang, Q., Xiang, S., Lv, X., Ma, L., ... Wang, L. (2017). Highly twisted bipolar emitter for efficient nondoped deep-blue electroluminescence. *Dyes and Pigments*, *140*, 328–336. doi:10.1016/j.dyepig.2017.01.028

Hua, R., & Jiang, H. (2006). Article. *Materials Letters*, *60*, 3822–3825. doi:10.1016/j.matlet.2006.03.121

Hua, R., Jia, Z., & Shi, C. (2007). Preparation of KMgF3 and Eu-doped KMgF3 nanocrystals in water-in-oil microemulsions. *Materials Research Bulletin*, *42*(2), 249–255. doi:10.1016/j.materresbull.2006.06.004

Hua, R., Jia, Z., Xie, D., & Shi, C. (2002). Solvothermal synthesis of the complex fluorides KMgF3 and KZnF3 with the Perovskite structures. *Materials Research Bulletin*, *37*(6), 1189–1195. doi:10.1016/S0025-5408(02)00732-8

Hua, R., Zang, C., Shao, C., Xie, D., & Shi, C. (2003). Synthesis of barium fluoride nanoparticles from microemulsion. *Nanotech.*, *14*(6), 588–591. doi:10.1088/0957-4484/14/6/304

Hughes, G., & Bryce, M. R. (2005). Electron-transporting materials for organic electroluminescent and electrophosphorescent devices. *Journal of Materials Chemistry*, *15*(1), 94. doi:10.1039/b413249c

Hu, H., & Bai, Y. (2012). Upconversion white luminescence of TeO_2:Tm^{3+}/Er^{3+}/Yb^{3+} nanoparticles. *Journal of Alloys and Compounds*, *527*, 25–29. doi:10.1016/j.jallcom.2012.02.129

Hung, W. Y., Chi, L. C., Chen, W. J., Chen, Y. M., Chou, S. H., & Wong, K. T. (2010). A new benzimidazole/carbazole hybrid bipolar material for highly efficient deep-blue electrofluorescence, yellow–green electrophosphorescence, and two-color-based white OLEDs. *Journal of Materials Chemistry*, *20*(45), 10113. doi:10.1039/c0jm02143a

Ichikawa, M., Wakabayashi, K., Hayashi, S., Yokoyama, N., Koyama, T., & Taniguchi, Y. (2010). Bi- or ter-pyridine tris-substituted benzenes as electron-transporting materials for organic light-emitting devices. *Organic Electronics*, *11*(12), 1966–1973. doi:10.1016/j.orgel.2010.09.004

Idris, N. M., Gnanasammandhan, M. K., Zhang, J., Ho, P. C., Mahendran, R., & Zhang, Y. (2012). In vivo photodynamic therapy using upconversion nanoparticles as remote-controlled nanotransducers. *Nature Medicine*, *18*(10), 1580–1585. doi:10.1038/nm.2933 PMID:22983397

Im, W. B., Brinkley, S., Hu, J., Mikhailovsky, A., DenBaars, S. P., & Seshadri, R. (2010). $Sr_{2.975-x}Ba_xCe_{0.025}AlO_4F$: A Highly Efficient Green-Emitting Oxyfluoride Phosphor for Solid State White Lighting. *Chemistry of Materials*, *22*(9), 2842–2849. doi:10.1021/cm100010z

Inomata, H., Goushi, K., Masuko, T., Konno, T., Imai, T., Sasabe, H., ... Adachi, C. (2004). High-Efficiency Organic Electrophosphorescent Diodes Using 1,3,5-Triazine Electron Transport Materials. *Chemistry of Materials*, *16*(7), 1285–1291. doi:10.1021/cm034689t

Islam, A., Wang, Q., Zhang, L., Lei, T., Hong, L., Yang, R., ... Ge, Z. (2017). Efficient non-doped deep blue organic light emitting diodes with high external quantum efficiency and a low efficiency roll-off based on donor-acceptor molecules. *Dyes and Pigments*, *142*, 499–506. doi:10.1016/j.dyepig.2017.03.061

Ivanova, S., & Strong, P. F. (2009). 153 μm to NIR-VIS-UV upconversion in Er-doped fluoride glass for high-efficiency solar cells. *J. Opt. Soc. B*, *29*, 1930.

Ivanova, S., Pellé, F., Tkachuk, A., Joubert, M. F., Guyot, Y., & Gapontzev, V. P. (2008). Upconversion luminescence dynamics ofEr-doped fluoride crystals for optical converters. *Journal of Luminescence*, *128*(5-6), 914–917. doi:10.1016/j.jlumin.2007.11.031

Ivaturi, A., MacDougall, S. K., Martín-Rodríguez, R., Quintanilla, M., Marques-Hueso, J., Krämer, K. W., ... Richards, B. S. (2013). Optimizing infrared to near infrared upconversion quantum yield of β-$NaYF_4$: Er^{3+} in fluoropolymer matrix for photovoltaic devices. *Journal of Applied Physics*, *114*(1), 013505. doi:10.1063/1.4812578

Ivey. (1963). *Electroluminescence and related effects*. New York: Academic Press.

Jain, I. P., & Agarwal, G. (2011). Ion beam induced surface and interface engineering. *Surface Science Reports*, *66*(3-4), 77–172. doi:10.1016/j.surfrep.2010.11.001

Jain, S. R., Adiga, K. C., & Vernerker, V. R. P. (1981). A new approach to thermochemical calculations of condensed fuel-oxidizer mixtures. *Combustion and Flame*, *40*, 71–79. doi:10.1016/0010-2180(81)90111-5

Jain, V. K., & Mitra, S. (1977). Thermoluminescence studies of some silicate minerals. *Thermochimica Acta*, *18*(2), 241–244. doi:10.1016/0040-6031(77)80024-5

Jairam, T., Ulla, H., Kajjam, A. B., Satyanarayan, M. N., & Vaidyanathan, S. (2017). Star-Shaped Phenanthroimidazole-Triphenylamine-Based Yellow Organic Emitter for Organic Light Emitting Diodes. *ChemistrySelect*, *2*(8), 2611–2620. doi:10.1002lct.201700056

Jayabharathi, J., Prabhakaran, A., Thanikachalam, V., Jeeva, P., & Sundharesan, M. (2016). Air Stable HyLEDs Using Efficient Electron Injection and Emitting Materials. *Industrial & Engineering Chemistry Research*, *55*(29), 8087–8095. doi:10.1021/acs.iecr.6b01151

Jayabharathi, J., Prabhakaran, A., Thanikachalam, V., & Sundharesan, M. (2016). Highly efficient non-doped blue electroluminescent materials for organic lightemitting devices. *RSC Advances*, *6*(67), 62208–62217. doi:10.1039/C6RA11439C

Jayasudha, S., Madhukumar, K., Nair, C. M. K., Nair, R. G., Rajesh, S., Elias, T. S., ... Gopakumar, N. (2017). TL dosimetric characterization of gamma irradiated $SrSO_4$:Eu phosphors. *Journal of Luminescence*, *183*, 259–265. doi:10.1016/j.jlumin.2016.11.044

Jeong, J. H., Moon, B. K., Seo, H. J., Bae, J. S., Yi, S. S., Kim, I. W., & Park, H. L. (2003). Stark-shift modulation absorption spectroscopy of single quantum dots. *Applied Physics Letters*, *83*, 1346. doi:10.1063/1.1602169

Jeon, S. O., Jang, S. E., Son, H. S., & Lee, J. Y. (2011). External Quantum Efficiency Above 20% in Deep Blue Phosphorescent Organic Light-Emitting Diodes. *Advanced Materials*, *23*(12), 1436–1441. doi:10.1002/adma.201004372 PMID:21433109

Jeon, S. O., & Lee, J. Y. (2012). Phosphine oxide derivatives for organic light emitting diodes. *Journal of Materials Chemistry*, *22*(10), 4233–4243. doi:10.1039/C1JM14832J

Jeon, S. O., Yook, K. S., Jogo, C. W., & Lee, J. Y. (2010). High-Efficiency Deep-Blue-Phosphorescent Organic Light-Emitting Diodes Using a Phosphine Oxide and a Phosphine Sulfide High-Triplet-Energy Host Material with Bipolar Charge-Transport Properties. *Advanced Materials*, *22*(16), 1872–1876. doi:10.1002/adma.200903321 PMID:20512965

Jia, G., Zheng, Y. H., Liu, K., Song, Y. H., You, H. P., & Zhang, H. J. (2009). Article. *Journal of Physical Chemistry*, *C113*, 153–158.

Jianbei, Q. I. U., Qing, J. I. A. O., Dacheng, Z. H. O. U., & Zhengwen, Y. A. N. G. (2016)... *Journal of Rare Earths*, *34*(4), 341. doi:10.1016/S1002-0721(16)60034-0

Jiang, W., Duan, L., Qiao, J., Dong, G., Wang, L., & Qiu, Y. (2011). Tuning of Charge Balance in Bipolar Host Materials for Highly Efficient Solution-Processed Phosphorescent Devices. *Organic Letters*, *13*(12), 3146–3149. doi:10.1021/ol201039n PMID:21591625

Jiao, H., & Wang, Y. (2009). $Ca_2Al_2SiO_7$:Ce^{3+},Tb^{3+} : A White-Light Phosphor Suitable for White-Light-Emitting Diodes. *Journal of the Electrochemical Society*, *5*(5), J117–J120. doi:10.1149/1.3097191

Jiao, M. M., Jia, Y. C., Lü, W., Lv, W. Z., Zhao, Q., Shao, B. Q., & You, H. P. (2014). $Sr_3GdNa(PO_4)_3F:Eu^{2+}$, Mn^{2+}: a potential color tunable phosphor for white LEDs. *Journal of Materials Chemistry. C, Materials for Optical and Electronic Devices*, *2*(1), 90–97. doi:10.1039/C3TC31837K

Jiao, M., Xu, Q., Yang, C., & You, H. (2017). Cation substitution induced novel gehlenite $Ca_2GaAlSiO_7:Eu^{2+}/Ce^{3+}$ phosphor with green/blue emission for UV-WLEDs. *RSC Advances*, *7*(46), 28647–28654. doi:10.1039/C7RA04105E

Jia, Y., Pang, R., Li, H., Sun, W., Fu, J., Jiang, L., ... Liu, R. S. (2015). Single-phased white-light-emitting $Ca_4(PO_4)_2O:Ce^{3+},Eu^{2+}$ phosphors based on energy transfer. *Dalton Transactions (Cambridge, England)*, *44*(25), 11399–11407. doi:10.1039/C5DT01018G PMID:26017804

Jiayue, S., Jianbo, X., & Zhang, X. (2011). Du Haiyan. *Journal of Rare Earths*, *29*(1), 32–38.

Ji, H., Huang, Z., Xia, Z., Molokeev, M. S., Atuchin, V. V., Fang, M., & Huang, S. (2014). New Yellow-Emitting Whitlockite-type Structure $Sr_{1.75}Ca_{1.25}(PO_4)_2:Eu^{2+}$ Phosphor for Near-UV Pumped White Light-Emitting Devices. *Inorganic Chemistry*, *53*(10), 5129–5135. doi:10.1021/ic500230v PMID:24773050

Jingwei, X., Xiaoran, L., & Younan, X. (2008). Putting Electrospun Nanofibers to Work for Biomedical Research. *Macromolecular Rapid Communications*, *9*(22), 1775–1792. PMID:20011452

Joshi, C., Kumar, K., & Rai, S. B. (2009). Upconversion and anomalous power dependence in $Ca_{12}Al_{14}O_{33}$: Er^{3+}/Yb^{3+} single phase nanophosphor. *Journal of Applied Physics*, *105*(12), 123103. doi:10.1063/1.3148294

Judd, B. R. (1962). Optical Absorption Intensities of Rare-Earth Ions. *Physical Review*, *127*(3), 750–761. doi:10.1103/PhysRev.127.750

Ju, G. F., Hu, Y. H., Chen, L., & Wang, X. J. (2012). Persistent luminescence and its mechanism of $Ba_5(PO_4)_3Cl:Ce^{3+},Eu^{2+}$. *Journal of Applied Physics*, *111*(11), 113508. doi:10.1063/1.4723646

Ju, G., Hu, Y., Chen, L., Wang, X., Mu, Z., Wu, H., & Kang, F. (2012). Luminescence properties of $Y_2O_3:Bi^{3+}$, Ln^{3+} (Ln=Sm, Eu, Dy, Er, Ho) and the sensitization of Ln^{3+} by Bi^{3+}. *Journal of Luminescence*, *132*(8), 1853–1859. doi:10.1016/j.jlumin.2012.03.020

Kadono, K., Takahashi, M., Higuchi, H., Shojiya, M., Ohtsuki, T., Kawamoto, Y., & Peyghambarian, N. (1997). Frequency upconversion of Er3+ in Ga2S3-GeS2-La2S3 glasses. *SPIE Proc.*, *2996*, 25.

Kahng, D. (1968). Electroluminescence of rare earth and transition metal molecules in II-VI compound via impact excitation. *Applied Physics Letters*, *13*(6), 210–212. doi:10.1063/1.1652574

Kalaji, A., Mikami, M., & Cheetham, A. K. (2014). Ce^{3+}-Activated γ- Ca_2SiO_4 and Other Olivine-Type $ABXO_4$ Phosphors for Solid-State Lighting. *Chemistry of Materials*, *26*(13), 3966–3975. doi:10.1021/cm501516m

Kamtekar, K. T., Monkman, A. P., & Bryce, M. R. (2010). Recent Advances in White Organic Light-Emitting Materials and Devices (WOLEDs). *Advanced Materials*, *22*(5), 572–58. doi:10.1002/adma.200902148 PMID:20217752

Kang, F., Peng, M., Yang, X., Dong, G., Nie, G., Liang, W., ... Qiu, J. (2014). Broadly tuning Bi^{3+} emission via crystal field modulation in solid solution compounds $(Y,Lu,Sc)VO_4$:Bi for ultraviolet converted white LEDs. *Journal of Materials Chemistry. C, Materials for Optical and Electronic Devices*, *2*(30), 6068–6076. doi:10.1039/C4TC00238E

Kang, J. W., Lee, D. S., Park, H. D., Park, Y. S., Kim, J. W., Jeong, W. I., ... Kim, J. J. (2007). Silane- and triazine-containing hole and exciton blocking material for high-efficiency phosphorescent organic light emitting diodes. *Journal of Materials Chemistry*, *17*(35), 3714. doi:10.1039/b705741e

Kang, Y. C., Chung, Y. S., & Park, S. B. (1999). Preparation of YAG:Europium Red Phosphors by Spray Pyrolysis Using a Filter-Expansion Aerosol Generator. *Journal of the American Ceramic Society*, *82*(8), 2056–2060. doi:10.1111/j.1151-2916.1999.tb02040.x

Kang, Z. J., Wang, Y. X., You, F. T., & Lin, J. H. (2001). Hydrothermal Syntheses and Crystal Structure of NH4Ln3F10 (Ln=Dy, Ho, Y, Er, Tm). *Journal of Solid State Chemistry*, *158*(2), 358–362. doi:10.1006/jssc.2001.9131

Kang, Z., Barta, M., Nadler, J., Wagner, B., Rosson, R., & Kahn, B. (2010). Synthesis of BaF2:Ce nanophosphor and epoxy encapsulated transparent nanocomposite. *Journal of Luminescence*, *131*(10), 2140–2143. doi:10.1016/j.jlumin.2011.05.030

Karbowiak, M., Mech, A., Bednarkiewicz, A., Strek, W., & Kepinski, L. (2005). Comparison of different NaGdF4:Eu3+ synthesis routes and their influence on its structural and luminescent properties. *Journal of Physics and Chemistry of Solids*, *66*(6), 1008–1019. doi:10.1016/j.jpcs.2005.01.002

Kasturi, S., & Sivakumar, V. (2016). Eu^{2+} luminescence in $Ca_3Si_2O_7$ and spectral widening and tuning of Eu^{2+} emission color (orangish-red to green) by crystal chemical substitution. *RSC Advances*, *6*(101), 98652–98662. doi:10.1039/C6RA24258H

Kaur, G., Singh, S. K., & Rai, S. B. (2010). Eu^{3+} and Yb^{3+} codoped Gd_2O_3 single phase nanophosphor: An enhanced monochromatic red emission through cooperative upconversion and downconversion. *Journal of Applied Physics*, *107*(7), 073514. doi:10.1063/1.3373628

Kaur, J., Dubey, V., & Suryanarayana, N. S. (2015). Comparative study of ML and PL spectra of different impurity-doped (Zn, Cd) S mixed phosphors. *Research on Chemical Intermediates*, *39*(9), 4337–4349. doi:10.100711164-012-0950-x

Kaur, J., Suryanarayana, N. S., & Dubey, V. (2010a). Effect of temperature on the ML of Au doped (Zn, Cd) S mixed phosphors. *Chinese Chemical Letters*, *22*(6), 709–712. doi:10.1016/j.cclet.2010.12.006

Kaur, J., Suryanarayana, N. S., & Dubey, V. (2010b). ML spectra of gold doped (Zn, Cd) S mixed phosphors. *Chalcogenide Letters*, *7*(5), 329–334.

Kaur, J., Suryanarayana, N. S., Dubey, V., Agrawal, S., Sriwastawa, R., Jaykumar, B., ... Murthy, K. V. R. (2012). TL glow curve analysis of beta irradiated Eu3+ doped Y4Al2O9 phosphor. *Int. Jour. Lumi. Appl.*, *2*(I), 60–62.

Kawamoto, Y., Kanno, R., Shojiaya, M., & Kadono, K. (1995). Structural study of ion-conducting MX-K (or Rb) X-CsX-BaX_2halideglasses (M=Li, Cu or Ag) with single X species (X=Cl, Br or I). *Journal of Non-Crystalline Solids*, *184*, 155–159. doi:10.1016/0022-3093(94)00608-3

Kawamoto, Y., Kanno, R., Yokota, R., Takahashi, M., Tanabe, S., & Hirao, K. (1993). Compositional dependence of upconversion luminescence of Er^{3+} in ZrF_4 based glasses III. anion substitution effect. *Journal of Solid State Chemistry*, *103*(2), 334–340. doi:10.1006/jssc.1993.1108

Keating, P. N. (1961). Thermally Stimulated Emission and Conductivity Peaks in the Case of Temperature Dependent Trapping Cross Sections. *Proceedings of the Physical Society*, *78*(6), 1408–1415. doi:10.1088/0370-1328/78/6/343

Kellerer, A. (1985). Fundamentals of microdosimetry. In K. R. Kase, B. E. Bjarngard, & F. H. Attix (Eds.), *The Dosimetry of lonuinsRadiation* (Vol. 1). New York: Academic Press.

Kelley, T. W., Diaz, A. L., & Dang, T. A. (2006). Quantitative Determination of Nonradiative Host-to-Activator Energy Transfer Efficiencies in YBO_3:Eu^{3+} and Y_2O_3:Eu^{3+} under Vacuum Ultraviolet Excitation. *Chemistry of Materials*, *18*(13), 3130–3136. doi:10.1021/cm060383l

Kennedy, G. C. (1950). Pressure-volume-temperature relations in water at elevated temperatures and pressures. *American Journal of Science*, *248*(8), 540–564. doi:10.2475/ajs.248.8.540

Khare, A., & Bhusan, S. (2006). Electroluminescence studies of chemically deposited (Zn-Cd)S:Cu,F films. *Crystal research tech., 41*(7), 689-697.

Kijima, Y., Asai, N., & Tamura, S. (1999). A Blue Organic Light Emitting Diode. *Japanese Journal of Applied Physics, 38*(Part 1, No. 9A), 5274–5277. doi:10.1143/JJAP.38.5274

Kim, Y. H., Arunkumar, P., Kim, B. Y., Unithrattil, S., Kim, E., Moon, S. H., . . . Lee, J.-S. Im, … W. B. (2017). A zero-thermal-quenching phosphor. Nature Materials, 16, 543–550.

Kimata, M., Ii, N. I., & Ibaraki, N. (1982). The structural property of synthetic gehlenite, $Ca_2Al_2SiO_7$. *Neues Jahrbuch für Mineralogie. Abhandlungen, 3*, 254–267.

Kim, B., Park, Y., Lee, J., Yokoyama, D., Lee, J. H., Kido, J., & Park, J. (2013). Synthesis and electroluminescence properties of highly efficient blue fluorescence emitters using dual core chromophores. *Journal of Materials Chemistry. C, Materials for Optical and Electronic Devices, 1*(3), 432–440. doi:10.1039/C2TC00185C

Kim, C. H., Kwon, I. E., Park, C. H., Hwang, Y. J., Bae, H. S., Yu, B. Y., ... Hong, G. Y. (2000). Phosphors for plasma display panels. *Journal of Alloys and Compounds, 311*(1), 33–39. doi:10.1016/S0925-8388(00)00856-2

Kim, D. H., Cho, N. S., Oh, H. Y., Yang, J. H., Jeon, W. S., Park, J. S., ... Kwon, J. H. (2013). Highly Efficient Red Phosphorescent Dopants in Organic Light-Emitting Devices. *Advanced Materials, 23*(24), 2721–2726. doi:10.1002/adma.201100405 PMID:21495090

Kim, D., Kim, S. C., Bae, J. S., Kim, S., Kim, S. J., & Park, J. C. (2016). Eu^{2+}-Activated Alkaline-Earth Halophosphates, $M_5(PO_4)_3X:Eu^{2+}$ (M = Ca, Sr, Ba; X = F, Cl, Br) for NUV-LEDs: Site-Selective Crystal Field Effect. *Inorganic Chemistry, 55*(17), 8359–8370. doi:10.1021/acs.inorgchem.6b00637 PMID:27494550

Kimizuka, N. (2008). Self-Assembled Nanomaterials. *Advances in Polymer Science, 219*, 1–26.

Kim, J. S., Jeon, P. E., Choi, J. C., & Park, H. L. (2005). Emission color variation of $M_2SiO_4:Eu^{2+}$ (M=Ba, Sr, Ca) phosphors for light-emitting diode. *Solid State Communications, 133*(3), 187–190. doi:10.1016/j.ssc.2004.10.017

Kim, J. S., Jeon, P. E., Choi, J. C., Park, H. L., Mho, S. I., & Kim, G. C. (2004). Warm-white-light emitting diode utilizing a single-phase full-color $Ba_3MgSi_2O_8:Eu^{2+}$,$Ba_3MgSi_2O_8:Eu^{2+}$, Mn^{2+} phosphor. *Applied Physics Letters, 84*(15), 2931–2933. doi:10.1063/1.1695441

Kim, J.-Y., Bae, M. J., Park, S. H., Jeong, T., Song, S., Lee, J., ... Yu, S. G. (2011). Electroluminescence enhancement of the phosphor dispersed in a polymer matrix using the tandem structure. *Organic Electronics, 12*(3), 529–533. doi:10.1016/j.orgel.2010.12.020

Kim, S. H., Jang, J., & Lee, J. Y. (2007). Triplet host engineering for triplet exciton management in phosphorescent organic light-emitting diodes. *Applied Physics Letters, 91*, 083511. doi:10.1063/1.2773941

Kim, S. Y., Jeong, W. I., Mayr, C., Park, Y. S., Kim, K. H., Lee, J. H., ... Kim, J. J. (2013). Organic Light-Emitting Diodes with 30% External Quantum Efficiency Based on a Horizontally Oriented Emitter. *Advanced Functional Materials, 23*(31), 3896–3900. doi:10.1002/adfm.201300104

Kim, T. G., Kim, T., Kim, J., Kim, S. J., & Im, S. J. (2014). Interplay between Crystal Structure and Photoluminescence Properties of β-$Ca_3SiO_4Cl_2:Eu^{2+}$. *The Journal of Physical Chemistry C, 118*(23), 12428–12435. doi:10.1021/jp5002379

Kimura, T., Takahashi, T., & Yamaguchi, T. (1980). Preparation and Characteristics of Ni- Ferrite Powders obtained in the presence of Fused Salts. *Journal of Materials Science, 15*(6), 1491–1497. doi:10.1007/BF00752130

Kingsley, J. J., & Pederson, L. R. (1993). Combustion synthesis of perovskite LnCrO3 powders using ammonium dichromate. *Materials Letters*, *18*(1-2), 89–96. doi:10.1016/0167-577X(93)90063-4

Kingsley, J. J., & Pederson, L. R. (1993). Energetic Materials in Ceramics Synthesis. *Proceedings of the Materials Research Society*, *296*, 361. doi:10.1557/PROC-296-361

Kitai, H. (1993). *Solid State Luminescence: Theory, Materials & Devices*. London: Chapman & Hall. doi:10.1007/978-94-011-1522-3

Kitis, G., Chen, R., & Pagonis, V. (2008). Thermoluminescence glow-peak shape methods based on mixed order kinetics. [a]. *Physica Status Solidi*, *205*(5), 1181–1189. doi:10.1002/pssa.200723470

Kitis, G., Gomez-Ros, J. M., & Tuyn, J. W. N. (1998). Thermoluminescence glow-curve deconvolution functions for first, second and general orders of kinetics. *Journal of Physics. D, Applied Physics*, *31*(19), 2636–2641. doi:10.1088/0022-3727/31/19/037

Kitis, G., & Pagonis, V. (2007). Peak shape methods for general order thermoluminescence glow-peaks: A reappraisal. *Nucl. Instm. Methods B*, *262*(2), 313–322. doi:10.1016/j.nimb.2007.05.027

Kitis, G., & Vlachos, N. D. (2013). General semi-analytical expressions for TL, OSL and other luminescence stimulation modes derived from the OTOR model using the Lambert W-function. *Radiation Measurements*, *48*, 47–54. doi:10.1016/j.radmeas.2012.09.006

Kodaira, C. A., Brito, H. F., Malta, O. L., & Serra, O. A. (2003). Luminescence and energy transfer of the europium (III) tungstate obtained via the Pechini method. *Journal of Luminescence*, *101*(1-2), 11–21. doi:10.1016/S0022-2313(02)00384-8

Kodama, N., & Oishi, S. (2005). Visible quantum cutting through downconversion in KLiGdF5:Eu3+ crystals. *Journal of Applied Physics*, *98*(10), 103515–103520. doi:10.1063/1.2135893

Komarneni, S. (2003). Article. *Current Science*, *85*(12), 1730–1734.

Kondakova, M. E., Pawlik, T. D., Young, R. H., Giesen, D. J., Kondakov, D. Y., Brown, C. T., ... Klubek, K. P. (2008). High-efficiency, low-voltage phosphorescent organic light-emitting diode devices with mixed host. *Journal of Applied Physics*, *104*(9), 094501. doi:10.1063/1.3000046

Kore, B. P., Kumar, A., Pandey, A., Kroon, R. E., Terblans, J. J., Dhoble, S. J., & Swart, H. C. (2017). Spectroscopic Investigation of Up-Conversion Properties in Green Emitting $BaMgF_4$: Yb^{3+}, Tb^{3+} Phosphor. *Inorganic Chemistry*, *56*(9), 4996–5005. doi:10.1021/acs.inorgchem.7b00044 PMID:28406623

Kraft, A., Grimsdale, A. C., & Holmes, A. B. (1998). Electroluminescent Conjugated Polymers-Seeing Polymers in a New Light. *Angewandte Chemie International Edition*, *37*(4), 402–428. doi:

Kraft, S., Schattat, B., Bolse, W., Klaumünzer, S., Harbsmeier, F., Kulinska, A., & Löffl, A. (2002). Ion beam mixing of ZnO/SiO_2 and Sb/Ni/Si interfaces under swift heavy ion irradiation. *Journal of Applied Physics*, *91*(3), 1129–1134. doi:10.1063/1.1425439

Krames, M. R., Shchekin, O. B., Mueller-Mach, R., Mueller, G. O., Zhou, L., Harbers, G., & Craford, M. G. (2007). Status and Future of High-Power Light-Emitting Diodes for Solid-State Lighting. *Journal of Display Technology*, *3*(2), 160–175. doi:10.1109/JDT.2007.895339

Krauser, J., Zollondz, J. H., Weidinger, A., & Trautmann, C. (2003). Conductivity of nanometer-sized ion tracks in diamond-like carbon films. *Journal of Applied Physics*, *94*(3), 1959–1964. doi:10.1063/1.1587263

Krüger, H. (2010). $Ca_{5.45}Li_{3.55}[SiO_4]_3O_{0.45}F_{1.55}$ and $Ca_7K[SiO_4]_3F_3$: single-crystal synthesis and structures of two trigonal oxyfluorides. *Zeitschrift für Kristallographie – Crystalline Materials, 225,* 418-424.

Kuck, S., Sokolska, I., & Henke, M. (2003). Article. *Journal of Luminescence, 102,* 176–181.

Kulkarni, A. P., Tonzola, C. J., Babel, A., & Jenekhe, S. A. (2004). Electron Transport Materials for Organic Light-Emitting Diodes. *Chemistry of Materials, 16*(23), 4556–4573. doi:10.1021/cm0494731

Kumar, A., Tiwari, S. P., Singh, A. K., & Kumar, K. (2016). Synthesis of Gd_2O_3: Ho^{3+}/ Yb^{3+} upconversion nanoparticles for latent fingermark detection. *Applied Physics. B, Lasers and Optics, 122*(7), 1–10. doi:10.100700340-016-6468-y

Kumar, G. A., Pokhrel, M., Martinez, A., Dennis, R. C., Villegas, I. L., & Sardar, D. K. (2012a). Synthesis and spectroscopy of color tunable Y_2O_2S:Yb^{3+},Er^{3+} phosphors with intense emission. *Journal of Alloys and Compounds, 513,* 559–565. doi:10.1016/j.jallcom.2011.11.006

Kumar, G. A., Pokhrel, M., & Sardar, D. K. (2012b). Intense visible and near infrared upconversion in M_2O_2S: Er (M= Y, Gd, La) phosphor under 1550nm excitation. *Materials Letters, 68,* 395–398. doi:10.1016/j.matlet.2011.10.087

Kumari, A., Pandey, A., Dey, R., & Rai, V. K. (2014a). Simultaneous influence of Zn^{2+}/Mg^{2+} on the luminescent behaviour of La_2O_3: Tm^{3+}–Yb^{3+} phosphors. *RSC Advances, 4*(42), 21844–21851. doi:10.1039/C4RA01400F

Kumari, A., Rai, V. K., & Kumar, K. (2014b). Yellow–orange upconversion emission in Eu^{3+}/Yb^{3+} codoped $BaTiO_3$ phosphor. *Spectrochimica Acta. Part A: Molecular and Biomolecular Spectroscopy, 127,* 98–101. doi:10.1016/j.saa.2014.02.023 PMID:24632162

Kumari, A., Soni, A. K., & Rai, V. K. (2016). Rare Earth Doped Upconverting Phosphors for Display and Temperature Sensing Applications. *Materials Focus, 5*(3), 187–194. doi:10.1166/mat.2016.1328

Kumari, A., Soni, A. K., & Rai, V. K. (2017). Near infrared to blue upconverting Tm^{3+}/Yb^{3+}/Li^+: $Gd_2(MoO_4)_3$ phosphors for light emitting display devices. *Infrared Physics & Technology, 81,* 313–319. doi:10.1016/j.infrared.2017.01.024

Kumar, P., Rodrigues, G., Lakshmy, P. S., Kanjilal, D., & Kumar, R. (2008). Charge-state distributions of metallic electron cyclotron resonance plasmas. *Journal of Vacuum Science & Technology. A, Vacuum, Surfaces, and Films, 26*(1), 97–102. doi:10.1116/1.2823486

Kumar, R., Nyk, M., Ohulchanskyy, T. Y., Flask, C. A., & Prasad, P. N. (2009). Combined Optical and MR Bioimaging Using Rare Earth Ion-Doped NaYF4 Nanocrystals. *Advanced Functional Materials, 19*(6), 853–859. doi:10.1002/adfm.200800765

Kumar, S., Gathania, A. K., Vij, A., & Kumar, R. (2016). Gamma induced thermoluminescence and color centers study of Dy doped LiF micro-cubes. *Ceramics International, 42*(13), 14511–14517. doi:10.1016/j.ceramint.2016.06.062

Kuo, C. J., Li, T. Y., Lien, C. C., Liu, C. H., Wub, F. I., & Huang, M. J. (2009). Bis(phenanthroimidazolyl)biphenyl derivatives as saturated blue emitters for electroluminescent devices. *Journal of Materials Chemistry, 19*(13), 1865–1871. doi:10.1039/b816327h

Kurosaki, K., Tanaka, T., Maekawa, T., & Yamanaka, S. (2005). Thermal properties of polycrystalline sintered SrY_2O_4. *Journal of Alloys and Compounds, 395*(1-2), 318–321. doi:10.1016/j.jallcom.2004.11.041

Kurosaki, K., Tanaka, T., Maekawa, T., & Yamanaka, S. (2005). Thermophysical properties of SrY_2O_4. *Journal of Alloys and Compounds, 398*(1-2), 304–308. doi:10.1016/j.jallcom.2005.02.034

Kuwahara, H., Sato, Y., Kato, H., Kobayashi, M., Masaki, T., & Kakihana, M. (2016). Effect of Site Occupancies on Deep-red Emission from Eu^{2+}-activated Ca_2SiO_4 Phosphor. *Chemistry Letters, 45*(3), 321–323. doi:10.1246/cl.151145

Kwon, J. E., Park, S., & Park, S. Y. (2013). Realizing Molecular Pixel System for Full-Color Fluorescence Reproduction: RGB-Emitting Molecular Mixture Free from Energy Transfer Crosstalk. *Journal of the American Chemical Society, 135*(30), 11239–11246. doi:10.1021/ja404256s PMID:23876082

Lahariya, V. (2017). Study of electroluminescence study in cadmium sulfide polymer nanocomposite films. *Journal of Nano Research, 49*, 181-189.

Lai, M. Y., Chen, C. H., Huang, W. S., Lin, J. T., Ke, T. H., Chen, L. Y., ... Wu, C. C. (2008). Benzimidazole/Amine-Based Compounds Capable of Ambipolar Transport for Application in Single-Layer Blue-Emitting OLEDs and as Hosts for Phosphorescent Emitters. *Angewandte Chemie International Edition, 47*(3), 581–585. doi:10.1002/anie.200704113 PMID:18058878

Lakshmanan, A. R. (2001) A new high sensitive $CaSO_4$: Dy thermo stimulated luminescence phosphor. *Physica Status Solidi (A), 186*, 153–166.

Lakshmanan, A. R., Jose, M. T., & Annalakshmi, O. (2008). High-sensitive $CaSO_4$: Dy thermoluminescent phosphor synthesis by co-precipitation technique. *Radiation Protection Dosimetry, 132*(1), 42–50. doi:10.1093/rpd/ncn215 PMID:18716066

Lakshminarasimhan, N., & Varadaraju, U. V. (2005). White-Light Generation in Sr_2SiO_4:Eu^{2+},Ce^{3+} under Near-UV Excitation A Novel Phosphor for Solid-State Lighting. *Journal of the Electrochemical Society, 152*(9), H152–H156. doi:10.1149/1.1997169

Lau, K. T., Gu, C., & Hui, D. (2006). A critical review on nanotube and nanotube/nanoclay related polymer composite materials. *Composites. Part B, Engineering, 37*(6), 425–436. doi:10.1016/j.compositesb.2006.02.020

Laversenne, L., Guyot, Y., Goutaudier, C., Cohen-Adad, M. T., & Boulon, G. (2001). Optimization of spectroscopic properties of Yb^{3+}-doped refractory sesquioxides: Cubic Y_2O_3, Lu_2O_3 and monoclinic Gd_2O_3. *Optical Materials, 16*(4), 475–483. doi:10.1016/S0925-3467(00)00095-1

Le Masson, N. J. M., Bos, A. J. J., Winkelman, A. J. M., & van Eijk, C. W. E. (2001). Optically stimulated luminescence in KMgF/sub 3/:Ce/sup 3+/ comparison of dosimetric characteristics with Al/sub 2/O/sub 3/:C. *IEEE Transactions on Nuclear Science, 48*(4), 1143–1147. doi:10.1109/23.958739

Le Toquin, R., & Cheetham, A. K. (2006). Red-emitting cerium-based phosphor materials for solid-state lighting applications. *Chemical Physics Letters, 423*(4), 352–356. doi:10.1016/j.cplett.2006.03.056

Lee, C. W., & Lee, J. Y. (2013). Above 30% External Quantum Efficiency in Blue Phosphorescent Organic Light-Emitting Diodes Using Pyrido[2,3-*b*]indole Derivatives as Host Materials. *Advanced Materials, 25*(38), 5450–5454. doi:10.1002/adma.201301091 PMID:23788128

Lee, J. H., Cheng, S. H., Yoo, S. J., Shin, H., Chang, J. H., Wu, C. I., ... Kim, J. J. (2015). An Exciplex Forming Host for Highly Efficient Blue Organic Light Emitting Diodes with Low Driving Voltage. *Advanced Functional Materials, 25*(3), 361–366. doi:10.1002/adfm.201402707

Lee, J. H., Kang, Y. C., Park, S. B., & Park, H. D. (2001). Luminescence Characteristics of Eu-doped Strontium Halophosphate Phosphor Particles Prepared by Spray Pyrolysis. *Japanese Journal of Applied Physics, 40*(Part 1), 3222–3226. doi:10.1143/JJAP.40.3222

Lee, K.-G., Yu, B.-Y., Pyun, C.-H., & Mho, S.-I. (2002). Vacuum ultraviolet excitation and photoluminescence characteristics of (Y,Gd)Al3(BO3)4/Eu3+. *Solid State Communications, 122*(9), 485–488. doi:10.1016/S0038-1098(02)00195-3

Lee, L. S., Rand, S. C., & Schawlow, A. L. (1984). Cooperative energy transfer among Pr^{3+} ions in LaF_3. *Physical Review B: Condensed Matter and Materials Physics*, *29*(12), 6901–6906. doi:10.1103/PhysRevB.29.6901

Lee, S., Kim, B., Jung, H., Shin, H., Lee, H., Lee, J., & Park, J. (2017). Synthesis and electroluminescence properties of new blue dual-core OLED emitters using bulky side chromophores. *Dyes Pigm*, *136*, 255–261. doi:10.1016/j.dyepig.2016.08.010

Lee, S., & Park, S. (2013). Preparation and luminescent properties of Tb^{3+} and Tb^{3+}–Ce^{3+} doped $Ba_9Y_2Si_6O_{24}$ phosphors. *Journal of Luminescence*, *143*, 215–218. doi:10.1016/j.jlumin.2013.05.008

Lee, S., & Sohn, K. S. (2010). Effect of inhomogeneous broadening on time-resolved photoluminescence in $CaAlSiN_3$:Eu^{2+}. *Optics Letters*, *35*(7), 1004–1006. doi:10.1364/OL.35.001004 PMID:20364198

Lee, T. W., Noh, T., Shin, H. W., Kwon, O., Park, J. J., Choi, B. K., ... Kim, Y. R. (2009). Characteristics of Solution-Processed Small-Molecule Organic Films and Light-Emitting Diodes Compared with their Vacuum-Deposited Counterparts. *Advanced Functional Materials*, *19*(10), 1625–1630. doi:10.1002/adfm.200801045

Lee, Y. H., Wu, T. C., Liaw, C. W., Wen, T. C., Feng, S. W., Lee, J. J., ... Guo, T. F. (2013). Non-doped active layer, benzo[k]fluoranthene-based linear acenes, for deep blue- to green-emissive organic light-emitting diodes. *Organic Electronics*, *14*(4), 1064–1072. doi:10.1016/j.orgel.2013.01.021

Lemanski, K., Walerczyk, W., & Deren, P. J. (2016). Luminescent properties of europium ions in $CaAl_2SiO_6$. *Journal of Alloys and Compounds*, *672*, 595–599. doi:10.1016/j.jallcom.2016.02.208

Leung, M., Yang, C. C., Lee, J. H., Tsai, H. H., Lin, C. F., Huang, C. Y., ... Chiu, C. F. (2007). The Unusual Electrochemical and Photophysical Behavior of 2,2'-Bis(1,3,4-oxadiazol-2-yl)biphenyls, Effective Electron Transport Hosts for Phosphorescent Organic Light Emitting Diodes. *Organic Letters*, *9*(2), 235–238. doi:10.1021/ol062668+ PMID:17217273

Levy, P. W. (1985). Thermoluminescence kinetics in materials exposed to the low doses applicable to dating and dosimetry. *Nuclear Tracks*, *10*, 547.

Liang, B., Wang, L., Xu, Y. H., Shi, H. H., & Cao, Y. (2007). High-Efficiency Red Phosphorescent Iridium Dendrimers with Charge-Transporting Dendrons: Synthesis and Electroluminescent Properties. *Advanced Functional Materials*, *17*(17), 3580–3589. doi:10.1002/adfm.200700222

Liang, H. B., Tao, Y., Xu, J. H., He, H., Wu, H., Chen, W. X., ... Su, Q. (2004). Photoluminescence of Ce^{3+},Pr^{3+} and Tb^{3+} activated $Sr_3Ln(PO_4)_3$ under VUV-UV excitation. *Journal of Solid State Chemistry*, *177*(3), 901–908. doi:10.1016/j.jssc.2003.09.037

Lian, H., Liu, J., Ye, Z., & Shi, C. (2004). Synthesis and photoluminescence properties of erbium-doped BaF2 nanoparticles. *Chemical Physics Letters*, *386*(4-6), 291–294. doi:10.1016/j.cplett.2004.01.036

Li, C. L., Wei, J., Han, J., Li, Z., Song, X., Zhang, Z., ... Wang, Y. (2016). Efficient deep-blue OLEDs based on phenanthro[9,10-d]imidazole-containing emitters with AIE and bipolar transporting properties. *Journal of Materials Chemistry. C, Materials for Optical and Electronic Devices*, *4*(42), 10120–10129. doi:10.1039/C6TC03923E

Li, C., Li, Z., Yan, X., Zhang, Y., Zhang, Z., & Wang, Y. (2017). Structurally simple non-doped sky-blue OLEDs with high luminance and efficiencies at low driving voltages. *Journal of Materials Chemistry. C, Materials for Optical and Electronic Devices*, *5*(8), 1973–1980. doi:10.1039/C6TC05639C

Li, C., Wang, S., Chen, W., Wei, J., Yang, G., Ye, K., ... Wang, Y. (2015). High performance full color OLEDs based on a class of molecules with dual carrier transport channels and small singlet–triplet splitting. *Chemical Communications (Cambridge)*, *51*(53), 10632–10635. doi:10.1039/C5CC03492B PMID:26040328

Li, C., Wei, J., Song, X., Ye, K., Zhang, H., Zhang, J., & Wang, Y. (2016). Non-doped luminescent material based organic light-emitting devices displaying high brightness under very low driving voltage. *Journal of Materials Chemistry. C, Materials for Optical and Electronic Devices*, *4*(29), 7013–7019. doi:10.1039/C6TC01975G

Li, G. G., Geng, D. L., Shang, M. M., Peng, C., Cheng, Z. Y., & Lin, J. (2011). Tunable luminescence of Ce^{3+}/Mn^{2+}-coactivated $Ca_2Gd_8(SiO_4)_6O_2$ through energy transfer and modulation of excitation: Potential single-phase white/yellow-emitting phosphors. *Journal of Materials Chemistry*, *21*(35), 13334–13344. doi:10.1039/c1jm11650a

Li, G. G., Geng, D. L., Shang, M. M., Zhang, Y., Peng, C., Cheng, Z. Y., & Lin, J. (2011). Color Tuning Luminescence of $Ce^{3+}/Mn^{2+}/Tb^{3+}$-triactivated $Mg_2Y_8(SiO_4)_6O_2$ via Energy Transfer: Potential Single-Phase White-Light-Emitting Phosphors. *The Journal of Physical Chemistry C*, *115*(44), 21882–21897. doi:10.1021/jp204824d

Li, G. G., Zhang, Y., Geng, D. L., Shang, M. M., Peng, C., Cheng, Z. Y., & Lin, J. (2012). Single-Composition Trichromatic White-Emitting $Ca_4Y_6(SiO_4)_6O$: $Ce^{3+}/Mn^{2+}/Tb^{3+}$ Phosphor: Luminescence and Energy Transfer. *ACS Applied Materials & Interfaces*, *4*(1), 296–305. doi:10.1021/am201335d PMID:22136252

Li, G. H., Lan, S., Li, L. L., Li, M. M., Bao, W. W., Zou, H. F., ... Gan, S. C. (2012). Tunable luminescence properties of $NaLa(MoO_4)_2$:Ce^{3+},Tb^{3+} phosphors for near UV-excited white light-emitting-diodes. *Journal of Alloys and Compounds*, *513*, 145–149. doi:10.1016/j.jallcom.2011.10.008

Li, G., & Lin, J. (2014). Recent progress in low-voltage cathodoluminescent materials: Synthesis, improvement and emission properties. *Chemical Society Reviews*, *43*(20), 7099–7131. doi:10.1039/C4CS00109E PMID:24960634

Li, G., Tian, Y., Zhao, Y., & Lin, J. (2015). Recent progress in luminescence tuning of Ce^{3+} and Eu^{2+}-activated phosphors for pc-WLEDs. *Chemical Society Reviews*, *44*(23), 8688–8713. doi:10.1039/C4CS00446A PMID:26421319

Li, G., Wang, Y. H., Zeng, W., Chen, W. B., Han, S. C., Guo, H. J., & Wang, X. C. (2015). Luminescence properties of a new green afterglow phosphor $NaBaScSi_2O_7$:Eu^{2+}. *Dalton Transactions (Cambridge, England)*, *44*(40), 17572–17578. doi:10.1039/C5DT02748A PMID:26391314

Li, G., Wang, Y., Zeng, W., Chen, W., Han, S., Guo, H., & Li, Y. (2016). Photo-/cathodoluminescence and energy transfer properties of novel Ce^{3+} singly doped and Ce^{3+}/Tb^{3+} codoped $NaBaScSi_2O_7$ phosphors. *Journal of Materials Chemistry. C, Materials for Optical and Electronic Devices*, *4*(15), 3304–3312. doi:10.1039/C5TC04396D

Li, G., Zhao, J., Zhang, D., Zhu, J., Shi, Z., Tao, S., ... Tong, Q. (2017). Non-doped deep blue emitters based on twisted phenanthroimidazole derivatives for organic light-emitting devices (CIE y E 0.04). *New Journal of Chemistry*, *41*(12), 5191–5197. doi:10.1039/C7NJ00155J

Li, G., Zhao, Y., Wei, Y., Tian, Y., Quan, Z., & Lin, J. (2016). Novel yellowish-green light-emitting $Ca_{10}(PO_4)_6O$:Ce^{3+} phosphor: structural refinement, preferential site occupancy and color tuning. *ChemComm*, *52*, 3376–3379. PMID:26880115

Li, H., Jia, Z., & Shi, C. (2000). The Mild Hydrothermal Synthesis of Complex Fluorides of $AZnF_3$ (A = Na, K). *Chemistry Letters*, *9*(9), 1106–1110. doi:10.1246/cl.2000.1106

Li, J., Liu, D., Li, Y., Lee, C. S., Kwong, H. L., & Lee, S. (2005). A High Tg Carbazole-Based Hole-Transporting Material for Organic Light-Emitting Devices. *Chemistry of Materials*, *17*(5), 1208–1212. doi:10.1021/cm034731k

Li, J., Liu, J., & Yu, X. (2011). Synthesis and luminescence properties of Bi^{3+}-doped YVO_4 phosphors. *Journal of Alloys and Compounds*, *509*(41), 9897–9900. doi:10.1016/j.jallcom.2011.07.079

Li, K., Geng, D., Shang, M., Zhang, Y., Lian, H., & Lin, J. (2014). Color-Tunable Luminescence and Energy Transfer Properties of $Ca_9Mg(PO_4)_6F_2$:Eu^{2+}, Mn^{2+} Phosphors for UV-LEDs. *The Journal of Physical Chemistry C*, *118*(20), 11026–11034. doi:10.1021/jp501949m

Li, K., Shang, M., Lian, H., & Lin, J. (2016). Recent development in phosphors with different emitting colors via energy transfer. *Journal of Materials Chemistry. C, Materials for Optical and Electronic Devices*, *4*(24), 5507–5530. doi:10.1039/C6TC00436A

Li, K., Xu, J., Cai, X. C., Fan, J., Zhang, Y., Shang, M. M., ... Lin, J. (2015). An efficient green-emitting α-$Ca_{1.65}Sr_{0.35}SiO_4:Eu^{2+}$ phosphor for UV/n-UV w-LEDs: synthesis, luminescence and thermal properties. *Journal of Materials Chemistry. C, Materials for Optical and Electronic Devices*, *3*(24), 6341–6349. doi:10.1039/C5TC00796H

Li, N., Wang, P., Lai, S. L., Liu, W., Lee, C. S., Lee, S. T., & Liu, Z. (2010). Synthesis of multiaryl-substituted pyridine derivatives and applications in non-doped deep-blue OLEDs as electron-transporting layer with high hole-blocking ability. *Advanced Materials*, *22*(4), 527–530. doi:10.1002/adma.200902430 PMID:20217747

Lin, C. C., Xiao, Z. R., Guo, G. Y., Chan, T. S., & Liu, R. S. (2010). Versatile phosphate phosphors $ABPO_4$ in white light-emitting diodes: Collocated characteristic analysis and theoretical calculations. *Journal of the American Chemical Society*, *132*(9), 3020–3028. doi:10.1021/ja9092456 PMID:20155924

Lin, H., Meredith, G., Jiang, S., Peng, X., Luo, T., Peyghambarian, N., & Pun, E. Y. B. (2003). Optical transitions and visible upconversion in Er^{3+} doped niobic tellurite glass. *Journal of Applied Physics*, *93*(1), 186–191. doi:10.1063/1.1527209

Lin, J., Yu, M., Lin, C., & Liu, X. (2007). Multiform oxide optical materials via the versatile pechini-type sol– gel process: Synthesis and characteristics. *The Journal of Physical Chemistry C*, *111*(16), 5835–5845. doi:10.1021/jp070062c

Lin, M., Chi, L., Chang, H., Huang, Y., Tien, K., Chen, C., ... Chi, Y. (2012). A diarylboranesubstituted carbazole as a universal bipolar host material for highly efficient electrophosphorescence devices. *Journal of Materials Chemistry*, *22*(3), 870–876. doi:10.1039/C1JM13323C

Lin, N., Qiao, J., Duan, L., Xue, J., & Wang, L. (2014). Rational Design of Chelated Aluminum Complexes toward Highly Efficient and Thermally Stable Electron-Transporting Materials. *Chemistry of Materials*, *26*(12), 3693–3700. doi:10.1021/cm5011604

Liu, B., & Barbier, J. (1993). Structures of the Stuffed Tridymite Derivatives, $BaMSiO_4$ (M = Co, Zn, Mg). *Journal of Solid State Chemistry*, *102*(1), 115–125. doi:10.1006/jssc.1993.1013

Liu, B., Yuan, Y., He, D., Huang, D. Y., Luo, C. Y., Zhu, Z. L., ... Lee, C. S. (2016). High-Performance Blue OLEDs Based on Phenanthroimidazole Emitters via Substitutions at the C6- and C9-Positions for Improving Exciton Utilization. *Chemistry (Weinheim an der Bergstrasse, Germany)*, *22*, 1–9. PMID:27412733

Liu, B., Zhao, J., Luo, C., Lu, F., Tao, S., & Tong, Q. (2016). A novel bipolar phenanthroimidazole derivative host material for highly efficient green and orange-red phosphorescent OLEDs with low efficiency roll-off at high brightness. *Journal of Materials Chemistry. C, Materials for Optical and Electronic Devices*, *4*(10), 2003–2010. doi:10.1039/C5TC04393J

Liu, C. Y., Xia, Z. G., Lian, Z. P., Zhou, J., & Yan, Q. F. (2013). Structure and luminescence properties of green-emitting $NaBaScSi_2O_7:Eu^{2+}$ phosphors for near-UV-pumped light emitting diodes. *Journal of Materials Chemistry. C, Materials for Optical and Electronic Devices*, *1*(43), 7139–7147. doi:10.1039/C3TC31423E

Liu, D., Du, M., Chen, D., Ye, K., Zhang, Z., Liu, Y., & Wang, Y. (2015). A novel tetraphenylsilane–phenanthroimidazole hybrid host material for highly efficient blue fluorescent, green and red phosphorescent OLEDs. *Journal of Materials Chemistry. C, Materials for Optical and Electronic Devices*, *3*(17), 4394–4401. doi:10.1039/C5TC00333D

Liu, H. K., Luo, Y., Mao, Z. Y., Liao, L. B., & Xia, Z. G. (2014). A novel single-composition trichromatic white-emitting $Sr_{3.5}Y_{6.5}O_2(PO_4)_{1.5}(SiO_4)_{4.5}$[thin space(1/6-em)]:[thinspace(1/6-em)]$Ce^{3+}/Tb^{3+}/Mn^{2+}$ phosphor: synthesis, luminescent properties and applications for white LEDs. *Journal of Materials Chemistry. C, Materials for Optical and Electronic Devices*, *2*, 1619–1627. doi:10.1039/c3tc32003k

Liu, H., Chen, P., Hu, D. H., Tang, X. Y., Pan, Y. Y., Zhang, H. H., ... Ma, Y. G. (2014). Separation of Electrical and Optical Energy Gaps: Selectively Adjusting the Electrical and Optical Properties for a Highly Efficient Blue Emitter. *Chemistry (Weinheim an der Bergstrasse, Germany)*, *20*(8), 2149–2153. doi:10.1002/chem.201304544 PMID:24478181

Liu, J., Lian, H., Sun, J., & Shi, C. (2005). Characterization and Properties of Green Emitting $Ca_3SiO_4Cl_2$:Eu^{2+} Powder Phosphor for White Light-emitting Diodes. *Chemistry Letters*, *34*(10), 1340–1341. doi:10.1246/cl.2005.1340

Liu, J., Wu, Z., & Gong, M. (2008). Thermally stable luminescence of blue $LiSrPO_4$: Eu^{2+} phosphor for near-UV light-emitting diodes. *Applied Physics. B, Lasers and Optics*, *93*(2-3), 583–587. doi:10.100700340-008-3169-1

Liu, L., Zhang, B., Zhang, F., Pan, S., Zhang, F., Zhang, X., ... Yang, Z. (2015). $Pb_6Ba_2(BO_3)_5X$ (X = Cl, Br): New borate halides with strong predicted optical anisotropies derived from Pb^{2+} and $(BO_3)^{3-}$. *Dalton Transactions (Cambridge, England)*, *44*(15), 7041–7047. doi:10.1039/C5DT00641D PMID:25785911

Liu, M., Li, X. L., Chen, D. C., Xie, Z., Cai, X., Xie, G., ... Cao, Y. (2015). Study of Configuration Differentia and Highly Efficient, Deep-Blue, Organic Light-Emitting Diodes Based on Novel Naphtho[1,2-D]imidazole Derivatives. *Advanced Functional Materials*, *25*(32), 5190–5198. doi:10.1002/adfm.201502163

Liu, Q., Feng, W., & Li, F. (2014). Water-soluble lanthanide upconversion nanophosphors: Synthesis and bioimaging applications in vivo. *Coordination Chemistry Reviews*, *273*, 100–110. doi:10.1016/j.ccr.2014.01.004

Liu, Q., Liu, Y., Yang, Z., Han, Y., Li, X., & Fu, G. (2012). Multiwavelength excited white-emitting phosphor Dy3+-activated Ba3Bi(PO4)3. *Alloys Compd.*, *515*, 16–19. doi:10.1016/j.jallcom.2011.11.114

Liu, R. S., Liu, Y. H., Bagkar, N. C., & Hu, S. F. (2007). Enhanced luminescence of SrSi2O2N2:Eu2+ phosphors by codoping with Ce3+, Mn2+, and Dy3+ ions. *Applied Physics Letters*, *91*(6), 61119. doi:10.1063/1.2768916

Liu, S., & Su, Q. (1997). Sol-gel synthesis and luminescence of $Y_4Al_2O_9$:RE^{3+} (RE = Eu, Tb). *Journal of Alloys and Compounds*, *255*(1-2), 102–105. doi:10.1016/S0925-8388(96)02866-6

Liu, X. K., Zheng, C. J., Lo, M. F., Xiao, J., Lee, C. S., Fungc, M. K., & Zhang, X. H. (2014). A multifunctional phosphine oxide–diphenylamine hybrid compound as a high performance deep-blue fluorescent emitter and green phosphorescent host. *Chemical Communications (Cambridge)*, *50*(16), 2027. doi:10.1039/c3cc48531e PMID:24413567

Liu, X., Aidilibike, T., Guo, J., Li, Y., Di, W., & Qin, W. (2017). Upconversion luminescence of Sm^{2+} ions. *RSC Advances*, *7*(23), 14010–14014. doi:10.1039/C7RA00071E

Liu, X., Li, L., Noh, H. M., Moon, B. K., Choi, B. C., & Jeong, J. H. (2014). Chemical bond properties and charge transfer bands of O(2-)-Eu(3+), O(2-)-Mo(6+) and O(2-)-W(6+) in Eu(3+)-doped garnet hosts Ln3M5O12 and ABO4 molybdate and tungstate phosphors. *Dalton Transactions (Cambridge, England)*, *43*(23), 8814–8825. doi:10.1039/c4dt00674g PMID:24781975

Liu, Y. F., Zhang, X., Hao, Z. D., Luo, Y. S., Wang, X. J., & Zhang, J. H. (2011). Generating yellow and red emissions by co-doping Mn^{2+} to substitute for Ca^{2+} and Sc^{3+} sites in $Ca_3Sc_2Si_3O_{12}$:Ce^{3+} green emitting phosphor for white LED applications. *Journal of Materials Chemistry*, *21*(41), 16379–16384. doi:10.1039/c1jm11601k

Liu, Y., Tu, D., Zhu, H., & Chen, X. (2013). Lanthanide-doped luminescent nanoprobes: Controlled synthesis, optical spectroscopy, and bioapplications. *Chemical Society Reviews*, *42*(16), 6924–6958. doi:10.1039/c3cs60060b PMID:23775339

Liu, Y., Zhang, X., Hao, Z., Wang, X., & Zhang, J. (2011). Generation of broadband emission by incorporating N^{3-} into $Ca_3Sc_2Si_3O_{12}$:Ce^{3+} garnet for high rendering white LEDs. *Journal of Materials Chemistry*, *21*(17), 6354–6358. doi:10.1039/c0jm04404k

Liu, Y., Zhang, X., Hao, Z., Wang, X., & Zhang, J. (2011). Tunable full-color-emitting $Ca_3Sc_2Si_3O_{12}$:Ce^{3+}, Mn^{2+} phosphor via charge compensation and energy transfer. *ChemComm*, *47*, 10677–10679. PMID:21892443

Li, W., Liu, D., Shen, F., Ma, D., Wang, Z., Feng, T., ... Ma, Y. (2012). A Twisting Donor-Acceptor Molecule with an Intercrossed Excited State for Highly Efficient, Deep-Blue Electroluminescence. *Advanced Functional Materials*, *22*(13), 2797–2803. doi:10.1002/adfm.201200116

Li, W., Pan, Y., Yao, L., Liu, H., Zhang, S., Wang, C., ... Ma, Y. (2014). A Hybridized Local and Charge-Transfer Excited State for Highly Efficient Fluorescent OLEDs: Molecular Design, Spectral Character, and Full Exciton Utilization. *Adv. Opt. Mater*, *2*(9), 892–901. doi:10.1002/adom.201400154

Li, W., Yao, L., Liu, H., Wang, Z., Zhang, S., Xiao, R., ... Ma, Y. (2014). Highly efficient deep-blue OLED with an extraordinarily narrow FHWM of 35 nm and a y coordinate <0.05 based on a fully twisting donor–acceptor molecule. *Journal of Materials Chemistry. C, Materials for Optical and Electronic Devices*, *2*(24), 4733–4736. doi:10.1039/C4TC00487F

Lo, J. R., & Tseng, T. Y. (1998). Phase development and activation energy of the Y_2O_3-Al_2O_3 system by a modified sol-gel process. *Materials Chemistry and Physics*, *56*(1), 56–62. doi:10.1016/S0254-0584(98)00139-4

Lu, C. H., & Jagannathan, R. (2002). Cerium-ion-doped yttrium aluminium garnet nanophosphors prepared through sol-gel pyrolysis for luminescent lighting. *Applied Physics Letters*, *80*(19), 3608–3610. doi:10.1063/1.1475772

Lu, C., Huang, W., Ni, Y., & Xu, Z. (2011). Hydrothermal synthesis and luminescence properties of octahedral LiYbF4: Er3+ microcrystals. *Materials Research Bulletin*, *46*(2), 216–221. doi:10.1016/j.materresbull.2010.11.008

Lundin, N. J., Blackman, A. G., Gordon, K. C., & Officer, D. L. (2006). Cover Picture: Synthesis and Characterization of a Multicomponent Rhenium(I) Complex for Application as an OLED Dopant. *Angewandte Chemie International Edition*, *45*(16), 2582–2584. doi:10.1002/anie.200504532 PMID:16555347

Luo, J., Li, X., Hou, Q., Peng, J., Yang, W., & Cao, Y. (2007). High-efficiency white-light emission from a single copolymer: Fluorescent blue, green, and red chromophores on a conjugated polymer backbone. *Advanced Materials*, *19*(8), 1113–1117. doi:10.1002/adma.200601241

Lupei, V., Pavel, N., & Taira, T. (2002). Highly efficient continuous-wave 946-nm Nd:YAG laser emission under direct 885-nm pumping. *Applied Physics Letters*, *81*(15), 2677–2679. doi:10.1063/1.1511811

Luschik, C. B. (1956). The Investigation of Trapping Centres in Crystals by the Method of Thermal Bleaching. *Soviet Physics, JETP*, *3*, 390.

Lushchik, B. (1956). The Investigation of Trapping Centers in Crystals by the Method of Thermal Bleaching. *Soviet Physics, JETP*, *3*, 390–399.

Lu, W., Guo, N., Jia, Y. C., Zhao, Q., Lv, W. Z., Jiao, M. M., ... You, H. P. (2013). Tunable color of $Ce^{3+}/Tb^{3+}/Mn^{2+}$-coactivated $CaScAlSiO_6$ via energy transfer: A single-component red/white-emitting phosphor. *Inorganic Chemistry*, *52*(6), 3007–3012. doi:10.1021/ic3024666 PMID:23445353

Lü, W., Jia, Y. C., Lv, W. Z., Zhao, Q., & You, H. P. (2013). An orange-emitting phosphor via the efficient Ce^{3+}–Mn^{2+} and Eu^{2+}–Mn^{2+} energy transfers in $La_{9.33}(SiO_4)_6O_2$ for UV or near-UV LEDs. *New Journal of Chemistry*, *37*(11), 3701–3705. doi:10.1039/c3nj00758h

Lv, W., Jia, Y., Zhao, Q., Jiao, M., Shao, B., Lü, W., & You, H. (2014). Crystal Structure and Luminescence Properties of $Ca_8Mg_3Al_2Si_7O_{28}$:Eu^{2+} for WLEDs. *Advanced Optical Materials*, *2*(2), 183–188. doi:10.1002/adom.201300448

Lv, W., Jia, Y., Zhao, Q., Lü, W., Jiao, M., Shao, B., & You, H. P. (2014). Synthesis, Structure, and Luminescence Properties of $K_2Ba_7Si_{16}O_{40}$:Eu^{2+} for White Light Emitting Diodes. *The Journal of Physical Chemistry C, 118*(9), 4649–4655. doi:10.1021/jp500662a

Lv, Y., Jin, Y., Zhang, S., Wang, C., Ju, G., Xue, F., ... Hu, Y. (2017). $Sr_3GdLi(PO_4)_3F$:Eu^{2+}, Mn^{2+}: A tunable blue-white color emitting phosphor via energy transfer for near-UV white LEDs. *Ceramics International, 43*(12), 8824–8830. doi:10.1016/j.ceramint.2017.04.015

Maciel, G. S., de Araújo, C. B., Messaddeq, Y., & Aegerter, M. A. (1997). Frequency upconversion in Er 3 + -doped fluoroindate glasses pumped at 1.48 μm. *Physical Review B: Condensed Matter and Materials Physics, 55*(10), 6335–6342. doi:10.1103/PhysRevB.55.6335

Maciel, G. S., Guimaraes, R. B., Barreto, P. G., Carvalho, I. C. S., & Rakov, N. (2009). The influence of Yb 3+ doping on the upconversion luminescence of Pr^{3+} in aluminum oxide based powders prepared by combustion synthesis. *Optical Materials, 31*(11), 1735–1740. doi:10.1016/j.optmat.2009.05.007

Maciel, G. S., Rakov, N., Fokine, M., Carvalho, I. C. S., & Pinheiro, C. B. (2006). Strong upconversion from $Er_3Al_5O_{12}$ceramic powders prepared by low temperature direct combustion synthesis. *Applied Physics Letters, 89*(8), 109. doi:10.1063/1.2338558

Mader, H. S., Kele, P., Saleh, S. M., & Wolfbeis, O. S. (2010). Upconverting luminescent nanoparticles for use in bioconjugation and bioimaging. *Current Opinion in Chemical Biology, 14*(5), 582–596. doi:10.1016/j.cbpa.2010.08.014 PMID:20829098

Madhusoodanan, U., Jose, M. T., Tomita, A., Hoffmann, W., & Lakshmanan, A. R. (1999). A new thermostimulated luminescence phosphor based on CaSO4:Ag,Tm for applications in radiation dosimetry. *Journal of Luminescence, 82*(3), 221–232. doi:10.1016/S0022-2313(99)00044-7

Maeda, S., & Armes, S. (1995). Preparation and characterisation of polypyrrole – Tin (IV) oxide nanocomposite colloids. *Chemistry of Materials, 7*(1), 171–178. doi:10.1021/cm00049a026

Maggay, I. V. B., Lin, P. C., & Liu, W. R. (2014). Enhanced luminescence intensity of novel red-emitting phosphor -$Sr_3Lu_2(BO_3)_4$:Bi^{3+},Eu^{3+} via energy transfer. *Journal of Solid State Lighting, 1*(1), 13. doi:10.118640539-014-0013-6

Mai, H. X., Zhang, Y. W., Si, R., Yan, Z. G., Sun, L. D., You, L. P., & Yan, C. H. (2006). High-quality sodium rare-earth fluoride nanocrystals: Controlled synthesis and optical properties. *Journal of the American Chemical Society, 128*(19), 6426–6436. doi:10.1021/ja060212h PMID:16683808

Mai, H.-X., Zhang, Y.-W., Sun, L.-D., & Yan, C.-H. (2007). Highly Efficient Multicolor Up-Conversion Emissions and Their Mechanisms of Monodisperse NaYF4:Yb,Er Core and Core/Shell-Structured Nanocrystals. *The Journal of Physical Chemistry C, 111*(37), 13721–13729. doi:10.1021/jp073920d

Maldiney, T., Viana, B., Gourier, D., Richard, C., Scherman, D., Bessodes, M., ... Smet, P. F. (2012). *In vivo* optical imaging with rare earth doped $Ca_2Si_5N_8$ persistent luminescence nanoparticles. *Optical Materials Express, 2*(3), 261–268. doi:10.1364/OME.2.000261

Malik, M. A., Wani, M. Y., & Hashim, M. A. (2012). Microemulsion method: A novel route to synthesize organic and inorganic nanomaterials. *Arabian Journal of Chemistry, 5*(4), 397–417. doi:10.1016/j.arabjc.2010.09.027

Manam, J., & Das, S. (2008). Thermally stimulated luminescence studies of undoped, Cu- and Mn-doped $CaSO_4$ compounds. *Radiation Effects and Defects in Solids, 163*(12), 955–965. doi:10.1080/10420150802163869

Manam, J., & Das, S. (2010). Characterization and TSL dosimetric properties of Mn doped $BaSO_4$ phosphor prepared by recrystallisation method. *Journal of Alloys and Compounds*, *489*(1), 84–90. doi:10.1016/j.jallcom.2009.09.018

Mancic, L., Del Rosario, G., Stanojevic, Z., & Milosevic, O. (2007). Phase evolution in Ce-doped yttrium–aluminum-based particles derived from aerosol. *Journal of the European Ceramic Society*, *27*(13-15), 4329–4332. doi:10.1016/j.jeurceramsoc.2007.02.153

Manjunatha, C., Sunitha, D. V., Nagabhushana, H., Singh, F., Sharma, S. C., Chakradhar, R. P. S., & Nagabhushana, B. M. (2013). Thermoluminescence properties of 100 MeV Si^{7+} swift heavy ions and UV irradiated $CdSiO_3$: Ce^{3+} nano phosphor. *Journal of Luminescence*, *134*, 358–368. doi:10.1016/j.jlumin.2012.08.020

Manzoor, K., Aditya, V., Vadera, S., Kumar, N., & Kutty, T. (2005). Enhanced electroluminescence properties of doped ZnS nanorods formed by the self-assembly of colloidal nanocrystals. *Solid State Communications*, *135*(1-2), 16–20. doi:10.1016/j.ssc.2005.04.001

Mao, Y., Banerjee, S., & Wong, S. S. (2003). Large Scale Synthesis of Single – Crystalline Perovskite Nanostructures. *Journal of the American Chemical Society*, *125*(51), 15718–15719. doi:10.1021/ja038192w PMID:14677942

Mao, Y., Park, T. J., Zhang, F., Zhou, H., & Wong, S. (2007). S., "Environmentally Friendly Methodologies of Nano-structure Synthesis. *Small*, *3*(7), 1122–1139. doi:10.1002mll.200700048 PMID:17554768

Ma, P., Yuan, B., Sheng, Y., Zheng, K., Wang, Y., Xu, C., ... Song, Y. (2017). Tunable emission, thermal stability and energy-transfer properties of $SrAl_2Si_2O_8$: Ce^{3+}/Tb^{3+} phosphors for w-LEDs. *Journal of Alloys and Compounds*, *714*, 627–635. doi:10.1016/j.jallcom.2017.04.296

Ma, R., Bullock, E., Maynard, P., Reedy, B., Shimmon, R., Lennard, C., ... McDonagh, A. (2011). Fingermark detection on non-porous and semi-porous surfaces using $NaYF_4$: Er, Yb up-converter particles. *Forensic Science International*, *207*(1), 145–149. doi:10.1016/j.forsciint.2010.09.020 PMID:20980110

Maria, W. W., Uwe, K., & Ekkehart, T. (2010). The Crystal Structures of three new complex silicates of Scandium. *Canadian Mineralogist*, *48*(1), 51–68. doi:10.3749/canmin.48.1.51

Marrale, M., Longo, A., Bartolotta, A., Basile, S., D'Oca, M. C., Tomarchio, E., ... Di. (2012). Thermoluminescence response of sodalime glass irradiated with proton and neutron beams. *Nuclear Instruments and Methods in Physics Research Section B*, *292*, 55–58. doi:10.1016/j.nimb.2012.10.003

Marshall, C. D., Payne, S. A., Spaeth, J. A., Krupke, W. F., Quarles, G. J., Castillo, V., & Chai, B. H. T. (1994). Ultra-violet laser emission properties of Ce^3+-doped LiSrAlF_6 and LiCaAlF_6. *Journal of the Optical Society of America. B, Optical Physics*, *11*(10), 2054–2065. doi:10.1364/JOSAB.11.002054

Martinez, A., Morales, J., Diaz-Torres, L. A., Salas, P., De la Rosa, E., Oliva, J., & Desirena, H. (2010). Green and red upconverted emission of hydrothermal synthesized Y_2O_3: Er^{3+}–Yb^{3+} nanophosphors using different solvent ratio conditions. *Materials Science and Engineering B*, *174*(1), 164–168. doi:10.1016/j.mseb.2010.04.028

Martini, M., Meinardi, F., & Scacco, A. (1988). Impurity-induced thermally stimulated luminescence of KMgF3:Ce3+ crystals. *Chemical Physics Letters*, *293*(1-2), 43–46. doi:10.1016/S0009-2614(98)00764-7

Mattoussi, H., Radzilowski, L. H., Dabbousi, B. O., Thomas, E. L., Bawendi, M. G., & Rubner, M. F. (1998). Electro-luminescence from heterostructures of poly(phenylenevinylene) and inorganic CdSe nanocrystals. *Journal of Applied Physics*, *83*(12), 7965–7974. doi:10.1063/1.367978

May, E., & Partridge, J. A. (1964). Thermoluminescent Kinetics of Alpha-Irradiated Alkali Halides. *The Journal of Chemical Physics*, *40*(5), 1401–1408. doi:10.1063/1.1725324

Mayer, I., & Cohen, S. (1983). The crystal structure of $Ca_6Eu_2Na_2(PO_4)_6F_2$. *Journal of Solid State Chemistry*, *48*(1), 17–20. doi:10.1016/0022-4596(83)90054-3

Mayer, I., Roth, R. S., & Brown, W. E. (1974). Rare earth substituted fluoride-phosphate apatites. *Journal of Solid State Chemistry*, *11*(1), 33–37. doi:10.1016/0022-4596(74)90143-1

Mazumdar, P. S., Singh, S. J., & Gartia, R. K. (1988). The determination of activation energy from the shape of a thermoluminescence peak. *Journal of Physics. D, Applied Physics*, *21*(5), 815–819. doi:10.1088/0022-3727/21/5/023

Mckeever, S. W. S. (1985). *Thermoluminescence in solids*. Cambridge, UK: Cambridge University Press. doi:10.1017/CBO9780511564994

McKinlay, F. (1981). *Thermoluminescence Dosimetry. Mishawaka*. CRC Press.

Melo, A. P., Valerio, M. E. G., & Caldas, L. V. E. (2004). Thermoluminescent characteristics of mineral samples acquired as jade. *Nucl. Instm. Methods B*, *218*, 198–201. doi:10.1016/j.nimb.2003.12.058

Melo, A. P., Valerio, M. E. G., & Caldas, L. V. E. (2007). Thermoluminescent characteristics of Actinolite-Teflon composites for gamma high-doses. *Physica Status Solidi*, *4*(c), 1052–1055. doi:10.1002/pssc.200673854

Meng, X., Qiu, K., Tian, Z., Shi, X., You, J., Wang, Z., ... Yang, Z. (2017). Tunable-emission single-phase phosphors $Ba_3Ca_2(PO_4)_3F$:M (M=Ce^{3+},Eu^{2+},Mn^{2+}): Crystal structure, luminescence and energy transfer. *Journal of Alloys and Compounds*, *719*, 322–330. doi:10.1016/j.jallcom.2017.05.197

Mialon, G., Türkcan, S., Dantelle, G., Collins, D. P., Hadjipanayi, M., Taylor, R. A., ... Boilot, J. P. (2010). High up-conversion efficiency of YVO_4: Yb, Er nanoparticles in water down to the single-particle level. *The Journal of Physical Chemistry C*, *114*(51), 22449–22454. doi:10.1021/jp107900z

Miao, S., Xia, Z., Zhang, J., & Liu, Q. (2014). Increased Eu^{2+} Content and Codoping Mn^{2+} Induced Tunable Full-Color Emitting Phosphor $Ba_{1.55}Ca_{0.45}SiO_4$:Eu^{2+}, Mn^{2+}. *Inorganic Chemistry*, *53*(19), 10386–10393. doi:10.1021/ic501450f PMID:25254295

Minsung, K., Kobayashi, M., Kato, H., Yamane, H., Sato, Y., & Kakihana, M. (2015). Crystal structures and luminescence properties of Eu2+-activated new $NaBa_{0.5}Ca_{0.5}PO_4$ and $Na_3Ba_2Ca(PO_4)_3$. *Dalton Transactions (Cambridge, England)*, *44*(4), 1900–1904. doi:10.1039/C4DT03024A

Mishra, Y. K., Avasthi, D. K., Kulriya, P. K., Singh, F., Kabiraj, D., & Tripathi, A. (2007), Controlled growth of gold nanoparticles induced by ion irradiation: An in situ x-ray diffraction study. *Appl. Phys. Lett.*, *90*, 073110 (1-3).

Mishra, K., Giri, N. K., & Rai, S. B. (2011). Preparation and characterization of upconversion luminescent Tm^{3+}/Yb^{3+} co-doped Y_2O_3 nanophosphor. *Applied Physics. B, Lasers and Optics*, *103*(4), 863–875. doi:10.100700340-011-4379-5

Mishra, Y. K., Kabiraj, D., Avasthi, D. K., & Pivin, J. C. (2007). Swift heavy ion-induced dissolution of gold nanoparticles in silica matrix. *Radiation Effects and Defects in Solids*, *162*(3-4), 207–213. doi:10.1080/10420150601132883

Misra, R., Jadhav, T., Dhokale, B., & Mobin, S. M. (2014). Reversible mechanochromism and enhanced AIE in tetraphenylethene substituted phenanthroimidazoles. *Chemical Communications (Cambridge)*, *50*(65), 9076. doi:10.1039/C4CC02824D PMID:24983656

Moghe, A. K., & Gupta, B. S. (2008). Co-axial Electrospinning for Nanofiber structures: Preparation and Applications. *Polymer Reviews (Philadelphia, Pa.)*, *48*(2), 353–377. doi:10.1080/15583720802022257

Moghe, A. K., & Gupta, B. S. (2009). Hybrid Nanofiber Structures for Tissue Engineering. *AATCC Review*, *9*(10), 43–47.

Mohanta, D., Mishra, N. C., & Choudhury, A. (2004). SHI-induced grain growth and grain fragmentation effects in polymer-embedded CdS quantum dot systems. *Materials Letters, 58*(29), 3694–3699. doi:10.1016/j.matlet.2004.06.061

Mokoena, T. P., Linganiso, E. C., Swart, H. C., Kumar, V., & Ntwaeaborwa, O. M. (2017). Cooperative luminescence from low temperature synthesized α-Al_2O_3: Yb^{3+} phosphor by using solution combustion. *Ceramics International, 43*(1), 174–181. doi:10.1016/j.ceramint.2016.09.130

Molina, P., Vasyliev, V., Villora, E. G., & Shimamura, K. (2011). Tb^{3+}-Yb^{3+} cooperative down and up conversion processes in $Tb_{0.81}Ca_{0.19}F_{2.81}$: Yb^{3+} single crystals. *Journal of Applied Physics, 110*(12), 123527. doi:10.1063/1.3671406

Mondragon, G., Mendoza-Anaya, D., Nicho-Diaz, M. E., Garcia, R., & Reyes-Gasga, J. (2008). Thermoluminescence response of previously heated NaCl: Eu crystals to UV radiation, *Journal of Applied Physics D. Applied Physics (Berlin), 41*, 8.

Mondragón-Galicia, G., Mendoza-Anaya, D., Nicho-Diaz, M. E., García, R., & Reyes-Gasga, J. (2008). Thermoluminescence response of previously heated NaCl: Eu crystals to UV radiation. *Journal of Physics. D, Applied Physics, 41*(4), 045103. doi:10.1088/0022-3727/41/4/045103

Morey, G. W., & Niggli, P. (1913). The Hydrothermal Formation of Silicates, A Review. *Journal of the American Chemical Society, 35*(9), 1086–1130. doi:10.1021/ja02198a600

Morozov, V. A., Belik, A. A., Stefanovich, S. Y., Grebenev, V. V., Lebedev, O. I., Van Tendeloo, G., & Lazoryak, B. I. (2002). High-Temperature Phase Transition in the Whitlockite-Type Phosphate $Ca_9In(PO_4)_7$. *Journal of Solid State Chemistry, 165*(2), 278–288. doi:10.1006/jssc.2001.9521

Murray, C. B., Kagan, C. R., & Bawendi, M. G. (1995). Self-Organization of CdSe Nanocrystallites into Three-Dimensional Quantum Dot Superlattices. *Science, 270*(5240), 1335–1338. doi:10.1126cience.270.5240.1335

Murray, C. B., Norris, D. J., & Bawendi, M. G. (1993). Synthesis and characterization of nearly monodisperse CdE (E = sulfur, selenium, tellurium) semiconductor nanocrystallites. *Journal of the American Chemical Society, 115*(19), 8706–8715. doi:10.1021/ja00072a025

Nagpure, I. M., Shinde, K. N., Dhoble, S. J., & Kumar, A. (2009). Photoluminescence characterization of Dy^{3+} and Eu^{2+} ion in $M_5(PO_4)_3F$ (M = Ba, Sr, Ca) phosphors. *Journal of Alloys and Compounds, 481*(1-2), 632–638. doi:10.1016/j.jallcom.2009.03.069

Nain, A., Amon, C., & Sitti, M. (2005). Proximal Probes based Nanorobotic Drawing of Polymer Micro/Nanofibers. *IEEE Transactions on Nanotechnology, 5*(5), 499–510. doi:10.1109/TNANO.2006.880453

Nakajima, T., Zemva, B., & Tressaud, T. (2000). *Adv. Inorganic Fluorides: Synthesis.* Characterization & Applications, Elsevier Series S.A.

Nakanotani, H., Higuchi, T., & Furukawa, T. Masuil, K.,Morimoto, K.,Numata, M.,Tanaka, H.,Sagara, Y., Yasuda,T., &Adachi, C.(2014).High-efficiency organic light-emitting diodes with fluorescent emitters. *Nature Communications.* doi:10.1038/ncomms5016

Naka, S., Okada, H., Onnagawa, H., & Tsutsui, T. (2000). High electron mobility in bathophenanthroline. *Applied Physics Letters, 76*(2), 197–199. doi:10.1063/1.125701

Nakazawa, E., & Shionoya, S. (1970). Cooperative Luminescence in $YbPO_4$. *Physical Review Letters, 25*(25), 1710–1712. doi:10.1103/PhysRevLett.25.1710

Nastasi, M., Mayer, J. W., & Hirvonen, J. K. (1996). *Ion-Solid Interactions.* Cambridge, UK: Cambridge University Press. doi:10.1017/CBO9780511565007

Nayak, P. K., Agarwal, N., Ali, F., Patankar, M. P., Narasimhan, K. L., & Periasamy, N. (2010). Blue and white light electroluminescence in a multilayer OLED using a new aluminium complex. *Journal of Chemical Sciences*, *122*(6), 847–855. doi:10.100712039-010-0073-0

Neha, T., & Dubey, V. (2016). Luminescence studies and infrared emission of erbium-doped calcium zirconate phosphor. *Luminescence*, *31*(3), 837–842. doi:10.1002/bio.3039 PMID:26456020

Ni, S., Chou, L., & Chang, J. (2007). Preparation and characterization of forsterite (Mg2SiO4) bioceramics. *Ceram. Interna.*, *33*(1), 83–88. doi:10.1016/j.ceramint.2005.07.021

Northup, M. A., & Lee, O. I. (1940). Experiments on the Thermoluminescence of Some Common and Unusual Minerals. *Journal of the Optical Society of America*, *30*(5), 206. doi:10.1364/JOSA.30.000206

Nyk, M., Kumar, R., Ohulchanskyy, T. Y., Bergey, E. J., & Prasad, P. N. (2008). High Contrast in Vitro and in Vivo Photoluminescence Bioimaging Using Near Infrared to Near Infrared Up-Conversion in Tm3+ and Yb^{3+} Doped Fluoride Nanophosphors. *Nano Letters*, *8*(11), 3834–3838. doi:10.1021/nl802223f PMID:18928324

O'Brien, D. F., Baldo, M. A., Thompson, M. E., & Forrest, S. R. (1999). Improved Energy Transfer in Electrophosphorescent Devices. *Applied Physics Letters*, *74*(3), 442–444. doi:10.1063/1.123055

Ofelt, G. S. (1962). Intensities of Crystal Spectra of Rare-Earth Ions. *The Journal of Chemical Physics*, *37*(3), 511–520. doi:10.1063/1.1701366

Ogasawara, K., Watanabe, S., Sakai, Y., Toyoshima, H., Ishii, T., Brik, M. G., & Tanaka, I. (2004). Calculations of Complete $4f^n$ and $4f^{n-1}5d^1$ Energy Level Schemes of Free Trivalent Rare-Earth Ions. *Japanese Journal of Applied Physics*, *43*(5A), L611–L613. doi:10.1143/JJAP.43.L611

Okazaki, C., Shiiki, M., Suzuki, T., & Suzuki, K. (2000). Luminescence saturation properties of PDP phosphors. *Journal of Luminescence*, *87-89*, 1280–1282. doi:10.1016/S0022-2313(99)00575-X

Opanowicz, A. (2000). Determination of the recombination coefficient and the density of deep traps from the simultaneously measured thermally stimulated conductivity and luminescence. *Journal of Physics. D, Applied Physics*, *33*(14), 1635–1642. doi:10.1088/0022-3727/33/14/301

Opanowicz, A. (2007). Analysis of thermally stimulated luminescence and conductivity without quasi-equilibrium approximation. *Journal of Physics. D, Applied Physics*, *40*(16), 4980–4990. doi:10.1088/0022-3727/40/16/034

Opanowicz, A., & Pietrucha, P. (2003). Heating-rate method for determination of kinetic parameters from thermally stimulated conductivity and luminescence. *Journal of Applied Physics*, *93*(2), 957–967. doi:10.1063/1.1530713

Ouyang, X., Li, X. L., Ai, L., Mi, D., Ge, Z., & Su, S. J. (2015). Novel Hot Exciton Blue Fluorophores for High Performance Fluorescent/Phosphorescent Hybrid White Organic Light-Emitting Diodes with Superhigh Phosphorescent Dopant Concentration and Improved Efficiency Roll-Off. *ACS Applied Materials & Interfaces*, *7*(15), 7869–7877. doi:10.1021/am5081106 PMID:25826288

Ouyang, X., Li, X. L., Zhang, X., Islam, A., Ge, Z., & Su, S. J. (2015). Effective management of intramolecular charge transfer to obtain from blue to violet-blue OLEDs based on a couple of Phenanthrene isomers. *Dyes and Pigments*, *122*, 264–271. doi:10.1016/j.dyepig.2015.06.036

Padmaperuma, A. B., Sapochak, L. S., & Burrows, P. E. (2006). New Charge Transporting Host Material for Short Wavelength Organic Electrophosphorescence: 2,7-Bis(diphenylphosphineoxide)-9,9-dimethylfluorene. *Chemistry of Materials*, *18*(9), 2389–2396. doi:10.1021/cm0600677

Pagonis, V., Kitis, G., & Furetta, C. (2006). *Numerical and Practical Exercises in Thermoluminescence*. New York: Springer.

Pagonis, V., Kitis, G., & Furetta, C. (2006). *Numerical and practical exercises in thermoluminescence*. Springer Science & Business Media.

Pan, A. C., Del Cañizo, C., Cánovas, E., Santos, N. M., Leitão, J. P., & Luque, A. (2010). Enhancement of up-conversion efficiency by combining rare earth-doped phosphors with PbS quantum dots. *Solar Energy Materials and Solar Cells*, *94*(11), 1923–1926. doi:10.1016/j.solmat.2010.06.028

Pandey, A., Dey, R., Pandey, S., & Rai, V. K. (2014a). Visible Light-Emitting Er^{3+}-Tm^{3+}–Yb^{3+} Codoped ZrO_2 Phosphor for Sensing and Photonics Applications. *Journal of Display Technology*, *10*(6), 494–499. doi:10.1109/JDT.2014.2305773

Pandey, A., Dey, R., & Rai, V. K. (2013a). Sensitization Effect of Yb^{3+} in Upconversion Luminescence of Eu^{3+} Codoped Y_2O_3 Phosphor. *Journal of Physical Chemistry & Biophysics*, *3*(5), 1. doi:10.4172/2161-0398.1000129

Pandey, A., Kumar, V., Som, S., Yousif, A., Kroon, R. E., Coetsee, E., & Swart, H. C. (2017). Photon and electron beam pumped luminescence of Ho^{3+} activated $CaMoO_4$ phosphor. *Applied Surface Science*, *423*, 1169–1175. doi:10.1016/j.apsusc.2017.06.326

Pandey, A., & Rai, V. K. (2012). Colour emission tunability in Ho^{3+}–Tm^{3+}–Yb^{3+} co-doped Y_2O_3 upconverted phosphor. *Applied Physics. B, Lasers and Optics*, *109*(4), 611–616. doi:10.100700340-012-5224-1

Pandey, A., & Rai, V. K. (2013). Improved luminescence and temperature sensing performance of Ho^{3+}– Yb^{3+}–Zn^{2+}: Y_2O_3 phosphor. *Dalton Transactions (Cambridge, England)*, *42*(30), 11005–11011. doi:10.1039/c3dt50592h PMID:23797536

Pandey, A., & Rai, V. K. (2014a). Pr^{3+}–Yb^{3+} codoped Y_2O_3 phosphor for display devices. *Materials Research Bulletin*, *57*, 156–161. doi:10.1016/j.materresbull.2014.04.071

Pandey, A., & Rai, V. K. (2014b). Rare earth doped materials for temperature sensors. In Y. Dwivedi, S. B. Rai, & J. P. Singh (Eds.), *Spectroscopic techniques for security forensic and environmental applications* (pp. 279–292). Nova Publisher.

Pandey, A., Rai, V. K., Dey, R., & Kumar, K. (2013b). Enriched green upconversion emission in combustion synthesized Y_2O_3: Ho^{3+}– Yb^{3+} phosphor. *Materials Chemistry and Physics*, *139*(2), 483–488. doi:10.1016/j.matchemphys.2013.01.043

Pandey, A., Rai, V. K., & Kumar, K. (2014b). Influence of Li^+ codoping on visible emission of Y_2O_3:Tb^{3+},Yb^{3+} phosphor. *Spectrochimica Acta. Part A: Molecular and Biomolecular Spectroscopy*, *118*, 619–623. doi:10.1016/j.saa.2013.08.109 PMID:24095771

Pandey, A., Rai, V. K., Kumar, V., Kumar, V., & Swart, H. C. (2015). Upconversion based temperature sensing ability of Er^{3+}–Yb^{3+} codoped $SrWO_4$: An optical heating phosphor. *Sensors and Actuators. B, Chemical*, *209*, 352–358. doi:10.1016/j.snb.2014.11.126

Pandey, S., Pandey, A., & Rai, V. K. (2013). Synthesis and Characterization of Sm^{3+}–Yb^{3+} Codoped Y_2O_3 Phosphor. *Journal of Display Technology*, *9*(12), 989–994. doi:10.1109/JDT.2013.2275964

Pandurangappa, C., & Lakshminarasappa, B. N. (2012). Optical studies on lanthanum-doped calcium fluoride. *Journal of Materials Science*, *47*(2), 892–897. doi:10.100710853-011-5869-x

Pang, X. L., Jia, C. H., Li, G. Q., & Zhang, W. F. (2011). Bright white upconversion luminescence from Er^{3+}–Tm^{3+}–Yb^{3+} doped $CaSnO_3$ powders. *Optical Materials*, *34*(1), 234–238. doi:10.1016/j.optmat.2011.08.019

Pan, Y. X., Wu, M. M., & Su, Q. (2003). Synthesis of Eu3+-doped calcium and strontium carbonate phosphors at room temperature. *Materials Research Bulletin*, *38*(9-10), 1537–1544. doi:10.1016/S0025-5408(03)00174-0

Parchur, A. K., & Ningthoujam, R. S. (2012). Behaviour of electric and magnetic dipole transitions of Eu^{3+}, $^5D_0 \rightarrow {}^7F_0$ and Eu–O charge transfer band in Li^+ co-doped $YPO_4:Eu^{3+}$. *RSC Advances*, *2*(29), 10859–10868. doi:10.1039/c2ra22144f

Parganiha, Y., Kaur, J., Dubey, V., & Chandrakar, D. (2015). Synthesis, characterization, thermoluminescence and optical studies of Eu 3+ doped Y 2 SiO 5 phosphor. *Superlattices and Microstructures*, *77*, 152–161. doi:10.1016/j.spmi.2014.11.010

Park, C., Park, O., Lim, J., & Kim, H. (2001). The fabrication of syndiotactic polystyrene/organophilic clay nanocomposites and their properties. *Polymer*, *42*(17), 7465–7475. doi:10.1016/S0032-3861(01)00213-0

Park, J. K., Lim, M. A., Kim, C. H., Park, H. D., Park, J. T., & Choi, S. Y. (2003). White light-emitting diodes of GaN-based $Sr_2SiO_4:EuSr_2SiO_4:Eu$ and the luminescent properties. *Applied Physics Letters*, *82*(5), 683–685. doi:10.1063/1.1544055

Park, S. (2014). Ce^{3+}–Mn^{2+} cooperative $Ba_9Y_2Si_6O_{24}$ orthosilicate phosphors. *Materials Letters*, *135*, 59–62. doi:10.1016/j.matlet.2014.07.134

Park, S. G., Park, C. H., Yu, B. Y., Bae, H. S., Kim, C. H., & Pyun, C. H. (1999). Structure and luminescence of $SrY_2O_4:Eu$. *Journal of the Electrochemical Society*, *146*(10), 3903. doi:10.1149/1.1392570

Park, S., & Vogt, T. (2009). Luminescent phosphors, based on rare earth substituted oxyfluorides in the $A(1)_{3-x}A(2)_xMO_4F$ family with A(1)/A(2) = Sr, Ca, Ba and M=Al, Ga. *Journal of Luminescence*, *129*(9), 952–957. doi:10.1016/j.jlumin.2009.04.005

Park, Y. I., Kim, J. H., Lee, K. T., Jeon, K. S., Na, H. B., Yu, J. H., ... Lee, N. (2009). Article. *Advanced Materials*, *21*, 4467–4471. doi:10.1002/adma.200901356

Park, Y. R., Jeong, H. Y., Soo Seo, Y., Choi, W. K., & Hong, Y. J. (2017). Quantum dots Light emitting diodes with nitrogen doped carbon nanodot hole transport and electronic energy transfer layer. *Scientific Reports*, *7*, 46422. doi:10.1038rep46422 PMID:28401912

Patra, et al. (2015). Development of Thermal Neutron Detector based on Lithium Tetraborate (LTB) Single Crystal. *BARC News Letter*, 9-13.

Patra, A., Friend, C. S., Kapoor, R., & Prasad, P. N. (2002). Upconversion in $Er^{3+}:ZrO_2$ nanocrystals. *The Journal of Physical Chemistry B*, *106*(8), 1909–1912. doi:10.1021/jp013576z

Patra, A., Friend, C. S., Kapoor, R., & Prasad, P. N. (2003). Effect of crystal nature on upconversion luminescence in $Er^{3+}:ZrO_2$ nanocrystals. *Applied Physics Letters*, *83*(2), 284–286. doi:10.1063/1.1592891

Patro, L. N., & Hariharan, K. (2011). Influence of synthesis methodology on the ionic transport properties of BaSnF4. *Materials Research Bulletin*, *46*(5), 732–737. doi:10.1016/j.materresbull.2011.01.010

Pavitra, E., Raju, G. S. R., Park, W., & Yu, J. S. (2005). Concentration and penetration depth dependent tunable emissions from Eu3+ co-doped $SrY_2O_4:Dy^{3+}$nanocrystalline phosphor. *Journal of Chemistry*, *38*, 163.

Peacock, R. D. (1975). The intensities of lanthanide f ↔ f transitions. *Structure and Bonding*, *22*, 83–122. doi:10.1007/BFb0116556

Peng, B., & Izumitani, T. (1994). The Fluorescence Properties of Eu3+ in Various Glasses and the Energy Transfer Between Eu3+ and Sm3+ in Borosilico-phosphate Glass. *Rev. Laser Eng.*, *22*(1), 16–27. doi:10.2184/lsj.22.16

Peng, J. H., Binner, J., & Bradshaw, S. (2002). Microwave initiated self-propagating high temperature synthesis of materials. *Materials Science and Technology*, *18*(12), 1419–1427. doi:10.1179/026708302225006142

Peng, J., Ye, K., Zhang, G., Zhan, Y., Jia, J., Xue, P., & Lu, R. (2014). Synthesis, photophysical and electroluminescent properties of phenanthroimidazole derivatives with terminal carbazole or pyrene. *Synthetic Metals*, *193*, 94–101. doi:10.1016/j.synthmet.2014.04.004

Peng, L., Huang, M., Cao, S., Liu, B., Han, T., & Zhao, C. (2016). Enhanced upconversion in Dy^{3+}, Yb^{3+} co-doped Gd_2O_3 monodisperse nanocrystals. *Journal of Sol-Gel Science and Technology*, *78*(2), 307–312. doi:10.100710971-015-3952-5

Peng, X. G., Manna, L., Yang, W. D., Wickham, J., Scher, E., Kadavanich, A., & Alivisatos, A. P. (2000). Shape control of CdSe nanocrystals. *Nature*, *404*(6773), 59–61. doi:10.1038/35003535 PMID:10716439

Phillips, J. M., Coltrin, M. E., Crawford, M. H., Fischer, A. J., Krames, M. R., Mueller-Mach, R., ... Tsao, J. Y. (2007). Research challenges to ultra-efficient inorganic solid-state lighting. *Laser & Photonics Reviews*, *1*(4), 307–333. doi:10.1002/lpor.200710019

Poort, S. H. M., Van Krevel, J. W. H., Stomphorst, R., Vink, A. P., & Blasse, G. (1996). Luminescence of Eu^{2+}in Host Lattices with Three Alkaline Earth Ions in a Row. *Journal of Solid State Chemistry*, *122*(2), 432–435. doi:10.1006/jssc.1996.0137

Prasad, P. N. (2004). *Nanophotonics*. John Wiley & Sons. doi:10.1002/0471670251

Prashantha, S. C., Lakshminarasappa, B. N., & Nagabhushana, B. M. (2011). Photoluminescence and thermoluminescence studies of Mg_2SiO_4:Eu^{3+} nano phosphor. *Journal of Alloys and Compounds*, *509*(42), 10185–10189. doi:10.1016/j.jallcom.2011.03.148

Prodjosantoso, A. K., Kennedy, B. J., Vogt, T., & Woodward, P. M. (2003). Cation and anion ordering in the layered oxyfluorides $Sr_{3-x}A_xAlO_4F$ (A=Ba, Ca). *Journal of Solid State Chemistry*, *172*(1), 89–94. doi:10.1016/S0022-4596(02)00128-7

Puchalska, M., & Zych, E. (2012). Effect of Na^+ co-dopant and activator concentration on luminescent properties of $CaGa_4O_7$:Eu^{3+}. *Journal of Luminescence*, *132*(11), 2879–2884. doi:10.1016/j.jlumin.2012.06.007

Purwanto, A., Wang, W. N., Lenggoro, I. W., & Okuyama, K. (2007). Formation and Luminescence Enhancement of Agglomerate-Free YAG:Ce[sup 3+] Submicrometer Particles by Flame-Assisted Spray Pyrolysis. *Journal of the Electrochemical Society*, *154*(3), 91–96. doi:10.1149/1.2426874

Qin, G. S., Qin, W. P., Huang, S. H., Wu, C. F., Zhao, D., Chen, B. J., ... Shulin, E. (2002). Infrared-to-violet upconversion from Yb^{3+} and Er^{3+}codoped amorphous fluoride film prepared by pulsed laser deposition. *Journal of Applied Physics*, *92*(11), 6936–6938. doi:10.1063/1.1521801

Qin, W., Yang, Z., Jiang, Y., Lam, J. W. Y., Liang, G., Kwok, H. S., & Tang, B. Z. (2015). Construction of Efficient Deep Blue Aggregation-Induced Emission Luminogen from Triphenylethene for Nondoped Organic Light-Emitting Diodes. *Chemistry of Materials*, *27*(11), 3892–3901. doi:10.1021/acs.chemmater.5b00568

Qi, S. Y., Huang, Y. L., Li, Y. D., Cai, P. Q., Kim, S. I. I., & Seo, H. J. (2014). Probe spectrum measurements of Eu^{3+} ions as a relevant tool for monitoring *in-vitro* hydroxyapatite formation in a new borate biomaterial. *Journal of Materials Chemistry. B, Materials for Biology and Medicine*, *2*(37), 6387–6396. doi:10.1039/C4TB00871E

Qi, S. Y., Huang, Y. L., Tsuboi, T., Huang, W., & Seo, H. J. (2014). Versatile luminescence of $Eu^{2+, 3+}$-activated fluorosilicate apatites $M_2Y_3[SiO_4]_3F$ (M= Sr, Ba) suitable for white light emitting diodes. *Optical Materials Express*, *4*(2), 396–402. doi:10.1364/OME.4.000396

Que, M. D., Ci, Z. P., Wang, Y. H., Zhu, G., Xin, S. Y., Shi, Y. R., & Wang, Q. (2013). Crystal structure and luminescence properties of a cyan emitting $Ca_{10}(SiO_4)_3(SO_4)_3F_2$:$Eu^{2+}$ phosphor. *CrystEngComm*, *15*(32), 6389–6394. doi:10.1039/c3ce40482j

Que, W., Zhou, Y., Lam, Y. L., Chan, Y. C., Kam, C. H., Liu, B., ... Mendis, F. V. C. (1998). Photoluminescence and Electroluminescence from copper doped Zinc sulfide nanocrystals/Polymer composites. *Applied Physics Letters*, *73*(19), 2727–2729. doi:10.1063/1.122571

Rabenau, A., & Rau, H. (1969). Article. *Philips Tech. Rev.*, *30*, 89–96.

Rabinovitch, Y., Moune, O. K., Tetard, D., & Faucher, M. D. (2004). Synthesis, Optical Absorption, and Site-Selective Excitation of the $^3P_0$ Levels in $Y_4Al_2O_9$:Pr^{3+}. *The Journal of Physical Chemistry A*, *108*(40), 8244–8255. doi:10.1021/jp048123j

Rack, P. D., & Holloway, P. H. (1998). The structure, device physics, and material properties of thin film electroluminescent displays. *Materials Science and Engineering R Reports*, *21*(4), 171–219. doi:10.1016/S0927-796X(97)00010-7

Rahaman. (2003). *Ceramic Processing and sintering* (2nd ed.). Marcel Dekker.

Rai, M., Mishra, K., Singh, S. K., Verma, R. K., & Rai, S. B. (2012). Infrared to visible upconversion in Ho^{3+}/Yb^{3+} co-doped Y_2O_3 phosphor: Effect of laser input power and external temperature. *Spectrochimica Acta. Part A: Molecular and Biomolecular Spectroscopy*, *97*, 825–829. doi:10.1016/j.saa.2012.07.071 PMID:22902580

Rai, V. K. (2012). Frequency Upconversion in Rare earth ions. In A. H. Al-Khursan (Ed.), *Solid State Laser* (pp. 209–224). Intech Publisher.

Rai, V. K., Dey, R., & Kumar, K. (2013a). White upconversion emission in Y_2O_3: Er^{3+}–Tm^{3+}–Yb^{3+} phosphor. *Materials Research Bulletin*, *48*(6), 2232–2236. doi:10.1016/j.materresbull.2013.02.064

Rai, V. K., & Pandey, A. (2016). Efficient color tunable $ZnWO_4$: Er^{3+}-Yb^{3+} phosphor for high temperature sensing. *Journal of Display Technology*, *12*(11), 1472–1477. doi:10.1109/JDT.2016.2602503

Rai, V. K., Pandey, A., & Dey, R. (2013b). Photoluminescence study of Y_2O_3: Er^{3+}–Eu^{3+}–Yb^{3+} phosphor for lighting and sensing applications. *Journal of Applied Physics*, *113*(8), 083104. doi:10.1063/1.4793265

Rajurkar, N., Jha, A., & Bhadane, R. (1994). Instrumental neutron activation analysis of some ancient punchmark coins from India. *Journal of Radioanalytical and Nuclear Chemistry*, *186*(5), 393–400. doi:10.1007/BF02162623

Ramachandra Naik, S. C. (2014). Article. *Biological Chemistry*, *195*, 140–149.

Ramakrishna, S., Fujihara, K., Teo, W. E., Lim, C. T., & Ma, Z. (2005). *An introduction to Electrospinning and Nanofibers, Basics related to Electrospinning*. Singapore: World Scientific Publishing Co. Pvt. Ltd. doi:10.1142/5894

Ramakrishna, S., Fujihara, K., Teo, W.-E., Yong, T., Ma, Z., & Ramaseshan, R. (2006). Electrospun nanofibers: Solving global issues. *Materials Today*, *9*(3), 40–50. doi:10.1016/S1369-7021(06)71389-X

Ramrakhiani, M., & Nogriya, V. (2012). Chapter X Synthesis and Characterization of Zinc Sulfide nanocrystals and ZnS/PVA nanocomposite for luminescence application Polymer Processing and Characterization Apple. Academic Press Publisher.

Ramrakhiani, M., & Nogriya, V. (2013). Photo and Electroluminescence of Cadmium selenide nanocrystals and Nonocomposites. *Journal of Luminescence*, *133*, 129–133. doi:10.1016/j.jlumin.2011.09.046

Randall, J. T., & Wilkins, M. H. F. (1945a). Phosphorescence and Electron Traps. I. The Study of Trap Distributions. *Proc. Roy. Soc.*, *184*(999), 365–389. doi:10.1098/rspa.1945.0024

Randall, J. T., & Wilkins, M. H. F. (1945b). Phosphorescence and Electron Traps. II. The Study of Trap Distributions. *Proc. Roy. Soc.*, *184*(999), 390–407. doi:10.1098/rspa.1945.0025

Rao, R. P., & Devine, D. J. (2000). RE-activated lanthanide phosphate phosphors for PDP applications. *J. Lumin. 87-89*, 1260–1263.

Rao, C. N. R. (1994). *Chemical Approaches to the Synthesis of Inorganic Materials*. New York: John Wiley.

Reddy, A. A., Das, S., Ahmad, S., Babu, S. S., Ferreira, J. M. F., & Prakash, G. V. (2012). Influence of the annealing temperatures on the photoluminescence of $KCaBO_3:Eu^{3+}$ phosphor. *RSC Advances*, *2*(23), 8768–8776. doi:10.1039/c2ra20866k

Reddy, S. S., Nagabhushana, K. R., & Singh, F. (2016). Thermoluminescence studies of γ-irradiated $Al_2O_3:Ce^{3+}$ phosphor. *Nuclear Instruments and Methods in Physics Research Section B*, *379*, 146–151. doi:10.1016/j.nimb.2016.03.051

Redhammer, G. R., & Roth, G. (2003). Lithium and sodium yttrium orthosilicate oxyapatite, $LiY_9(SiO_4)_6O_2$ and $NaY_9(SiO_4)_6O_2$, at both 100 K and near room temperature. *Acta Crystallographica. Section C, Crystal Structure Communications*, *59*(12), i120–i124. doi:10.1107/S0108270103023321 PMID:14671334

Reid, A. F., & Ringwood, A. E. (1969). Six-coordinate silicon: High pressure strontium and barium aluminosilicates with the hollandite structure. *Journal of Solid State Chemistry*, *1*(1), 6–9. doi:10.1016/0022-4596(69)90002-4

Reis, K., Ramanan, A., & Whittingham, M. (1990). Hydrothermal synthesis of sodium tungstates. *Chemistry of Materials*, *2*(3), 219–221. doi:10.1021/cm00009a003

Reneker, D. H., & Chun, I. (1996). Nanometre diameter fibres of polymer produced by electrospinning. *Nanotechnology*, *7*(3), 216–223. doi:10.1088/0957-4484/7/3/009

Reneker, D. H., & Yarin, A. L. (2008). Electrospinning jets and polymer nanofibers. *Polymer*, *49*(10), 2387–2425. doi:10.1016/j.polymer.2008.02.002

Ren, X. F., Li, J., Holmes, R. J., Djurovich, P. I., Forrest, S. R., & Thompson, M. E. (2004). Ultrahigh Energy Gap Hosts in Deep Blue Organic Electrophosphorescent Devices. *Chemistry of Materials*, *16*(23), 4743–4747. doi:10.1021/cm049402m

Retricevic, V., & Gayen, S. K. (1998). Article. *Applied Physics Letters*, *52*, 1040.

Revathi, V. (2014). Article. *Chin Shu Hsueh Pao*, *27*, 557–562.

Riseberg, L. A., & Moos, H. W. (1968). Multiphonon orbit-lattice relaxation of excited states of rare-earth ions in crystals. *Physical Review*, *174*(2), 429–438. doi:10.1103/PhysRev.174.429

Robertson, J. M., Van Tol, M. W., Smits, W. H., & Heynen, J. P. H. (1981). Color shift of the Ce^{3+} emission in monocrystalline epitaxially grown garnet layers. *Philips Journal of Research*, *36*, 15–30.

Roco, M. C. (1999). Nanoparticles and nanotechnology research. *Journal of Nanoparticle Research*, *1*(1), 1–6. doi:10.1023/A:1010093308079

Ronda, C. (2008). *Luminescence: from theory to application*. Weinheim: Wiley-VCH publication.

Ronda, C. R. (1997). Recent achievements in research on phosphors for lamps and displays. *Journal of Luminescence*, *49*, 72.

Russ, M. J., & Kennedy, D. I. (1967). The Effects of Double Insulating Layers on the Electroluminescence of Evaporated ZnS:Mn Films. *Journal of the Electrochemical Society*, *114*(10), 1066–1071. doi:10.1149/1.2424188

Ryan, F. M., Warren, R. W., Hopkins, R. H., & Murphy, J. (1978). Selective Site Laser Excitation and ESR Studies of Nd^{3+} Ions in $Ca_5(PO_4)_3F$. *Journal of the Electrochemical Society*, *125*(9), 1493–1498. doi:10.1149/1.2131702

Sahare, P. D., Bakare, J. S., Dhole, S. D., Ingale, N. B., & Rupasov, A. A. (2010). Synthesis and luminescence properties of nanocrystalline LiF:Mg,Cu,P phosphor. *Journal of Luminescence*, *130*(2), 258–265. doi:10.1016/j.jlumin.2009.08.018

Salah, N. (2008). Carbon ions irradiation on nano- and microcrystalline $CaSO_4$: Dy *Journal of Applied Physics D. Applied Physics (Berlin)*, *41*, 155302–155308.

Salah, N. (2011). Nanocrystalline materials for the dosimetry of heavy charged particles: A review. *Radiation Physics and Chemistry*, *80*(1), 1–10. doi:10.1016/j.radphyschem.2010.08.003

Salah, N., Sahare, P. D., Lochab, S. P., & Kumar, P. (2006). TL and PL studies on $CaSO_4$: Dy nanoparticles. *Radiation Measurements*, *41*(1), 40–47. doi:10.1016/j.radmeas.2005.07.026

Salah, N., Sahare, P. D., & Rupasov, A. A. (2007). Thermoluminescence of nanocrystalline LiF: Mg, Cu, P. *Journal of Luminescence*, *124*(2), 357–364. doi:10.1016/j.jlumin.2006.04.004

Saluja, J. K., Parganiha, Y., Tiwari, N., Dubey, V., Tiwari, R., & Prabhath, A. (2016). Mechano and photoluminescence spectra of cadmium sulphide and cadmium selenide doped phosphors. *Optik-International Journal for Light and Electron Optics*, *127*(19), 7958–7966. doi:10.1016/j.ijleo.2016.05.011

Santhanam, R., & Rambabu, B. (2010). Research Progress in High Voltage Spinel $LiNi_{0.5}Mn_{1.5}O_4$ Material. *Journal of Power Sources*, *195*(17), 5442–5451. doi:10.1016/j.jpowsour.2010.03.067

Saradhi, M. P., & Varadaraju, U. V. (2006). Photoluminescence studies on Eu^{2+}-activated Li_2SrSiO_4 a potential orange-yellow phosphor for solid-state lighting. *Chemistry of Materials*, *18*(22), 5267–5272. doi:10.1021/cm061362u

Saravanapavan & Hench. (2003). J. Non-Crys. *Soli.*, *318*, 14.

Sasabe, H., Chiba, T., Su, S. J., Pu, Y. J., Nakayama, K. I., & Kido, J. (2008). 2-Phenylpyrimidine skeleton-based electron-transport materials for extremely efficient green organic light-emitting devices. *Chemical Communications (Cambridge)*, (44): 5821. doi:10.1039/b812270a PMID:19009093

Sasabe, H., & Kido, J. (2011). Multifunctional Materials in High-Performance OLEDs: Challenges for Solid-State Lighting. *Chemistry of Materials*, *23*(3), 621–630. doi:10.1021/cm1024052

Sato, M., Tanaka, T., & Ohta, M. (1994). Photostimulated Luminescence and Structural Characterization of $Ba_5(PO_4)_3Cl{:}Eu^{2+}$ Phosphors. *Journal of the Electrochemical Society*, *141*(7), 1851–1855. doi:10.1149/1.2055016

Sato, Y., Kuwahara, H., Kato, H., Kobayashi, M., Masaki, T., & Kakihana, M. (2015). Large redshifts in emission and excitation from Eu^{2+}-activated Sr_2SiO_4 and Ba_2SiO_4 phosphors induced by controlling Eu^{2+} occupancy on the basis on crystal-site engineering. *Optics and Photonics Journal*, *5*(11), 326–333. doi:10.4236/opj.2015.511031

Schafhäutl, G. A. (1845). Article. *Royal Bavarian Academy of Sciences*, *20*, 557.

Schiwietz, G., Staufenbiel, F., Roth, M., Czerski, K., & Grande, P. L. (2005). *Electrons as a tool for short-time track characterization*. Aschaenburg.

Schubert, E. F., & Kim, J. K. (2005). Solid-state light sources getting smart. *Science*, *308*(5726), 1274–1278. doi:10.1126cience.1108712 PMID:15919985

Schwartz, G., Reineke, S., Rosenow, T. C., Walzer, K., & Leo, K. (2009). Triplet Harvesting in Hybrid White Organic Light-Emitting Diodes. *Advanced Functional Materials*, *19*(9), 1319–1333. doi:10.1002/adfm.200801503

Setlur, A. A., Heward, W. J., Gao, Y., Srivastava, A. M., Gopi Chandran, R., & Shankar, M. (2006). Crystal Chemistry and Luminescence of Ce^{3+}-Doped $Lu_2CaMg_2(Si,Ge)_3O_{12}$ and Its Use in LED Based Lighting. *Chemistry of Materials*, *18*(14), 3314–3322. doi:10.1021/cm060898c

Shalav, A., Richards, B. S., Trupke, T., Kramer, K. W., & Güdel, H. U. (2005). Application of NaYF4:Er^{3+} up-converting phosphors for enhanced near-infrared silicon solar cell response. *Applied Physics Letters*, *86*(1), 013505. doi:10.1063/1.1844592

Shan, G. B., & Demopoulos, G. P. (2010). Near-Infrared Sunlight Harvesting in Dye-Sensitized Solar Cells Via the Insertion of an Upconverter-TiO_2 Nanocomposite Layer. *Advanced Materials*, *22*(39), 4373–4377. doi:10.1002/adma.201001816 PMID:20809511

Shani, G. (2000). *Radiation Dosimetry-Instrumentation and Methods* (2nd ed.). CRC Press.

Shan, T., Gao, Z., Tang, X., He, X., Gao, Y., Li, J., ... Ma, Y. (2017). Highly efficient and stable pure blue nondoped organic light-emitting diodes at high luminance based on phenanthroimidazole-pyrene derivative enabled by triplei-triplet annihilation. *Dyes and Pigments*, *142*, 189–197. doi:10.1016/j.dyepig.2017.03.032

Shan, T., Liu, Y., Tang, X., Bai, Q., Gao, Y., Gao, Z., ... Ma, Y. (2016). Highly Efficient Deep Blue Organic Light-Emitting Diodes Based on Imidazole: Significantly Enhanced Performance by Effective Energy Transfer with Negligible Efficiency Roll-off. *ACS Applied Materials & Interfaces*, *8*(42), 28771–28779. doi:10.1021/acsami.6b10004 PMID:27723979

Sharma, V., Shukla, A., & Gopalakrishnan, K. (1992). Effect of aliovalent-cation substitution on the oxygen-ion conductivity of Bi4V2O11. *J. Solid State Ionics*, *58*(3-4), 359–362. doi:10.1016/0167-2738(92)90141-B

Shi, H., Xin, D., Dong, X., Dai, J., Wu, X., Miao, Y., ... Choi, M. (2014). A star-shaped bipolar host material based on carbazole and dimesitylboron moieties for fabrication of highly efficient red, green and blue electrophosphorescent devices. *Journal of Materials Chemistry. C, Materials for Optical and Electronic Devices*, *2*(12), 2160–2168. doi:10.1039/C3TC32236J

Shih, P. I., Chien, C. H., Wu, F. I., & Shu, C. F. (2007). A Novel Fluorene-Triphenylamine Hybrid That is a Highly Efficient Host Material for Blue-, Green-, and Red-Light-Emitting Electrophosphorescent Devices. *Advanced Functional Materials*, *17*(17), 3514–3520. doi:10.1002/adfm.200700191

Shih, P. I., Chuang, C. Y., Chien, C. H., Diau, E. W. G., & Shu, C. F. (2007). Highly Efficient Non-Doped Blue-Light-Emitting Diodes Based on an Anthrancene Derivative End-Capped with Tetraphenylethylene Groups. *Advanced Functional Materials*, *17*(16), 3141–3146. doi:10.1002/adfm.200700355

Shi, L. A., Huang, Y. L., & Seo, H. J. (2010). Emission Red Shift and Unusual Band Narrowing of Mn^{2+} in $NaCaPO_4$ Phosphor. *The Journal of Physical Chemistry A*, *114*(26), 6927–6934. doi:10.1021/jp101772z PMID:20540584

Shinsho, K., Harada, K., Yamamoto, Y., & Urushiyama, A. (2008). Differences in glow curves structure of nano- and microcrystals of $CaSO_4$:Dy measured at a low heating rate. *Radiation Measurements*, *43*(2-6), 236–240. doi:10.1016/j.radmeas.2007.11.037

Shockley, W., & Queisser, H. J. (1961). Detailed balance limit of efficiency of p-n junction solar cells. *Journal of Applied Physics*, *32*(3), 510–519. doi:10.1063/1.1736034

Shojiya, M., Takahashi, M., Kanno, R., Kawamoto, Y., & Kadono, K. (1994). Upconversion luminescence of Er^{3+}in chloride glasses based on $ZnCl_2$ or $CdCl_2$. *Applied Physics Letters*, *65*(15), 1874–1876. doi:10.1063/1.112874

Shojiya, M., Takahashi, M., Kanno, R., Kawamoto, Y., & Kadono, K. (1995). Upconversion luminescence of Er^{3+} in CdX_2system glasses (X=Cl, Br, I). *Applied Physics Letters*, *67*(17), 2453–2455. doi:10.1063/1.114605

Shojiya, M., Takahashi, M., Kanno, R., Kawamoto, Y., & Kadono, K. (1997). Optical transitions of Er^{3+} ions in $ZnCl_2$-based glass. *Journal of Applied Physics*, *82*(12), 6259–6266. doi:10.1063/1.366545

Shrivastava, R., Kaur, J., Dubey, V., & Jaykumar, B. (2014). Photoluminescence, trap states and thermoluminescence decay process study of Ca2MgSi2O7: Eu2+, Dy3+ phosphor. *Bulletin of Materials Science*, *37*(4), 925–929. doi:10.100712034-014-0027-0

Shrivastava, R., Kaur, J., Dubey, V., Jaykumar, B., & Loreti, S. (2015). Photoluminescence and Thermoluminescence Investigation of Europium- and Dysprosium-Doped Dibarium Magnesium Silicate Phosphor. *Spectroscopy Letters*, *48*(3), 179–183. doi:10.1080/00387010.2013.872666

Singh, T. S. C. (1989). *Thermally stimulated luminescence of some solids* (Unpublished doctoral dissertation). Manipur University, Imphal, India.

Singh, I. R., Ningthoujam, R. S., Sudarsan, V., Shrivastav, I., Singh, S. D., Dey, G. K., & Kulshreshtha, S. K. (2008). Luminescence study on Eu^{3+} doped Y_2O_3 nanoparticles: Particle size, concentration and core–shell formation effects. *Nano Tech.*, *19*, 1.

Singh, J., Manam, J., & Singh, F. (2017). Synthesis and thermoluminescence studies of γ-irradiated Dy^{3+} doped $SrGd_2O_4$ phosphor. *Materials Research Bulletin*, *94*, 113–121. doi:10.1016/j.materresbull.2017.05.052

Singh, S. J., Karmakar, M., Bhattacharya, M., Singh, S. D., Singh, W. S., & Azharuddin, S. (2012). On the reappraisal of the peak shape method for general order thermoluminescence. *Indian Journal of Physics*, *86*(2), 113–116. doi:10.100712648-012-0013-6

Singh, S. J., Karmakar, M., & Singh, S. D. (2013). On the determination of the order of kinetics in thermoluminescence by peak-shape method. *Rad. Eff. Def. Sol.*, *168*(5), 352–357. doi:10.1080/10420150.2013.771358

Singh, S. J., Karmakar, M., Singh, W. S., & Singh, S. D. (2012). Modified peak shape method for the determination of activation energy in thermoluminescence. *Physica Scripta*, *86*(3), 035702. doi:10.1088/0031-8949/86/03/035702

Singh, S. K., Kumar, K., & Rai, S. B. (2009). Multifunctional Er^{3+}–Yb^{3}codoped Gd_2O_3 nanocrystalline phosphor synthesized through optimized combustion route. *Applied Physics. B, Lasers and Optics*, *94*(1), 165–173. doi:10.100700340-008-3261-6

Singh, W. S., Singh, S. D., & Mazumdar, P. S. (1998). On the mixed order kinetics of the thermoluminescence glow peak. *Journal of Physics Condensed Matter*, *10*(22), 4937–4946. doi:10.1088/0953-8984/10/22/018

Sivakumar, V., & Kasturi, S. (2017). Eu3+ based orange - red emitting inorganic colour convertors: An overview. In Synthesis of phosphors and their applications. Singapore: Pan Stanford Publishing Pte Ltd.

Sivakumar, V., Lakshmanan, A., Satheesh Kumar, R., Kalpana, S., Sangeetha Rani, R., & Jose, M. T. (2012).. . *Indian Journal of Pure and Applied Physics*, *50*, 123–128.

Sivakumar, V., & Varadaraju, U. V. (2007). Ce^{3+} \ Eu^{2+} Energy Transfer Studies on $BaMgSiO_4$ A Green Phosphor for Three Band White LEDs. *Journal of the Electrochemical Society*, *154*(5), J167–J171. doi:10.1149/1.2714303

Sivakumar, V., & Varadaraju, U. V. (2009). Eu^{2+}, Ce^{3+} Luminescence and $Ce^{3+} \rightarrow Eu^{2+}$ Energy-Transfer Studies on Sr_2LiSiO_4F: A White Light-Emitting Phosphor. *Journal of the Electrochemical Society*, *156*(7), J179–J184. doi:10.1149/1.3122663

Snitzer, E., & Woodcock, R. (1965). Yb^{5+}-Er^{3+}-Ybw glass laser. *Applied Physics Letters*, *6*(3), 5. doi:10.1063/1.1754157

Sohn, K. S., Kim, S., & Park, H. D. (2002). Luminescence quenching in thermally-treated barium magnesium aluminate phosphor. *Applied Physics Letters*, *81*(10), 1759–1761. doi:10.1063/1.1504866

Somiya, S., Hirano, S., Yoshimura, M., & Yanagisawa, K. (1981). Hydrothermal crystal growth of perovskite-type fluorides. *Journal of Materials Science*, *16*(3), 813–816. doi:10.1007/BF02402800

Som, S., Choubey, A., & Sharma, S. K. (2015). Spectral and trapping parameters of Eu in Gd_2O_2S nanophosphor. *Journal of Experimental Nanoscience*, *10*(5), 350–370. doi:10.1080/17458080.2013.837972

Som, S., Chowdhury, M., & Sharma, S. K. (2015). Kinetic parameters of γ-irradiated Y2O3 phosphors: Effect of doping/codoping and heating rate. *Radiation Physics and Chemistry*, *110*, 51–58. doi:10.1016/j.radphyschem.2015.01.015

Som, S., Dutta, S., Chowdhury, M., Kumar, V., Kumar, V., Swart, H. C., & Sharma, S. K. (2014). A comparative investigation on ion impact parameters and TL response of Y_2O_3: Tb^{3+}nano phosphor exposed to swift heavy ions for space dosimetry. *Journal of Alloys and Compounds*, *589*, 5–18. doi:10.1016/j.jallcom.2013.11.182

Som, S., Kunti, A. K., Kumar, V., Kumar, V., Dutta, S., Chowdhury, M., ... Swart, H. C. (2014). Defect correlated fluorescent quenching and electron-phonon coupling in the spectral transition of Eu^{3+} in $CaTiO_3$ for red emission in display application. *Journal of Applied Physics*, *115*(19), 193101. doi:10.1063/1.4876316

Som, S., Mitra, P., Kumar, V., Kumar, V., Terblans, J. J., Swart, H. C., & Sharma, S. K. (2014). The energy transfer phenomena and color tunability in Y_2O_2S: Eu^{3+}/ Dy^{3+} micro-fibers for white emission in solid-state lighting applications. *Dalton Transactions (Cambridge, England)*, *43*(26), 9860–9871. doi:10.1039/C4DT00349G PMID:24710323

Song, H., Sun, B., Wang, T., Lu, S., Yang, L., Chen, B., ... Kong, X. (2004). Three-photon upconversion luminescence phenomenon for the green levels in Er3+/Yb3+ codoped cubic nanocrystalline yttria. *Solid State Communications*, *132*(6), 409–413. doi:10.1016/j.ssc.2004.07.044

Stenberg, L., Sellar, J. R., & Hyde, B. G. (1986). Incommensurately modulated α′-Sr_2SiO_4. *Nature*, *320*(6061), 428–429. doi:10.1038/320428a0

Stouwdam Jan, W., Hebbink Gerald, A., Jurriaan, H., & van Veggel Frank, C. J. M. (2003). Article. *Chemistry of Materials*, *15*, 4604–4616. doi:10.1021/cm034495d

Stouwdam Jan, W., & van Veggel Frank, C. J. M. (2002). Article. *Nano Letters*, *2*, 733–737. doi:10.1021/nl025562q

Subeesh, M. S., Shanmugasundaram, K., Sunesh, C. D., Nguyen, T. P., & Choe, Y. (2015). Phenanthroimidazole Derivative as an Easily Accessible Emitter for Non-Doped Light-Emitting Electrochemical Cells. *The Journal of Physical Chemistry C*, *119*(41), 23676–23684. doi:10.1021/acs.jpcc.5b07871

Sugita, M., Tsuji, M., & Abe, M. (1990). Synthetic Inorganic Ion-Exchange Materials. LVIII. Hydrothermal Synthesis of a New Layered Lithium Titanate and Its Alkali Ion Exchange. *Bulletin of the Chemical Society of Japan*, *63*(7), 1978–1984. doi:10.1246/bcsj.63.1978

Sukul, P. P., Mahata, M. K., & Kumar, K. (2017). NIR optimized dual mode photoluminescence in Nd doped Y_2O_3 ceramic phosphor. *Journal of Luminescence*, *185*, 92–98. doi:10.1016/j.jlumin.2017.01.008

Su, M. Y., Zhou, Y. F., Wang, K., Huang, D. C., Xu, W. T., & Cao, Y. G. (2015). Effect of Yb^{3+} concentration on upconversion luminescence of AlON:Er^{3+} phosphors. *Journal of Rare Earths*, *33*(3), 227–230. doi:10.1016/S1002-0721(14)60407-5

Sun, D. M., Zhou, X. K., Li, H. H., Sun, X. L., Zheng, Y. H., Ren, Z. J., ... Yan, S. K. (2014). A versatile hybrid polyphenylsilane host for highly efficient solution-processed blue and deep blue electrophosphorescence. *Journal of Materials Chemistry. C, Materials for Optical and Electronic Devices*, *2*(39), 8277–8284. doi:10.1039/C4TC01467G

Sunitha, D. V., Manjunatha, C., Shilpa, C. J., Nagabhushana, H., Sharma, S. C., Nagabhushana, B. M., & Chakradhar, R. P. S. (2012). $CdSiO_3$: Pr^{3+} nanophosphor: Synthesis, characterization and thermoluminescence studies. *Spectrochimica Acta. Part A: Molecular and Biomolecular Spectroscopy*, *99*, 279–287. doi:10.1016/j.saa.2012.08.057 PMID:23085285

Sunitha, D. V., Nagabhushana, H., & Fouran Singh, B. M. (2012). Thermo, Iono and photoluminescence properties of 100 MeV Si^{7+} ions bombarded $CaSiO_3$:Eu^{3+} nanophosphor. *Journal of Luminescence*, *132*(8), 2065–2071. doi:10.1016/j.jlumin.2012.03.019

Sunta, M., Ayta, W. E. F., Chubaci, J. F. D., & Watanabe, S. (2001). Dependence of quasi-equilibrium on heating rate and its use in the study of thermoluminescent materials. *Journal of Physics. D, Applied Physics*, *34*(22), 3285–3295. doi:10.1088/0022-3727/34/22/313

Sun, W., Jia, Y., Pang, R., Li, H., Ma, T., Li, D., ... Li, C. (2015). $Sr_9Mg_{1.5}(PO_4)_7$:Eu^{2+}: A Novel Broadband Orange-Yellow-Emitting Phosphor for Blue Light-Excited Warm White LEDs. *ACS Applied Materials & Interfaces*, *7*(45), 25219–25226. doi:10.1021/acsami.5b06961 PMID:26503633

Sun, X.-Y., Liu, Y., Liu, X.-L., Cao, R.-P., Li, Y.-N., & Lin, L.-W. (2014). Substitution of Y^{3+} for Gd^{3+} on the luminescent properties of $BaGd_2O_4$:Eu^{3+} scintillating phosphors. *Optical Materials*, *36*(9), 1478–1483. doi:10.1016/j.optmat.2014.03.044

Sun, Y., Duan, L., Zhang, D., Qiao, J., Dong, G., Wang, L., & Qiu, Y. (2011). A Pyridine-Containing Anthracene Derivative with High Electron and Hole Mobilities for Highly Efficient and Stable Fluorescent Organic Light-Emitting Diodes. *Advanced Functional Materials*, *21*(10), 1881–1886. doi:10.1002/adfm.201002691

Suo, H., Hu, F., Zhao, X., Zhang, Z., Li, T., Duan, C., ... Guo, C. (2017). All-in-one thermometer-heater up-converting platform YF_3:Yb^{3+},Tm^{3+} operating in the first biological window. *Journal of Materials Chemistry. C, Materials for Optical and Electronic Devices*, *5*(6), 1501–1507. doi:10.1039/C6TC05449H

Su, S. J., Cai, C., & Kido, J. (2011). RGB Phosphorescent Organic Light-Emitting Diodes by Using Host Materials with Heterocyclic Cores: Effect of Nitrogen Atom Orientations. *Chemistry of Materials*, *23*(2), 274–284. doi:10.1021/cm102975d

Su, S. J., Sasabe, H., Pu, Y. J., Nakayama, K., & Kido, J. (2010). Pyridine-Containing Bipolar Host Materials for Highly Efficient Blue Phosphorescent OLEDs. *Advanced Materials*, *22*, 3311. doi:10.1002/adma.200904249 PMID:20552601

Su, S. J., Takahashi, Y., Chiba, T., Takeda, T., & Kido, J. (2009). Relationship Structure–Property of Pyridine-Containing Triphenyl Benzene Electron-Transport Materials for Highly Efficient Blue Phosphorescent OLEDs. *Advanced Functional Materials*, *19*, 1260. doi:10.1002/adfm.200800809

Su, S. J., Tanaka, D., Li, Y. J., Sasabe, H., Takeda, T., & Kido, J. (2008). Novel Four-Pyridylbenzene-Armed Biphenyls as Electron-Transport Materials for Phosphorescent OLEDs. *Organic Letters*, *10*(5), 941–944. doi:10.1021/ol7030872 PMID:18237184

Takahashi, M., Kanno, R., Kawamoto, Y., Tanabe, S., & Hirao, K. (1994). Compositional dependence of Er^{3+}upconversion luminescence in $MFLiFZrF_4$ glasses (M: Alkali metals). *Journal of Non-Crystalline Solids*, *168*(1), 137–143. doi:10.1016/0022-3093(94)90129-5

Takizawa, S., Montes, V. A., & Anzenbacher, P. (2009). Phenylbenzimidazole-Based New Bipolar Host Materials for Efficient Phosphorescent Organic Light-Emitting Diodes. *Jr. Chemistry of Materials*, *21*(12), 2452–2458. doi:10.1021/cm9004954

Tamao, K., Uchida, M., Izumizawa, T., Furukawa, K., & Yamaguchi, S. (1996). Silole Derivatives as Efficient Electron Transporting Materials. *Journal of the American Chemical Society*, *118*(47), 11974–11975. doi:10.1021/ja962829c

Tamrakar, R. K., & Bisen, D. P. (2015). Thermoluminescence studies of ultraviolet and gamma irradiated erbium (III)- and ytterbium (III)-doped gadolinium oxide phosphors. *Materials Science in Semiconductor Processing*, *33*, 169–188. doi:10.1016/j.mssp.2015.01.044

Tam, T. T. H., Du, N. V., Kien, N. D. T., Thang, C. X., Cuong, N. D., Huy, P. T., ... Nguyen, D. H. (2014). Co-precipitation synthesis and optical properties of green-emitting $Ba_2MgSi_2O_7$: Eu^{2+} phosphor. *Journal of Luminescence*, *147*, 358–362. doi:10.1016/j.jlumin.2013.11.066

Tang, C. W., & VanSlyke, S. A. (1987). Organic electroluminescent diodes. *Applied Physics Letters*, *51*(12), 913–915. doi:10.1063/1.98799

Tang, C., Bi, R., Tao, Y., Wang, F., Cao, X., Wang, S., ... Huang, W. (2015). A versatile efficient one-step approach for carbazole–pyridine hybrid molecules: Highly efficient host materials for blue phosphorescent OLEDs. *Chemical Communications (Cambridge)*, *51*(9), 1650–1653. doi:10.1039/C4CC08335K PMID:25500635

Tang, X., Bai, Q., Peng, Q., Gao, Y., Li, J., Liu, Y., ... Ma, Y. (2015). Efficient Deep Blue Electroluminescence with an External Quantum Efficiency of 6.8% and CIEy < 0.08 Based on a Phenanthroimidazole–Sulfone Hybrid Donor–Acceptor Molecule. *Chemistry of Materials*, *27*(20), 7050–7057. doi:10.1021/acs.chemmater.5b02685

Tang, X., Shan, T., Bai, Q., Ma, H., He, X., & Lu, P. (2017). Efficient Deep Blue Electroluminescence Based on Phenanthroimidazole-Dibenzothiophene Derivatives with Different Oxidation States of Sulfur Atom. *Chemistry, an Asian Journal*, *12*(5), 552–560. doi:10.1002/asia.201601626 PMID:28070976

Tang, Y. S., Hu, S. F., Lin, C. C., Bagkar, N. C., & Liu, R. S. (2007). Thermally stable luminescence of $KSrPO_4$:Eu^{2+} phosphor for white light UV light-emitting diodes. *Applied Physics Letters*, *90*(15), 151108. doi:10.1063/1.2721846

Tang, Z. B., Wang, D. Y., Khan, W. U., Du, S. S., Wang, X. C., & Wang, Y. H. (2016). Novel zirconium silicate phosphor $K_2ZrSi_2O_7$:Eu^{2+} for white light-emitting diodes and field emission displays. *Journal of Materials Chemistry. C, Materials for Optical and Electronic Devices*, *4*(23), 5307–5313. doi:10.1039/C6TC01449F

Tao, L., Li, J., Yu, Y., Jiang, Y., Zhou, C. S., Lee, S. T., ... Kwon, O. (2009). Bipolar Molecule as an Excellent Hole-Transporter for Organic-Light Emitting Devices. *Chemistry of Materials*, *21*(7), 1284–1287. doi:10.1021/cm803087c

Tao, Y., Wang, Q., Ao, L., Zhong, C., Qin, J., Yang, C., & Ma, D. (2010). Molecular design of host materials based on triphenylamine/oxadiazole hybrids for excellent deep-red phosphorescent organic light-emitting diodes. *Journal of Materials Chemistry*, *20*(9), 1759. doi:10.1039/b920227g

Tao, Y., Wang, Q., Ao, L., Zhong, C., Yang, C., Qin, J., & Ma, D. (2010). Highly Efficient Phosphorescent Organic Light-Emitting Diodes Hosted by 1,2,4-Triazole-Cored Triphenylamine Derivatives: Relationship between Structure and Optoelectronic Properties. *The Journal of Physical Chemistry C*, *114*(1), 601–609. doi:10.1021/jp908886d

Tao, Y., Wang, Q., Shang, Y., Yang, C. L., Ao, L., Qin, J., ... Shuai, Z. (2009). Multifunctional bipolar triphenylamine/oxadiazole derivatives: Highly efficient blue fluorescence, red phosphorescence host and two-color based white OLEDs. *Chemical Communications (Cambridge)*, (1): 77–79. doi:10.1039/B816264F PMID:19082004

Tao, Y., Wang, Q., Yang, C., Wang, Q., Zhang, Z., Zou, T., ... Ma, D. (2008). A Simple Carbazole/Oxadiazole Hybrid Molecule: An Excellent Bipolar Host for Green and Red Phosphorescent OLEDs. *Angewandte Chemie International Edition*, *47*(42), 8104–8107. doi:10.1002/anie.200803396 PMID:18798180

Tao, Y., Yang, C., & Qin, J. (2011). Organic host materials for phosphorescent organic light-emitting diodes. *Chemical Society Reviews*, *40*(5), 2943–2970. doi:10.1039/c0cs00160k PMID:21369622

Tao, Z. X., Huang, Y. L., & Seo, H. J. (2013). Blue luminescence and structural properties of Ce^{3+}-activated phosphosilicate apatite $Sr_5(PO_4)_2(SiO_4)$. *Dalton Transactions (Cambridge, England)*, *42*(6), 2121–2129. doi:10.1039/C2DT32117C PMID:23187727

Tavgeniene, D., Krucaite, G., Baranauskyte, U., Wu, J. Z., Su, H. Y., Huang, C. W., ... Grigalevicius, S. (2017). Phenanthro[9,10-d]imidazole based new host materials for efficient red phosphorescent OLEDs. *Dyes and Pigments*, *137*, 615–621. doi:10.1016/j.dyepig.2016.11.003

Taylor, G. (1969). Electrically driven jets. *Proceedings of the Royal Society of London. Series A, Mathematical and Physical Sciences*, *313*(1515), 453–475. doi:10.1098/rspa.1969.0205

Teo, W. E., & Ramakrishna, S. (2006). A review on electrospinning design and nanofibre assemblies. *Nanotechnology*, *17*(14), R89–R106. doi:10.1088/0957-4484/17/14/R01 PMID:19661572

Thornton, W. A. (1961). Direct Current Electroluminescence at Low Voltages. *Physical Review*, *122*(1), 58–59. doi:10.1103/PhysRev.122.58

Tian, L., Xu, Z., Zhao, S., Cui, Y., Liang, Z., Zhang, J., & Xu, X. (2014). The Upconversion Luminescence of $Er^{3+}/Yb^{3+}/Nd^{3+}$ Triply-Doped β-$NaYF_4$ Nanocrystals under 808-nm Excitation. *Materials (Basel)*, *7*(11), 7289–7303. doi:10.3390/ma7117289 PMID:28788246

Tian, Z., Liang, H., Lin, H., Su, Q., Guo, B., Zhang, G., & Fu, Y. (2006). Luminescence of $NaGdFPO_4$:Ln^{3+} after VUV excitation: A comparison with $GdPO_4$:Ln^{3+} (Ln=Ce, Tb). *Journal of Solid State Chemistry*, *179*(5), 1356–1362. doi:10.1016/j.jssc.2006.01.044

Tian, Z., Liu, J., Voigt, J. A., Xu, H., & Mcddermott, M. J. (2003). Dendritic Growth of Cubically Ordered Nanoporous Materials through Self-Assembly. *Nano Letters*, *3*(1), 89–92. doi:10.1021/nl025828t

Tikhomirov, V. K., Driesen, K., Rodriguez, V. D., Gredin, P., Mortier, M., & Moshchalkov, V. V. (2009). Optical nanoheater based on the Yb^{3+}-Er^{3+} co-doped nanoparticles. *Optics Express*, *17*(14), 11794–11798. doi:10.1364/OE.17.011794 PMID:19582094

Tiwari, S. P., Kumar, A., &Kumar,K. (2016). Upconversion phosphor materials for beginners: synthesis and applications. *Research Frontiers in Sciences,* 23-42.

Tiwari, N., Dubey, V., & Kuraria, R. K. (2015). Mechanoluminescence Study of Europium Doped CaZrO3 Phosphor. *Journal of Fluorescence*, *26*(4), 1309–1315. doi:10.100710895-016-1817-0 PMID:27184974

Tiwari, R., Dubey, V., & Chandra, B. P. (2014). Exact model for the elastico mechanoluminescence of II-VI phosphors. *Materials Physics and Mechanics*, *19*(1), 25–38.

Tiwari, S. P., Kumar, K., & Rai, V. K. (2015a). Plasmonic enhancement in upconversion emission of La_2O_3: Er^{3+}/Yb^{3+} phosphor via introducing silver metal nanoparticles. *Applied Physics. B, Lasers and Optics*, *121*(2), 221–228. doi:10.100700340-015-6223-9

Tiwari, S. P., Kumar, K., & Rai, V. K. (2015b). Latent fingermarks detection for La_2O_3: Er^{3+}/Yb^{3+} phosphor material in upconversion emission mode: A comparative study. *Journal of Applied Physics*, *118*(18), 183109. doi:10.1063/1.4935279

Togashi, K., Nomura, S., Yokoyama, N., Yasuda, T., & Adachi, C. (2012). Low driving voltage characteristics of triphenylene derivatives as electron transport materials in organic light-emitting diodes. *Journal of Materials Chemistry*, *22*(38), 20689. doi:10.1039/c2jm33669c

Triolo, A., Marrale, M., & Brai, M. (2007). Neutron–gamma mixed field measurements by means of MCP–TLD600 dosimeter pair. *Nuclear Instruments and Methods in Physics Research Section B*, *264*(1), 183–188. doi:10.1016/j.nimb.2007.08.009

Trupke, T., Green, M. A., & Würfel, P. (2002). Improving solar cell efficiencies by up-conversion of sub-band-gap light. *Journal of Applied Physics*, *92*(7), 4117–4122. doi:10.1063/1.1505677

Tsai, H., Lin, H. W., Su, H. C., Ke, T. H., Wu, C. C., Fang, F. C., ... Wu, C. I. (2006). Incorporation of a CN group into mCP: A new bipolar host material for highly efficient blue and white electrophosphorescent devices. *Advanced Materials, 18*, 1216–1220. doi:10.1002/adma.200502283

Tsai, M. T. (2002). Synthesis of nanocrystalline forsterite fiber via a chemical route. *Materials Research Bulletin, 37*(13), 2213–2226. doi:10.1016/S0025-5408(02)00926-1

Ulla, H., Garudachari, B., Satyanarayan, M. N., Umesh, G., & Isloor, A. M. (2014). Blue organic light emitting materials: Synthesis and characterization of novel 1,8-naphthalimide derivatives. *Optical Materials, 36*(3), 704–711. doi:10.1016/j.optmat.2013.11.017

Ulla, H., Raveendra Kiran, M., Garudachari, B., Satyanarayan, M. N., Umesh, G., & Isloor, A. M. (2014). Blue emitting halogen–phenoxy substituted 1,8-naphthalimides for potential organic light emitting diode applications. *Optical Materials, 37*, 311–321. doi:10.1016/j.optmat.2014.06.016

Vahlenkamp, T., & Wegner, G. (1994). Poly(2,5 dialkoxy-p-phenylene)S- synthesis and properties. *Macromolecular Chemistry and Physics, 195*(6), 1933–1952. doi:10.1002/macp.1994.021950605

Vaidhyanathan, B., Balaji, K., & Rao, K. J. (1998). Microwave-Assisted Solid-State Synthesis of Oxide Ion Conducting Stabilized Bismuth Vanadate Phases. *Chemistry of Materials, 10*(11), 3400–3401. doi:10.1021/cm980092f

Vecht, A., & Werring, N. J. (1970). Direct Current Electroluminescence in ZnS. *J. Phys. D, 3*(2), 105–120. doi:10.1088/0022-3727/3/2/202

Venkatachari, K. R., Huang, D., Ortrander, S. P., Schulze, W. A., & Stangle, G. C. (1995). A combustion synthesis process for synthesizing nanocrystalline zirconia powders. *Journal of Materials Research, 10*(03), 748–755. doi:10.1557/JMR.1995.0748

Venugopal, T., Jeeva, P., Jayabharathi, J., Ramanathan, P., Prabhakaran, A., & Purani, E. S. (2016). Nondoped Blue Organic Light Emitting Devices with Donor−π−Acceptor Derivatives as the Emissive Material. *Industrial & Engineering Chemistry Research, 55*(36), 9639–9647. doi:10.1021/acs.iecr.6b02549

Venugopal, T., Palanivel, J., & Jayaraman, J. (2017). Nondoped blue fluorescent OLED based on cyanophenanthrimidazole-styryl-triphenylamine/carbazole materials. *Journal of Physical Organic Chemistry, 30*(12), 3695. doi:10.1002/poc.3695

Venugopal, T., Sarojpurani, E., Jayabharathi, J., & Jeeva, P. (2017). Efficient phenanthroimidazole-styryltriphenylamine derivatives for blue OLEDs: A combined experimental and theoretical study. *New Journal of Chemistry, 41*(6), 2443–2457. doi:10.1039/C6NJ03801H

Verma, D., Saetre, T. O., & Midtgård, O. M. (2012, June). Review on up/down conversion materials for solar cell application. In *Photovoltaic Specialists Conference (PVSC), 2012 38th IEEE* (pp. 002608-002613). IEEE. 10.1109/PVSC.2012.6318129

Verma, R. K., Singh, S. K., & Rai, S. B. (2012). Upconversion, avalanche effect and controlled optical switching in Yb^{3+}, Ho^{3+} co-doped $Ca_{12}Al_{14}O_{33}$ phosphor. *Current Applied Physics, 12*(6), 1481–1484. doi:10.1016/j.cap.2012.04.018

Vetrone, F., Boyer, J. C., Capobianco, J. A., Speghini, A., & Bettinelli, M. (2003). Concentration-dependent near-infrared to visible upconversion in nanocrystalline and bulk Y_2O_3: Er^{3+}. *Chemistry of Materials, 15*(14), 2737–2743. doi:10.1021/cm0301930

Vetrone, F., Boyer, J. C., Capobianco, J. A., Speghini, A., & Bettinelli, M. (2003). Effect of Yb^{3+} Codoping on the Upconversion Emission in Nanocrystalline Y_2O_3:Er^{3+}. *The Journal of Physical Chemistry B, 107*(5), 1107–1112. doi:10.1021/jp0218692

Vetrone, F., Boyer, J. C., Capobianco, J. A., Speghini, A., & Bettinelli, M. (2004). Significance of Yb^{3+} concentration on the upconversion mechanisms in codoped Y_2O_3:Er^{3+}, Yb^{3+} nanocrystals. *Journal of Applied Physics, 96*(1), 661–667. doi:10.1063/1.1739523

Vij, A., Kumar, R., Chawla, A. K., Lochab, S. P., Chandra, R., & Singh, N. (2010). Swift heavy ion induced synthesis and enhanced photoluminescence of SrS: Ce nanoparticles. *Optical Materials, 33*(1), 58–62. doi:10.1016/j.optmat.2010.07.024

Vij, D. (Ed.). (2004). *Handbook of Electroluminescent Materials*. Bristol, UK: Institute of Physics. doi:10.1201/9781420033410

Vishwakarma, P., Ramrakhiani, M., Singh, P., & Bisen, D. P. (2011). Synthesis and electroluminescence studies of manganese doped cadmium sulfide nanoparticles. *Open Nanoscience Journal, 5*, 34-40.

Vodopyanov, K. L., Shori, R., & Stafsudd, O. M. (1998). Generation of **Q**-switched Er:YAG laser pulses using evanescent wave absorption in ethanol. *Applied Physics Letters, 72*(18), 2211–2213. doi:10.1063/1.121325

Vogt, T., Woodward, P. M., Hunter, B. A., Prodjosantoso, A. K., & Kennedy, B. J. (1999). Sr_3MO_4F (M=Al, Ga) - A New Family of Ordered Oxyfluorides. *Journal of Solid State Chemistry, 144*(1), 228–231. doi:10.1006/jssc.1999.8172

Wada, A., Yasuda, T., Zhang, Q., Yang, Y. S., Takasu, I., Enomoto, S., & Adachi, C. (2013). A host material consisting of a phosphinic amide directly linked donor–acceptor structure for efficient blue phosphorescent organic light-emitting diodes. *Journal of Materials Chemistry. C, Materials for Optical and Electronic Devices, 1*(13), 2404. doi:10.1039/c3tc00939d

Wade, S. A., Collins, S. F., & Baxter, G. W. (2003). Fluorescence intensity ratio technique for optical fiber point temperature sensing. *Journal of Applied Physics, 94*(8), 4743–4756. doi:10.1063/1.1606526

Wagiran, H., Hossain, I., Bradley, D., Yaakob, A. N. H., & Ramli, T. (2012). Thermoluminescence Responses of Photon and Electron Irradiated Ge- and Al-Doped SiO_2 Optical Fibres. *Chinese Physics Letters, 29*(2), 027802. doi:10.1088/0256-307X/29/2/027802

Wakamiya, A., Taniguchi, T., & Yamaguchi, S. (2006). Intramolecular B–N coordination as a scaffold for electron-transporting materials: Synthesis and properties of boryl-substituted thienylthiazoles. *Angewandte Chemie International Edition, 45*(19), 3170–3173. doi:10.1002/anie.200504391 PMID:16570334

Wang, X-J., Jia, Z.Y., Strek, W., Pazik, R., Hreniak, D., & Yen, W. M. (2005). Crystal size dependence of the persistent phosphorescence in $Sr_2ZnSi_2O_7$: Eu^{2+}, Dy^{3+}. *Microelectronics Journal, 36*, 546–548.

Wang, B., Lv, X., Tan, J., Zhang, Q., Huang, Z., Yi, W., & Wang, L. (2016). Bipolar phenanthroimidazole–diazacarbazole hybrids with appropriate bandgaps for highly efficient and low roll-off red, green and blue electroluminescent devices. *Journal of Materials Chemistry. C, Materials for Optical and Electronic Devices, 4*(36), 8473–8482. doi:10.1039/C6TC02683D

Wang, B., Mu, G., Tan, J., Lei, Z., Jin, J., & Wang, L. (2015). Pyridine-containing phenanthroimidazole electron-transport materials with electron mobility/energy-level trade-off optimization for highly efficient and low roll-off sky blue fluorescent OLEDs. *Journal of Materials Chemistry. C, Materials for Optical and Electronic Devices, 3*(29), 7709–7719. doi:10.1039/C5TC01436K

Wang, D. W., Huang, S. H., You, F. T., Qi, S. Q., Fu, Y. B., Zhang, G. B., ... Huang, Y. (2007). Vacuum ultraviolet spectroscopic properties of Pr3+ in MYF4 (M=Li, Na, and K) and LiLuF4. *Journal of Luminescence, 122*, 450–456. doi:10.1016/j.jlumin.2006.01.203

Wang, D. Y., & Wang, Y. H. (2006). Photoluminescence of $Y_4Al_2O_9$:Re (Re= Tb^{3+}, Eu^{3+}) under VUV excitation. *Journal of Alloys and Compounds*, *425*(1-2), L5–L7. doi:10.1016/j.jallcom.2006.01.033

Wang, F., & Liu, X. (2009). Recent advances in the chemistry of lanthanide-doped upconversion nanocrystals. *Chemical Society Reviews*, *38*(4), 976–989. doi:10.1039/b809132n PMID:19421576

Wang, G., Peng, Q., & Li, Y. (2011a). Lanthanide-doped nanocrystals: Synthesis, optical-magnetic properties, and applications. *Accounts of Chemical Research*, *44*(5), 322–332. doi:10.1021/ar100129p PMID:21395256

Wang, H. Q., Batentschuk, M., Osvet, A., Pinna, L., & Brabec, C. J. (2011b). Rare-Earth Ion Doped Up-Conversion Materials for Photovoltaic Applications. *Advanced Materials*, *23*(22-23), 2675–2680. doi:10.1002/adma.201100511 PMID:21823249

Wang, H. Q., Stubhan, T., Osvet, A., Litzov, I., & Brabec, C. J. (2012). Up-conversion semiconducting MoO 3: Yb/Er nanocomposites as buffer layer in organic solar cells. *Solar Energy Materials and Solar Cells*, *105*, 196–201. doi:10.1016/j.solmat.2012.06.005

Wang, H., Duan, C. K., & Tanner, P. A. (2008). Visible upconversion luminescence from Y_2O_3: Eu^{3+}, Yb^{3+}. *The Journal of Physical Chemistry C*, *112*(42), 16651–16654. doi:10.1021/jp8046505

Wang, H., Qian, C., Yi, Z., Rao, L., Liu, H., & Zeng, S. (2013). Hydrothermal Synthesis and Tunable Multicolor Upconversion Emission of Cubic Phase Y_2O_3 Nanoparticles. *Advances in Condensed Matter Physics.*

Wang, J., Lin, H., Huang, Q., Xiao, G., Xu, J., Wang, B., ... Wang, Y. (2017). Structure and luminescence behavior of a single-ion activated single-phased $Ba_2Y_3(SiO_4)_3F$:Eu white-light phosphor. *Journal of Materials Chemistry. C, Materials for Optical and Electronic Devices*, *5*(7), 1789–1797. doi:10.1039/C6TC04350J

Wang, J., Lou, X., Liu, Y., Zhao, G., Islam, A., Wang, S., & Ge, Z. (2015). Controllable molecular configuration for significant improvement of blue OLEDs based on novel twisted anthracene derivatives. *Dyes and Pigments*, *118*, 137–144. doi:10.1016/j.dyepig.2015.03.005

Wang, J., & Xu, Y. (2009). Mirrabos Hojamberdiev, Jianhong Peng, Gangqiang Zhu, A facile route to synthesize luminescent YVO_4:Eu^{3+}. *J. Non-Cryst. Solid.*, *355*, 903–907.

Wang, J., Yue, Y., Yao, J., & Hu, Z. (2011). Growth and characterization of Ba(Al,Ga)BO3F2 crystal. *Journal of Crystal Growth*, *318*(1), 962–965. doi:10.1016/j.jcrysgro.2010.10.223

Wang, K., Wang, S., Wei, J., Chen, S., Liu, D., Liu, Y., & Wang, Y. (2014). New multifunctional phenanthroimidazole–phosphine oxide hybrids for high-performance red, green and blue electroluminescent devices. *Journal of Materials Chemistry. C, Materials for Optical and Electronic Devices*, *2*(33), 6817–6826. doi:10.1039/C4TC00749B

Wang, K., Wang, S., Wei, J., Miao, Y., Liu, Y., & Wang, Y. (2014). Novel diarylborane–phenanthroimidazole hybrid bipolar host materials for high-performance red, yellow and green electrophosphorescent devices. *Organic Electronics*, *15*(11), 3211–3220. doi:10.1016/j.orgel.2014.08.062

Wang, L., & Li, Y. (2006). Green upconversion nanocrystals for DNA detection. *Chemical Communications*, (24): 2557–2559. doi:10.1039/b604871d PMID:16779476

Wang, Q., & Ma, D. (2010). Management of charges and excitons for high-performance white organic light-emitting diodes. *Chemical Society Reviews*, *39*(7), 2387–2398. doi:10.1039/b909057f PMID:20358089

Wang, Q., Oswald, I. W. H., Yang, X., Zhou, G., Jia, H., Qiao, Q., ... Gnade, B. E. (2014). A Non-Doped Phosphorescent Organic Light-Emitting Device with Above 31% External Quantum Efficiency. *Advanced Materials*, *26*(48), 8107–8113. doi:10.1002/adma.201402947 PMID:25219957

Wang, Q., Zhu, G., Xin, S., Ding, X., Xu, J., Wang, Y., & Wang, Y. (2015). A blue-emitting Sc silicate phosphor for ultraviolet excited light-emitting diodes. *Physical Chemistry Chemical Physics*, *17*(41), 27292–27299. doi:10.1039/C5CP02899J PMID:26242881

Wang, S. S., Chen, W. T., Li, Y., Wang, J., Sheu, H. S., & Liu, R. S. (2013). Neighboring-Cation Substitution Tuning of Photoluminescence by Remote-Controlled Activator in Phosphor Lattice. *Journal of the American Chemical Society*, *135*(34), 12504–12507. doi:10.1021/ja404510v PMID:23941075

Wang, X., Liu, Q., Bu, Y., Liu, C. S., Liu, T., & Yan, X. (2015). Optical temperature sensing of rare-earth ion doped phosphors. *RSC Advances*, *5*(105), 86219–86236. doi:10.1039/C5RA16986K

Wang, X., Zhao, Z., Wu, Q., Wang, C., Wang, Q., Yanyan, L., & Wang, Y. (2016). Structure, photoluminescence and abnormal thermal quenching behavior of Eu^{2+}-doped $Na_3Sc_2(PO_4)_3$: A novel blue-emitting phosphor for n-UV LEDs. *Journal of Materials Chemistry. C, Materials for Optical and Electronic Devices*, *4*(37), 8795–8801. doi:10.1039/C6TC01049K

Wang, Y., Huang, C., Chen, L., & Fang, Z. (1996). Crystal growth of Cr3+:LiCaAlF6 by Bridgman technique. *Journal of Crystal Growth*, *167*(1-2), 176–179. doi:10.1016/0022-0248(96)00250-3

Wang, Y., Zhang, X., Han, B., Peng, J., Shiyou, H. H., Huang, Y., ... Lu, Z. (2010). The synthesis and photoluminescence characteristics of novel blue light-emitting naphthalimide derivatives. *Dyes and Pigments*, *86*(2), 190–196. doi:10.1016/j.dyepig.2010.01.003

Wang, Y., Zheng, M., Lu, H., Feng, S., Ji, G., & Cao, J. (2010). Template Synthesis of Carbon Nanofibers Containing Linear Mesocage Arrays. *Nanoscale Research Letters*, *5*(6), 913–916. doi:10.100711671-010-9562-9 PMID:20671793

Wang, Z. M., Song, X. H., Gao, Z., Yu, D. W., Zhang, X. J., Lu, P., ... Ma, Y. G. (2012). Tuning of the electronic and optical properties of 4,4′-bis(1-phenylphenanthro[9,10-d]imidazol-2-yl)biphenyl via cyano substitution in un-conjugated phenyl. *RSC Advances*, *2*(25), 9635–9642. doi:10.1039/c2ra21054a

Wang, Z. Y., Xia, Z. G., Molokeev, M. S., Atuchin, V. V., & Liu, Q. L. (2014). Blue-shift of Eu^{2+} emission in $(Ba,Sr)_3Lu(PO_4)_3$:Eu^{2+} eulytite solid-solution phosphors resulting from release of neighbouring-cation-induced stress. *Dalton Transactions (Cambridge, England)*, *43*(44), 16800–16804. doi:10.1039/C4DT02319F PMID:25287659

Wang, Z., Feng, Y., Li, H., Gao, Z., Zhang, X., Lu, P., ... Liu, S. (2014). Dimeric phenanthroimidazole for blue electroluminescent materials: The effect of substituted position attached to biphenyl center. *Physical Chemistry Chemical Physics*, *16*(22), 10837–10843. doi:10.1039/C4CP00209A PMID:24759956

Wang, Z., Li, X., Xue, K., Li, H., Zhang, X., Liu, Y., ... Chen, P. (2016). Towards stable deep-blue emission and low efficiency roll-off in OLEDs based on phenanthroimidazole dimers. *Journal of Materials Chemistry. C, Materials for Optical and Electronic Devices*, *4*(9), 1886–1894. doi:10.1039/C5TC04048E

Wang, Z., Lu, P., Chen, S., Gao, Z., Shen, F., Zhang, W., ... Ma, Y. (2011). Phenanthro[9,10-d]imidazole as a new building block for blue light emitting materials. *Journal of Materials Chemistry*, *21*(14), 5451. doi:10.1039/c1jm10321k

Wang, Z., Xu, C., Wang, W., Dong, X., Zhao, B., & Ji, B. (2011). Novel pyrene derivatives: Synthesis, properties and highly efficient non-doped deep-blue electroluminescent device. *Dyes and Pigments*, *92*(1), 732–736. doi:10.1016/j.dyepig.2011.06.025

Watnabe, S. (1951). Properties of a $CaSO_4$-Mn Phosphor under Vacuum Ultraviolet Excitation. *Physical Review*, *83*(4), 785–791. doi:10.1103/PhysRev.83.785

Weber, M. J. (1967). Probabilities for Radiative and Nonradiative Decay of Er 3 + in La F 3. *Physical Review*, *157*(2), 262–271. doi:10.1103/PhysRev.157.262

Wegh, R. T., Donker, H., Meijerink, A., Lamminmaki, R. J., & Holsa, J. (1997). Vacuum-ultraviolet spectroscopy and quantum cutting for Gd 3 + in LiYF 4. *Physical Review B: Condensed Matter and Materials Physics*, *56*(21), 13841–13848. doi:10.1103/PhysRevB.56.13841

Wegh, R. T., Donker, H., Oskam, K. D., & Meijerink, A. (1999). Visible Quantum Cutting in LiGdF $_4$:Eu $^{3+}$ Through Downconversion. *Science*, *283*(5402), 663–666. doi:10.1126cience.283.5402.663 PMID:9924020

Wen, D., Dong, Z., Shi, J., Gong, M., & Wu, M. (2013). Standard White-Emitting $Ca_8MgY(PO_4)_7$:Eu^{2+},Mn^{2+} Phosphor for White-Light-Emitting LEDs. *ECS Journal of Solid State Science and Technology*, *2*(9), R178–R185. doi:10.1149/2.019309jss

Wen, D., Kuwahara, H., Kato, H., Kobayashi, M., Sato, Y., Masaki, T., & Kakihana, M. (2016). Anomalous Orange Light-Emitting $(Sr,Ba)_2SiO_4$: Eu^{2+} Phosphors for Warm White LEDs. *ACS Applied Materials & Interfaces*, *8*(18), 11615–11620. doi:10.1021/acsami.6b02237 PMID:27090069

West, A. R. (2007). *Solid state chemistry and its applications*. John Wiley & Sons.

Whittingham, M., Guo, J. D., Chen, R., Chirayil, T., Janauer, G., & Aavalij, P. (1995). The hydrothermal synthesis of new oxide materials. *Solid State Ionics*, *75*, 257–268. doi:10.1016/0167-2738(94)00220-M

Wickham. (1971). The Preparation of Ferrites with the Aid of Fused Salts. *Proceedings of the International Conference*, 105-107.

Wickleder, M. S. (2002). Inorganic Lanthanide Compounds with Complex Anions. *Chemical Reviews*, *102*(6), 2011–2088. doi:10.1021/cr010308o PMID:12059261

Wierzbicka-Wieczorek, M., Kolitsch, U., & Tillmanns, E. (2010). Synthesis and structural study of five new trisilicates, $BaREE_2Si_3O_{10}$ (REE = Gd, Er, Yb, Sc) and $SrY_2Si_3O_{10}$, including a review on the geometry of the Si_3O_{10} unit. *European Journal of Mineralogy*, *22*, 245–258. doi:10.1127/0935-1221/2010/0022-1969

Wilson, J., & Hawkes, J. F. B. (1999). *Optoelectronics: An Introduction* (2nd ed.). Prentice-Hall of India Pvt. Ltd.

Wu, C. A., Chou, H. H., Shih, C. H., Wu, F. I., Cheng, C. H., Huang, H. L., ... Tseng, M. R. (2012). Synthesis and physical properties of meta-terphenyloxadiazole derivatives and their application as electron transporting materials for blue phosphorescent and fluorescent devices. *Journal of Materials Chemistry*, *22*(34), 17792–17799. doi:10.1039/c2jm33376g

Wu, K. C., Ku, P. J., Lin, C. S., Shih, H. T., Wu, F. I., Huang, M. J., ... Cheng, H. (2008). The Photophysical Properties of Dipyrenylbenzenes and Their Application as Exceedingly Efficient Blue Emitters for Electroluminescent Devices. *Advanced Functional Materials*, *18*(1), 67–75. doi:10.1002/adfm.200700803

Wulf, H. C., & Gniadecka, M. (1996). CaF_2:Dy and CaF_2 crystal-based UV dosimeters. *Skin Research and Technology*, *2*(3), 108–113. doi:10.1111/j.1600-0846.1996.tb00071.x PMID:27327385

Wu, M. F., Yeh, S. J., Chen, C. T., Murayama, H., Tsuboi, T., Li, W. S., & Wang, J. K. (2007). The Quest for High-Performance Host Materials for Electrophosphorescent Blue Dopants. *Advanced Functional Materials*, *17*(12), 1887–1895. doi:10.1002/adfm.200600800

Wu, W. C., Yeh, H. C., Chan, L. H., & Chen, C. T. (2002). Red Organic Light-Emitting Diodes with a Non-doping Amorphous Red Emitter. *Advanced Materials*, *14*(15), 1072–1075. doi:

Wu, W., & Xia, Z. G. (2013). Synthesis and color-tunable luminescence properties of Eu^{2+} and Mn^{2+}-activated $Ca_3Mg_3(PO_4)_4$ phosphor for solid state lighting. *RSC Advances*, *3*(17), 6051–6057. doi:10.1039/c3ra40313k

Wu, X., & Wang, L. (2013). Dual-Functional Upconverter-Doped TiO_2 Hollow Shells for Light Scattering and Near-Infrared Sunlight Harvesting in Dye-Sensitized Solar Cells. *Advanced Energy Materials*, *3*(6), 704–707. doi:10.1002/aenm.201200933

Wyszecki, G., & Stiles, W. S. (1967). *Color Science* (1st ed.). John Wiley & Sons, Inc.

Xia, G., Zhou, S., Zhang, J., Wang, S., Hong, W., & Jun, X. (2005). Sol–gel combustion synthesis and luminescence of Y4Al2O9:Eu3+ nanocrystal. *Journal of Non-Crystalline Solids*, *351*(37-39), 2979–2982. doi:10.1016/j.jnoncrysol.2005.07.007

Xia, H., Li, M., Lu, D., Zhang, C., Xie, W., Liu, X., ... Ma, Y. (2007). Host selection and configuration design of electrophosphorescent devices. *Advanced Functional Materials*, *17*(11), 1757–1764. doi:10.1002/adfm.200600882

Xiao, L., Su, S. J., Agata, Y., Lan, H., & Kido, J. (2009). Nearly 100% Internal Quantum Efficiency in an Organic Blue-Light Electrophosphorescent Device Using a Weak Electron Transporting Material with a Wide Energy Gap. *Advanced Materials*, *21*(12), 1271–1274. doi:10.1002/adma.200802034

Xiao, W. G., Zhang, X., Hao, Z. D., Pan, G. H., Luo, Y. S., Zhang, L. G., & Zhang, J. H. (2015). Blue-emitting $K_2Al_2B_2O_7$:Eu^{2+} phosphor with high thermal stability and high color purity for near-UV-pumped white light-emitting diodes. *Inorganic Chemistry*, *54*(7), 3189–3195. doi:10.1021/ic502773t PMID:25751730

Xiao, Y., Hao, Z., Zhang, L., Xiao, W., Wu, D., Zhang, X., ... Zhang, J. (2017). Highly efficient green-emitting phosphors $Ba_2Y_5B_5O_{17}$ with low thermal quenching due to fast energy transfer from Ce^{3+} to Tb^{3+}. *Inorganic Chemistry*, *56*(8), 4539–4545. doi:10.1021/acs.inorgchem.7b00085 PMID:28358516

Xia, W., & Chang, J. (2008). Preparation and the phase transformation behavior of amorphous mesoporous calcium silicate. *Microporous and Mesoporous Materials*, *108*(1-3), 345–351. doi:10.1016/j.micromeso.2007.04.013

Xia, Y., Chen, J., Liu, Y., Molokeev, M. S., Guan, M., Huang, Z., & Fang, M. (2016). Crystal structure evolution and luminescence properties of color tunable solid solution phosphors $Ca^{2+}xLa_8-x(SiO_4)_6-x(PO_4)xO_2$:$Eu^{2+}$. *Dalton Transactions (Cambridge, England)*, *45*(3), 1007–1015. doi:10.1039/C5DT03786G PMID:26648132

Xia, Z. G., Li, Q., & Sun, J. Y. (2007). Luminescence properties of $Ba_5SiO_4(F,Cl)_6$:Eu^{2+} phosphor. *Materials Letters*, *61*(8-9), 1885–1888. doi:10.1016/j.matlet.2006.07.149

Xia, Z. G., & Liu, Q. L. (2016). Progress in discovery and structural design of color conversion phosphors for LEDs. *Progress in Materials Science*, *84*, 59–117. doi:10.1016/j.pmatsci.2016.09.007

Xia, Z. G., & Liu, R. S. (2012). Tunable Blue-Green Color Emission and Energy Transfer of $Ca_2Al_3O_6F$:Ce^{3+}, Tb^{3+} Phosphors for Near-UV White LEDs. *The Journal of Physical Chemistry C*, *116*(29), 15604–15609. doi:10.1021/jp304722z

Xia, Z. G., Sun, J. Y., Du, H. Y., & Zhou, W. (2006). Luminescence of Eu^{2+} in alkali earth chlorosilicate phosphor and their color-tunable properties. *Optical Materials*, *28*(5), 524–529. doi:10.1016/j.optmat.2005.01.029

Xia, Z., Molokeev, M. S., Im, W. B., Unithrattil, S., & Liu, Q. (2015). Crystal Structure and Photoluminescence Evolution of $La_5(Si_2+xB1-x)(O_{13}-xNx)$:$Ce^{3+}$ Solid Solution Phosphors. *The Journal of Physical Chemistry C*, *119*(17), 9488–9495. doi:10.1021/acs.jpcc.5b01211

Xia, Z., Zhang, Y., Molokeev, M. S., & Atuchin, V. V. (2013). Structural and Luminescence Properties of Yellow-Emitting $NaScSi_2O_6$:Eu^{2+} Phosphors: Eu^{2+} Site Preference Analysis and Generation of Red Emission by Codoping Mn^{2+} for White-Light-Emitting Diode Applications. *The Journal of Physical Chemistry C*, *117*(40), 20847–20854. doi:10.1021/jp4062225

Xie, M., Li, D., Pan, R., Zhou, X., & Zhu, G. (2015). Crystallographic sites of Eu2+ and Eu^{2+}–Tb^{3+} energy transfer in Sr_2LiSiO_4F. *RSC Advances*, *5*(29), 22856–22862. doi:10.1039/C5RA00585J

Xie, R. J., Hirosaki, N., Mitomo, M., Yamamoto, Y., Suehiro, T., & Sakuma, K. (2004). Optical Properties of Eu^{2+} in α-SiAlON. *The Journal of Physical Chemistry B, 108*(32), 12027–12031. doi:10.1021/jp048295g

Xing, M. M., Ma, Y. B., Luo, X. X., Fu, Y., Jiang, T., Wang, H., & Duan, X. L. (2014). Design and achieving of multicolor upconversion emission based on rare-earth-doped tellurite. *Journal of Rare Earths, 32*(5), 394–398. doi:10.1016/S1002-0721(14)60083-1

Xin, S. Y., Wang, Y. H., Zhu, G., Zhang, F., Gong, Y., Wen, Y., & Liu, B. T. (2013). Tunable white light emitting from mono Ce^{3+} doped $Sr_5(PO_4)_2SiO_4$ phosphors for light emitting diodes. *Materials Research Bulletin, 48*(4), 1627–1631. doi:10.1016/j.materresbull.2013.01.008

Xu, H., Liu, Q., Huang, L., He, Z., Wang, S., Jiang, C., & Peng, Q. (2017). Luminescence and energy transfer of color-tunable $Na_6Ca_3Si_6O_{18}$:Ce^{3+},Tb^{3+} phosphors for WLED. *Optics & Laser Technology, 89*, 151–155. doi:10.1016/j.optlastec.2016.09.035

Xu, J., Zhang, H., Xiong, Q., Liang, G., Wang, L., Shen, X., ... Xu, W. (2011). Large Enhancement in short-circuit current densities of dye-sensitized solar cells via photon conversion. *Journal of Physics: Conference Series, 276*(1), 012195. doi:10.1088/1742-6596/276/1/012195

Xu, W., Jia, W. Y., Revira, I., Monge, K., & Liu, H. M. (2001). Optical properties of multiple sites of Eu^{3+} in SrY_2O_4 single-crystal fibers. *Journal of the Electrochemical Society, 148*(12), 179. doi:10.1149/1.1415724

Xu, X. G., Chen, J., Deng, S. Z., Xu, N. S., & Lin, J. (2010). Cathodoluminescent properties of nanocrystalline Lu 3 Ga 5 O 12: Tb 3+ phosphor for field emission display application a. *Journal of Vacuum Science & Technology B. Nanotechnology and Microelectronics: Materials, Processing, Measurement, and Phenomena, 28*(3), 490–494.

Yadav, R. S., Verma, R. K., Bahadur, A., & Rai, S. B. (2015). Infrared to infrared upconversion emission in Pr 3+/Yb 3+ co-doped La 2 O 3 and La (OH) 3 nano-phosphors: A comparative study. *Spectrochimica Acta. Part A: Molecular and Biomolecular Spectroscopy, 142*, 324–330. doi:10.1016/j.saa.2015.01.095 PMID:25706603

Yadav, R., Khan Atif, F., Yadav, A., Chander, H., Haranath, D., Gupta, B. K., ... Chawla, S. (2009). Intense red-emitting $Y_4Al_2O_9$:Eu^{3+}phosphor with short decay time and high color purity for advanced plasma display panel. *Optics Express, 17*(24), 22023. doi:10.1364/OE.17.022023 PMID:19997447

Yamane, H., Nagasawa, T., Murakami, Y., Kamata, T., Shindo, D., Shimada, M., & Endo, T. (1998b). Crystal Structure and Thermal Expansion of Ca2Y2Si2O9. *Materials Research Bulletin, 33*(6), 845–853. doi:10.1016/S0025-5408(98)00059-2

Yamane, H., Shimada, M., & Hunter, B. A. (1998a). High-Temperature Neutron Diffraction Study of Y4Al2O9. *Journal of Solid State Chemistry, 141*(2), 466–474. doi:10.1006/jssc.1998.7998

Yamashita, T., Nada, N., Onishi, H., & Kitamura, S. (1971). Calcium sulfate activated by thulium or dysprosium for thermoluminescence dosimetry. *Health Physics, 21*(2), 295–300. doi:10.1097/00004032-197108000-00016 PMID:5094199

Yan, B., & Lei, F. (2010). Molten Salt Synthesis, Characterization and Luminescence of $ZnWO_4$:Eu^{3+} Nanoparticles. *Journal of Alloys and Compounds, 507*(2), 460–464. doi:10.1016/j.jallcom.2010.07.203

Yang, B., & Ma, Y. G. (2013). Progress in next-generation organic electroluminescent materials: Material design beyond exciton statistics. *Science in China. Series B, Chemistry, 43*, 1457.

Yang, C. H., Pan, Y. X., Zhang, Q. Y., & Jiang, Z. H. (2007). Cooperative energy transfer and frequency upconversion in Yb^{3+}–Tb^{3+} and Nd^{3+}–Yb^{3+}–Tb^{3+} Codoped $GdAl_3(BO_3)_4$ Phosphors. *Journal of Fluorescence, 17*(5), 500–504. doi:10.100710895-007-0200-6 PMID:17609865

Yang, H. M., Shi, J. X., Liang, H. B., & Gong, M. L. (2006). Novel red phosphor Mg_2GeO_4 doped with Eu^{3+} for PDP applications. *J. Mater. Sci. Eng. B*, *127*(2-3), 276–279. doi:10.1016/j.mseb.2005.10.014

Yang, J. K., Xiao, S. G., Ding, J. W., Yang, X. L., & Wang, X. F. (2009). Preparation and photoluminescence properties of SrY_2O_4:Yb^{3+},Er^{3+} powders. *Journal of Alloys and Compounds*, *474*(1-2), 424–427. doi:10.1016/j.jallcom.2008.06.158

Yang, J., Huang, J., Li, Q., & Li, Z. (2016). Blue AIEgens: Approaches to control the intramolecular conjugation and the optimized performance of OLED devices. *Journal of Materials Chemistry. C, Materials for Optical and Electronic Devices*, *4*(14), 2663–2684. doi:10.1039/C5TC03262H

Yang, P., Yu, X., Yu, H., Jiang, T., Zhou, D., & Qiu, J. (2012). Effects of crystal field on photoluminescence properties of $Ca_2Al_2SiO_7$:Eu^{2+} phosphors. *Journal of Rare Earths*, *12*(12), 1208–1212. doi:10.1016/S1002-0721(12)60207-5

Yang, Z. W., Zhu, K., Song, Z. G., Zhou, D. C., Yin, Z. Y., Yan, L., & Qiu, J. B. (2011). Significant reduction of upconversion emission in $CaTiO_3$:Yb,Er inverse opals. *Thin Solid Films*, *519*(16), 5696–5699. doi:10.1016/j.tsf.2011.03.027

Yang, Z., Lin, P. C., Guo, C. F., & Liu, W. R. (2015). Color-tunable luminescence and energy transfer properties of Eu^{2+}- and Mn^{2+}-activated $BaCa_2MgSi_2O_8$ phosphor for ultraviolet light-emitting diodes. *RSC Advances*, *5*(17), 13184–13191. doi:10.1039/C4RA14848G

Yan, R. X., & Li, Y. D. (2005). Down/Up Conversion in Ln3+-Doped YF3 Nanocrystals. *Advanced Functional Materials*, *15*(5), 763–770. doi:10.1002/adfm.200305044

Yan, Z. Y., Yan, B., & Jia, L. P. (2013). Article. *Materials Research Bulletin*, *1*, 39.

Yazici, A. N., Chen, R., Solak, S., & Yegingil, Z. (2002). The analysis of thermoluminescent glow peaks of CaF_2: Dy (TLD-200) after β-irradiation. *Journal of Applied Physics D Applied Physics (Berlin)*, *35*, 2526–2535.

Ye, D., Chen, M., Liu, S., Su, J., Wang, Y. F., Lo, C. C., ... Kido, J. (2014). Pyridine-Containing Electron-Transport Materials for Highly Efficient Blue Phosphorescent OLEDs with Ultralow Operating Voltage and Reduced Efficiency Roll-Off. *Advanced Functional Materials*, *24*(21), 3268–3275. doi:10.1002/adfm.201303785

Yeh, C. L., & Sung, W. Y. (2004). Synthesis of NiTi intermetallics by self-propagating combustion. *Journal of Alloys and Compounds*, *376*(1-2), 79–88. doi:10.1016/j.jallcom.2003.12.016

Yeh, K. Y., Lin, C. H., Maggay, I. V. B., & Liu, W. R. (2016). Novel green-emitting $K_2Ba_5Si_{12}O_{30}$:Eu^{2+} phosphors with excellent thermal quenching for white light-emitting diodes. *Optical Materials*, *59*, 8–14. doi:10.1016/j.optmat.2016.04.026

Ye, J., Chen, Z., Fung, M. K., Zheng, C., Ou, X., Zhang, X., ... Lee, C. S. (2013). Carbazole/Sulfone Hybrid D-π-A-Structured Bipolar Fluorophores for High-Efficiency Blue-Violet Electroluminescence. *Chemistry of Materials*, *25*(13), 2630–2637. doi:10.1021/cm400945h

Ye, S., Xiao, F., Pan, Y. X., Ma, Y. Y., & Zhang, Q. Y. (2010). Phosphors in phosphor-converted white light-emitting diodes: Recent advances in materials, techniques and properties. *Materials Science and Engineering R Reports*, *71*(1), 1–34. doi:10.1016/j.mser.2010.07.001

Ye, X., Collins, J. E., Kang, Y., Chen, J., Chen, D. T. N., Yodh, A. G., & Murray, C. B. (2010). Morphologically controlled synthesis of colloidal upconversion nanophosphors and their shape-directed self-assembly. *Proceedings of the National Academy of Sciences of the United States of America*, *107*(52), 22430–22435. doi:10.1073/pnas.1008958107 PMID:21148771

Ye, Y. S., Jiang, Z. H., Wang, Q. Z., Zhu, Z. S., Wang, X., Sui, Z. L., ... Ding, Z. J. (2014). Upconversion luminescence of NaYF4:Yb,Er nanocrystals with high uniformity. *Journal of Rare Earths*, *32*(9), 802–805. doi:10.1016/S1002-0721(14)60144-7

Yi, G. S., Lu, H. C., Zhao, S. Y., Yue, G., & Yang, W. J. (2004). Article. *Nano Letters*, *4*, 2191–2196. doi:10.1021/nl048680h

Yi, H., Wu, L., Wu, L., Zhao, L., Xia, Z., Zhang, Y., ... Xu, J. (2016). Crystal. Structure of High-Temperature Phase β-$NaSrBO_3$ and Photoluminescence of β-$NaSrBO_3$:Ce^{3+} I*norgan.ic. Chemistry*, *55*, 6487–6495. PMID:27299933

Yook, K. S., & Lee, J. Y. (2012). Organic Materials for Deep Blue Phosphorescent Organic Light-Emitting Diodes. *Advanced Materials*, *24*(24), 3169–3190. doi:10.1002/adma.201200627 PMID:22641461

Yoon, K. H., Cho, Y. S., & Kang, D. H. (1998). Molten Salt Synthesis of Lead-based Relaxors. *Journal of Materials Science*, *33*(12), 2977–2984. doi:10.1023/A:1004310931643

Yoshimura, M., & Byrappa, K. (2008). Hydrothermal processing of materials: Past, present and future. *Journal of Materials Science*, *43*(7), 2085–2103. doi:10.100710853-007-1853-x

Yoshimura, M., Suchanek, W. L., & Byrappa, K. (2000). Soft Solution Processing: A Strategy for One-Step Processing of Advanced Inorganic Materials. *Materials Research Bulletin*, *25*(09), 17–25. doi:10.1557/mrs2000.175

Yossian, D., & Horowitz, Y. S. (1997). Mixed-order and general-order kinetics applied to synthetic glow peaks and to peak 5 in Lif:Mg,Ti (TLD-100). *Radiation Measurements*, *27*(3), 465–471. doi:10.1016/S1350-4487(97)00007-3

You, F., Wang, Y., Lin, J., & Tao, Y. (2002). Hydrothermal synthesis and luminescence properties of NaGdF4:Eu. *Journal of Alloys and Compounds*, *343*(1-2), 151–155. doi:10.1016/S0925-8388(02)00203-7

Yuan, B., Huang, Y., Yu, Y. M., & Seo, H. J. (2012). Luminescence and structure of Eu^{2+}-doped $Ba_2CaMg_2Si_6O_{17.}$ *Ceramics International*, *38*(3), 2219–2223. doi:10.1016/j.ceramint.2011.10.070

Yuan, W. Z., Gong, Y., Chen, S., Shen, X. Y., Lam, J. W. Y., Lu, P., ... Tang, B. Z. (2012). Efficient Solid Emitters with Aggregation-Induced Emission and Intramolecular Charge Transfer Characteristics: Molecular Design, Synthesis, Photophysical Behaviors, and OLED Application. *Chemistry of Materials*, *24*(8), 1518–1528. doi:10.1021/cm300416y

Yuan, Y., Chen, J. X., Lu, F., Tong, Q. X., Yang, Q. D., Mo, H. W., ... Lee, C.-S. (2013). Bipolar Phenanthroimidazole Derivatives Containing Bulky Polyaromatic Hydrocarbons for Nondoped Blue Electroluminescence Devices with High Efficiency and Low Efficiency Roll-Off. *Chemistry of Materials*, *25*(24), 4957–4965. doi:10.1021/cm4030414

Yuan, Y., Li, D., Zhang, X., Zhao, X., Liu, Y., Zhang, J., & Wang, Y. (2011). Phenanthroimidazole-derivative semiconductors as functional layer in high performance OLEDs. *New Journal of Chemistry*, *35*(7), 1534–1540. doi:10.1039/c1nj20072k

Yu, D., Zhao, F., Han, C., Xu, H., Li, J., Zhang, Z., ... Yan, P. (2012). Ternary Ambipolar Phosphine Oxide Hosts Based on Indirect Linkage for Highly Efficient Blue Electrophosphorescence: Towards High Triplet Energy, Low Driving Voltage and Stable Efficiencies. *Advanced Materials*, *24*(4), 509–514. doi:10.1002/adma.201104214 PMID:22213078

Yu, D., Zhao, F., Zhang, Z., Han, C., Xu, H., Li, J., ... Yan, P. (2012). Insulated donor–p–acceptor systems based on fluorene-phosphine oxide hybrids for non-doped deep-blue electroluminescent devices. *Chemical Communications (Cambridge)*, *48*(49), 6157. doi:10.1039/c2cc31066j PMID:22590706

Yu, R. J., Li, H. J., Ma, H. L., Wang, C. F., & Wang, H. (2014). A New Blue-Emitting Phosphor of Ce^{3+}-Activated Fluorosilicate Apatite $Ba_2Y_3[SiO_4]_3F$. *Journal of the American Ceramic Society*, *97*(4), 1151–1156. doi:10.1111/jace.12765

Yu, R., Zhong, S., Xue, N., Lia, H., & Ma, H. (2014). Synthesis, structure, and peculiar green emission of $NaBaBO_3$:Ce^{3+} phosphors. *Dalton Transactions (Cambridge, England)*, *43*(28), 10969–10976. doi:10.1039/C4DT00859F PMID:24905051

Zahedifar, M., Sadeghi, E., & Mohebbi, Z. (2012). Synthesis and thermoluminescence characteristics of Mn doped CaF_2 nanoparticles. *Nuclear Instruments and Methods in Physics Research Section B, 274*, 162–166. doi:10.1016/j.nimb.2011.12.015

Zeng, L., Lee, T. Y. H., Merkel, P. B., & Chen, S. H. (2009). A new class of non-conjugated bipolar hybrid hosts for phosphorescent organic light-emitting diodes. *Journal of Materials Chemistry, 19*(46), 8772. doi:10.1039/b909787b

Zeng, Q., Tanno, H., Egoshi, K., Tanamachi, N., & Zhang, S. (2006). Ba_5SiO_4:Eu^{2+}: An intense blue emission phosphor under vacumm ultraviolet and near-ultraviolet excitation. *Applied Physics Letters, 88*(5), 051906. doi:10.1063/1.2168492

Zhai, D. Y., Ning, L. X., Huang, Y. C., & Liu, G. K. (2014). Ce–O Covalence in Silicate Oxyapatites and Its Influence on Luminescence Dynamics. *The Journal of Physical Chemistry C, 118*(29), 16051–16059. doi:10.1021/jp5049293

Zhang, S., Hou, Y., Fujii, H., Onishi, T., Kokubu, M., Obata, M., Tanno, H., Kono, T., & Uchiike, H. (2003). Effect of Nonstoichiometry on the Deterioration of Eu^{2+} -Doped Hexagonal Aluminate Phosphor for Plasma Display Applications. *Jpn. J. Appl. Phys., 42*(Part 1, No. 2A), 477–480.

Zhang, C. M., Huang, S. S., Yang, D. M., Kang, X. J., Shang, M. M., Peng, C., & Lin, J. (2010). Tunable luminescence in Ce^{3+}, Mn^{2+}-codoped calcium fluorapatite through combining emissions and modulation of excitation: A novel strategy to white light emission. *Journal of Materials Chemistry, 20*(32), 6674–6680. doi:10.1039/c0jm01036g

Zhang, C., Sun, L., Zhang, Y., & Yan, C. (2010). Rare earth upconversion nanophosphors: Synthesis, functionalization and application as biolabels and energy transfer donors. *Journal of Rare Earths, 28*(6), 807–819. doi:10.1016/S1002-0721(09)60206-4

Zhang, F., Li, W., Wei, D., Wei, X., Li, Z., Zhang, S., ... Zhai, B. (2016). Synthesis, photophysical and electroluminescent properties of blue-emitting dual core imidazole–anthracene/pyrene derivatives. *RSC Advances, 6*(65), 60264–60270. doi:10.1039/C6RA04958C

Zhang, F., Wan, Y., Shi, Y., Tu, B., & Zhao, D. (2008). Ordered Mesostructured Rare-Earth Fluoride Nanowire Arrays with Upconversion Fluorescence. *Chemistry of Materials, 20*(11), 3778–3784. doi:10.1021/cm800489e

Zhang, H., Yang, D., Ji, Y., Ma, X., Xu, J., & Que, D. (2004). Low temperature synthesis of flowerlike ZnO nanostructures by cetyltrimethylammonium bromide-assisted hydrothermal process. *The Journal of Physical Chemistry B, 108*(13), 3955–3958. doi:10.1021/jp036826f

Zhang, J., Hao, Z., Zhang, X., Luo, Y., Ren, X., Wang, X., & Zhang, J. (2009). Color tunable phosphorescence in KY3F10:Tb3+ for x-ray or cathode-ray tubes. *Journal of Applied Physics, 106*(3), 034915–034916. doi:10.1063/1.3190511

Zhang, J., Wang, Y., Zhang, F., & Huang, Y. (2011). Single-Phase White-Emitting $Ca_8MgGd(PO_4)_7$:Ln^{3+}, Mn^{2+} ($Ln^{3+}=Ce^{3+}$, Tb^{3+}, and Dy^{3+}) for Mercury-Free Lamps. *Journal of the Electrochemical Society, 158*(4), J110–J114. doi:10.1149/1.3551508

Zhang, Q. Y., & Huang, X. Y. (2010). Recent progress in quantum cutting phosphors. *Progress in Materials Science, 55*(5), 353–427. doi:10.1016/j.pmatsci.2009.10.001

Zhang, Q. Y., Yang, G. F., & Jiang, Z. H. (2007). Cooperative downconversion in GdAl3(BO3)4:RE3+,Yb3+ (RE=Pr, Tb, and Tm). *Applied Physics Letters, 91*(5), 051903. doi:10.1063/1.2757595

Zhang, Q., Li, J., Shizu, K., Huang, S., Hirata, S., Miyazaki, H., & Adachi, C. (2012). Design of Efficient Thermally Activated Delayed Fluorescence Materials for Pure Blue Organic Light Emitting Diodes. *Journal of the American Chemical Society, 134*(36), 14706–14709. doi:10.1021/ja306538w PMID:22931361

Zhang, S. Y., Huang, Y. L., Nakai, Y., Tsuboi, T., & Seo, H. J. (2011). The Luminescence Characterization and Thermal Stability of Eu^{2+} Ions-Doped $NaBaPO_4$ Phosphor. *Journal of the American Ceramic Society*, *94*(9), 2987–2992. doi:10.1111/j.1551-2916.2011.04481.x

Zhang, S., Gelainb, F., & Zhao, X. (2005). Designer self-assembling peptide nanofiber scaffolds for 3D tissue cell cultures. *Seminars in Cancer Biology*, *15*(5), 413–420. doi:10.1016/j.semcancer.2005.05.007 PMID:16061392

Zhang, S., Hu, Y., Chen, L., Fu, Y., & Ju, G. (2016). $La_3GaGe_5O_{16}$:Cr^{3+} phosphor: the near-infrared persistent luminescence. *Optical Materials Express*, *4*(4), 1247–1255. doi:10.1364/OME.6.001247

Zhang, S., Li, W., Yao, L., Pan, Y., Shen, F., Xiao, R., ... Ma, Y. (2013). Enhanced proportion of radiative excitons in non-doped electro-fluorescence generated from an imidazole derivative with an orthogonal donor–acceptor structure. *Chemical Communications (Cambridge)*, *49*(96), 11302. doi:10.1039/c3cc47130f PMID:24158601

Zhang, W. P., Xie, P. B., Duan, C. K., Yan, K., Yin, M., Lou, L. R., ... Krupa, J. C. (1998). Preparation and size effect on concentration quenching of nanocrystalline Y2SiO5:Eu. *Chemical Physics Letters*, *292*(1-2), 133–136. doi:10.1016/S0009-2614(98)00656-3

Zhang, X. X., Bass, M., Villaverde, A. B., Lefaucheur, J., Pham, A., & Chai, B. H. T. (1993). Efficient laser performance of Nd:$GdLiF_4$: A new laser crystal. *Applied Physics Letters*, *62*(11), 1197–1199. doi:10.1063/1.108732

Zhang, X. X., Hong, P., Bass, M., & Chai, B. H. T. (1993). Ho^{3+} to Yb^{3+}back transfer and thermal quenching of upconversion green emission in fluoride crystals. *Applied Physics Letters*, *63*(19), 2606–2608. doi:10.1063/1.110445

Zhang, X., Lin, J., Ouyang, X., Liu, Y., Liu, X., & Ge, Z. (2013). Novel host materials based on phenanthroimidazole derivatives for highly efficient green phosphorescent OLEDs. *Journal of Photochemistry and Photobiology A Chemistry*, *268*, 37–43. doi:10.1016/j.jphotochem.2013.06.012

Zhang, X., Quan, Z., Yang, J., Yang, P., Lian, H., & Lin, J. (2008). Solvothermal synthesis of well-dispersed MF_2 (M = Ca,Sr,Ba) nanocrystals and their optical properties. *Nanotech.*, *19*(7), 075603–075608. doi:10.1088/0957-4484/19/7/075603

Zhang, Y. (1994). Article. *Journal of Materials Research*, *9*, 1997. doi:10.1557/JMR.1994.1997

Zhang, Y. W., Sun, X., Si, R., You, L. P., & Yan, C. H. (2005). Single-crystalline and monodisperse LaF_3 triangular nanoplates from a single-source precursor. *Journal of the American Chemical Society*, *127*(10), 3260–3261. doi:10.1021/ja042801y PMID:15755126

Zhang, Y., Lai, S. L., Tong, Q. X., Chan, M. Y., Ng, T. W., Wen, Z. C., ... Lee, C. S. (2011). Synthesis and characterization of phenanthroimidazole derivatives for applications in organic electroluminescent devices. *Journal of Materials Chemistry*, *21*(22), 8206. doi:10.1039/c1jm10326a

Zhang, Y., Lai, S. L., Tong, Q. X., Lo, M. F., Ng, T. W., Chan, M. Y., ... Lee, C. S. (2012). High Efficiency Nondoped Deep-Blue Organic Light Emitting Devices Based on Imidazole-π-triphenylamine Derivatives. *Chemistry of Materials*, *24*(1), 61–70. doi:10.1021/cm201789u

Zhang, Y., Li, G. G., Geng, D. L., Shang, M. M., Peng, C., & Lin, J. (2012). Color-tunable emission and energy transfer in $Ca_3Gd_7(PO_4)(SiO_4)_5O_2$:$Ce^{3+}$/$Tb^{3+}$/$Mn^{2+}$ phosphors. *Inorganic Chemistry*, *51*(21), 11655–11664. doi:10.1021/ic3015578 PMID:23075230

Zhang, Y., Ng, T. W., Lua, F., Tong, Q. X., Lai, S. L., Chan, M. Y., ... Lee, C. S. (2013). A pyrene-phenanthroimidazole derivative for non-doped blue organic light-emitting devices. *Dyes and Pigments*, *98*(2), 190–194. doi:10.1016/j.dyepig.2013.01.022

Zhang, Y., Wang, J. H., Han, G., Lu, F., & Tong, Q. X. (2016). Phenanthroimidazole derivatives as emitters for non-doped deep-blue organic light emitting devices. *RSC Advances*, *6*(75), 70800–70809. doi:10.1039/C6RA13605B

Zhao, C., Feng, S., Chao, Z., Shi, C., Xu, R., & Ni, J. (1996). Hydrothermal synthesis of the complex fluorides LiBaF3 and KMgF3 with perovskite structures under mild conditions. *Chemical Communications (Cambridge)*, (14): 1641–1642. doi:10.1039/cc9960001641

Zhao, D., Feng, F., Quo, Q., Melosh, N., Fredrickson, G. H., Chmelka, B. F., & Stucky, G. D. (1999). Triblock Copolymer Syntheses of Mesoporous Silica with Periodic 50 to 300 Angstrom Pores. *Science*, *279*(5350), 548–552. doi:10.1126cience.279.5350.548 PMID:9438845

Zhao, J., Han, W., Chen, H., Tu, M., Zeng, R., Shi, Y., ... Zhou, C. (2001). Preparation, structure and crystallinity of chitosan nano-fibers by a solid–liquid phase separation technique. *Carbohydrate Polymers*, *83*(4), 1541–1546. doi:10.1016/j.carbpol.2010.10.009

Zhao, J., Xie, G., Yin, C., Xie, L., Han, C., Chen, R., ... Huang, W. (2011). Harmonizing Triplet Level and Ambipolar Characteristics of Wide-Gap Phosphine Oxide Hosts toward Highly Efficient and Low Driving Voltage Blue and Green PHOLEDs: An Effective Strategy Based on Spiro-Systems. *Chemistry of Materials*, *23*(24), 5331–5339. doi:10.1021/cm201654c

Zhao, X., Roy, R., Cherian, K., & Badzian, A. (1997). Article. *Science*, *385*, 513–515.

Zheng, Z., & Tang, W. (2016). Efficient sensitization of Eu^{2+}/Mn^{2+} emissions by Ce^{3+} doping in NaMgPO4 host under UV excitation. *Applied Physics. A, Materials Science & Processing*, *122*(3), 229. doi:10.100700339-016-9784-5

Zhong, G. Y., Xu, Z., He, J., Zhang, S. T., Zhan, Y. Q., Wang, X. J., ... Hou, X. Y. (2002). Aggregation and permeation of 4-(dicyanomethylene)-2-methyl-6-(p-dimethylaminostyryl)-4H-pyran molecules in Alq_3. *Applied Physics Letters*, *81*(6), 1122–1124. doi:10.1063/1.1497438

Zhong, J., Chen, D., Yuan, Y., Chen, L., Yu, H., & Ji, Z. (2017). Synthesis and spectroscopic investigation of $Ba_3La_6(SiO_4)_6$:Eu^{2+} green phosphors for white light-emitting diodes. *Chemical Engineering Journal*, *309*, 795–801. doi:10.1016/j.cej.2016.10.109

Zhong, J., Zhuang, W., Xing, X., Liu, R., Li, Y., Liu, Y., & Hu, Y. (2015). Synthesis, Crystal Structures, and Photoluminescence Properties of Ce^{3+}-Doped $Ca_2LaZr_2Ga_3O_{12}$: New Garnet Green-Emitting Phosphors for White LEDs. *The Journal of Physical Chemistry C*, *119*(10), 5562–5569. doi:10.1021/jp508409r

Zhou, D. C., Song, Z. G., Chi, G. W., & Qiu, J. B. (2009). NIR broadband luminescence and energy transfer in Er^{3+}-Tm^{3+}-co-doped tellurite glasses. *Journal of Alloys and Compounds*, *481*(1), 881–884. doi:10.1016/j.jallcom.2009.03.141

Zhou, D. L., Hu, Y. B., Qiu, J. B., Yang, Z. W., & Song, Z. G. (2010). Characteristics and mechanism of upconversion luminescence in $Er^{3+}/Yb^{3+}/Tb^{3+}$co-doped oxyfluorogermanateglasses. *Journal of Inorganic Materials*, *25*, 551–556.

Zhou, J., Liu, Q., Feng, W., Sun, Y., & Li, F. (2014). Upconversion luminescent materials: Advances and applications. *Chemical Reviews*, *115*(1), 395–465. doi:10.1021/cr400478f PMID:25492128

Zhou, J., Liu, Z., & Li, F. (2012). Upconversion nanophosphors for small-animal imaging. *Chemical Society Reviews*, *41*(3), 1323–1349. doi:10.1039/C1CS15187H PMID:22008740

Zhou, J., & Xia, Z. (2014). Synthesis, luminescence properties and energy transfer behavior of $Na_2CaMg(PO_4)_2$:Eu^{2+}, Mn^{2+} phosphors. *Journal of Luminescence*, *146*, 22–26. doi:10.1016/j.jlumin.2013.09.031

Zhou, J., Xia, Z., Yang, M., & Shen, K. (2012). High efficiency blue-emitting phosphor: Ce^{3+}-doped $Ca_{5.45}Li_{3.55}(SiO_4)_3O_{0.45}F_{1.55}$ for near UV-pumped light-emitting diodes. *Journal of Materials Chemistry*, *22*(41), 21935–21941. doi:10.1039/c2jm34146h

Zhou, T. M., Zhang, Y. Q., Wu, Z. L., & Chen, B. J. (2015). Concentration effect and temperature quenching of up-conversion luminescence in $BaGd_2ZnO_5$:Er^{3+}/Yb^{3+} phosphor. *Journal of Rare Earths*, *33*(7), 686–692. doi:10.1016/S1002-0721(14)60471-3

Zhou, W., Pan, F., Zhou, L., Hou, D., Huang, Y., Tao, Y., & Liang, H. (2016). Site Occupancies, Luminescence, and Thermometric Properties of $LiY_9(SiO_4)_6O_2$:Ce^{3+} Phosphors. *Inorganic Chemistry*, *55*(20), 10415–10424. doi:10.1021/acs.inorgchem.6b01656 PMID:27700068

Zhou, Y. H., Lin, J., Yu, M., Wang, S. B., & Zhang, H. J. (2002). Synthesis-dependent luminescence properties of Y3Al5O12:Re3+ (Re=Ce, Sm, Tb) phosphors. *Materials Letters*, *56*(5), 628–636. doi:10.1016/S0167-577X(02)00567-0

Zhuang, S.,Shangguan, R., Huang, H.,Tu, G., Wang, L., &Zhu, X. (2014).Synthesis, characterization, physical properties, and blue electroluminescent device applications of phenanthroimidazole derivatives containing anthracene or pyrene moiety. *Dyes and Pigments,101,* 93-102.

Zhuang, S., Shangguan, R., Jin, J., Tu, G., Wanga, L., Chen, J., ... Zhu, X. (2012). Efficient nondoped blue organic light-emitting diodes based on phenanthroimidazole-substituted anthracene derivatives. *Organic Electronics*, *13*(12), 3050–3059. doi:10.1016/j.orgel.2012.08.032

Zhu, G., Shi, Y. R., Mikami, M., Shimomura, Y., & Wang, Y. H. (2014). Electronic structure and photo/cathodoluminescence properties investigation of green emission phosphor $NaBaScSi_2O_7$:Eu^{2+} with high thermal stability. *CrystEngComm*, *16*(27), 6089–6097. doi:10.1039/C4CE00523F

Zhu, M., & Yang, C. (2013). Blue fluorescent emitters: Design tactics and applications in organic light-emitting diodes. *Chemical Society Reviews*, *42*(12), 4963. doi:10.1039/c3cs35440g PMID:23436004

Zhu, X. H., Peng, J., Cao, Y., & Roncali, J. (2011). Solution-processable single-material molecular emitters for organic light-emitting devices. *Chemical Society Reviews*, *40*(7), 3509–3524. doi:10.1039/c1cs15016b PMID:21483933

Zhu, Y. C., Zhou, L., Li, H. Y., Xu, Q. L., Teng, M. Y., Zheng, Y. X., ... You, X. Z. (2011). Highly Efficient Green and Blue-Green Phosphorescent OLEDs Based on Iridium Complexes with the Tetraphenylimidodiphosphinate Ligand. *Advanced Materials*, *23*(35), 4041–4046. doi:10.1002/adma.201101792 PMID:21800382

Zhu, Z. L., Chen, M., Chen, W. C., Ni, S. F., Peng, Y. Y., Zhang, C., ... Lee, C. S. (2016). Removing shortcomings of linear molecules to develop high efficiencies deep-blue organic electroluminescent materials. *Organic Electronics*, *38*, 323–329. doi:10.1016/j.orgel.2016.09.004

Zhu, Z. L., Chen, W. C., Zhang, L. D., Liu, X. L., Tong, Q. X., Wong, F. L., ... Lee, C. S. (2016). A pyridine based meta-linking deep-blue emitter with high conjugation extent and electroluminescence efficiencies. *Journal of Materials Chemistry. C, Materials for Optical and Electronic Devices*, *4*(26), 6249–6255. doi:10.1039/C6TC01689H

Ziegler, J. F., Biersack, J. P., & Ziegler, M. D. (2008). *SRIM- The Stopping and Range of Ions in Matter*. Lulu Press Co.

Zou, W., Visser, C., Maduro, J. A., Pshenichnikov, M. S., & Hummelen, J. C. (2012). Broadband dye-sensitized upconversion of near-infrared light. *Nature Photonics*, *6*(8), 560–564. doi:10.1038/nphoton.2012.158

About the Contributors

Sukhamoy Bhattacharyya has done B. Sc. (Physics Hons.) and M. Sc. (Physics) from Jadavpur University, India. He has done his Ph. D in 2009 from Indian Association for the Cultivation of Science, Kolkata, India in the field of Theoretical Atomic Physics. He was awarded Fulbright-Nehru award for pursuing post doctoral research at Auburn University, Alabama, USA. At present he is an Assistant Professor of Physics in Acharya Prafulla Chandra College, Kolkata, India. Along with his research interest in Atomic Physics, he is also involved in the phenomenological research of thermoluminescence. He has more than 30 research papers in many international journals of repute. He has also served as reviewer of several journals.

Neha Dubey completed Ph.D. from RDVV Jabalpur, Having more than 25 articles in reputed Scopus index journals, including two book published by her. 1 book chapter in Springerlink. She has got young scientist award and so many academic award

Somrita Dutta is a Postdoctoral Fellow at National Chiao Tung University, Taiwan. She has achieved the doctorate degree from the Department of Applied Physics, Indian School of Mines, Dhanbad in 2015. She has also Postdoctoral experience from the Department of Chemistry, University of the Free State, South Africa. Her major research interests are in luminescence and spectroscopic properties of different phosphor materials for LED and display. Recently she is involved in the research for white light emitting diode fabrication.

Vijay Kumar is an Assistant Professor at Chandigarh University, Gharuan, Mohali, India. He was a postdoc fellow in Professor Swart's group at the University of the Free State, South Africa from 2013-2016. He has done his Ph.D. (Physics/Material Science) from Sant Longowal Institute of Engineering and Technology, Longowal (Deemed to be University) and in Collaboration with Inter University Accelerator Center (Formerly known as Nuclear Science Center), New Delhi. During the last six years of his research career, he has published a reasonable and significant number of publications to his credit on diverse topics/areas. So far, he has been published more than 70 research papers in many of the reputed international journals, which attracted more than 1300 citations with an h-index of 21. His current research involves the synthesis and spectroscopic investigations of rare earth/transitional metal ions doped nanomaterials, nanocomposites, and hybrid materials to make color tunable emission in solid-state lighting and white light LEDs. He is also working on the synthesis and characterization of a biomaterial with electro-conductive properties that could be used in biomedical applications with better biocompatibility. He has received Young Scientist Award under the fast track scheme of Department of Science and Technology (Ministry of Science and Technology, Government of India), New Delhi. He has also received an Indo – US R&D Joint Networked Center Research Project grant from The Indo-US Science and Technology Forum (IUSSTF) in collaboration with Harvard-MIT Division of Health Sciences and Technology, Massachusetts Institute of Technology (MIT), USA; Texas A&M University, USA and Indian Institute of Technology Kanpur.

Vikas Lahariya is working as an Assistant Professor in department of Physics, Amity School of Applied Science, Amity University Haryana since 2013. She earned his Ph.D. degree from Rani Durgavati University Jabalpur in 2011 on Electroluminescence in Nanoparticles. She has gained excellent teaching and research experience of about 10 years. Dr. V. Lahariya has published a number of research articles in peer review highly reputed national/international journals and presented research/scientific work in national and international conference/workshop. She has been a reviewer of national/ international journal. In addition Dr. V. Lahariya have been a member of reviewer committee of international conference and organising member of national conference

Partha Sarathi Majumdar has done B. Sc. (Physics Hons.) and M. Sc. (Physics) from University of Calcutta, India. He has done his Ph. D in 1984 from Indian Association for the Cultivation of Science, Kolkata, India in the field of Atomic Collision theory. He served as an Associate Professor in the Dept. of Physics, Manipur University, India before joining Acharya Prafulla Chandra College, Kolkata. Along with his research in Atomic Physics, he developed strong interest in the Study of Thermoluminescence. He has more than 100 research papers in many international journals of repute. He has also acted as reviewer of several journals.

Anurag Pandey is working as Postdoctoral fellow at Departments of Physics and Electronics, Rhodes University (RU), Grahamstown, South Africa. Before joined RU, he was a postdoctoral fellow at University of the Free State, Bloemfontein, South Africa. He obtained his M.Sc. (Physics) and PhD (Applied Physics) degrees from U. P. Autonomous College Varanasi and Indian Institute of Technology (ISM) Dhanbad in 2010 and 2014, respectively. His current research interest in Lanthanides activated luminescent materials for multifunctional applications. He has published more than 30 articles in reputed international journals and presented his work in several international conferences.

Jyoti Singh is an Assistant Professor at Galgotias University, Greater Noida, Uttar Pradesh, India and currently working on the topic "Luminescence studies of rare-earth doped lanthanide based alkaline earth inter-oxides".

Sudipta Som, as a postdoctoral fellow in Chemical Engineering at National Taiwan University, is noted for his research in the areas of physics, materials physics, and luminescence spectroscopy. His ultimate goal is to contribute to the field of solid state research and light emitting materials. Dr. Som has been at National Taiwan University since 2014. Prior to that, he spent a year as a postdoctoral fellow in physics at the University of the Free State in South Africa. Dr. Som's strong educational foundation is a major factor behind his success in physics research. Born in India, he attended the University of Calcutta and graduated with first class honors with a Bachelor of Science in physics in 2008. He went on to study at the Indian School of Mines (now IIT, Dhanbad), from which he received a Master of Science in applied physics in 2010 and a Ph.D. in applied physics in 2014. His recent research interests are in spectroscopic and luminescence behavior of different materials for LED appliances.

Hendrik C. Swart is a B1 NRF rated researcher and currently a senior professor in the Department of Physics at the University of the Free State. Over the past 20 years he has led research in the area of the degradation of phosphors for field emission displays, as well as developing materials for nano solid state lighting. He has been key in the development of processes to synthesise and deposit thin films of various types of semiconductor nano-particles which will enhance the colour, luminescent intensity and lifetime of such displays. He has more than 500 publications in international peer reviewed journals, 100 peer reviewed conference proceedings and editor/author or co-editor/author of 12 book chapters and or books with more than 4700 cited author references and more than 530 national and international conference contributions (authored and co-authored). He has an ISI H-index of 33 (rid=g-2696-2012). He is a reviewer for about 50 international and national professional journals in his field (or in related fields), and a member of the editorial board of the high impact factor journal 'Critical Reviews in Solid State and Materials Sciences. He is on the editorial board of Applied Surface Science and one of the newly appointed editors of the Journal of Nano Research. He has received the South African National Science and Technology Forum (NSTF) award in 2009 for research capacity development of students in the niche area of nanophysics. His commitment to the next generation of scientists is also reflected by the awards he received from the Faculty of Natural and Agricultural Sciences at the University of the Free State, South Africa for excellence (2012), research (2014), mentorship (2008) and for academic entrepreneurship (2009). The International Association of Advanced Materials (IAAM) honored him with the prestigious IAAM Scientist award for 2014 at the SETCOR International Conference on Smart Materials and Surfaces, Bangkok - Thailand. He received honorary membership of the Golden Key Association (2012). He was chair of national and international conferences. He has supervised 60 PhD and MSc students successfully in the past with another 17 in progress and has established a National Nano Surface Characterization Facility (NNSCF) containing state of the art surface characterization equipment. Interdisciplinary results obtained with these systems were recently selected for the cover of Federation of European Microbiological Societies (FEMS) Yeast research journal for 2013. A research chair in Solid State Luminescent and Advanced Materials was awarded to him from the South African Research Chairs Initiative (SARChI) at the end of 2012. The main focus of his research group will be the improvement of luminescent materials for applications in Flat Panel Displays and Solar Cells; the development of Organic light emitting diodes (OLED) materials as well as materials for power saving solid state lighting devices.

Nilesh S. Ugemuge is currently working as Assistant Professor in Department of Applied Physics, Priyadarshini College of Engineering, Nagpur. He obtained his doctoral degree(Ph.D.) from Department of Physics, RTM Nagpur University, Nagpur on the topic entitled "Synthesis and Characterization of Fluorides by Solid State Metathesis". Currently pursuing second Ph.D. In Management from RTM Nagpur University, Nagpur. He was awarded masters degree(M.Sc.) in Physics from Department of Physics, RTM Nagpur University, Nagpur. He has passed UGC-NET(HRM), UGC-NET(IR), UGC-NET(PM), UGC-NET(LW),UGC-NET(LSW),UGC-NET(SW) and SET(SW). Also he did PGDHRM from MGLI Ahmadabad, MA(Economics), MA(Public Administration),Master in Social Work(MSW),MIRPM and

B.Ed. from RTM Nagpur University Nagpur. Also completed Certificate and Diploma course in Vedic Mathematics from Kavi Kulguru Sanskrit Vidyapeeth, Ramtek. He has published several research papers in various International/National Journals and Conferences and many of them in a reputed high impact journals. His areas of research interest are Condensed Matter Physics, Luminescence, Organization Behaviour and Human Resource Management. He has published 11 books on various subjects and edited 30 books with HPH. He is involved in teaching Applied Physics since last 9 years. He is associated with various professional bodies viz. ISTE, LSI, IAEng, SSUN, etc.

Sivakumar Vaidyanathan received his Bachelor (2000) and Master (2002) degree in Chemistry from Muthurangam Government Arts College, Vellore (affiliated to University of Madras) and Ph.D. in Solid State Chemistry (2007) from Indian Institute of Technology Madras with Prof. U. V. Varadaraju. Then he moved to Prof. Duk Young Jeon group at Korea Advanced Institute of Science and Technology, the Republic of Korea as a Brain Korea 21 postdoctoral research fellow (2007 – 2009). He also worked as a post-doctoral fellow at Commissariat à l'Energie Atomique (CEA), Saclay, France for a year. After spending a year in NIT Calicut as a Faculty in the Department of Chemistry, He has moved to Department of Chemistry, National Institute of Technology Rourkela. His research focuses on the Lanthanide photochemistry, Molecular Materials for OLEDs, Phosphors for LEDs (Solid State Lighting) and organic fluorophores for diverge applications.

Index

M

O

P

R

S

T

U

www.ingramcontent.com/pod-product-compliance
Lightning Source LLC
LaVergne TN
LVHW080530300926
849162LV00007B/1580